Speciation Studies in Soil, Sediment and Environmental Samples

Speciation Studies in Soil, Sediment and Environmental Samples

Editor

Dr. Sezgin Bakirdere
Department of Chemistry
Yildiz Technical University
Davutpasa Campus
Esenler, 34210, Istanbul, Turkey

CRC Press
Taylor & Francis Group
Boca Raton   London   New York

CRC Press is an imprint of the
Taylor & Francis Group, an **informa** business

A SCIENCE PUBLISHERS BOOK

CRC Press
Taylor & Francis Group
6000 Broken Sound Parkway NW, Suite 300
Boca Raton, FL 33487-2742

© 2014 Copyright reserved
CRC Press is an imprint of Taylor & Francis Group, an Informa business

Cover illustrations: Background illustration reproduced from Wikimedia Commons. NASA image courtesy Norman Kuring, Ocean Color Team.
Images of structures reproduced by kind courtesy of Sezgin Bakırdere

Library of Congress Cataloging-in-Publication Data

Speciation studies in soil, sediment, and environmental samples /
editor, Sezgin Bakirdere.
       pages cm
  "A CRC title."
  Includes bibliographical references and index.
  ISBN 978-1-4665-9484-5 (hardcover : alk. paper) 1. Speciation
(Chemistry) 2. Soil chemistry. 3. Sedimentation analysis. I.
Bakirdere, Sezgin, 1980- editor of compilation.

  QD75.3.S64 2014
  631.4'1--dc23
                                                      2013023189

Visit the Taylor & Francis Web site at
http://www.taylorandfrancis.com

CRC Press Web site at
http://www.crcpress.com

Science Publishers Web site at
http://www.scipub.net

Preface

Recent developments in scientific community have clearly indicated that speciation studies in biological and environmental matrices are much more remarkable than the total element determination due to the tremendous difference in bioavailability and toxicity of various chemical forms of a particular element. There has long been evidence that transport, distribution and bioavailability of the elements in any systems directly correspond to their chemical forms including different organometallic compounds and oxidation states. Hence, the number of scientists working on the topic of speciation has grown expeditiously, which provides substantial literature, to figure out the probable effects on human health, and to enhance the awareness of the importance of speciation analysis. Range of highly sensitive separation-detection techniques as well as hyphenated systems have been developed for the identification and quantification of the species present in a particular system at ultra-trace levels. Each technique has its own advantages and disadvantages with respect to precision, sensitivity and detection limit.

This book aims to evaluate speciation analysis in depth and present a comprehensive review of state-of-the-art analytical approaches and methodologies used for the speciation of elements in environmental samples. The book presents a survey of speciation of many elements and developments in the speciation area to date. Particular emphasis has been given to soil, sediment and environmental samples within the book. Most of the chapters are dedicated to various extraction, separation and detection methods as well as the applications of elemental speciation in variety of matrices. Throughout the chapters, examples of chromatographic and non-chromatographic analytical methods are presented for the speciation of different analytes.

The book, covering fifteen chapters, includes various topics such as aspects of speciation; sample pre-treatment methods for organometallic species determination; separation techniques for elemental speciation; role and importance of hyphenated techniques in speciation analysis; speciation of selenium, chromium, vanadium, thallium, antimony, arsenic, mercury, zinc, tin, iodine and tellurium in variety of matrices; trace elements, and human health. Encompassing the extensive coverage, this book will be of interest to researchers working on or entering the field of speciation.

I would like to thank my wife, Gülhan, and my small princess, Bengisu Ece, for their indefatigable moral support while preparing this book. I wish to thank Dr. M. Ferdi Fellah for the structures he drew for the cover page. I would also like to extend my most heartfelt thanks to Yildiz Technical University for supporting me in completing this book.

30th June 2013   **Sezgin Bakirdere**

Istanbul
Turkey

Contents

Preface   v

1. **Aspects of Speciation**   1
   Emrah Yıldırım and *Lütfiye Sezen Yıldırım*

2. **Sample Pre-treatment Methods for Organometallic**   19
   Species Determination
   Antonio Moreda-Piñeiro, Jorge Moreda-Piñeiro and
   Pilar Bermejo-Barrera

3. **Separation Techniques for Elemental Speciation in Soil,**   202
   Sediments, and Environmental Samples
   Márcia F. Mesko, Carla A. Hartwig, Cezar A. Bizzi,
   Edson I. Müller, Fábio A. Duarte and *Paola A. Mello*

4. **Role and Importance of Hyphenated Techniques in**   242
   Speciation Analysis
   Rajmund Michalski, Magdalena Jabłońska and *Sebastian Szopa*

5. **Selenium Speciation in the Environment**   263
   Rodolfo G. Wuilloud and *Paula Berton*

6. **Speciation of Chromium and Vanadium in Soil Matrices**   306
   Khakhathi L. Mandiwana and *Nikolay Panichev*

7. **Speciation and Solubility of Thallium in Low**   325
   Temperature Systems: Additional Aqueous and Solid
   Thallium Species Potentially Important in Soil
   Environments
   Yongliang Xiong

8. **Fractionation and Speciation Analysis of Antimony in**   341
   Atmospheric Aerosols and Related Matrices
   Patricia Smichowski

 9. **Speciation of Arsenic in Soil, Sediment and**   363
    Environmental Samples
    Selin Bora, Işıl Aydın, Ersin Kılınç and Fırat Aydın

10. **Methods for Mercury Speciation in Environmental**   390
    Samples
    Zhenli Zhu, Qian He and Zhifu Liu

11. **Zinc Speciation Studies in Soil, Sediment and**   433
    Environmental Samples
    Todd P. Luxton, Bradley W. Miller and Kirk G. Scheckel

12. **Speciation Analysis of Tin in Environmental Samples**   478
    Valderi Luiz Dressler, Clarissa Marques Moreira dos Santos,
    Fabiane Goldschmidt Antes, Erico Marlon de Moraes Flores and
    Dirce Pozebon

13. **Speciation and Bioavailability of Iodine in Edible Seaweed** 513
    Vanessa Romaris Hortas, Antonio Moreda Piñeiro and
    Pilar Bermejo Barrera

14. **Speciation and Determination of Tellurium in Water,**   527
    Soil, Sediment and other Environmental Samples
    M.S. El-Shahawi, H.M. Al-Saidi, E.A. Al-Harbi,
    A.S. Bashammakh and A.A. Alsibbai

15. **Trace Elements and Human Health**   545
    Mehrdad Gholami and Hojatollah Yamini

Index   599

Color Plate Section   603

Aspects of Speciation

Emrah Yıldırım[1],* and Lütfiye Sezen Yıldırım[2]

In this chapter, the authors aim to make a brief introduction to trace element analysis, history and development followed by a discussion on the need for speciation analysis and an introduction to definitions and terms. Basic principles and operations of chemical analysis process are discussed from the speciation point of view.

History of Spectroscopy and Trace Element Determination

Many people believe that the roots of spectroscopy date back to the 1600s, from the time of Sir Isaac Newton (Koirtyohann 1991). With his famous experimental setup, which was no more than a piece of black paper having a hole at the center and a simple glass prism, Newton was able to show that light is composed of waves. Later on, William Herschel discovered the infrared region of the spectrum in 1800 by the heating effect and a year later ultraviolet region was discovered by Johann Wilhelm Ritter. A careful examination of the sun's spectrum by the English scientist William Wollaston revealed that the sun's spectrum itself is not continuous but has some dark lines at certain positions. Thirteen years later, Jospeh von Fraunhofer carefully mapped the dark lines and named them. He realized

[1]Department of Chemistry, Middle East Technical University, 06800, Ankara/TURKEY.
Email: yemrah@metu.edu.tr
[2]Central Laboratory, Middle East Technical University, 06800, Ankara/TURKEY.
Email: ksezen@metu.edu.tr
*Corresponding author

that the reason for black lines was the absence of the corresponding waves but the cause was still unknown (Rust et al. 2005). The correct explanation was stated by Kirchhoff and Bunsen in the papers published in 1859 and 1860. Kirchhoff and Bunsen were able to see the interaction between matter and light by performing a series of careful experiments indicating the theory of absorption and emission (Koirtyohann 1991). The last component of the modern spectroscopy, namely scattering, was known after the works of Indian physicist Sir ChandrasekharaVenkata Raman in the 1920s (Gilbert 1999).

Among the spectroscopic techniques, atomic emission, atomic absorption and atomic fluorescence spectroscopy were the three that gave rise to the trace element analysis. The first practical AAS instrument was designed by Walsh and a team of Australian researchers in 1955 and announced to the world by the publication *"The application of atomic absorption spectra to chemical analysis"* (Walsh 1955). During the mid-20th century, arc and sparc spectroscopy were popular for the quantitative applications but had the drawback of requiring sample preparation and availability of the matched standards. Atomic absorption was gaining increasing popularity until the successful application of inductively coupled plasma atomic emission spectroscopy by Stanley Greenfield in 1962 (Mermet 2005). In 1980 Houk and his team took the plasma source one step further and coupled it to a mass spectrometer (Houk et al. 1980, Taylor 2001). This way lower detection limits were achieved with a simpler mass spectrum.

With the help of these three leading instruments, namely FAAS, ICP-OES and ICP-MS, trace element research gained more importance, numerous papers were published, and still continue to be published.

Trace Element Speciation

In the early stages of trace element analysis development, researchers mainly focused on the total amount of elements rather than the individual fraction of each form. This is partly because they thought that elements behave in a definite way at certain matrices regardless of their chemical form. The other reason is that most atomic techniques are destructive and they do not preserve the oxidation state or any other molecular information (Sanz-Medel 1998). This means, that in order to determine specific forms of an element, one must separate it from the others before it reaches an atomic detector.

As more information became available about the elements and their interaction in environmental and biological research, it was realized that the distribution, mobility and biological availability of chemical elements depends not simply on their concentration but, critically, on the chemical and physical association which they undergo in natural systems.

An interesting example which demonstrates the importance of speciation is the case of arsenic. It is well known that As is toxic even at trace levels. However, marine creatures, basically fish, contain total As at 1–70 mg/kg levels (dry weight) (Wrobel et al. 2002, Serafimovski et al. 2006, Shah et al. 2009). Without chemical speciation, one may easily draw a misleading conclusion that As content of marine organisms is more than the permissible limit, and hence consumption is detrimental to health. However, it is well known that the marine organisms have the ability to convert inorganic arsenic forms to organic ones, such as arsenobetaine, arsenosugars, arsenoribosides and arsenocholine, and these organic forms are essentially nontoxic compared to inorganic arsenic species (Maher and Butler 1988). LD_{50} values of some arsenic species are given in Table 1.1 (Morrison et al. 1989). LD_{50} is defined as the concentration leading to the death of 50 percent of a population in a study group consisted of mostly rats.

Another good example is chromium species. It is well known that trace amount of Cr(III) is essential for sugar and lipid metabolism, whereas Cr(VI) is very toxic and has no use in the body (Anderson 1998). According to World Health Organization (WHO), the maximum recommended concentration of Cr(VI) is 0.05 mg/L in drinking water (WHO 2003). If toxicity is the main concern, the amount of trivalent and hexavalent chromium species must be defined to make a more reliable judgment of a situation.

These examples make it evident that for molecular, as well as oxidation state, information is crucial for trace element determination. As this fact is realized in time, more effort will be expended for the speciation issue.

Table 1.1 LD_{50} values of some arsenic species of a selected rat population.

Species	LD_{50} value, mg/kg
Arsenite	14
Arsenate	20
Arsine	3
Monomethyl arsenic acid	700–1800
Dimethyl arsenic acid	700–2600
Arsenocholine	> 10000
Arsenobetaine	> 10000

Term and Definitions

In the literature various definitions are identified for the speciation analysis. In order to eliminate the confusion, in 2000, IUPAC published a guideline (explaining) leading to definitions and differences between speciation and fractionation (Templeton et al. 2000). According to the Guidelines, speciation analysis is defined as:

> *"the analytical activity of identifying and/or measuring the quantities of one or more individual chemical species in a sample."*

And speciation and fractionation are defined as:

> *"The **speciation** of an element is the distribution of an element amongst defined chemical species in a system."*

> *"**Fractionation** is the process of classification of an analyte or a group of analytes from a certain sample according to physical (e.g., size, solubility) or chemical (e.g., bonding, reactivity) properties."*

These definitions cover different cases such as isotopic abundances, different oxidation states, molecular and complex forms and macromolecular species, where the terms speciation and fractionation are applied.

In the book "Chemical Speciation in the Environment", Ure emphasized that the definitions given above are well applicable to the solutions but do not cover the studies dealing with soil, sediments, geochemical and biological samples (Ure and Davidson 2002). Ure suggested that a definition should cover the speciation:

a) functionally
b) operationally and
c) Different chemical compounds or oxidation states. He proposed the term for speciation as:

1. The process of identifying and quantifying the different, defined species forms or phases presents in a material
   or
2. The description of the amounts or kinds of these species forms or phases present.

Quality Control Procedures and Certified Reference Materials

Quality control and quality assurance protocols are the necessary components of an analytical procedure in order to provide valid and reliable data. For total element determination, analytes are usually converted to the most stable and easily analyzed form, called isoformation, in order to minimize losses due to precipitation, volatilization or adsorption of metastable forms. Isoformation also minimizes the uncertainty caused by different behavior of species in the analytical system that includes difference in nebulization efficiencies of organic-inorganic species, different transport and atomization profiles. The development of such protocols is especially difficult in the case of chemical speciation (Rosenberg and Ariese 2001) since such an approach is not valid for speciation work where retention of chemical diversity is needed. As a result, species should be conserved "as they are". This necessity makes the quality control and quality assurance more challenging due to low stability of target species, possibility of contamination/analyte losses during the whole process and over/under estimation of the quantity because of improper separation/detection. There is no generally accepted quality control-quality assurance procedure for speciation process since there are too many variables affecting the final result of a speciation process for different species and matrices. Therefore, each laboratory should develop its own way of validation and try to find comparable data with the others having different approaches to the same problem.

Certified reference materials (CRM) are the most accepted validation materials in analytical chemistry and widely used by many laboratories. There are many different CRM available for different matrices on total element determination. But in the case of speciation work, few are found in the market. There are several challenges for the production of CRM for speciation. Mainly, extraction efficiency of a method rarely reaches 100 percent and creates an uncertainty in the total amount present. Secondly, stability of the analyte is critical and largely depends on the storage conditions. And lastly, real world concentrations of analytes are usually low; hence uncertainty in the quantity is high. CRM's of Hg, As, Sn and Se in some matrices are available but these are very specific to analyte. In the following paragraphs, some of the difficulties and advice during a speciation process will be discussed.

Sampling and Storage

Speciation analysis often requires advanced techniques or a combination of them. As a result, *in vivo* analysis is not possible for most cases. Samples need to be collected and transported to the properly equipped laboratories. The first step in an *in vitro* speciation process is the collection of the sample containing the analyte species from the sampling site. This stage is followed by proper storage of the sample till the next analytical process or between any analytical/physical processes. These processes are extremely important in a speciation study since improper sampling and storage may destroy the whole effort.

Sampling

A key parameter in the sampling is not to disturb the conditions critical for interspecies-intraspecies conversion. Critical conditions in the analyte equilibrium should be defined before the sampling process and sample collection must be designed accordingly. Among many parameters, temperature, pH, ionic strength, presence of oxidizing-reducing agents or any bacterial action should be considered. It is always advised to read the literature and the properties of the sample in order to find optimal sampling conditions.

Prior to collection

Sampling time is a critical factor by itself. For some cases, sampling area is so dynamic that composition may vary within seasons, days or even hours (Gonçalves et al. 1994, Li and Zhang 2010, Okuda et al. 2004, Kulshrestha et al. 2009, Lillebø et al. 2010). This is especially true for water bodies and air samples. The flow of a river, for instance, is continuously changing depending on the season, weather conditions and precipitation profile of the area. These parameters change the analytical composition as well as other environmental parameters that have an impact on the interconversion or intraconversion of species, such as pH or ionic strength. The detailed study of the changes going on in the sampling site is necessary in order not to over- or under-estimate the amount of specific species.

The sampling site should be selected according to the purpose of the conducted study. The distance of the local or global sources, including natural and anthropogenic ones, transport efficiencies between the sampling site and sources, impact of the local variables determine the sampling site and sampling frequency. For an uncontrolled environmental study, sampling frequency and number of sampling points will be as much as possible. This is mostly determined by the availability of the staff, budget of the study and the location of the sampling site.

To assess the homogeneity of the samples, representativeness must be high. Representativeness of a sample refers to how closely a sample reflects the characteristic of a certain population (Kahneman and Tversky 1972). A good sampling strategy greatly increases the representativeness of the sample to be collected. It is especially important for the solid samples since homogeneity is difficult to obtain. Increasing the amount of sample and collecting samples from different points of a large body greatly increase the representativeness.

Collection

When all the parameters regarding the sampling time, place and frequency have been determined, it is time to collect the samples. The sampling apparatus has to be cleaned beforehand. For most cases, this can be done by simply soaking in a suitable cleaning solution for at least 24 hr. PTFE, PEEK, high density polyethylene, polypropylene and glass materials can be cleaned with dilute (5 to 10% (v/v)) nitric acid, rinsed with plenty of deionized water and dried preferably in a clean room or under a laminar flow (Reimann et al. 1999). When the sample apparatus is dry it should be kept in airtight containers till the sample collection. If any biological study is conducted or there is a risk for biological contamination, all the apparatus must be autoclaved and treated with alcohol. Any metal device must not be treated with corrosive acids and contact of metal devices with speciation samples should be avoided. Solid samples, such as soil or sediment, may be collected by means of high density plastic or PTFE shovels or spoons. If this kind of material is not available, metal apparatus coated with plastic layers may be used. This layer should be strong enough and durable throughout the sampling process. Sometimes, it is necessary to eliminate the large particles such as stones or plant remnant, by hand or by the use of a sieve with a mesh size usually smaller than 2 mm. This step is crucial since grounding of stones or other materials change the composition of the sample. One of the biggest challenges with the soil samples is homogeneity. This can be minimized by collecting a large amount of heterogeneous sample, mixing thoroughly and random sampling from the mixture. As the number of rehomogenation and sampling steps increase, representativeness of the system is greatly increased. Sampling should be done without excessive visualization of the sample. This technique minimizes the operator's preference about the sample.

Water samples are collected by using polycarbonate, polypropylene, PTFE and high density polyethylene containers. The contamination or loss of analyte due to adsorption on sampling vessels is more severe for liquid samples since they have a large contact area with vessel walls. It is a good practice to note the physical conditions of the sample including clarity, cloudiness, color, smell, foam, temperature and pH at the time of sampling for further studies. For the total element determination, acidification of the sample is advised but it must not be done for speciation work because species distribution largely depends on the pH of the medium. If the sample is collected from a running water stream, the container mouth should be faced upstream and it should be filled with the sample and discharged away from the sampling site several times in order to remove any residue in the sample containers. Moreover, the containers should be filled completely; otherwise volatile forms will be transferred to the gas phase and may be lost when the container is opened. For some cases, especially water bodies with dynamic flows, sampling depth is critical and should be precisely controlled and recorded. For rivers and shallow waters, debris, sediment or other suspended particles may be present in the water. These particles may absorb analyte atoms or desorb interfering components to the sample. The collection of these large particles should be avoided in these cases. If necessary, inert filters can be capped to the sample container so as to keep these particles out.

Storage

Storage is another key parameter towards the reliable analysis of a sample. It covers the time after sampling till the analysis or between any mechanical or chemical process. The main aim is to minimize the loss of any analyte species, prevent contamination and avoid conversion of species to other forms especially for speciation studies or keeping the sample *"as it is"*. There are general precautions that one should keep in mind but the overall process is highly sample- and matrix-dependent. The ideal conditions for sample storage should be optimized and the integrity of the procedure should be validated.

Liquid samples are put into amber containers or colored containers and stored in the dark in order to prevent any photochemical process. Another advantage is the prevention of the formation of algae or other light-sensitive organisms. Suspended particles may dissolve and give ions or they may adsorb analyte and settle down to create a homogeneity problem. In order to remove suspended solids, deposit, algae and other microorganisms, water samples are usually filtered through 0.20 or 0.45 μm membrane filters (Heumann et al. 2004). Suspended particles may be analyzed separately if desired. Liquid samples are dynamic and very susceptible to transformation. Therefore, storage periods should be kept as short as possible. For some cases, analyte species may be extracted from the sample and stored as extracted. This approach is useful if the interconversion of species is relatively easy.

The formation of algae or other bacterial actions is also possible for solid samples. Wet storage of solid samples, such as soil or sediment, leads to microbial actions and conditions change from oxidizing to reducing (Ure and Davidson 2002), which may lead to various element transformation. Sterilization by γ irradiation may be used to minimize microbial based changes (Bartlett and James 1980). Bartlett and James studied the storage conditions of soil samples and found that air-drying of soil caused major changes in the soil composition. Also, remoisturing the sample did not return it back to the metastable form or it caused unpredictable changes. If samples are only to be stored for a short time (for a week), they are best stored at the initial moisture condition. This is especially important for speciation work since volatile or semi volatile species may be lost during the drying stage.

Another key parameter during storage is the temperature. The use of low temperature seems to be the safest method since low temperature decreases kinetics of a process. Liquid samples are mostly stored at cold but above freezing point (2–5°C) for a short while, or below freezing temperatures (–20°C or –80°C) for a relatively long time (in the order of weeks). Deep freezing seems to be feasible but the recovery and rehomogenization of the sample should be considered beforehand and organized accordingly. The sample may freeze inhomogeneously resulting in different concentration throughout the whole body; hence the sample should be recovered completely rather than small pieces. The homogeneity problem is more severe for solid samples, which should be homogenized before their storage at low temperatures. Preferably, they should be stored as separate homogeneous numbers for each individual process and once recovered; they should not be deep frozen again.

Biological or solid samples can be freeze-dried and stored at cold temperatures. Freeze-drying is a useful process that removes the water molecules leaving a water free sample, the life time of many species are extended by this way and any water based reactions are avoided. Freeze-dried samples are hydroscopic, hence they need to be stored at humidity controlled conditions (Heumann et al. 2004). However, some volatile species may be lost during the freeze-drying process. Hjorth studied the effect of freeze drying on the composition of major elements and trace metals in lake sediments. The results indicated that freeze drying does not preserve the chemical speciation pattern of major elements, and even Si and Al, which are considered as immobile, were affected by the process (Hjorth 2004).

Containers

There are hundreds of materials that one can use as a sample container, but not all of them are suitable for storing a sample for speciation analysis. Identification of the sample and the container material should be analyzed and possible reactions should be considered. Sample size is to be selected to minimize the contact area between the sample and the container surface. Apparently, a metal container is not a good choice for the speciation study since metals are usually more reactive towards chemicals, especially acidic or basic ones. Even though there are less reactive metallic alloys present, stainless steel is still susceptible to corrosion at an atomic level. The corrosion of the metal may lead to transfer of metal ions to the sample or the sample may react with the metal resulting in a change of composition. This situation is much more severe for samples collected for speciation studies.

Glass is an alternative to metal. Regarding the reactivity, glass is much less reactive than metals. It is composed mostly of silicon, aluminum and some dopants added for specific purposes. There are hundreds of types of glass available in the market. Among them, borosilicate glass is the most frequently used one in the field of chemistry due to its resistance to high temperature (softening point of about 800°C), good chemical durability and thermal shock resistance (Smallman and Ngan 2007). However, it is still not as inert as one may think. Glass surface has an active silanol group that can act as an ion exchanger transferring ions with the sample (Gardiner 1993).

Plastic containers are more inert compared to others; hence, they are widely used as sample containers. High density polymers have low porosity, high resistance to chemicals and mechanical stress. However, some metals are used in the production process of such materials, such as Cr or Ni. It was recently found that some high density polyethylene (HDPE) containers found in the market released a significant amount of Ba, Sr and Zn to the sample solution (Reimann et al. 2007). Therefore, the leaching test using different solvents should be performed beforehand to evaluate any possible contamination.

Sample Preparation

Preparation of the sample for the analysis is another critical point. The number of preparation steps must be as low as possible in order to decrease the possibility of sample loss and contamination. If possible, direct analysis is advised. But this is not feasible for most cases. After recovery of the sample from the storage conditions, analyte species are extracted to another medium, mainly water based or organic solvents. Depending on the nature of the analyte as well as the sample, there are several extraction procedures mentioned in the literature. Detailed extraction procedures for specific analytes and media will be discussed in detail in the following chapters, basically liquid-liquid extraction (El-Shahawi et al. 2007, Ying et al. 2011, Pena-Pereira et al. 2009), liquid solid extraction (Narin et al. 2008, Erdoğan et al. 2011) and enzymatic extraction (Pardo-Martinez 2001).

Selection of an extraction solvent strongly depends on the analyte. Some organic molecules may be extracted with high efficiency using DI water (Huerta et al. 2005, Ochsenkühn-Petropoulou et al. 2003), whereas ionic species mainly require acidic (Mir et al. 2007, Zhang and Frankenberger Jr. 2001) or alkaline (Séby et al. 1997, Mir et al. 2007) solution. Molecules bonded to proteins require enzymatic extraction (Moreda-Piñeiro et al. 2010, Pardo-Martínez et al. 2001, Reyes et al. 2009). In order to shorten the extraction time and increase efficiency, these methods are usually combined with microwave irradiation (García Salgado et al. 2006, Wang et al. 2008,

Jamali et al. 2009, Arain et al. 2008), ultrasonic or mechanical shaking (Huerga et al. 2005, Montes-Bayón et al. 2006, López et al. 2010, Sanz et al. 2005, Pérez-Cid et al. 1998), pressurized liquid (Moscoso-Pérez et al. 2008, Moreda-Piñeiro et al. 2010, Alonso-Rodríguez et al. 2006, Chiron et al. 2000) or supercritical fluids (Bayona 2000, Bayona and Cai 1994, Kumar et al. 1993, Foy and Pacey 2000, Lorenzo et al. 1999).

Here the main aim is to extract target species as much as possible while minimizing the extraction of others and interferents (Rosenberg and Ariese 2001). Moreover, it is necessary to control the process in order to keep the species as they are. This is the most difficult of all. If one cannot extract the whole analyte in a certain matrix, which is the usual case, the determination of the extraction efficiency is vital for valid quantitative data. There are several methods to estimate extraction efficiency, but none of them are ideal with certain weaknesses.

The use of certified reference materials is the recommended approach for efficiency calculations, but very few numbers of certified materials related with speciation analysis are available to resemble the concentration levels of real world samples.

The spiking experiment is the most widely used alternative. It includes the injection of the target species to the matrix at a high concentration and extracting it back. This approach is particularly useful for simple matrices, such as environmental water samples. However, spiked species may not bond to the more complex matrices as the naturally occurring ones do. This is especially true for biological matrices where the interaction schemes are complicated. In this case, erroneous efficiencies may be calculated.

Quantification

Quantification of the analyte molecules or ions can be done either by chromatographic separation followed by detection or non-chromatographic separation. Non-chromatographic speciation makes use of the different physical, chemical or kinetic behavior of individual species. One popular way is to make use of difference in hydride generation efficiency of different forms of an element. Hydride forming efficiency of Te(IV), for instance, is much higher compared to Te(VI) at the same conditions. For the determination of Te(IV), samples were acidified and analyzed directly without applying a reduction procedure. For the total Te determination, a reduction procedure was applied. The difference between the total Te and Te(IV) gave the concentration of Te(VI) species in the sample, since Te(IV) and Te(VI) forms are known to be present as the dominant inorganic Te species in the nature (D'Ulivo 1997).

Another non-chromatographic speciation approach is the selective sorption of a species on a solid phase sorbent. This method, similar to the hydride generation case mentioned above, relies on the different absorption behavior of chemical forms. Garbos et al. (1997) used Polyorgs 31 complexing sorbent to quantify Sb(III) and Sb(V). By controlling pH, they were able to separate and quantify Sb(III) and Sb (V) (Garbos et al. 1997). Detection is either conducted online or offline. In both cases, no modification in the detection system is needed hence compatibility is not a big issue when compared with chromatographic speciation with online detection.

The non-chromatographic speciation approach is a fast, reliable and low-cost method, but this approach is strongly limited to certain forms of elements, mostly inorganic forms, and they do not yield comprehensive information about all species for most of the cases. Selenium, for instance, has many different organic forms as well as inorganic ones (D'Ulivo 1997). The non-chromatographic speciation approach fails to give information about all species hence the whole picture cannot be visualized. They are useful when the sample medium is not complex and certain forms dominate the system. This is particularly true for natural water samples where inorganic forms are mostly present (Kumar and Riyazuddin 2007).

Chromatographic separation makes use of a separation unit, namely gas chromatography (GC), liquid chromatography (LC or HPLC), capillary electrophoresis (CE), supercritical fluid chromatography (SCF) and thin layer chromatography (TLC), in combination with a detection system. As detectors, mainly element specific detectors such as atomic absorption spectrometry (AAS), atomic fluorescence spectrometry (AFS), inductively coupled plasma optical emission spectrometry (ICP-OES), inductively coupled plasma-mass spectrometry (ICP-MS) and microwave induced plasma spectrometry (MIPS) are used.

There are different possible combinations among these instruments. Some are readily compatible with each other while others need special connections. GC is mostly used for volatiling species or species that can be derivatized to volatile and thermally stable forms. The transfer line between the column outlet and detector inlet should be heated homogeneously (Ellis and Roberts 1997). The formation of any cooled areas leads to sample deposition and disturbed peak shapes. Furthermore, the connection has to be airtight and designed to have minimum dead volume. GC-AAS (Dirkx et al. 1995, Ritsema et al. 1998), GC-AES (Dirkx et al. 1995, Cui et al. 2011) and GC-ICP-MS (Mishra et al. 2005, Uveges et al. 2007, Krystek et al. 2012, Buenno and Pannier 2009) have been readily applied in the literature. Even though LC is more popular in speciation works, sample transport efficiency, which is usually 5–10 percent for conventional nebulizers, may reach to 100 percent when the gas outlet of GC is directly connected to the atomizer bypassing the nebulizer unit. This is particularly useful for AAS since sensitivity is low for speciation work.

Among different combinations, coupling of liquid chromatography with ICP-MS is the most popular system since both use compatible flow rates and no special interface is necessary. A PTFE, PEEK or tygon tubing is placed between the outlet of the analytical column and sample introduction system of the detection unit. The connection diameter and the length of the tubing have a great influence on the resolution of the hyphenated system. Due to the design of ICP plasma, water based mobile phases with low dissolved salt content (usually below 1 percent) should be used for the separation process. As the organic content and dissolved solid amount of the mobile phase increase the plasma instability, carbon build-up, clogging of sampler, skimmer cones and plasma overload are observed (Gettar et al. 2000). The introduction of oxygen to the plasma, cooling spray chamber and increasing RF power are used to decrease the negative effect of the mobile phase (de Leon et al. 2002).

Among the different modes, reversed phase, ion chromatography and ion pairing chromatography are the most widely used (Montes-Bayon et al. 2003). Ion pairing chromatography has the advantage of separating both the ionic and molecular species at a single run whereas the reversed phase is mainly used for molecular forms and ion chromatography is used for ionic species.

Another sample introduction technique gaining more popularity is the LC separation followed by post column chemical vapor generation (CVG) (Arslan et al. 2011). This technique has the advantage of separating the analyte species from the sample matrix and mobile phase; hence, it allows the use of high organic and salt containing mobile phases. Moreover, analyte transport efficiency is much higher compared to conventional nebulization techniques. The sensitivity of the system can be increased by this way. Thus, use of low cost AAS and AFS systems as detectors becomes feasible. However, It is known that only certain forms of elements are capable of forming volatile species. For instance, among other organic and inorganic forms, only As(III) is capable of forming volatile species on the reduction with a reducing agent with high efficiency. As a result, organic forms must be converted to As(III) before the CVG unit. In this case, online oxidation/ reduction systems are used. The most popular approaches include UV irradiation, microwave assisted oxidation/reduction and use of strong oxidants. The details of the process are given elsewhere (Arslan et al. 2011).

The application of each system to specific element will be discussed in detail in the proceeding chapters.

Legislation

Legislation concentrates more on toxicological issues regarding the speciation issue. The current situation in the legislation is basically on total element concentration rather than specific forms of elements. As discussed below, there are several factors affecting this situation.

First of all, the necessary data regarding the chemistry, toxicology and transport of different forms of an element in short-, mid- and long-term interval is not well detailed. Many researchers are studying these effects trying to create a database regarding these issues. Documenting every possibility in detail is a rather long process. After the creation of reliable and multi-dimensional information, reviewing this data set will be very beneficial for setting values for specific species.

Another issue is the technological drawbacks. Speciation is a rather time-consuming process that requires advanced instrumentation and expert staff. For most cases, use of a chromatographic separation unit hyphenated with an element specific detector is necessary. Operation of such a system requires experience and knowledge and the development of easily applicable and a species specific method is required together with the validation of the analytical methods.

The stability of the species is another issue. For cases where inter-conversion of species is relatively easy, parameters such as pH, temperature, reducing/oxidizing environment or ionic strength should not be changed or should be limited to certain safe levels during sampling, storage and sample preparation stages. This can be greatly enhanced by the development of *in situ* methods.

References

Alonso-Rodríguez, E., J. Moreda-Piñeiro, P. López-Mahía, S. Muniategui-Lorenzo, E. Fernández-Fernández, D. Prada-Rodríguez, A. Moreda-Piñeiro, A. Bermejo-Barrera and P. Bermejo-Barrera. 2006. Pressurized liquid extraction of organometals and its feasibility for total metal extraction. Trac-Trend. Anal. Chem. 25: 511–519.

Anderson, R.A. 1998. Chromium, glucose intolerance and diabetes. J. Am. Coll. Nutr. 17: 548–555.

Arain, M.B., T.G. Kazi, M.K. Jamali, N. Jalbani, H.I. Afridi and J.A. Baig. 2008. Speciation of heavy metals in sediment by conventional, ultrasound and microwave assisted single extraction methods: A comparison with modified sequential extraction procedure. J. Hazard. Mater. 154 : 998–1006.

Arslan Y., E. Yildirim, M. Gholami and S. Bakirdere. 2011. Lower limits of detection in speciation analysis by coupling high-performance liquid chromatography and chemical-vapor generation. Trac-Trend. Anal. Chem. 30: 569–585

Bartlett, R. and B. James. 1980. Studying dried, stored soil samples—some pitfalls. Soil Sci. Soc. Am. J. 44: 721–724.

Bayona, J.M. 2000. Supercritical fluid extraction in speciation studies. Trac-Trend. Anal. Chem. 19: 107–112.

Bayona, J.M. and Y. Cai. 1994. The role of supercritical fluid extraction and chromatography in organotin speciation studies. Trac-Trend. Anal. Chem. 13: 327–332.

Bueno, M. and F. Pannier. 2009. Quantitative analysis of volatile selenium metabolites in normal urine by headspace solid phase microextraction gas chromatography–inductively coupled plasma mass spectrometry. Talanta. 78. 759–763.

Chiron, S., S. Roy, R. Cottier and R. Jeannot. 2000. Speciation of butyl- and phenyltin compounds in sediments using pressurized liquid extraction and liquid chromatography–inductively coupled plasma mass spectrometry. J. Chromatogr. A 879: 137–145.

Cui, Z., K. Zhang, Q. Zhou, J. Liu and G. Jiang. 2011. Determination of methyltin compounds in urine of occupationally exposed and general population by in situ ethylation and headspace SPME coupled with GC-FPD. Talanta. 85: 1028–1033.

Dirkx, W.M.R., R. Lobinski and F.C. Adams. 1995. Speciation analysis of organotin by GC-AAS and GC-AES after extraction and derivatization. Techniques and Instrumentation in Analytical Chemistry. 11: 357–409.

Ellis, L.A. and D.J. Roberts. 1997. Chromatographic and hyphenated methods for elemental speciation analysis in environmental media. J. Chromatogr. A 774: 3–19.

D'Ulivo, A. 1997. Determination of selenium and tellurium in environmental samples. Analyst. 122: 117R–144R.

El-Shahawi, M.S., A.S. Bashammakh and S.O. Bahaffi. 2007. Chemical speciation and recovery of gold(I, III) from wastewater and silver by liquid–liquid extraction with the ion-pair reagent amiloride mono hydrochloride and AAS determination. Talanta. 72: 1494–1499.

Erdoğan, H., Ö. Yalçınkaya and A.R. Türker. 2011. Determination of inorganic arsenic species by hydride generation atomic absorption spectrometry in water samples after preconcentration/separation on nano ZrO_2/B_2O_3 by solid phase extraction. Desalination. 280: 391–396.

Foy, G.P. and G.E. Pacey. 2000. Specific extraction of chromium(VI) using supercritical fluid extraction. Talanta. 51: 339–347.

Garbos, S., E. Bulska, A. Hulanicki, N.I. Shcherbinina and E.M. Sedykh. 1997. Preconcentration of inorganic species of antimony by sorption on Polyorgs 31 followed by atomic absorption spectrometry detection. Anal. Chim. Acta. 342: 167–174.

García Salgado, S., M.A. Quijano Nieto and M.M. Bonilla Simón. 2006. Determination of soluble toxic arsenic species in alga samples by microwave-assisted extraction and high performance liquid chromatography-hydride generation-inductively coupled plasma-atomic emission spectrometry. J. Chromatogr. A 1129: 54–60.

Gardiner, P.E. 1993. Considerations in the preparation of biological and environmental reference materials for use in the study of the chemical speciation of trace-elements. Fresenius' J. Anal. Chem. 345: 287–290.

Gettar, R.T., R.N. Garavaglia, E.A. Gautier and D.A. Batistoni. 2000. Determination of inorganic and organic anionic arsenic species in water by ion chromatography coupled to hydride generation–inductively coupled plasma atomic emission spectrometry. J. Chromatogr. A 884: 211–221.

Gilbert, A.S. 1999. Vibrational, Rotational and Raman Spectrocopy, Historical Perspective. *In:* J.C. Lindon, J.L. Holmes and G.E. Tranter [eds.]. Encyclopedia of Spectroscopy and Spectrometry. 2nd edn. Academic Press. Oxford. pp. 2938–2948.

Gonçalves, E.P.R., H.M.V.M. Soares, R.A.R. Boaventura, A. A.S.C. Machado and J.C.G. Esteves da Silva. 1994. Seasonal variations of heavy metals in sediments and aquatic mosses from the Cávado river basin (Portugal). Sci. Total Environ. 142: 143–156.

Heumann, K., R. Cornelis, J. Caruso and H. Crews. 2004. Handbook of Elemental Speciation —Techniques and Methodology. John Wiley & Sons. Chichester.

Helmers, E. 1997. Sampling of sea and fresh water for the analysis of trace elements. *In:* M. Stoeppler [ed.]. Sampling and Sample Preparation. Springer, Berlin, Germany, pp. 26–42.

Hjorth, T. 2004. Effects of freeze-drying on partitioning patterns of major elements and trace metals in lake sediments. Anal. Chim. Acta. 526: 95–102.

Houk, R.S., V.A. Fassel, G.D. Flesch, H.J. Svec, A.L. Gray and C.E. Taylor. 1980. Inductively coupled argon plasma as an ion-source for mass-spectrometric determination of trace-elements. Anal. Chem. 52: 2283–2289.

Huerga, A., I. Lavilla and C. Bendicho. 2005. Speciation of the immediately mobilisable As(III), As(V), MMA and DMA in river sediments by high performance liquid chromatography-hydride generation-atomic fluorescence spectrometry following ultrasonic extraction. Anal. Chim. Acta. 534: 121–128.

Huerta, V.D., M.L.F. Sánchez and A. Sanz-Medel. 2005. Qualitative and quantitative speciation analysis of water soluble selenium in three edible wild mushrooms species by liquid chromatography using post-column isotope dilution ICP-MS. Anal. Chim. Acta. 538: 99–105.

Jamali, M.K., T.G. Kazi, M.B. Arain, H.I. Afridi, N. Jalbani, G.A. Kandhro, A.Q. Shah and J.A. Baig. 2009. Speciation of heavy metals in untreated sewage sludge by using microwave assisted sequential extraction procedure. J. Hazard. Mater. 163: 1157–1164.

Kahneman, D. and A. Tversky. 1972. Subjective probability: A judgment of representativeness. Cogn. Psychol. 3: 430–454.

Koirtyohann, S.R. 1991. A history of atomic-absorption spectrometry from an academic perspective. Anal. Chem. 63: A1024–A1031.

Krystek, P., P. Favaro, P. Bode and R. Ritsema. 2012. Methyl mercury in nail clippings in relation to fish consumption analysis with gas chromatography coupled to inductively coupled plasma mass spectrometry: A first orientation. Talanta. 97: 83–86.

Kulshrestha, A., P.G. Satsangi, J. Masih and A. Taneja. 2009. Metal concentration of PM2.5 and PM10 particles and seasonal variations in urban and rural environment of Agra, India. Sci. Total Environ. 407: 6196–6204.

Kumar, A.R. and P. Riyazuddin. 2007. Non-chromatographic hydride generation atomic spectrometric techniques for the speciation analysis of arsenic, antimony, selenium, and tellurium in water samples—a review. Intern. J. Environ. Anal. Chem. 87: 469–500.

Kumar, U.T., N.P. Vela, J.G. Dorsey and J.A. Caruso. 1993. Supercritical fluid extraction of organotins from biological samples and speciation by liquid chromatography and inductively coupled plasma mass spectrometry. J. Chromatogr. A 655: 340–345.

de León, C.A.P., M. Montes-Bayon and J.A. Caruso. 2002. Elemental speciation by chromatographic separation with inductively coupled plasma mass spectrometry detection. J. Chromatogr. A 974: 1–21.

Li, S. and Q.F. Zhang. 2010. Risk assessment and seasonal variations of dissolved trace elements and heavy metals in the Upper Han River, China. J. Hazard. Mater. 181: 1051–1058.

Lillebø, A.I., M. Válega, M. Otero, M.A. Pardal, E. Pereira and A.C. Duarte. 2010. Daily and inter-tidal variations of Fe, Mn and Hg in the water column of a contaminated salt marsh: Halophytes effect. Estuarine, Coastal Shelf Sci. 88: 91–98.

López, I., S. Cuello, C. Cámara and Y. Madrid. 2010. Approach for rapid extraction and speciation of mercury using a microtip ultrasonic probe followed by LC-ICP-MS. Talanta. 82: 594–599.

Lorenzo, R.A., M.J. Vázquez, A.M. Carro and R. Cela. 1999. Methylmercury extraction from aquatic sediments: A comparison between manual, supercritical fluid and microwave-assisted techniques. Trac-Trend. Anal. Chem. 18: 410–416.

Maher, W. and E. Butler. 1988. Arsenic in the marine environment. Appl. Organomet. Chem. 2: 191–214.

Mermet, J.M. 2005. Atomic Emission Spectrometry Inductively Coupled Plasma. *In:* P.T. Alan and P. Colin [eds.]. Encyclopedia of Analytical Science. 2nd edn. Elsevier. Oxford. pp. 210–215.

Mir, K.A., A. Rutter, I. Koch, P. Smith, K.J. Reimer and J.S. Poland. 2007. Extraction and speciation of arsenic in plants grown on arsenic contaminated soils. Talanta. 72: 1507–1518.

Mishra, S., R.M. Tripathi, S. Bhalke, V.K. Shukla and V.D. Puranik. 2005. Determination of methylmercury and mercury(II) in a marine ecosystem using solid-phase microextraction gas chromatography-mass spectrometry. Anal. Chim. Acta. 551: 192–198.

Montes-Bayón, A., K. DeNicola and J.A. Caruso. 2003. Liquid chromatography–inductively coupled plasma mass spectrometry. J. Chromatog. A 1000: 457–476.

Montes-Bayón, M., M.J.D. Molet, E.B. González and A. Sanz-Medel. 2006. Evaluation of different sample extraction strategies for selenium determination in selenium-enriched plants (Allium sativum and Brassica juncea) and Se speciation by HPLC-ICP-MS. Talanta. 68: 1287–1293.

Moreda-Piñeiro, J., E. Alonso-Rodríguez, A. Moreda-Piñeiro, C. Moscoso-Pérez, S. Muniategui-Lorenzo, P. López-Mahía, D. Prada-Rodríguez and P. Bermejo-Barrera. 2010. Simultaneous pressurized enzymatic hydrolysis extraction and clean up for arsenic speciation in seafood samples before high performance liquid chromatography–inductively coupled plasma-mass spectrometry determination. Anal. Chim. Acta. 679: 63–73.

Morrison, G.M.P., G.E. Batley and T.M. Florence. 1989. Metal speciation and toxicity. Chem. Br. 25: 791–&.

Moscoso-Pérez, C., J. Moreda-Piñeiro, P. López-Mahía, S. Muniategui-Lorenzo, E. Fernández-Fernández and D. Prada-Rodríguez. 2008. Pressurized liquid extraction followed by high performance liquid chromatography coupled to hydride generation atomic fluorescence spectrometry for arsenic and selenium speciation in atmospheric particulate matter. J. Chrom. A 1215: 15–20.

Narin, I., A. Kars and M. Soylak. 2008. A novel solid phase extraction procedure on Amberlite XAD-1180 for speciation of Cr(III), Cr(VI) and total chromium in environmental and pharmaceutical samples. J. Hazard. Mater. 150: 453–458.

Ochsenkühn-Petropoulou, M., B. Michalke, D. Kavouras and P. Schramel. 2003. Selenium speciation analysis in a sediment using strong anion exchange and reversed phase chromatography coupled with inductively coupled plasma-mass spectrometry. Anal. Chim. Acta. 478: 219–227.

Okuda, T., J. Kato, J. Mori, M. Tenmoku, Y. Suda, S. Tanaka, K. He, Y.L Ma, F. Yang, X.C. Yu, F.K. Duan and Y. Lei. 2004. Daily concentrations of trace metals in aerosols in Beijing, China, determined by using inductively coupled plasma mass spectrometry equipped with laser ablation analysis, and source identification of aerosols. Sci. Total Environ. 330: 145–158.

Pardo-Martınez, M., P. Viñas, A. Fisher and S.J. Hill. 2001. Comparison of enzymatic extraction procedures for use with directly coupled high performance liquid chromatography-inductively coupled plasma mass spectrometry for the speciation of arsenic in baby foods. Anal. Chim. Acta. 441(1): 29–36.

Pena-Pereira, F., I. Lavilla and C. Bendicho. 2009. Miniaturized preconcentration methods based on liquid–liquid extraction and their application in inorganic ultratrace analysis and speciation: A review. Spectrochim. Acta, Part B. 64: 1–15.

Pérez-Cid, B., I. Lavilla and C. Bendicho. 1998. Speeding up of a three-stage sequential extraction method for metal speciation using focused ultrasound. Anal. Chim. Acta. 360: 35–41.

Reimann, C., A. Grimstvedt, B. Frengstad and T.E. Finne. 2007. White HDPE bottles as source of serious contamination of water samples with Ba and Zn. Sci. Total Environ. 374: 292–296.

Reimann, C., U. Siewers, H. Skarphagen and D. Banks. 1999. Does bottle type and acid-washing influence trace element analyses by ICP-MS on water samples? A test covering 62 elements and four bottle types: high density polyethene (HDPE), polypropene (PP), fluorinated ethene propene copolymer (FEP) and perfluoroalkoxy polymer (PFA). Sci. Total Environ. 239: 111–130.

Reyes, L.H., J.L. Guzmán Mar, G.M.M. Rahman, B. Seybert, T. Fahrenholz and H.M.S. Kingston. 2009. Simultaneous determination of arsenic and selenium species in fish tissues using microwave-assisted enzymatic extraction and ion chromatography–inductively coupled plasma mass spectrometry. Talanta. 78: 983–990.

Ritsema, R., T. de Smaele, L. Moens, A.S. de Jong and O.F.X. Donard.1998. Determination of butyltins in harbour sediment and water by aqueous phase ethylation GC-ICP-MS and hydride generation GC-AAS. Environ. Pollut. 99: 271–277.

Rosenberg E. and F. Ariese. 2001. Quality control in speciation analysis. *In:* L. Ebdon, L. Pitts, R. Cornelis, H. Crews, O.F.X. Donard and P. Quevauviller [eds.]. Trace Element Speciation for Environment, Food and Health. Royal Society of Chemistry, Cambridge, UK, pp. 17–50.

Rust, J.A., J.A. Nobrega, C.P. Calloway and B.T. Jones. 2005. Fraunhofer effect atomic absorption spectrometry. Anal. Chem. 77: 1060–1067.

Sanz, E., R. Muñoz-Olivas and C. Cámara. 2005. A rapid and novel alternative to conventional sample treatment for arsenic speciation in rice using enzymatic ultrasonic probe. Anal. Chim. Acta. 535(1–2): 227–235.

Sanz-Medel, A. 1998. Trace element analytical speciation in biological systems: importance, challenges and trends. Spectrochim. Acta, Part B. 53: 197–211.

Serafimovski, I., I.B. Karadjova, T. Stafilov and D.L. Tsalev. 2006. Determination of total arsenic and toxicologically relevant arsenic species in fish by using electrothermal and hydride generation atomic absorption spectrometry. Microchem. J. 83: 55–60.

Shah, A.Q., T.G. Kazi, M.B. Arain, M.K. Jamali, H.I Afridi, N. Jalbani, J.A. Baig and G.A. Kandhro. 2009. Accumulation of arsenic in different fresh water fish species— potential contribution to high arsenic intakes. Food Chem. 112: 520–524.

Smallman, R.E. and A.H.W. Ngan. 2007. Non-metallics I—Ceramics, glass, glass-ceramics. *In:* Physical Metallurgy and Advanced Materials Engineering. 7th edn. Butterworth-Heinemann. Oxford. pp. 513–548.

Séby, F., M. Potin Gautier, G. Lespés and M. Astruc. 1997. Selenium speciation in soils after alkaline extraction. Sci. Total Environ. 207(2–3): 81–90.

Taylor, H.E. 2001. Inductively Coupled Plasmas. *In:* Inductively Coupled Plasma-Mass Spectrometry. Academic Press. San Diego. pp. 15–27.

Templeton, D.M., F. Ariese, R. Cornelis, L.G. Danielsson, H. Muntau, H.P. Van Leeuwen and R. Lobinski. 2000. Guidelines for terms related to chemical speciation and fractionation of elements. Definitions, structural aspects, and methodological approaches. Pure Appl. Chem. 72: 1453–1470.

Ure, A. and C. Davidson. 2002. Chemical Speciation in the environment. 2nd edition. Wiley-Blackwell Science. Oxford, pp. 1–4.

Uveges, M., L. Abranko and P. Fodor. 2007. Optimization of GC–ICPMS system parameters for the determination of butyltin compounds in Hungarian freshwater origin sediment and mussel samples. Talanta. 73: 490–497.

Walsh, A. 1955. The Application of Atomic Absorption Spectra to Chemical Analysis. Spectrochim. Acta. 7: 108–117.

Wang, X.P., H.Y. Jin, L. Ding, H.R. Zhang, H.Q. Zhang, C.L. Qu and A.M. Yu. 2008. Organotin speciation in textile and plastics by microwave-assisted extraction HPLC–ESI-MS. Talanta. 75: 556–563.

WHO. 2003. Guidelines for Drinking water quality, Chromium.

Wrobel, K., K. Wrobel, B. Parker, S.S. Kannamkumarath and J.A. Caruso. 2002. Determination of As(III), As(V), monomethylarsonic acid, dimethylarsinic acid and arsenobetaine by HPLC–ICP–MS: analysis of reference materials, fish tissues and urine. Talanta. 58: 899–907.

Ying, L.Y., H.L. Jiang, S.C. Zhou and Y. Zhou. 2011. Ionic liquid as a complexation and extraction medium combined with high-performance liquid chromatography in the evaluation of chromium(VI) and chromium(III) speciation in wastewater samples. Microchem. J. 98: 200–203.

Zhang, Y.Q. and W.T. Frankenberger Jr. 2001. Speciation of selenium in plant water extracts by ion exchange chromatography-hydride generation atomic absorption spectrometry. Sci. Total Environ. 269: 39–47.

CHAPTER

2

Sample Pre-treatment Methods for Organometallic Species Determination

Antonio Moreda-Piñeiro,[1,]* Jorge Moreda-Piñeiro[2] and Pilar Bermejo-Barrera[1]

Introduction

Speciation analysis of metal and organometallic compounds is an important topic because the toxicity, bioavailability and mobility of elements in the environment and biosphere are critically dependent on their chemical species. Elemental speciation in real-world samples is challenging work because of the sample matrix concomitants and because the elemental species usually occur at low concentration levels. Therefore, sample pre-treatments are usually required for enriching the target species, and for target isolation from the sample matrix.

[1]Department of Analytical Chemistry, Nutrition and Bromatology, Faculty of Chemistry, University of Santiago de Compostela, Avenida das Ciencias, s/n, 15782 Santiago de Compostela, Spain.
[2]Department of Analytical Chemistry, Faculty of Sciences, University of A Coruña, Campus da Zapateira, s/n, 15071 A Coruña, Spain.
*Corresponding author

List of abbreviations after the text.

Sample pre-treatment procedures are necessary to isolate the desired components from complex liquid and solid matrices, and also because most of the analytical instruments cannot directly handle the liquid or the solid materials. The basic concept of sample pre-treatment procedure is to convert a real matrix sample into a sample suitable for analysis. This process almost inevitably changes the interactions of analytes with the matrix. These interactions are determined by the physical and chemical properties of both analytes and matrices, and they affect the applicability of different sample pre-treatment techniques and analytical methods, as well as their efficiency and repeatability.

In many analytical procedures, sample preparation is the most time-consuming and cost-determining step. In addition, large volumes of samples are often required for trace analysis, and handling it can be extremely time-consuming and tedious.

Historically, the sample preparation step has been considered as the most polluting step of the whole analytical procedure. The extensive use of organic solvents in analytical laboratories is no longer desirable because of environmental and health concerns. Since the acceptance of the philosophy and ideas of Analytical Green Chemistry (AGC) in analytical laboratories, sample preparation techniques that minimize solvent consumption have substituted the more solvent consuming techniques such as liquid–liquid extraction, solid-phase extraction or Soxhlet extraction.

AGC is a sustainable development concept which proposes removing or minimizing the environmental impact caused by analytical methodologies (Armenta et al. 2008). From the 12 principles proposed by the Green Chemistry (GC), the following are those more closely related to analytical chemistry and also to sample pre-treatment (Anastas and Warner 1998):

Principle 1: prevention of the waste generation;

Principle 5: safer solvents and auxiliaries;

Principle 6: design of energy efficiency;

Principle 8: to diminish derivatives;

Principle 11: real-time analysis for pollution prevention; and,

Principle 12: inherently safer chemistry for accident prevention.

Therefore, several new sample pre-treatments comprise some or all of the six AGC principles referred to above. Some examples when treating liquid materials are the following: micro-extraction based on sorbent enrichment (solid phase micro-extraction and stir bar sorptive extraction); liquid phase micro-extraction (liquid-liquid micro-extraction, liquid liquid liquid micro-extraction, single drop micro-extraction, dispersive liquid-liquid micro-extraction and solidified drop liquid phase micro-extraction); membrane-based micro-extraction (hollow fiber liquid phase micro-extraction, solvent bar micro-extraction, phase transfer membrane supported liquid-liquid-liquid micro-extraction, microporous membrane liquid liquid extraction and supported liquid membrane extraction probe). For solid samples, procedures based on pressurized liquid extraction, enzymatic hydrolysis extraction and matrix solid phase dispersion can also be considered. New pre-treatment techniques also miniaturize sample preparation and, thus, reduce organic solvent consumption. In addition, on-line extraction and analysis coupling results in higher sensitivity, negligible analyte losses, and lower sample amount for analysis.

Sample pre-treatments can be achieved by using very different techniques; some of them are simple and inexpensive, and do not require specialized equipment. This is the case of distillation and liquid-liquid extraction procedures. However, other techniques can be sophisticated, such as pressurized liquid extraction, which requires a high cost for acquisition and mantainance. Independent of the degree of sophistication, sample pre-treatment methods have the same goals:

1) High selectivity by removing potential interferences; e.g., high salinity matrices such as seawater samples; high concentrations of ions from river waters and groundwater; organic matter from biological samples or mineral matrices from sediments, soils or atmospheric particulate matter. In this sense, removal of the matrix components can also avoid the potential deterioration of the analytical instruments, such as blockage of ICP torch injectors, etc.

2) Improvements in sensitivity by increasing the analyte concentration (pre-concentration methods); e.g., emerging contaminants concentrations in open seawaters and atmospheric particulate matter are in the range of ng L^{-1} and ng m^{-3}, respectively.

3) If necessary, the analyte conversion into a more suitable form; e.g., metals/organometals derivatization to organic chelate by using adequate ligands in liquid-liquid extraction, liquid-liquid micro-extraction or solid phase extraction methodologies; organometallic compounds derivatization into apolar volatile species by Grignard reaction (ethylation, propylation or pentylation) in liquid-liquid extraction or solid phase micro-extraction methodologies; etc. Derivatization is a process of chemically modifying a species to produce a new species which has properties more suitable for analysis. The main advantages of derivatization are: (i) it increases the volatility of the species; (ii) it increases the stability of the species; (iii) it improves the chromatographic behavior or detectability of the complex and (iv) it enhances sensitivity. However, derivatization procedures have some drawbacks such as the time needed for converting the analyte into the derivatized form, the lack of suitable reagents for reacting with certain

functional groups of the analyte, and the occurrence of interferences due to the excess of the derivatization reagent.

4) The transfer of the analyte to a compatible solvent for a certain analytical technique to allow the analytical determination; e.g., back-extraction liquid-liquid extraction procedures into aqueous/acid solution for further metal quantification by atomic absorption or emission spectrometric methods, or a change from a polar to a non-polar solvent or vice versa for organic compounds determination by chromatographic techniques.

5) If possible, the isolation of certain species (speciation studies). This goal is less important when chromatographic techniques are used.

6) The establishment of a robust and reliable method that is not affected by variations in the sample matrix.

Although, many traditional sample pre-treatment methods are still in use (e.g., precipitation, co-precipitation, liquid-liquid extraction, solid phase extraction, acid extraction or acid digestion), there have been trends in recent year towards:

1) Use of small initial sample sizes; e.g., sample volumes around a few milliliters are enough for single-drop micro-extraction or dispersive liquid-liquid micro-extraction procedures.

2) Ease of automation, which minimizes the sample and extract handling, and avoids analyte losses and sample contamination; e.g., microwave or ultrasound assisted extraction techniques and pressurized liquid extraction.

3) Miniaturization; e.g., miniaturized devices available for operations such as liquid-liquid extraction and pre-concentration into solid supports by flow injection analysis (FIA), and new miniaturized liquid-liquid extraction procedures such as single-drop micro-extraction, dispersive liquid-liquid micro-extraction or hollow fiber liquid phase micro-extraction.

4) Use of non exhaustive extraction procedures in which only a small fraction of the analyte is extracted/pre-concentrated for further analysis; e.g., the analyte extraction is incomplete when performing single drop micro-extraction and solid phase micro-extraction pre-treatments. The high repeatability of measurement of time in the sequence of operations (as a result of automation) ensures that the sample and the patterns are exactly the same process and exactly in the same period of time, achieving a high reproducibility.

5) High selectivity in the extraction; e.g., the development of new solid phases for solid phase extraction such as molecularly/ionic imprinted polymers for organic compounds and metal pre-concentration,

respectively; carbon nanotubes, nanometer-sized materials, ordered mesoporus materials, etc.

6) Reduction of the pre-treatment time; e.g., microwave assisted extraction and pressurized liquid extraction dramatically shorten the extraction times, (completed extractions after 1 to 5 min).

7) Increase in the extraction efficiency without degradation or losses of the target, and decrease in the number of extraction cycles; e.g., two–three extraction steps are typical for liquid-liquid extraction methodologies, while one extraction step is enough for single drop micro-extraction or dispersive liquid-liquid micro-extraction. The extraction cycles for organic compounds isolation by Soxhlet are typically around 24 cycles, whereas pressurized liquid extraction only requires one to three cycles.

8) High throughput performance, which implies that the number of samples that could be pre-treated per unit of time should be similar to the number of samples per unit of time that modern analytical detectors can analyze. Microwave assisted extraction and pressurized liquid extraction procedures allow the pre-treatment of high number of sample per unit of time; thus, these procedures could be coupled on-line with analytical instruments.

9) The on-line coupling with analytical instruments, transferring the treated sample to the detection systems; e.g., metals determination from solid samples by using continuous digestion/extraction devices coupled to spectrometric instruments; or analyte pre-concentration by solid phase micro-extraction coupled to GC.

10) Increase in the operator safety and decrease in the environmental impact of the analytical methodologies. Four top priorities concerning the elimination or reduction of organic solvents and/or highly toxic or ecotoxic reagents, the prevention of waste generation and the reduction of energy consumption, were identified in the analytical laboratory, which connects with the principles established by the AGC. The use of procedures that reduce or eliminate the use or generation of feedstock, products, by-products, solvents, reagents, etc., that are hazardous to human health and to the environment.

A description of the theoretical aspects, experimental parameters and advantages and disadvantages of these sample pre-treatment methods will be developed in the following sections. In addition, the application of those techniques to the isolation and pre-concentration of organometallic compounds from environmental and biological samples will also be discussed.

Sample Pre-treatment for Liquid Samples: Classical Approaches

Several pre-treatment techniques, including precipitation and co-precipitation, solvent extraction, adsorption on solid sorbent, electrolysis, purge and trap or Soxhlet extraction (called classical pre-treatment procedures) are still used for analyte extraction and pre-concentration. A great number of different standard and official analytical methods, which are available for the determination of several trace analytes in environmental and biological samples, use some of these classical approaches.

Precipitation and co-precipitation

General aspects

Precipitation and co-precipitation procedures are based on a selective or specific precipitation of the analyte by the addition of different precipitating and/or co-precipitating agents under certain conditions (pH, temperature, stirring, etc.). The precipitate obtained is then separated from the aqueous medium by filtration, centrifugation or flotation, and is re-dissolved in a small volume of mineral acids or organic solvents for further analysis. Solid particles could also be introduced as a solid or as a slurry into atomic spectrometric atomizers, or irradiated with thermal neutrons flux (neutron activation analysis (NAA)). Special care must be taken to ensure that no precipitate is lost during either filtration or centrifugation steps.

Precipitation procedures are mainly used for isolating analytes which are present in the sample at high concentrations. Because of the relative high concentration of the targets, precipitation is achieved by directly adding an organic or an inorganic precipitant. However, co-precipitation approaches are preferred when the target analyte are at low concentrations. In these cases, a carrier element and an organic co-precipitant are added to sample. Carrier elements are metals which easily form insoluble hydroxides, oxides or which can easily be reduced. Some examples of carrier elements are copper, zinc, iron, lanthanum, indium and magnesium. As organic co-precipitants, reagents such as dithiocarbamate, diethyldithiophosphate, heterocyclic azo-dyes, quinoline, among others, are used. In this case, co-precipitation involves two steps: (1) the trace analyte reacts with an organic/inorganic compound and form a solid phase; and (2) the major precipitate reacts with other metals to form chelates.

A quantitative precipitation of the analyte from the solution must be reached. Colloidal suspensions (particle diameters from 10^{-7} to 10^{-4} cm) due to poorly soluble substances must be avoided. The relative super-saturation affects the particle size and it is expressed as:

$$\frac{Q-S}{S}$$

Eqn. 1

where Q is the instantaneous concentration of the added species and S is the equilibrium solubility of the compound that precipitates. Particle size seems to be inversely proportional to relative super-saturation. The electric double layer formed during precipitation keeps the colloidal precipitate particles from coming into contact with each other, thus preventing further coagulation. The formation of a colloidal suspension can be minimized or prevented by several methods; for instance, under constant stirring or by performing the precipitation at a temperature close to the boiling point of water (heating increases overall thermal motion, affecting the mobility of adsorbed ions and colloidal precipitate particles). The final effect of temperature is that there are collisions among particles that increase the particle size as a result of coagulation.

As stated, temperature is a critical parameter. However, precipitation and co-precipitation procedures are affected by other experimental parameters such as the stirring time and the stirring rate, the standing time, the pH, the amount of precipitant and/or co-precipitant and the sample volume. In addition, centrifugation rate and time (for those procedures that use a centrifugation step for isolating the precipitate from aqueous sample) are also important factors.

Advantages and disadvantages

The disadvantages of these procedures are that they can lead to sample contamination because they require the addition of large amounts of reagents, which result in high concentrations when compared with trace element concentrations. In addition, these procedures require a high volume of sample and they are tedious and time-consuming. The use of thermo-responsive polymers can reduce or remove some of these drawbacks. In addition, the implementation of rapid, convenient and sensitive on-line procedures by flow injection systems is desirable, thereby reducing sample and reagents consumption while enhancing throughput.

Precipitation and co-precipitation methods for inorganic species pre-concentration

Co-precipitation methods are focused on inorganic As, Sb and Se speciation. Inorganic species of As, Sb and Se have been preconcentrated by co-precipitation using inorganic and organic precipitants. These methods are based on the selective precipitation of an inorganic species exhibiting a certain oxidation state at a fixed pH. Then, a reduction or oxidation step allows the quantitative precipitation of the total target and the concentration of the element with the other oxidation state is calculated by difference. As inorganic precipitants, Mo(VI) (Van Elteren et al. 1991) and $Zr(OH)_4$ (Yalei et al. 1993) have been proposed for arsenate precipitation; whereas, $Ce(OH)_4$ (Elçi et al. 2008) and Na_2SO_4 plus $BaCl_2$ (Okamoto et al. 2010) have been used for arsenite precipitation. Selective selenite co-precipitation was achieved by using $Mg(NO_3)_2$ (Tuzen et al. 2007) and $La(OH)_3$ (Wu et al. 2007).

Similarly, dibenzyldithiocarbamate (DBDTC) (Van Elteren et al. 1991, Johannes et al. 2001, 2002) has also been used for arsenite. The combination of a chelating agent and a metal as a carrier element has also been used. Therefore, ammonium pyrrolidinedithiocarbamate (APDC) (Zhang et al. 2007) and APDC after Pb(IV) addition has been used as carriers for arsenite, antimonite and selenite preconcentration from natural waters (Sun and Yang 1999).

Flow injection on-line preconcentration systems co-precipitation

One of the major advances in analytical chemistry in recent decades has been the development of automated systems for analysis, which provide analytical data with minimal operator intervention. Thus, several separation methods based on ion exchange, liquid-liquid extraction and precipitation and co-precipitation have been successfully adapted mainly to flow injection (FI) on line pre-concentration. This has led to numerous rapid, convenient, sensitive and reliable spectroscopic methods.

There are two types of flow injection method for on- line co-precipitation; the methods which use a filtration stage, and those procedures without filtration steps. The first method type often has a stainless steel high performance liquid chromatography (HPLC) screen or a polytetrafluoroethylene (PTFE) membrane on a polypropylene support as filtering devices. After complete co-precipitation and retention of the precipitate, a small volume of an acid is used to dissolve and flush the analyte to the detector. The use of filters in these on-line procedures can be troublesome when large amounts of precipitate are formed (filter clogging). In these cases, large volume filters can be used, although large volume filters impair the attainable enrichment factor through dispersion and retard the dissolution of the precipitate.

The second method (without filtration stage) uses knotted reactors (KR) for precipitate collection. KRs are generally made from PTFE tubing. They present a 100–150 cm length and 0.5–1.5 mm inner diameter and are made of tubing by tying interlaced knots. The KR produced from PTFE tubes was able to retain the complexes under appropriate experimental conditions. Two factors are responsible for the retention of the complex molecules on the wall of the KR. The first factor is related to the fact that the molecular species are launched at the KR's inner walls by centrifugal forces generated by secondary flows in three dimensionally disordered systems. The second factor is attributed to the nature of the KR's material and the nature of the complex. After precipitate adsorption onto the walls of the KR, an elution/dissolution step with an acid solution is needed to sweep the target to the detector.

KRs show several advantages for collecting the precipitate: (a) they offer a relatively high capacity due to the large internal surface area; (b) they offer a low back pressure (even for high flow rates) due to the absence of any filter; (c) contamination is avoided due to the inert nature of tube material; (d) the reactor is easy and cheap to make; and, (e) the lifetime is unlimited and the reactor requires no maintenance (Welz and Sperling 1999).

Since the first application of an on-line co-precipitation system with KR by Fang et al. (Fang et al. 1991) for lead determination, several methods have been developed for the preconcentration of trace elements from waters. In the speciation field, a selenite FI co-precipitation system using KR was performed by using $La(OH)_3$ as a carrier element (Wu et al. 2007).

Liquid-liquid extraction

Liquid-liquid extraction (LLE) is the oldest and the most widely used sample pre-treatment method for liquid samples. Extraction of low concentrations of inorganic and organic analytes from complex aqueous samples into organic solvents is the aim of this approach.

General aspects

LLE procedures are based on the relative solubility of the target compound in two immiscible phases (Fig. 2.1), which leads to an improvement in the selectivity by isolating the analyte and an increase in sensitivity by transferring the analyte to a small solvent volume. The isolation and pre-concentration is attained at the same time making the solvent more suitable for the analytical signal acquisition. An ideal LLE procedure must guarantee a quantitative analyte isolation from the aqueous matrix sample whilst interference species remain in the aqueous phase.

A quantitative measure of a target analyte distribution between an aqueous and an organic phases is referred to as a distribution or partition coefficient. It is the ratio, K, of the solubility of analyte dissolved in the organic phase ($S_{organic}$) to the solubility of analyte dissolved in the aqueous phase (S_{water}). Note that K is independent of the amounts of the two solvents mixed.

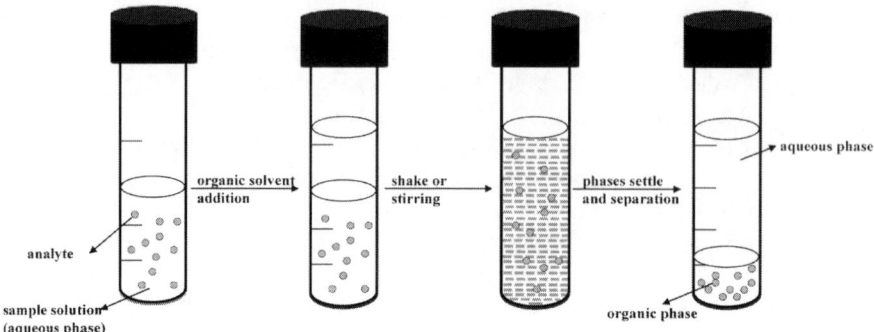

Figure 2.1 Schematic liquid-liquid extraction (LLE) set up.

$$K = \frac{S_{organic}}{S_{water}}$$ Eqn. 2

The constant K, is essentially the ratio of the concentrations of the solute in the two different solvents once the system reaches equilibrium. At equilibrium the molecules naturally distribute themselves in the solvent where they are more soluble.

By using the appropriate solvent system, analyte can be selected specifically and extracted from one solvent to another. The distribution coefficient is a ratio, and unless K is very large, not all of a solute will transfer to the organic phase in a single extraction. Usually two, three, or four extractions of the aqueous phase with an organic solvent are performed in sequence to remove as much of the analyte from the aqueous phase as possible. The effectiveness of multiple small volume extractions versus one large volume extraction can be demonstrated by a simple calculation. If one extraction can recover 60 percent of the analyte, a second extraction with the same solvent and under similar operating conditions may be able to extract 60 percent of the remaining analyte. Therefore, an 84 percent would have been recovered after two successive extractions. Similarly, a third extraction on the remaining 16 percent would extract a further 9.6 percent, and the total extraction accounts for 93.6 percent. In contrast, one extraction with large solvent volumes would have only recovered 60 percent. This fact can mathematically be proved by the following equation:

$$FE_{organic} = \left(\frac{1}{1 + \dfrac{V_{organic}}{V_{water}} nK} \right)^{n}$$ Eqn. 3

This equation provides the fraction of material extracted by the organic solvent ($FE_{organic}$) where n is the number of extractions performed, K is the distribution coefficient, V_{water} is the volume water and $V_{organic}$ is the volume of organic solvent.

The pre-concentration of trace metals and organometallic compounds by LLE procedures is commonly achieved by target derivatization with a complexing or derivatizing agent in a buffered aqueous sample, followed by the extraction of the formed complexes into an organic phase. Complexation agents are organic compounds with functional groups containing nitrogen (as amines, amides or nitriles), oxygen (carboxylic, hydroxyl or ether functional groups) or sulfur (thiols, thiocarbamides) atoms, which are capable of chelating trace metals. Derivatizing agents are mainly Grignard reagents or tetralkylborate reagent such as sodium tetraethyl- or tetrapropylborate directly in the aqueous matrix used for mercury, lead and tin species derivatization.

Derivatization agents used for LLE procedures should offer two characteristics. First, these substances must guarantee the complexation/extraction of most of the target; and second, the extraction of these complexes must not be affected over a fairly wide range of pH so that error when adjusting the pH of the solution can be allowed. This last task is not always possible because most of the derivatization agents show a strong pH dependence. Therefore, a pH adjustment to within one pH unit or less is required which can result in serious errors in routine applications.

The use of buffer solutions to fix the pH is recommended in routine LLE work. The buffer solution must offer the following characteristics: (1) buffer solution must be stable, (2) buffer solution must offer a high buffering capacity, and (3) buffer solution must not participate in any reaction. Citrate- and acetate- based buffer solutions are the most useful for fixing the pH before LLE.

In addition, the solvent used to extract the analyte/derivatized analyte must also show several desirable characteristics:

1) Solvents must extract the desired derivatized compound by using one or few extraction steps; i.e., two extraction steps are generally required for quantitative recovery of most common trace analytes.
2) Solvents must be immiscible with the aqueous solution; i.e., the water solubility of 4-methyl-2-pentanone (IBMK) is relatively high, which allows salt carry-over and matrix interferences when this organic phase is used.
3) Solvents must not form emulsions because emulsion formation hinders phase separation.
4) Solvents must be compatible with the analytical technique used for determination; i.e., solvent must have good burning characteristics

or must not create problems during the determination. For example some common analytical techniques for determination are flame atomic absorption spectrometry (FAAS), electrothermal atomic absorption spectrometry (ETAAS), inductively coupled plasma atomic emission spectrometry (ICP-OES) and ICP-mass spectrometry (ICP-MS). Some organic solvents may produce smoke and other particulate carbon matter which may cause interferences, extinguish the plasma or may coat the ICP-MS interface region with carbon. Ideally, the solvent should also have good chromatographic properties.

5) Solvents must enhance rather than suppress the signal sensitivity as compared to analyte sensitivity in water.

6) Finally, the extracted analyte must be stable in the organic solvent.

As stated, the efficiency of LLE process depends on the affinity of analytes to the extracting solvent, the pH of the aqueous phase and the stability of the derivatized compound formed. Several variables such as the nature of the derivatization agent, the concentration of derivatization agent, the nature of the organic phase, the organic phase/aqueous phase ratio, the pH of aqueous phase (selection of the appropriate buffer solution), the number of successive extractions, and the shaking and standing times, must be carefully selected.

Advantages and disadvantages

Although LLE procedures have proven to be reliable and efficient techniques, these methodologies are time-consuming, use large volumes of reagent and are labor-intensive. They also have a tendency for emulsion formation and have poor potential for automation. As LLE is a multi-step process, it often results in analyte losses. In addition, LLE requires a large amount of high purity solvents, which are expensive and toxic, and the procedure generates hazardous laboratory wastes, for which disposal or treatment can be harmful to the environment and the public health. Moreover, a derivatization step previous to extraction into the organic phase is required when dealing with metals and organometallic compounds; and back-extractions of the preconcentrate complex into the aqueous phase must be performed before detection/quantification by the most common spectrometric techniques.

The implementation of LLE techniques in on-line mode coupled with atomic spectrometry instruments has been developed to minimize many of these drawbacks. Most of the drawbacks of conventional LLE procedures can also be overcome by miniaturizing the whole procedure, mainly by drastically reducing the extractant phase volume. This has led to new extractive techniques based on liquid phase micro-extractions. Some examples of these methodologies are the liquid-liquid micro-extraction, the single-drop micro-extraction, the dispersive liquid-liquid micro-extraction or the hollow fiber liquid-phase micro-extraction. These new methodologies will be described in the next sections.

Liquid-liquid extraction methods for inorganic and organometallic species preconcentration

Several chelating/derivatizing agents have been used for inorganic As, Sb, Se and Te speciation by non-chromatographic methods. These methods are based on the selective complexation/derivatization of the target (As, Sb, Se or Te) exhibiting a certain oxidation state. APDC is the most popular chelating agent for these purposes. APDC solutions are stable in acidic conditions, and they can be used at room temperature and at different pHs without decomposition. APDC (Menéndez-García et al. 1995, Yusof et al. 1999, Serafimovska et al. 2011) forms dithiocarbamate complexes with As(III), Sb(III) and Se(IV). 4-chloro-o-phenylendiamine (Gómez-Ariza et al. 1999a) and 4,4'-dichloro (3-mercapto-1,5-diphenylformazan) (Cl_2H_2DZ) (El-Shahawi et al. 2006) derivatizes Se(IV) to form piazselenols (Gómez-Ariza et al. 1999a), or forms a yellow-red colored complex with Se(IV) (El-Shahawi et al. 2006). 4-chloro-o-phenylenediamine requires a high temperature of reaction (75°C). However, a shorter reaction time was sufficient to obtain higher reaction yields. Rhodamine B (Trivelin et al. 2006) forms an ion pair with Sb(V); side reactions could occur and a lack of stability of the ion pair was usually verified when using rhodamine B, which results in poor precision and inadequate accuracy of the procedure. Finally, N-n-octylaniline (Sargar and Anuse 2001) forms a complex with Te(IV). Total concentration of the inorganic target is obtained by repeating the LLE procedure after reduction (Menéndez-García et al. 1995, Yusof et al. 1999, Gómez-Ariza et al. 1999a, El-Shahawi et al. 2006, Serafimovska et al. 2011), or oxidation (Trivelin et al. 2006). The target which exhibits the other oxidation state is then calculated by the difference.

Similarly, 2,3-dimercaptopropanol has been used as a derivatizing agent for As species (As(III); As(V); monomethylarsenic acid, MMA; and dimethylarsonic acid, DMA) (Huang et al. 2008a), although the derivatization reaction is not completely successful. In general, derivatizing agents with thiol groups, such as BAL, are not recommended because of their reducing properties. Ethylchloroformate (ECF) forms N-ethoxycarbonyl-O-ethyl ester derivatives with selenomethionine (SeMet) and selenocysteine (SeCys) (Devos et al. 2002, Işcioğlu and Henden 2004). This compound reacts in a one step reaction with both the amino and the carboxylic acid residues in the amino acids, and offers the advantage of fast kinetics (< 15 min). Tetraalkylborate reagents such as sodium tetraethylborate

(NaBEt$_4$) (Michel and Averty 1991, Stäb et al. 1993, Kuballa et al. 1995, Li et al. 1995, Lespes et al. 1996, Carlier-Pinasseau et al. 1996a, b, 1997, Tao et al. 1999, Colombini et al. 2004, Zachariadis and Rosenberg 2009a), and sodium tetrakis (4-fluorophenyl)borate (NaB(FC$_6$H$_4$)$_4$) (Tsunoi et al. 2004) form volatile tin tetra-substituted species from monobutyltin, MBT; dibutyltin, DBT; tributyltin, TBT; monophenyltin, MPhT; diphenyltin, DPhT; and triphenyltin, TPhT. Sodium tetrahydroborate (NaBH$_4$) (Tolosa et al. 1996) can be employed to produce alkyl-stannanes which are also used for tin speciation, although they are less frequently used due to their limited thermal stability. The use of Grignard reagents avoids the co-extraction of various high molecular weight components from complex samples into the organic phase. Therefore, the formation of gels or foams that damages the two phase separation is minimized. Finally, CH$_3$MgCl was also used as a derivatizing agent for tin species extraction/preconcentration (Tolosa et al. 1991).

Many types of non-polar solvents such as chloroform (Yusof et al. 1999, Devos et al. 2002), IBMK (Menéndez-García et al. 1995), xylene (Sargar and Anuse 2001, Serafimovska et al. 2011), toluene (Gómez-Ariza et al. 1999a, Trivelin et al. 2006), hexane (Tsunoi et al. 2004, El-Shahawi et al. 2006, Zachariadis and Rosenberg 2009a), dichloromethane (DCM) (Kuballa et al. 1995, Huang et al. 2008a), and isooctane (Stäb et al. 1993, Carlier-Pinasseau et al. 1996a, b, 1997, Tao et al. 1999, Colombini et al. 2004) are commonly used. All these solvents offer a relatively low solubility in water, high extraction efficiencies and poor emulsion formation.

Several procedures without a derivatization step have also been proposed. Mono-ethylmercury (MEM) and mono-phenylmercury (MPhM) can be acid leached from soils and extracted with toluene, re-extracted then with cysteine, and injected as an aqueous phase into HPLC-UV (Gaona and Valiente 2003). TBT and TPhT have been extracted from seawater with hexane-diethyl ether and gas chromatographed by a capillary doped with a dilute HBr–methanolic solution (Mizuishi et al. 1998). Finally, organotin compounds have also been extracted from seawater with DCM/tropolone (Yu et al. 2009).

After LLE, the organic phases could be introduced into conventional atomizers (Trivelin et al. 2006, Serafimovska et al. 2011). However, the direct analysis of organic phases causes several solvent effects arising from changes in the density, surface tension and viscosity which have a decisive effect on the droplet sizes of the aerosol (ICP-OES/MS), or which affect the drop deposition into the graphite furnace by the autosampler (ETAAS). Other solvents such chloroform tend to evaporate quickly leaving the solid complex behind, which clogs the spray chamber or the injector of the ICP torch.

Thus, direct analysis of organic extracts by atomic spectrometry, and also in liquid chromatography or capillary electrophoresis instruments is not advisable because of poor repeatability and sensitivity, and an organic solvent evaporation step followed by aqueous dissolution (Huang et al. 2008a, Yu et al. 2009) or a back extraction step into aqueous medium (Gaona and Valiente 2003) are usually performed. However, the direct injection of organic phases into GC is not a problem, and organic solvent dryness, re-dissolution or back-extractions are avoided (Michel and Averty 1991, Tolosa et al. 1991, Lespes et al. 1996, Carlier-Pinasseau et al. 1996a, b, Tolosa et al. 1996, Mizuishi et al. 1998, Tao et al. 1999, Gómez-Ariza et al. 1999a, Devos et al. 2002, İşcioğlu and Henden 2004, Colombini et al. 2004, Tsunoi et al. 2004, Zachariadis and Rosenberg 2009a).

Flow injection on-line pre-concentration systems by LLE

The introduction of flow injection liquid–liquid extraction (FI-LLE) for metal pre-concentration in flame atomic absorption spectrometry (FAAS) by Nord and Karlberg (Nord and Karlberg 1981) led to several developments for on-line preconcentration systems. The extraction process in FI mode consists of three steps: (1) derivatization, (2) extraction, and (3) phase separation (Lespes et al. 1996). In the first step, the acidified samples containing the target analytes (butyl- and phenyltin compounds) and rhodamine B (derivatizating reagent) are pumped to the extraction chamber. Toluene is then added, and the sample/reagent/solvent mixture inside the extraction chamber is stirred for a period of 100 sec (formation and extraction of the ion-pair). After the extraction of the ion-pair, the agitation is stopped and a time of 80 sec is needed to allow phase separation inside the extraction chamber. Finally, the carrier stream fills the extraction chamber and transfers the less dense organic phase containing the ion-pair towards the detector.

New developments: Liquid-liquid micro-extraction

1) *General aspects.* Liquid-liquid micro-extraction (LLME) is a solvent micro-extraction technique in which the phase ratio values are higher than 100 (Majors 1996). The Murray flask and its following adaptations were first described for the pre-concentration of organic pollutants (organochloride and organophosphorus pesticides) (Murray 1979, Zapf et al. 1995, Barrio et al. 1996) using LLME from aqueous matrices. The Murray flask consists of a 1000 mL flask with a capillary tube at the top and a lateral arm near the base. The capillary tube has the function of collecting the small volume of organic solvent after agitation of the sample. The lateral arm is used to move the organic extraction

solution to the capillary tube using either water or air. The use of air allows multiple extractions of the sample while the aqueous volume is maintained.

2) *Advantages and disadvantages*. The technique is faster and simpler than conventional LLE methods. It is also inexpensive (the use of large volumes of toxic and expensive solvents and complex equipment are avoided). It is a sensitive and selective technique, and minimizes waste. Compared with conventional LLE, LLME may provide poorer analyte recoveries, but the concentration in the organic phase is greatly enhanced. In addition, the amount of organic solvent used is reduced and pre-concentration is performed in a single step, which reduces problems derived from contamination and analyte loss.

3) *Liquid-liquid micro-extraction for organometallic species pre-concentration.* Despite the advantages of this technique, few developments have been reported using LLME and/or micro-volume back-extraction for the pre-concentration of organometallic compounds. Only one non-chromatographic development has been addressed for inorganic arsenic (As(III) and As(V)) and organic arsenic (MMA and DMA) species from environmental samples using the ionic liquid (IL) tetradecyl(trihexyl)phosphonium chloride, CYPHOS® IL 101, as an ion pairing reagent (Monasterio and Wuilloud 2010). As(V) species was selectively separated by forming As(V)-molybdate heteropoly acid [As(V)-MHPA] complex with molybdenum, followed by ion-pairing reaction with CYPHOS® IL 101 and micro-extraction in chloroform. Total inorganic arsenic and total arsenic were evaluated after oxidation with $K_2S_2O_8$ and potassium peroxydisulfate (boiling for 30 min), respectively. Organic arsenic was calculated by the difference.

Purge and Trap

General aspects

Volatile anlytes can be concentrated by either static headspace or dynamic headspace (purge and trap) sampling. When using the static headspace concentration mode, the sample is placed in a closed sample chamber. Molecules of the volatile compounds in the sample migrate to the headspace above the sample and equilibrium is established between the concentration of the compounds in the vapor and liquid phase. Once equilibrium is reached, an aliquot of the headspace above the sample is injected onto the GC column. A major problem with the static headspace techniques is derived from the sample matrix which can affect the equilibrium significantly, and hence, determinations of compounds exhibiting high solubility in the sample matrix often yield low sensitivity.

Purge and trap (PT) (Fig. 2.2) is a technique frequently used to isolate volatile analytes from liquid samples and solid samples by gas chromatography (GC) analysis. PT-gas chromatography (PT-GC), as first described by Swinnerton et al. (Swinnerton et al. 1962), has become a valuable and widely accepted method for the determination of volatile organic compounds (VOCs).

Samples are introduced into a purge vessel and a flow of inert gas is passed through the sample at a constant flow rate for a fixed time. Volatile compounds, which are purged from the sample into the headspace above the sample, are diluted in the extractant gas and must be focused in a trap before being introduced into the chromatographic column. This focalization can be performed in a cold trap, by cryofocusing or on cartridges packed with certain sorbent materials. Next, the analytes are released into a GC column by heating the cryo- or sorbent trap.

Purge and trap techniques involve the following series of steps that must be followed to ensure accurate and reproducible results:

1) Standby: during the standby mode, the purge gas flow is stopped, the trap is cooled and the system is made ready for starting an analysis. The desorption gas bypasses the trap and is directed onto the column as the carrier gas flow. The gas flow rate through the column can be measured.

2) Wet purge: during the wet purge, the purge gas flow passes through the purge vessel, removes volatile analytes from the sample, and

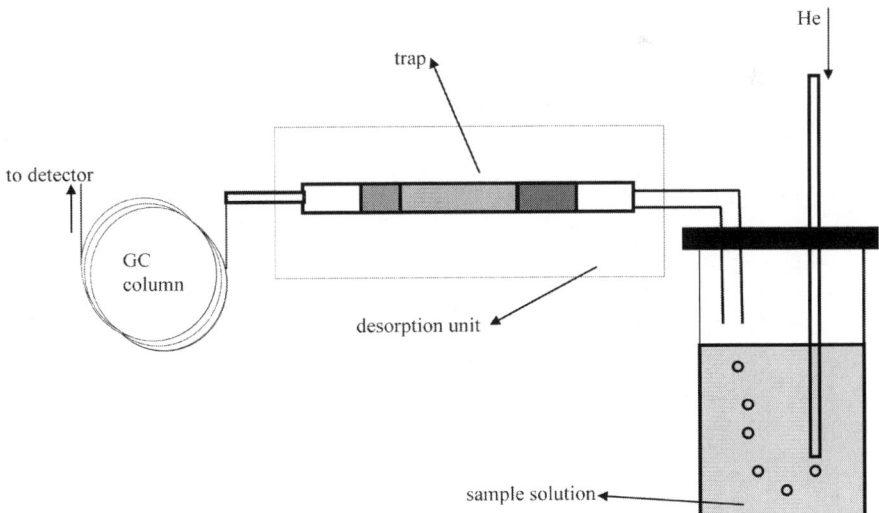

Figure 2.2 Schematic of purge and trap (PT) set-up.

sweeps the analytes through the heated valve onto the adsorbent trap. The analytes are collected on the trap and the purge gas exits through the purge vent. The purge gas flow is typically set at 30–50 mL min^{-1}. Samples are usually purged for 10–15 min. During the purge mode the desorb (carrier) gas is directed onto the column.

3) Dry purge: during the wet purge, a large amount of water is removed from the sample and is collected on the trap. The excess water accumulated in the trap is therefore removed during the dry purge step. At this point the purge gas bypasses the purge vessel and is directed to the trap, while the desorb (carrier) gas is directed onto the column. Only traps that incorporate hydrophobic adsorbents can be dry purged.

4) Desorb preheat: once the analytes have been trapped and excess water removed, the purge gas flow is stopped. During this static period the trap is rapidly heated to ~ 5°C below the desorb temperature of the adsorbent material used. Without a desorb preheat step the peaks would be tailing, resulting in poor chromatography. During the desorb pre-heat step the desorb (carrier) gas is directed onto the column.

5) Desorb: once the desorb preheat temperature is reached, the purge and trap unit valve is rotated. This directs the desorb (carrier) gas flow to backflush the adsorbent trap and carries the analytes in a narrow band to the GC system. While the sample transfer occurs, the trap is heated to its final desorb temperature. Desorb temperatures range from 180 to 250°C according to the nature of the adsorbent material and to the model of concentrator. The desorb flow rate is extremely important. It must be high enough to ensure that the sample remains in a narrow band during the transfer to the GC column. The optimum desorb flow rate for a purge and trap system is > 20 mL min^{-1}; however, this flow rate is too high to be used with capillary columns and must be reduced to retain column efficiency. Cryofocusing (i.e., cold trapping) by installing a secondary cold trap or by cooling the GC column to subambient temperatures can be used to reduce band broadening. The desorb time is inversely proportional to flow rate and trap temperature, so that as the flow rate/trap temperature increases, the desorb time decreases because of the analytes flush off the trap at a higher rate.

6) Trap bake: the trap is baked with gas flow to remove any remaining sample components and contaminants from the trap after the desorb step. This step generally lasts 6–10 min at typical temperatures of 10–20°C above the desorb temperature. The maximum trap temperature must not be exceeded to prevent damage to the adsorbent materials.

Published PT procedures encompass different purge vessel designs, sorbents for trapping, valves and trap types, and modes for refocusing analyte bands desorbed from traps before their introduction into the GC.

Three types of purge vessels (i.e., spargers) are commonly used in purge and trap systems. (1) Frit spargers are mainly used for water samples. The frit creates many small bubbles that travel through the sample to increase the purging efficiency. (2) Fritless spargers are used for samples containing high amounts of particulate matter, or for industrial wastewater samples that may foam. They create fewer bubbles (low purging efficiency) but eliminate plugged frits and they reduce foaming problems. (3) Needle spargers are used when purging soil, sludge or solid samples. A narrow gauge needle is inserted into the sample and is used to release a small stream of purge gas. Common sizes of spargers are 5 mL and 25 mL.

The trapping material (adsorbents) efficiency is greatly affected by the temperature and desorption time. The adsorbent must be able to retain compounds during the entire purging sequence and must allow a fast release of target analytes during the desorption step. Each adsorbent has a unique trapping capability for a specific class or classes of compounds. Therefore, a trap may have several different beds of adsorbents. The weakest adsorbent material is placed at the inlet end of the trap, then the next strongest adsorbent, and so on. The more volatile compounds pass through the weaker adsorbents and are retained by the stronger adsorbents, while the less volatile compounds are retained on the weaker adsorbents and never reach the stronger adsorbents. The purged analytes trapped onto a solid sorbent must be desorbed by heating, and must be transferred to the GC using a desorption gas (the GC carrier gas is usually used for this purpose).

The most widely used adsorbent traps are:

a) Tenax® TA: Tenax adsorbent is commercially available in high purity and is usually recommended for thermal desorption applications. The manufacturer's operating temperature is 230°C but, the material performs best when kept below 200°C.

b) Silica Gel (surface area: 200–800 $m^2 g^{-1}$): Silica gel is a stronger adsorbent than Tenax® adsorbent. Silica gel is frequently used in conjunction with Tenax® adsorbent as a trap for volatile organic pollutants. It is an excellent trapping material for polar and highly volatile compounds that are gases at room temperature; however, silica gel is extremely hydrophilic and will retain large amounts of water. However it should be kept in mind that if a trap contains silica gel, dry purging will not reduce the water content.

c) Charcoal (surface area: 900 m^2 g^{-1}): Charcoal is another strong adsorbent material. It is usually used in series after silica gel for trapping very volatile compounds that might break through the gel. Charcoal is hydrophobic, and does not retain significant amounts of water.

d) Graphitized carbon black or Carbopack® adsorbent (surface area: 10–100 m^2 g^{-1}): Graphitized carbon black (GCB) is an alternative to Tenax® adsorbent. GCB is available in many pore sizes and is effective at trapping volatile organics in the same range as Tenax® adsorbent. GCB is hydrophobic and offers excellent thermal stability. Highly volatile compounds are not well retained on GCB and must be trapped on stronger adsorbent materials such as carbon molecular sieves.

e) Carbon molecular sieves (surface area: 50–800 m^2 g^{-1}): Carbon molecular sieves such as Carbosieve™-SIII offer high surface areas and are recommended for trapping highly volatile compounds. They are commonly used in series after GCB because they retain compounds that break through the GCB. Carbon molecular sieves are hydrophobic and have excellent thermal stability.

f) Carboxen™-1000 adsorbent (surface area: 1200 m^{2}g^{-1}): Carboxen®-1000 adsorbent is a strong adsorbent designed to be used as the inner most adsorbent bed in the trap. This material traps Freon® compounds, permanent gases and light hydrocarbons. The characteristics of the material are very similar to those offered by Carbosieve® S-III packing material, and it is stable up to temperatures of 300°C.

Chromosorb W and Supelcoport SP 2100 are also used for the trapping of organometallic compounds.

The purge and desorb flows are controlled by an automated switching valve. The valve is contained in a heated compartment to prevent sample condensation. By rotating the valve, the purge and desorb flow paths can be changed during the purge and trap sequence.

Several trap types are available: (1) Type "K" trap (Vocarb™ 3000 Trap): this trap has exceptional ability to retain highly volatile compounds with minimal bleed, activity, or breakdown, yet it works well for trapping higher boiling compounds. The trap resists adsorption of water and methanol, and virtually eliminates the need for moisture control systems and the dry purge step on the concentrator. This trap contains Carboxen™ 1000 adsorbent. (2) Type "J" trap (BTEXTRAP™ Trap): this trap retains less water and methanol compared to the "K" trap, and can withstand higher temperatures than the Tenax®/silica gel trap. The disadvantage of the "J" trap is the limited ability to retain more polar analytes. (3) Type "B" trap (Tenax®/silica gel traps): Tenax®/silica gel traps are used for a variety of methods. These traps offer better recoveries for polar analytes than those obtained with the "K" trap, but the silica gel layer adsorbs water, methanol and carbon dioxide. traps. (4) Type "F" trap (OV®-1/Tenax®/silica gel traps): Although these traps are recommended in many EPA methods, they exhibit more bleed and activity than the Tenax®/silica gel trap, with no significant improvement in performance. (5) Type "I" trap (Vocarb 4000™): The "I" trap is used for increasing the response of less volatile compounds. It is used only for applications involving larger molecular size targets. Common desorb times of 2 to 4 min should be increased when using the "I" trap for compounds of high boiling points. The Tenax®/silica gel trap also has better lot-to-lot reproducibility compared with the "K" or "I".

The purge efficiency of an analyte is based on vapor pressure, solubility, temperature of sample and purge volume. Higher analyte vapor pressure involves quick analyte vaporization. Polar organic compounds have poor purge efficiencies because they are quite soluble in water (strong dipole-dipole interaction and hydrogen bonds). Increasing the temperature of the sample puts enough thermal energy into the molecule to break the dipole-dipole interaction thus increasing the purge efficiency. The total amount of analyte removed from the sample is directly proportional to the purge volume. The purge volume is a product of the purge flow rate and the purge time. The recommended gas for both purging and for desorption is helium. The purge gas flow typically is set at 30–50 mL min^{-1}. The desorb gas flow ranges from 10–80 mL min^{-1}, depending on the column type and GC equipment used.

Advantages and disadvantages

Due to the high sensitivity and recoveries, PT still remains the most frequently used pre-concentration method for the determination of volatile compounds in water. In addition to high sensitivity, PT has the advantages of good precision and the possibility of automation.

One drawback associated with this methodology is excessive water vapor that is purged with the volatiles by the stream of the inert gas. This gives rise to peak distortion, especially in the early part of the chromatogram. To avoid this problem, a device combining a solvent elimination system, consisting of a Nafion desiccator (whose walls are permeable to water vapor but not to organic compounds) and a double-trap system with different sorbents have been developed. Moreover, the water problem is even more prominent when cryogenic traps are used because the trap may become blocked by ice plugging. Therefore, these traps are usually combined with a "drying step" in which the water vapor is removed prior to cryogenic trapping.

Purge and trap methods for organometallic species pre-concentration

Arsenic, antimony and tellurium species. The first application using derivatization by hydride generation (HG) followed by purge and trap-gas chromatography (PT-GC) for arsenic compounds (As(III), As(V), MMA and DMA) was developed in 1973 (Braman and Foreback 1973). This method comprised borohydride reduction of As(III) to arsine at pH 6, and reduction of As(V), MMA and DMA to their respective arsines, monomethyl arsine and dimethyl arsine, at pH 1. The arsines were then trapped and volatilized by either flash heating followed by GC (or by successive thermal desorption from the GC material in the trap). Although high sensitivity was achieved, this technique was limited to the determination of those hydride forming species.

PT-GC interfaced with atomic fluorescence spectrometry (AFS) has been developed for quantification of arsenic species (As(III); As(V); MMA; DMA; trimethylarsine oxide, TMAO; arsenobetaine, AsB; arsenocholine, AsC; and tetramethylarsonium ion, TMAI). The purge vessel was constructed from Pyrex glass and had a sodium borohydride introduction port with a septum. A helium flow purged the sample via a capillary tube and stripped the arsines from the solution and swept them into a liquid nitrogen-cooled U-tube. The U-tube was constructed from Pyrex glass and the upstream leg was packed (height, 6.5 cm) with Chromosorb W (0.15–0.18 mm) (Šlejkovec et al. 1998).

After derivatization with $NaBH_4$, PT-GC coupled with inductively coupled plasma-mass spectrometry (ICP-MS) has also been applied to study the formation of ionic and volatile arsenic compounds (As(III); As(V); MMA; DMA; trimethylarsenic, TMA; monomethylarsonic acid, MMAA $(CH_3)As(O)(OH)_2$; dimethylarsinic acid DMAA $(CH_3)_2$, As(O) OH; and TMAO) produced in a bath culture of the anaerobic methanogen *Methanobacterium formicicum* (Wickenheiser et al. 1998); and also for arsine (AsH_3), methylarsine $((CH_3)AsH_2)$, dimethylarsine $((CH_3)_2AsH)$, and trimethylarsine $((CH_3)_3As)$ released from bacteria cultures (Yuan et al. 2008). Trimethylantimony (TMSb) produced by the filamentous fungus *Scopulariopsis brevicaulis* culture grown in an antimony(III)-rich medium under aerobic conditions was also determined using PT-GC-ICP-MS (Andrewes et al. 1999). The determination of inorganic and methylated species of As (MMA and DMA) and Sb (TMSb and dimethylantimony (DMSb)) in marine and freshwater samples (Ellwood and Maher 2002) and mono-, di- and trimethylated species of arsenic, antimony and tin in urban soils (Duester et al. 2005) have been undertaken using PT-GC-ICP-MS. Volatile species produced were initially trapped on the pretrap (–196°C) and then they can be cryofocused on the column (Wickenheiser et al. 1998), or trapped in a chromatographic column immersed in liquid nitrogen (Yuan et al. 2008) or in a U-shaped tube containing Supelcoport SP 2100 at $-78°C$ (Andrewes et al. 1999). On other occasions, a Teflon cryogenic trap packed with Chromosorb W was used for trapping the volatile species (Ellwood and Maher 2002).

PT-GC coupled with MS has also been used for arsenic (diarsine, H_2As-AsH_2; monomethyl diarsine, $(CH_3)_2As$-AsH_2; and dimethylarsenomercaptane, $(CH_3)_2As$-$S(CH_3)$) determination in hot springs waters (Kösters et al. 2003). The application also included the assessment of antimony (TMSb) and tellurium (dimethyltellurium, DMTe) species. After $NaBH_4$ derivatization, volatile analytes were purged with helium into a Tenax® trap at room temperature. The trap is put into an automatic desorbing unit and the trapped analytes are desorbed at a temperature of $250°C$ and transferred to the GC/MS instrument.

Methylated antimony compounds (TMSb, DMSb, monomethylantimony (MMSb)) and inorganic antimony Sb(III) have been determined in plants using PT coupled with quartz furnace atomic absorption spectrometry (QF-AAS) after $NaBH_4$ derivatization (Craig et al. 1999). The reaction solution was purged with helium and the antimony compounds were trapped in a glass column using liquid nitrogen ($-196°C$). The species were then selectively eluted according to their boiling points by removing the liquid nitrogen and heating.

Mercury species. For Hg(II) and MMM, the Bloom's method (Bloom 1989), which is the basis for EPA method 1630 (EPA Method 1630 2001), is usually used when analyzing aqueous samples or diluted extracts from solid samples. Targets are derivatized with $NaBEt_4$ forming the volatile (CH_3) $(CH_3CH_2)Hg$ and $(CH_3CH_2)_2Hg$ species, which are then purged from solution onto a Tenax® trap, followed by thermal desorption and transfer to a packed GC column. After separation on the column, alkylated Hg species are pyrolyzed and detected using AFS.

Several adaptations of this method have been reported, one of the most powerful being the use of ICP-MS to replace AFS as the detector. ICP-MS can be used with isotope dilution (Hintelmann et al. 1995, Hintelmann and Evans 1997, Hintelmann et al. 1997, Rodríguez Martin-Doimeadios et al. 2003, Lambertsson and Björn 2004, Perna et al. 2005, Dzurko et al. 2009, Jackson et al. 2009, Baxter et al. 2011), although the former operation mode has also been applied (Amouroux et al. 1998, Lambertsson et al. 2001, Mao et al. 2008, Jae-Sung et al. 2010, Mao et al. 2010). Isotope dilution offers an advantage during the quantification of inorganic mercury, which is prohibited by external calibration because of the decomposition of MEM to Hg^0 during trap heating. A drawback of ethylation is that both inorganic mercury and MEM present in the sample are ethylated to form Et_2Hg, making these species indistinguishable. The use of propylating or phenylating reagents allows determination of EtHg in the sample, and these reagents are robust over a larger range of pH, and form more stable derivatives than NaBEt$_4$.

Developments include the determination of MMM in sediments (Hintelmann et al. 1995, Hintelmann and Evans 1997, Mao et al. 2008, Jae-Sung et al. 2010), biological samples (Mao et al. 2008, Jae-Sung et al. 2010), lake water (Jackson et al. 2009), plasma and serum samples (Baxter et al. 2011), mire, fresh and seawater (Lambertsson et al. 2001), soil (Mao et al. 2010). Inorganic mercury assessment (Hintelmann and Evans 1997), diethylmercury (DEM) (Amouroux et al. 1998, Mao et al. 2008), and simultaneous dimethylselenium (DMSe), dimethyldiselenium (DMDSe$_2$), dimethylmercury (DMM), tetramethyltin (TeMT), tetraethyltin (TeET), trimethyllead (TML), tetraethyllead (TeEL) determination (Amouroux et al. 1998, Mao et al. 2008) in natural waters have been described. Volatile compounds generation was achieved after using NaBH$_4$ (Hintelmann et al. 1995), NaBEt$_4$ (Lambertsson et al. 2001, Amouroux et al. 1998, Jackson et al. 2009, Jae-Sung et al. 2010, Baxter et al. 2011), NaBPh$_4$ (Fernández et al. 2000, Grinberg et al. 2003a, Mao et al. 2008, Mao et al. 2010), or NaBPr$_4$ (Fernández et al. 2000, Demuth and Heumann 2001, Grinberg et al. 2003a) for derivatization. PT pre-collection on quartz tube/liners packed with Tenax® adsorbent has been proposed (Hintelmann et al. 1995, Hintelmann and Evans 1997, Lambertsson et al. 2001, Mao et al. 2008, Jackson et al. 2009, Jae-Sung et al. 2010, Mao et al. 2010, Baxter et al. 2011). Thermodesorption was typically performed at 200°C, while the cryogenic trap operated at –196°C with flash desorption at 300°C (Hintelmann et al. 1995, Hintelmann and Evans 1997, Lambertsson et al. 2001, Mao et al. 2008, Jae-Sung et al. 2010, Mao et al. 2010, Baxter et al. 2011).

Recently, a new automated system for MMM determination using PT-GC with AFS or ICP-MS was developed. Ethylated Hg species were transported to three Tenax® traps in the PT unit by a N$_2$ stream. Hg species were thermally desorbed by IR radiation and then passed to the GC/pyrolysis unit where they were isothermally separated (Taylor et al. 2011).

PT-GC in line with a Fourier transform infrared spectrometer (FTIR) (Filippelli et al. 1992, Filippelli and Baldi 1993, Filippelli 1994) after NBH$_4$ derivatization has also been developed for monomethylmercury (MMM) and dimethylmercury (DMM) determination in fish and sediment extracts and natural waters. The sample was purged with nitrogen, and volatile compounds were concentrated in a cold trap (–120°C). The trap was heated (250°C), and targets were automatically injected into the GC column hyphenated to the FTIR spectrometer (Filippelli et al. 1992, Filippelli and Baldi 1993).

Quartz furnace-atomic absorption spectrometry (QFAAS) has also been coupled with PT-GC after hydride generation ($NaBH_4$) for Hg(II), MMM, DMM and DEM determination in estuarine waters (Puk and Weber 1994) and in marine biological samples (Välimäki and Perämäki 2001). Quartz tube trap are typically filled with 200 mg of Tenax® and the trapped mercury species are thermally desorbed at 200°C.

Several applications using MIP-AED as a detection system coupled with PT-GC have also been used for methylated species of Hg (Hg(II), MMM and DMM), Sn (Sn(IV), monomethyltin (MMT), dimethyltin (DMT), trimethyltin (TrMT), tetramethyltin (TeMT) and MBT) and Pb (TeEL) in river and soil run-off water and fish extracts (Ceulemans and Adams 1996, Gerbersmann et al. 1997, Reuther et al. 1999) after ethylation (Ceulemans and Adams 1996, Gerbersmann et al. 1997) or hydride generation ($NaBH_4$) (Reuther et al. 1999). Pre-concentrated species on a capillary cryogenic trap (fused-silica liner at –100°C (Ceulemans and Adams 1996), or –170°C (Gerbersmann et al. 1997)), or on Chromosorb (at –160°C) (Reuther et al. 1999) were desorbed by heating at 200°C. Recently, a new instrument, automated speciation analyser (ASA) based on a combination of a PT, a multicapillary column and a miniature MIP-ED was developed for the pre-concentration (cooled fused silica capillary coated with a 5 µm CP-Sil 5CB layer), separation and quantification of Hg(II) and MMM from environmental samples (Slaets and Adams 2000).

PT-GC coupled with furnace atomization plasma emission spectrometry (FAPES) was also used for Hg(II) and MMM determination in biological tissue extracts after ethylation ($NaBEt_4$), purging (with He), trapping on Tenax®, and thermal desorption onto an isothermal (90°C) GC column (Jiménez and Sturgeon 1997).

PT-GC coupled with AFS was applied to MMM and DMM determination in seawater, sediment and tissue samples after ethylation, trapping on Carbotrap™ (Pongratz and Heumann 1998, Bloom et al. 2005) or Tenax® (Bowman and Hammerschmidt 2011) and thermal desorption. Carbotrap™ was found to be the best choice as a sampling media, whereas Tenax® was found to be inadequate due to high breakthrough (> 70%) (Bloom et al. 2005).

Finally, substitution of the packed column GC with a capillary GC has been demonstrated to improve resolution and peak shape, particularly for heavier organometallic species where the capillary GC offers a much larger number of theoretical plates, providing higher resolution. Additionally, gradient heating can be used to decrease retention times of later eluting compounds and provide further improvement in the peak shape. However, the major drawback associated with capillary GC is a much lower column capacity than packed column GC. This fact limits the amount of sample that can be loaded and makes it incompatible with purge and trap preconcentration, and hinders achievement of low detection limits. Multicapillary gas chromatography (MCGC) hyphenated to ICP-MS (Slaets et al. 1999, Leenaers et al. 2002, Jitaru et al. 2003) or MIP-AED (Rodríguez-Pereiro et al. 1998) have also been coupled to PT for Hg(II) and MMM determination in ice, urine and fish samples. After *in situ* ethylation, targets are trapped in a narrow bore capillary coated with a 5 µm layer of 5 percent phenyl–95 percent dimethylpolysiloxane at –75°C (Leenaers et al. 2002, Jitaru et al. 2003) or cryotrapped (–100°C) in a fused silica capillary coated with CP-Sil 8 CB layer, followed by their flash desorption (Rodríguez-Pereiro et al. 1998, Slaets et al. 1999).

Selenium species. The most common approach for selenium species preconcentration by PT involves cryogenic trapping followed by thermal desorption (Tanzer and Heumann 1990, Masscheleyn et al. 1991, Cai et al. 1995, Amouroux et al. 1998, de la Calle-Guntiñas et al. 1995, 1999, Pécheyran et al. 1998, Zhang et al. 2000, Tessier et al. 2002). DMSe and DMDSe from natural water and plants (Tanzer and Heumann 1990, de la Calle-Guntiñas et al. 1995, Amouroux et al. 1998, Pécheyran et al. 1998, Tessier et al. 2002, Campillo et al. 2005) and DMSe and other species such allylmethyl selenide, methanesulfenoselenoic acid methyl ester and 2-propenesulfenoselenoic acid methyl ester (Cai et al. 1995) from human breath have been preconcentrated by PT followed by GC separation and MIP-AED (de la Calle-Guntiñas et al. 1995), AFS (Pécheyran et al. 1998), ICP-MS (Amouroux et al. 1998, Tessier et al. 2002), AED (Cai et al. 1995) and FPD (de la Calle-Guntiñas et al. 1999) determination. In addition, several non-chromatographic procedures were also proposed (Masscheleyn et al. 1991, Zhang et al. 2000). After purging by passing a helium flow through the aqueous solution (maintained at 35°C), Se species were trapped in a packed column immersed in liquid N_2 (–196°C) (Tanzer and Heumann 1990, Masscheleyn et al. 1991, de la Calle-Guntiñas et al. 1995, 1999, Cai et al. 1995, Amouroux et al. 1998, Pécheyran et al. 1998, Zhang et al. 2000, Tessier et al. 2002) or in a column coated with Tenax®, silica gel and activated carbon (at 40°C) (Campillo et al. 2005). Once concentrated, the volatile selenium compounds were desorbed by heating the trap at 210°C. The PT system was directly coupled with the gas chromatograph.

PT coupled with QFAAS (Masscheleyn et al. 1991) and HGAAS (Zhang et al. 2000) without chromatographic separation have been applied for the determination of inorganic selenium species (Se(IV) and Se(VI)) and organic selenium forms (DMSe and dimethylselenoxide (DMSeO), methylselenomethionine (MSeMet) and dimethylselenoniopropionate (DMSeP)) in sediment, soils and water samples. After DMSe purge out and trapping on a U-tube immersed in liquid N_2, Se(IV) and oxidized methylated Se-compounds are reduced using $NaBH_4$ to H_2Se and DMSe, respectively (Masscheleyn et al. 1991). Entrained by a He carrier gas the H_2Se and DMSe are trapped in a liquid N_2 cooled U-tube, and the separation is accomplished by controlled heating of the sample trap (Masscheleyn et al. 1991). The second non-chromatographic method involved the reaction with $NaBH_4$ followed by a purging stage of the derivatized species into alkaline H_2O_2 traps (Zhang et al. 2000). The alkaline H_2O_2 solutions were heated in a hot water bath (50–70°C) until all H_2O_2 was decomposed and all Se was oxidized to Se[VI]. Se(VI) was then reduced to Se(IV) and determined by HGAAS (Zhang et al. 2000).

Lead species. Tetramethyllead (TeML) and triethyllead (TEL) (Amouroux et al. 1998, Reuther et al. 1999); TML (Craig et al. 1995, Wasik et al. 1998); and TeML, dimethyllead (DML), monomethyllead (MML) and Pb(IV) (Wasik et al. 1998) in rain water and road dust have been determined by PT-GC coupled with ICP-MS and MIP-AED after ethylation. Organolead compounds were trapped at –100°C in a thick film-coated capillary tube, followed by electrically heating for desorption before injection of the desorbed analytes onto the GC (Wasik et al. 1998).

Tin species. Butyl-, methyl- and phenyltin compounds in freshwater, seawater and sediment have been assessed by different PT-GC methods coupled with flame photometric detector (FPD) (Jo-Anne et al. 1982), AAS (Dowson et al. 1992, Cai et al. 1993), ICP-MS (Amouroux et al. 1998), MS (Eiden et al. 1998, Kösters et al. 2003), QF-AAS (Bowles et al. 2004), and MIP-AED (Reuther et al. 1999, Campillo et al. 2004). Hydridization with $NaBH_4$ (Jo-Anne et al. 1982) and ethylation with $NaBEt_4$ (Cai et al. 1993, Eiden et al. 1998, Bowles et al. 2004, Campillo et al. 2004) has been usually performed. Traps filled with adsorbents such as Tenax® (Jo-Anne et al. 1982, Eiden et al. 1998, Bowles et al. 2004, Campillo et al. 2004) or Chromosorb cooled with liquid N_2 (Cai et al. 1993), followed by thermal desorption have been used.

Finally, a PT non-chromatographic method for organotin compounds (TPhT, DPhT, MPhT, TBT, DBT, MBT and inorganic tin) in seawater has been developed (Sato et al. 1996). After volatilization by hydride generation and trapping on Chromosorb, organotin compounds were separated on the basis of the different boiling points before ICP-MS detection.

Solid phase extraction

General aspects

Solid phase extraction (SPE) is based on the partitioning of compounds between a liquid phase (sample) and a solid phase (ion exchange resin, polymeric resin, activated carbon, silica, etc.) whereby the intermolecular forces between the phases influence retention and elution.

There are two SPE modes, the batch and column methods. The batch method consists of mixing the sample, at a fixed pH, with a few grams of resin, the mixture is then shaken or stirred for several minutes to facilitate analyte adsorption onto the solid phase. After filtration or centrifugation, desorption is achieved by stirring the solid phase with a few milliliters of an adequate solvent. Analytes are finally determined in this eluate solution. In the column method (Fig. 2.3), a PTFE or glass column or a syringe is packed with a few milligrams of solid phase. A small amount of glasswool, or a frit made of suitable materials, is placed at both ends to prevent loss of the particles during sample loading. The use of commercial cartridges, disks or SEP-PAK devices (various adsorbents within a plastic syringe) has also been used for this pre-concentration mode. Before use, several solvents and doubly distilled de-ionized water must be passed successively through the column to equilibrate, clean and neutralize the sorbent material. The sample, at the desired pH value, is then passed through the column at a fixed flow rate by using a peristaltic pump. Afterwards, the analytes retained on column could be eluted with the appropriate solvent and the analytes in the eluate solution determined.

Retention on the solid phases may involve non-polar, polar or ionic interactions and Van der Waals forces, or hydrophobic interactions, which are not very strong. On other occasions, strong interactions can be attained. In this case, the addition of chelating agents to the sample, prior to the extraction, or the use of solid phases with immobilized chelating groups is required.

There are different solid phase materials that can be used for metal and organometallic compound pre-concentration. These materials must fulfill several requirements to be used as solid-phase extraction adsorbents. First,

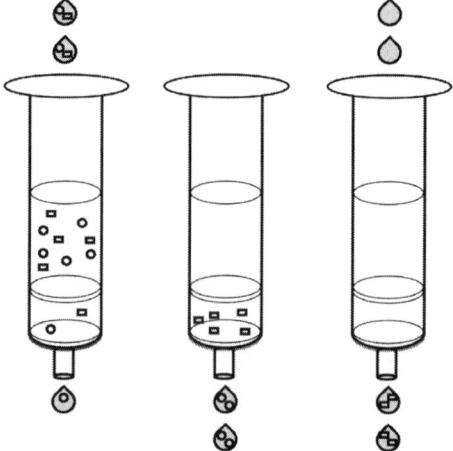

Figure 2.3 Schematic of solid phase extraction (SPE) set-up.

these supports must allow the extraction of analytes in a wide pH range, and they must provide a high sorption capacity, selectivity and enrichment factors. In addition, a fast and quantitative adsorption of the target analytes and easy desorption of the adsorbed compounds is desirable. Finally, the solid supports should have a long life time so that repeated adsorption/desorption cycles can be performed with the same sorbent.

The wide range of sorbents available for SPE pre-concentration, including synthetic and natural materials provides a wide range of interactions. The current trends in SPE are mainly focused on the development of new sorbents such as nanometer-sized materials, egg-shell membranes, modified silica beads, molecularly/ion imprinted polymers, or mesoporous materials. These new solid supports are prepared mainly to improve selectivity.

Optimization of the SPE process involves the study of several factors, such as: (1) the choice of sorbent; (2) the volume of the sample flushed through the column (dynamic mode); (3) the derivatizing agent; (4) the mass of the sorbent and breakthrough volume of the sorbent (volume of sample per gram of adsorbent resin which causes the analyte molecules to migrate from the front of the adsorbent bed to the back of the adsorbent bed); (5) the elution profile; (6) the sample pH; and, (7) the retention (adsorption) and elution flow rates. The choice of the sorbent is a key factor in SPE because it can control parameters such as selectivity, affinity and capacity. The choice depends largely on the analyte and the interactions with the chosen sorbent. However, it also depends on the kind of sample matrix and its interactions with both the sorbent and the analyte.

Another important consideration is related to the washing and conditioning of the column before the pre-concentration step. After the sorbent material has been packed into the column or micro-column, organic solvents, dilute acids or high purity de-ionized water must be passed through the column in sequence to clean the solid support. Then, the column must be conditioned to the desired pH by passing through the column the same buffer solution used for fixing the pH of samples.

Advantages and disadvantages

SPE has gradually replaced classical LLE and is becoming the most common sample pre-treatment technique in environmental areas. SPE overcomes the problems arising from the use of organic solvents and offers several major advantages that include (1) higher enrichment factors, (2) simple operation, (3) lower cost and less time, (4) improved selectivity, specificity and repeatability, (5) safety with respect to hazardous reagents, (6) less volume requirement and the use of environmentally friendly reagents, (7) absence of foaming and emulsions, (8) the ability for coupling with different modern detection techniques in on-line and off-line modes, and (9) ease of automation.

However, some disadvantages such as (1) low recovery of analyte, (2) low selectivity, (3) incomplete interferences removal, (4) high variability (RSDs), (5) high cost (time and materials), and (6) expensive instrumentation for off-line and on-line SPE; are usually related to conventional SPE.

Solid phase extraction for organometallic species pre-concentration

Several methods for inorganic As, Sb, Se and Te species and for organic species of As, Hg, Se and Sn in liquid samples (surface water, groundwater, seawater, waste water, tap and drinking water, pool water) and extracts from solid materials (hair, nail, blood, fish tissue and vitamin tablets) have been developed by using very different solid phase materials.

Inorganic As, Sb, Se and Te species. Different substances such as synthetic resins (ion exchange resins, polymeric resins, chelating resins) and biological substances (micro-organisms) have been proposed and applied as solid-phase extraction sorbents for inorganic As, Sb, Se and Te species.

Anion exchange resins such as Dowex 1-X8 (Smichowski et al. 2002, Stripeikis et al. 2004), strong anion exchange cartridges (500 mg silica-based, anion exchange sorbent) (Gong et al. 2006), strong anion exchange SAX resin (silica-based chloride-form) with trimethylaminopropyl functional groups (Sigrist et al. 2011), Amberlite IRA 910 (Calvo-Fornieles et al. 2011), and Muromac® anion exchange resin (chloride-form) (Jitmanee et al. 2007) have been used as solid phase extractors for As(V) (Smichowski et al. 2002, Gong et al. 2006, Sigrist et al. 2011), Sb(V) (Calvo-Fornieles et al. 2011), Se(IV) (Stripeikis et al. 2004) and Se(VI) (Stripeikis et al. 2004, Jitmanee et al. 2007). Activated alumina (Al_2O_3) has also been used for Se(IV) and Se(VI) pre-concentration (Pyrzynska et al. 1998, Wójcik et al. 2003). Alumina offers many desirable ion exchange properties, e.g., it undergoes little swelling or shrinking in electrolyte solution, is stable under strong oxidizing and reducing conditions, and it can operate as an anion and as a cation exchanger depending on the working pH.

Polymeric resins have also been proposed for inorganic As, Sb, Se and Te species pre-concentration. There are two ways for analyte pre-concentration by using these solid supports. In the first method, chelates of the analyte must be formed by adding the specific chelating reagent to the sample, and then passing the sample through the column filled with the resin (column mode). When performing the SPE batch mode, the sample must be stirred with the resin for retaining the chelates. In the second method, chelating agents are first immobilized on the solid phase (chelating resins) and then, the sample is passed through the column containing the modified solid phase (the sample is stirred with the modified chelating resin when performing SPE batch modes). The use of chelating resins is especially appealing because of the high selectivity offered. Therefore, different applications have been developed, such as the use of CTAB-modified alkyl silica sorbent for As(V) and Se(VI) pre-concentration (Xiong et al. 2008), chelating celluloses (imino diacetic acid–ethyl cellulose, IDAEC) for Sb(III) pre-concentration, and 2,2'-diaminodiethylaminecellulose (DENC) for Sb(III) and Sb(V) pre-concentration (Zhi-Perényi et al. 2008). On other occasions chelating reagents such as [1,5-bis(2-pyridyl)-3-sulfophenyl methylene] thiocarbonohydrazide immobilized on aminopropyl-controlled pore glass (PSTH-cpg) for Sb(III) pre-concentration (Calvo-Fornieles et al. 2011), polyurethane foams (PUFs) loaded with the chromogenic reagent 4,4-dichlorodithizone (Cl_2H_2DZ) for Se(IV) pre-concentration (El-Shahawi and El-Sonbati 2005), and iron(III)-loaded Chelex-100 resin (Ferri and Sangiorgio 1996) have been used. Non-polar sorbents such as C_{18} or C_8 have also been used after treatment with APDC (Yu et al. 2002, 2003a, 2004, Mulugeta et al. 2010) for As(III), Sb(III), Se(IV) and Te(IV) pre-concentration. Diaion HP-2MG (hydrophilic nature methacrylic ester copolymer) has also been proposed for Se(IV) pre-concentration after APDC derivatization (Saygi et al. 2007).

Living or non-living micro-organisms can accumulate metals and metalloids from aqueous solutions by different chemical and biological mechanisms without pre-concentrating the matrix. This ability is attributed to stable complexes formed between the analytes and any functional groups present in the cell walls of the micro-organisms. In this technique, micro-organisms are immobilized on a solid support or are used as freely suspended cells. The use of loaded cell systems on a support has many advantages over the use of freely suspended cells. These include better capability of reusing the biomass, better precision for recovery, easy separation of cells from the reaction mixture, high biomass loadings, and minimal plugging in continuous flow systems. As an example, immobilization of *Streptococcus pyogenes* on Sepabeads SP 70 has been used for As(III) pre-concentration (Uluozlu et al. 2010).

All the procedures commented above, except for the work of Wójcik (Wójcik et al. 2003), are non-chromatographic approaches. These methods are therefore based on the selective complexation/adsorption of As, Sb, Se or Te at one specific oxidation state. After reduction or oxidation, the SPE can be repeated for assessing the total concentration of the inorganic target (the amount of analyte exhibiting the other oxidation state is then calculated by the difference). Finally, some of these SPE methods (Smichowski et al. 2002, Yu et al. 2002, 2003a, 2004, Gong et al. 2006, Sigrist et al. 2011) are used as selective separation procedures without pre-concentration purposes.

Organic As, Hg, Se and Sn species. Several anion and cation resins, activated carbon, polymeric resins, chelating resins and micro-organisms have been proposed and applied as solid-phase extraction sorbents for assessing organic As, Hg, Sb, Se and Sn species.

Anion/cation exchange resins

Sep-pak VAC ($-C(O)NH(CH_2)_3N(CH_3)_3^+$ Cl$^-$) has been used for As(V), MMA and DMA adsorption before target analyte elution with H_3PO_4 (Gómez et al. 1997); whereas, selective roxarsone (3-nitro-4-hydroxyphenylarsonic acid, 3-NHPAA) adsorption was achieved on strong anion exchangers (HAX cartridges) (Jaafar et al. 2007). On other occasions, As(V) and MMA adsorption onto SAX-3 cartridges has been proposed, and selective elution with acetic acid was reported (Yu et al. 2003b).

Other applications involved DMA, AsB, AsC, TMAI and TMAO adsorption on SCX-3 cartridges (ethylbenzene sulfonic acid-based strong cation exchange resin) followed by a sequential elution with 1.0 M HNO$_3$ (Yu et al. 2003b). In this development As(III) was not retained and remained in solution (Yu et al. 2003b). Similarly, strong anion (SAX) and cation exchange (SCX) columns in tandem have been proposed for retaining As(V), MMA and DMA (As(III) remained in the aqueous sample) (Menéndez-Sánchez et al. 2009, Watts et al. 2010). Selective MMA elution was achieved with an acetate buffer at pH 3.4 (Menéndez-Sánchez et al. 2009) or with 80 mM acetic acid (Watts et al. 2010); whereas, As(V) and DMA elution was performed with 1.0 M HNO$_3$ (Menéndez-Sánchez et al. 2009, Watts et al. 2010).

Anion exchange resins based on a styrene-divinylbenzene copolymer (Lewatit Mono-Plus M 500) have also been used for As(V), MMA and DMA adsorption, allowing the direct As(III) (not retained) determination using ICP-MS (Issa et al. 2011). In this application a hybrid resin HY-Fe (hybrid macroporous monodispersed polystyrene-based resin) was used for retaining As(III), As(V) and MMA, while the unretained DMA in the effluent was determined using ICP-MS (Issa et al. 2011). Similarly, As(III) and As(V) can be retained on a HY-AgCl resin (silver loaded ion exchange), allowing the ICP-MS determination of the unretained MMA and DMA (Issa et al. 2011).

A similar approach has been adopted by Voice et al. (Voice et al. 2011) using Dowex 50W cation exchanger. Under optimized conditions the DMA was adsorbed on the cation exchange resin while unretained As(III), As(V) and MMA were collected. DMA species were then eluted with 1.0 M HCl (Voice et al. 2011). The procedure continued by retaining the As(V) and MMA species on cation exchange resin (Chromosorb W, silanized diatomaceous earth modified with dioctyltin dichloride), allowing As(III) species to remain in the effluent. As(V) and MMA were then eluted with 0.2 M HCl (0.2M) and 1.0 M HCl, respectively (Voice et al. 2011).

A similar application was developed by Shi et al. (Shi et al. 2001), which involved the use of Dowex 50W-X8-200 for retaining AsB and DMA (unretained As(III), As(V) and MMA were collected in the effluent). AsB and DMA were then eluted with 1.0 M NH$_3$ before HPLC measurement. The unretained As(III), As(V) and MMA in the effluent were then also separated and quantified using HPLC (Shi et al. 2001).

Hg(II) adsorption can be performed on a column packed with the Dowex 1X-8 anion exchange resin after complexation with methylthymol blue (MTB) and elution with 3.0M HNO$_3$ (de Wuilloud et al. 2002). The MMM species adsorption was achieved after oxidation to inorganic mercury and MTB complexation (total mercury determination). The MMM concentration was then calculated by the difference (de Wuilloud et al. 2002).

Other developments are based on Hg(II), MMM, MEM and MPhM adsorption on anion exchange (quaternary amine type resin) with on-column formation of Hg complexes with mercaptopropionic acid (MPA) and elution with 100 mM KClO$_4$ (Houserová et al. 2006). Hg(II), MMM, MEM and MPhM adsorption can also be performed on strong cation exchangers (SCX, benzenesulfonic acid-type) and L-cysteine elution (Delgado et al. 2007). Finally, Hg(II) adsorption on an anion exchange membrane disk (Anion-SR) after derivatization to a tetra-chloride complex has been also reported (Serra et al. 2009). In this development MMM did not interact with the adsorbent and a subsequent reduction with 2.0% (w/v) SnCl$_2$ 2.0% w/v) and UV irradiation was therefore needed before total mercury could be pre-concentrated (Serra et al. 2009).

Se(IV), Se(VI), DMSe, DMDSe, diethylselenium (DESe) and diethyldiselenium (DEDSe) adsorption have been performed on strong anion (SAX) and octadecylsiloxane C$_{18}$ cartridges in tandem. Selective Se(IV) and Se(VI) elution from a SAX column was achieved using 1.0 M acetic acid and 1.0 M HCl, respectively. Organic Se elution from a C$_{18}$ cartridge was then performed with CS$_2$ (Gómez-Ariza et al. 1999b). Other applications used an Amberlite IRA-743 resin column for retaining Se(IV), Se(VI), SeMet and SeCys before 1.0M HClO$_4$ elution (Bueno and Potin-Gautier 2002); whereas strong cation-exchange (SCX) cartridges can be used for SeMet, MSeMet and methylselenocysteine (MSeCys) adsorption before 8.0 mM pyridinium formate elution (Ketavarapu-Yathavakilla et al. 2005). A last application of SCX cation exchanger is focused on adsorbing dextro selenomethionine (D-SeMet) and levo selenomethionine (L-SeMet) (Gómez-Ariza et al. 2007).

Some metal-loaded activated charcoal (MC*) applications have also been reported by Latva et al. (Latva et al. 1999, 2000). In these developments phenylarsonic acid (PAS) species were adsorbed on VC* or CeC*, while after filtration, As(V) species were collected onto LaC*, and As(III) species were separated by APDC coprecipitation (Fe(III) as a carrier, and the precipitate was bound onto activated charcoal). Finally, DMA in the filtrate was collected onto ZrC*. Arsenic species concentrations in the MC* were directly measured by energy dispersive X-ray fluorescence (EDXRF) (Latva et al. 1999) or GFAAS (Latva et al. 2000).

Finally, MBT, DBT and TBT adsorption on ENVI-Carb (non-porous carbon) before elution with tropolone has been also proposed (Schwarz et al. 1995).

Polymeric resins

Most of the developments are focused on mercury speciation. Applications using C18 Sep-pak cartridges modified with 2- mercaptoethanol were developed for Hg(II) and MMM pre-concentration (Aizpún et al. 1994, Cairns et al. 2008). Target analyte elution was achieved using acetonitrile (ACN) (Aizpún et al. 1994) or L-cysteine (0.5% w/v) (Cairns et al. 2008). On other occasions, Hg(II), MMM, MEM, and MPhM species were adsorbed on RP C_{18} (ODS Hypersil) (column mode) after target analyte derivatization with sodium pyrrolidinedithiocarbamate (SPDC) followed by deionized water/acetonitrile elution (Falter and Schöler 1995, Falter and Ilgen 1997). In the same way, Hg(II), MMM, MEM and MPhM can be adsorbed on C_{18} micro-columns after on-line APDC derivatization (Yin et al. 1998, Houserová et al. 2007a, Río-Segade and Tyson 2007). Elution was achieved using methanol/ACN/DIW mixtures (Yin et al. 1998) or methanol (Houserová et al. 2007a), although the slurry sampling techique coupled with cold vapor atomic absorption spectrometry (CV-AAS) can also be performed (Río-Segade and Tyson 2007). Dithizone (DZ) immobilized on a reversed-phase C_{18} cartridge was also used to retain Hg(II), MMM and MPhM (Sánchez et al. 2000, Margetinová et al. 2008, Yong-Guang et al. 2010). Methanol (Sánchez et al. 2000, Margetinová et al. 2008) or $Na_2S_2O_3$ solution (Yong-Guang et al. 2010) were used for elution. Similarly, octadecyl silica C_{18} extraction disks modified with 1,3-bis(2-cyanobenzene)triazene (CBT) and ACN elution was also proposed for Hg(II), MMM, and MPhM pre-concentration (Hashempur et al. 2008).

Hg(II), MMM, MEM, and the tin species Sn(IV), MBT, DBT and TBT were also adsorbed on C_{60} fullerene (column mode) after complexation with sodium diethyldithiocarbamate (SDDC) before eluting with ethyl acetate containing $NaBPr_4$ as derivatizing reagent (Muñoz et al. 2005a).

Other developments consisted of using a column packed with PTFE for adsorbing Hg(II) after APDC derivatization allowing the determination of the unretained MMM complex using CV-AAS (Zachariadis et al. 2005). Adsorption onto the inner walls of KRs (150 cm) after Hg(IIcomplexation with ammonium diethyl dithiophosphate (ADDP) and MMM complexation with DZ has also been described (Wu et al. 2006). In this application, 15% (v/v) HCl was used as an eluting solution. Finally, Hg(II), MMM, MEM and MPhM species have been adsorbed on a column of PUFs loaded with SDDC before n-butanol elution (dos Santos et al. 2009). Another application involved MMM and MEM adsorption onto sulfydryl cotton fibre (SCF) adsorbent packed in a screening column. Elution was performed with a KBr/CuSO$_4$ mixture followed by a DCM back-extraction (Cai et al. 1996).

Concerning selenium species, C$_{18}$ cartridges modified with hexanesulfonic acid were proposed for retaining SeMet, MSeMet, trimethylselonium (TMSe), selenoethionine (SeEt) and selenoadenosylmethionine (AdoSeMeT) before elution with methanol (Wrobel et al. 2003a). Finally, arsenic species (As(III), As(V), MMA and DMA) were adsorbed on MnO$_2$ (140–160 mg) mini-columns. The process involved As(III) oxidation to As(V) followed by elution with tetramethylammonium hydroxide (TMAH) (Tian et al. 2011).

Chelating resins

A dithiocarbamate resin has been proposed for retaining Hg(II) and MMM. After shaking (for 8–22 hr) and filtration, Hg species were eluted with an acidic thiourea solution, followed by Hg species extraction with toluene (500µL) and butylation with a Grignard reagent (butylmagnesium chloride in tetrahydrofuran (THF)). The butylated mercury forms were then injected (13 µl) into GC-MIP-AES (Emteborg et al. 1995, Hänström et al. 1996).

A chloromethylated Merrifield resin containing a 1,2-bis (o-aminophenylthio) ethane moiety (column mode) was also used for Hg(II) and MMM adsorption. Elution was performed with ethanol (75% v/v) and thiourea (10% w/v) in 1.0 M HCl for Hg(II) and MMM, respectively (Mondal and Das 2003).

Hg(II) and MMM on column adsorption was also carried out with polyaniline (PANI) (Balarama-Krishna et al. 2005, 2010) followed by sequential elution with HCl (0.3% v/v) for MMM and 0.3% HCl (containing 0.02% w/v thiourea for Hg(II)) (Balarama-Krishna et al. 2005, 2010).

Finally, Hg(II) adsorption was also achieved on a column filled with aminopropyl-controlled pore glass functionalized with [1,5-bis (2 pyridyl)-3-sulfophenyl methylene thiocarbonohydrazide] and elution with thiourea (6.0% w/v) (Vereda-Alonso et al. 2008). MMM adsorption was performed after oxidation to Hg(II) (total mercury determination), and the MMM concentration was then calculated by the difference.

Micro-organisms

Applications are mainly focused on using micro-columns packed with *Chlorella vulgaris* immobilized on silica gel for Hg(II) and MMM adsorption followed by sequential elution with HCl (30 mM) for MMM and HCl (1.5 M) for Hg(II) (Tajes-Martínez et al. 2006). On other occasions, SeMet and selenourea (SeU) adsorption was achieved on a *Pseudomonas putida* culture medium. In this application an aliquot of the culture was injected directly as a slurry into the graphite furnace (GF) (Aller and Robles 1998a, b).

Some of the SPE procedures discussed previously have been developed for selective retention of a particular species (without pre-concentration purposes) followed by selective elution and quantification by non-chromatographic techniques (Latva et al. 1999, 2000, Shi et al. 2001, Yu et al. 2003b, Jaafar et al. 2007, Menéndez-Sánchez et al. 2009, Watts et al. 2010, Issa et al. 2011, Voice et al. 2011). On other occasions, SPE was used as a sampling and storage method for organometallic compounds from natural waters (Tajes-Martínez et al. 2006, Hashempur et al. 2008, Voice et al. 2011). The use of micro-columns filled with suitable resins allows the storage of species until analysis, and avoids problems derived from maintaining species integrity in the aqueous solution. Finally, these procedures have also been used for sample clean-up purposes (Gómez-Ariza et al. 2007), and also for minimizing artifact problems during speciation analysis (Delgado et al. 2007).

Flow injection on-line pre-concentration systems using SPE

As stated earlier, flow injection on-line SPE techniques provide improvements in the detection limits and precision of measurements, and also reduces the interference from matrix sample concomitants. In comparison with off-line batch modes, these systems offer a number of significant advantages for target analyte determination, such as a higher efficiency, and lower sample and reagents consumption. In addition, the whole pre-concentration procedure can be performed in closed systems that minimizes airborne contamination.

After the first development by Olsen et al. (Olsen et al. 1983) for FI sorbent extraction using the chelation exchanger Chelex-100 as column packing material for Cd, Cu, Pb, and Zn pre-concentration, this promising technique has been extensively applied by using numerous packing materials for target analyte pre-concentration from environmental and biological samples. Regarding mercury and tin species, FI on-line pre-concentration methods are based on using micro-columns packed with ion exchange resins such as Dowex 1X-8 (de Wuilloud et al. 2002) and Anion-SR (Serra et al. 2009). Polymeric C_{18} after derivatization with SPDC (Falter and Schöler 1995), APDC (Falter and Ilgen 1997), or 2-mercaptoethanol (Houserová et al. 2007a) have also been used. In addition, C_{60} fullerene (Muñoz et al. 2005a), and chelating resins such as PANI (Balarama-Krishna et al. 2010), or aminopropyl-controlled pore glass functionalized with [1,5-bis (2 pyridyl)-3-sulfophenyl methylene thiocarbonohydrazide] (Vereda-Alonso et al. 2008), and PTFE after complexation with APDC (Zachariadis et al. 2005), ADDP or DZ (Wu et al. 2006) have also been proposed. The FI systems have been coupled to CV-AAS (de Wuilloud et al. 2002, Zachariadis et al. 2005), CV-AFS (Wu et al. 2006, Serra et al. 2009), CV-ETAAS (Vereda-Alonso et al. 2008) CV-ICP-MS (Balarama-Krishna et al. 2010), HPLC-CV-AAS (Falter and Ilgen 1997), HPLC-ICP-MS (Falter and Schöler 1995, Houserová et al. 2007a) and GC-MS (Muñoz et al. 2005a).

FI sorbent extraction consists of two steps. In the first stage, the sample is aspirated by a pump (A) through a valve (load position) into the column and then to waste (this allows the sample volume taken to be adjusted according to the analyte concentration). In the second step, the valve is switched to another position (elute position) and a pump (B) propels the eluting solvent through the column. Analytes are rapidly released from the column and led to the detector.

New developments in solid phase extraction: nanotubes, nanoparticles, mesoporous silica, molecularly/ionic impression polymers

Conventional sorbents used in typical SPE procedures for both organic compounds and trace elements pre-concentration can be affected by interfering concomitants from the matrix which can be co-extracted with the target analytes. More selective systems for separation are therefore required, and the development or synthesis of new sorbents has increased in recent years.

High selectivity can be expected from sorbents such as immunosorbents (ISs) and molecularly imprinted polymers (MIPs), although MIPs are advantageous over ISs because of the fast and less expensive synthesis, and the high degree of molecular recognition. Carbon nanotubes, nanometer-sized materials and mesoporous materials have been explored extensively as new adsorbents in the SPE technique. Some of these, have been applied to the pre-concentration of organometallic compounds and will be described here.

Nanometer-size materials

After the introduction of carbon nanotubes (CNTs) in 1991 (Iijima 1991), the application of this material for SPE purposes has increased due to the novel structure characteristics and unique properties. New mechanical and electronic properties, a large specific surface area and a high thermal stability indicate the tremendous potential of CNTs for engineering applications such as hydrogen storage, field emission, quantum nanowires, catalyst supports, chemical sensors and as a packing material for gas chromatography. CNTs can be visualized as a graphite sheet rolled up into a cylinder, and they are classified into single-walled carbon nanotubes (SWNTs) and multi-walled carbon nanotubes (MWNTs) according to the carbon atom layers in the wall of the nanotubes. The hexagonal arrays of carbon atoms in graphite sheets of CNT's surface have a strong interaction with other molecules or atoms, which make CNTs a promising adsorbent material. All the facts mentioned above reveal that CNTs may have great analytical potential as a solid phase extraction adsorbent for metal ions and organometallic compounds.

SWNTs have been developed for the pre-concentration and separation of inorganic arsenic and antimony species. As(III) and Sb(III) species are selectively sorbed on the micro-column packed with SWNTs in the presence of APDC, while As(V) and Sb(V) species are not retained (Chen et al. 2009a, Zhu et al. 2009, Wu et al. 2011, López-García et al. 2011). Total As and total Sb were determined by the same protocol after As(V) and Sb(V) reduction with thiourea or ammonium iodide, and pentavalent arsenic and antimony species were then calculated by the difference (Wu et al. 2011, López-García et al. 2011). The proposed method was applied for the assessment of inorganic arsenic species in groundwater and lake water.

MWNTs has been also applied for the determination of TML, dimethyllead (DML), TEL and diethyllead (DEL); and also for mercury species (MMM, MEM, Hg(II)), and tin species (MBT, DBT, TBT, Sn(IV)) in environmental samples (water and coastal sediment samples) (Muñoz et al. 2005b).

Nanoparticles

Nanometer-sized materials have attracted substantial interest in the scientific community because of their special properties (Henglein 1989). The size range of nanoparticles is from 1.0 to almost 100 nm, which falls between the classical fields of chemistry and solid-state physics. The relatively large surface area and highly active surface sites of nanoparticles enable them to have a wide range of potential applications, including shape-selective catalysis, chromatographic separations, enzyme encapsulation, DNA transfection, drug delivery and sorption of metal ions.

One of the properties of nanoparticles is that most of the atoms are on the surface. The surface atoms are unsaturated and can therefore bind with other atoms that possess highly chemical activity. Consequently, the nanometer material can adsorb metal ions with a high capacity. Investigations of the surface chemistry of highly dispersed oxides, i.e., TiO_2, Al_2O_3, ZrO_2, CeO_2 and ZnO, show that these materials have very high sorptive capacities. The sorption properties of many oxides, including nanometer-sized TiO_2, mainly depend on the characteristics of the solid, i.e., crystal structure, morphology, defects, specific surface area, hydroxyl coverage, surface impurities and modifiers. Since these nanometer-sized metal oxides are not target-selective, the change of the characteristics of the solid by using a suitable coating has proven to be one of the most efficient ways to improve the selectivity.

Therefore, TiO_2 nanoparticles (Xiao et al. 2007) and nanometer titanium dioxide immobilized on silica gel (immobilized nanometer TiO_2) prepared by a sol–gel method (Liang and Liu 2007) have been used as an SPE adsorbent for separating and pre-concentrating arsenite and arsenate. The adsorption of arsenite and arsenate were greater than 98 percent within the 5 to 7.5 pH range, whereas only As(III) could be quantitatively retained on the sorbent when working at pHs within the 9.5–10.5 range (Liang and Liu 2007). Based on this fact, a speciation scheme for inorganic arsenic was established, which involved determining total As at pH 6.0 and As(III) at pH 10 (As(V) was obtained by difference) (Liang and Liu 2007). Similarly, dimercaptosuccinic acid chemically modified TiO_2 was employed as a micro-column packing material for the simultaneous separation/pre-concentration of inorganic arsenic and antimony species. Both trivalent and pentavalent inorganic As and Sb species could be quantitatively adsorbed on dimercaptosuccinic acid modified TiO_2 at pHs from 4 to 7; whereas, only trivalent species (As(III) and Sb(III)) were quantitatively retained at pHs ranging from 10 to 11. These findings were applied for a flow injection on-line micro-column separation/pre-concentration coupled with ICP-OES for the simultaneous speciation of inorganic arsenic and antimony in natural waters (Huang et al. 2007).

Nanometer Al_2O_3 packed micro-columns were also used for direct speciation of dissolved inorganic (Huang et al. 2008b, Liu 2010) and organic (Huang et al. 2008b) selenium species in environmental and biological samples. Ion chromatography coupled with ICP-MS (Liu 2010) and FI dual-column pre-concentration/separation on-line coupled with ICP-MS (Huang et al. 2008b) were used for the determination of ultra-trace amounts of selenium species in aqueous samples. Inorganic selenium species were successfully extracted on a nano-Al_2O_3 solid phase column and then quantitatively eluted with NaOH (0.1–0.2 M). On the other hand, a second column packed with TiO_2 chemically modified with dimercaptosuccinic acid adsorbs selectively Se(IV) and selenocystine ($SeCys_2$) (Huang et al. 2008b).

Palladium nanoparticles (PdNP) (Sounderajan et al. 2009) and Nano ZrO_2/B_2O_3 particles (Erdoğan et al. 2011) have also been applied for inorganic arsenic speciation. PdNP obtained by sodium borohydride reduction gave quantitative recoveries for As(V) and As(III); whereas, the PdNP obtained by hydrazine treatment could only account for As(III) species. Reduced palladium particles are collected, dissolved in the minimum amount of nitric acid and quantified using ETAAS (Sounderajan et al. 2009). The hybrid sorbent (ZrO_2/B_2O_3) pre-concentrates As(V), and As(III) was then oxidized to As(V) using 0.1 M $KMnO_4$ in acidic medium for total As determination (As(III) concentration is calculated by the difference) (Erdoğan et al. 2011).

Finally, magnetic nanoparticles (MNPs) such as 1,5-diphenylcarbazide doped magnetic Fe_3O_4 nanoparticles (Zhai et al. 2010) and amino-modified silica-coated magnetic nanoparticles (Huang et al. 2011) have also been used for Hg(II) (Zhai et al. 2010) and As(V) pre-concentration (Huang et al. 2011). Maximum adsorption when using Fe_3O_4 nanoparticles occurred at pHs higher than 6, and equilibrium was achieved within 5.0 min. For amino-modified silica-coated magnetic nanoparticles, As(V) was selectively adsorbed at pHs from 3 to 8, while As(III) is not retained. The adsorbed As(V) was quantitatively recovered from the MNPs using 1.0 M nitric acid. Total inorganic As was extracted after the permanganate oxidation of As(III) to As(V). Without filtration or centrifugation, these mercury/arsenic loaded nanoparticles could be separated easily from the aqueous solution by simply applying an external magnetic field.

Ordered mesoporous silica

Mesoporous materials have developed quickly during the last decade and have attracted much attention in various scientific areas of physics, chemistry and material science. These materials offer high performances, with a large surface area, high porosity, well-defined pore size and an ordered pore arrangement, excellent mechanical resistance properties, no swelling, good chemical stability and well modified surface properties (Kresage et al. 1992). Therefore, ordered mesoporous silica has become an ideal adsorbing material for SPE procedures. Modified mesoporous adsorbents are synthesized in the presence of surfactants, which are used as templates, and can later be functionalized with different organic groups. The development of functionalized mesoporous materials for adsorption applications has generated considerable interest. In particular, materials whose surfaces have been functionalized with groups containing sulfur and nitrogen as active donor atoms offer high selectivity. Therefore, micro-columns packed with 3-(2-aminoethylamino) propyltrimethoxysilane modified ordered mesoporous silica has been applied to inorganic arsenic speciation in natural waters (Chen et al. 2009b). As(V) can be selectively adsorbed on the micro-column within pHs of 3–9, while As(III) is not retained. Total inorganic arsenic was extracted after the oxidation of As(III) to As(V) with $KMnO_4$.

Molecularly imprinted polymers and ionic imprinted polymers

Molecularly imprinted polymers (MIPs) are crosslinked polymers with specific binding sites for a particular analyte. These binding sites are tailor-made *in situ* by the copolymerization of crosslinking monomers and functional monomers in the presence of the print molecule, called the template. After polymerization, the template is removed from the polymer. This leaves recognition sites that, in terms of size, shape and functionality, are complementary to the print molecule. Thus, MIPs are materials with high selectivity for a target molecule. This selectivity arises from the synthetic procedure followed to prepare the MIP.

MIPs can be synthesized following three different imprinting approaches: the non-covalent, the covalent and the semi-covalent. In all of these protocols, a template molecule (with amino, carboxyl, or keto groups) interacts with an appropriate functional monomer to establish specific interactions. Once the polymer is obtained, the template is removed from the polymer; consequently, the template leaves its imprint in the polymer structure and this imprint is responsible for the recognition properties of the MIP. These imprints or cavities are also known as binding sites.

Addressing molecularly imprinted polymerization in the presence of ions as templates, instead of molecules, new supports for recognizing ions (ionic imprinted polymers, IIPs) can be synthesized. These polymers offer all the benefits derived from MIPs and have a high capacity for recognizing ions. Moreover bulk polymerization, different approaches for synthesizing MIPs and IIPs (suspension, emulsion, dispersion and precipitation methods) have been proposed by several authors.

In contrast to MIPs for which the organic compound target (template) can react with different vinylated monomers, when using ions as templates stable ion-vinylated monomer complexes are not formed. Therefore, only certain complexing agents which can react with the ion and which can polymerize through vinyl groups can be used. As reviewed by Rao et al. (Rao et al. 2004, 2006), bifunctional reagents, showing vynil groups and functional groups to interact with the dissolved ions, can be used following three main approaches: (i) linear chain polymers carrying metal-binding groups being cross-linked with a bifunctional reagent; (ii) chemical immobilization by preparation of binary complexes of metal ions with ligands having vinyl groups, isolation and then polymerization with matrix-forming monomers; and (iii) surface imprinting conducted on an aqueous-organic interface.

Polymerization under these conditions leads to complexing ligands being chemically immobilized in the polymeric matrix. However, the main drawback of these approaches is that vinylated ligands, which act as a complexing reagent for the target ion and as a monomer for polymerizing, are not commercially available and they must be synthesized beforehand. Additionally, a different approach consisting of trapping a non-vinylated chelating ligand via imprinting of binary/ternary mixed ligand complexes of metal ions with non-vinylated chelating agent and vinylated ligand is also possible. In this case, the use of specially synthesized vinylated ligands is avoided. After polymerization, the non-vinylated ligand is not chemically bonded to the polymer chains, but instead it is trapped inside the polymeric matrix.

The selectivity of MIPs/IIPs has been exploited in several applications, such as sensors, capillary electrochromatography, enantiomeric separations, antibody and receptor mimics in immunoassay type analyses and as enzyme mimics in catalytic applications and site-mediated synthesis, and solid-phase extraction (SPE). The first study of MIPs for SPE purposes was made by Sellergren in 1994 (Sellergren 1994). This author prepared a MIP that could selectively extract pentanamide from diluted human urine samples. The high selectivity provided by this polymer meant that pentanamide could be detected directly at low concentrations, thus making a subsequent chromatographic analysis unnecessary. Since then, several MIPs have been prepared for use as SPE materials (Masqué et al. 2001, Caro et al. 2006) for extracting organic compounds from several matrices. However the number of studies for organometallic compounds is limited.

The use of MIPs in SPE is advantageous mainly when a selective extraction must be performed and the commonly used sorbents lack selectivity. Then, MIPs for SPE not only allow the analyte to be pre-concentrated but other compounds present in the sample matrix can also be removed.

For organotin compounds, several MIPs have been synthesized by the non-covalent free radical approach using sodium methacrylate (NaMA) or 4-vinylpyridine (4-VP) as monomers in the presence of TBT as template molecule in three different polymerization media (toluene, acetonitrile and methanol/water) (Gallego-Gallegos et al. 2005, 2006a). The results clearly showed the presence of cavities within the polymeric matrix allowing specific recognition of TBT. This procedure has been applied to seawater (Gallego-Gallegos et al. 2005) and extracts from seafood tissues (Gallego-Gallegos et al. 2006a). MIPs synthesized by the covalent imprinting approach have also been used for recognizing organotin species. The synthesis has been accomplished by co-polymerization of the complex Bu_2SnO-*m*-vinylbenzoin as the imprinting template plus co-monomer sodium methacrylate, and ethylene glycol dimethacrylate as a cross-linker. The imprinting effect depends of the pH (Gallego-Gallegos et al. 2006b).

For organomercury compounds, an ion-imprinting polymer has been prepared by pre-organizing the template (methylmercury ion) and the monomer (methacryloyl-(l)-cysteine methylester, MAC) as a ternary complex in the pre-polymerization mixture before polymerization. Methylmercury-imprinted beads were produced by a dispersion polymerization technique by using methylmercury–methacryloyl-(l)-cysteine (MMM–MAC) complex monomer and ethylene glycoldimethacrylate (EDMA). After removal of methylmercury ions, methylmercury-imprinted beads were used for solid-phase extraction and determination of mercury compounds (Büyüktiryaki et al. 2007).

Finally, molecular imprinting technology has also been employed to prepare a specific affinity chromatographic stationary phase for tin speciation (MBT, DBT, TBT and TPhT) purposes (Gallego-Gallegos et al. 2010). Tributyltin was chosen as the template molecule and the non-covalent approach was applied. Three different polymerization methods were evaluated: (i) a composite material, (ii) a polymer prepared via-Iniferter grafting; (iii) an emulsion polymer. The main advantage of this proposed stationary phase is that good recovery is obtained for all species, including MBT. Baseline resolution for TBT and TPhT has also been obtained. The high selectivity of this column prevents matrix interferences.

Sample Pre-treatments for Liquid Samples: Cloud Point Extraction

Cloud point extraction (CPE) methodology is based on the property of non-ionic and zwitterionic surfactants of separating into two liquid phases when their aqueous solutions are heated above a certain temperature. Any hydrophobic species (hydrophobic organic compounds or metal ions after reaction with a suitable hydrophobic chelating agent form sparingly water-soluble complexes) in solution are able to react with and bind to micelles and become concentrated in a small volume of the surfactant rich phase. CPE was initially introduced for the pre-concentration of metals, in the form of their hydrophobic complexes, by Miura et al. (Miura et al. 1976), as an alternative method for avoiding the use of organic solvents. It was then extensively exploited for isolation of inorganic compounds (Watanabe 1982, Pelizzetti and Pramauro 1985, Hinze and Pramauro 1993, Stalikas 2002, Burguera and Burguera 2004, Bezerra et al. 2005, Silva et al. 2006) and organic substances (Watanabe 1982, Pelizzetti and Pramauro 1985, Hinze and Pramauro 1993, Carabias-Martínez et al. 2000, Sosa-Ferrera et al. 2004). Current trends and new CPE developments are mainly related to automation and to the use of ionic liquids and micellar formations (Burguera and Burguera 2004, Paleólogos et al. 2005, Baghdadi and Shemirani 2008).

General aspects

Aqueous solutions of certain surface-active agents (such as non-ionic surfactants) display the so-called cloud point phenomenon in which the aqueous surfactant solution (surfactant above the critical micellar concentration, CMC) suddenly becomes turbid because of a decrease in the solubility of the surfactant in water (Hinze and Pramauro 1993). Surfactant agents are amphiphilic molecules with distinct hydrophobic and hydrophilic moieties; a polar or ionic group connected to a long hydrocarbon tail. At low concentrations, surfactant molecules are present mainly as monomers. When their concentration increases above a certain threshold (CMC) surfactant monomers spontaneously accumulate to form colloidal-sized clusters, known as micelles.

CPE is based on certain properties of non-ionic surfactants in aqueous solutions to form micelles and become turbid at levels above their CMC. If some condition, such as temperature or pressure, is appropriately altered or, if an appropriate substance (electrolyte) is added to the solution when the non-ionic surfactant is above the CMC, the system composed of a unique phase is separated into two isotropic phases. One of these phases contains a surfactant at a concentration below, or equal to CMC, and the other phase is a surfactant rich phase of a small volume, in which the surfactant concentration is close to the CMC (Bezerra et al. 2005). Above the "cloud point", hydrophobic species (hydrophobic organic compounds and metal ions or organometallic compounds after reaction with a suitable hydrophobic ligand) present in sample can be entrapped in the surfactant micelles (Fig. 2.4). They can therefore be separated from the bulk sample

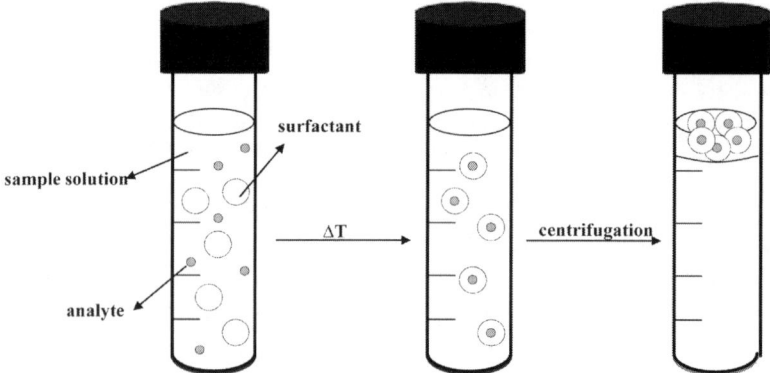

Figure 2.4 Schematic of cloud point extraction (CPE) set-up.

matrix and concentrated in the small volume of the surfactant-rich phase (Attwood and Florence 1985).

The pre-concentration factor in CPE is defined as:

$$f_c = \frac{C_S}{C_O}$$ Eqn. 4

where C_s is the analyte concentration in the surfactant-rich phase after phase separation and C_0 is the initial analyte concentration in the bulk solution before phase separation. The equilibrium partition coefficient, K_p, of analyte in the CPE process is described as:

$$K_p = \frac{C_S}{C_W}$$ Eqn. 5

where C_w is the analyte concentration in the water phase after phase separation. The recovery efficiency, R, can be characterized as the percentage of analyte extracted from the bulk solution into the surfactant-rich phase:

$$R = \frac{C_S V_S}{C_O V_t} x100\%$$ Eqn. 6

$$R = f_c \left(\frac{R_V}{1 + R_V} \right) x100\%$$ Eqn. 7

where, V_t is the total volume of the solution, V_s is the volume of the surfactant-rich phase and R_V is the phase-volume ratio (defined as the phase volume ratio of the surfactant-rich phase V_S to that of the coexisting water phase V_W, i.e.. $R_V = V_S/V_W$) (Kun-Chih et al. 2007).

   In the micellar structure, surfactant aggregates orientate their hydrocarbon tail towards the center of the formation, creating a non-polar core. Hydrophobic and covalent compounds initially present in the aqueous solution are favorably partitioned in the non-polar micro-environment. Although the exact mechanism via which this phenomenon occurs is yet to be defined, several studies have shown that such phase separations result from the competition between entropy (which favors miscibility of micelles in water) and enthalpy (which favors separation), so the clouding and phase-separation procedure is reversible (Paleólogos et al. 2005). This behavior, shown by numerous hydrophilic groupings, is especially observed with polyoxyethylene surfactants, and it can be attributed to the two ethylene oxide segments in the micelle that repel each other at low temperature when they are hydrated, and that attract each other as the temperature increases due to dehydration. This effect causes a decrease in the effective area occupied by the polar group on the micelle surface, increasing the size of the micelle, which can become infinite at the cloud point, leading to the phase separation. Any hydrophobic species remain preferentially in the surfactant-rich phase, and they can be extracted or pre-concentrated. More detailed discussion on CPE phenomenon can be consulted in specialized literature (Gullickson et al. 1989, Hinze and Pramauro 1993, Paleólogos et al. 2005).

The separation of inorganic and organic As, Hg, Sb, Se and Sn species by CPE involves the prior formation of a complex with sufficient hydrophobicity to be extracted in to the small volume of surfactant-rich phase, thereby obtaining the desired pre-concentration. This usually requires the optimization of several parameters such as pH at which complex formation occurs, and ionic strength. In general, the following factors must be optimized to achieve a successful CPE procedure:

Selection of the appropriate chelating agent

The ligand is selected with the requirement that the derived complex is sufficiently hydrophobic, possesses a high partition coefficient and is formed quickly and quantitatively with the least possible excess. Based on their reactivity and formation constants with the target metal species, some of the most used reagents are carbamates, pyridylazo, quinoline and naphthol derivatives. Other reagents, such as O,O-diethyldithiophosphate, have been used for more specific applications.

Concentration of chelating agent

The concentration of the chelating agent has to compensate sufficiently for any consumption of the reagent by other species present. Additionally, chelating agents with lower partition coefficients than others have to be in sufficiently large excess to extract efficiently.

pH

Although the chelate formation is strongly dependent on the pH, this variable offers little influence on the extraction efficiency of the complexes formed.

Ionic strength

Although ionic strength has proved to have a negligible effect on the performance of CPE, increasing ionic strength enhances phase separation. Moreover, the addition of salt can markedly facilitate the phase-separation process.

Surfactant selection

The non-ionic surfactant must offer high extraction efficiency for the analyte. In addition, the two phases must be easy to separate after centrifugal settling and cooling stages should be omitted. Moreover, the surfactant must be inexpensive and it must offer a low cloud point temperature, which simplifies the procedure because the incubation step is avoided. The Triton X series of non-ionic surfactants are most frequently used for the formation of the surfactant-rich phase because they offer several advantages such as commercial availability, low toxicity, low cloud point temperature, and high density of the surfactant-rich phase, which facilitates phase separation by centrifugation. Triton X-114 and Triton X-100 are the most commonly used surfactants. The cloud point achieved with these reagents is low, being 23 and 65°C, respectively.

Surfactant concentration

There is a narrow range within which easy phase separation, maximum extraction efficiency and analytical signal are accomplished. Outside of this optimal range, the pre-concentration factor decreases. However, if the surfactant concentration is lower than the recommended value, poor repeatability is achieved.

Incubation time

Metal reaction with chelating agents and their transportation inside the micelle are kinetically controlled. It is therefore essential to maintain the reaction time above a minimum threshold for quantitative extraction. The use of the shortest incubation time and the lowest possible equilibration temperature is desirable. This compromises completion of the reaction and efficient separation of the phases. A reaction time of up to 10 min is usually used for most applications.

Incubation temperature

An optimum incubation temperature is required for achieving an easy phase separation. Temperature also seems to play an additional role in enhancing pre-concentration efficiencies and enhancement factors. The application of high temperatures leads to dehydration of the micelle, and an increase of the phase-volume ratio, and thus the signal enhancement by a factor higher than 3.

Centrifugation time

Centrifugation time hardly ever affects micelle formation but it can accelerate phase separation similar to conventional solid phase (precipitate) –liquid separations. Centrifugation times of around 5–10 min have been found to be adequate in most of the CPE procedures.

Advantages and disadvantages

The use of CPE procedures offers an interesting green alternative to the conventional extraction systems. The use of small amounts of non-ionic surfactants eliminates handling large volumes of volatile and flammable organic solvents used in conventional solvent extraction. The surfactant-rich phase can be burned in the presence of acetone or ethanol, providing a good solution for waste disposal. The small volume of the surfactant-rich phase obtained with this methodology permits the design of extraction schemes that are simple, fast, cheap and safe.

CPE offers high capacity to concentrate analytes with high quantitative recoveries and high pre-concentration factors. Concentration factors of 10–100 are easily obtained with good recoveries, resulting in highly sensitive analyses (Stalikas 2002). Pre-concentration factors can also be modified on demand by varying the amount of surfactant. Another feature is the small amount of aqueous sample required, typically below 50–100 ml (conventional LLE requires sample volumes up to 1000 ml for achieving the same pre-concentration factors as those achieved by micelle-mediated extraction methods). In addition, the presence of surfactants can minimize analytes losses arising from their adsorption onto the container.

CPE can be used to separate and/or pre-concentrate analytes in a step prior to their determination using hydrodynamic analytical systems such as HPLC or flow injection analysis (FIA). In addition, the small volume of the surfactant-rich phase obtained from CPE is compatible with the hydroorganic phases frequently used in the reversed phase chromatographic mode. Finally, CPE can be applied for extracting thermally sensitive analytes because of the mild conditions inherent to CPE techniques.

There are, of course, several limitations of CPE procedures, mainly attributed to the manipulation of the surfactant-rich phase obtained. As this phase is viscous, it cannot be injected directly into atomic spectrometry instruments. In addition, the adsorption of surfactant in the surfactant-rich phase onto the chromatographic stationary phase or the inner wall of electrophoretic capillary may interfere with the chromatography/ electrophoresis injection and separation resulting in poor reproducibility and efficiency. For these reasons, the surfactant-rich phase must be diluted with an aqueous or organic solvent to reduce the viscosity. This decreases the theoretical pre-concentration factors. To overcome this drawback, the dual-cloud point extraction technique (dCPE) was developed. In dCPE, the cloud point procedure is carried out twice during a single sample pretreatment process. The first part of dCPE procedure is done like traditional CPE. The surfactant is added into the solution containing the analytes that are hydrophobic or can form hydrophobic complexes with suitable ligands. Following the thermostatic bath and centrifugation processes, the analytes of interest and other hydrophobic interfering species are extracted into the surfactant-rich phase. However, instead of the direct analysis, dCPE performs another round of the cloud point procedure, in which the surfactant-rich phase is treated with another aqueous solution containing a special ligand which can form new hydrophilic complexes with the analytes. After the thermostatic bath and centrifugation, the targets are back-extracted into the aqueous phase. The obtained aqueous extract is directly injected into conventional instruments (Xue-Bo 2007).

In addition, non-ionic surfactants present another important drawback: a high background absorbance in the ultraviolet region and high fluorescence signals due to the presence of an aromatic moiety in their structure. One possible way of overcoming this pitfall is the use of surfactants that do not absorb at the working wavelengths normally used in chromatography such as alkylammoniosulfate zwitterionic surfactants (Saitoh and Hinze 1991). Another approach consists of using electrochemical detection (García-Pinto et al. 1992) (although the approach is applicable only when the pre-concentrated analytes are electroactive). Otherwise, when using atomic detectors the use of non-ionic surfactants is not a problem.

Cloud point extraction for organometallic species pre-concentration

Inorganic As, Sb, Se and Sn species

Due to the selective reactivity of several chelating agents or ion-pairing forming ligands with arsenite or with arsenate at certain pHs, these reagents could be used before a CPE procedure for inorganic arsenic speciation without chromatographic separation. Thus, As(III) forms complexes with APDC (at pH 4.2–4.5) (An-na et al. 2005, Baig et al. 2009, 2011, Shah and Kandhro 2012) or O,O-diethyldithiophosphate (DDTP) (Amjadi et al. 2010) and an ion-pairing complex with Pyromine B in the presence of dodecyl sulfate (SDS) (at pH 10) (Ulusoy et al. 2011a). The complexes are then extracted into non-ionic surfactant-rich phases (Triton X-114) (An-na et al. 2005, Baig et al. 2009, 2011, Shah and Kandhro 2012) or into a reverse micelle-based coacervative phase (decanoic acid reverse micelles in DIW/ THF mixture) (Amjadi et al. 2010). Arsenate could be extracted using CPE after prior reduction to arsenite by using potassium iodide/ascorbic acid (Amjadi et al. 2010) or sodium thiosulfate (Ulusoy et al. 2011a). Total inorganic arsenic could therefore be determined using CPE because the arsenate concentration was then calculated by the difference. Another alternative could be the total arsenic pre-concentration using SPE (TiO_2 adsorbent) followed by direct slurry sampling for ETAAS determination (arsenate concentration is then obtained by the difference) (Baig et al. 2009, Shah and Kandhro 2012).

Similarly, arsenate reacts with molybdate forming a yellow heteropoly acid complex in sulfuric acid medium (Baig et al. 2011, Shemirani et al. 2005) or with Pyronine B in the presence of cetyl pyridinium chloride (CPC) at pH 8 (Ulusoy et al. 2011b). Arsenate is quantitatively extracted to the non-ionic surfactant-rich phase (Triton X-114) after centrifugation. Total inorganic arsenic was extracted after oxidation of As(III) to As(V) with $KMnO_4$, and As(III) is obtained by the difference.

Several non-chromatographic methods for inorganic Sb, Se and Sn speciation have been proposed. Antimonite reacts selectively with N-benzoyl-N-phenyhydroxylamine (BPHA) at pH 2 (Fan 2005), with APDC at pH 5.5 (Li et al. 2006, Hagarova et al. 2008), with DDTC at pH 6.0 (Li et al. 2008a), and with DDTP (Oliveira-Souza and Teixeira-Tarley 2008). Similarly, selenite selectively reacts with DDTC at pH 6.0 (Li et al. 2008a), with APDC at pH 3.0 (Chen et al. 2006), with 3,3-diaminobenzidine (DAB) at pH 1.9 (Suvarna-Sounderajan et al. 2010), and with 2,3-diaminonaphthalene (DAN) at pH 1.5 (Güler et al. 2011). Tin species such as stannate reacts with 1-(2-pyridylazo)-2-naphthol (PAN) (Zhu et al. 2006). Stannite reacts with α-polyoxometalate in acidic medium (pH 1.2) and forms Keggin-type complexes; whereas, both stannite and stannate species form those complexes at pH 3.7 (Gholivand et al. 2008). In those conditions, when the system temperature is higher than the cloud point extraction temperature, the formed complexes with antimonite, selenite or stannate can enter the surfactant-rich phase Triton X-114 (Fan 2005, Li et al. 2006, 2008a, Chen et al. 2006, Hagarova et al. 2008, Oliveira-Souza and Teixeira-Tarley 2008, Suvarna-Sounderajan et al. 2010, Güler et al. 2011), Triton X-100–SDS–NaCl

(Zhu et al. 2006), or Triton X-100–cetyltrimethylammonium bromide (CTAB) (Gholivand et al. 2008)); whereas, the antimonite or selenate or stannite (Zhu et al. 2006) or stannate (Gholivand et al. 2008) remain in the aqueous phase. Antimonate can be then be extracted using CPE after prior reduction to antimonite by using L-cysteine (Fan 2005, Li et al. 2006, 2008a, Oliveira-Souza and Teixeira-Tarley 2008). In the same way, selenate is reduced to selenite by L-cysteine (Li et al. 2008a), by boiling in HCl medium (Chen et al. 2006), or by microwave heating in HCl medium (Suvarna-Sounderajan et al. 2010, Güler et al. 2011) before total inorganic selenium CPE. In these applications antimonate and selenate concentrations are finally obtained by calculating the differences between total concentrations of antimony/ selenium and the concentrations of the antimonite and selenite.

The developed CPE procedures have been used mainly in combination with ETAAS (An-na et al. 2005, Shemirani et al. 2005, Zhu et al. 2006, Baig et al. 2009, 2011, Gholivand et al. 2008, Oliveira-Souza and Teixeira-Tarley 2008, Amjadi et al. 2010, Suvarna-Sounderajan et al. 2010), HG-AAS (Ulusoy et al. 2011a), FAAS (Fan 2005, Ulusoy et al. 2011b), electrothermal vaporization (ETV) coupled with ICP-OES (Li et al. 2006) or ICP-MS (Chen et al. 2006, Li et al. 2008a), and spectrofluorimetry (Güler et al. 2011).

Organic Hg and Sn species

Organic mercury (MMM, MEM and MPhM) and tin species (TBT) have also been extracted by CPE followed by several non-chromatographic and chromatographic methods, mainly HPLC and capillary electrophoresis (CE).

A non-chromatographic procedure for Hg(II) and MMM speciation based on Hg(II) complexation with I^- and methyl green (MG) has been proposed. Hg(II) reacts with I^- and forms HgI_4^{2-}, the reaction of HgI_4^{2-} with MG cation forms a hydrophobic ion-associated complex which is extracted into the surfactant-rich phase of the non-ionic surfactant Triton X-114. MMM remains in the supernatant phase. The surfactant-rich phase containing Hg(II) was diluted with HNO_3 and determined by ICP-OES. The supernatant is also subjected to the similar CPE procedure for the pre-concentration of MMM by the addition of APDC. The MMM in the micelles was directly analyzed after disposal as describe above (Li and Hu 2007). The method allows the fast and accurate direct determination of inorganic mercury and methylmercury in food samples (after acid leaching), thereby avoiding the uncertainty of the differential approach.

Another non-chromatographic procedure involves the use of ultrasound to assist the extraction into the micelles of both non-ionic (Triton X-114) and anionic (sodium dodecane sulfonic acid SDSA) surfactants for the isolation of TBT from fish and mussel tissues prior to ETAAS determination. The application of ultrasound accelerates the CPE. Isolation and subsequent determination of TBT is achieved by ultrasound-assisted back extraction into a water immiscible solvent, utilizing the selective partition of TBT from an alkaline environment (Louppis et al. 2010).

Several chromatographic methods have been developed for mercury speciation. Thus, mercury species (Hg(II), MMM, MEM and MPhM) from natural waters; fish and hair samples (after acid leaching) were taken into complexes with APDC (Li-Pimg 2005) or DDTC (Chen et al. 2009c, 2009d) in aqueous non-ionic surfactant Triton X-114 medium and concentrated in the surfactant-rich phase (by bringing the solution to the temperature of 40°C). The use of APDC as chelating agent is preferable to the use of DDTC because the MEM complex is subjected to partial decomposition during the extraction process (Chen et al. 2009c).

Finally, a novel dual-cloud point extraction (dCPE) technique has been also proposed for mercury species pre-concentration from natural water and fish tissues (Xue-Bo 2007). In dCPE, the cloud point was carried out twice during the sample pretreatment. First, Hg(II), MMM, MDM and MPhM formed hydrophobic complexes with PAN. After heating and centrifuging, the complexes were extracted into the formed Triton X-114 surfactant-rich phase. Instead of the direct injection or analysis, the surfactant-rich phase containing the Hg species was treated with L-cysteine aqueous solution. The Hg species were then transferred back into aqueous phase by forming hydrophilic Hg-L-cysteine complexes. After dCPE, the aqueous phase containing the Hg-L-cysteine complexes was subjected to CE with UV detection for mercury speciation analysis (Xue-Bo 2007). This procedure offers several advantages: because the concentration of Triton X-114 in the extract after dCPE was only around critical micelle concentration, the adsorption of surfactant on the capillary wall and its possible influence on the sample injection and separation in traditional CPE were minimized. In addition, the hydrophobic interfering species were removed completely by using dCPE, leading to a significant improvement in analysis selectivity.

Flow injection on-line pre-concentration systems by CPE

As stated, one of the major advances in analytical chemistry in recent decades has been the development of automated systems for analysis, i.e., batch analyzers and continuous analyzers (mainly FIA and continuous flow (CFA)), which provide analytical data with minimal operator intervention. Several pre-treatment procedures have been successfully adapted to FI and CF on line pre-concentration, and rapid, convenient, sensitive and reliable methods for trace analysis have been developed.

Fang et al. (Fang et al. 2001) proposed the on-line incorporation of CPE to flow injection analysis for the first time. The authors evaluated the analytical performance of the on-line CPE–FIA system by using hematoporphyrin as a model test compound. After this first development, few studies have exploited the analytical advantage offered by the on-line application of this technique for trace analyte pre-concentration.

On-line CPE-FIA is based on the on-line mixing of the sample with a surfactant rich solution, which allows phase separation in the flow on a high surface area material (a glass tube packed with a suitable filtering material such as cotton, nylon or glass-wool) that can quantitatively intercept the surfactant aggregates. Subsequently, the surfactant-rich phase containing the analytes is desorbed by a suitable elution agent and is transported on-line towards the measuring device.

Li et al. (Li et al. 2006) have applied this approach for inorganic antimony speciation in environmental and biological samples by using APDC/Triton X-114. When the solution is clouded, it was loaded on-line into a cotton column. Under these conditions, the surfactant rich phase was trapped inside the column and the retained chelates were eluted using acetonitrile and directly introduced into the graphite furnace for ETV-ICP-OES determination.

Sample Pre-treatment for Liquid Samples: Micro-extraction Techniques

As stated above, classical sample pre-treatment methods consume huge amounts of hazardous solvents leading to the generation of toxic wastes which pose a serious health risk. Unlike LLE, SPE offers many advantages such as simplicity of operation and high enrichment factor; however, the amounts of solvents consumed in the elution process are still relatively high. CPE is proven to be comparatively simple, cost effective and poses lower toxic threats to the environment. However, the CPE exhibits several limitations such as the viscous surfactant-rich phase that prevents a smooth and instant injection to conventional analytical instruments. In addition, the surfactant bears chromophores that can interfere with UV absorption by overlapping with the analyte signal.

Miniaturization of the extraction procedure is one of the recent trends in sample pre-treatment techniques which involve a great reduction of solvent and leads to new solvent micro-extraction methodologies. One of the techniques that evolved from this approach is the single drop micro-extraction (SDME), in which the extraction phase is a micro-drop of water immiscible organic solvent suspended in the aqueous sample.

Recently, several new techniques have been developed for pre-concentration and clean-up purposes. These techniques include solid-phase micro-extraction (SPME), stir bar sorptive extraction (SBSE), and liquid-phase micro-extraction (LPME). These extractive procedures are appealing methodologies for trace analysis because of the low amounts of organic solvent used (solvent-free) and because they are easy to use.

Micro-extraction is an extraction technique where the volume of the extracting phase is very small in relation to the volume of the sample. In contrast with classical LLE and the widely applicable SPE, which are exhaustive processes, micro-extraction is not intended to be an exhaustive extraction procedure, so only a fraction of the initial analyte is likely to be extracted for subsequent analysis (Stalikas and Fiamegos 2008). In micro-extraction, the extraction yield hinges on the partition coefficient of analyte(s) between the bulk (sample or donor) phase and the deprived (extractant or acceptor) phase. Since partitioning does not depend on analyte concentration, quantification of sample concentration may be done from the absolute amount extracted. As micro-extraction is a non exhaustive extraction, care should be taken to ensure the distribution constant is equal in all experiments including calibration and sample extraction. Thus, standards have to undergo the same extraction procedure. In addition, the use of surrogate standard could be useful to improve the precision.

Depending on the type of the extracting phase, micro-extraction techniques may be divided into three broad groups: (A) micro-extraction based on sorbent enrichment; (B) membrane micro-extraction and (C) liquid-phase micro-extraction (Stalikas and Fiamegos 2008).

Micro-extraction based on sorbent enrichment

Micro-extraction based on sorbent enrichment or sorbent micro-extraction comprises solid phase micro-extraction and stir-bar sorptive extraction.

Solid phase micro-extraction

The most popular and successful of the sorbent-based, solvent-free micro-extraction techniques is solid phase micro-extraction (SPME). SPME is a technique whereby an analyte is sorbed onto the surface of the coated silica fiber. This is followed by desorption of the analytes into a suitable instrument such as GC or HPLC for the separation which is attached to a suitable detector for quantification. Sorption of analyte onto a suitably coated silica fiber or stationary phase is the most important stage. Operationally, SPME encompasses non-exhaustive, equilibrium and non-equilibrium, batch and flow through micro-extraction techniques.

General aspects

Solid phase micro-extraction (SPME) is a solvent free sample preparation technique that uses a fused silica fiber coated with an appropriate stationary phase on the outside surface of a needle of a modified micro-syringe (Fig. 2.5). An important feature of this technique is that extraction and injection are incorporated into the same device, thus minimizing analysis time. In addition, the direct desorption of analytes into the analytical instrument is possible. It was originally developed by Arthur and Pawliszyn (Arthur and Pawliszyn 1990) and subsequently a number of books have been written on the technique (Wercinski 1999). SPME is essentially a two step process, first the partitioning of analytes between the sample matrix and the fiber coating, and then the desorption of the (concentrated) extract from the fiber into the analytical instrument, usually a GC or HPLC. In the case of SPME-GC, the analytes are thermally desorbed into the injector of the chromatograph. It is however, generally limited to volatile and thermally stable compounds. Some of the applications involve the derivatization

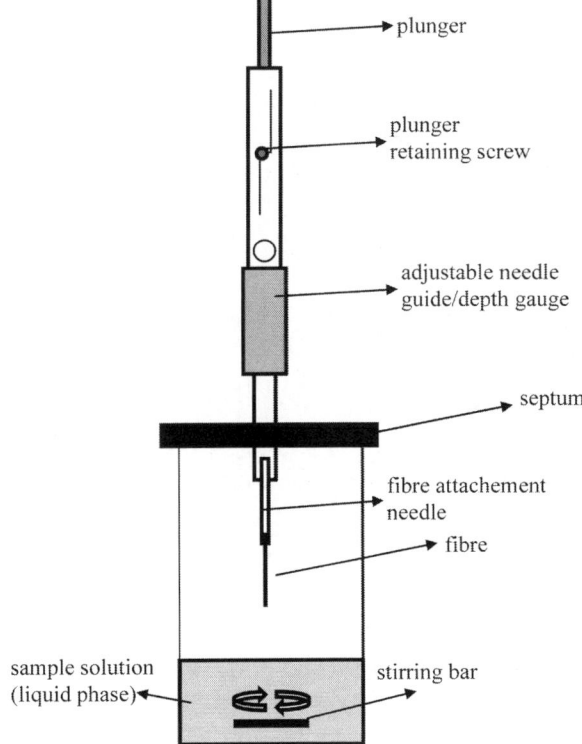

Figure 2.5 Schematic of solid phase micro-extraction (SPME) set-up.

in the sample matrix, in the injection port and on the fiber derivatization after and/or during SPME to overcome the problem of its limited use. More recently, SPME was applied to non-volatile and thermally unstable compounds by interfacing with HPLC. In SPME-GC, the fiber is introduced into the injector port and analytes are thermally desorbed from the coating. But in SPME–HPLC, desorption is carried out in an appropriate interface. It consists of six port injector with a special fiber desorption chamber, installed in place of a sample loop. Desorption is carried out by using an organic solvent or mobile phase because the thermal desorption at high temperature leads to degradation of the polymer and incomplete desorption of many non-volatile compounds from the fiber.

SPME is not an exhaustive extraction technique but is an equilibrium technique. The maximum sensitivity is obtained at the equilibrium point; however, it is not necessary to reach this point and the extractions can instead be performed for a defined period of time.

The SPME technique relies upon the equilibration of analytes between the liquid, and or the headspace gas and the stationary phase coating of the fiber. Equilibration, therefore, depends on the dissociation constant of the analyte and the thickness of the stationary phase. The amount of analyte adsorbed by the fiber is directly proportional to the concentration of the analyte in the sample when the system is at equilibrium. The extraction efficiency in SPME can be evaluated from the amount of analyte extracted by the coating. For coatings that extract analytes based on absorption, the amount of analyte extracted can be expressed as:

$$n_A = \frac{K_A V_f V_s C_A^0}{K_A V_f + V_s}$$ Eqn. 9

where n_A is the amount of analyte A extracted by the coating at equilibrium, V_s and V_f are the volumes of the sample solution and coating, respectively, C_A^0 is the initial concentration of the analyte in the sample, and K_A is the partition coefficient (Pawliszyn 1997).

For porous coatings that extract analytes by adsorption, the amount of analyte extracted by the coating can be expressed as follows:

$$n_A = \frac{K_A V_f V_s C_A^0 \left(C_{f\,max} - C_{fA}^\infty \right)}{\left[V_S + \left(K_A V_f \left(C_{f\,max} - C_{fA}^\infty \right) \right) \right]}$$ Eqn. 10

where $C_{f\,max}$ is the maximum concentration of active sites on the coating, C_{fA}^∞ is the equilibrium concentration of the analyte on the coating, and K_A is the adsorption equilibrium constant (Górecki et al. 1999).

However, the use of the above equations to obtain n_A is difficult because some of the terms such as K_A, $C_{f\,max}$, C_{fA}^∞ and V_f are often unknown or difficult to measure. Fortunately, the amount of analyte extracted (n_A) by a SPME coating can be easily obtained by experimental measurements with the following equation:

$$n_A = FxA = \left(\frac{m}{A_d}\right)xA$$

Eqn. 11

where n_A is the amount (mass) of analyte extracted by SPME, F is the detector response factor which can be calculated by comparing the amount of analyte (m) injected and the area counts (A_d) obtained by liquid injection ($F = m/A_d$), and A is the response (area counts) obtained by SPME. Therefore, practically, the extraction efficiencies of different coatings for the same analyte can be evaluated by comparing the n_A values obtained by SPME experiments under the same extraction conditions.

There are two different techniques for the SPME method: fiber SPME and in-tube SPME.

Fiber SPME is a modified syringe-like instrument which consists of a fiber holder and fiber assembly with built-in fiber inside the needle. The fused silica fiber is coated with a relatively thin film of several polymeric phases. Because of its small physical diameter, cylindrical geometry and stability at higher temperatures, it can be incorporated into a syringe-like holder. The SPME holder provides protection to the fiber and allows piercing of a rubber septum of the GC injector. The fused silica fiber is retracted within the needle of the SPME holder when it is not in use. During operation, the silica fiber is exposed to the sample.

There are two modes of fiber SPME for extraction of analytes: Headspace SPME (HS-SPME) and direct immersion SPME (DI-SPME) (Fig. 2.6). HS-SPME involves exposure of the fiber to the vapor phase above the gaseous, liquid or solid sample. In this case, the fiber is not in direct contact with the sample. The analytes need to be transported through a layer of air before they can reach the coating. HS is used when GC is employed for final analysis. HS eliminates the need for filtration or centrifugation of the sample. In DI-SPME the coated fiber is immersed in the liquid sample and the analytes are transported directly to the extraction phase. For volatile compounds, fiber HS-SPME is preferred over DI-SPME. The fiber coating can be damaged by high molecular weight species and other non-volatile contaminants present in the liquid sample matrix in case of DI-SPME as the fiber is directly immersed into it. It is also found that HS is more selective than DI.

There are seven different types of fibers commercially available: polydimethylsiloxane (PDMS), polydimethylsiloxane/divinylbenzene (PDMS/DVB), stableflex polydimethylsiloxane/ divinylbenzene (PDMS/DVB), polyacrylate (PA), carboxen/polydimethylsiloxane (CAR/PDMS),

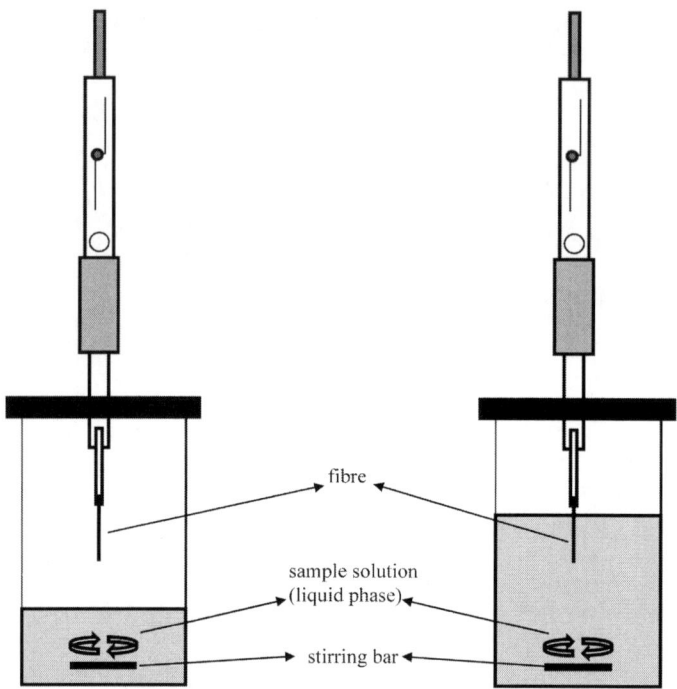

Figure 2.6 Schematic of direct immersion solid phase micro-extraction (DI-SPME) and headspace solid phase micro-extraction (HS-SPME) set-up.

carbowax/divinylbenzene (CW/DVB) and stableflex divinylbenzene/ carboxen/polydimethylsiloxane (DVB/CAR/PDMS). Despite development of new micro-extraction sorbents, the dearth of SPME sorbent coatings for polar compounds (the most well known being polyacrylate) and their lack of stability is still important. Selection of the type of polymer used for the extraction depends upon the chemical nature of the analyte, i.e., polarity and volatility. In general, polar fibers are used for polar analytes and non-polar ones for non-polar analytes. Extraction efficiency can be improved by modification of matrix, target analytes and fiber chemistry. It is reported that the extraction yields for DVB/CAR/PDMS fibers is much higher than that observed for PDMS fibers when extracting organometallic compounds. PMDS is the most useful liquid type coating. The coating should be thin to reduce the extraction times because thick coatings require more time for equilibration. Coatings can be attached to the fused-silica core by various methods. Commercially available fibers can be damaged in strong organic solvents (SPME-HPLC), strong acids and alkali solutions. To overcome this problem, fibers are prepared by sol-gel technology which are stable even in strong organic solvents (xylene and methylene chloride) as well as in acidic and basic solutions. Because the sol-gel coating is chemically bonded to the surface of the fused-silica fiber, these fibers can be used up to 320°C, whereas commercial PDMS fibers begin to bleed at lower temperatures (200°C). The high porosity of sol-gel fibers results in higher sensitivity and faster extraction times than for commercial fibers (Kaur et al. 2006).

Several variables such as those involved in the analyte adsorption (fiber cooling, sample heating, extraction time, sample agitation, salting out effect, sample pH, derivatizing agent concentration) and those related to desorption (desorption temperature in GC or desorption solvent in HPLC) must be optimized for each application.

Analyte equilibrium concentration in the HS can be increased by heating the sample and by cooling the fiber. This is applied when assessing very volatile components in heavily contaminated liquid and solid samples.

Agitation also speeds up the transfer of analytes from matrix to coating of the fiber. Magnetic stirring, sonication, intrusive stirring or an elliptical shaker are typically used. Another way to speed up the extraction consists of fiber vibration and rotation, which also increases precision.

The addition of electrolytes (salting out effect) improves the extraction time by increasing the ionic strength and reducing analyte solubility. Electrolyte are usually NaCl, $NaHCO_3$, K_2SO_4 and $(NH_4)_2SO_4$. The salting out effect makes HS-SPME more effective.

When dealing with ionic organometallic compounds, a derivatization step prior to extraction (direct derivatization) or after the extraction (injector port derivatization), is required. In direct derivatization, the reagent is added directly to the sample. The derivatized analytes can be extracted onto the SPME fiber and analyzed using GC with a suitable detector. In injector port derivatization, a fiber is placed in the aqueous phase to isolate the analytes. After the desired extraction time, the fiber is transferred into the hot GC injector port for desorption, derivatization, separation and quantification. The use of tetraalkylborates such as $NaBEt_4$, and $NaBPh_4$, tetra(n-propylborate) ($NaBPr_4$) and tetraammonium tetrabutylborate for the determination of organometallic species has been used extensively. Deuterated sodium ethylborates hydride generation and Grignard derivatization have also been proposed. Grignard derivatization requires aprotic media and, therefore, numerous handling steps. It is a complicated process and involves consumption of reagents and time. Fiber derivatization is another alternative: the fiber is immersed in reagent solution and then in a sample. The analyte is extracted and converted to a derivative in the coating.

In SPME coupled with GC, analytes are thermally desorbed into the injector of a GC instrument. This method is used for volatile and thermally stable compounds. After analyte extraction, the fiber is introduced into the injector port for analyte thermal desorption from the coating. Thermal desorption has the disadvantage of the degradation of the polymer coating at high temperatures. In addition, many non-volatile compounds cannot be completely desorbed from the fiber at high temperatures. These compounds therefore need to be derivatized before GC separation to increase their volatility. SPME coupled to HPLC is performed in a specially designed six port injector with a desorption chamber installed in place of a sample loop. Mobile phase is used for analytes desorption. Two modes of desorption are used in HPLC: (1) dynamic desorption mode, and (2) static desorption mode. Dynamic desorption involves the removal of analytes by a moving stream of mobile phase and static desorption involves the soaking of fiber in the mobile phase for a specified time for desorption of strongly adsorbed analytes. The dynamic mode of desorption is sufficient if the analyte is not strongly adsorbed onto the fiber. In this case, the analyte can be removed by a stream of mobile phase. However, if the analytes are more strongly adsorbed on the fiber, it is dipped into the mobile phase or another strong solvent for a specified time. Desorption performed in this way is known as static desorption. Each type of desorption should be undertaken using a minimum quantity of solvent (Kaur et al. 2006).

Advantages and disadvantages

SPME offers numerous advantages in sample preparation for speciation analysis, such as the combination of sampling and extraction into one step, the ability to examine smaller sample sizes, the application for polar and non-polar analytes in several matrices. SPME is easy to combine with virtually any detection system, is quick to achieve sample-matrix separation, is solvent free and enables miniaturization of manipulation in sample pretreatment techniques. Thus, SPME is an increasingly common method of sample isolation and enhancement for chromatographic analysis and is a useful alternative to LLE and SPE. SPME has been widely used in environmental analysis because it is a fairly safe method when dealing with highly toxic chemicals.

However, the main drawbacks of SPME are the relatively few stationary phases available and their limited life time (as they tend to degrade with the number of samplings). As a solution when dealing with dirty samples, the fiber can be placed inside a hollow cellulose membrane. SPME fiber coatings are fragile and relatively expensive. In addition, the carry-over between extractions cannot be avoided completely. A recent development is that of "superelastic SPME" where the fiber is a metal alloy with elastic properties and can be coated with PDMS/DVB, carboxen/PDMS and DVB/carboxen-PDMS as well as PDMS. This improves robustness and overcomes problems with the breaking of fibers due to misalignment with injection ports or in viscous matrices. In addition, the small volume of the stationary phase coating used (≤ 0.5 µL), which implies a large (sample/

stationary phase) phase ratio, which may also lead to incomplete extraction and limits the sample enrichment capabilities. Matrix effects can be an issue and quantitation generally requires matrix matched standards or the method of standard addition. The use of an isotopically labeled internal standard should be considered. The presence of high concentrations of matrix components or other compounds can result in competitive binding and displacement and potentially large errors can occur. Finally, the fiber should be cleaned before analyzing any sample as the contaminants are responsible for the background in the chromatogram. It is done in the desorption chamber of the HPLC system by a flowing solvent.

Recent developments in SPME

In-tube SPME developed by Eisert and Pawliszyn (Eisert and Pawliszyn 1997, Mester and Pawliszyn 1999) is an automated version of SPME that can easily be coupled to conventional HPLC autosamplers for on-line sample extraction/pre-concentration, separation and quantitation. It has been termed "in-tube" SPME because the extraction phase is coated inside a section of fused-silica tubing rather than coated on the surface of a fused-silica rod as in the conventional syringe-like SPME device. This new solventless and miniature extraction technique offers a reduced pressure drop during extraction and desorption under flow conditions. The trapped analytes are then desorbed or eluted by a solvent. This technique was developed because of the difficulties of interfacing SPME with LC systems. Automation of in-tube SPME not only shortens the analysis time, but also provides better accuracy, precision and sensitivity relative to off-line manual techniques. This technology has become increasingly popular because it is simple, fast, solvent-free and inexpensive. It also has the ability to combine with different detection techniques in on-line modes such as HPLC, GC and ICP-MS. However, commercial capillaries that are currently used for in-tube SPME do not show high extraction efficiencies for polar and ionic species. Therefore, it is necessary to develop new coating materials for SPME to extend the applications.

Solid phase micro-extraction for organometallic species pre-concentration

Arsenic species. Several fibers have been proposed for arsenic species pre-concentration. Different commercial SPME fibers (PA, PDMS/DVB, PDMS, CAR) were tested for the extraction of methylarsenic compounds (MMA and DMA) present in human urine. The best results were obtained using the PDMS coating (Mester and Pawliszyn 2000). Direct extraction of the methylarsenic compounds by SPME after thioglycol methylate derivatization was done. GC-MS was used as a separation and quantification system (Mester and Pawliszyn 2000). In the same way, PDMS fibers (100 μm) and a Carboxen/PDMS fibre (75 μm) were compared for the extraction of As species (triphenylarsine, TPhA and triethylarsine, TEA) from aqueous samples. It was found that the amount of extractable analytes is greatly increased for the PDMS fiber (Mothes and Wennrich 2000). These species were successfully determined by GC-AED after NaBEt$_4$ derivatization (Mothes and Wennrich 2000).

PDMS (Killelea and Aldstadt 2001) and PDMS–DVB (Roerdink and Aldstadt 2004) fibers were also used for DMA, MMA and TPhA (Killelea and Aldstadt 2001) and roxarsone, MMA and DMA (Roerdink and Aldstadt 2004) extraction and CGC-MS or CGC-pulsed flame photometric detection (PFPD) determination. Benzothiophene (BTP) and 1,3-propanedithiol (PDT) were used as derivatizing agents (Killelea and Aldstadt 2001, Roerdink and Aldstadt 2004).

Conductive polymer films (poly(3-dodecylthiophene), poly(3-octylthiophene) and poly(3-hexadecylthiophene)) were used as SPME elements for the direct and specific extraction of trace levels of AsB. HPLC coupled to ICP-MS was used as a separation and detection system (Yates et al. 2002, Tamer et al. 2003). Desorption was performed by using 0.1 M ammonium carbonate in deionized water (Yates et al. 2002) and 30 mM $(NH_4)_2CO_3$ (pH 8.70) (Tamer et al. 2003).

In addition, sol–gel-prepared SPME fibers (Gbatu et al. 1999) were also applied to extract organoarsenic (TPhA) and also organomercury (diphenylmercury, DPhM) and organotin (trimethylphenyltin, TMPhT) compounds from aqueous solutions prior to separation using HPLC with UV absorbance detection. The detection limits were comparable or slightly better than those obtained by using commercial SPME fibers. Analyte desorption was performed with 80/20 acetonitrile/H$_2$O at a flow rate of 1.0 mL min^{-1} as a mobile phase (Gbatu et al. 1999).

In tube-SPME coupled with liquid chromatography-electrospray ionization mass spectrometry (LC-ESI-MS) (Wu et al. 2000) and in-tube SPME on line ICP-MS (Hu et al. 2008) have also been applied to organoarsenic compounds extraction. A polypyrrole (PPY) coated capillary and several commercially available capillary stationary phases (PPY coated capillary, Omegawax 250, Supel-Q PLOT, SPB-1 and SPB-5) were used to examine MMA, DMA AsB and AsC extraction efficiencies (Wu et al. 2000). Compared with commercial capillaries of the time that were used for in-tube SPME, the PPY coated capillary showed better extraction efficiencies, especially for anionic species (Wu et al. 2000). Ordered mesoporous Al$_2$O$_3$ coating prepared by sol-gel technology was also used as coating material for inorganic arsenic (As(III) and As(V)) and also for Cr(III) and Cr(VI) pre-concentration from natural waters (Hu et al. 2008). The ordered mesoporous

Al_2O_3 coated capillary showed an excellent solvent and thermal stability and could be re-used for more than 30 times without decreasing extraction efficiency. Finally, analytes were desorbed by using the mobile phase (30% methanol + 70% aqueous solution, 100 mM ammonium acetate, 0.6% acetic acid) (Wu et al. 2000) or eluted on-line with 10 mM NaOH at a flow rate of 0.1 mLmin⁻¹ (Hu et al. 2008).

Mercury species. Mercury species were mainly extracted by using PDMS-coated silica fibers (Cai and Bayona 1995, Snell et al. 1996, Moens et al. 1997, Guidotti and Vitali 1998, Cai et al. 1998a, Bin et al. 1998, De Smaele et al. 1999, Mothes and Wennrich 1999, 2000, Beichert et al. 2000, Carro et al. 2002, Carpinteiro-Botana et al. 2002a, Díez and Bayona 2002, Rodil et al. 2002, Yang et al. 2003a, b, Grinberg et al. 2003b, Fragueiro et al. 2004, Davis et al. 2004, Mishra et al. 2005, Jókai et al. 2005, Zachariadis and Kapsimali 2008, Bravo-Sánchez et al. 2004, Geerdink et al. 2007, Carrasco et al. 2007). In addition, other fibers such as CAR/PDMS- (Mothes and Wennrich 2000, Grinberg et al. 2003b, Jitaru et al. 2004, Jitaru and Adams 2004, Geerdink et al. 2007), PDMS/DVB- (Barshick et al. 1998, Mester et al. 2000, Grinberg et al. 2003b, Fragueiro et al. 2004, Geerdink et al. 2007, Yeuk-Ki et al. 2010), CAR-DVB- (Grinberg et al. 2003b), DVB/CAR/PDMS- (Grinberg et al. 2003b, Centineo et al. 2004, 2006) and PA-coated silica fibers (Grinberg et al. 2003b) were also used. For the PDMS and PDMS-DVB fibers, phase equilibration could be reached after 10 min for both MMM and Hg(II) derivatized species but, when using PDMS-DVB fibers, a very high Hg⁰ peak was observed (Grinberg et al. 2003b). For CAR-DVB, equilibrium was achieved after 7 min for MMM but only after 20 min for Hg(II). A longer extraction time, greater than 20 min, is necessary when using the DVB-CAR-PDMS and PA. For CAR-PDMS fibers equilibrium was not achieved even after 30 min and a severe tailing from Hg⁰ was observed for all three fibers (Grinberg et al. 2003b). PDMS and PDMS-DVB fibers showed the best performances (Geerdink et al. 2007). Although the extraction efficiency for MMM derivative of the PDMS-CAR fiber is similar to the other fibers, desorption of MMM derivative from a PDMS-CAR fiber is poor (Geerdink et al. 2007).

As derivatizing agents, NaBEt₄ is usually used (Cai and Bayona 1995, Moens et al. 1997, Guidotti and Vitali 1998, Cai et al. 1998, De Smaele et al. 1999, Beichert et al. 2000, Mothes and Wennrich 2000, Carpinteiro-Botana et al. 2002a, Díez and Bayona 2002, Grinberg et al. 2003b, Davis et al. 2004, Jitaru et al. 2004, Jitaru and Adams 2004, Centineo et al. 2004, Zachariadis and Kapsimali 2008). Derivatization with NaBPh₄ (Carro et al. 2002, Rodil et al. 2002, Yang et al. 2003a, Grinberg et al. 2003b, Davis et al. 2004, Bravo-Sánchez et al. 2004, Mishra et al. 2005, Jókai et al. 2005) and NaBPr₄ (Yang et al. 2003b, Grinberg et al. 2003b, Centineo et al. 2006, Geerdink et al. 2007, Carrasco et al. 2007) have been proposed. Other reagents such as methylpentacyanocobaltate(III) ($K_3[Co(CN)_5CH_3]$) and methylbis-(dimethylglyoximato)pyridinecobalt(III) ($CH_3Co(dmgH)_2Py$) (Barshick et al. 1998), and butylmagnesium chloride (Snell et al. 1996), have also been described in the literature. In general, propylation was proved to be more sensitive, robust and faster than ethylation or phenylation (Grinberg et al. 2003b). Other derivatizations based on using KBH_4 (Bin et al. 1998) or $NaBH_4$ (Fragueiro et al. 2004), and also chloride derivatives by heating and stirring with HCl (MMM determination) (Fragueiro et al. 2004) have also been proposed.

The analytes were desorbed in the GC and subsequently analyzed using MS (Cai and Bayona 1995, Guidotti and Vitali 1998, Barshick et al. 1998, Yang et al. 2003a, Centineo et al. 2004, 2006, Mishra et al. 2005, Zachariadis and Kapsimali 2008), ICP-MS (Moens et al. 1997, De Smaele et al. 1999, Mester et al. 2000a, Yang et al. 2003b, Davis et al. 2004, Bravo-Sánchez et al. 2004, Jitaru et al. 2004, Jitaru and Adams 2004), MIP-AED (Mothes and Wennrich 1999, Carro et al. 2002, Carpinteiro-Botana et al. 2002, Rodil et al. 2002, Geerdink et al. 2007), AFS (Cai et al. 1998, Díez and Bayona 2002, Jókai et al. 2005, Carrasco et al. 2007), QF-AAS (Bin et al. 1998, Fragueiro et al. 2004) or FAPES (Grinberg et al. 2003b, Davis et al. 2004). HS-SPME and DI-SPME were used. Compared with the DI-SPME sampling from aqueous phase, the HS-SPME sampling procedure is more suitable since it eliminates the memory effects of Hg(II) (Moens et al. 1997, Guidotti and Vitali 1998, Mothes and Wennrich 1999, Cai et al. 1998, Bin et al. 1998, De Smaele et al. 1999, Beichert et al. 2000, Díez and Bayona 2002, Rodil et al. 2002, Grinberg et al. 2003b, Yang et al. 2003b, Fragueiro et al. 2004, Bravo-Sánchez et al. 2004, Davis et al. 2004, Jitaru et al. 2004, Jitaru and Adams 2004, Centineo et al. 2004, 2006, Jókai et al. 2005, Geerdink et al. 2007, Carrasco et al. 2007, Zachariadis and Kapsimali 2008, Yeuk-Ki et al. 2010). When dealing with clean samples, DI-SPME provided good sensitivity and linearity over two orders of magnitude; whereas, HS-SPME showed lower sensitivity, but a linear range of more than three orders of magnitude (Mester et al. 2000a).

SPME coupled with HPLC-ICP-MS has been also used for the extraction and speciation of MMM and MEM in urine. Without any tedious sample pre-treatments or derivatization procedures, headspace SPME offers a clean and convenient one-step extraction followed by direct desorption into HPLC minimizing the introduction of artifacts (Yeuk-Ki et al. 2010). A mobile phase of 0.4% (w/v) L-cysteine, 0.05% (v/v) 2-mercaptoethanol, 5% (v/v) methanol in water from the LC pump was directed into the desorption chamber for analyte desorption (Yeuk-Ki et al. 2010).

Capillary fused-silica fiber coated with polyimide was used for extraction of organomercuric species (MMM, MEM and MPhM) from soils by CGC-QF-AAS. After *in situ* hydride generation (KBH$_4$), HS-SPME was performed (Bin and Guibin 1999).

In addition, sol–gel-prepared SPME fibers (Gbatu et al. 1999) were also applied to extract organomercury (diphenylmercury, DPhM) compounds from aqueous solutions prior to separation using HPLC with UV absorbance detection. The detection limits were comparable or slightly better than those obtained by using commercial SPME fibers (Gbatu et al. 1999).

Finally, combined SPME and electrochemically aided extraction was performed for determining mercury species. An SPME/EC fiber was made of a carbon steel wire with a 10-μm gold coating. Hg(II) ions were electrochemically extracted from the aqueous solution, desorbed with a dedicated desorption system, and then detected by ion trap GC-MS. Hg(II) ions were detected in aqueous solution and mercury vapor in gas (Guo et al. 1996).

Lead compounds. PDMS (Tutschku et al. 1996, Górecki and Pawliszyn 1996, Moens et al. 1997, De Smaele et al. 1999, Mothes and Wennrich 2000, Fragueiro et al. 2000, Yu and Pawliszyn 2000, Crnoja et al. 2001, Peñalver et al. 2011), CAR/PDMS, PDMS/DVB (Jitaru et al. 2004), and DVB/CAR/ PDMS (Centineo et al. 2004) fibers have been used for the extraction of Pb species (TeEL, TEL, TML, DML) from extracts from environmental and biological samples, gasoline and natural water. It was found that the amount of extractable analytes is greatly increased when using PDMS fibers (Mothes and Wennrich 2000). The HS-SPME mode was usually used (Górecki and Pawliszyn 1996, Moens et al. 1997, De Smaele et al. 1999, Fragueiro et al. 2000, Yu and Pawliszyn 2000, Crnoja et al. 2001, Jitaru et al. 2004, Centineo et al. 2004, Peñalver et al. 2011).

These species were successfully determined using GC coupled with an atomic emission detector (AED) (Tutschku et al. 1996, Mothes and Wennrich 2000, Crnoja et al. 2001, Peñalver et al. 2011), ICP-MS (Moens et al. 1997, De Smaele et al. 1999, Jitaru et al. 2004), MS (Górecki and Pawliszyn 1996, Yu and Pawliszyn 2000, Centineo et al. 2004), QF-AAS (Fragueiro et al. 2000), and flame ionization detector (FID) (Yu and Pawliszyn 2000). Derivatization by NaBEt$_4$ was usually performed (Tutschku et al. 1996, Górecki and Pawliszyn 1996, Moens et al. 1997, De Smaele et al. 1999, Mothes and Wennrich 2000, Centineo et al. 2004). Other derivatization techniques based on deuterium-labeled sodium tetraethylborate (NaB(C$_2$D$_5$)$_4$ (DSTEB)) (Yu and Pawliszyn 2000), NaBPr$_4$ (Crnoja et al. 2001, Peñalver et al. 2011), NaBPh$_4$ (Peñalver et al. 2011), and hydride generation (NaBH$_4$) (Jitaru et al. 2004) were proposed. Because the isotope labeled ethyl group does not occur in environmental samples, DSTEB can be used for distinguishing the original ethyl group from the introduced ethyl group. The discrimination among inorganic lead, triethyllead, diethyllead, tetraethyllead, methyllead, and mixed methylethyllead species is therefore possible (Yu and Pawliszyn 2000). Higher sensitivity was attained with propylation derivatization (Peñalver et al. 2011). A shorter SPME pre-concentration step can be selected when using $NaBPh_4$, because the derivative compounds were unstable at high adsorption, although this procedure can be used if lower sensitivity is required (Peñalver et al. 2011).

The design of different thermal desorption units (spherical, vial-based and tube-shaped volatilizers) and their influence in the thermal desorption of some lead species (e.g., TEL) was studied. Low detection limits were achieved by using Vial-based and Tube-shaped designs (Fragueiro et al. 2000). Finally, several in-tube SPME applications for organolead species pre-concentration have been developed (Mester and Pawliszyn 1999, Mester and Pawliszyn 2000). In-tube SPME on line coupled with HPLC–MS was applied to TML and TEL speciation (Mester and Pawliszyn 1999, Mester et al. 2000b). For the extraction, a Supel-Q-Plot gas chromatography column coated with a porous divinylbenzene polymer (Mester and Pawliszyn 1999, Mester et al. 2000b), OmegaWax 250 a bonded poly(ethylene glycol) coating (Mester et al. 2000b) and Nukol a poly(ethylene glycol) modified with a nitroterephthalic acid coating (Mester et al. 2000b) were used. For the SupelQ Plott and Omegawax capillaries the extractions are based on hydrophobic interactions between the compound, the mobile phase and the capillary coating. For the Nukol capillary, the interaction is based upon a light cation-exchange effect due to the presence of carboxylic functional groups in the modification agent. Because of this, the best results were obtained from the Nukol column (Mester et al. 2000b). Desorption was carried out by using the mobile phase 0.1% (w/v) trifluoroacetic acid (TFA) + 12% (v/v) methanol (Mester and Pawliszyn 1999, Mester et al. 2000b).

Manganese compounds. Direct determination of organomanganese compounds (cyclopentadienyl manganese tricarbonyl and methylcyclopentadienyl manganese tricarbonyl) from seawater was performed by using HS-SPME with PDMS fiber and GC-AED after $NaBPr_4$ and $NaBPh_4$ derivatization (Peñalver et al. 2011).

Se compounds. SPME by using silica fibers with PDMS coatings prepared by the sol–gel process was also used as a sample preparation strategy for the analysis of seleno amino acids (SeMet, SeEt, and SeCys) by GC-ICP-MS (Vonderheide et al. 2002). Acylation of the amino group and esterification of the carboxylic group in these compounds was performed with isobutylchloroformate to increase volatility (Vonderheide et al. 2002). DMSe and DMDSe contents in garlic samples (Dietz et al. 2003) and urine (Bueno and Pannier 2009); and DMSe and DESe from gastric digestion of selenized yeast (Sanz-Landaluze et al. 2004) and enriched yeast samples (Dietz et al. 2004) have been also extracted by HS-SPME (CAR/PDMS fiber) and thermal desorption and detection by AAS or ICP-MS (Vonderheide et al. 2002, Bueno and Pannier 2009, Dietz et al. 2004), GC-MIP-AED (Sanz-Landaluze et al. 2004, Dietz et al. 2004) or AFS (Dietz et al. 2004).

Inorganic selenium species from selenized yeast material (Dimitrakakis et al. 2004) and human urine (Kapsimali and Zachariadis 2010) were also pre-concentrated by HS-SPME using PDMS-coated fiber. GC–MIP-AED (Dietz et al. 2004) and GC-MS (Kapsimali and Zachariadis 2010) were used for separation and quantification of selenite after ethylation, propylation and phenylation derivatization. Ethylation over phenylation was preferable for the headspace extraction because of the higher volatility of the diethyl-derivative of selenites (Kapsimali and Zachariadis 2010). Selenite, DMSe and DMDSe were extracted from natural water using a DVB/CAR/PDMS fiber, and separated and quantified by GC-AED (Campillo et al. 2007). Sodium tetraethylborate and 4,5-dichloro-1,2-phenylenediamine were used as derivatizing agents. DI-SPME and HS-SPME were tested (Campillo et al. 2007).

Antimony compounds. Direct determination of antimony compounds (stibine (SbH_3), MMSb, DMSb and TMSb) from gases from *Cryptococcus humicolus* cultures was performed by using HS-SPME with PDMS fiber and GC-MS (Smith et al. 2002).

Tellurium compounds. Volatile organotellurium compounds (methanetellurol, CH_3TeH; DMTe; dimethyl tellurenyl sulfide, CH_3TeSCH_3; dimethylditelluride, DMDTe), which were released into the headspace gas above liquid cultures of *Escherichia coli* when amended with tellurite anions in micromolar amounts, was determined by using HS-SPME with CAR/PDMS fiber and GC-MS (Swearingen et al. 2004).

Organotin Compounds. HS-SPME was extensively applied for assessing inorganic tin and organotin compounds (tetramethyltin (TeMT), trimethyltin (TMT), DMT, MMT); tetraethyltin (TeET); butyltin (MBT, DBT, TBT and tetrabutyltin (TeBT)); phenyltin (MPhT, DPhT and TPhT); and octyltins (monooctyltin (MOcT), dioctyltin (DOcT) and trioctyltin (TOcT)) in environmental solid and liquid samples. *In situ* derivatization with $NaBEt_4$ was usually performed (Morcillo et al. 1995, Tutschku et al. 1996, Moens et al. 1997, Guidotti and Vitali 1997, Lespes et al. 1998, De Smaele et al. 1999, Vercauteren et al. 2000, Aguerre et al. 2000, Jiang and Liu 2000, Millán and Pawliszyn 2000, Cardellicchio et al. 2001, Aguerre et al. 2001a, Aguerre et al. 2001b, Crnoja et al. 2001, Carpinteiro-Botana et al. 2002a, Azenha and Vasconcelos 2002, Bancon-Montigny et al. 2002, Carpinteiro-Botana et al. 2002b, Aguerre et al. 2003, Le Gac et al. 2003, Colombini et al. 2004, Centineo et al. 2004, Bravo et al. 2004, Chi-Chi and Maw-Rong 2005, Devos et al. 2005, Zuliani et al. 2006, Zachariadis and Rosenberg 2009b,

Nikolaos and Zachariadis 2010). Derivatization with NaBPr$_4$ (Crnoja et al. 2001) or hydride generation (Gui-Bin et al. 2000, Ji-Yan et al. 2001, Ji-Yan and Gui-Bin 2002) were also proposed. Commercial PDMS-fibers were usually used (Morcillo et al. 1995, Tutschku et al. 1996, Moens et al. 1997, Guidotti and Vitali 1997, Lespes et al. 1998, De Smaele et al. 1999, Vercauteren et al. 2000, Aguerre et al. 2000, Millán and Pawliszyn 2000, Gui-Bin et al. 2000, Ji-Yan et al. 2001, Cardellicchio et al. 2001, Aguerre et al. 2001a, Aguerre et al. 2001b, Ji-Yan and Gui-Bin 2002, Carpinteiro-Botana et al. 2002a, Azenha and Vasconcelos 2002, Bancon-Montigny et al. 2002, Carpinteiro-Botana et al. 2002b, Aguerre et al. 2003, Colombini et al. 2004, Jitaru et al. 2004, Centineo et al. 2004, Bravo et al. 2004, Chi-Chi and Maw-Rong 2005, Devos et al. 2005, Zuliani et al. 2006, Zachariadis and Rosenberg 2009b, Nikolaos and Zachariadis 2010). Fibers based on CAR/PDMS- (Lespes et al. 1998, Le Gac et al. 2003, Jitaru et al. 2004), or DVB/CAR/PDMS-coated fused silica fiber were also used.

Different quantification techniques such as ICP-MS (Moens et al. 1997, De Smaele et al. 1999, Jitaru et al. 2004, Vercauteren et al. 2000, Aguerre et al. 2001), MIP-AES (Carpinteiro-Botana et al. 2002a, Aguerre et al. 2001a, Zachariadis and Rosenberg 2009b), MS (Guidotti and Vitali 1997, Cardellicchio et al. 2001, Azenha and Vasconcelos 2002, Bancon-Montigny et al. 2002, Colombini et al. 2004, Centineo et al. 2004, Chi-Chi and Maw-Rong 2005, Devos et al. 2005, Nikolaos and Zachariadis 2010), AED (Tutschku et al. 1996, Crnoja et al. 2001, Carpinteiro-Botana et al. 2002b), FPD (Lespes et al. 1998, Aguerre et al. 2000, Jiang and Liu 2000, Millán and Pawliszyn 2000, Aguerre et al. 2001a, Ji-Yan and Gui-Bin 2002), QSIL-FPD (modified flame photometric detector using quartz surface-induced tin emission) (Gui-Bin et al. 2000, Ji-Yan et al. 2001), PFPD (Aguerre et al. 2001a, Le Gac et al. 2003, Bravo et al. 2004, Zuliani et al. 2006), ICP-OES (Aguerre et al. 2001b, 2003) have been used.

HS-SPME (by using PDMS-coated fiber) has also been applied for extracting fentabutatin oxide (FBTO) and two FBTO by-products (bis(2-methyl-2-phenylpropyl) tin and mono(2-methyl-2-phenylpropyl) tin) in wines. Selective determination of FBTO was achieved using GC–AED (Montes et al. 2009).

SPME methods coupled with GC are extensively used for tin species pre-concentration and separation; however, few applications describe SPME based extraction of organotins with LC analysis. TMPhT compounds from aqueous solutions were separated using HPLC with UV absorbance detection after adsorption on sol–gel-prepared SPME fibers (Gbatu et al. 1999). DI-SPME methods coupled to HPLC and the subsequent detection using ICP-MS have also been used for TBT, TPhT, TMT and tripropyltin

(TPrT) determination in freshwater and seawater samples (Ugarte et al. 2009). These organometallic compounds were extracted on commercially available coated fibers (PDMS, PA, PDMS/DVB and carbowax/templated resin (CAR/TPR)). The results showed that PDMS/DVB fibers gave the best extraction efficiency for TMT, TPrT and TPhT, while the most polar CAR/TPR exhibited slightly better extraction for TBT (Ugarte et al. 2009). Analytes were desorbed in static mode in a SPME/HPLC interface (by soaking the fiber in mobile phase inside the desorption chamber of the interface for several minutes) (Ugarte et al. 2009).

In-tube SPME combined with LC-ESI-MS has been used for the sampling and determination of TBT in environmental samples (Wu et al. 2001). A commercially available capillary, Supel-Q PLOT, was used. Desorption was carried out by using 80% methanol + 20% aqueous solution (0.1% TFA) (Wu et al. 2001).

Another approach was the use of SPME without chromatographic separation. Tributyltin was collected from the sample headspace above various matrices, by passive sampling using a PDMS/DVB-coated SPME fiber, after TBT derivatization by halide generation (Mester et al. 2001).

Stir bar sorptive extraction

General aspects

Stir bar sorptive extraction (SBSE), introduced by Baltussen et al. is a solventless sample preparation technique based upon sorptive extraction (Baltussen et al. 1999). In sorptive extraction the analytes are extracted from the matrix into a non-miscible liquid phase. A stir bar (with length from 1 to 4 cm) coated with a relatively thick layer of PDMS (0.1–1.0 mm, resulting in PDMS volumes varying from 50 to 300 μL) is employed for extracting analytes from a variety of matrices (Fig. 2.7). PDMS is a well known stationary phase in gas chromatography (GC), is thermo-stable, can be used in a wide range of temperatures (220–320°C), and has adequate diffusion properties.

SBSE is a sample preparation technique derived from SPME (stirrer variation of SPME), and its extraction mechanism and advantages are similar to those of SPME. Generally, the volume of the extraction phase of SBSE is 50–250 times larger than that of a SPME fiber. It provides better reproducibility and higher sensitivity than SPME. SBSE is especially suitable for trace/ultra-trace analysis. The SBSE extraction process and its analytical applications have been thoroughly described in recent reviews (Baltussen et al. 2002, Kawaguchi et al. 2006, David and Sandra 2007, Sánchez-Rojas et al. 2009, Lancas et al. 2009, Prieto et al. 2010).

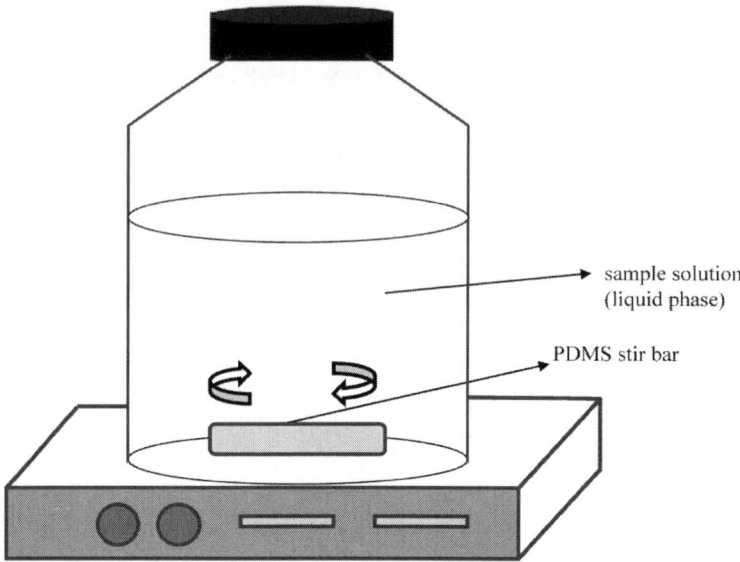

Figure 2.7 Schematic of stir bar sorptive extraction (SBSE) set-up.

The sorptive extraction is kinetically governed until a steady state is attained, where the extraction efficiency is governed by the distribution or partition coefficient of the target analyte between both phases ($K_{PDMS,w}$) and their respective volumes.

Starting from the kinetic step and considering a first-order one compartment model, the following equation can be used (Prieto et al. 2010):

$$C_{PDMS}(t) = C_{w,0} x \frac{K_1}{K_2} x \left(1 - e^{-k_2 x t}\right)$$ Eqn. 12

where $C_{PDMS}(t)$ is the concentration of the target analyte in the stir bar as a function of time, t, $C_{w,0}$ the initial concentration of the target analyte in the aqueous phase, and k_1 and k_2 are the uptake and the elimination rate constants, respectively.

When chemical equilibrium is attained, the yield of the extraction can be estimated from the mass-balance equation and the partition coefficient as follows:

$$m_{w,0} = m_{PDMS} + m_w$$ Eqn. 13

$$K_{PDMS,w} = \frac{C_{PDMS}}{C_w} = \frac{m_{PDMS}}{m_w} x \frac{V_w}{V_{PDMS}} = \frac{m_{PDMS}}{m_w} x \beta$$ Eqn. 14

where $m_{w,0}$ is the initial mass of the target analyte in the aqueous phase that is distributed between PDMS (m_{PDMS}) and water (m_w). Additionally, the partition coefficient ($K_{PDMS,w}$) is defined as the ratio of the concentrations of the target analyte between the PDMS phase (C_{PDMS}) and the aqueous phase (C_w). Once the phase ratio ($\beta = V_w/V_{PDMS}$) is included, the volumes of each phase are considered as well.

Combining equations 13 and 14, the theoretical recovery (R%) of a given SBSE setup can be calculated as follows:

$$R = \frac{m_{PDMS}}{m_{w,0}} = \frac{K_{PDMS,w}}{K_{PDMS,w} + \beta} \qquad \text{Eqn. 15}$$

In contrast to SPME, the use of SBSE as equilibrium sampling devices requires more stringent conditions (i.e., higher phase ratios and only polar analytes).

Similar to SPME, SBSE extraction can be performed in two modes: direct-SBSE (immersion) and headspace sorptive extraction (HS-SBSE). The process of direct SBSE begins by introducing the stir bar directly into the aqueous sample for stirring and extraction simultaneously in a given time, avoiding the competitive adsorptions of the stirring magnet in SPME. In HS-SBSE, the sorptive bar is exposed in the headspace above the sample solution to extract volatile analytes. Regardless of the increase in equilibrium time (several hours) compared with direct-SBSE, the headspace extraction is very advantageous in reducing the risk of contamination and prolonging the lifetime of the sorptive bar coatings because of the avoidance of direct contact with the complex sample–matrix.

After extraction, the stir bar is removed from the solution or from the head space and can then be thermally desorbed into a gas chromatograph or ultrasonically desorbed with few µL of a desorption solution.

Several factors such as extraction time, extraction temperature, stirring speed, pH of the sample, the addition of an inert salt, addition of an organic modifier, can all affect the extraction step. Some of these variables, such as sample pH or addition of an inert salt, modify the analytes or sample conditions and affect the equilibrium. Another group of variables accelerates the process affecting its kinetics, such as stirring speed. In any case, these variables must be optimized in SBSE procedures.

Extraction time. The extraction time (typically between 30 and 240 min) is controlled by parameters such as sample volume, stirring speed and stir bar dimensions. Working under equilibrium guarantees maximum sensitivity but, above all, better precision. However, sometimes, to minimize the analysis time, scientists sacrifice sensitivity and precision and work under non-equilibrium conditions.

Extraction temperature. At elevated temperatures (40–60°C) the extraction equilibrium is reached faster, but at temperatures higher than 70°C the sorption distribution coefficient of the analytes decreases and, thus, the extraction efficiencies become lower. In addition, the lifetime of the PDMS extraction phase can be reduced significantly at temperatures above 40°C.

Stirring speed. The stirring rate (up to 750 rpm) accelerates the extraction, increasing the extraction efficiency at a fixed extraction time. The decrease of the thickness of the boundary layer between the stir-bar and the solution bulk explain this fact. For higher stirring rate (> 750 rpm) no effect on extraction efficiency was observed. In addition, very high stirring rates may cause physical damage to the extraction phase arising from the direct contact of the stir-bar with the bottom of the sample vial.

Sample pH. Sample pH must be adjusted in order to obtain the solute partially or totally in the non-ionic form leading to the maximum extraction efficiency. However, too acidic (pH < 2) or too basic (pH > 9) conditions are not recommended to avoid PDMS-phase degradation and extend PDMS-coated stir-bar lifetime.

Salt addition. The addition of sodium chloride modifies the ionic strength of the sample solution. In general, it has been observed that for hydrophobic analytes the addition of an inert salt does not improve, but even reduces, the extraction efficiency. On the contrary, for polar analytes the response increases with the addition of inert salts.

Organic modifier addition. The addition of organic modifiers (MeOH or ACN) minimizes analyte adsorption to the glass walls. However, the addition of such modifiers can also increase the solubility of the solutes in the water phase and can therefore minimize the extraction efficiency.

Advantages and disadvantages

The proposed method has many practical advantages such as the use of small sample volumes and simplicity of extraction. It is also a solvent-free procedure and offers high sensitivity. In addition, efficiency is better than that offered by SPME because of the much larger volume of the PDMS-phase extraction.

Operations such as removing the stir-bar from the sample, rinsing and drying (optionally liquid desorption, if applied) are usually performed manually, which is laborious and can lead to errors. Moreover, for some applications, a special and complex thermal desorption unit for the stirring bar is required.

In addition, the commercially available SBSE coatings are limited, which discourages the selectivity and applicability of SBSE as a sample preparation technique. At present, only PDMS-coated stir-bars are commercially available and this represents one of the main drawbacks of SBSE, because of the non-polarity of the PDMS polymer, polar compounds are poorly extracted.

To improve the extraction performance of SBSE for polar compounds, some novel extraction coatings with good capability have been explored, including dual-phase stir bar (PDMS/activated carbons), restricted access material alkyl-diol-silica, polyurethane foams, PDMS/β-cyclodextrin (β-CD), polyphthalazine ether sulfone ketone, nylon-6 polymer imprinted with L-glutamine, vinylpyrrolidine/divinylbenzene monolithic material, PDMS/polyvinylalcohol (PVA), titania immobilized polypropylene hollow fibre (TiO$_2$-PPHF) or carbowax (CAR)-PDMS-PVA.

Stir bar sorptive extraction for organometallic species pre-concentration

Although SBSE has been successfully developed to analyze organic compounds in environmental, food and biological samples, very few reports on the application of SBSE to the trace elements and their species analysis have been published. This is mainly due to the lack of suitable coatings and coating techniques.

Thus, direct-SBSE (Ito et al. 2008, Ito et al. 2009, Vercauteren et al. 2001) and HS-SBSE (Prieto et al. 2008) used a commercial PDMS coated stir bar and thermal desorption coupled with GC for mercury speciation (Hg(II), MMM, MEM, DEM) (Ito et al. 2008, Prieto et al. 2008, Ito et al. 2009), and for the determination of tin species (MBT, DBT, TBT, TPhT) (Vercauteren et al. 2001, Prieto et al. 2008). The method involves an *in situ* alkylation (propyl- (Ito et al. 2008) or ethyl- (Vercauteren et al. 2001, Prieto et al. 2008, Ito et al. 2009) derivatization) of the species. Thermal desorption step requires that the stir bar is transferred into a glass thermal desorption tube operating at 200–290°C, and the desorbed compounds are then cryofocused at –150°C.

As stated above, further developments of SBSE, especially in the preparation of novel coatings for stir bars, mainly for polar compounds such as organometallic compounds, are needed. Several high polarity extraction phases have been prepared (by sol–gel immersion and low temperature hydrothermal process) for arsenic (Mao et al. 2011) and selenium (Duan et al. 2009) compounds.

TiO$_2$-PPHF was prepared for the extraction of inorganic arsenic species, MMA, DMA and phenyl arsenic compounds and their possible transformation products (4-hydroxyphenylarsonic acid, 3-nitro-4-

hydroxyphenylarsonic acid, phenylarsonic acid, 4-aminophenylarsonic acid and 4-nitrophenylarsonic acid) in chicken tissues (Mao et al. 2011). The obtained TiO_2-PPHF inherits the adsorption properties of TiO_2 and the toughness of PPHF. Before the arsenic compounds were separated and quantified using HPLC-ICP-MS, they were desorbed by sonication with KH_2PO_4 (desorption solution). The TiO_2-PPHF coating was demonstrated to be a highly selective coating for arsenic species, and could easily be prepared at a low cost. In addition, with the disposable coating, the carry-over effect commonly encountered in conventional SBSE was minimized. To avoid the carry-over effect, the TiO_2-PPHF stir bar coating was discarded after use.

Several sorptive bar coatings, including PDMS, PDMS-poly(vinylalcohol) (PVA), carbowax (CAR)-PDMS-PVA and PDMS-β-cyclodextrin (β-CD) have been prepared for the extraction (headspace sorptive extraction) of volatile organo seleno compounds (DMSe and DMDSe) in garlic, onion and their juices (Duan et al. 2009). The best extraction efficiencies were obtained with PDMS-PVA sorptive bar coating. The DMSe and DMDSe retained on the sorptive bar were desorbed by stirring with methanol and were then introduced into the CGC coupled with ICP-MS for separation and determination.

Liquid phase micro-extraction

Liquid phase micro-extraction (LPME) is a miniaturized sample pre-treatment technique. It is simple, inexpensive and environmentally friendly, and has gained increasing popularity since its introduction in the mid to late 1990s (Pena-Pereira et al. 2010, Dadfarnia and Shabani 2010). In LPME, analytes are extracted from a donor phase (aqueous phase) into the acceptor phase (organic phase) previous complexation or derivatization. This is a non-exhaustive extraction approach.

Extraction efficiency is defined as the fraction of the analyte extracted into the acceptor phase of the total amount of analyte in the sample. It can be experimentally measured either by analyzing the contents of the acceptor phase after extraction of a sample containing a known amount of analyte.

Equations describing the effects of several parameters that affect the efficiency of LPME method are as follows:

$$C_{o,f} = KC_{aq,f} = \frac{KC_{aq,i}}{1 + KV_o / V_{aq}}$$
Eqn. 16

$$\frac{dC_o}{dt} = \frac{A_{i\beta}}{V_o \left(KC_{aq} - C_o \right)}$$
Eqn. 17

where $C_{o,f}$ is the final concentration of the analyte in the organic phase; $C_{aq,f}$ and $C_{aq,i}$ are the final and initial analyte concentrations in the aqueous phase, respectively; V_o and V_{aq} are the organic and aqueous phase volumes, respectively; K is the distribution coefficient; C_o and C_{aq} are the analyte concentrations in the organic and aqueous phases at the time t, respectively; A_i is the interfacial area and β is the overall mass transfer coefficient with respect to the organic phase (He and Lee 1997).

The pre-concentration factor (PF) and enhancement factor (EF) were used to evaluate the extraction efficiency. PF is defined as the ratio between the analyte concentration in the organic phase $(C_{o,f})$ and the initial concentration of analyte $(C_{aq,i})$ within the sample:

$$PF = \frac{C_{o,f}}{C_{aq,i}} x100\% \qquad \text{Eqn. 18}$$

$C_{o,f}$ was calculated from a suitable calibration curve obtained from the direct injection of the standards into an analytical instrument.

EF is defined as the slopes ratio of two calibration curves for analyte species with or without the pre-concentration procedure:

$$EF = \frac{m_{LPME}}{m} x100\% \qquad \text{Eqn. 19}$$

m_{LPME} is the slope of the calibration curve after the LPME procedure and m is the slope of the calibration curve obtained by direct injection into the instrument of the standards.

Finally, the extraction efficiency (EE_{eq}) of LPME is represented as:

$$EE_{eq} = \frac{V_{eq,a}C_{eq,a}}{V_i C_i} x100\% \qquad \text{Eqn. 20}$$

where C_i is the initial concentration of analyte in the donor phase, and $C_{eq'a}$ is the analyte concentration at equilibrium in the acceptor phase; V_i is the volume of the donor phase, and $V_{eq,a}$ is the volume of the acceptor phase at equilibrium.

LPME can be divided into four main categories:

1) dispersive liquid–liquid micro-extraction,
2) single drop micro-extraction,
3) solidified floating organic drop micro-extraction, and
4) membrane-based micro-extraction.

Basic concepts and principles of these techniques, involving parameters, advantages and disadvantages and main application to the extraction and pre-concentration of organometals will be commented on in the later.

Despite the novelty of these techniques, LPME has been used extensively for organic compounds isolation. However, applications for organometallic compounds (and also metals) pre-concentration are scarce. Three main reasons may be responsible for this (Li et al. 2011):

1) The main principle of LPME is based on the diffusion effect; however, with the exception of organomercury and organotin species, most of the metal species could not be directly extracted in LPME systems based on diffusion alone, because their ionic/polar nature would result in poor partition coefficients.

2) To increase partition coefficients for different metals and their species, several strategies including complexation and chemical derivatization (pre-extraction, *in situ*, and post-extraction) have been provided. These approaches, however, offer some disadvantages such as the reduced number of derivatization reagents, and the loss of certain species during the deribatization process, mainly when using alkyl- or arylborates derivatizing reagents.

3) The extraction phase containing the analyte must be compatible with the mobile phase or the electrolyte buffer used in HPLC or CE. Only three-phase LPME mode (liquid–liquid–liquid micro-extraction, LLLME) meets this requirement. However, the application of LLLME for speciation analysis is limited to those compounds which could be converted by reactions (protonation or complexation) into species that have a high affinity for the aqueous phase.

Recently, a novel LPME concept of phase transfer/membrane supported based liquid–liquid–liquid micro-extraction (PT/MS-LLLME) was developed to expand the applications of LPME in element speciation, avoiding the above mentioned drawbacks (Li et al. 2011).

Single-drop micro-extraction

General aspects

Single-drop micro-extraction (SDME) is a miniaturization of the traditional LLE. Classically, this pre-concentration technique is based on the use of a micro-drop of water immiscible solvent exposed to the aqueous sample donor solution (direct-SDME or continuous flow micro-extraction (CFME)) (Fig. 2.8). In Direct-SDME, analytes are extracted from the bulk aqueous phase onto a micro-drop of extractant phase by immersion of the drop in a stirred aqueous sample solution (Jeannot and Cantwell 1997) (Fig. 2.9A,B).

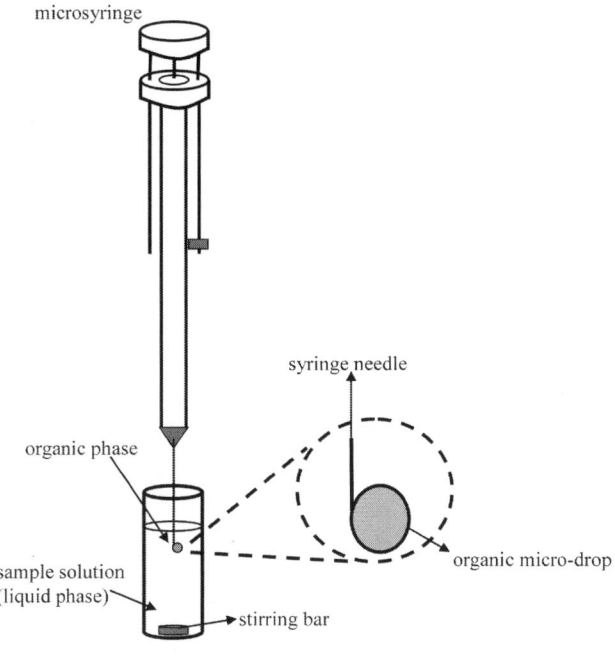

Direct-SDME

Figure 2.8 Schematic of single-drop micro-extraction (SDME) set-up.

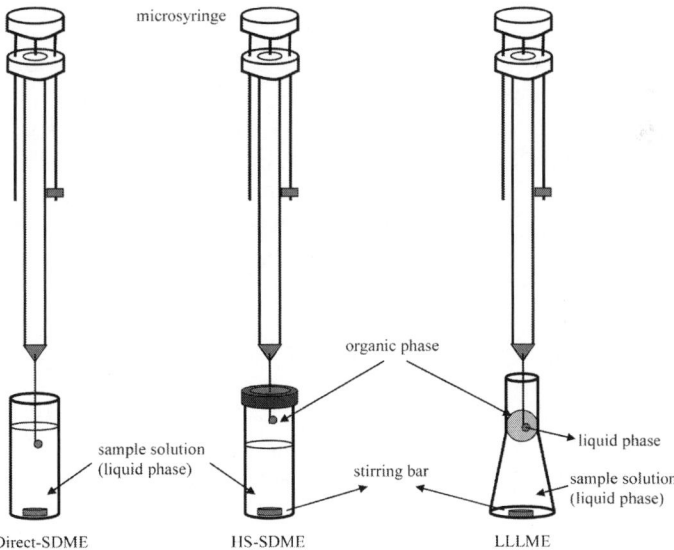

Figure 2.9 Schematic of direct single-drop micro-extraction (Direct-SDME), headspace single-drop microextraction (HS-SDME) and liquid–liquid–liquid micro-extraction (LLLME) set-up.

Another variation of the technique consists of thin water immiscible organic extraction phase (of lower density than water) layered over the aqueous sample (liquid–liquid–liquid micro-extraction, LLLME) (Ma and Cantwell 1999) (Fig. 2.9C). The LLLME system is a three phase micro-extraction technique, the donor solution, the organic solvent phase and the acceptor solution. In this system, the analytes are extracted from the donor solution into the organic solvent phase and extracted back simultaneously into the acceptor phase with the help of stirring. The acceptor phase is an aqueous microdrop suspended from a syringe and immersed into the organic phase. After extraction the microdrop was withdrawn back into the syringe for further detection (Ma and Cantwell 1998, Ma and Cantwell 1999). In addition, when analytes are volatile or semi-volatile, pre-concentration can be achieved by exposing the drop to the headspace above the sample (Fig. 2.9), which leads to the headspace single-drop micro-extraction (HS-SDME) (Theis et al. 2001). CFME is another mode of SDME (Liu and Lee 2000), wherein the extraction is performed in a glass extraction chamber instead of a vial. The sample, instead of being stirred, is pumped continuously at a constant flow rate and, when the extraction chamber is full of sample, a drop is formed at the tip of a microsyringe needle. In contrast to the other SDME micro-extraction modes, a solvent drop makes contact with a fresh and flowing sample solution. Recently, a modification of this micro-extraction mode, called cycle-flow micro-extraction was developed (Xiu et al. 2004). SDME is not exhaustive, and only a small fraction of analyte is extracted/pre-concentrated for analysis.

Although SDME was originally developed for organic analytes, its potential for pre-concentration of trace metals, elemental species and organometals has been recognized. The first publication of this methodology for inorganic species dates back to 2003, and is attributed to Chamsaz et al. (Chamsaz et al. 2003) who used the headspace mode to determine arsenic using ETAAS. Since this pioneering development, the number of applications for metal and organometallic compounds pre-concentration has significantly grown. A summary of the different modes of SDME, including CFME, and their combination with different analytical techniques, can be found in recent reviews (Xu et al. 2007, Pena-Pereira et al. 2009a).

To develop SDME methodology for extracting trace species, several parameters controlling optimum performance, such as the selection of organic solvent, solvent drop size, ionic strength of the sample, magnetic stirring rate, and stirring and sorption times were assessed. In addition, variables affecting the formation of metal complexes (chelating agent selection/concentration and pH of the complex formation and extraction) must be taken into account. Sample flow rate must be also considered when CFME is used.

Solvent selection. The choice of an appropriate water immiscible solvent should be based on selectivity, extraction efficiency, incidence of drop loss, rate of drop dissolution and level of toxicity. Candidate solvents were selected based on high boiling point and low vapor pressure to minimize any evaporation during the extraction process. Moreover, the extractant phase should be compatible with the analytical technique used. Several organic solvents and ILs have been selected according to the polarity of the species. Recently, the use of ILs has been increasing because of their particular physicochemical properties, being environmentally friendly extractant phases. Reagents such as 1-alkyl-3-methylimidazolium hexafluorophosphates ([C_nMIM][PF$_6$], n = 4, 6, 8) have been used for SDME applications in both direct immersion and headspace modes for extraction.

Drop size. Drop size shows an important influence on the extraction efficiency. Therefore, a more efficient mass transfer from aqueous solution to the organic drop can be expected when the superficial area increases. However, larger drops are difficult to manipulate and they are easy to dissolve or displace into aqueous solutions. Larger drops cannot tolerate the high sample stirring rate and long extraction times. Under these conditions, the micro-drop easily falls off the needle of the microsyringe.

Extraction time. Since SDME is not an exhaustive extraction technique, the mass transfer is a time-dependent process. The maximum sensitivity is attained at equilibrium. The selection of an exposure time that guarantees an acceptable compromise between the amount of analyte extracted and the time for analysis is needed. Therefore, the time for the transfer of analyte into the drop solvent increases the extraction efficiency until equilibrium is obtained (around 15 min) and levelling off at higher extraction times. Thus, SDME obtains maximum sensitivity after the equilibrium between aqueous and organic phase has been achieved. However, large extraction times may result in drop dissolution and can produce drop losses.

Magnetic stirring. Magnetic stirring was used to allow the mass transfer process. An increase in the stirring rate of the aqueous sample causes an enhancement of the extraction efficiency in SDME, thereby reducing the time required to reach thermodynamic equilibrium. However, high stirring rates usually lead to drop instability, drop displacement or drop dissolution. These problems are attributed to the relatively large vortex formed in the lower region of the drop, and manifest themselves as poor repeatability for measurements.

Ionic strength of the sample. Addition of salt to the aqueous sample is usually performed to improve the extraction of several analytes when classical liquid–liquid extraction is used because the increase in ionic strength brings a reduction in the solubility of the hydrophobic analytes in the water solution. Nevertheless, in SDME the addition of salt to the sample can cause two opposite effects: (1) a positive effect owing to the salting out effect and, (2) a negative effect due to the modification of the nature of the Nernst diffusion film when the ionic strength of the aqueous phase is modified, reducing the diffusion rates of analytes into the microdrop, and consequently diminishing the analytical signals.

Sample flow rate. The flow rate of sample solution affects extraction dynamics remarkably since the thickness of the interfacial layer surrounding the microdroplet will vary with the change of flow rate. However, for high flow rates, although the extraction efficiency was better, air bubble formation occurred frequently, leading to quantification problems.

Advantages and disadvantages

SDME is a simple, very inexpensive (compared with sorbent-based approaches such as SPME or SPE), fast, effective, virtually solvent-free (environmentally friendly) sample preparation procedure based on a high reduction of the extractant phase-to-sample volume ratio. There is a clear advantage in applying headspace sampling for compounds exhibiting high vapor pressure (such as derivatized organotin compounds) as the extraction process can be significantly faster in this mode. However, SDME is not very robust, as it requires special solvent handling equipment and the droplets can be lost from the needle tip of the microsyringe during extraction. This is especially the case when samples are stirred vigorously to speed up the extraction process. More reliable methods are obtained when working for short times and at low stirring rates. This increases precision of the SDME procedures, although it decreases the sensitivity.

Single-drop micro-extraction for organometallic species pre-concentration

Few analytical applications of SDME in speciation studies have been reported, although all SDME modes (direct-SDME, HS-SDME, CFME and LLLME) have been applied to organometallic species extraction from waters.

Few pre-concentration methods have been proposed for antimony species while for arsenic, no data has been found in the literature. After complexation by N-benzoyl-N-phenylhydroxylamine (BPHA), a single drop of chloroform has been used (direct-SDME mode) for antimonite pre-concentration from river water and seawater (Fan 2007). Total Sb was then determined after Sb(V) reduction to Sb(III) by L-cysteine and the concentration of antimonate was calculated by subtracting the Sb(III) concentration from the total Sb concentration (Fan 2007).

In recent studies, the organic single drop is replaced by an aqueous single drop containing Pd(II) (Pena-Pereira et al. 2009b, Gil et al. 2005) or Pt(IV) (Gil et al. 2005). In these applications HS-SDME mode was used after species derivatization by adding reducing agents (hydride generation/ cold vapor). The hydrogen evolved in the headspace after decomposition of sodium tetrahydroborate (III) injected in the system caused the formation of finely dispersed Pd(0) or Pt(0) in the drop, which in turn, were responsible for the sequestration of hydrides formed. Pd(II) or Pt(IV) are trapping agents contained in the aqueous drop. The enriched drop with hydrides is subsequently injected into a graphite tube for determination using electrothermal atomic absorption spectrometry (ETAAS). Thus, a Pd(II)-containing aqueous single drop has been used to pre-concentrate antimonite and total antimony (after Sb(V) reduction to Sb(III) with L-cysteine) from spring and seawaters (Pena-Pereira et al. 2009b). Also, Pd(II) and Pt(IV)-containing aqueous single drop have been used for MMM pre-concentration after generation of methylmercury hydride (Gil et al. 2005). In contrast to other SDME methods, no toxic organic solvents are required, since the sequestration mechanism lies in the catalytic decomposition of the hydrides onto an aqueous drop containing Pd or Pt.

Continuous-flow micro-extraction and cycle-flow micro-extraction modes have also been used for the extraction of Se(IV) and total inorganic selenium (after reduction of selenate to selenite by gentle boiling in HCl) from natural waters (Xia et al. 2006). The CFME system consisted of a PFA tubing (0.30-mm i.d.) connected to the extraction chamber (~ 0.2-mL volume), a pump and a microsyringe (10 μL) used to introduce the extracting solution (CCl$_4$). For the cycle-flow micro-extraction system, the waste outlet of tubing was put into the sample reservoir.

LLLME coupled with CE was developed for MMM and MPhM determination in water samples (Fan and Liu 2008). In the method developed MMM and MPhM were complexed with PAN to form hydrophobic complexes. When the sample solution was stirred, analytes were extracted into the organic layer (200 μL toluene) and back-extracted simultaneously into an L-cysteine microdrop (4.0 μL).

The use of ILs has replaced typical organic solvents used in SDME techniques. ILs are environmentally benign solvents and are compatible with chromatographic systems:

1) A microdrop of 1-hexyl-3-methylimidazolium hexafluorophosphate ([C$_6$MIM][PF$_6$]) was used for the micro-extraction of neutral mercury chelates obtained after the formation of dithizonate derivatives of several Hg species (MMM, MEM, MPhM and Hg(II)) (Pena-Pereira

et al. 2009c). Afterwards, the separation and determination was performed using HPLC with a photodiode array detector.

2) A microdrop of tetradecyl(trihexyl)phosphonium chloride (CYPHOS IL 101™) was used (in head space mode) for pre-concentrating inorganic (InHg) and organomercury (OrgHg) species after *in situ* cold vapor generation (CV). Stannous chloride was used to reduce Hg(II) to volatile Hg^0, while oxidation of OrHg species by UV irradiation permitted the determination of total Hg. OrgHg species concentration was evaluated based on the difference between total Hg and InHg concentration (Martinis and Wuilloud 2010). The IL microdrops were injected directly into the graphite furnace.

3) Microdrops of 1-butyl-3-methylimidazolium hexafluorophosphate ($[C_4MIM][PF_6]$) and 1-octyl-3-methylimidazolium hexafluorophosphate ($[C_8MIM][PF_6]$) were also used for several mercury (MMM, MEM, MPhM, DMM and DEM) and tin species (MMT, DMT, TMT, MBT, DBT,TBT, TPhT and DOcT) extraction (Jing-Fu et al. 2005). GC-MS was used for further quantification.

Finally, tin species (MBT, DBT, TBT and TPhT) have also been pre-concentrated by using a microdrop of decane (Colombini et al. 2004) or $α,α,α$-trifluorotoluene (Shioji et al. 2004) (in HS-SDME mode), after ethylation with $NaBEt_4$ (Colombini et al. 2004) or phenylation with tetrakis(4-fluorophenyl) borate (Shioji et al. 2004). Pre-concentrated species were then determined using gas chromatography-mass spectrometry (GC-MS).

Dispersive liquid-liquid micro-extraction

General aspects

A more recent technique, introduced by Rezaee et al., which does not involve the use of either a fiber or a syringe has been termed as dispersive liquid–liquid micro-extraction (DLLME) (Rezaee et al. 2006). As the name suggests, it is based on a ternary component solvent system similar to homogeneous liquid–liquid extraction and cloud point extraction. In DLLME a cloudy solution is formed when an appropriate mixture of extraction solvent (i.e., a few microlitres of an organic solvent with high density such as tetrachloromethane, chloroform, carbon disulfide, nitrobenzene, bromobenzene, chlorobenzene or 1,2-dichlorobenzene) and disperser solvent (i.e., methanol, ethanol, acetonitrile or acetone) is quickly injected (using a micropipette or a syringe) into the aqueous sample. Thus a high turbulence is produced. This turbulent regimen gives rise to the formation of small droplets, which are dispersed throughout the aqueous sample. Emulsified droplets have a large interfacial area. The resulting fine droplets of extraction solvent provide very efficient extraction, due in part, to their extreme surface area (the surface area between extraction solvent and aqueous sample is very large). Thus, the analytes of interest in the sample solution are quickly transferred into the fine droplets of extraction solvent. After centrifuging the cloudy solution, the sedimented phase at the bottom of a conical tube is recovered and analyzed using the most appropriate analytical technique (Fig. 2.10).

Only water-immiscible extractant solvents with higher density than water are used to ease their collection as they settle below the aqueous phase after centrifuging. The disperser solvent should be miscible in the extractant solvent and the aqueous sample. The nature of the emulsifier (disperser solvent) can also have an influence on droplet size distribution, the mean droplet size, and also on the emulsion viscosity (Schramm 2005). Liquid–liquid dispersions play an essential role in separation processes and reaction systems. This is because the large interfacial area caused but the dispersion facilitates mass transfer and reaction rate (Takahiko et al. 2000).

Several variables, such as the extraction solvent and disperser types, the volumes for both extraction solvent and disperser, the extraction time and the ionic strength, must be optimized carefully.

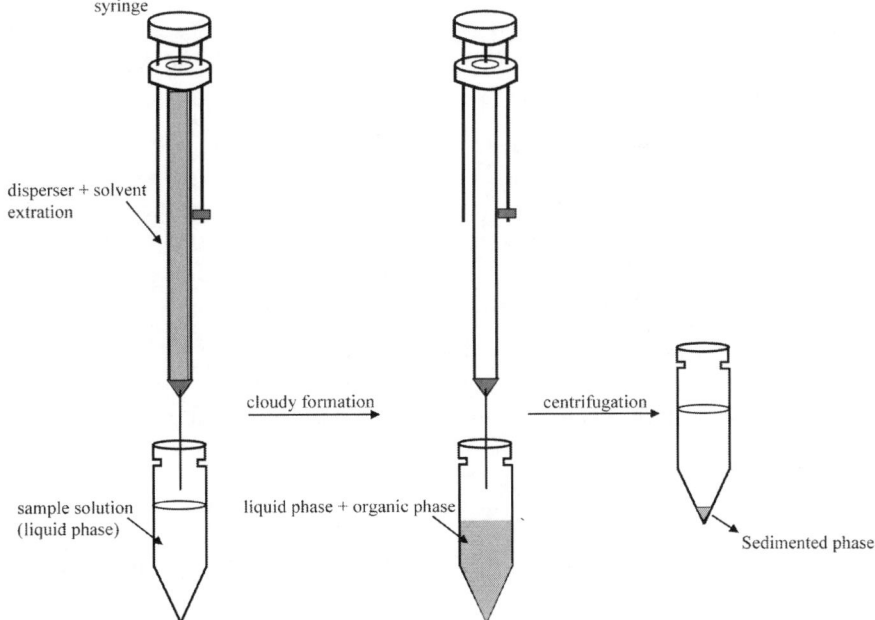

Figure 2.10 Schematic of the dispersive liquid–liquid micro-extraction (DLLME) set-up.

Solvent extraction selection. The selection of the extraction solvent deserves special attention. Higher density solvents are preferred because of quick accumulation at the bottom of the test tube. Higher extraction capacity for the analytes of interest and lower solubility in water are also preferred. Finally, good chromatographic behavior is also recommended. A variety of water immiscible organic solvents such as chlorobenzene, tetrachloroethylene and carbon tetrachloride have the above-mentioned characteristics.

*Extraction solvent volume.*The use of low extraction solvent volumes decreases the volume of the sedimented phase; which makes the removal of the sedimented phase from the test tube difficult. On the other hand, large volumes of extraction solvent may recover more analyte, in absolute terms, but it will be at a lower concentration. Thus, the enrichment factor decreases for high volumes of solvent because the volume of the sedimented phase increases. Subsequently, high enrichment factors can be obtained when using low volumes of the extraction solvent.

Disperser selection. The selection of the disperser solvent (acetone, acetonitrile, methanol or ethanol) is an important factor. Disperser solvent should have good dispersive ability. It should be chosen according to the miscibility properties of the extraction solvent and the aqueous sample. Also the disperser toxicity could be a useful criterion of choice.

Disperser volume. It is clear that at low volumes of the disperser, dispersion was incomplete, while for volumes exceeding a few milliliters, the enrichment factor decreased.

Extraction time. Extraction time is defined as the time elapsing from the incorporation of the mixture of extraction and disperser agents to the moment centrifugation begins. This variable is not significant for the extraction efficiency. This is because of the very large contact surface between the organic and aqueous phases after the formation of the cloudy mixture, which result in a very rapid extraction process.

Ionic strength. Ionic strength has no significant effect on the enrichment factor. This is possibly as a result of two opposing effects of salt addition. One of them is the increase in the volume of sedimented phase and this decreases the enrichment factor. The other is the salting-out effect that increases the enrichment factor. Therefore, the enrichment factor remains nearly constant when changing this parameter.

Centrifugation time. In DLLME centrifugation of the cloudy solution is the most time-consuming step, a process that takes approximately 5 min.

This method has been applied successfully for the determination of organic pollutants and metallic ion compounds (Theis et al. 2001, Nuhu et al. 2011) in environmental samples.

Recently, DLLME was combined with ultrasound-assistance (US-DLLME) to improve the extraction efficiency (Saleh et al. 2009a, Wei-Xun et al. 2011). The application of ultrasound improved the formation of the emulsion when the mixed solution containing disperser solvent and extraction solvent was injected into the sample. This procedure requires times for emulsification in the range of 5–10 min. These times are significantly higher than the time needed to disperse an organic solvent in conventional DLLME.

Organic solvents (such as carbon tetrachloride, chloroform, nitrobenzene, bromobenzene, chlorobenzene or 1,2-dichlorobenzene) are generally used as the extractants in DLLME and are toxic. Therefore, green solvents (ILs) have gradually taken the place of classical organic solvents. ILs solvents exhibit excellent properties, including low volatility, good chemical and thermal stability and good solubility with most organic solvents. Thus, IL-DLLME has been successfully applied to the enrichment of various organic compounds and inorganic metals in environmental and biological samples (Liu et al. 2009, Zhou et al. 2009, Mallah et al. 2009, Berton et al. 2009, Berton and Wuilloud 2010, Jia et al. 2011a).

Advantages and disadvantages

From the commercial, economical and environmental point of view, DLLME offers several important advantages over conventional solvent extraction methods: faster operation, short time and lower cost, easier manipulation, low amounts of organic extraction solvents (volumes at µL level), low sample volume, high recovery and enrichment factors, good repeatability and less stringent requirements for separation. DLLME does not require a syringe as a drop holder during the extraction process, and problems such as drop dislodgment are avoided. Thus, DLLME could be considered as an alternative technique to fit these purposes because of its simplicity and applicability in almost all analytical laboratories.

However, this micro-extraction technique appears to be difficult to automate (Theis et al. 2001). Also because of the difficulty in collecting micro-volumes of floating organic solvents, the selected extracting solvent must be denser than the aqueous samples. There is, therefore, a small number of extractants, which can efficiently extract the analytes and which have higher density than water, must form a stable cloudy solution and must be easily removed. However, the use of organic solvents with densities of less than 1.0 g mL^{-1} (i.e., toluene, 1-octanol, 1-undecanol, 1-dodecane and 1-dodecanol) have been reported in the literature (Saleh et al. 2009a).

Dispersive liquid-liquid micro-extraction for organometallic species pre-concentration

Although DLLME has been recently described, several methods for speciation studies in liquid samples (surface water, seawater, waste water, tap and drinking water, garlic extracts, urine, blood, edible oils and liquid cosmestics) have been developed.

Inorganic As, Sb, Se and Te species. Non-chromatographic methods based on the selective derivatization of arsenite, antimonite, selenite and tellurite with complexing agents such as APDC (Rivas et al. 2009, Najafi et al. 2010, Martinis et al. 2011), BPHA (Yousefi et al. 2010, Shuai and Zhefeng 2011) or 4-nitro-o-phenylendiamine (Bidari et al. 2008) (in acid media) have been proposed for inorganic As speciation (Rivas et al. 2009), for inorganic Sb (Rivas et al. 2009, Yousefi et al. 2010, Shuai and Zhefeng 2011), inorganic Se (Bidari et al. 2008, Martinis et al. 2011) and inorganic Te (Najafi et al. 2010). The complexing agent is added to the sample before adding the mixture of the extractant and the disperser solvent. The fact that asrsenate, antimoniate, selenate and tellurate do not react with complexing agent allows speciation of the inorganic forms of these elements. Then, arsenate, antimonate, selenate and tellurate were reduced to arsenite, antimonite, selenite and tellurite by the addition of sodium thiosulfate (Rivas et al. 2009) or by gentle boiling in acidic media (Bidari et al. 2008, Najafi et al. 2010, Martinis et al. 2011) and the derivatization and extraction steps were repeated for total target analyte determination. The arsenate, antimonate, selenate and tellurate concentrations were obtained by the difference.

Carbon tetrachloride (Rivas et al. 2009, Najafi et al. 2010), chloroform (Yousefi et al. 2010), chlorobenzene (Bidari et al. 2008) and IL (CYPHOS® IL 101)) (Martinis et al. 2011) were used as extracting solvents, whereas methanol (Rivas et al. 2009) and ethanol (Yousefi et al. 2010, Bidari et al. 2008, Najafi et al. 2010) were used as disperser solvents.

DLLME has mainly been used in combination with ETAAS for inorganic As, Sb, Se and Te species quantitation (Rivas et al. 2009, Yousefi et al. 2010, Najafi et al. 2010, Shuai and Zhefeng 2011, Martinis et al. 2011) CG-AED was also used for Se(IV) determination (Bidari et al. 2008).

Although DLLME automation is difficult, a semi-automatic method based on the use of on-line IL-DLLME coupled with ETAAS has been proposed for Se(IV) and Se(VI) speciation. Retention and separation of the IL phase was achieved with a Florisil® packed micro-column after DLLME with CYPHOS® IL 101. Se(IV) was selectively separated by the forming Se-APDC complex followed by extraction with CYPHOS® IL 101. Se(VI) was reduced and then determined indirectly. The on-line IL-DLLME approach makes the use of extraction solvents with lower density than aqueous media feasible, which represents a considerable advantage over formal DLLME procedures (Martinis et al. 2011).

Organic As, Hg and Sn species. DLLME was also used for organic As (Wei-Xun et al. 2011), Hg (Jia et al. 2011a, Gao and Ma 2011, Jia et al. 2011b) and Sn (Birjandi et al. 2008) species extraction.

Thus, trace amounts of organoarsenic compounds such as MMA, DMA and Roxarsone were extracted from edible oils (sunflower oil, frying oil, soybean oil, palm oil, olive oil and grape oil) using ultrasound assisted-DLLME (UA-DLLME) (Wei-Xun et al. 2011). Extracts were analyzed using HPLC–MS. A homogeneous dispersion of the extraction solvent using a disperser solvent with shake-assist was difficult when dealing with oil samples. To overcome this problem, ultrasound could be used, which improved the homogeneity of the mixture when the solution containing disperser solvent (hexane) and extraction solvent (ammonium formate buffer solution) was injected into the oil sample. Thus, the extraction efficiency was enhanced up to 1.5 (DMA) and 2.7 (MMA).

This technique was also used for mercury species extraction from natural water and liquid cosmetics. Hg(II) and MMM were extracted from tap and lake water, snow and seawater after complexation with DDTC (Jia et al. 2011b). IL-DLLME by using 1-hexyl-3-methylimidazolium hexafluorophosphate hexafluorophosphate ([C6MIM][PF6]) was used for trace amounts of Hg(II) (Gao and Ma 2011, Jia et al. 2011a), MMM (Gao and Ma 2011, Jia et al. 2011a), MPhM (Gao and Ma 2011) and MEM (Jia et al. 2011a) pre-concentration from natural waters. Complexation with DZ for waters (Gao and Ma 2011) and with APDC (Jia et al. 2011a) for liquid cosmetic (skin refresheners and hand moisturizing lotion samples) was reported. The sedimented phases were analyzed using HPLC with diode array detection (DAD) (Gao and Ma 2011) and HPLC-ICP-MS (Jia et al. 2011b, Jia et al. 2011a).

Finally, ultra-trace concentrations of butyltin (MBT, DBT and TBT) and phenyltin (MPhT, DPhT and TPhT) compounds were extracted from natural water samples after derivatization with sodium tetraethylborate (NaBEt$_4$). High enrichment factors (825–1036) and low detection limits (0.2–1.0 ng L^{-1}) were obtained under the optimum conditions (Birjandi et al. 2008).

Solidified drop liquid phase micro-extraction

General aspects

Recently, Zanjani et al. (Zanjani et al. 2007) have reported a new liquid–liquid micro-extraction method based on solidification of floating organic drop which they termed solidified drop liquid phase micro-extraction

(SDLPME). They used this successfully for the extraction and determination of polycyclic aromatic hydrocarbons. In this method, a small volume of an organic solvent with a melting point near room temperature (in the range of 10–30°C) was floated on the surface of an aqueous solution (Fig. 2.11). The aqueous phase was stirred for a prescribed period of time, and the sample was then transferred into an ice bath. When the organic solvent was solidified (about 5 min) it was transferred into a small conical vial and the melted organic solvent was used for analyte determination.

Several variables, such as the extraction solvent, extraction solvent volume, extraction temperature, stirring rate, the extraction time and the ionic strength, must be optimized carefully.

Extraction solvent selection. The selection of an appropriate extraction solvent is of great importance for the optimization of the SDLPME process. As discussed previously, for other pre-concentration techniques, the selected solvent should be immiscible with water and exhibit a high boiling point with low vapor pressure to reduce the evaporation risk during the extraction time. The solvent should exhibit good chromatographic behavior and should, also exhibit appropriate extraction efficiencies to yield high pre-concentration factors (Tor and Aydin 2006, Zanjani et al. 2007, Ganjali et al. 2010). Finally, the solvent must offer a melting point near room temperature (in the range of 10–30°C). Several extracting solvents, such as 1-undecanol (13–15°C), 1-dodecanol (22–24°C), 2-dodecanol (17–18°C),

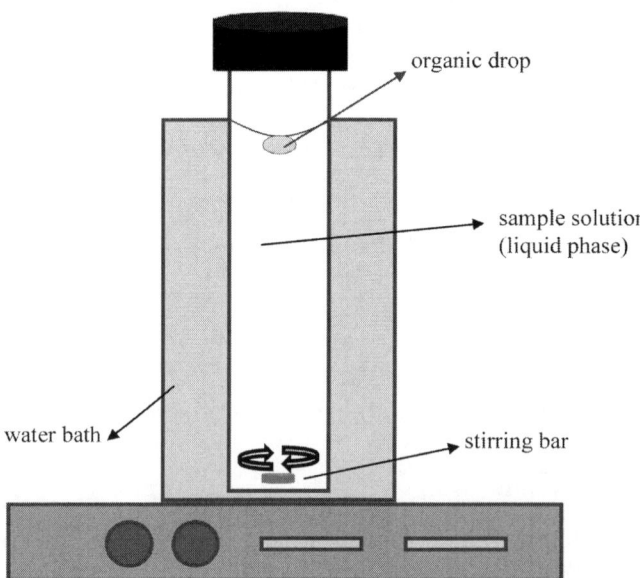

Figure 2.11 Schematic of solid drop based liquid-phase micro-extraction (SDLPME) set-up.

1-bromohexadecane (17-18°C), n-hexadecane (18°C), 1,10-dichlorodecane (14–16°C) and 1-chlorooctadecane (20–23°C) have a melting point near room temperature (Zanjani et al. 2007). Among those, 1-undecanol and 1-dodecanol have been usually used.

Organic solvent volume. In a similar way to that for other liquid-liquid micro-extraction techniques, the rate of analyte transport into the microdrop is directly related to the interfacial area between the two liquid phases and is inversely related to the organic-phase volume. Thus, by increasing the drop volume, the effect of the interfacial area predominate and the analytical signals are increased. By further increasing the microdrop volume, the effect of the solvent volume predominates and the analytical signals are decreased. Thus, at equilibrium conditions similar results for different microdrop volumes are expected. Solvent volumes < 10 µL are typical in this technique (Zanjani et al. 2007, Ganjali et al. 2010).

Sample solution temperature. Generally the increase of temperature facilitates mass transfer rate and the partition coefficient of the analyte from the sample to the organic solvent and thus increases the efficiency of the extraction. Therefore, it affects the extraction kinetics and consequently, the time required to reach equilibrium diminishes (He and Kang 2006). The effect of sample solution temperature on the extraction efficiency is usually studied in the range of 20–70°C by floating the extracting solvent microdrop during the extraction time. It is clear that by increasing the temperature, the extraction efficiency of the analyte increases. However, at higher temperatures (> 60°C) the over-pressurization of the sample vial makes the extraction system unstable. Thus, in this method the sample solution temperature should not exceed 60°C (Zanjani et al. 2007, Ganjali et al. 2010).

Stirring rate. Sample agitation is another important parameter which enhances extraction efficiency and reduces extraction time. It also facilitates the mass transfer process and thus improves the extraction efficiency (Zanjani et al. 2007). Based on the penetration theory of mass transfer of solute, the aqueous-phase mass-transfer coefficient expands with increasing stirring rate (Zhao et al. 2004a). In SDLPME, since a specific holder is not required for supporting the organic microdrop, stirring of the sample solution at high speeds is feasible. The stirring rate, however, should not be excessively high otherwise instability of the organic drop results. Another advantage is that by using a multi-stirrer, parallel extraction of many samples is possible (Ganjali et al. 2010).

Extraction time. The amount of analyte extracted at a given time depends upon the mass transfer of analyte from the aqueous phase to the organic phase (Batlle and Nerín 2004). To improve the repeatability of extraction it is necessary to choose an extraction time during which the equilibrium between the aqueous and organic phase is reached. However, it is not practical to wait for equilibrium to occur. Instead, the extraction time should be just long enough for the extraction rate to slow down so that an improved precision is obtained. Usually the extraction time for the method varies from 20 to 35 min. Therefore, the method is not exhaustive; i.e., not all of the analyte is extracted from the sample solution (Zanjani et al. 2007, Ganjali et al. 2010).

Salt addition. The salting-out effect has been widely applied to LLE and SPME. However in SDLPME and LPME in general, some contradictory results have been reported (Psillakis and Kalogerakis 2001, Zhao et al. 2004b). One of them is an observed decrease in extraction efficiency at higher salt concentrations. This can be explained by the fact that the addition of salt can restrict the transport of the analyte to the extracting solvent drop because of an increase in sample viscosity. By increasing the salt concentration, diffusion of analytes towards the organic solvent becomes more and more difficult. On the other hand, the addition of salt is reported to promote the transfer of the analyte into the extracting solvent. This fact can be explained by assuming that water molecules form hydration spheres around the salt ions. These hydration spheres reduce the concentration of water available to dissolve analyte molecules; thus it was expected that this would drive additional analytes into the extracting solvent (Zanjani et al. 2007, Ganjali et al. 2010).

Advantages and disadvantages

The proposed method has advantages such as simplicity, good accuracy and precision, short extraction time, low cost, and minimum organic solvent consumption. Since a fresh portion of organic solvent is used for each extraction, there is no memory effect, and because the volume of the organic phase is ~ 10 μL, large pre-concentration factors are achievable. A special microdrop holder is not necessary as the sample solution can be stirred at high speeds (Zanjani et al. 2007, Ganjali et al. 2010). The main drawback of the proposed method is the solvent which may lead to some analyte signals overlapping during the analysis stage. The use of selective detection systems such as GC–MS can decrease this limitation (Zanjani et al. 2007, Ganjali et al. 2010).

Solidified drop liquid phase micro-extraction for organometallic species pre-concentration

Despite the great advantages of this micro-extraction technique, only a few applications have been developed for organometallic compounds.

Thus, arsenite (Ghambarian et al. 2010) and selenite (Wang et al. 2011) were extracted by selective formation of the As(III)-APDC/Se(IV)-APDC complexes at acidic pH into 15 µL (Ghambarian et al. 2010) or 40 µL (Wang et al. 2011) of 1-undecanol. Arsenate and selenate form weak complexes with the ligand under the same pH conditions. Total inorganic As was extracted in a similar way after reduction of As(V) to As(III) with potassium iodide and sodium thiosulfate. The As(V) concentration was calculated by the difference (Ghambarian et al. 2010).

Membrane-base micro-extraction

Membrane base micro-extraction is based on the utilization of suitable polymeric membranes which allow the separation of the sample from the extractant. Several membranes can be used such as flat membranes, known as supported-liquid membrane extraction, hollow fiber membranes, and miniaturized liquid membrane probes. A hollow fiber membrane (HF) has been used to replace planar supported-liquid membrane extraction systems for a simpler operation at a lower cost. It also provides better enrichment factors, yielding an improvement in limits of detection. In HF membranes, the pores of a disposable polymer can contain the trapped extractant, so the requirement for immiscibility of both extractant and water is not as restrictive as with SDME techniques. The miniaturized liquid membrane probe is made of a glass tube closed with a porous filter (pore diameter 0.2µm), which is soaked with organic solvent. The configuration is simple and inexpensive, since no pumps are needed and, depending on the matrix, it can be used for multiple analyses after regeneration. These techniques are most often used together with HPLC or CE techniques because of the aqueous acceptor phase. Nonetheless, introducing a phase-switching step to the procedure allows application of the supported-liquid membrane extraction technique in combination with GC. Overall, membrane techniques are characterized by the distinct advantages of simplicity, low consumption of organic solvent, low cost and far-reaching possibilities for modifying membrane surface, porous properties and capacity for extraction, in several modes (Stalikas and Fiamegos 2008).

Hollow fiber liquid phase micro-extraction, solvent bar micro-extraction, phase transfer/membrane supported based liquid–liquid–liquid micro-extraction, supported liquid membrane extraction probe and microporous membrane liquid-liquid extraction are all versions of the technique and will be described next.

Hollow fiber liquid phase micro-extraction

General aspects

To overcome drawbacks associated with SDME, Pedersen-Bjergaard and Rasmussen (Pedersen-Bjergaard and Rasmussen 1999), introduced a new inexpensive solvent-minimized liquid–liquid extraction technique based on the application of a supported liquid membrane, named hollow fiber liquid-phase micro-extraction (HF-LPME), where a hollow fiber is used to accommodate the acceptor-solution volumes (typically, 5.0–25 µL).

The hollow fiber membrane is made of polypropylene with a 0.2 µm pore size, 600–1200 µm internal diameter and 200 µm wall thickness. The volume of aqueous sample is typically in the 0.1–4.0 mL range, and the length of the hollow fiber is normally 1.5–10 cm. Polypropylene was selected because it is highly compatible with a broad range of organic solvents. In addition, polypropylene, with a pore size of approximately 0.2 µm, strongly immobilizes the organic solvents used in LPME. This strong immobilization is important for ensuring that the organic phase does not leak during extraction, as that may alter extraction performance and the characteristics of the system. This is especially critical because extraction devices are often vigorously stirred to improve extraction speed, and because many samples, such as human blood plasma, may emulsify substantial amounts of water-immiscible organic solvents (Rasmussen and Pedersen-Bjergaard 2004).

In this concept, the analytes of interest are extracted from aqueous samples through a thin layer of organic solvent immobilized within the pores of a porous hollow fiber, and into an acceptor solution inside the lumen of the hollow fiber (Fig. 2.12). The disposable nature of the hollow fiber totally eliminates the possibility of sample carry-over and ensures reproducibility. In addition, the small pore size prevents large molecules and particles present in the donor solution from entering the accepting phase, thus yielding very clean extracts. HF-LPME is a technique which allows extraction and pre-concentration of analytes from complex samples in both a simple and inexpensive way. Two reviews described this appealing technique in detail (Rasmussen and Pedersen-Bjergaard 2004, Psillakis and Kalogerakis 2003).

In HF-LPME, the organic phase is protected by the hollow fiber and the process of organic solvent dissolution into bulk solution was thus retarded. This fact allows faster stirring and longer extraction times. Another factor which contributes to improvement in sensitivity is that the surface area for the rod-like configuration of the two-phase HF-LPME system is larger than the spherical surface adopted by the drop-based SDME methods.

Solvent impregnation of the fiber is essential since the extraction occurs on the surface of the immobilized solvent. The pores of a porous hydrophobic

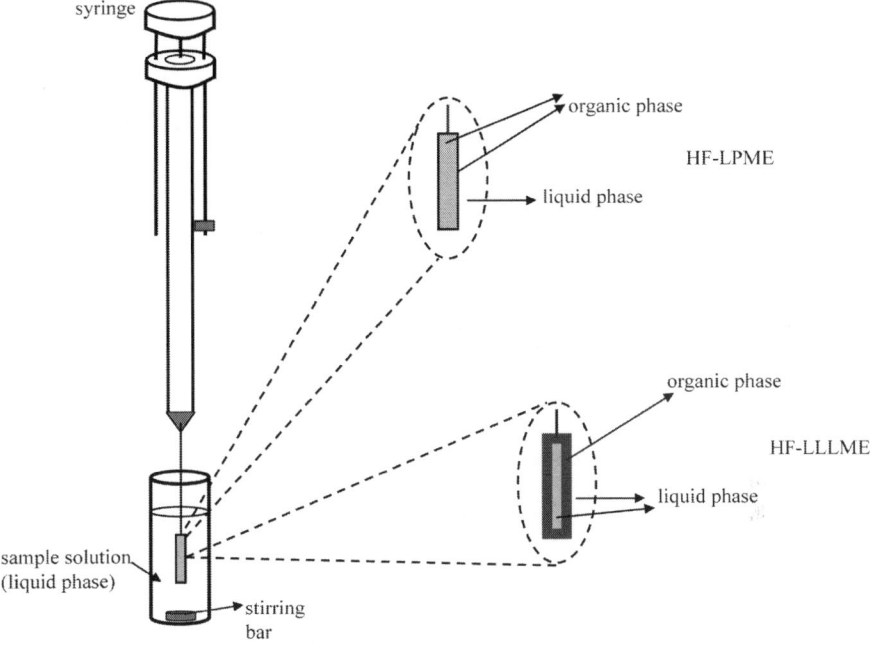

Figure 2.12 Schematic of hollow fiber liquid-phase micro-extraction (HF-LPME) and and hollow fiber liquid–liquid–liquid micro-extraction (HF-LLLME) set-ups.

polymer membrane are filled with an organic liquid, which is held in place by capillary forces. The extractant must exhibit a polarity matching that of the hollow fiber to be easily immobilized within the pores.

The extraction modes of HF-LPME mainly consist of either a two phase system or a three-phase system, according to the difference of the acceptor solution:

A) In the two phase LPME sampling mode (HF-LPME), the analyte is extracted from the aqueous sample into the same organic phase immobilized in the pores and lumen of the hollow fiber (typically made of polypropylene) and supported by a microsyringe. In this sampling mode, the acceptor phase is organic and the extraction process occurs in the pores of the hollow fiber, where the solvent is immobilized.

B) In the three-phase sampling mode, hollow fiber liquid–liquid–liquid micro-extraction (HF-LLLME), the analyte is extracted from an aqueous sample (donor phase) through the water-immiscible extractant (organic phase) immobilized in the pores of the hollow fiber and then extracted back into another aqueous phase (acceptor phase) inside the lumen of the hollow fiber. In HF-LLLME, the hollow fiber membrane was used to separate the three liquid phases. The organic phase is impregnated in the pores of the polypropylene hollow fiber and is sandwiched between the two aqueous phases. One aqueous phase is outside the fiber and the other phase is inside the lumen of the fiber. Three-phase micro-extraction was developed to extract ionizable and chargeable compounds from different aqueous samples.

Compared with HF-LPME, HF-LLLME provided a higher enrichment factor. In addition, HF-LLLME is more suitable to extract those compounds that could be converted by reactions, such as protonation or complexation, to species that will have very good solubility for the acceptor phase.

Several factors, such as the nature of the extraction solvent, the sample stirring speed and the extraction time, must be optimized in LPME procedures. In addition, the stripping solution concentration must also be considered in the three-phase sampling mode.

Organic solvent selection. The choice of the organic solvent must be based on several considerations: (1) it should be immiscible with water to prevent leakage and its dissolution into the aqueous phase; (2) it should exhibit a low volatility, which will restrict solvent evaporation during extraction; (3) it should be strongly immobilized in the pores of the hollow fiber to prevent leakage; and (4) it should provide an appropriate extraction selectivity to leading to high extraction recoveries. For the three-phase mode, 1-octanol and dihexyl ether have been the most popular, but hexane, octane, nonane, dichloromethane, butyl acetate, 2-octanone and diamyl ether have also been used. For the two-phase mode, 1-octanol has been the most popular candidate, but most solvents mentioned for three-phase LPME have also been used in two-phase systems (Rasmussen and Pedersen-Bjergaard 2004).

Sample stirring speed. Agitation of the sample is routinely applied to accelerate the extraction kinetics and reduces the time required to reach the thermodynamic equilibrium. Increasing the agitation rate of the donor solution enhances extraction, as the diffusion of analytes through the interfacial layer of the hollow fiber is facilitated, and improves the repeatability of the extraction method. In LPME, the acceptor solution is confined within the fiber, and it can tolerate very high stirring speeds (around 1000 rpm). However, high stirring rates can cause bubbles which can become attached to the surface of the hollow fiber and can decrease analyte transfer and promote solvent evaporation. This leads to bad precision in measurements (Psillakis and Kalogerakis 2003).

Extraction time. The extraction efficiency of HF-LPME is based on the analyte's partitioning between the aqueous sample and the organic solvent. This equilibrium is attained only after exposing the acceptor solution to the sample for a long period of time for which solvent loss through dissolution could occur. Although longer exposure times of the acceptor solution generally result in increased extraction efficiency, long extraction times are not always practical.

Volumes of donor and acceptor solutions. In general, sensitivity of two-phase and the three-phase LPME systems can be increased by decreasing the volume ratio of the acceptor-to-donor phase. However, the volume of the acceptor solution used for extraction may also be adjusted, depending on the analytical technique coupled to LPME (Psillakis and Kalogerakis 2003).

Salt addition. Depending on the nature of the target analytes, addition of salt to the sample solution can decrease their solubility and therefore enhance extraction because of the salting-out effect (Psillakis and Kalogerakis 2003).

Advantages and disadvantages

This miniaturized separation and pre-concentration technique offers many advantages over traditional LLE, such as low cost, extreme simplicity, safety and environmental friendliness. Very small amounts of the solvent are used, and exposure of the operator to toxic organic solvents is therefore minimized. In general, the extraction efficiency achieved with HF-LPME is higher than with Direct-SDME. This is because the organic solvent is protected by the hollow fiber, which allows the use of vigorous stirring rates (to accelerate the extraction kinetics) and high extraction time. In addition, the contact area between the aqueous sample and the extractant phase is higher than in the case of SDME, which increases the mass transfer rate. The disposable nature of the hollow fiber helps to eliminate sample carry-over and cross-contamination, enhancing the repeatability and reproducibility. Moreover, the small pore size allows micro-filtration of the sample. Large molecules and particles are prevented from entering the accepting phase and this results in a very clean extraction. Thus, the analysis of complex samples (biological fluids such as urine and plasma) is feasible. At the same time, the technique combines extraction, concentration, clean-up and sample introduction into one step.

However, it should be noted that this technique suffers from some drawbacks such as the manipulation of the hollow fiber at the time of placing it at the tip of the needle of the micro-syringe. This operation is carried out before the micro-extraction process and it could be a source of contamination. In addition, before use, each hollow fiber must be sonicated for several minutes in acetone to remove any possible contaminants.

Hollow fiber liquid-phase micro-extraction for organometallic species pre-concentration

Inorganic arsenic (Jiang et al. 2009), antimony (Zeng et al. 2011), selenium (Xia et al. 2006, Saleh et al. 2009b, Ghasmi et al. 2010) and tellurium (Ghasmi et al. 2010) species have been pre-concentrated using the two phase LPME sampling mode using APDC (Xia et al. 2006, Jiang et al. 2009, Ghasmi et al. 2010), DDTC (Zeng et al. 2011) or o-phenylenediamine (Saleh et al. 2009b) as extractants combined with ETAAS (Jiang et al. 2009, Ghasmi et al. 2010), FAAS (Zeng et al. 2011), ETV-ICP-MS (Xia et al. 2006) or HPLC (Saleh et al. 2009b). The complexes are formed at pH values within the 3.0–4.0 range for arsenic, around 5.0 for antimony, and within the 2.0–4.0 range for selenium and tellurium. As(III)-APDC, Sb(III)-DDTC, Se(IV)-APDC or piazselenol (Se(IV)-o-phenylenediamine) and Te(IV)-APDC complexes are extracted with toluene (Jiang et al. 2009, Ghasmi et al. 2010), 1-octanol (Zeng et al. 2011, Saleh et al. 2009b) or CCl_4 (Xia et al. 2006) immobilized in the pores of the polypropylene hollow fiber (1.5–12 cm) as liquid membrane, and As(V), Sb(V), Se(VI) and Te(VI) species remain in the aqueous layer. For total inorganic target analyte determination, As(V) and Sb(V) forms must be reduced to the trivalent state by L-cysteine addition, while Se(VI) and Te(VI) must be reduced to tetravalent state by boiling with HCl. The concentration of the As(V), Sb(V), Se(VI) and Te(VI) species are then obtained by the difference between the concentration of the total inorganic target and the concentration of As(III), Sb(III), Se(IV), and Te(IV).

Similarly, MMM was also extracted, using the two phase LPME sampling mode, into the organic phase (toluene) prior to its determination using ETAAS, while inorganic mercury remained as a free species in the sample solution. The total mercury was determined using ICP-MS, and the levels of inorganic mercury were obtained by subtracting MeHg from total mercury (Jiang et al. 2008).

Two phase LPME sampling mode was also applied to selenoamino acids (MSeCys, SeMet and SeEt) pre-concentration from biological sample extracts after derivatization with ethyl chloroformate (ECF) to improve the volatility of selenoamino acids (Duan and Hu 2009). A toluene/chloroform (3:1 v/v) mixture was used as organic phase. GC–ICP-MS was used for selenoamino acid separation and detection.

Three-phase sampling mode (HF-LLLME) has been used for mercury species extraction from environmental water, seafood and biological samples (Xia et al. 2007, Jiang et al. 2008, Li et al. 2008b). Toluene (Xia et al. 2007, Jiang et al. 2008) or bromobenzene (Li et al. 2008b) impregnated in the pores of the hollow fiber wall were used as organic phases; and thiourea (Jiang et al. 2008), $Na_2S_2O_3$ (Xia et al. 2007) or L-cysteine (Li et al. 2008b) aqueous solution in the lumen of hollow fiber were used as acceptor phases. The simultaneous pre-concentration of MMM, MEM and MPhM could be performed, and the mercury species in the post-extraction acceptor phase were analyzed directly using HPLC (Xia et al. 2007) or CE (Li et al. 2008b) with UV detection at 200–254 nm. ETAAS was also used for MMM determination after HF-LLLME (Jiang et al. 2008).

Several modifications of conventional HF-LPME have been developed. Volatile selenium compounds (DMSe and DMDSe) were extracted by using a headspace hollow fiber protected liquid-phase micro-extraction (HS-HF-LPME) in which the organic solvent (1-decanol) was impregnated in the pores and filled inside the porous hollow fiber membrane (Ghasemi et al. 2011). In HS-HF-LPME the surface area of the extraction phase in contact with the headspace is dramatically higher than in SPME or HS-SDME.

A continuous on-line extraction using a conventional hollow fiber membrane was developed for arsenate extraction via chelation with dibutyl butylphosphonate (DBBP) and tributyl phosphate (TBP) (Hylton and Mitra 2008). The design of the system is simple, offered high active surface area per unit volume, and allowed fast flow rates leading to enhanced mass transfer. A polypropylene hollow fiber (19 cm) was soaked in a DBBP/TBP (90:10, v/v) mixture for a few minutes. It was then placed in the channel of the microfluidic extractor, and the ends were attached to the respective inlets and outlets. A syringe filled with acceptor solution (NaCl) was attached to the inlet of the fiber lumen which was flushed to remove excess organic extractant. The donor solution was pumped through the microfluidic channel around the hollow fiber using the micro-syringe pump. The acceptor solution was also pumped with a syringe pump. In a typical extraction, the solutions were allowed to flow until 10mL of the extract had been collected.

Solvent bar micro-extraction

In 2004, Jiang and Lee reported another new mode of LPME called solvent bar micro-extraction (SBME) (Jiang and Lee 2004). In this approach, the organic extractant solvent was confined to a short length of a hollow fiber membrane (HFM), sealed at both ends, that was placed in a stirred aqueous sample solution (Fig. 2.13). The solvent bar tumbles freely in the sample solution during extraction, which effectively increases the extraction efficiency. Compared with HF-LPME, SBME has a higher enrichment factor and faster extraction kinetics. In comparison with stir bar sorptive extraction (SBSE), the procedure is very easy to operate and no extra desorption unit is needed. The results show that this technique is stable, reproducible and efficient.

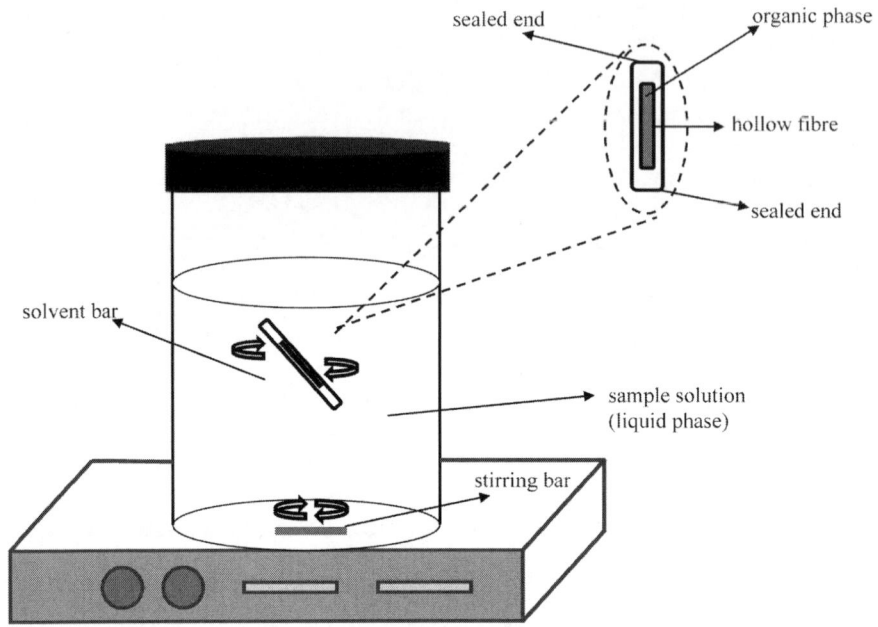

Figure 2.13 Schematic of solvent bar micro-extraction (SBME) set-up.

Only one application of SBME was reported for inorganic arsenic species extraction from natural waters (Pu et al. 2009). The method is based on the chelation of As(III) with APDC under the selected conditions, and the extraction of the As(III)-APDC complex into the organic phase, while As(V) remained in aqueous solution. The post-extraction organic phase was directly injected into ETV-ICP-MS for determination of As(III). As(V) was reduced to As(III) by L-cysteine and was then subjected to SBME prior to total As determination. The assay of As(V) was based on subtracting As(III) from total As.

Phase transfer membrane supported liquid liquid liquid micro-extraction

Recently, Yang and Ying (Yang and Ying 2009) reported a general phase transfer liquid–liquid extraction (PT-LLE) protocol to extract metal ions from an aqueous solution to an organic medium. However, LLE is well-known to have some shortcomings such as emulsion formation, the use of large sample volumes and toxic organic solvents, which make LLE environmentally unfriendly as a sample pre-concentration method. Recently, Li et al. (Li et al. 2011) developed a novel sample pre-treatment technique termed phase transfer membrane supported liquid–liquid–liquid micro-extraction

(PT/MS-LLLME) for the simultaneous extraction of inorganic and organic mercury species which avoided drawbacks associated with LLE.

In PT/MS-LLLME analytes are extracted from the donor phase, through the organic phase, and further into the acceptor phase by complexation with the help of an intermediate solvent (Fig. 2.14). The mechanism of PT/MS-LLLME is similar to that of the other three phase LPME techniques, which are non-exhaustive extraction approaches. In this procedure the volume of the organic phase and the acceptor phase is not constant. This is because the diffusion of the intermediate solvent from the donor phase, through the organic phase, and into the acceptor phase, changes the volumes of the phases. In addition, the introduction of intermediate solvent into the donor phase, results in changes of the donor phase properties.

At present, one application of PT/MS-LLLME has been reported for the determination of organometallic species. Thus, PT/MS-LLLME was used for mercury species (Hg(II), MMM, MEM and MPhM) pre-concentration from biological extracts and environmental water samples. Acetonitrile as an intermediate solvent was added into the donor phase to improve the contact between mercury species and the complexing reagent (dodecylamine (DDA)). A nylon membrane supporting carrier, containing 50 μL of acceptor

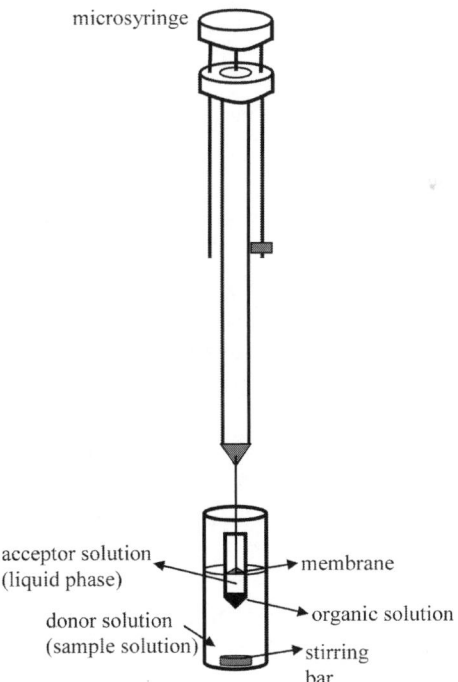

Figure 2.14 Schematic of phase transfer membrane supported liquid–liquid–liquid micro-extraction (PT/MS-LLLME) set-up.

solution (L-cysteine) was hung up. The acceptor solutions were directly analyzed by large volume sample stacking capillary electrophoresis/ ultraviolet detection (LVSS-CE/UV) (Li et al. 2011).

Microporous membrane liquid liquid extraction

Microporous membrane liquid–liquid extraction (MMLLE) systems have mainly been restricted to industrial scale operations such as extraction of phenols and volatile organic compounds or metals. However, Shen et al. (Shen et al. 1998) recently reported a fully automated, on-line MMLLE system for the determination of local anaesthetics in blood plasma by capillary GC. MMLLE is a two phase extraction technique where the sample solution is pumped through the donor channel of a membrane device holding a hydrophobic microporous membrane. The pores of the membrane and acceptor channel of the device are filled with an organic solvent. The analytes in a small volume of sample (< 1.0 mL) were extracted into the organic acceptor phase. The method has good potential for clean-up and enrichment of biological tissue, sediment digest samples and contaminated waters. It also offers high selectivity and low carry-over effects (Shen et al. 1998).

One application of MMLLE was reported for ionic and nonionic organotin compound (MBT, DBT, TBT, TPrT, TeBT, DPhT and TPhT) extraction from natural water after derivatizing with $NaBEt_4$. Isooctane was used as the acceptor phase (Ndungu and Mathiasson 2000).

Supported liquid membrane extraction probe

Supported liquid membrane (SLM) extraction was first introduced by Audunsson (Audunsson 1986) using a flat unit and has been utilized to extract and enrich many trace metals. Many different designs and sizes of membrane extraction units have been described. The liquid membrane extraction probe (SLMP) is a miniaturized device made from a glass tube of dimensions 13 mm i.d., 16 mm o.d and 92 mm length. In this configuration, one end of the glass tube was closed with a porous membrane sealed with PTFE tape and soaked with organic solvent. The inside of this probe served as the acceptor. The probe was then immersed in a vial containing the stirred sample (Fig. 2.15). The configuration is simple and inexpensive because no pumps are needed, and it is possible to perform many probe extractions simultaneously, which is an important aspect in routine analysis. Stirring the sample allows faster mass transport, and offers high extraction efficiency. Depending on the sample matrix, the LMP can be used many times, and it is also very easy to regenerate (Soko et al. 2003).

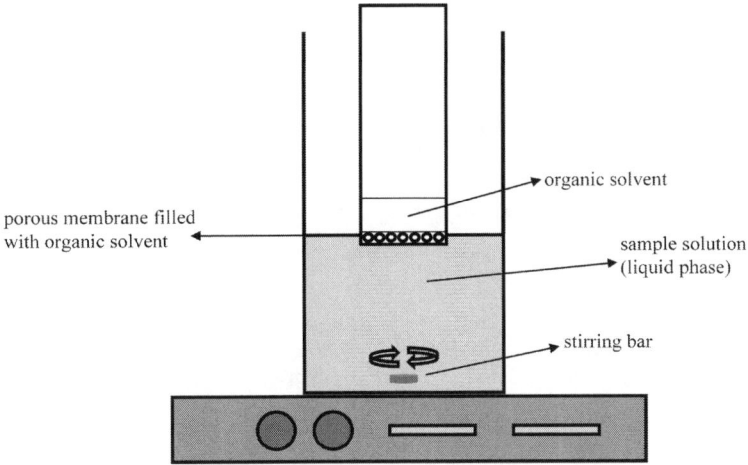

Figure 2.15 Schematic of supported liquid membrane probe (SLMP) set-up.

Few applications of SLMP have been reported for organometallic species analysis. LMP has been applied to the extraction and pre-concentration of the organotin (MBT, DBT, TBT and TPhT) (Cukrowska et al. 2004) and organolead (TeML, TML, TeEL, TEL) (Cukrowska et al. 2007) species from aqueous samples to an organic acceptor solution (isooctane) by ethylation (Cukrowska et al. 2004) or phenylation (Cukrowska et al. 2007) derivatization.

Sample Pre-treatments for Solid Samples

Solid-liquid extraction—Introduction

Solid-liquid extraction methods are based on target desorption, solvation and diffusion from the sample matrix (solid phase) allowing them to transfer into the liquid phase (solvent). The solvent must first enter through the pores of the particles (Fig. 2.16) and desorp the target analytes bound to specific points in the matrix. Desorbed analytes will then diffuse into the solvent contained in the pore (or through the organic matter to the solvent contained in the pore); and finally, from the solvent contained in the pore to the liquid phase (solvent). In addition to the solvent nature, which will govern the extraction efficiency, other different variables (mainly heat and pressure) also affect these mechanisms. The solid-liquid extraction procedures developed are thus performed under different conditions to enhance the extraction efficiency. Optimization of such environmental variables will lead to the various techniques for solid-liquid extraction.

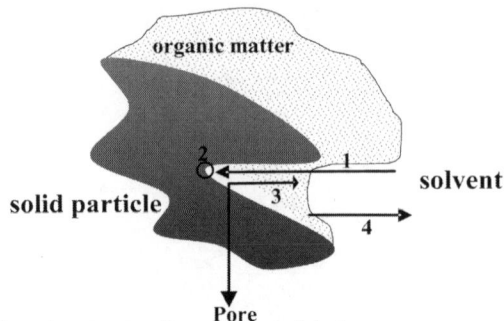

1: **The solvent enters into solid particles' pores**
2: **Targets desorption**
3: **Targets diffusion into the solvent contained in the pore**
4: **Targets diffusion from the solvent contained in the pore to the solvent**

Figure 2.16 Schematic representation of the different mechanisms involved in solid-liquid extraction.

The solvent nature will be dependent on the sample matrix and the elemental species of interest. In some cases, dilute acids such as hydrochloric acid, hydrobromic acid, carboxylic acids or organic solvents such as methanol, dichloromethane or hexane are adequate for target analyte isolation. However, especially for biological matrices, some elemental species can be integrated into macromolecules such as proteins (as for iodinated amino acids and selenium containing amino acids) and extraction efficiencies with conventional solvents can be poor. As will be discussed here, polar solvents such as methanol and water (and methanol-water mixtures) are commonly used for extracting arsenic species; while organotin compounds are usually extracted with acetic acid-methanol solvents. However, mercury species extraction, mainly methyl-mercury and inorganic mercury, can be performed with different solvents such as sodium hydroxide or potassium hydroxide in methanol, tetramethyl ammonium hydroxide (TMAH), hydrochloric acid or acetic acid, combined with organic solvents such as methanol or toluene.

Conventional mechanical shaking procedures

The simplest solid-liquid extraction procedures can be performed by mechanical agitation or by magnetic stirring. These methods consist of performing the solid-liquid extraction by continuous contact between the sample (solid) and the extracting solution (liquid). Since temperature and pressure are not varied during the process, the extraction efficiency is generally low and, low yields can be obtained even after large extraction times (several hours). The long extraction times lead to time-consuming procedures. In addition, the risk of species degradation/inter-conversion and/or sample contamination can be important. Once the extraction procedure has finished, the supernatant (extract) must be separated from the solid (pellet) by filtration or centrifugation (a preferable option as it is a fast procedure, it is economical because no costly membranes or other disposables are used, and offers a consistent performance due to no mid-run degradation as filters clog).

Tin species

Butyl-tin compounds (MBT, DBT, and TBT) have been extracted from marine sediments with acetic acid/methanol (3/1) by mechanical shaking for 12 hr (Ruiz Encinar et al. 2002a, García Alonso et al. 2002), or from sewage sludge with glacial acetic acid (16 hr) (Zuliani et al. 2010). Quantitative recoveries were found for DBT and TBT, but the method showed a low efficiency for extracting MBT which is strongly bound to the sample's matrix. However, the use of ethanol as an extracting solution for isolating MBT, DBT, TBT, MPhT, DPhT, TPhT, MOcT, DOcT and TOcT from soils gave extraction yields similar to those obtained after ultrasound and microwave assisted extraction protocols, and even after pressurized conditions (pressurized liquid extraction, PLE) (Heroult et al. 2008). Species degradation/inter-conversion was not observed under the mechanical stirring conditions used. In a similar study developed by Inagaki et al. (Inagaki et al. 2007), no degradation of MBT, DBT, TBT, MPhT, DPhT and TPhT was observed when shaking marine sediments with an extracting solution consisting of tropolone dissolved in toluene plus hydrochloric acid in methanol (Inagaki et al. 2007).

Arsenic species

Inorganic arsenic (As(III) and As(V)) and organic arsenic species (MMA and DMA) have also been extracted from plants by mechanical agitation (extracting time of 14 hr) using different extracting solutions (water and methanol/water mixtures ranging from 10/90 to 90/10) (Bohari et al. 2002). Low extractable ratios were found for inorganic arsenic species when comparing with those offered by ultrasound and microwave-assisted methods. Similarly, Giral et al. (Giral et al. 2010) have reported inefficient inorganic arsenic yields from soils when using mechanical shaking (160 osc/min for 30 min or 300 osc/min for 60 min) when using hydrochloric acid or orthophosphoric acid, respectively, as extracting solutions. In addition to the low recoveries found, As(III) oxidation to As(V) was observed even when using short shaking times. These results agree with those reported by Al-Assaf et al. (Al-Assaf et al. 2009) who observed As(III) conversion into As(V) (approximately 5 percent) when shaking soils with diluted phosphoric acid and sodium hydroxide solutions. Similar results were reported by Montperrus et al. (Montperrus et al. 2002) when dealing with soil, river sediment and sewage sludge, although As(III) conversion to As(V) could be minimized under shaking when using water as an extracting solution. Mechanical shaking has been used for extracting water-soluble arsenic species, including arsenosugars, from edible seaweeds (Llorente-Mirandes et al. 2011), and from fish (Sloth et al. 2003). Different methanol/water mixtures when treating submerged freshwater plants in Moira River, Ontari, Canada (Zheng and Hintelmann 2009), and methanol/water or ethanol/water solutions for land plants (Yuan et al. 2005) have also been evaluated with the assistance of continuous shaking.

Antimony species

Mechanical agitation was also tested for extracting antimony species from land plants (Miravet et al. 2005). Best performances were obtained when using water-methanol (1/9) with 0.1 M citric acid as a solvent (mechanical shaking for 14 hr). However, low recoveries of Sb(III), Sb(V) and methyl-antimony (MeSb) were observed. These yields were increased slightly after a combined method consisting of a first stage of mechanical shaking (4 hr) followed by a second extraction step involving sonication for 1 hr. Similarly, soluble-water Sb(III) and Sb(V) from atmospheric particulate matter were also measured after mechanical shaking for 4 hr (Zheng et al. 2001), and when extracting inorganic arsenic, antimony and selenium species from fly ash following a sequential five-step procedure to assess the exchangeable fraction, and the fractions bound to carbonate, to Fe-Mn oxides and to organic matter (Narukawa et al. 2005).

Other speciation studies

Concerning mercury species, satisfactory results have been obtained when extracting MMM and inorganic mercury from fish tissues under continuous shaking with methanolic KOH (Grinberg et al. 2003a, b, Yang et al. 2003a, b), concentrated hydrochloric acid (Tu et al. 2000), or dilute hydrochloric acid plus 2-mercaptoethanol (Meng et al. 2007). In addition, the simultaneous extraction of MMM, MBT, DBT and TBT from biota (mechanical agitation with TMAH for 3 hr), and from sediments (mechanical agitation with aqueous EDTA/NaDDTC for 1 hr) has also been described (Poperechna and Heumann 2005).

Quantitative recoveries for selenium-methyl-selenocysteine (Se-Met-Se-Cys) and selenomethionine (SeMet) from plants were also reported after mechanical shaking with hydrochloric acid or ammonium acetate (37°C for 20 hr) (Montes-Bayón et al. 2006).

Pt(II) and Pt(IV) extraction from road dust was achieved by mechanical shaking with 100 mM EDTA in 50 percent methanol for 30 min (Nischwitz et al. 2004).

Mechanical agitation was also used for isolating iodine water soluble species as well as iodine bound to proteins from iodine-enriched microalgae (water and TRIS/HCl, pH 9.0 as extracting solutions) by Gómez-Jacinto et al. (Gómez-Jacinto et al. 2010); whereas, TRIS (pH 7.4) was also used for extracting MIT, DIT, rT3 and T4 from biota (Simon et al. 2002).

A recent application dealing with mercury species in fish involved the extraction at 4°C with a solution consisting of 0.075 percent PAN and 0.08 percent Triton X-114 for 1 hr with shaking every 10 min. The extracted species were then pre-concentrated in the Triton X-114 phase after heating (cloud point extraction) (Yin 2007). Other applications involved the use of Soxhlet extraction (methanol/water mixture) for extracting arsenic species from marine biota (Gómez-Ariza et al. 2000), or water/methanol and water/ ethanol when treating land plants (Yuan et al. 2005).

Finally, temperature-controlled procedures, where the sample is extracted by heating at low temperatures have also been described as sample pre-treatments for speciation studies. As examples, phenylarsenic acids have been extracted from plants with water as a solvent by slight heating at 20 and 37°C for 20 hr (Schmidt et al. 2008), while MMM and inorganic mercury were extracted from fish with dilute nitric acid at 55°C for 16 hr (Taylor et al. 2008).

Microwave assisted extraction (MAE)

General aspects

The use of microwave energy for assisting sample preparation procedures is a current practice in analytical chemistry. Well-established methodologies for total sample decomposition when determining trace and ultratrace elements, as well as for assisting solid-liquid extraction procedures when isolating organic compounds have been proposed, and MAE can be considered as the chosen sample pre-treatment for routine analysis.

Microwaves are non-ionizing radiation with frequencies between 300 and 300000 MHz (wavelengths within the 0.005–1 m range). The interaction between materials and microwaves does not involve changes in the molecular structure of the material, but the absorption of the energy associated with the microwave radiation. The rate of absorbed energy depends upon the dissipation factor (tan δ) of the material (parameter defined as a ratio of the sample's dielectric loss, ε'', to its dielectric constant, ε', as tan $\delta = \varepsilon''/\varepsilon'$) (Kingston and Jassie, 1988). The dielectric constant (ε') can therefore be considered as a measure of the material to minimize the penetration of microwave energy; however, the dielectric loss (ε'') offers a measure of the material's ability to dissipate microwave energy as heat. Materials which exhibit low dissipation factors, such as quartz (0.6) or PTFE (1.5), can be regarded as transparent materials to microwave radiation (insulators). However, materials showing high dissipation factors, such as water (1570) and several organic solvent of polar nature, offer absorptive properties to microwave radiation and are referred to as dielectric materials.

As reported, the rapid heating achieved by microwave irradiation is attributed to two different mechanisms: molecular motion by migration of ions (dissolved salts in the solvent), and rotation of dipoles (molecular solvents). Migration of ions, also referred to as ionic conduction, is the conductive migration of dissolved ions under the electromagnetic field associated with the microwave radiation. Since the solution has resistance to the ionic movement, energy dissipation as heat is produced. Rotation of dipoles is a consequence of the alignment of molecules (permanent or induced dipole moments) in the electric field supplied by microwaves. As the electric field increases, it aligns the polarized molecules; however, the thermally induced disorder is restored when the electric field decreases or disappears, and thermal energy is then released.

Heat produced by microwave absorption (microwave heating, Fig. 2.17A) is more efficient than conventional heating (conductive heating, Fig. 2.17B) provided by a hot plate or multi-sampler heating devices. This is because microwaves are directly absorbed by the sample-solvent mixture (the vessel containing the sample-solvent mixtures is transparent to microwaves). In contrast, a certain time is needed to heat the vessel under conductive heating, and then to transfer the heating energy to the sample-solvent mixture. In addition, solvent vaporization at the surface of the liquid generates a thermal gradient (convection currents), which means that only a small portion of the sample-solvent mixture is at the programmed temperature. Therefore, solid-liquid extraction procedures based on conductive heating require long extraction times, and the poor heat transfer between the vessel, the solvent and the solid samples leads to the use of large solvent volumes.

Dilute acids, such as hydrochloric acid and carboxylic acids; as well as polar organic solvents such as acetone, methanol or acetonitrile are excellent dielectric materials and they release heat as a consequence of their interaction with microwaves. These polar solvents are adequate for extracting polar compounds such as organonetallic species (the presence of a metal such as As, Hg, Pb, Sb, Se, Sn, or a non-metal as Br, I or P, increases the polarity of these compounds). However, some MAE developments using toluene (solvent exhibits low polarity) as an extracting solution for methyl-mercury from fish tissues have also been reported (Carbonell et al. 2009). In this case, the addition of Wefflon, a material which interacts

A

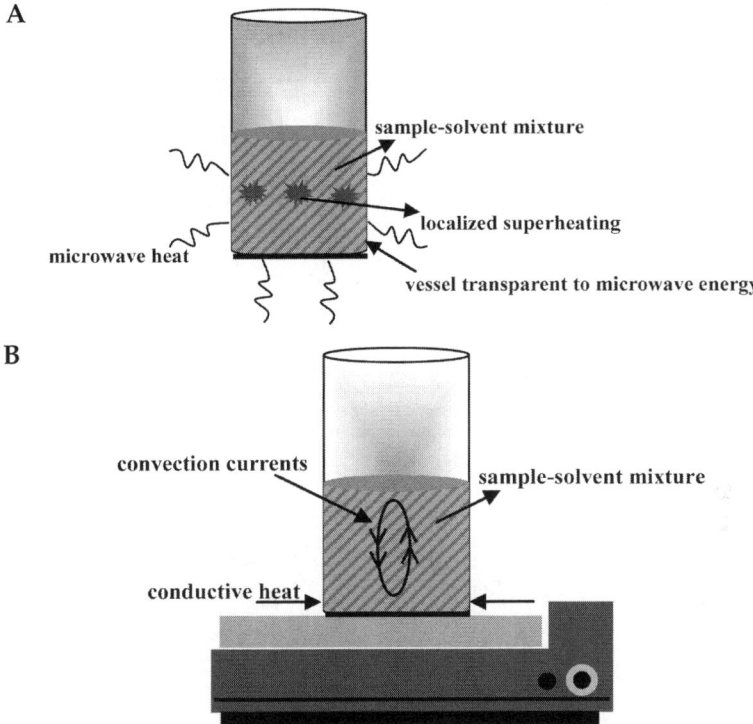

B

Figure 2.17 Microwave heating (A) and conventional heating (B) of solvents.

with microwaves very strongly, is needed. The material is heated and the heat is then transferred to the solvent. As stated previously, heat enhances target analytes desorption from the sample matrix as well as their diffusion towards the solvent. In addition, pressure is also increased under microwave irradiation. This is mainly important when treating biological materials: the rapid heating attributed to microwaves evaporates moisture inside cells and it generates high pressure on the cell membranes/walls. This can lead to subsequent cell disruption.

Microwave instrumentation

Closed-vessel and open-focused type microwave ovens are commercially available. The former operates under pressurized conditions. The reactors (closed-vessels) are equipped with a safety relief valve which opens when the internal pressure exceeds the target pressure. In this case, the safety relief valve works by lifting of the vessel cap to release excess pressure and then immediately re-sealing to prevent loss of sample. Typically, closed-vessels are designed to operate at internal pressures up to 120 psi. Reactors are usually made of materials transparent to the microwave radiation, such as PTFE (dissipation factor of 1.5), polystyrene (3.3) and quartz (0.6).

A closed-vessel microwave instrument for laboratory purposes consists of a microwave generator (magnetron), the waveguide, microwave cavity, mode stirrer, a circulator and turntable (Fig. 2.18A). The magnetron generates the microwave radiation which is propagated through the waveguide to the oven cavity, where it is dispersed via the mode stirrer. Most of the commercial microwave ovens have a fixed-tuned magnetron

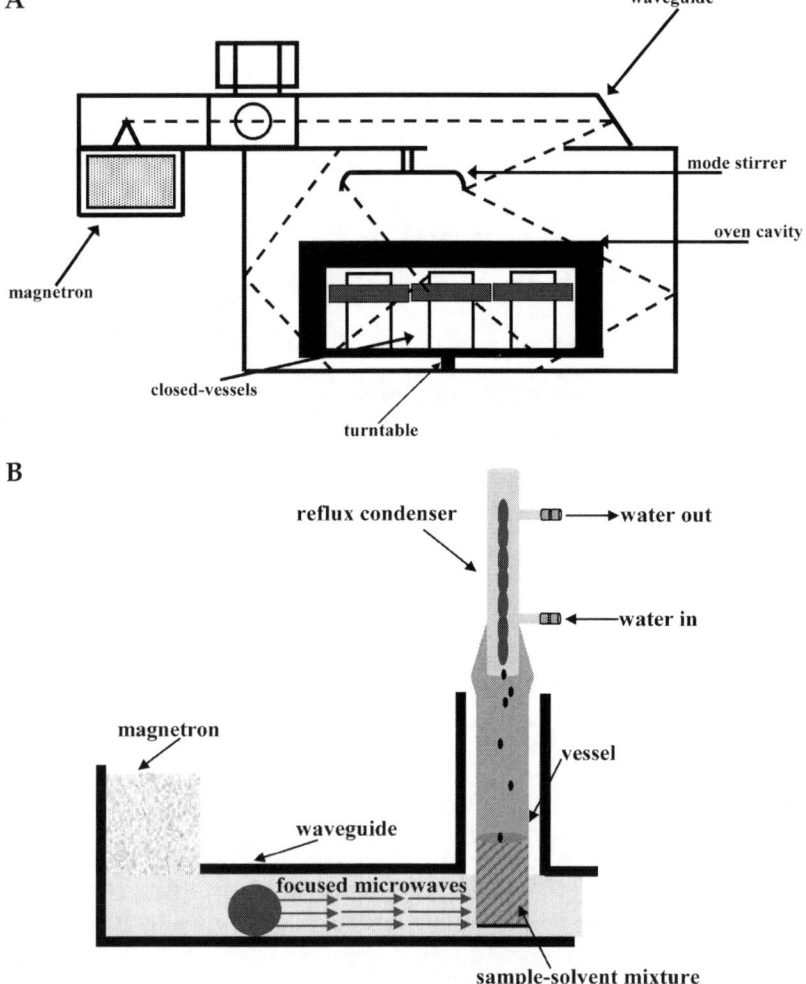

Figure 2.18 Schematic diagram of a closed vessel microwave instrument (A) and an open-focused type microwave (B).

(output frequency of 2450 MHz) which transforms approximately 1200 W of electrical line power to 600 W of electromagnetic energy. A turntable (carousel) is needed in microwave ovens which use more than one extraction vessel so that all vessels can be exposed to microwaves uniformly. Turntables are available that rotate 360° continuously, or that alternate back and forth at 180°. In addition, other features include a function for monitoring both pressure and temperature.

MAE in open-focused type microwave ovens occur under atmospheric pressure. These ovens consist of a large test tube (open vessel) made of glass or quartz (Fig. 2.18B) fitted to a water-cooled reflux condenser which causes the solvent to condense and return to the vessel, and which minimizes loss of volatile compounds. Microwave radiation from a magnetron is focused via the waveguide on the test tube containing the sample and solvent, and heating is precisely controlled (temperatures up to the boiling point of the solvent). The solvent is therefore heated to boil in the vessel, and condensed by the water-cooled reflux several times thus enabling desorption of target analytes from solid particles and diffusion into the solvent.

Closed-vessel microwave ovens offer the advantage of avoiding the loss of volatile compounds during the extraction, and also that airborne contamination is minimized. However, the whole extraction time can be longer than those obtained when using open-focused microwave systems because a certain cooling time after extraction and before handling of vessels is needed (the cooling time is minimal for open-focused microwave ovens).

MAE for speciation studies

Mercury compounds. MMM has been commonly extracted from solid environmental materials by MAE. Special care must be taken to avoid species degradation under microwave irradiation. Some studies have reported monoethyl-mercury (MEM) losses after irradiation at 40 W for times higher than 2 min, while MMM is not degraded/lost even after 8 min at 120 W (Abrankó et al. 2005). Recent studies have shown that the degree of mercury species transformation under MAE conditions is dependent on the solvent type (Reyes et al. 2008), and the highest levels of transformation were observed when using acid solvents such as nitric acid and acetic acid. In contrast, MAE with alkaline solvents offered quantitative mercury species recoveries and negligible species inter-conversion. Other studies have pointed out that the sample's matrix also affects the degree of species degradation. Sediments exhibiting high contents of carbon and sulfur are reported to generate MMM artifacts from inorganic mercury (Rodríguez-Martín-Doimeadios et al. 2003).

MMM and inorganic mercury are commonly extracted from soils, sediments and biological environmental materials using alkaline solvents such as methanolic sodium hydroxide (Abrankó et al. 2007), methanolic potassium hydroxide (Grinberg et al. 2003a, b, Cabañero Ortiz et al. 2002), TMAH alone (Berzas Nevado et al. 2011, Castillo et al. 2010, Chen et al. 2004, Monperrus et al. 2003a, Jimenez Moreno et al. 2006, Berzas Nevado et al. 2005, Vidler et al. 2007) or combined with methanol (Cabañero Ortiz et al. 2002). On other occasions, mixtures consisting of TMAH, methanolic potassium hydroxide and hydrochloric acid have also been proposed (Serafimovski et al. 2008). Acid extracting solutions combined or not with organic solvents such as toluene, have also been used for isolating mercury species. Most of the developments have used glacial acetic acid (Abuín et al. 2000, Davis et al. 2004), dilute acetic acid solutions (Tutschku et al. 2002), and concentrated (Carbonell et al. 2009) or dilute (Houserová et al. 2007b, Margetínová et al. 2008, Park et al. 2011, Hinojosa Reyes et al. 2009, Rodil et al. 2002) hydrochloric acid, and nitric acid (Rodríguez Martín-Doimeadios et al. 2003, Berzas Nevado et al. 2008). As described previously, some applications required the addition of Wefflon™ heating bottom or Wefflon™ stir bars for achieving a constant temperature in the vessel (Carbonell et al. 2009). Other extracting solutions used consisted of 2-mercaptoethanol plus L-cysteine (Chiou et al. 2001), or 2-mercaptoethanol plus EDTA, which allows the simultaneous extraction of MMM and organolead compounds (TMT and TET) (Chang et al. 2007).

GC separation methods for mercury species, and also for tin and lead species, require the conversion of these compounds into volatile derivatives. As reported, this derivatization stage is usually performed via ethylation or phenylation in an aqueous medium, or butylation with a Grignard reagent (Rodríguez Pereiro and Carro Díaz 2002). This step is usually carried out after target analyte extraction, but procedures involving the simultaneous extraction and derivatization stages have been reported, which reduce the sample pre-treatment time (Abuín et al. 2000). Derivatization ensures a better volatilization of the target analytes in the GC injector as well as a clean-up of the extract. The possibility of using SPME for pre-concentrating the phenyl- or ethyl- derivatives of the target analytes before GC analysis has also been explored, and different SPME methods have been developed after mercury species extraction from solid environmental materials and after derivatization (Abrankó et al. 2007, Davis et al. 2004, Rodil et al. 2002, Tutschku et al. 2002). On other occasions, clean-up stages based on LLE with solutions containing L-cysteine (Carbonell et al. 2009) have been used before analysis.

Tin compounds. Organotin compounds (OTC), mainly butyl derivatives, have been extracted from marine sediments and marine biota by MAE methods. Butyl-tin species degradation (TBT and DBT to MBT) (García Alonso et al. 2002, Ruiz Encinar et al. 2002a), and from DBT to MBT (Inagaki et al. 2007) has been reported when assisting the extraction with microwave energy. Methanol/acetic acid (1/3) mixtures were used in two of these studies (García Alonso et al. 2002, Ruiz Encinar et al. 2002a), while an extracting solution containing tropolone in toluene plus acetic acid in methanol was used in the latter study (Inagaki et al. 2007). A recent study on phenyl-tin compounds extraction from biological matrices has shown that degradation of TPhT to DPhT or of DPhT to MPhT can be attributed mainly to the presence of tropolone in the extracting solutions, and significant degradation occurs when assisting the extraction either with microwaves or with ultrasound (Van et al. 2008). However, Yu et al. (Yu et al. 2010) have reported that DPhT and DBT are degraded when using tropolone-free extracting solutions and ultrasound energy for assisting the organotin extraction.

Extractions are often performed with acetic acid/methanol solutions when dealing with MBT, DBT and TBT compounds (García Alonso et al. 2002, Ruiz Encinar et al. 2002a, Jimenez Moreno et al. 2006, Rodríguez-González et al. 2004, Rodríguez-González et al. 2005), and with phenyl-tin species (Van et al. 2008). However, other mixtures such as acetic acid plus sodium 1-pentasulfonate (Chao and Jiang 1998), tartaric acid plus methanol (Flores et al. 2011), or glacial acetic acid (Yang and Lam 2001) have also been proposed. Alkaline extractants such as TMAH (Monperrus et al. 2003a) have also been proposed for extracting butyl-tin compounds from marine biota. The simultaneous MAE of butyl-, phenyl- and octyl-tin compounds from soils has also been performed with glacial acetic acid (Heroult et al. 2008, Monperrus et al. 2003b, Tutschku et al. 2002), while extracting solutions consisting of acetic acid in methanol plus tropolone in toluene were also used for the simultaneous extraction of butyl- and phenyl-tin species (Inagaki et al. 2007). Finally, methanol was also used for extracting triorganotin compounds (TMT, TET, TPT and TBT) from marine biota (Yang et al. 2010).

In a similar way to mercury species, derivatization by ethylation or phenylation is often performed when separating OTCs using GC methods (Yang and Lam 2001, García Alonso et al. 2002, Monperrus et al. 2003b, Flores et al. 2011) and, SPME techniques have also been used to pre-concentrate the phenyl- and ethyl-OTC derivatives (Tutschku et al. 2002). Other clean-up stages for OTC speciation were based on SPE with Florisil or C_{18} supports (Inagaki et al. 2007).

Arsenic compounds. Developments based on MAE for arsenic speciation studies have shown that the presence of reducing agents such as ascorbic acid or hydroxylammonium hydrochloride in the extracting solutions (diluted phosphoric acid) minimizes inorganic arsenic inter-conversion when treating soils, and also increases the stability of inorganic arsenic, MMA and DMA in the extracts over time (Ruiz-Chancho et al. 2005). However, oxidation of As(III) to As(V) has been reported in strong alkaline medium (extraction with TMAH) when analyzing freshwater plants (Zheng and Hintelmann 2009).

Methanol/water mixtures are the extracting solution of choice when isolating arsenic species from environmental materials, although ethanol/water has been also proposed when treating land plants (Yuan et al. 2005). These mixtures have been used for extracting inorganic arsenic and organic arsenic (mainly MMA, DMA, AsB and AsC) from marine biota (Dagnac et al. 1998, Dagnac et al. 1999, Ackley et al. 1999, Gómez-Ariza et al. 2000, Vilanó and Rubio 2001, Bohari et al. 2002, Kirby and Maher 2002, Tukai et al. 2002, Karthikeyan and Hirata 2004, Yeh and Jiang 2005, Dufailly et al. 2007, Wang et al. 2007, Han et al. 2009, Williams et al. 2009, Nam et al. 2010), although dilute orthophosphoric acid has also been used as an extractant (Bohari et al. 2002). On other occasions, water has been used as an extracting solution (Hirata and Toshimitsu 2005, García Salgado et al. 2006, Leufroy et al. 2011, Llorente-Mirandes et al. 2011), and water was also adequate for isolating arsenosugars from seaweed (Llorente-Mirandes et al. 2011). Rahman et al. (Rahman et al. 2009a) have recently proposed a solution containing sucrose, MES, EDTA, L-ascorbate; ammonium dihydrogenphosphate (pH 5.6) plus methanol–water (1:1) for isolating inorganic arsenic plus MMA and DMA from land plants. In addition, alkaline extracting solutions such as TMAH and ethanolic sodium hydroxide have also been used for arsenic speciation studies in land plants (Monperrus et al. 2003b) and marine biota (Sloth et al. 2005).

Some authors have reported the need for a clean-up stage before HPLC analysis mainly when treating biological matrices. Fat removal from the extract with hexane based-LLE (Gómez-Ariza et al. 2000) or C_{18} based-SPE (Gómez-Ariza et al. 2000, Vilanó and Rubio 2001) has been recommended. Finally, solutions of orthophosphoric acid or phosphate buffers have also been used for extracting arsenic species, mainly inorganic arsenic, from soils, sediments, sewage sludge, and atmospheric particulate matter (Montperrus et al. 2002, Rattanachongkiat et al. 2004, Ruiz-Chancho et al. 2005, Oliveira et al. 2005, Yuan et al. 2007, Rahman et al. 2009b, Giral et al. 2010); and hydroxylamine hydrochloride and diammonium oxalate solutions have also been proposed for arsenic speciation studies when treating inorganic environmental samples (Montperrus et al. 2002, Ruiz-Chancho et al. 2005, Oliveira et al. 2005).

Chromium, vanadium, antimony, lead, iodine and bromine compounds. Some MAE protocols have been reported for isolating inorganic chromium from environmental matrices. It must be pointed out that over-estimations of up to 35 percent have been reported for the Cr(VI) species: Cr(III) is converted to Cr(VI)) in soil and sediment extracts after MAE (95°C for 1 hr) (Rahman et al. 2005). In this study the EPA Method 6800 was used with a microwave-assisted extraction technique (EPA Method 3060A), which requires the use of 2.5 M sodium hydroxide plus sodium carbonate (0.75 g). The same standard method (EPA Method 3060A) was used by Priego-Capote and Luque de Castro for Cr(III) and C(VI) determination in inorganic reference materials (Priego-Capote and Luque de Castro 2006). Finally, other MAE applications were based on using dilute hydrofluoric acid when extracting soils and plants for chromium and also for vanadium (V(IV) and V(V)) isolation (Kuo et al. 2007).

Concerning antimony, significant conversion of Sb(III) to Sb(V) ($\approx 36\%$) has been reported when extracting inorganic antimony species from soils with citric acid solutions (Telford et al. 2008). Land plants were also treated with methanol/water under microwave irradiation for extracting inorganic antimony and MeSb (Miravet et al. 2005), while Foster et al. (Foster et al. 2005) have investigated the use of EDTA, dilute nitric acid and dilute sodium hydroxide as extracting solutions for the speciation of inorganic antimony in aquatic plants and marine biota.

Finally, focused MAE (EDTA solution or EDTA plus 2-mercaptoethanol as extractants) has been applied for lead speciation studies (inorganic lead, trimethyl-lead and triethyl-lead) in marine biota (Lee and Jiang 2005, Chang et al. 2007); whereas, dilute TMAH has been chosen for inorganic iodine (iodide and iodate) and bromine (bromide and bromate) assessment in seaweed (Chen et al. 2007).

Ultrasound assisted extraction (UAE)

General aspects

Ultrasound is often defined as a cyclic sound pressure with frequencies greater than the upper limit of human hearing (approximately 20 kHz). The ultrasound frequency varies when ultrasound passes through a liquid and, as a consequence, repeated expansion and compression cycles occur. The successive expansion and compression cycles of ultrasound when interacting with a liquid are responsible for the cavitation phenomenon, which is macroscopically confirmed by an increase in temperature of the irradiated liquid. The increase in temperature is attributed to the negative pressure (expansion cycle) combined with positive pressure (compression cycle) in the liquid. Bubble nucleation occurs because molecules of vapor and gases migrate towards the cavities during the expansion cycles followed by their expulsion from the bubbles during the compression cycles. The successive expansion and compression cycles lead to collapse of the bubbles and the generation of high local temperatures and pressure inside the liquid (Mason 1999).

The cavitation process occurs by asymmetric collapse when a solid is suspended in a liquid. This leads to the generation of high-speed microjets toward the solid surface, which provides solid erosion and cleavage. As a consequence, the analyte desorption rate and the analyte transport from the solid particles to the liquid phase is more effective (Luque-García and Luque de Castro 2003, Santos Júnior et al. 2006), and quantitative extraction can then be achieved in short time periods.

Ultrasound instrumentation

Two ultrasound generator devices are usually available: ultrasound water-baths (ultrasonic baths, ultrasonic cleaning devices) and ultrasonic probes. The former consists of a tank in which the ultrasound transducer is placed at the bottom, although there are devices in which an additional transducer is placed at the side of the tank. This last design offers a more uniform distribution of ultrasound. The ultrasound transducer transfers the ultrasound energy (commonly between 1 and 5 Wcm^{-2}) to the liquid (ultrapure water) contained in the tank. As a consequence, ultrasound dissipation through the water contained in the tank and through the reaction vessel (containing the sample and the extracting solution) occurs, and the ultrasound energy which reaches the sample-solvent mixture is at a lower energy than the ultrasound intensity supplied by the transducer. In addition, the ultrasound energy which interacts with the sample-solvent mixture in the vessel depends on several factors, mainly the position in which the sample vessel is placed inside the bath (vertical and horizontal position) and the number of vessels. Other parameters which influence the amount of the effective ultrasound energy are the vessels' wall thickness and the vessels' material, the bath dimensions, water volume contained in the tank, and the presence of certain detergents in the water (better ultrasonic transmission through the liquid is achieved). The intensity of the ultrasound energy is therefore uniform in neither the tank nor inside the reaction vessels (Mason and Lorimer 2002). The so-called "aluminium foil test" (Standards Association of Australia 1999) is frequently performed to find the best place inside the bath in which the ultrasound intensity is the highest (Mason 1999). Taking all these facts into account and because of the different ultrasound frequencies and powers supplied by the transducers in the different commercial ultrasonic baths, attempts for comparing experimental conditions used in different applications are difficult (Santos Júnior et al. 2006).

Better performances are achieved when using ultrasonic probes. In these devices, the ultrasound transducer (ultrasound energy within the 50–750 Wcm^{-2} range) is coupled to a detachable metal probe or tip, usually made of a titanium alloy (Fig. 2.19). The tip is immersed into the vessel containing the sample and the solvent, and the ultrasound energy is directly transmitted to the sample-solvent mixture. Because of the higher ultrasound energy supplied by ultrasonic probes, the sonication efficiency using then is therefore higher than that supplied by ultrasonic water-baths. Lower extraction times and better extraction efficiencies are therefore obtained. However, because of the high ultrasound transmission, important increases in the temperature of the sample-solvent mixture can occur. This can diminish the cavitation phenomenon, and can also damage/alter the sample or the target analytes. The pulse working mode is therefore recommended when requiring high sonication times. Periods of time without ultrasonic irradiation allow cooling of the sample-extractant mixture, minimizing loss of analyte/degradation by heat. In addition, the vessel containing the sample-solvent mixtures can be placed in an ice-bath to minimize the increase of the bulk temperature.

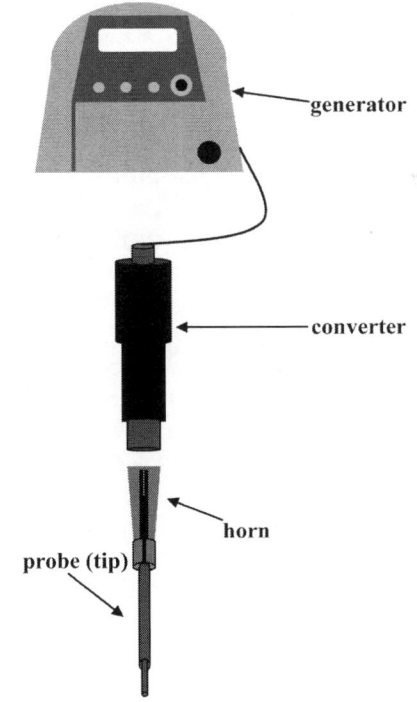

Figure 2.19 Schematic diagram of an ultrasonic probe.

There are different detachable tip (probe) sizes that may be used as a function of the volume of the sample-extracting liquid mixture to be sonicated, and they work at a certain amplitude values. Therefore, the volume of the sample-solvent mixtures conditions the tip size and also the working amplitude. The sonication efficiency is also dependent on the shape of the tip (Santos and Capelo 2007). Although titanium alloys tips are used most often, tips made of silica glass or pure titanium are also available, and they are highly recommended to minimize metal contamination (Capelo et al. 2005, Santos and Capelo 2007). In addition to silica glass probes, there are other commercially available probes such as spiral tips, in which the ultrasonic power is distributed across the entire surface; and multiple tips, which are manufactured allowing the use of two or more probes at the same time. In addition to the tip (Fig. 2.19), an ultrasonic probe also consists of a generator which converts the main voltage to high frequencies (typically 20 kHz); an ultrasonic converter, which transforms electrical energy into mechanical vibrations of a certain frequency (typically 20 kHz); and a horn which increases the sonication amplitude. Finally, the dimensions and shape of the vessels are important parameters for achieving good performance. In this sense, conical-type vessels with the diameter as small as possible in order to rise up the liquid level are recommended to avoid the formation of aerosols and foams (Santos and Capelo 2007).

UAE for speciation studies

Mercury species. Ultrasound assisted extraction can be performed with an ultrasonic water-bath but ultrasonic probes are used more often. As reported previously, Reyes et al. (Reyes et al. 2008) showed that both Hg^{2+} methylation and MMM demethylation occur when using acid extracting solutions with microwave and ultrasound assistance, and the degree of mercury species transformation is more obvious when both species are in a similar concentration range. The authors concluded that alkaline extractants (methanolic potassium hydroxide and TMAH) and assistance with ultrasound (water-bath, 30 min) or microwaves (temperature increase from room temperature to 180°C in 10 min followed by heating at 180°C for 10 min) offer quantitative mercury species recoveries and maintenance of mercury species integrity.

Extracting solutions consisting of 2-mercaptoethanol/L-cysteine/hydrochloric acid have also been used when treating marine biota (ultrasonic water-bath, 15 min) (Lemos Batista et al. 2011). Similarly, extractants such as perchloric acid/L-cysteine/methanol/toluene (Maldonado Santoyo et al. 2009) or TMAH plus dilute potassium hydroxide (Zabaljauregui et al. 2007) have been proposed. On other occasions, USE procedures followed by LLE for extract clean-up have also been implemented (Gao and Liu 2011).

In this case, after USE (ultrasonic water-bath, 70°C, 30 min) with a mixture of acidic potassium bromide/copper sulfate plus hydrochloric acid and TMAH (or methanolic potassium hydroxide), MMM is isolated from the extract by LLE with dichloromethane and extracted back into the aqueous dilute sodium thiosulfate. Other, simple methods which use ultrasonic probes require dilute hydrochloric acid and sonication for 2 min (20 percent amplitude) (Fragueiro et al. 2004), or 5 min (30 percent amplitude) (López et al. 2010).

Tin species. Recent literature shows that tin species degradation occurs when assisting the extraction procedures with microwaves and with ultrasound. This is more significant for phenyl-tin compounds, and the presence of tropolone in the extracting solution has been shown to increase TPhT conversion to DPhT, and DPhT conversion to MPhT when assisting with either microwaves or ultrasound (Van et al. 2008). Degradation of DPhT and DBT has also been reported when sonicating marine biota with a mixture consisting of acetonitrile/water/acetic acid/triethylamine (Yu et al. 2010), and also when treating marine sediments with glacial acetic acid (TPhT degradation increases when sonicating for times more than 4 min) (Carpinteiro et al. 2001).

In addition to the use of ultrasonic water-baths for extracting phenyl-tin and butyl-tin compounds under acid (Carpinteiro et al. 2001, Van et al. 2008, Yu et al. 2010) and alkaline conditions (Zabaljauregui et al. 2007, Van et al. 2008), and with organic solvents such as ethanol (Heroult et al. 2008), ultrasonic probes have also been used. A mixture of acetic acid/methanol (9/1) has been proposed by Campillo et al. (Campillo et al. 2004) for isolating butyl-tin species (MBT, DBT, TBT), methyl-tin compounds (DMT and TeMT), and MPhT from sediments (sonication for 30 sec, 60 percent amplitude). Similarly, butyl-tin compounds have been extracted from marine biota using acetic acid/methanol (1/1) as an extracting solution, or dilute hydrochloric acid in methanol when treating sediments (sonication for 30 sec, 20–25 percent amplitude) (Gallego-Gallegos et al. 2006a). These authors have proposed a clean-up stage based on SPE with molecularly imprinted polymers (MIPs) as solid supports. MIP cartridges are conditioned with toluene or with methanol/dichloromethane (4/1) depending on the MIP used (covalent or non-covalent polymer, respectively), and target analyte elution is performed with dilute hydrochloric acid in methanol (Gallego-Gallegos et al. 2006a).

Finally, for mercury determination using GC, tin speciation using GC needs target analyte derivatization (via ethylation) into volatile species (Carpinteiro et al. 2001, Campillo et al. 2004, Gallego-Gallegos et al. 2006a, Zabaljauregui et al. 2007).

Arsenic species. As stated previously a low rate conversion of As(III) to As(V) was observed when isolating arsenic species from soils by mechanical shaking with dilute phosphoric acid followed by extraction with dilute sodium hydroxide (Al-Assaf et al. 2009). However, the same authors have reported that a sonication probe is not suitable for arsenic speciation in soils as it accelerates the oxidization of As(III) to As(V) (Al-Assaf et al. 2009).

A sonication probe has been used for extracting water-soluble arsenic species and exchangeable arsenic species from river sediments (Huerga et al. 2005). Water containing 1 M phosphate buffer was the extracting solution used to assess the water-soluble and exchangeable fractions, under sonication for 1 min and 30 percent amplitude. Similar conditions (sonication for 1 min, 30 percent amplitude) were used by Sanz et al. for isolating arsenic species (mainly AsB) from fish with water (Sanz et al. 2005a) and from soils (As(III), As(V), MMA and DMA) with phosphoric acid (Sanz et al. 2005a, Sanz et al. 2007). Finally, an extracting solution consisting of methanol/water (1/1) was used by Karthikeyan et al. (Karthikeyan et al. 2003) for arsenic speciation studies in algae and shrimps. In this application, sonication was performed for 15 min in an ultrasonic water-bath.

Chromium, antimony, selenium, sulfur and iodine species. The EPA Method 3060A (0.75 g sodium carbonate plus 10 mL of 2.5 M sodium hydroxide as extracting solution) was used under ultrasound for isolating Cr(III) and Cr(VI) from inorganic reference materials (Priego-Capote and Luque de Castro 2006). Ten sonication cycles of 3 min (90 percent amplitude) were used for achieving quantitative results. Inorganic antimony was also isolated from airborne particulate matter under probe sonication for 3 min with dilute hydroxylammonium hydrochloride (Bellido-Martín et al. 2009). In addition, Sb(III), Sb(V) and TMSb were extracted from land plants by sonication for 1hr (ultrasonic water-bath) with a water/methanol (9/1) mixture containing 0.1 M citric acid (Miravet et al. 2005). Higher yields for the different antimony species were obtained than those obtained after mechanical shaking for 4 hr. Non-protein amino acids such as SeMet and Se-methyl-Se-Cys have been isolated from selenium-enriched plants by use of an ultrasonic probe for 3 min with dilute hydrochloric acid or ammonium acetate buffer (pH 5.6) (Montes-Bayón et al. 2006). The extraction time is reduced when comparing with mechanical shaking procedures under controlled temperature (37°C) performed for 24 hr. In addition, the authors reported negligible species degradation under the optimized conditions. Probe sonication was also used for extracting sulfate and elemental sulfur from zinc sulfide (Dash et al. 2005). The solid sample can be treated directly with chloroform (sonication for 5 min, 40 percent amplitude), or after effervescence ceased through the addition of 5 mL of 15 percent hydrochloric acid. Finally, other applications of ultrasound water-bath assisted extractions focused on isolating iodide and iodate from atmospheric particulate matter (filters treated with water as an extractant under irradiation at 35°C for 60 min) (Xu et al. 2010).

Supercritical fluid extraction (SFE)

General aspects

A supercritical fluid (SF) is any substance which is above its critical temperature and pressure. The triple point shown in the pressure-temperature phase diagram (Fig. 2.20) corresponds to the values for temperature and pressure at which the three phases (solid, liquid and gas) co-exist in equilibrium. The critical point, however, is the value of temperature and pressure at which no gas is formed when increasing the temperature, and at which no liquefaction occurs when raising the pressure. At this point the liquid and gas phases disappear to become a single supercritical phase. A substance under these conditions (SF) will exhibit properties between those of a gas and of a liquid. The density and the solvating power of a supercritical fluid are close to those of liquids, which facilitates target analyte solubilization. In addition, the viscosity is close to that offered by a gas, which provides good flow characteristics and mass transfer properties. The diffusivity of a supercritical fluid is also high (as that observed in a gas) and the surface tension is low. These properties facilitate penetration of solid samples, even low porosity materials, and accelerates target analyte desorption and then mass transfer.

Carbon dioxide is the most common supercritical fluid used for analytical applications. The advantages of carbon dioxide are that it offers low toxicity, is a non-flammable gas, and can be obtained at high

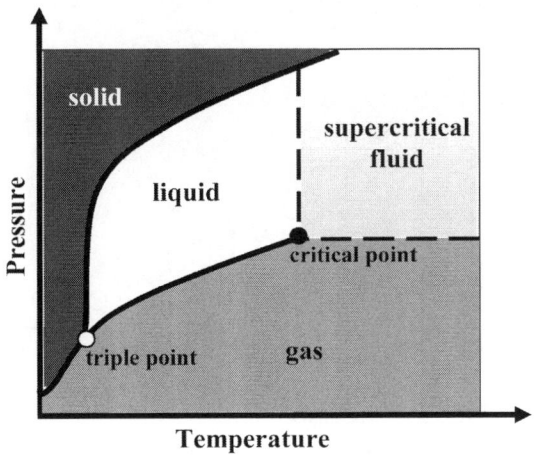

Figure 2.20 Schematic phase diagram for a pure substance.

purity and at low cost. In addition, the critical temperature (31.1°C) and pressure (73.8 bar) of carbon dioxide are moderate, which is useful when dealing with thermally labile targets. Carbon dioxide is a gas at room temperature, and coupling supercritical fluid equipment with GC is not a problem. Finally, the reactivity of carbon dioxide is low, which minimizes inter-conversion/degradation of the extracted species. Extractions based on supercritical carbon dioxide are adequate when isolating non-polar compounds. However, yields for moderately polar and polar substances are low. This drawback can be overcome by adding small amounts of polar organic solvents, referred to as modifiers, which varies the polarity of the supercritical fluid. Nevertheless other modifiers' mechanisms have been proposed, such as a higher affinity of the modifier to the specific points in the matrix where the targets are bonded. This interaction is mainly through hydrogen bonds (organic solvents such as methanol) or through π-interactions (organic solvents such as toluene). On other occasions, certain complexing agents can be added. These substances react with the target analytes and the complexes formed can be more soluble in the fluid. The addition of the modifier, usually methanol, is often performed by mixing with the sample in the extraction cell. However, it can be added by in-line mixing with carbon dioxide.

Supercritical fluid extraction instrumentation

Figure 2.21 shows the major components of an SFE system. Basically, it consists of a carbon dioxide cylinder equipped with a dip tube for liquefying the carbon dioxide before pumping. The liquid carbon dioxide is propelled by the pump to the extraction cell. As shown, the pump must be cooled

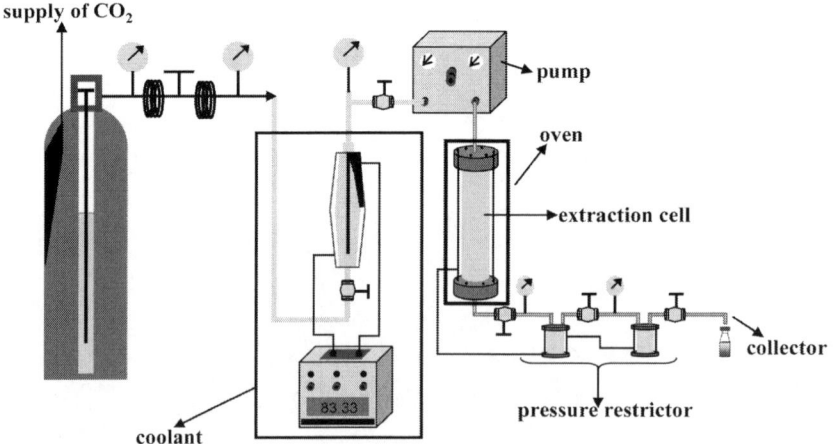

Figure 2.21 Schematic diagram of an SFE system.

to avoid liquid carbon dioxide cavitation. This can easily be performed by using a re-circulating water-bath or a Peltier device. The liquid carbon dioxide enters the extraction vessel containing the sample which is located in an oven for achieving the critical temperature. In addition, the extraction cell is connected to electronically controlled pressure restrictors which vary the pressure inside the cell for achieving the critical pressure. The extraction vessels are therefore designed for working at high pressures, and they are often made of stainless steel. The extract is finally collected in an open vial containing the collection solvent. However, studies using C_{18}-bonded silica traps or SPE cartridges have been reported for on-line extract clean-up/pre-concentration purposes.

SFE for speciation studies

Tin species. The polar/ionic nature of organometallic species and the strong electrostatic/hydrophobic interactions with the matrix, mean that low extraction yields are expected when using SFE (Bayona 2000). Therefore, according to Dietz et al. (Dietz et al. 2007), SFE applications in speciation studies are scarce. As reported, two different strategies can be performed to extract polar compounds using supercritical CO_2; i.e., the use of modifiers to increase the polarity of supercritical CO_2 (soluble organometallic species in the modified supercritical CO_2), or complexation/chelation or derivatization of target analytes to obtain less polar compounds (more soluble compounds in the supercritical CO_2).

Methanol has frequently been used as a modifier when extracting OTCs from sediments (Chau et al. 1995, Liu et al. 1994) and biota (Kumar et al. 1993). However, as concluded by Lopez-Avila et al. (Lopez-Avila et al. 1997), supercritical CO_2 modified with 5 percent methanol for treating complex samples such as soils and sediments works satisfactorily for tri- and tetra-substituted OTCs but falls when isolating mono- and di-substituted compounds. Alternatively, acid modifiers such as formic acid have also been proposed when treating foodstuffs (Oudsema and Poole 1992) and marine sediments (Cai et al. 1998), although a higher proportion of the acid modifier is needed when treating sediments. Acetic acid alone (Fernández-Escobar and Bayona 1997) or combined with tropolone (Morabito et al. 1999) has also been proposed as a modifier for the SFE of butyl- and phenyl-tin compounds from marine biota. In the former case, tropolone was added as a complexing agent while, in the latter, it was mixed with the supercritical CO_2-acetic acid mixture. On other occasions, high acidity of the supercritical fluid is obtained by mixing methanol with small amounts of hydrochloric acid, and the modified supercritical CO_2-methanol-hydrochloric acid was found to be adequate when extracting butyl-tin compounds from sediments (Dachs et al. 1994, Quevauviller et al. 1994a, b).

As reported by several authors (Bayona 2000, Chau et al. 1995, Liu et al. 1994, Cai et al. 1998b), the use complexing agents in the presence of acid modifiers does not improve OTC extraction efficiency, and acidic supercritical CO_2 can be used directly. However, the presence of complexing agents is advantageous when using supercritical CO_2 modified with methanol, as shown recently by Liu et al. (Liu et al. 2011) for the extraction of TMT and butyl- and phenyl-tin compounds from clams. The authors concluded that the presence of complexing agents such as APDC, and especially DDC, increases the extraction yields of some OTCs, mainly for MPT and TBT.

Finally, certain derivatizing agents such as hexylmagnesium bromide have been shown to be adequate for the extraction of butyl- and phenyl-tin compounds as their hexylated derivatives. In these cases, derivatized organotin compounds can be extracted directly with unmodified supercritical CO_2 (Cai et al. 1994).

Mercury, lead and chromium species. First attempts dealing with SFE of mercury species (inorganic mercury, MMM and dimethyl-mercury) used spiked experiments with cellulose and the use of methanol as a modifier and lithium bis(trifluoroethyl)-dithiocarbamate (LiFDDC) as a complexing agent (Wai et al. 1993). Similarly, SFE with methanol as a modifier and LIFDDC as a chelator was tested by Foy and Pacey (Foy and Pacey 2003) for extracting inorganic mercury, MMM and MPhM spiked onto celite. Application of SFE conditions to real samples such as sediments shows moderate MMM recoveries when using unmodified supercritical CO_2 (Emteborg et al. 1996). In addition, Lorenzo et al. (Lorenzo et al. 1999) have reported MMM losses during the clean-up stage just after rinsing out of the trap with toluene (two 1 mL aliquots). These authors concluded that the extraction yields depend on the sediment matrix and that the SFE method was inefficient when treating marine sediment with high carbonate contents.

Concerning lead, first experiments were conducted with spiked (trimethyl-lead, triethyl-lead and diethyl-lead) sediment and urban dust samples (Johansson et al. 1995). Modified CO_2-methanol under optimum pressure and temperature conditions led to quantitative recoveries for trimethyl-lead and triethyl-lead. Recently, Zúñiga et al. (Zúñiga et al. 2009) developed an SFE method (modified CO_2-methanol, *in situ* NaDDTC complexation) for isolating triethyl-lead, trimethyl-lead and inorganic lead from sand, urban dust and river sediments. Quantitative recoveries were obtained for trimethyl-lead and inorganic lead (although the addition of a chelator was needed for achieving quantitative results for inorganic lead), but extraction yields around 70 percent were obtained for triethyl-lead.

Finally, Foy and Pacey (Foy and Pacey 2000) investigated the applicability of SFE for the selective extraction of Cr(VI). The targets (Cr(VI) and Cr(III)) were loaded onto a solid support (celite) and selective extraction conditions for Cr(VI) were found using fluorinated dithiocarbamate as a chelator, and methanol and water as modifiers. As shown, the efficiency of extraction was mostly affected by the presence of water. Wang and Chiu (Wang and Chiu 2007) also assessed Cr(III) and Cr(VI) in sand and wood waste samples under SFE conditions: LiFDDC as a chelating agent was effective in removing chromium species in methanol-modified CO_2 via *in-situ* chelation/SFE. The authors concluded that Cr(III) and Cr(VI) extraction efficiencies from solid matrices increased to a maximum of 92 percent in the presence of a small amount of water.

Pressurized liquid extraction (PLE)

General aspects

Pressurized liquid extraction (PLE), also referred to as pressurized fluid extraction (PFE), pressurized solvent extraction (PSE) or accelerated solvent extraction (ASE) (Fisher et al. 1997, Giergielewicz-Mozajska et al. 2001), consists of using solvents at a high pressure and/or high temperature without reaching the critical point. When using water as a solvent, the technique is often referred as to pressurized hot water extraction (PHWE) (Kronholm et al. 2007), but the instrumentation as well as critical parameters controlling the extraction process is similar to PLE. As shown in Fig. 2.22 when a liquid (solvent) is heated and subjected to high pressures but where the temperature and pressure remain below the critical temperature and critical pressure, the solvent remains in liquid phase (compressible

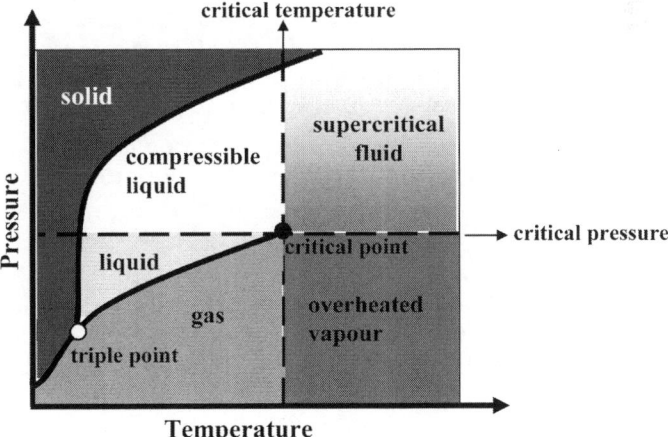

Figure 2.22 Schematic phase diagram for a pure substance showing conditions for pressurized liquids.

or pressurized liquid). Under these conditions several properties of the solvent change dramatically; i.e., the dielectric constant, density, viscosity and surface tension of the pressurized solvent decrease. The decrease in the solvent viscosity and surface tension affect the penetration of the solvent within the sample matrix, and an improved solvent-sample contact is achieved. In addition, disruption of target analyte-matrix bonds is assisted by the high temperature inside the extraction cells. Since most organometallic species are bonded to the matrix by electrostatic interactions (hydrogen bonding and van der Waals forces), the high temperature as well as the improved solvent properties of the pressurized solvents allow a fast analyte desorption, an increase in the solubility of targets in the extracting solution, and hence, an efficient mass transfer from the solid particles towards the extracting solvent.

The main advantage of PLE procedures is the short extraction time required for quantitative extractions, typically from 2 to 10 min. In addition, the technique usually requires low volumes of extracting solution (within the 15–25 mL range) and is suitable when extracting thermolabile compounds. The high level of automation and the possibility of performing in cell clean-up stages are other advantages inherent to PLE. In contrast, the high cost of equipment is perhaps the main disadvantage

Pressurized liquid extraction instrumentation

The basic components of PLE equipment are a pump for propelling the solvent and an oven where the sample and the solvent inside the extraction cell is heated and pressurized (Fig. 2.23). The pump can operate with one or several solvent lines, and extractions with solvent mixtures are possible. In addition, the system can be operated by performing successive extractions with different solvents. Along with SFE devices, PLE takes place in stainless-steel extraction cells, in which the sample, previously mixed/dispersed in an inert support, is placed between two cellulose filters. The cells consist of cylinders of a variable volume (e.g., one commercial supplier produces cells of volume 5, 11 and 30 mL) which are fitted with two finger-tight end caps. Once the extraction cell is inside the oven, a programmable solvent volume is pumped, and the cell is heated and pressurized at the fixed settings. The solvent volume is dependent on the extraction cell volume, and also on the amount of sample and dispersing support in the cell. Some commercial equipment use a sensor system to ensure that the extraction cell is full of solvent. A certain time for heating the extraction cell is required when working at temperatures higher than room temperature. This heating time depends on the programmable temperature (up to 200°C). The applied pressure can be varied within the 500–3000 psi (35–200 bar) range. After a fixed static time (extraction time) the static valve is released and a small

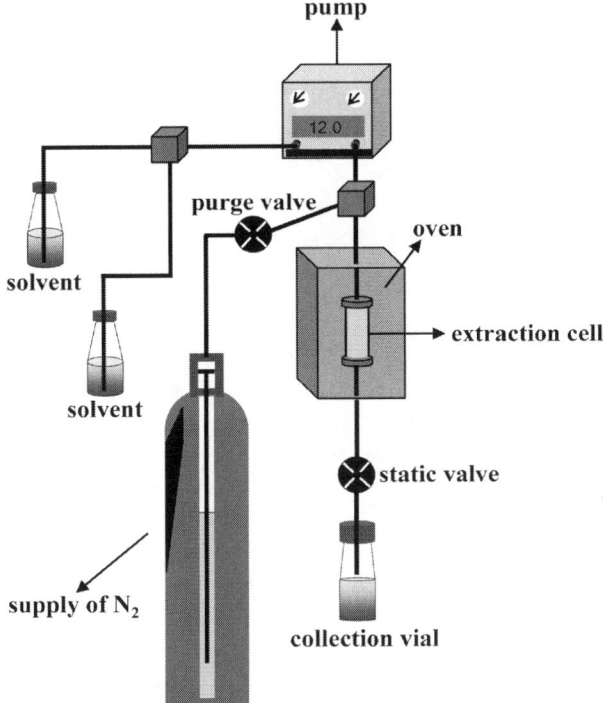

Figure 2.23 Schematic diagram of a PLE system.

volume of fresh solvent is passed through the system, expelling the volume inside the cell to the collector vial. This volume (flush volume) is usually set at 60 percent of the volume of the pressurized solvent used for the extraction. Finally, the system is purged by passing N_2 gas and the extract is transferred to a closed collection vial whose cap contains a solvent-resistant septum. The extract transfer is via stainless-steel tubing ending in a needle, which punctures the septum located in the cap of the collection vial. The extract volume can vary among repeated extractions under the same fixed solvent volume used (according to the extraction cell volume) and flush volume, and so dilution to a fixed volume is required unless the extract was subjected to dryness evaporation before dilution.

As proved by Wahlen et al. (Wahlen et al. 2004) approximately 95 percent of the AsB and DMA was extracted from the dispersed sample (fish tissue) in the first cycle, and subsequent cycles with fresh solvent could be used to complete the extraction. Similar studies when dealing with butyl-tin species in sediments showed that all of the TBT could be extracted in the first two 1-min cycles but the complete extraction of DBT required three steps, which demonstrates the greater polarity of this compound and hence greater affinity to the matrix compared with TBT (Wahlen and Wolff-Briche 2003).

As mentioned previously, sample dispersion in an inert support before filling the extraction cells is required. This ensures good solvent sample contact by avoiding the aggregation of sample particles thereby increasing the sample surface area. Different dispersing supports have been proposed for several applications, but functionalized C_{18}, diatomaceous earth, sea sand, alumina, and Florisil are used most frequently. In certain applications, such as when working with wet samples, the addition and/ or dispersion with drying substances is needed. Anhydrous sodium sulfate is usually recommended when using organic solvents as extractants. However, applications using high amounts of diatomaceous earth have been developed for treating wet samples (mussels) with aqueous solvents (Santiago-Rivas et al. 2007). On other occasions, such as for samples containing high levels of sulfur the use of copper or tetrabutylammonium sulfite powder is recommended to prevent system blockage.

An additional advantage of PLE is the possibility of performing *in situ* or in-cell clean up of the extract. In these cases, a fixed amount of the clean-up material (commonly C_{18} or Florisil) is loaded at the bottom of the extraction cell, and the sample-dispersing agent mixture is then loaded. After completion of the extraction, the extract passes through the clean-up material bed during the N_2 gas purge stage just before being transferred to the collection vial.

PLE for speciation studies

Arsenic species. In addition to the high degree of automation of PLE systems, an additional advantage of using pressurized solvents is that the integrity of chemical species is expected to be maintained. Schmidt et al. (Schmidt et al. 2000) proved that several PLE conditions, including the use of high temperatures, had little influence on the stability/integrity of arsenic species. In addition, as shown by Gallagher et al. (Gallagher et al. 2002) there are no interactions between the stainless-steel components of the PLE system and the arsenic species, and negligible blanks are obtained. Inorganic (As(III) and As(V)) and organic arsenic (MMA and DMA) have been extracted from different land plants using pressurized water (1450 psi and 120°C) as a solvent, obtaining quantitative yields (Schmidt et al. 2000, Schmidt et al. 2004). Methanol/water mixtures have been used when treating certified reference materials of marine origin (McKiernan et al. 1999, Gallagher et al. 2002), marine biota (Mato-Fernández et al. 2007), and land plants (Heitkemper et al. 2001). These developments were performed at room temperature and high pressure (1500 psi) in three cycles of 1 or 5 min (McKiernan et al. 1999, Heitkemper et al. 2001, Gallagher et al. 2002) or moderate pressure (500 psi) in one cycle of 2 min (Mato-Fernández et al. 2007). Arsenic speciation studies have also been performed in vegetables (carrots) and infant food products after PLE with methanol/water (Vela et al. 2001, Vela and Heitkemper 2004), and this solvent was also pressurized for isolating arsenosugars, as well as inorganic arsenic, DMA and AsB, from seaweed (Gallagher et al. 1999, Gallagher et al. 2001). Finally, methanol alone or combined with acetic acid has been used by Wahlen et al. (Wahlen 2004, Wahlen and Catterick 2004, Wahlen et al. 2004) for achieving arsenic speciation in marine biota. One of these applications led to the determination of compromise extraction conditions for the simultaneous isolation of arsenic (MMA, DMA and AsB) and tin (DBT and TBT) compounds as well as the mercury compound MMM (Wahlen and Catterick 2004). Results showed quantitative results for arsenic and tin species and yields of approximately 75 percent for MMM. PLE has also been applied for extracting inorganic arsenic and selenium species from atmospheric particulate matter (Moscoso-Pérez et al. 2008). A solvent consisting of aqueous 40 mM EDTA was pressurized at 100°C and 1000 psi, and quantitative recoveries were obtained by using a static time of 5 min (one cycle).

Teflon beads, glass beads and diatomaceous earth have been used most often as dispersing agents. The use of sand when treating vegetables (Vela et al. 2001) required a previous acid-wash process of the sand to remove inorganic arsenic contamination. In addition, moderate values for inorganic arsenic have also been found when using diatomaceous earth for dispersing the sample (Mato-Fernández et al. 2007). Finally, sample dispersion was not required in some applications such as when analyzing atmospheric particulate matter (Moscoso-Pérez et al. 2008). In this case, the sample (PM10) was uniformly dispersed in the quartz filter used for sampling and the addition of an extra solid support was not advantageous.

Tin species. As reported by Arnold et al. (Arnold et al. 1998) for the analysis of sediments and sewage sludge, the use of acid extractants under pressurized conditions can partially digest the sample matrix, with the increase in hydrolysis of certain alkyl and aryl ligands of OTCs. These authors showed that DBT was largely degraded to MBT and to inorganic tin when using oxalic acid in methanol as a solvent, even in the presence of tropolone as a chelating agent. Best performance and maintenance of OTC integrity was obtained when using acetic acid/sodium acetate buffer solutions in methanol (Arnold et al. 1998). The authors suggested that dissolved sodium acetate enhances the extractions because the acetate acts as a chelating agent by complexing OTCs, and sodium acts as a cationic exchanger for the OTCs linked to the clay fraction of sediments. A combined study of the effect of the temperature and the solvent composition (concentration of acetic acid and sodium acetate in methanol) showed that OTC degradation is not affected by temperature but high concentrations of acetic acid in the solvent can degrade DBT and MBT. The authors proposed 0.25 M for acetic acid and sodium acetate in methanol as an extractant for isolating butyl- and phenyl-tin compounds. In addition, recoveries for tricyclohexyl-tin (TCyT) were moderate under the proposed PLE conditions.

Ruiz-Encinar et al. (Ruiz Encinar et al. 2002b) have, however, found that temperature is a critical parameter to ensure the integrity of OTCs, mainly DBT. Their results confirm that DBT is stable up to 50°C when using a methanol/acetic acid (9/1) as an extracting solution (pressure of 1750 psi), while TBT can be extracted from sediments without degradation even working at temperatures up to 140°C. A similar extractant (methanol/ acetic acid, 8.5/1.5) was successfully used by Heroult et al. (Heroult et al. 2008) for butyl-, phenyl- and octyl-tin compounds in soils. Other authors, such as Marcic et al. (Marcic et al. 2005), have reported OTC degradation when using high pressures, i.e., TBT was degraded at pressures of above 100 bars when using a methanol/ethyl acetate (1/1) mixture as a solvent (extracting temperature of 90°C) for treating vegetables.

An extracting solvent consisting of acetic acid/sodium acetate in methanol, as proposed by Arnold et al. (Arnold et al. 1998), was also used by Wahlen and Wolff-Briche for butyl-tin isolation from sediments (Wahlen and Wolff-Briche 2003). Similarly, acetic acid/sodium acetate in methanol was the solvent of choice for extracting butyl- and phenyl-tin compounds from biota (Looser et al. 2000, Wahlen and Catterick 2003). The same extractant modified with a small amount of tropolone as a complexing agent was used when treating biological samples with high lipid contents (Wasik and Ciesielski 2004) and sulfur-rich sediments (Wasik et al. 2007). The results showed that the presence of tropolone improved MBT recovery considerably (approximately 60 percent of MBT was obtained in the absence of this chelating agent). Further applications were therefore based on the use of tropolone as a chelating agent such as in the extracting solution (acetic acid/tropolone in methanol) proposed for isolating butyl- and phenyl-tin species from sediments (Chiron et al. 2000, Inagaki et al. 2007), tropolone in hexane (Konieczka et al. 2007), or the hydrochloric acid/tropolone/ hexane mixture proposed in an international intercomparison exercise for TBT determination in sediments (Sturgeon et al. 2003).

After PLE extraction different clean-up strategies were performed before analysis. As previously mentioned, OTC derivatization by ethylation is a common practise when using GC as a separation technique. Ethylation was the method of choice for most of the applications that used PLE as a sample pre-treatment and GC hyphenated with AES, MS or ICP-MS (Arnold et al. 1998, Looser et al. 2000, Ruiz Encinar et al. 2002b, Wasik and Ciesielski 2004, Marcic et al. 2005, Wasik et al. 2007, Inagaki et al. 2007, Heroult et al. 2008), although derivatizations with pentyl magnesium bromide have also been proposed (Konieczka et al. 2007). Some authors have also reported extract clean-up before HPLC separations. In this case, C_{18}-SPE methods were proposed (Chiron et al. 2000).

Selenium species. Most of the PLE procedures proposed for selenium species aimed at isolating water-soluble selenium species. Water containing phenylmethanesulfonylfluoride (PMSF) as a protease inhibitor, and dithiothreitol (DTT) as an antioxidant, at low concentrations have been used for extracting inorganic selenium (Se(IV) and Se(VI)) and organic selenium species (SeCys, SeMet and SeCys$_2$) from selenized yeast (Goenaga et al. 2004). The same PLE procedure combined with an enzymatic hydrolysis method was also applied for assessing the same target analytes in addition to soluble γ-glutamyl-semethylselenocysteine in selenized yeast (Goenaga Infante et al. 2005) and vegetables (Peachey et al. 2009, Goenaga Infante et al. 2009). The application of the developed method also allowed the identification of methylseleninic acid in watercress (Peachey et al. 2009). Other applications involved the use of water/methanol mixtures for assessing water-soluble selenium from selenized and native yeast (Goenaga et al. 2004). These authors concluded that recoveries achieved after PLE are quite similar to those obtained after sample pre-treatments based on mechanical shaking, Soxhlet extraction and even sonication.

Enzymatic hydrolysis methods

General aspects

Enzymatic hydrolysis methods encompass a range of procedures consisting of hydrolyzing biomolecules by breaking down certain bonds and allowing the release of target analytes from the sample matrix. The procedures are very useful when the organometallic species are integrated in macromolecules such as proteins. This is the case of some selenium species (SeCys and SeMet) and iodine compounds (MIT and DIT). Enzymes hydrolyze/digest macromolecules contained in biological samples under mild temperature and pH conditions and in the absence of polluting or toxic reagents. These procedures are therefore considered to be environmentally friendly methods (Moreda-Piñeiro et al. 2009). In addition, because of the moderate temperatures and pHs used, target analyte degradation/inter-conversion phenomena are minimized, resulting in being appealing procedures for speciation studies.

The enzyme must be selected carefully based on the nature of sample (high protein, fat or carbohydrate content). Some enzymes can be of proteolytic nature (proteases), and they attack the peptidic bonds of proteins and peptides. As examples, Pronase E, pepsin, pancreatin, trypsin, and mainly protease XIV are frequently used for the release of organometallic species. The latter protease is able to digest long protein chains into the individual amino acids. However, the enzyme can be chosen to hydrolyze other biomolecules such as lipids and carbohydrates. Enzymes such as lipases are used in the former application and they hydrolyze fats into long-chain fatty acids and glycerol. The latter enzymes, mainly amylase types, hydrolyze starch and glycogen to maltose and to residual polysaccharides. In some cases, mixtures of different enzymes can be used simultaneously for the hydrolysis of different biomolecules. However, this approach is not always possible because the different optimum pH values inherent to the different enzymes; i.e., pepsin normally works at pH around 2 while lipase or α-amylase work at a pH of around 7 (Peña-Farfal et al. 2004a).

However, it must be pointed out that species alteration or co-elution of some species can occur when using certain enzymes. As reported by Pardo-Martínez et al. (Pardo-Martínez et al. 2001) when assessing arsenic species in baby foods, pancreatin offered a good overall extraction efficiency, but was found to either cause species alteration or to disrupt the chromatography, thus causing co-elution of some species.

In addition to temperature and pH, there are other parameters that affect the enzymatic activity. These include the ionic strength, the sample to enzyme ratio, and the presence of certain substances that can inhibit the enzyme activity (inhibitors) or that can increase the enzyme action (activators). The sample to enzyme ratio needs to be low enough to guarantee total sample degradation and target analyte release. In this sense, after evaluating several enzymatic methods based on a 24-hr incubation time for selenium speciation in yeast, Yang et al. (Yang et al. 2004), concluded that in addition to the enzyme type, the optimization of the amount of enzyme as a function of the incubation time is quite important. They found that quantitative SeMet yields can be achieved when using 200 mg of protease XIV and a hydrolysis times of 72 hr; and also when using a higher amount of protease (400 mg) with incubation for 24 hr. However, it must be mentioned that large amounts of hydrolyzed matrix components will be present in the extract/digest when using large amounts of enzymes, which can diminish the adequate response of the analytical instruments, and which can involve exhaustive clean-up stages before analysis (Vale et al. 2008). Concerning the use of activators, 1,4-dithiothreitol (DTT) is frequently used before Pronase E enzymatic hydrolysis of hair when assessing drugs of abuse (Chiarotti 1993). In a similar manner, urea and DTT/iodoacetamide pre-treatments before enzymatic hydrolysis have been proposed when assessing selenium species in biological materials (Pedrero et al. 2011, Ruiz Encinar et al. 2004, Bierla et al. 2008): DTT is reported to reduce Se–S and Se–Se bridges in the protein structure, and then allows a fast action of the protease (Ruiz Encinar et al. 2004, Bierla et al. 2008).

Enzymatic hydrolysis procedures require homogenization of the enzyme/buffer solution mixed with the sample. Different homogenization methods, such as the use of Potter homogenizer, are often used to bring enzyme/buffer solution and sample into intimate contact (Pardo-Martínez et al. 2001).

Assisted enzymatic hydrolysis methods

Despite the advantages of the enzymatic hydrolysis in speciation studies, their main drawback is the long time required for hydrolysis completion. As will be discussed later, it has been reported that between 4 to 24 hr of time required for conventional enzymatic hydrolysis procedures. This long time is attributed to the high stability of cell membranes/walls, which must first be hydrolyzed by the enzyme before digesting the cytosolic content. Cell membranes/walls are mainly constituted of proteins and polar lipids in which phospholipids and sterols form the lipid bilayer and the proteins are embedded at irregular intervals. This structure, called fluid mosaic model, leads to a very stable and strong array.

The limitation of the long time necessary for hydrolysis has recently been overcome by assisting or accelerating the enzymatic process through several different approaches. The use of ultrasound energy supplied by either ultrasonic probes or ultrasonic water-baths was the first approach for reducing the enzymatic hydrolysis time from several hours to minutes (ultrasound water-baths) or even seconds (ultrasounds probes). The reduction of the enzymatic hydrolysis time when using ultrasound is attributed to the rapid and efficient cell membranes/walls disruption by the action of ultrasound which allows a direct contact between the enzyme and the cytosolic structures (Peña-Farfal et al. 2004b). As reviewed by Vale et al. (Vale et al. 2008), the enhancement of enzymatic kinetics by ultrasonication depends on the intensity of sonication rather than on the frequency of sonication (Sakakibara et al. 1996), and enzyme denaturization (inactive enzyme) by ultrasonication is negligible when using short irradiation times (Talukder et al. 2006). In addition, the enzymatic reaction constant (Michaelis constant, K_M) appears to be unchanged when sonicating; whereas, the reaction rate (maximum speed rate, V_{max}) increases (Talukder et al. 2006).

In addition to ultrasound, other recent approaches for accelerating the enzymatic hydrolysis are based on pressurization (Moreda-Piñeiro et al. 2010) and microwave irradiation (Peachey et al. 2008, Guzmán Mar et al. 2009). In a way similar to ultrasound, pressurization and microwaves also disrupt cell membranes/walls so that a rapid attack by the enzyme on the cytosolic components occurs.

Enzymatic hydrolysis methods for speciation studies

Selenium species. As mentioned above, the adequate selection of the enzyme and also the optimization of the sample to enzyme ratio and the extracting time is important for achieving quantitative results. As an example, Yang et al. (Yang et al. 2004) have reported quantitative SeMet concentrations in yeast after applying an enzymatic hydrolysis method based on protease XIV (200 mg, 37°C, 72 hr; or 400 mg, 37°C, 24 hr). SeMet levels found using enzymatic hydrolysis were comparable to those obtained when using the methanesulfonic acid reflux reference method. However, other authors have reported lower SeMet recoveries from yeast and Brazil nuts when using enzyme–based sample pre-treatments when compared with those concentrations found after refluxing with methanesulfonic acid (Wrobel et al. 2003b). This latter application used proteinase K (20 mg, 50°C, 18 hr) followed by a second stage with proteinase K (20 mg, 50°C, 6 hr), and a third step with protease XIV (20 mg, 37°C, 6 hr) (Wrobel et al. 2003b).

Concerning SeMet integrity during the enzyme-based procedures Larsen et al. (Larsen et al. 2003) found that the degree of oxidation (formation of selenomethionine-Se-oxide, SeOMet) is large and variable when using β-glucosidase (30 mg, 25°C, 3 hr) followed by hydrolysis with endo- and exopeptidases (15 mg, 50°C, 24 hr). However, SeMet oxidation was low when using protease XIV (20mg, 37°C, 24 hr). The integrity of SeMet when performing ultrasound-assisted enzymatic procedures was also evaluated. Fang et al. (Fang et al. 2009) proved that SeMet was efficiently extracted without oxidization from selenium-enriched rice by using an ultrasonic water-bath assisted enzymatic hydrolysis. A two-stage procedure based on using α-amylase (37°C, 30 min. without stirring, and 2 hr with constant stirring) plus protease XIV (45°C, 2 hr with constant stirring) was used. Pedrero et al. (Pedrero et al. 2007), however, found that SeMet oxidation was significant when assisting the enzymatic hydrolysis with an ultrasonic probe; whereas the integrity of SeMet was preserved under conventional incubation procedures. In addition, these authors also showed that the stability of the enzymatic digest from yeast after 24 hr storage at different temperatures was also affected more when the enzymatic extracts were obtained by ultrasonication. These findings can be attributed to the large concentrations of matrix components in the enzymatic extract when assisting the procedure by ultrasound, and it agrees with those obtained by Dumont et al. (Dumont et al. 2004) who concluded that the degree of transformation of SeMet during storage is affected by the sample matrix components present in the extracts.

Stability of SeMet and trimethylselenonium ion (TMSe$^+$) in enzymatic extracts from oyster was also evaluated by Moreno et al. (Moreno et al. 2002). SeMet degradation was not observed when storing the enzymatic extracts

(Subtilsim was used as a protease) at 4°C and –18°C in both polyethylene and Pyrex containers for at least 10 d. Under the same conditions, TMSe$^+$ was stable for at least 15 d. Concerning total selenium, variations in the enzymatic extracts were not observed for at least 30 d when using Pyrex containers and storage at 4°C. Both polyethylene and Pyrex containers (4°C, 30 d) could be used for storing the enzymatic extracts from the soluble fraction from oyster tissue (water-soluble selenium). Therefore, in addition to the container material, the presence of matrix components in the extract appears to play an important role in the stability of selenium (Dumont et al. 2004, Pedrero et al. 2007). However, Mazej et al. (Mazej et al. 2006) have reported that the degree of transformation of SeMet can also be attributed to the activity of the enzyme rather than the sample matrix. These authors performed different experiments with SeMet solutions and SeMet solutions incubated with protease XIV at 37°C for 24 hr, and showed that oxidation of SeMet occured only in the presence of the enzyme. Extracts with the enzyme had to be stored a –20°C and analyzed as soon as possible; whereas SeMet solution without enzyme could be stored at –20°C or at 4°C for at least 2 wk.

Recently, Cuderman and Stibilj (Cuderman and Stibilj 2010) studied the stability of selenium species in enzymatic extracts from land plants rich in phenolic substances. Studies proved that organic selenium species (SeMet, SeCys$_2$, and SeMeSeCys) as well as Se(VI) were stable. However, the presence of rutin and tannin at high concentrations caused a strong decrease in Se(IV) response (60 percent) in aqueous extracts, whereas, the decrease was approximately 20 percent in the enzymatic digests (conventional enzymatic hydrolysis with protease XIV). In addition, the response for Se(IV) was not stable over 4 d at 4°C.

Conventional enzymatic hydrolysis with protease XIV under controlled temperature and shaking for 24 hr are procedures most frequently reported for selenium speciation studies. Conditions based on protease dissolved in water and shaking at room temperature were applied for selenium speciation studies in selenized dietary supplements (Amoako et al. 2007, Amoako et al. 2009) and in selenium-enriched yeast (Bird et al. 1997). Other studies performed the extraction at 37°C such as when treating land plants (Cuderman and Stibilj 2010, Mazej et al. 2006, Mazej et al. 2008, Montes-Bayón et al. 2002, Vogrinčič et al. 2009) and selenized yeast (Larsen et al. 2003). On other occasions, the enzymatic hydrolysis was performed using protease XIV combined with other enzymes such as α-amylase or β-amylase (Cuderman et al. 2008), or with lipase VII (Gómez-Ariza et al. 2002, Díaz Huerta et al. 2003, Li et al. 2010a, Stiboller et al. 2011). The development by Stiboller et al. (Stiboller et al. 2011) deserves special attention. In contrast to other methods, this application used a microtiter plate format which allowed a reduction of the required sample volumes to 1 mL per extract, and hence a reduction in the amount of enzyme. The enzymatic hydrolysis (protease XIV plus lipase VII) was performed in a parallel reaction platform made out of sintered silicon carbide, fitted with standard disposable glass HPLC/GC vials. Due to the high thermal conductivity of silicon carbide, this set-up can be placed on a standard hotplate to accurately maintain the desired extraction conditions (37°C, 20 hr) for all positions of the microtiter plate (5 × 4 = 20 positions).

Since pH is an important parameter controlling the enzyme activity, several conventional enzymatic procedures perform the process at a fixed pH, normally the optimum working pH of the enzyme. A pH within the 7.0–7.5 range is often used when using protease XIV, and different buffer solutions consisting of phosphate buffer or Tris/HCl buffer solution have been proposed (Smrkolj et al. 2005, Hinojosa Reyes et al. 2006, Smrkolj et al. 2006, Pedrero et al. 2007, Zheng and Hintelmann 2009, Seppänen et al. 2010, Pedrero et al. 2011). However, some applications used a pH of 6.8 fixed with ammonium acetate buffer (Egressy-Molnár et al. 2011). Similarly, procedures at a fixed pH using protease XIV together with cellulase or α-amylase (Bryszewska et al. 2005), with lipase VII (Goenaga Infante et al. 2005, Peachey et al. 2009, Goenaga Infante et al. 2009), or with driselase (Hinojosa Reyes et al. 2006) have also been developed.

Selenium speciation studies have also been developed using enzymatic hydrolysis based on other proteases such as proteinase K to assess chiral speciation of DL-SeMet in selenized yeast (Day et al. 2002) or SeMet in plants (Montes-Bayón et al. 2002), Subtilisim for selenium speciation in oyster (Moreno et al. 2002), pronase E when treating Antarctic krill (Siwek et al. 2005) and cereals (Stadlober et al. 2001), and Subtilisim plus pronase E for studies in marine biota (Quijano et al. 2000). In addition, sequential enzymatic hydrolysis based on using pepsin at pH 2.1 (37°C, shaking at 200 rpm, 20 hr) followed by trypsin treatment at pH 7.6 (37°C, shaking at 200 rpm, 20 hr) has also been proposed for selenium speciation in mushroom samples (Stefánka et al. 2001).

As mentioned, ultrasound energy has frequently been used for assisting/accelerating the enzymatic hydrolysis procedures. Most of the developments are based on ultrasonic probes which allows the completion of the enzymatic hydrolysis process in a few minutes. An example is selenized yeast that was treated with protease XIV dissolved in water, with assistance from an ultrasonic probe (20 W) for 30 sec. This allowed quantitative SeMet recoveries (Capelo et al. 2004). A further application involved the extraction in a buffered medium (Tris/HCl, pH 7.5) and sonication for 2 min (Pedrero et al. 2007). Similarly, inorganic selenium plus Se(Cys)$_2$, SeMet and TMSe$^+$ were extracted from animal-based food (chicken) with protease XIV in a Tris/HCl buffer by ultrasonication (20W) for 120 sec (Cabañero et al. 2005). Ultrasonic probe—assisted enzymatic hydrolysis was also applied in selenium speciation studies in potatoes (protease XIV, 50 percent amplitude, 3 min) (Cuderman et al. 2008), plants (protease XIV, 40 percent amplitude, 2 min) (Seppänen et al. 2010), food supplements (protease XIV, 50 percent amplitude, 2 min) (Vale et al. 2010), and fish and Antarctic krill (pronase E, 50 percent amplitude, 15 min, sample-enzyme mixture in an ice-bath) (Egressy-Molnár et al. 2011).

Some applications based on the use of ultrasonic water-baths have also been proposed. Ultrasound assisted α-amylase (37°C for 30 min plus 37°C plus 2 hr with constant stirring) followed by protease XIV (45°C for 2 hr with constant stirring) was used for selenium speciation in rice (Fang et al. 2009). Similarly, protease XIV with lipase VII (Tris/HCl buffer, pH 7.5) was used for treating selenized supplements (sonication for 2 hr at 25°C) (Stiboller et al. 2011).

Enzymatic hydrolysis has also been assisted by microwave energy. Peachey et al. (Peachey et al. 2008) proved the effectiveness of focused microwaves for accelerating the enzymatic hydrolysis of selenized yeast with a protease XIV-lipase VII mixture dissolved in Tris/HCl solution (pH 7.5). Quantitative SeMet yields were obtained after two consecutive extraction cycles, each of 15 min (total extraction time of 30 min) at 37°C (microwave power of 60 W). Guzmán Mar et al. (Guzmán Mar et al. 2009) have also applied microwave energy (closed-vessel microwave instrument) for the enzymatic hydrolysis of rice (protease XIV plus α-amylase dissolved in water) for extracting selenium and arsenic species simultaneously. Operating conditions consisted of a temperature ramp from room temperature to 37°C over 5 min followed by irradiation at 37°C for 40 min.

Arsenic species. Zheng and Hintelmann (Zheng and Hintelmann 2009) have compared different sample pre-treatments for isolating arsenic species (As(III), As(V), MMA, DMA and TMAO) from land plants, including a protease XIV based enzymatic hydrolysis procedure. The procedure consisted of incubating the sample with the enzyme (30 mg) in a phosphate buffer (20 mM, pH 7.5) for 16 hr at 37°C in a shaker. Low extraction efficiencies (approximately 48 percent of the total arsenic) were obtained. These yields were lower than those obtained when using shaking with ultrapure water. The low efficiency is probably due to an inappropriate selection of the enzyme and, as mentioned above for selenium speciation studies, a mixture of a protease and cellulase or α-amylase may potentially offer better results. Conventional enzymatic hydrolysis procedures based on the use of trypsin in a buffered solution of ammonium hydrogen carbonate (pH 8.0) were applied by Rattanachongkiat et al. for assessing inorganic arsenic, MMA, DMA and AsB in fish and crustacean samples (Rattanachongkiat et al. 2004). The hydrolysis was performed at 37°C for

12 hr and quantitative yields were reported. Trypsin dissolved in 0.1 M ammonium hydrogen carbonate solution was also used by Pardo-Martínez et al. for arsenic speciation in baby foods under conventional conditions (37°C, overnight) (Pardo-Martínez et al. 2001). These authors also proved interconversion of arsenic species attributed to the use of pancreatin.

Ultrasonic probe-assisted enzymatic hydrolysis has also been proposed by Sanz et al. for assessing arsenic speciation in biological samples (chicken and fish) (Sanz et al. 2005a) and rice (Sanz et al. 2005a, b). The authors reported a quantitative extraction for total arsenic (> 95%) and approximately 90 percent recovery as a sum of the arsenic species from rice when applying an enzymatic treatment based on using aqueous α-amylase (30 percent amplitude, 1 min) followed by aqueous protease XIV (30 percent amplitude, 2 min) (Sanz et al. 2005a, b). In contrast to As(V), DMA, and MMA, As(III) species appeared to be bonded to the hydrolyzed matrix, because the concentration found in the enzymatic extracts was higher than that found when using other extractants. Aqueous protease XIV was used when treating fish (30 percent amplitude, 1 min), yielding recoveries > 90% for total As content in all cases, while sonication time had to be increased to 4 min for quantitative arsenic recoveries from chicken tissues (Sanz et al. 2005a).

Marine biota have also been treated using ultrasonic water-bath assisted enzymatic hydrolysis by pepsin (pH 3.0 adjusted with dilute hydrochloric acid) under ultrasound irradiation (35 kHz, 40°C) for 5 min (Moreda-Piñeiro et al. 2011). Quantitative recoveries for AsB (the major arsenic species) and MMA, DMA, As(III) and As(V) were obtained. The authors reported that a clean-up stage was needed before analysis, and so an Envicarb-based SPE was used. Similarly, the assistance of the enzymatic hydrolysis of marine biota using pepsin was also performed by pressurization (Moreda-Piñeiro et al. 2010). In this application, the sample/pepsin mixture was dispersed with C_{18} and the enzymatic extraction was performed with water (pH 4.0 adjusted with dilute hydrochloric acid) under pressurization (50°C, 1500 psi) with two cycles of 3 min each. The main advantage of the procedure was the possibility of performing an in-cell clean-up procedure by using C_{18} (optimum amount of 2 g) as a solid support loaded at the bottom of the extraction cell. Once enzymatic hydrolysis had occurred, the extract was passed through the sorbent and it was then collected ready for analysis.

Finally, as mentioned earlier, microwave energy (closed-vessel microwave instrument) was used for assisting the enzymatic hydrolysis of rice (aqueous protease XVI/α-amylase, temperature ramp of 5 min from room temperature to 37°C, and 37°C for 40 min) for assessing selenium and arsenic speciation (Guzmán Mar et al. 2009).

Mercury species. Conventional enzymatic hydrolysis methods based on using protease XIV in buffered solutions (phosphate buffer, pH 7.5) have been proposed by Rai et al. for isolating MMM and inorganic mercury from fish (Rai et al. 2002). The authors reported the addition of a small amount of cysteine to the enzymatic reaction medium for complexing the extracted mercury species, and performed the enzymatic hydrolysis at 37°C for 2 hr under continuous rotation (20 rpm). Quantitative recoveries were obtained when analyzing different certified reference materials (total mercury recoveries within the 92–107 percent range). Conventional procedures based on using trypsin as a protease (ammonium acetate buffer at pH 8.0, 37°C, shaking at 20 rpm for 24 hr) were also developed for assessing mercury species in fish (Lemes and Wang 2009) and rice (Li et al. 2010b). Experiments based on using MeHg-cysteine and MeHg-glutathione complexes have demonstrated the presence and dominance of MeHg-cysteine in fish muscle (Lemes and Wang 2009), and in uncooked rice (Li et al. 2010b). However, although cooking does not change the total mercury or total MeHg concentration in rice, no MeHg-cysteine was measurable after cooking, suggesting that most, if not all, of the MeHg-cysteine is converted to other forms of MeHg (Li et al. 2010b). These findings are quite important because the MeHg, binding with L-cysteine is thought to be the main pathway of MeHg transport across the blood–brain barrier.

Ultrasound probe assisted enzymatic hydrolysis has also been proposed for mercury speciation in marine biota (López et al. 2010). Experimental conditions based on using aqueous protease XIV plus 2.5% (v/v) 2-mercaptoethanol for complexing the extracted mercury species under ultrasonic irradiation (40 percent amplitude) for 2 min were finally chosen. The authors reported that no transformations between mercury species were detected, and the method was mainly applied for treating small size samples.

Tin and antimony species. Rodríguez-González et al. (Rodríguez-González et al. 2004) have compared different sample pre-treatments for assessing butyl-tin compounds (MBT, DBT and TBT) in mussel tissue. As mentioned earlier, the authors found that MAE with methanol/acetic acid (1/3) led to quantitative extractions with negligible inter-conversion of target analytes. Similarly, quantitative recoveries with no degradation of the species were obtained when using enzymatic hydrolysis (protease XIV plus lipase VII in a buffered medium at pH 7.5, 37°C, 12 hr). However, the authors have reported long isotope equilibration times (up to 12 hr) when using enzymatic digestions, which is a limiting factor when applying isotope dilution mass spectrometry strategies for speciation studies.

Other applications of enzymatic hydrolysis procedures with protease XIV/lipase VII mixtures were developed for butyl- and phenyl-tin speciation in marine biota (Carlier-Pinasseau et al. 1996b, Pannier et al. 1996). The enzymatic reaction was performed in a buffered medium (phosphate buffer at pH 7.5) at 37°C for 4 hr. However, quantitative recoveries were only achieved for butyltin species, and a solid-liquid extraction procedure by shaking with methanol/hydrochloric acid as extractant was preferable when dealing with phenyltin species (Carlier-Pinasseau et al. 1996b).

Finally, Foster et al. (Foster et al. 2005) compared different enzymatic hydrolysis methods and MAE procedures when isolating inorganic antimony species from algae, land plants and animal tissues. After lipid removal with a chloroform/methanol mixture, algae and plants were treated with cellulose dissolved in dilute EDTA solutions (pH 5.0); whereas, protease dissolved in a phosphate buffer (pH 7.5) and in the presence of EDTA was chosen for treating animal tissues. Enzymatic hydrolysis for both cases was performed at 37°C for 12 hr. The authors reported low inorganic antimony recoveries from animal tissues (protease-based enzymatic hydrolysis) and from algae (cellulase as an enzyme); whereas, similar antimony yields to those obtained when using solid-liquid extraction with EDTA or sodium hydroxide were obtained from plants (moss).

Matrix solid phase dispersion

Matrix solid phase dispersion (MSPD), first introduced by Barker et al. (Barker et al. 1989), is a fast and efficient alternative technique for extracting organic compounds from solid, semi-solid and/or highly viscous samples. The large number of MSPD applications (Barker 2000, Kristenson et al. 2006, Barker 2007) is due to its flexibility, selectivity, and the possibility of performing the extraction and the clean-up stage simultaneously, resulting in rapid pre-treatments with minimal sample manipulation.

MSPD is based on sample disruption by mechanical blending with a solid support bonded-phase which disrupts the sample architecture and provides a more finely divided material for extraction. After blending, the dispersed sample is transferred and packed into a column/syringe, and elution with an adequate solvent is then performed by gravity flow or by applying pressure or a vacuum. A practical advantage is that analyte elution occurs with relatively small solvent volumes, and the dispersing solid support can be chosen for retaining the interfering matrix components. On other occasions, a solid support for clean-up can be loaded at the bottom of the column, and the dispersed sample is loaded above the clean-up bed. Once the solvent passes through the dispersed samples, the extract is passed through the clean-up material before collection. Experimental parameters of MSPD procedures can be summarized as the nature of the solid dispersing support, and the polarity of the eluting solvents which must be chosen as a function of the target analytes' nature. Other experimental parameters are the average particle size; the possible chemical modification of the matrix or of the matrix solid support blend by the addition of chelating agents, acids, bases, etc; the elution sequence of solvents; and the elution volume.

Despite there being several applications of MSPD procedures for isolating organic compounds, at present there is only one application in the organometallic speciation field. Arsenic species (As(III), As(V), DMA, and AsB) were isolated from marine biota (fish and molluscs) by blending the dried sample thoroughly with diatomaceous earth (dispersing material mass to sample mass ratio of seven) in a glass mortar for 5 min (Moreda-Piñeiro et al. 2008). After a homogeneous mixture was obtained, it was quantitatively transferred by using a powder funnel to a 20 mL syringe containing C_{18} (2.0 g) between two polyethylene frits (*in situ* clean-up). Arsenic species were eluted from the syringes by gravity with methanol/ultrapure water (1/1) as an extracting solution. The mild operating conditions inherent to MSPD (room temperature, atmospheric pressure) maintained the integrity of all arsenic species as proved by analyzing different CRMs offering certified concentrations for some arsenic species, such as AsB and/or DMA. In addition, As(III) and DMA concentrations found in such CRMs agreed with those concentrations reported for such species by other researchers (Moreda-Piñeiro et al. 2008).

Abbreviations and Acronyms

The following abbreviations and acronyms are used in this chapter:

ACN	acetonitrile
ADDP	ammonium diethyldithiophosphate
AdoSeMet	selenoadenosylmethionine
AED	atomic emission detector
AFS	atomic fluorescence spectrometry
APDC	ammonium pyrrolidin dithiocarbamate
ASA	automated speciation analyzer
AsB	arsenobetaine
AsC	arsenocholine
ASE	accelerated solvent extraction
BPHA	N-benzoyl-N-phenyhydroxylamine
BTP	Benzothiophene
CAR	carboxene
CBT	1,3-bis(2-cyanobenzene)triazene
CE	capillary electrophoresis
CFA	continuous flow analysis

CGC	capillary gas chromatography
CFME	continuous flow micro-extraction
CMC	critical micellar concentration
[C_nMIM][PF6]	1-alkyl-3-methylimidazolium hexafluorophosphates
CNT	carbon nanotubes
CPC	cetyl pyridinium chloride
CPE	cloud point extraction
CTAB	cetyltrimethylammonium bromide
CV-AAS	cold vapor—atomic absorption spectrometry
CV-AFS	cold vapor—atomic fluorescence spectrometry
CV-ETAAS	cold vapor—electrothermal atomic absorption spectrometry
CV-ICP-MS	cold vapor—inductively coupled plasma-mass spectrometry
DAB	3,3-diaminobenzidine
DAD	diode array detector
DAN	2,3-diaminonaphthalene
DBBP	dibutyl butylphosphonate
DBDTC	dibenzyldithiocarbamate
DBT	dibutyltin
DCM	dichloromethane
dCPE	dual-cloud point extraction
DDA	dodecylamine
DDTP	O,O-diethyldithiophosphate
DEL	diethyllead
DEM	diethylmercury
DENC	2,2'-diaminodiethylaminecellulose
DESe	diethylselenium
DEDSe	diethyldiselenium
DI-SPME	direct inmersion-solid phase micro-extraction
DIT	diiodotyrosine
DIW	deionized water
DLLME	dispersive liquid–liquid micro-extraction
DMA	dimethylarsonic acid
DMAA	dimethylarsinic acid
DMDSe	dimethyldiselenium
DMDTe	dimethylditellurium
DML	dimethyllead
DMM	dimethylmercury
DMSb	dimethylantimony
DMSe	dimethylselenium
DMSeO	dimethylselenoxide
DMSeP	dimethylselenoniopropionate

DMT	dimethyltin
DMTe	dimethyltellurium
DOcT	dioctyltin
DPhM	diphenylmercury
DPhT	diphenyltin
DSTEB	deuterium-labeled sodium tetraethylborate
DTT	1,4-dithiothreitol
DVB	divinylbenzene
DZ	dithizone
ECF	ethylchloroformate
EDMA	ethylene glycoldimethacrylate
EDXRF	energy dispersive X-ray fluorescence
EPA	Environmental Protection Agency
ETAAS	electrothermal atomic absorption spectrometry
ETV	electrothermal vaporization
FAAS	flame atomic absorption spectrometry
FAPES	furnace atomization plasma emission spectrometry
FBTO	fentabutatin oxide
FI	flow injection
FIA	flow injection analysis
FID	flame ionization detector
FPD	flame photometric detector
FTI	Fourier transform infrared spectroscopy
GAC	green analytical chemistry
GC	gas chromatography
GC-MS	gas chromatography—mass spectrometry
GF	graphite furnace
HF	hollow fiber
HF-LLLME	hollow fiber–liquid-liquid-liquid micro-extraction
HF-LPME	hollow fiber–liquid phase micro-extraction
HFM	hollow fiber membrane
HG	hydride generation
HG-AAS	hydride generation—atomic absorption spectrometry
HPLC	high-performance liquid chromatography
HPLC-CV-AAS	high-performance liquid chromatography—cold vapor—atomic absorption spectrometry
HPLC-ICP-MS	high-performance liquid chromatography—inductively coupled plasma-mass spectrometry
HS	headspace
HS-HF-LPME	headspace-hollow fiber-liquid phase micro-extraction
HS-SBSE	headspace-stir bar sorptive extraction

HS-SDME	head space single-drop micro-extraction
HS-SPME	headspace-solid phase micro-extraction
IBMK	4-Methyl-2-pentanone
ICP-OES	inductively coupled plasma-optical emission spectrometry
ICP-MS	inductively coupled plasma-mass spectrometry
IDAEC	imino diacetic acid–ethyl cellulose
IIP	ionic imprinted polymer
ILs	ionic liquids
InHg	inorganic mercury
ISs	immunosorbents
KR	knotted reactor
LC-ESI-MS	liquid chromatography-electrospray ionization mass spectrometry
LiFDDC	lithium bis(trifluoroethyl)-dithiocarbamate
LLE	liquid-liquid extraction
LLLME	liquid-liquid-liquid micro-extraction
LLME	liquid–liquid micro-extraction
LOD	limit of detection
LPME	liquid phase micro-extraction
LVSS-CE-UV	large volume sample stacking capillary electrophoresis/ultraviolet detection
MAC	methacryloyl-(l)-cysteine methylester
MAE	microwave assisted extraction
MBT	monobutyltin
MCGC	multicapillary gas chromatography
MC*	metal-loaded activated charcoal
MEM	monoethylmercury
MES	2-(N-morpholino)ethanesulfonic acid
MG	methyl green
MPhM	monophenylmercury
MIPs	molecularly imprinted polymers
MIP-AED	microwave induced plasma-atomic emission detector
MIT	monoiodotyrosine
MMA	monomethylarsenic acid
MMAA	monomethylarsonic acid
MMLLE	microporous membrane liquid-liquid extraction
MMM	monomethylmercury
MMSb	monomethylantimony
MMT	monomethyltin
MNPs	magnetic nanoparticles
MPA	mercaptopropionic acid

MPhT	monophenyltin
MOcT	monooctyltin
MS	mass spectrometry
MSeCys	methylselenocysteine
MSeMet	methylselenomethionine
MSPD	matrix solid phase dispersion
MTB	methylthymol blue
MWCNT	multi-walled carbon nanotubes
3-NHPAA	3-nitro-4-hydroxyphenylarsonic acid,
OrgHg	organic mercury
OTC	organotin compounds
PA	polyacrylate
PAN	1-(2-pyridylazo)-2-naphthol
PANI	polyaniline
PDMS	polydimethylsiloxane
PdNP	palladium nanoparticles
PDT	1,3-propanedithiol
PFE	pressurized fluid extraction
PFPD	pulsed flame photometric detection
PPHF	polypropylene hollow fiber
PHWE	pressurized hot water extraction
PLE	pressurized liquid extraction
PMSF	phenylmethanesulfonylfluoride
PPY	polypyrrole
PSE	pressurized solvent extraction
psi	pounds per square inch
PSTH-cpg	[1,5-bis(2-pyridyl)-3-sulfophenyl methylene] thiocarbonohydrazide immobilized on aminopropyl-controlled pore glass
PT	purge and trap
PTFE	polytetrafluoroethylene
PT-LLE	phase transfer liquid–liquid extraction
PT/MS-LLLME	phase transfer membrane supported-liquid–liquid–liquid micro-extraction
PUF	polyurethane foam
QF-AAS	quartz furnace atomic absorption spectrometry
QSIL-FPD	quartz surface-induced tin emission-flame photometric detector
rT3	reversed-triiodothyronine
SBME	solvent bar micro-extraction
SBSE	stir bar sorptive extraction
SCF	sulfydryl cotton fiber
SDDC	sodium diethyldithiocarbamate

SDLPME	solidified drop liquid phase micro-extraction
SDME	single - drop micro-extraction
SDS	sodium dodecyl sulfate
SDSA	sodium dodecane sulfonic acid
SeCys	selenocysteine
Se(Cys)$_2$	selenocystine
SeEt	selenoethionine
SeMet	selenomethionine
SeU	selenourea
SFE	supercritical fluid extraction
SLM	supported liquid membrane
SLMP	supported liquid membrane extraction probe
SPDC	sodium pyrrolidine dithiocarbamate
SPE	solid phase extraction
SPME	solid phase micro-extraction
SWNT	single-walled carbon nanotubes
TBP	tributyl phosphate
TEA	triethylarsine
TeBT	tetrabutyltin
TCyT	tricyclohexyl-tin
TeEL	tetraethyllead
TeET	tetraethyltin
TEL	triethyllead
TeML	tetramethyllead
TeMT	tetramethyltin
TBT	tributyltin
TFA	trifluoroacetic acid
THF	tetrahydrofuran
TMAH	tetramethylammonium hydroxide
TMAI	tetramethylarsonium ion
TMAO	trimethylarsine oxide
TML	trimethyllead
TMPhT	trimethylphenyltin
TMSb	trimethylantimony
TMSe$^+$	trimethylselonium
TMT	trimethyltin
TOcT	trioctyltin
TPhA	triphenylarsine
TPhT	triphenyltin
TPrT	tripropyltin
TrMT	trimethyltin
T4	Thyroxine
UA-DLLME	ultrasound assisted-DLLME

UAE	ultrasond assisted extraction
US	ultrasound
UV	ultraviolet
VOCs	volatile organic compounds
Cl_2H_2DZ	4,4-dichlorodithizone

References

Abrankó, L., Z. Jókai and P. Fodor. 2005. Investigation of the species-specific degradation behaviour of methylmercury and ethylmercury under microwave irradiation. Anal. Bioanal. Chem. 383: 448–453.

Abrankó, L., B. Kmellár and P. Fodor. 2007. Comparison of extraction procedures for methylmercury determination by a SPME-GC-AFS system. Microchem. J. 85: 122–126.

Abuín, M., A.M. Carro and R.A. Lorenzo. 2000. Experimental design of a microwave-assisted extraction–derivatization method for the analysis of methylmercury. J. Chromatogr. A 889: 185–193.

Ackley, K.L., C. B'Hymer, K.L. Sutton and J.A. Caruso. 1999. Speciation of arsenic in fish tissue using microwave-assisted extraction followed by HPLC-ICP-MS. J. Anal. At. Spectrom. 14: 845–850.

Aguerre, S., C. Bancon-Montigny, G. Lespes and M. Potin-Gautier. 2000. Solid phase microextraction (SPME): a new procedure for the control of butyl- and phenyltin pollution in the environment by GC-FPD. Analyst. 125: 263–268.

Aguerre, S., G. Lespes, V. Desauziers and M. Potin-Gautier. 2001a. Speciation of organotins in environmental samples by SPME-GC: comparison of four specific detectors: FPD, PFPD, MIP-AES and ICP-MS. J. Anal. At. Spectrom. 16: 263–269.

Aguerre, S., C. Pécheyran, G. Lespes, E. Krupp, O.F.X. Donard and M. Potin-Gautier. 2001b. Optimization of the hyphenation between solid-phase microextraction, capillary gas chromatography and inductively coupled plasma atomic emission spectrometry for the routine speciation of organotin compounds in the environment. J. Anal. At. Spectrom. 16: 1429–1433.

Aguerre, S., C. Pécheyran and G. Lespes. 2003. Validation, using a chemometric approach, of gas chromatography–inductively coupled plasma–atomic emission spectrometry (GC–ICP–AES) for organotin determination. Anal. Bioanal. Chem. 376: 226–235.

Aizpún, B., M.L. Fernández, E. Blanco and A. Sanz-Medel. 1994. Speciation of inorganic mercury(II) and methylmercury by vesicle-mediated high-performance liquid chromatography coupled to cold vapour atomic absorption spectrometry. J. Anal. At. Spectrom. 9: 1279–1284.

Al–Assaf, K.H., J.F. Tyson and P.C. Uden. 2009. Determination of four arsenic species in soil by sequential extraction and high performance liquid chromatography with post-column hydride generation and inductively coupled plasma optical emission spectrometry detection. J. Anal. At. Spectrom. 24: 376–384.

Aller, A.J. and L.C. Robles. 1998a. Speciation of selenomethionine and selenourea using living bacterial cells. Analyst. 123: 919–927.

Aller, A.J. and L.C. Robles. 1998b. Determination of selenocystamine by slurry sampling electrothermal atomic absorption spectrometry after a selective preconcentration by living *Pseudomonas putida*. J. Anal. At. Spectrom. 13: 469–476.

Amjadi, M., J.L. Manzoori and Z. Taleb. 2010. Reverse micelle coacervate-based extraction combined with electrothermal atomic absorption spectrometry for the determination of arsenic in water and oyster tissue samples. Microchim. Acta. 169: 187–193.

Amoako, P.O., C.L. Kahakachchi, E.N. Dodova, P.C. Uden and J.F. Tyson. 2007. Speciation, quantification and stability of selenomethionine, S-(methylseleno)cysteine and selenomethionine Se-oxide in yeast-based nutritional supplements. J. Anal. At. Spectrom. 22: 938–946.

Amoako, P.O., P.C. Uden and J.F. Tyson. 2009. Speciation of selenium dietary supplements; formation of S-(methylseleno)cysteine and other selenium compounds. Anal. Chim. Acta. 652: 315–323.

Amouroux, D., E. Tessier, C. Pécheyran and O.X.F. Donard. 1998. Sampling and probing volatile metal(loid) species in natural waters by *in situ* purge and cryogenic trapping followed by gas chromatography and inductively coupled plasma mass spectrometry (P-CT-GC-ICP/MS). Anal. Chim. Acta. 377: 241–254.

Anastas, P.T. and J.C. Warner. 1998. Green Chemistry: Theory and Practice. Oxford University Press. New York, USA.

Andrewes, P., W.R. Cullen and E. Polishchuk. 1999. Confirmation of the aerobic production of trimethylstibine by *Scopulariopsis brevicaulis*. Appl. Organometal. Chem. 13: 659–664.

An-na, T., D. Guo-sheng and Y. Xiu-ping. 2005. Cloud point extraction for the determination of As(III) in water samples by electrothermal atomic absorption spectrometry. Talanta. 67: 942–946.

Armenta, S., S. Garrigues and M. de la Guardia. 2008. Green Analytical Chemistry. Trends Anal. Chem. 27: 497–511.

Arnold, C.G., M. Berg, S.R. Müller, U. Dommann and R.P. Schwarzenbach. 1998. Determination of organotin compounds in water, sediments, and sewage sludge using perdeuterated internal standards, accelerated solvent extraction, and large-volume-injection GC/MS. Anal. Chem. 70: 3094–3101.

Arthur, C.L. and J. Pawliszyn. 1990. Solid-phase microextraction with thermal-desorption using fused-silica optical fibers. Anal. Chem. 62: 2145–2148.

Attwood, D. and A.T. Florence. 1985. Surfactants Systems—Their Chemistry Pharmacy and Biology. Chapman and Hall. London, UK.

Audunsson, G. 1986. Aqueous/aqueous extraction by means of a liquid membrane for sample cleanup and preconcentration of amines in a flow system. Anal. Chem. 58: 2714–2723.

Australian Standard AS 2773-1999. Standards Association of Australia. Ultrasonic Cleaners For Health Care facilities. Australia.

Azenha, M. and M.T. Vasconcelos. 2002. Headspace solid-phase micro-extraction gas chromatography–mass detection method for the determination of butyltin compounds in wines. Anal. Chim. Acta. 458: 231–239.

Baghdadi, M. and F. Shemirani. 2008. Cold-induced aggregation microextraction: A novel sample preparation technique based on ionic liquids. Anal. Chim. Acta. 613: 56–63.

Baig, J.A., T.G. Kazi, A.Q. Shah, M.B. Arain, H.I. Afridi, G.A. Kandhro and S. Khan. 2009. Optimization of cloud point extraction and solid phase extraction methods for speciation of arsenic in natural water using multivariate technique. Anal. Chim. Acta. 651: 57–63.

Baig, J.A., T.G. Kazi, M.B. Arain, A.Q. Shah, G.A. Kandhro, H.I. Afridi, S. Khan, N.F. Kolachi and S.K. Wadhwa. 2011. Inorganic arsenic speciation in groundwater samples using electrothermal atomic spectrometry following selective separation and cloud point extraction. Anal. Sci. 27: 439–445.

Balarama-Krishna, M.V., D. Karunasagar, S.V. Rao and J. Arunachalam. 2005. Preconcentration and speciation of inorganic and methyl mercury in waters using polyaniline and gold trap-CVAAS. Talanta. 68: 329–335.

Balarama-Krishna, M.V., K. Chandrasekaran and D. Karunasagar. 2010. On-line speciation of inorganic and methyl mercury in waters and fish tissues using polyaniline micro-column and flow injection-chemical vapour generation-inductively coupled plasma mass spectrometry (FI-CVG-ICPMS). Talanta. 81: 462–472.

Baltussen, E., P. Sandra, F. David and C. Cramers. 1999. Stir bar sorptive extraction (SBSE), a novel extraction technique for aqueous samples: Theory and principles. J. Microcolumn. Sep. 11: 737–747.

Baltussen, E., C.A. Cramers and P.J.F. Sandra. 2002. Sorptive sample preparation—a review. Anal. Bioanal. Chem. 373: 3–22.

Bancon-Montigny, C., P. Maxwell, L. Yang, Z. Mester and R.E. Sturgeon. 2002. Improvement of measurement precision of SPME-GC/MS determination of tributyltin using isotope dilution calibration. Anal. Chem. 74: 5606–5613.

Barker, S.A. 2000. Matrix solid-phase dispersion. J. Chromatogr. A 885: 115–127.

Barker, S.A. 2007. Matrix solid phase dispersion (MSPD). J. Biochem. Biophys. Methods. 70: 151–162.

Barker, S.A., A.R. Long and C.R. Short. 1989. Isolation of drug residues from tissues by solid phase dispersion. J. Chromatogr. A 475: 353–361.

Barrio, C.S., E.R. Melgosa, J.S. Asensio and J.G. Bernal. 1996. Extraction of pesticides from aqueous samples: A comparative study. Mikrochim, Acta. 122: 267–277.

Barshick, C.M., S.-A. Barshick, P.F. Britt, D.A. Lake, M.A. Vance and E.B. Walsh. 1998. Development of a technique for the analysis of inorganic mercury salts in soils by gas chromatography/mass spectrometry. Int. J. Mass Spectrom. 178: 31–41.

Batlle, R. and C. Nerín. 2004. Application of single-drop microextraction to the determination of dialkyl phthalate esters in food simulants. J. Chromatogr. A 1045: 29–35.

Baxter, D.C., M. Faarinen, H. Österlund, I. Rodushkin and M. Christensen. 2011. Serum/plasma methylmercury determination by isotope dilution gas chromatography-inductively coupled plasma mass spectrometry. Anal. Chim. Acta. 701: 134–138.

Bayona, J.M. 2000. Supercritical fluid extraction in speciation studies. Trends Anal. Chem. 19: 107–112.

Beichert, A., S. Padberg and B.W. Wenclawiak. 2000. Selective determination of alkylmercury compounds in solid matrices after subcritical water extraction, followed by solid-phase microextraction and GC-MS. Appl. Organometal. Chem. 14: 493–498.

Bellido-Martín, A., J.L. Gómez-Ariza, P. Smichowsky and D. Sánchez-Rodas. 2009. Speciation of antimony in airborne particulate matter using ultrasound probe fast extraction and analysis by HPLC-HG-AFS. Anal. Chim. Acta. 649: 191–195.

Berton. P. and R.G. Wuilloud. 2010. Highly selective ionic liquid-based microextraction method for sensitive trace cobalt determination in environmental and biological samples. Anal. Chim. Acta. 662: 155–162.

Berton, P., E.M. Martinis, L.D. Martinez and R.G. Wuilloud. 2009. Room temperature ionic liquid-based microextraction for vanadium species separation and determination in water samples by electrothermal atomic absorption spectrometry. Anal. Chim. Acta. 640: 40–46.

Berzas Nevado, J.J., R.C. Rodríguez Martín-Doimeadios, F.J. Guzmán Bernardo and M. Jiménez Moreno. 2005. Determination of mercury species in fish reference materials by gas chromatography-atomic fluorescence detection after closed-vessel microwave-assisted extraction. J. Chromatogr. A 1093: 21–28.

Berzas Nevado, J.J., R.C. Rodríguez Martín-Doimeadios, F.J. Guzmán Bernardo and M. Jiménez Moreno. 2008. Determination of monomethylmercury in low- and high-polluted sediments by microwave extraction and gas chromatography with atomic fluorescence detection. Anal. Chim. Acta. 608: 30–37.

Berzas Nevado, J.J., R.C. Rodríguez Martín-Doimeadios, E.M. Krupp, F.J. Guzmán Bernardo, N. Rodríguez Fariñas, M. Jiménez Moreno, D. Wallace and M.J. Patiño Ropero. 2011. Comparison of gas chromatographic hyphenated techniques for mercury speciation analysis. J. Chromatogr. A 1218: 4545–4551.

Bezerra, M.D., M.A.Z. Arruda and S.L.C. Ferreira. 2005. Cloud point extraction as a procedure of separation and pre-concentration for metal determination using spectroanalytical techniques: A review. Appl. Spectros. Rev. 40: 269–299.

Bidari, A., P. Hemmatkhah, S. Jafarvand, M.R.M. Hosseini and Y. Assadi. 2008. Selenium analysis in water samples by dispersive liquid-liquid microextraction based on piazselenol formation and GC–ECD. Microchim. Acta. 163: 243–249.

Bierla, K., M. Dernovics, V. Vacchina, J. Szpunar, G. Bertin and R. Łobinski. 2008. Determination of selenocysteine and selenomethionine in edible animal tissues by 2D size-exclusion reversed-phase HPLC-ICP-MS following carbamidomethylation and proteolytic extraction. Anal. Bioanal. Chem. 390: 1789–1798.

Bin, H. and J. Guibin. 1999. Analysis of organomercuric species in soils from orchards and wheat fields by capillary gas chromatography on-line coupled with atomic absorption spectrometry after *in situ* hydride generation and headspace solid phase microextraction. Fresenius´ J. Anal. Chem. 365: 615–618.

Bin, H., J. Gui-Bin and N. Zhe-Ming. 1998. Determination of methylmercury in biological samples and sediments by capillary gas chromatography coupled with atomic absorption spectrometry after hydride derivatization and solid phase microextraction. J. Anal. At. Spectrom. 13: 1141–1144.

Bird, S.M., P.C. Uden, J.F. Tyson, E. Block and E. Denoyer. 1997. Speciation of selenoamino acids and organoselenium compounds in seleniumenriched yeast using high-performance liquid chromatography–inductively coupled plasma mass spectrometry. J. Anal. At. Spectrom. 12: 785–788.

Birjandi, A.P., A. Bidari, A.B.F. Rezaei, M.R.M. Hosseini and Y. Assadi. 2008. Speciation of butyl and phenyltin compounds using dispersive liquid–liquid microextraction and gas chromatography-flame photometric detection. J. Chromatog. A 1193: 19–25.

Bloom, N.S. 1989. Determination of picogram levels of methylmercury by aqueous phase ethylation, followed by cryogenic gas-chromatography with cold vapor atomic fluorescence detector. Can. J. Fish. Aquat. Sci. 46: 1131–1140.

Bloom, N.S., A.K. Grout and E.M. Prestbo. 2005. Development and complete validation of a method for the determination of dimethyl mercury in air and other media. Anal. Chim. Acta. 546: 92–101.

Bohari, Y., G. Lobos, H. Pinochet, F. Pannier, A. Astruc and M. Potin-Gautier. 2002. Speciation of arsenic in plants by HPLC-HG-AFS: extraction optimisation on CRM materials and application to cultivated samples. J. Environ. Monit. 4: 596–602.

Bowles, K.C., M.D. Tiltman, S.C. Apte, L.T. Hales and J. Kalman. 2004. Determination of butyltins in environmental samples using sodium tetraethylborate derivatisation: characterization and minimisation of interferences. Anal. Chim. Acta. 509: 127–135.

Bowman, K.L. and C.R. Hammerschmidt. 2011. Extraction of monomethylmercury from seawater for low-femtomolar determination. Limnol. Oceanogr.: Methods. 9: 121–128.

Braman, R.S. and C.C. Foreback. 1973. Methylated Forms of Arsenic in the Environment. Sci. 182: 1247–1249.

Bravo, M., G. Lespes, I. De Gregori, H. Pinochet and M. Potin-Gautier. 2004. Identification of sulfur interferences during organotin determination in harbour sediment samples by sodium tetraethyl borate ethylation and gas chromatography-pulsed flame photometric detection. J. Chromatog. A 1046: 217–224.

Bravo-Sánchez, L.R., J. Ruiz-Encinar, J.I. Fidalgo-Martínez and A. Sanz-Medel. 2004. Mercury speciation analysis in sea water by solid phase microextraction–gas chromatography–inductively coupled plasma mass spectrometry using ethyl and propyl derivatization. Matrix effects evaluation. Spectrochim. Acta Part B. 59: 59–66.

Bryszewska, M.A., W. Ambroziak, J. Rudzinski and D.J. Lewis. 2005. Characterisation of selenium compounds in rye seedling biomass using ^{75}Se-labelling/SDS-PAGE separation/gamma-scintillation counting, and HPLC-ICP-MS analysis of a range of enzymatic digests. Anal. Bioanal. Chem. 382: 1279–1287.

Bueno, M. and M. Potin-Gautier. 2002. Solid-phase extraction for the simultaneous preconcentration of organic (selenocystine) and inorganic [Se(IV), Se(VI)] selenium in natural waters. J. Chromatog. A. 963: 185–193.

Bueno, M. and F. Pannier. 2009. Quantitative analysis of volatile selenium metabolites in normal urine by headspace solid phase microextraction gas chromatography–inductively coupled plasma mass spectrometry. Talanta. 78: 759–763.

Burguera, J.L. and M. Burguera. 2004. Analytical applications of organized assemblies for on-line spectrometric determinations: present and future. Talanta. 64: 1099–1108.

Büyüktiryaki, S., R. Say, A. Denizli and A. Ersöz. 2007. Mimicking receptor for methylmercury preconcentration based on ion-imprinting. Talanta. 71: 699–705.

Cabañero Ortiz, A.I., Y. Madrid Albarrán and C. Cámara Rica. 2002. Evaluation of different sample pre-treatment and extraction procedures for mercury speciation in fish samples. J. Anal. At. Spectrom. 17: 1595–1601.

Cabañero, A.I., Y. Madrid and C. Cámara. 2005. Enzymatic probe sonication extraction of Se in animal-based food samples: a new perspective on sample preparation for total and Se speciation analysis. Anal. Bioanal. Chem. 381: 373–379.

Cai, X.J., E. Block, P.C. Uden, B.D. Quimby and J.J. Sullivan. 1995. Allium chemistry-Identification of natural abundance organoselenium compounds in human breath after ingestion of garlic using gas-chromatography with atomic emission detection. J. Agric. Food Chem. 43: 1751–1753.

Cai, Y. and J.M. Bayona. 1995. Determination of methylmercury in fish and river water samples using *in situ* sodium tetraethylborate derivatization following by solid-phase microextraction and gas chromatography-mass spectrometry. J. Chromatog. A 696: 113–122.

Cai, Y., S. Rapsomanikis and M.O. Andreae. 1993. Determination of butyltin compounds in river sediment samples by gas chromatography–atomic absorption spectrometry following *in situ* derivatization with sodium tetraethylborate. J. Anal. At. Spectrom. 8: 119–125.

Cai, Y., R. Alzaga and J.M. Bayona. 1994. *In situ* derivatization and supercritical fluid extraction for the simultaneous determination of butyltin and phenyltin compounds in sediment. Anal. Chem. 66: 1161–1167.

Cai, Y., R. Jaffé, Z. Alli and R.D. Jones. 1996. Determination of organomercury compounds in aqueous samples by capillary gas chromatography-atomic fluorescence spectrometry following solid-phase extraction. Anal. Chim. Acta. 334: 251–259.

Cai, Y., S. Monsalud, K.G. Furton, R. Jaffé and R.D. Jones. 1998a. Determination of methylmercury in fish and aqueous samples using solid-phase microextraction followed by gas chromatography-atomic fluorescence spectrometry. Appl. Organometal. Chem. 12: 565–569.

Cai, Y., M. Ábalos and J.M. Bayona. 1998b. Effects of complexing agents and acid modifiers on the supercritical fluid extraction of native phenyl- and butyl-tins from sediment. Appl. Organomet. Chem. 12: 577–584.

Cairns, W.R.L., M. Ranaldo, R. Hennebelle, C. Turetta, G. Capodaglio, C.F. Ferrari, A. Dommergue, P. Cescon and C. Barbante. 2008. Speciation analysis of mercury in seawater from the lagoon of Venice by on-line pre-concentration HPLC–ICP-MS. Anal. Chim. Acta. 622: 62–69.

Calvo-Fornieles, A., A. Garcıa de Torres, E. Vereda-Alonso, M.T. Siles-Cordero and J.M. Cano-Pavon. 2011. Speciation of antimony(III) and antimony(V) in seawater by flow injection solid phase extraction coupled with online hydride generation inductively coupled plasma mass spectrometry. J. Anal. At. Spectrom. 26: 1619–1626.

Campillo, N., N. Aguinaga, P. Viñas, I. López-García and M. Hernández-Córdoba. 2004. Speciation of organotin compounds in waters and marine sediments using purge-and-trap capillary gas chromatography with atomic emission detection. Anal. Chim. Acta. 525: 273–280.

Campillo, N., N. Aguinaga, P. Viñas, I. López-García and M. Hernández-Córdoba. 2005. Gas chromatography with atomic emission detection for dimethylselenide and dimethyldiselenide determination in waters and plant materials using a purge-and-trap preconcentration system. J. Chromatogr. A 1095: 138–144.

Campillo, N., R. Peñalver, M. Hernández-Córdoba, C. Pérez-Sirvent and M.J. Martínez-Sánchez. 2007. Comparison of two derivatizing agents for the simultaneous determination of selenite and organoselenium species by gas chromatography and atomic emission detection after preconcentration using solid-phase microextraction. J. Chromatog. A 1165: 191–199.

Capelo, J.L., P. Ximénez-Embún, Y. Madrid-Albarrán and C. Cámara. 2004. Enzymatic probe sonication: enhancement of protease-catalyzed hydrolysis of selenium bound to proteins in yeast. Anal. Chem. 76: 233–237.

Capelo, J.L., C. Maduro and C. Viena. 2005. Discussion of parameters associated with the ultrasonic solid-liquid extraction for elemental analysis (total content) by electrothermal atomic absorption spectrometry. An overview. Ultrasounds Sonochem. 12: 225–232.

Carabias-Martínez, R., E. Rodríguez-Gonzalo, B. Moreno-Cordero, J.L. Pérez-Pavón, C. García-Pinto and E. Fernández-Laespada. 2000. Surfactant cloud point extraction and preconcentration of organic compounds prior to chromatography and capillary electrophoresis. J. Chromatog. A 902: 251–265.

Carbonell, G., J.C. Bravo, C. Fernández and J.V. Tarazona. 2009. A new method for total mercury and methyl mercury analysis in muscle of seawater fish. Bull. Environ. Contam. Toxicol. 83: 210–213.

Cardellicchio, N., S. Giandomenico, A. Decataldo and A. Di Leo. 2001. Speciation of butyltin compounds in marine sediments with headspace solid phase microextraction and gas chromatography—mass spectrometry. Fresenius´ J. Anal. Chem. 369: 510–515.

Carlier-Pinasseau, C.C., G. Lespes and M. Astruc. 1996a. Determination of butyltin and phenyltin by GC-FPD following ethylation by NaBEt$_4$, Appl. Oraganometal. Chem. 10: 505–512.

Carlier-Pinasseau, C., A. Astruc, G. Lespes and M. Astruc. 1996b. Determination of butyl- and phenyltin compounds in biological material by gas chromatography-flame photometric detection after ethylation with sodium tetraethylborate. J. Chromatogr. A 750: 317–325.

Carlier-Pinasseau, C.C., G. Lespes and M. Astruc. 1997. Validation of organotin compound determination in environmental samples using NaBEt$_4$ ethylation and GC-FPD. Environ. Tech. 18: 1179–1186.

Caro, E., R.M. Marcé, F. Borrull, P.A.G. Cormack and D.C. Sherrington. 2006. Application of molecularly imprinted polymers to solid-phase extraction of compounds from environmental and biological samples. Trends Anal. Chem. 25: 143–154.

Carpinteiro-Botana, J., R. Rodil-Rodríguez, A.M. Carro-Díaz, R.A. Lorenzo-Ferreira, R. Cela-Torrijos and I. Rodríguez-Pereiro. 2002a. Fast and simultaneous determination of tin and mercury species using SPME, multicapillary gas chromatography and MIP-AES detection. J. Anal. At. Spectrom. 17: 904–907.

Carpinteiro-Botana, J., I. Rodríguez-Pereiro and R. Cela-Torrijos. 2002b. Rapid determination of butyltin species in water samples by multicapillary gas chromatography with atomic emission detection following headspace solid-phase microextraction. J. Chromatog. A 963: 195–203.

Carpinteiro, J., I. Rodríguez and R. Cela. 2001. Simultaneous determination of butyltin and phenyltin species in sediments using ultrasound-assisted leaching. Fresenius' J. Anal. Chem. 370: 872–877.

Carrasco, L., S. Díez and J.M. Bayona. 2007. Methylmercury determination in biota by solid-phase microextraction Matrix effect evaluation. J. Chromatog. A 1174: 2–6.

Carro, A.M., I. Neira, R. Rodil and R.A. Lorenzo. 2002. Speciation of mercury compounds by gas chromatography with atomic emission detection. Simultaneous optimization of a headspace solid-phase microextraction and derivatization procedure by use of chemometric techniques. Chromatographia. 56: 733–738.

Castillo, A., P. Rodríguez-González, G. Centineo, A.F. Roig-Navarro and J.I. García Alonso. 2010. Multiple spiking species-specific isotope dilution analysis by molecular mass spectrometry: simultaneous determination of inorganic mercury and methylmercury in fish tissues. Anal. Chem. 82: 2773–2783.

Centineo, G., E. Blanco-González and A. Sanz-Medel. 2004. Multielemental speciation analysis of organometallic compounds of mercury, lead and tin in natural water samples by headspace-solid phase microextraction followed by gas chromatography—mass spectrometry. J. Chromatog. A 1034: 191–197.

Centineo, G., E. Blanco-González, J.I. García-Alonso and A. Sanz-Medel. 2006. Isotope dilution SPME GC/MS for the determination of methylmercury in tuna fish samples. J. Mass Spectrom. 41: 77–83.

Ceulemans, M. and F.C. Adams. 1996. Integrated sample preparation and speciation analysis for the simultaneous determination of methylated species of tin, lead and mercury in water by purge-and-trap injection-capillary gas chromatography-atomic emission spectrometry. J. Anal. At. Spectrom. 11: 201–206.

Chamsaz, M., M.H. Arbab-Zavar and S. Nazari. 2003. Determination of arsenic by electrothermal atomic absorption spectrometry using headspace liquid phase microextraction after *in situ* hydride generation. J. Anal. At. Spectrom. 18: 1279–1282.

Chang, L.-F., S.-J. Jiang and A.C. Sahayam. 2007. Speciation analysis of mercury and lead in fish samples using liquid chromatography–inductively coupled plasma mass spectrometry. J. Chromatogr. A 1176: 143–148.

Chao, W.-S. and S.-J. Jiang. 1998. Determination of organotin compounds by liquid chromatography inductively coupled plasma mass spectrometry with a direct injection nebulizer. J. Anal. At. Spectrom. 13: 1337–1341.

Chau, Y.K., F. Yang and M. Brown. 1995. Supercritical fluid extraction of butyltin compounds from sediment. Anal. Chim. Acta. 304: 85–89.

Chen, B., B. Hu and M. He. 2006. Cloud point extraction combined with electrothermal vaporization inductively coupled plasma mass spectrometry for the speciation of inorganic selenium in environmental water samples. Rapid Commun. Mass Spectrom. 20: 2894–2900.

Chen, J.-H., K.-E. Wang and S.-J. Jiang. 2007. Determination of iodine and bromine compounds in foodstuffs by CE-inductively coupled plasma MS. Electrophoresis. 28: 4227–4232.

Chen, S.-S., S.-S. Chou and D.-F. Hwang. 2004. Determination of methylmercury in fish using focused microwave digestion following by Cu^{2+} addition, sodium tetrapropylborate derivatization, n-heptane extraction, and gas chromatography–mass spectrometry. J. Chromatogr. A 1024: 209–215.

Chen, S., X. Zhan, D. Lu, C. Liu and L. Zhu. 2009a. Speciation analysis of inorganic arsenic in natural water by carbon nanofibers separation and inductively coupled plasma mass spectrometry determination. Anal. Chim. Acta. 634: 192–196.

Chen, D., C. Huang, M. He and B. Hu. 2009b. Separation and preconcentration of inorganic arsenic species in natural water samples with 3-(2-aminoethylamino) propyltrimethoxysilane modified ordered mesoporous silica micro-column and their determination by inductively coupled plasma optical emission spectrometry. J. Hazard. Mat. 164: 1146–1151.

Chen, J., H. Chen, X. Jin and H. Chen. 2009c. Determination of ultra-trace amount methyl-, phenyl- and inorganic mercury in environmental and biological samples by liquid chromatography with inductively coupled plasma mass spectrometry after cloud point extraction preconcentration. Talanta. 77: 1381–1387.

Chen, H., J. Chen, X. Jin and D. Wei. 2009d. Determination of trace mercury species by high performance liquid chromatography–inductively coupled plasma mass spectrometry after cloud point extraction. J. Hazard. Mat. 172: 1282–1287.

Chiarotti, M. 1993. Overview on extraction procedures. Forensic Sci. Int. 63: 161–170.

Chi-Chi, C. and L. Maw-Rong. 2005. Determination of organotin compounds in water by headspace solid phase microextraction with gas chromatography—mass spectrometry. J. Chromatog. A 1064: 1–8.

Chiou, C.-S., S.-J. Jiang and K.S.K. Danadurai. 2001. Determination of mercury compounds in fish by microwave-assisted extraction and liquid chromatography-vapor generation-inductively coupled plasma mass spectrometry. Spectrochim. Acta. B 56: 1133–1142.

Chiron, S., S. Roy, R. Cottier and R. Jeannot. 2000. Speciation of butyl- and phenyltin compounds in sediments using pressurized liquid extraction and liquid chromatography–inductively coupled plasma mass spectrometry. J. Chromatogr. A 879: 137–145.

Colombini, V., C. Bancon-Montigny, L. Yang, P. Maxwell, R.E. Sturgeon and Z. Mester. 2004. Headspace single-drop microextraction for the detection of organotin compounds. Talanta. 63: 555–560.

Craig, P.J., R.J. Dewick and J.T. Elteren. 1995. Use of sodium tetraethylborate for the analysis of trimethyllead species in artificial rainwater and a natural road dust sample. Fresenius' J. Anal. Chem. 351: 467–470.

Craig, P.J., S.N. Forster, R.O. Jenkins and D. Miller. 1999. An analytical method for the detection of methylantimony species in environmental matrices: methylantimony levels in some UK plant material. Analyst. 124: 1243–1248.

Crnoja, M., C. Haberhauer-Troyer, E. Rosenberg and M. Grasserbauer. 2001. Determination of Sn- and Pb-organic compounds by solid-phase microextraction-gas chromatography-atomic emission detection (SPME-GC-AED) after *in situ* propylation with sodium tetrapropylborate. J. Anal. At. Spectrom. 16: 1160–1166.

Cuderman, P. and V. Stibilj. 2010. Stability of Se species in plant extracts rich in phenolic substances. Anal. Bioanal. Chem. 396: 1433–1439.

Cuderman, P., I. Kreft, M. Germ, M. Kovačevič and V. Stibilj. 2008. Selenium species in selenium-enriched and drought-exposed potatoes. J. Agric. Food Chem. 56: 9114–9120.

Cukrowska, E., L. Chimuka, H. Nsengimana and V. Kwaramba. 2004. Application of supported liquid membrane probe for extraction and preconcentration of organotin compounds from environmental water samples. Anal. Chim. Acta. 523: 141–147.

Cukrowska, E.M., H. Nsengimana and L. Chimuka. 2007. Speciation of alkyllead in aqueous samples with application of liquid membrane probe for extraction and preconcentration. J. Sep. Sci. 30: 2754–2759.

Dachs, J., R. Alzaga, J.M. Bayona and Ph. Quevauviller. 1994. Development of a supercritical fluid extraction procedure for tributyltin determination in sediments. Anal. Chim. Acta. 286: 319–327.

Dadfarnia, S. and A.M.H. Shabani. 2010. Recent development in liquid phase microextraction for determination of trace level concentration of metals—A review. Anal. Chim. Acta. 658: 107–119.

Dagnac, T., A. Padró, R. Rubio, G. Rauret. 1998. Optimisation of the extraction of arsenic species from mussels with low power focused microwaves by applying a Doehlert design. Anal. Chim. Acta. 364: 19–30.

Dagnac, T., A. Padró, R. Rubio G. Rauret. 1999. Speciation of arsenic in mussels by the coupled system liquid chromatography—UV irradiation—hydride generation-inductively coupled plasma mass spectrometry. Talanta. 48: 763–772.

Dash, K., S. Thangavel, N.V. Krishnamurthy, S.V. Rao, D. Karunasagar and J. Arunachalam. 2005. Ultrasound-assisted analyte extraction for the determination of sulfate and elemental sulfur in zinc sulfide by different liquid chromatography techniques. Analyst 130: 498–501.

David, F. and P. Sandra. 2007. Stir bar sorptive extraction for trace analysis. J. Chromatog. A 1152: 54–69.

Davis, W.C., S.S. Vander Pol, M.M. Schantz, S.E. Long, R.D. Day and S.J. Christopher. 2004. An accurate and sensitive method for the determination of methylmercury in biological specimens using GC-ICP-MS with solid phase microextraction. J. Anal. At. Spectrom. 19: 1546–1551.

Day, J.A., S.S. Kannamkumarath, E.G. Yanes, M. Montes-Bayón and J.A. Caruso. 2002. Chiral speciation of Marfey's derivatized DL-selenomethionine using capillary electrophoresis with UV and ICP-MS detection. J. Anal. At. Spectrom. 17: 27–31.

de la Calle-Guntiñas, M.B., M. Ceulemans, C. Witte, R. Łobiński and F.C. Adams. 1995. Evaluation of a purge-and-trap injection system for capillary gas chromatography-microwave induced plasma-atomic emission spectrometry for the determination of volatile selenium compounds in water. Microchim. Acta. 120: 73–82.

de la Calle-Guntiñas, M.B., F. Laturnus and F.C. Adams. 1999. Purge and trap thermal desorption device for the determination of dimethylselenide and dimethyldiselenide. Fresenius′ J. Anal. Chem. 364: 147–153.

de Smaele, T., L. Moens, P. Sandra and R. Dams. 1999. Determination of organometallic compounds in surface water and sediment samples with SPME-CGC-ICPMS. Mikrochim. Acta. 130: 241–251.

de Wuilloud, J.C.A., R.G. Wuilloud, R.A. Olsina and L.D. Martinez. 2002. Separation and preconcentration of inorganic and organomercury species in water samples using a selective reagent and an anion exchange resin and determination by flow injection-cold vapor atomic absorption spectrometry. J. Anal. At. Spectrom. 17: 389–394.

Delgado, A., A. Prieto, O. Zuloaga, A. de Diego and J.M. Madariaga. 2007. Production of artifact methylmercury during the analysis of certified reference sediments: Use of ionic exchange in the sample treatment step to minimise the problem. Anal. Chim. Acta. 582: 109–115.

Demuth, N. and K.G. Heumann. 2001. Validation of methylmercury determinations in aquatic systems by alkyl derivatization methods for GC analysis using ICP-IDMS. Anal. Chem. 73: 4020–4027.

Devos, C., K. Sandra and P. Sandra. 2002. Capillary gas chromatography inductively coupled plasma mass spectrometry (CGC-ICPMS) for the enantiomeric analysis of D,L-selenomethionine in food supplements and urine. J. Phar. Biomed. Anal. 27: 507–514.

Devos, C., M. Vliegen, B. Willaert, F. David, L. Moens and P. Sandra. 2005. Automated headspace-solid-phase micro extraction–retention time locked-isotope dilution gas chromatography–mass spectrometry for the analysis of organotin compounds in water and sediment samples. J. Chromatog. A 1079: 408–414.

Díaz Huerta, V., L. Hinojosa Reyes, J.M. Marchante-Gayón, M.L. Fernández Sánchez and A. Sanz-Medel. 2003. Total determination and quantitative speciation analysis of selenium in yeast and wheat flour by isotope dilution analysis ICP-MS. J. Anal. At. Spectrom. 18: 1243–1247.

Dietz, C., T. Pérez-Corona, Y. Madrid-Albarrán and C. Cámara. 2003. SPME for on-line volatile organo-selenium speciation. J. Anal. At. Spectrom. 18: 467–473.

Dietz, C., J. Sanz-Landaluze, P. Ximénez-Embún, Y. Madrid-Albarrán and C. Cámara. 2004. SPME–multicapillary GC coupled to different detection systems and applied to volatile organo-selenium speciation in yeast. J. Anal. At. Spectrom. 19: 260–266.

Dietz, C., J. Sanz, E. Sanz, R. Muñoz-Olivas and C. Cámara. 2007. Current perspectives in analyte extraction strategies for tin and arsenic speciation. J. Chromatogr. A 1153: 114–129.

Díez, S. and J.M. Bayona. 2002. Determination of methylmercury in human hair by ethylation followed by headspace solid-phase microextraction–gas chromatography–cold-vapour atomic fluorescence spectrometry. J. Chromatog. A. 963: 345–351.

Dimitrakakis, E., C. Haberhauer-Troyer, M. Ochsenkühn-Petropoulou and E. Rosenberg. 2004. Solid-phase microextraction–capillary gas chromatography combined with microwave-induced plasma atomic-emission spectrometry for selenite determination. Anal. Bioanal. Chem. 379: 842–848.

dos Santos, J.S., M. de la Guárdia, A. Pastor and M.L.P. dos Santos. 2009. Determination of organic and inorganic mercury species in water and sediment samples by HPLC on-line coupled with ICP-MS. Talanta. 80: 207–211.

Dowson, P.H., J.M. Bubb and J.N. Lester. 1992. Organotin distribution in sediments and waters of selected east coast estuaries in the UK. Marine Pollut. Bull. 24: 492–498.

Duan, J. and B. Hu. 2009. Separation and determination of seleno amino acids using gas chromatography hyphenated with inductively coupled plasma mass spectrometry after hollow fiber liquid phase microextraction. J. Mass. Spectrom. 44: 605–612.

Duan, J., X. Li, C. Yu and B. Hu. 2009. Headspace stir bar sorptive extraction combined with GC-ICP-MS for the speciation of dimethylselenide and dimethyldiselenide in biological samples. J. Anal. At. Spectrom. 24: 297–303.

Duester, L., R.A. Diaz-Bone, J. Kosters and A.V. Hirner. 2005. Methylated arsenic, antimony and tin species in soils. J. Environ. Monitor. 7: 1186–1193.

Dufailly, V., L. Noël, J.-M. Frémy, D. Beauchemin and T. Guérin. 2007. Optimisation by experimental design of an IEC/ICP-MS speciation method for arsenic in seafood following microwave assisted extraction. J. Anal. At. Spectrom. 22: 1168–1173.

Dumont, E., K. de Cremer, M. van Hulle, C.C. Chery, F. Vanhaecke and R. Cornelis. 2004. Separation and detection of Se-compounds by ion impairing assisted hydride generation atomic fluorescence spectrometry. J. Anal. At. Spectrom. 19: 167–171.

Dzurko, M., D. Foucher and H. Hintelmann. 2009. Determination of compound-specific Hg isotope ratios from transient signals using gas chromatography coupled to multicollector inductively coupled plasma mass spectrometry (MC-ICP/MS). Anal. Bioanal. Chem. 393: 345–355.

Egressy-Molnár, O., A. Vass, A. Németh, J.F. García-Reyes and M. Dernovics. 2011. Effect of sample preparation methods on the D,L-enantiomer ratio of extracted selenomethionine. Anal. Bioanal. Chem. 401: 373–380.

Eiden, R., H.F. Schöler and M. Gastner. 1998. *In situ* ethylation–purge and programmed-temperature-vaporizer cold trapping–gas chromatography–mass spectrometry as an automated technique for the determination of methyl- and butyltin compounds in aqueous samples. J. Chromatog. A 809: 151–157.

Eisert, R. and J. Pawliszyn. 1997. Automated in-tube solid-phase microextraction coupled to high-performance liquid chromatography. Anal. Chem. 69: 3140–3147.

Elçi, L., Ü. Divrikli and M. Soylak. 2008. Inorganic arsenic speciation in various water samples with GFAAS using coprecipitation. Int. J. Environ. Anal. Chem. 88: 711–723.

Ellwood, M.J. and W.A. Maher. 2002. An automated hydride generation-cryogenic trapping-ICP-MS system for measuring inorganic and methylated Ge, Sb and As species in marine and fresh waters. J. Anal. At. Spectrom. 17: 197–203.

El-Shahawi, M.S. and M.A. El-Sonbati. 2005. Retention profile, kinetics and sequential determination of selenium(IV) and (VI) employing 4,4'-dichlorodithizone immobilized-polyurethane foams. Talanta. 67: 806–815.

El-Shahawi, M.S., S. Mohammad, A.M. Othman, A.S. Bashammakh and M.A. El-Sonbati. 2006. Chemical equilibria and sequential extractive spectrophotometric determination of selenium(IV) and (VI) using the chromogenic reagent 4,4'-dichlorodithizone. Int. J. Environ. Anal. Chem. 86: 941–954.

Emteborg, H., D.C. Baxter, M. Sharp and W. Frech. 1995. Evaluation, mechanism and application of solid-phase extraction using a dithiocarbamate resin for the sampling and determination of mercury species in humic-rich natural waters. Analyst. 120: 69–77.

Emteborg, H., F. Odman, L. Karlsson, L. Mathiasson, W. Frech and D.C. Baxter. 1996. Determination of methylmercury in sediments using supercritical fluid extraction and gas chromatography coupled with microwave-induced plasma atomic emission spectrometry. Analyst 121: 19–29.

EPA Method 1630. 2001. Methyl mercury in water by distillation, aqueous ethylation, purge and trap, and CVAFS; U.S. Environmental Protection Agency. Washington, DC.

Erdoğan, H., O. Yalçınkaya and A.R. Türker. 2011. Determination of inorganic arsenic species by hydride generation atomic absorption spectrometry in water samples after preconcentration/separation on nano ZrO_2/B_2O_3 by solid phase extraction. Desalination. 280: 391–396.

Falter, R. and H.F. Schöler. 1995. Determination of mercury species in natural waters at picogram level with on-line RP C_{18} preconcentration and HPLC-UV-PCO-CVAAS. Fresenius' J. Anal. Chem. 353: 34–38.

Falter, R. and G. Ilgen. 1997. Determination of trace amounts of methylmercury in sediment and biological tissue by using water vapor distillation in combination with RP C_{18} preconcentration and HPLC-HPF/HHPN-ICP-MS. Fresenius' J. Anal. Chem. 358: 401–406.

Fang, Q., M. Du and C.W. Huie. 2001. Online incorporation of cloud point extraction to flow injection analysis. Anal. Chem. 73: 3502–3505.

Fang, Y., Y. Zhang, B. Catron, Q. Chan, Q. Hu and J.A. Caruso. 2009. Identification of selenium compounds using HPLC-ICPMS and nano-ESI-MS in selenium-enriched rice *via* foliar application. J. Anal. At. Spectrom. 24: 1657–1664.

Fang, Z., M. Sperling and B. Welz. 1991. Flame atomic-absorption spectrometric determination of lead in biological samples using a flow-injection system with online pre-concentration by co-precipitation without filtration. J. Anal. At. Spectrom. 6: 301–306.

Fan, Z. 2005. Speciation analysis of Antimony (III) and antimony (V) by flame atomic absorption spectrometry after separation/preconcentration with cloud point extraction. Microchim. Acta. 152: 29–33.

Fan, Z. 2007. Determination of antimony(III) and total antimony by single-drop microextraction combined with electrothermal atomic absorption spectrometry. Anal. Chim. Acta. 585: 300–304.

Fan, Z. and X. Liu. 2008. Determination of methylmercury and phenylmercury in water samples by liquid–liquid–liquid microextraction coupled with capillary electrophoresis. J. Chromatog. A 1180: 187–192.

Fernández, R.G., M.M. Bayón, J.I.G. Alonso and A. Sanz-Medel. 2000. Comparison of different derivatization approaches for mercury speciation in biological tissues by gas chromatography/inductively coupled plasma mass spectrometry. J. Mass Spectrom. 35: 639–646.

Fernández-Escobar, I. and J.M. Bayona. 1997. Supercritical fluid extraction of priority organotin contaminants from biological matrices. Anal. Chim. Acta. 355: 269–276.

Ferri, T. and P. Sangiorgio. 1996. Determination of selenium speciation in river waters by adsorption on iron(III)-Chelex-100 resin and differential pulse cathodic stripping voltammetry. Anal. Chim. Acta. 321: 185–193.

Filippelli, M. 1994. Methylmercury determination by purge and trap–GC–FTIR–AAS after NaBH$_4$ derivatization of an environmental thiosulfate extract. Appl. Organomet. Chem. 8: 687–691.

Filippelli, M. and F. Baldi. 1993. Alkylation of ionic mercury to methylmercury and dimethylmercury by methylcobalamin: Simultaneous determination by purge-and-trap GC in line with FTIR. Appl. Organomet. Chem. 7: 487–493.

Filippelli, M., F. Baldi, F.E. Brinckman and G.J. Olson. 1992. Methylmercury determination as volatlle methylmercury hydride by purge and trap gas chromatography in line with Fourier transform infrared spectroscopy. Environ. Sci. Technol. 26: 1457–1460.

Fisher, J.A., M.J. Scarlett and A.D. Stott. 1997. Accelerated solvent extraction: An evaluation for screening of soils for selected U.S. EPA semivolatile organic priority pollutants. Environ. Sci. Technol. 31: 1120–1127.

Flores, M., M. Bravo, H. Pinochet, P. Maxwell and Z. Mester. 2011. Tartaric acid extraction of organotin compounds from sediment samples Microchem. J. 98: 129–134.

Foster, S., W. Maher, F. Krikowa, K. Telford and M. Ellwood. 2005. Observations on the measurement of total antimony and antimony species in algae, plant and animal tissues. J. Environ. Monit. 7: 1214–1219.

Foy, G.P. and G.E. Pacey. 2000. Specific extraction of chromium(VI) using supercritical fluid extraction. Talanta. 51: 339–347.

Foy, G.P. and G.E. Pacey. 2003. Supercritical fluid extraction of mercury species. Talanta. 61: 849–853.

Fragueiro, M.S., F. Alava-Moreno, I. Lavilla and C. Bendicho. 2000. Determination of tetraethyllead by solid phase microextraction-thermal desorption-quartz furnace atomic absorption spectrometry. J. Anal. At. Spectrom. 15: 705–709.

Fragueiro, S., I. Lavilla and C. Bendicho. 2004. Direct coupling of solid phase microextraction and quartz tube-atomic absorption spectrometry for selective and sensitive determination of methylmercury in seafood: an assessment of chloride and hydride generation. J. Anal. At. Spectrom. 19: 250–254.

174 *Speciation Studies in Soil, Sediment and Environmental Samples*

Gallagher, P.A., X. Wei, J.A. Shoemaker, C.A. Brockhoff and J.T. Creed. 1999. Detection of arsenosugars from kelp extracts via IC-electrospray ionization-MS-MS and IC membrane hydride generation ICP-MS. J. Anal. At. Spectrom. 14: 1829–1834.

Gallagher, P.A., J.A. Shoemaker, X. Wei, C.A. Brockhoff-Schwegel and J.T. Creed. 2001. Extraction and detection of arsenicals in seaweed via accelerated solvent extraction with ion chromatographic separation and ICP–MS detection. Fresenius J. Anal. Chem. 369: 71–80.

Gallagher, P.A., S. Murray, X. Wei, C.A. Schwegel and J.T. Creed. 2002. An evaluation of sample dispersion media used with accelerated solvent extraction for the extraction and recovery of arsenicals from LFB and DORM-2. J. Anal. At. Spectrom. 17: 581–586.

Gallego-Gallegos, M., R. Muñoz-Olivas, A. Martin-Esteban and C. Cámara. 2005. Synthesis and evaluation of molecularly imprinted polymers for organotin compounds: a screening method for tributyltin detection in seawater. Anal. Chim. Acta. 531: 33–39.

Gallego-Gallegos, M., M. Liva, R. Muñoz-Olivas and C. Cámara. 2006a. Focused ultrasound and molecularly imprinted polymers: A new approach to organotin analysis in environmental samples. J. Chromatog. A 1114: 82–88.

Gallego-Gallegos, M., R. Muñoz-Olivas, C. Cámara, M.J. Mancheño and M.A. Sierra. 2006b. Synthesis of a pH dependent covalent imprinted polymer able to recognize organotin species. Analyst. 131: 98–105.

Gallego-Gallegos, M., M. Liva-Garrido, R. Muñoz-Olivas, P. Baravalle, C. Baggiani and C. Cámara. 2010. A new application of imprinted polymers: Speciation of organotin compounds. J. Chromatog. A 1217: 3400–3407.

Ganjali, M.R., H.R. Sobhi, H. Farahani, P. Norouzi, R. Dinarvand and A. Kashtiaray. 2010. Solid drop based liquid-phase microextraction. J. Chromatog. A 1217: 2337–2341.

Gao, E. and J. Liu. 2011. Rapid determination of mercury species in sewage sludge by high-performance liquid chromatography on-line coupled with cold-vapor atomic-fluorescence spectrometry after ultrasound-assisted extraction. Anal. Sci. 27: 637–641.

Gao, Z. and X. Ma. 2011. Speciation analysis of mercury in water samples using dispersive liquid–liquid microextraction combined with high-performance liquid chromatography. Anal. Chim. Acta. 702: 50–55.

Gaona, X. and M. Valiente. 2003. Stability study on a Westöö-based methodology to determine organomercury compounds in polluted soil samples. Anal. Chim. Acta. 480: 219–230.

García Alonso, J.I., J. Ruiz Encinar, P. Rodríguez González and A. Sanz-Medel. 2002. Determination of butyltin compounds in environmental samples by isotope-dilution GC–ICP–MS. Anal. Bioanal. Chem. 373: 432–440.

García Salgado, S., M.A. Quijano Nieto and M.M. Bonilla Simón. 2006. Determination of soluble toxic arsenic species in alga samples by microwave-assisted extraction and high performance liquid chromatography–hydride generation–inductively coupled plasma-atomic emission spectrometry. J. Chromatogr. A 1129: 54–60.

García-Pinto, C., J.L. Pérez-Pavón and B. Moreno-Cordero. 1992. Cloud point preconcentration and high-performance liquid chromatographic analysis with electrochemical detection. Anal. Chem. 64: 2334–2338.

Gbatu, T.P., K.L. Sutton and J.A. Caruso. 1999. Development of new SPME fibers by sol-gel technology for SPME-HPLC determination of organometals. Anal. Chim. Acta. 402: 67–79.

Geerdink, R.B., R. Breidenbach and O.J. Epema. 2007. Optimization of headspace solid-phase microextraction gas chromatography-atomic emission detection analysis of monomethylmercury. J. Chromatog. A 1174: 7–12.

Gerbersmann, C., M. Heisterkamp, F.C. Adams and J.A.C. Broekaert. 1997. Two methods for the speciation analysis of mercury in fish involving microwave-assisted digestion and gas chromatography-atomic emission spectrometry. Anal. Chim. Acta. 350: 273–285.

Ghambarian, M., M.R. Khalili-Zanjani, Y. Yamini, A. Esrafili and N. Yazdanfar. 2010. Preconcentration and speciation of arsenic in water specimens by the combination of solidification of floating drop microextraction and electrothermal atomic absorption spectrometry. Talanta. 81: 197–201.

Ghasemi, E., N.M. Najafi, F. Raofie and A. Ghassempour. 2010. Simultaneous speciation and preconcentration of ultra traces of inorganic tellurium and selenium in environmental samples by hollow fiber liquid phase microextraction prior to electrothermal atomic absorption spectroscopy determination. J. Hazard. Mat. 181: 491–496.

Ghasemi, E., M. Sillanpää and N.M. Najafi. 2011. Headspace hollow fiber protected liquid-phase microextraction combined with gas chromatography–mass spectroscopy for speciation and determination of volatile organic compounds of selenium in environmental and biological samples. J. Chromatog. A 1218: 380–386.

Gholivand, M.B., A. Babakhanian and E. Rafiee. 2008. Determination of Sn(II) and Sn(IV) after mixed micelle-mediated cloud point extraction using α-polyoxometalate as a complexing agent by flame atomic absorption spectrometry. Talanta. 76: 503–508.

Giergielewicz-Mozajska, H., L. Dabrowski and J. Namiesnik. 2001. Accelerated solvent extraction (ASE) in the analysis of environmental solid samples—Some aspects of theory and practice. Critical Rev. Anal. Chem. 31: 149–165.

Gil, S., S. Fragueiro, I. Lavilla and C. Bendicho. 2005. Determination of methylmercury by electrothermal atomic absorption spectrometry using headspace single-drop microextraction with *in situ* hydride generation. Spectrochim. Acta Part B. 60: 145–150.

Giral, M., G.J. Zagury, L. Deschênes and J.-P. Blouin. 2010. Comparison of four extraction procedures to assess arsenate and arsenite species in contaminated soils. Environ. Pollut. 158: 1890–1898.

Goenaga, H., G.O'Connor, M. Rayman, R. Wahlen, J. Entwisle, P. Norris, R. Hearn and T. Catterick. 2004. Selenium speciation analysis of selenium-enriched supplements by HPLC with ultrasonic nebulisation ICP-MS and electrospray MS/MS detection. J. Anal. At. Spectrom. 19: 1529–1538.

Goenaga Infante, H., G. O'Connor, M. Rayman, R. Wahlen, J.E. Spallholz, R. Hearn and T. Catterick. 2005. Identification of water-soluble gamma-glutamyl-Semethylselenocysteine in yeast-based selenium supplements by reversed-phase HPLC with ICP-MS and electrospray tandem MS detection. J. Anal. At. Spectrom. 20: 864–870.

Goenaga Infante, H., A. Arias Borrego, E. Peachey, R. Hearn, G. O'Connor, T. García Barrera and J.L. Gómez Ariza. 2009. Study of the effect of sample preparation and cooking on the selenium speciation of selenized potatoes by HPLC with ICP-MS and electrospray ionization MS/MS. J. Agric. Food Chem. 57: 38–45.

Gómez, M., C. Cámara, M.A. Palacios and A. López-Gonzálvez. 1997. Anionic cartridge preconcentrators for inorganic arsenic, monomethylarsonate and dimethylarsinate determination by on-line HPLC-HG-AAS. Fresenius´ J. Anal. Chem. 357: 844–849.

Gómez-Ariza, J.L., J.A. Pozas, I. Giráldez and E. Morales. 1999a. Comparison of three derivatization reagents for the analysis of Se(IV) based on piazselenol formation and gas chromatography-mass spectrometry. Talanta. 49: 285–292.

Gómez-Ariza, J.L., J.A. Pozas, I. Giráldez and E. Morales. 1999b. Use of solid phase extraction for speciation of selenium compounds in aqueous environmental samples. Analyst. 124: 75–78.

Gómez-Ariza, J.L., D. Sánchez-Rodas, I. Giráldez and E. Morales. 2000. Comparison of biota sample pretreatments for arsenic speciation with coupled HPLC-HG-ICP-MS. Analyst 125: 401–407.

Gómez-Ariza, J.L., M.A. Caro de la Torre, I. Giráldez, D. Sánchez-Rodas, A. Velasco and E. Morales. 2002. Pretreatment procedure for selenium speciation in shellfish using high-performance liquid chromatography-microwave-assisted digestion-hydride generation-atomic fluorescence spectrometry. Appl. Organometal. Chem. 16: 265–270.

Gómez-Ariza, J.L., V. Bernal-Daza and M.J. Villegas-Portero. 2007. First approach of a methodological set-up for selenomethionine chiral speciation in breast and formula milk using high-performance liquid chromatography coupled to atomic fluorescence spectroscopy. Appl. Organometal. Chem. 21: 434–440.

Gómez-Jacinto, V., A. Arias-Borrego, T. García-Barrera, I. Garbayo, C. Vilchez and J.L. Gómez-Ariza. 2010. Iodine speciation in iodine-enriched microalgae *Chlorella vulgaris*. Pure Appl. Chem. 82: 473–481.

Gong, Z., X. Lu, C. Watt, B. Wen, B. He, J. Mumford, Z. Ning, Y. Xia and X.C. Le. 2006. Speciation analysis of arsenic in groundwater from Inner Mongolia with an emphasis on acid-leachable particulate arsenic. Anal. Chim. Acta. 555: 181–187.

Górecki, T. and J. Pawliszyn. 1996. Determination of tetraethyllead and inorganic lead in water by solid phase microextraction/gas chromatography. Anal. Chem. 68: 3008–3014.

Górecki, T., X. Yu and J. Pawliszyn. 1999. Theory of analyte extraction by selected porous polymer SPME fibres. Analyst. 124: 643–649.

Grinberg, P., R.C. Campos, Z. Mester and R.E. Sturgeon. 2003a. A comparison of alkyl derivatization methods for speciation of mercury based on solid phase microextraction gas chromatography with furnace atomization plasma emission spectrometry detection. J. Anal. At. Spectrom. 18: 902–909.

Grinberg, P., R.C. Campos, Z. Mester and R.E. Sturgeon. 2003b. Solid phase microextraction capillary gas chromatography combined with furnace atomization plasma emission spectrometry for speciation of mercury in fish tissues. Spectrochim. Acta B 58: 427–441.

Gui-Bin, J., L. Ji-Yan and Y. Ke-Wu. 2000. Speciation analysis of butyltin compounds in Chinese seawater by capillary gas chromatography with flame photometric detection using *in situ* hydride derivatization followed by headspace solid-phase microextraction. Anal. Chim. Acta. 421: 67–74.

Guidotti, M. and M. Vitali. 1997. Determination of organotin compounds in water samples by solid phase microextraction (SPME) and GC/MS. Annal. Chim. (Rome). 87: 497–504.

Guidotti, M. and M. Vitali. 1998. Determination of urinary mercury and methylmercury by solid phase microextraction and GC/MS. J. High Res. Chromatogr. 21: 665–666.

Güler, N., M. Maden, S. Bakırdere, O. Yavuz-Ataman and M. Volkan. 2011. Speciation of selenium in vitamin tablets using spectrofluorometry following cloud point extraction. Food Chem. 129: 1793–1799.

Gullickson, N.D., J.F. Scamehom and J.H. Harwell. 1989. Surfactant-based separation processes. Marcel Dekker. New York.

Guo, F., T. Gorecki, D. Irish and J. Pawliszyn. 1996. Solid-phase microextraction combined with electrochemistry Anal. Commun. 33: 361–364.

Guzmán Mar, J.L., L. Hinojosa Reyes, G.M. Mizanur Rahman and H.M. Skip Kingston. 2009. Simultaneous extraction of arsenic and selenium species from rice products by microwave-assisted enzymatic extraction and analysis by ion chromatography-inductively coupled plasma-mass spectrometry. J. Agric. Food Chem. 57: 3005–3013.

Hagarova, I., J. Kubova, P. Matus and M. Bujdos. 2008. Speciation of inorganic antimony in natural waters by electrothermal atomic absorption spectrometry after selective separation and preconcentration of antimony(III) with cloud point extraction. Acta Chim. Slov. 55: 528–534.

Han, C., X. Cao, J.-J. Yu, X.-R Wang and Y. Shen. 2009. Arsenic speciation in *Sargassum fusiforme* by microwave-assisted extraction and LC-ICP-MS. Chromatogr. 69: 587–591.

Hänström, S., C. Briche, H. Emteborg and D.C. Baxter. 1996. Large-volume injections in capillary gas chromatography using a separately heated packed pre-column: application to mercury speciation in natural water. Analyst. 121: 1657–1663.

Hashempur, T., M.K. Rofouei and A.R. Khorrami. 2008. Speciation analysis of mercury contaminants in water samples by RP-HPLC after solid phase extraction on modified C_{18} extraction disks with 1,3-bis(2-cyanobenzene)triazene. Microchem. J. 89: 131–136.

He, Y. and H.K. Lee. 1997. Liquid phase microextraction in a single drop of organic solvent by using a conventional microsyringe. Anal. Chem. 69: 4634–4640.

He, Y. and Y.J. Kang. 2006. Single drop liquid-liquid-liquid microextraction of methamphetamine and amphetamine in urine. J. Chromatogr. A 1133: 35–40.

Heitkemper, D.T., N.P. Vela, K.R. Stewart and C.S. Westphal. 2001. Determination of total and speciated arsenic in rice by ion chromatography and inductively coupled plasma mass spectrometry. J. Anal. At. Spectrom. 16: 299–306.

Henglein, A. 1989. Small-particle research: physicochemical properties of extremely small colloidal metal and semiconductor particles. Chem. Rev. 89: 1861–1873.

Heroult, J., T. Zuliani, M. Bueno, L. Denaix and G. Lespes. 2008. Analytical advances in butyl-, phenyl- and octyltin speciation analysis in soil by GC-PFPD. Talanta. 75: 486–493.

Hinojosa Reyes, L., J.M. Marchante-Gayón, J.I. García Alonso and A. Sanz-Medel. 2006. Application of isotope dilution analysis for the evaluation of extraction conditions in the determination of total selenium and selenomethionine in yeast-based nutritional supplements. J. Agric. Food Chem. 54: 1557–1563.

Hinojosa Reyes, L., G.M. Mizanur Rahman and H.M. Skip Kingston. 2009. Robust microwave-assisted extraction protocol for determination of total mercury and methylmercury in fish tissues. Anal. Chim. Acta 631: 121–128.

Hintelmann, H. and R.D. Evans. 1997. Application of stable isotopes in environmental tracer studies-Measurement of monomethylmercury (CH_3Hg^+) by isotope dilution ICP-MS and detection of species transformation. Fresenius' J. Anal. Chem. 358: 378–385.

Hintelmann, H., R.D. Evans and J.Y. Villeneuve. 1995. Measurement of mercury methylation in sediments by using enriched stable mercury isotopes combined with methylmercury determination by gas chromatography–inductively coupled plasma mass spectrometry. J. Anal. At. Spectrom. 10: 619–624.

Hintelmann, H., R. Falter, G. Ilgen and R.D. Evans. 1997. Determination of artifactual formation of monomethylmercury (CH_3Hg^+) in environmental samples using stable Hg^{2+} isotopes with ICP-MS detection: Calculation of contents applying species specific isotope addition. Fresenius´ J. Anal. Chem. 358: 363–370.

Hinze, W. L. and E. Pramauro. 1993. A critical-review of surfactant-mediated phase separations (cloud-point extractions)—theory and applications. Critical Rev. Anal. Chem. 24: 133–177.

Hirata, S. and H. Toshimitsu. 2005. Determination of arsenic species and arsenosugars in marine samples by HPLC–ICP–MS. Anal. Bioanal. Chem. 383: 454–460.

Hou, L. and H.K. Lee. 2003. Dynamic three-phase microextraction as a sample preparation technique prior to capillary electrophoresis. Anal. Chem. 75: 2784–2789.

Houserová, P., P. Kubáň and V. Kubáň. 2006. Ion exchange preconcentration and separation of mercury species by CE with indirect contactless conductometric detection. Electrophoresis. 27: 4508–4515.

Houserová, P., D. Matějíček and V. Kubáň. 2007a. High-performance liquid chromatographic/ion-trap mass spectrometric speciation of aquatic mercury as its pyrrolidinedithiocarbamate complexes. Anal. Chim. Acta. 596: 242–250.

Houserová, P., V. Kubáň, S. Kráčmar and J. Sitko. 2007b. Total mercury and mercury species in birds and fish in an aquatic ecosystem in the Czech Republic. Environm. Pollut. 145: 185–194.

Hu, W., F. Zheng and B. Hu. 2008. Simultaneous separation and speciation of inorganic As(III)/As(V) and Cr(III)/Cr(VI) in natural waters utilizing capillary microextraction on ordered mesoporous Al_2O_3 prior to their on-line determination by ICP-MS. J. Hazard. Mat. 151: 58–64.

Huang, C., B. Hu and Z. Jiang. 2007. Simultaneous speciation of inorganic arsenic and antimony in natural waters by dimercaptosuccinic acid modified mesoporous titanium dioxide micro-column on-line separation and inductively coupled plasma optical emission spectrometry determination. Spectrochim. Acta Part B. 62: 454–460.

Huang, C., W. Xie, X. Li and J. Zhang. 2011. Speciation of inorganic arsenic in environmental waters using magnetic solid phase extraction and preconcentration followed by ICP-MS. Microchim. Acta. 173: 165–172.

Huang, M., Y.T. Wang and P.C. Ho. 2008a. Quantification of arsenic compounds using derivatization, solvent extraction and liquid chromatography electrospray ionization tandem mass spectrometry. J. Phar. Biomed. Anal. 48: 1381–1391.

Huang, C., B. Hu, M. He and J. Duan. 2008b. Organic and inorganic selenium speciation in environmental and biological samples by nanometer-sized materials packed dual-column separation/preconcentration on-line coupled with ICP-MS. J. Mass Spectrom. 43: 336–345.

Huerga, A., I. Lavilla and C. Bendicho. 2005. Speciation of the immediately mobilisable As(III), As(V), MMA and DMA in river sediments by high performance liquid chromatography–hydride generation–atomic fluorescence spectrometry following ultrasonic extraction. Anal. Chim. Acta. 534: 121–128.

Hylton, K. and S. Mitra. 2008. A microfluidic hollow fiber membrane extractor for arsenic(V) detection. Anal. Chim. Acta. 607: 45–49.

Iijima, S. 1991. Helical microtubules of graphitic carbon. Nature. 354: 56–58.

Inagaki, K., A. Takatsu, T. Watanabe, Y. Aoyagi, T. Yarita, K. Okamoto and K. Chiba. 2007. Certification of butyltins and phenyltins in marine sediment certified reference material by species-specific isotope-dilution mass spectrometric analysis using synthesized [118]Sn-enriched organotin compounds. Anal. Bioanal. Chem. 387: 2325–2334.

Işcioğlu, B. and E. Henden. 2004. Determination of selenoamino acids by gas chromatography–mass spectrometry. Anal. Chim. Acta. 505: 101–106.

Issa, N.B., V.N. Rajaković-Ognjanović, A.D. Marinković and L.V. Rajaković. 2011. Separation and determination of arsenic species in water by selective exchange and hybrid resins. Anal. Chim. Acta. 706: 191–198.

Ito, R., M. Kawaguchi, N. Sakui, H. Honda, N. Okanouchi, K. Saito and H. Nakazawa. 2008. Mercury speciation and analysis in drinking water by stir bar sorptive extraction with *in situ* propyl derivatization and thermal desorption–gas chromatography–mass spectrometry. J. Chromatog. A 1209: 267–270.

Ito, R., M. Kawaguchi, N. Sakui, H. Honda, N. Okanouchi, K. Saito, Y. Seto and H. Nakazawa. 2009. Stir bar sorptive extraction with *in situ* derivatization and thermal desorption–gas chromatography–mass spectrometry for trace analysis of methylmercury and mercury(II) in water sample. Talanta. 77: 1295–1298.

Jaafar, J., Z. Irwan, R. Ahamad, S. Terabe, T. Ikegami and N. Tanaka. 2007 Online preconcentration of arsenic compounds by dynamic pH junction-capillary electrophoresis. J. Sep. Sci. 30: 391–398.

Jackson, B., V.F. Taylor, A. Baker and E. Miller. 2009. Low-level mercury speciation in freshwaters by isotope dilution GC-ICP-MS. Environ. Sci. Technol. 43: 2463–2469.

Jae-Sung, P., L. Jung-Sub, K. Gun-Bae, C. Jun-Seok, S. Sun-Kyoung, K. Hak-Gu, H. Eun-Jin, C. Gi-Taeg and K. Young-Hee. 2010. Mercury and methylmercury in freshwater fish and sediments in South Korea using newly adopted purge and trap GC-MS detection method. Water Air Soil Pollut. 207: 391–401.

Jeannot, M.A. and F.F. Cantwell. 1997. Mass transfer characteristics of solvent extraction into a single drop at the tip of a syringe needle. Anal. Chem. 69: 235–239.

Jia, X., Y. Han, C. Wei, T. Duan and H. Chen. 2011a. Speciation of mercury in liquid cosmetic samples by ionic liquid based dispersive liquid–liquid microextraction combined with high-performance liquid chromatography-inductively coupled plasma mass spectrometry. J. Anal. At. Spectrom. 26: 1380–1386.

Jia, X., Y. Han, X. Liu, T. Duan and H. Chen. 2011b. Speciation of mercury in water samples by dispersive liquid–liquid microextraction combined with high performance liquid chromatography-inductively coupled plasma mass spectrometry. Spectrochim. Acta Part B. 66: 88–92.

Jiang, G.B. and J.Y. Liu. 2000. Determination of butyltin compounds in aqueous samples by gas chromatography with flame photometric detector and headspace solid-phase microextraction after *in situ* hydride derivatization. Anal. Sci. 16: 585–588.

Jiang, H., B. Hu, B. Chen and W. Zu. 2008. Hollow fiber liquid phase microextraction combined with graphite furnace atomic absorption spectrometry for the determination of methylmercury in human hair and sludge samples. Spectrochim. Acta Part B. 63: 770–776.

Jiang, H., B. Hu, B. Chen and L. Xia. 2009. Hollow fiber liquid phase microextraction combined with electrothermal atomic absorption spectrometry for the speciation of arsenic(III) and arsenic (V) in fresh waters and human hair extracts. Anal. Chim. Acta. 634: 15–21.

Jiang, X.M. and H.K. Lee. 2004. Solvent bar microextraction. Anal. Chem. 76: 5591–5596.

Jiménez, M.S. and R.E. Sturgeon. 1997. Speciation of methyl- and inorganic mercury in biological tissues using ethylation and gas chromatography with furnace atomization plasma emission spectrometric detection. J. Anal. At. Spectrom. 12: 597–601.

Jimenez Moreno, M., J. Pacheco-Arjona, P. Rodríguez-González, H. Preud'Homme, D. Amouroux and O.F.X. Donard. 2006. Simultaneous determination of monomethylmercury, monobutyltin, dibutyltin and tributyltin in environmental samples by multi-elemental species-specific isotope dilution analysis using electron ionisation GC-MS. J. Mass Spectrom. 41: 1491–1497.

Jing-Fu, L., C. Yu-Guang and J. Gui-Bin. 2005. Screening the extractability of some typical environmental pollutants by ionic liquids in liquid-phase microextraction. J. Sep. Sci. 28: 87–91.

Jitaru, P. and F.C. Adams. 2004. Speciation analysis of mercury by solid-phase microextraction and multicapillary gas chromatography hyphenated to inductively coupled plasma–time-of-flight-mass spectrometry. J. Chromatog. A 1055: 197–207.

Jitaru, P., H. Goenaga-Infante and F.C. Adams. 2003. Multicapillary gas chromatography coupled to inductively coupled plasma–time-of-flight mass spectrometry for rapid mercury speciation analysis. Anal. Chim. Acta. 489: 45–57.

Jitaru, P., H. Goenaga-Infante and F.C. Adams. 2004. Simultaneous multi-elemental speciation analysis of organometallic compounds by solid-phase microextraction and multicapillary gas chromatography hyphenated to inductively coupled plasma-time-of-flight-mass spectrometry. J. Anal. At. Spectrom. 19: 867–875.

Jitmanee, K., N. Teshima, T. Sakai and K. Grudpan. 2007. DRCTM ICP-MS coupled with automated flow injection system with anion exchange minicolumns for determination of selenium compounds in water samples. Talanta. 73: 352–357.

Ji-Yan, L. and J. Gui-Bin. 2002. Survey on the presence of butyltin compounds in Chinese alcoholic beverages, determined by using headspace solid-phase microextraction coupled with gas chromatography-flame photometric detection. J. Agric. Food Chem. 50: 6683–6687.

Ji-Yan, L., J. Gui-Bin, Z. Qun-Fang and Y. Ke-Wu. 2001. Headspace solid-phase microextraction of butyltin species in sediments and their gas chromatographic determination. J. Sep. Sci. 24: 459–464.

Jo-Anne, A., W.R. Jackson, F. Blair, E. Brinckman and W.P. Iverson. 1982. Gas-chromatographic speciation of methylstannanes in the Chesapeake Bay using purge and trap sampling with a tin-selective detector. Environ. Sci. Technol. 18: 110–119.

Johannes, T., J.T. van Elteren, Z. Šlejkovec, M. Svetina and J. Glinšek. 2001. Determination of ultratrace dissolved arsenite in water—selective coprecipitation in the field combined with HGAFS and ICP–MS measurement in the laboratory. Fresenius' J. Anal. Chem. 370: 408–412.

Johannes, T., J.T. van Elteren, V. Stibilj and Z. Šlejkovec. 2002. Speciation of inorganic arsenic in some bottled Slovene mineral waters using HPLC–HGAFS and selective coprecipitation combined with FI-HGAFS. Water Res. 36: 2967–2974.

Johansson, M., T. Bergloëf, D.C. Baxter and W. Frech. 1995. Supercritical fluid extraction of ionic alkyllead species from sediment and urban dust. Analyst. 120: 755–759.

Jókai, Z., L. Abrankó and P. Fodor. 2005. SPME-GC-Pyrolysis-AFS determination of methylmercury in marine fish products by alkaline sample preparation and aqueous phase phenylation derivatization. J. Agric. Food Chem. 53: 5499–5505.

Kapsimali, D.C. and G.A. Zachariadis. 2010. Comparison of tetraethylborate and tetraphenylborate for selenite determination in human urine by gas chromatography mass spectrometry, after headspace solid phase microextraction. Talanta. 80: 1311–1317.

Karthikeyan, S. and S. Hirata. 2004. Ion chromatography–inductively coupled plasma mass spectrometry determination of arsenic species in marine samples. Appl. Organometal. Chem. 18: 323–330.

Karthikeyan, S., K. Honda, O. Shikino and S. Hirata. 2003. Speciation of arsenic in marine algae and commercial shrimp using ion chromatography with ICP-MS detection. At. Spectrosc. 24: 79–88.

Kaur, V., A.K. Malik and N. Verma. 2006. Applications of solid phase microextraction for the determination of metallic and organometallic species. J. Sep. Sci. 29: 333–345.

Kawaguchi, M., R. Ito, K. Saito and H. Nakazawa. 2006. Novel stir bar sorptive extraction methods for environmental and biomedical analysis. J. Pharm. Biomed. Anal. 40: 500–508.

Ketavarapu-Yathavakilla, S.V., M. Shah, S. Mounicou and J.A. Caruso. 2005. Speciation of cationic selenium compounds in Brassica juncea leaves by strong cation-exchange chromatography with inductively coupled plasma mass spectrometry. J. Chromatog. A 1100: 153–159.

Killelea, D.R. and J.H. Aldstadt III. 2001. Solid-phase microextraction method for gas chromatography with mass spectrometric and pulsed flame photometric detection: studies of organoarsenical speciation. J. Chromatog. A 918: 169–175.

Kingston, H.M. and L.B. Jassie. 1988. Introduction to microwave sample preparation. Theory and practice. ACS Professional Reference Book, Washington, USA.

Kirby, J. and W. Maher. 2002. Measurement of water-soluble arsenic species in freeze-dried marine animal tissues by microwave-assisted extraction and HPLC-ICP-MS. J. Anal. At. Spectrom. 17: 838–843.

Konieczka, P., B. Sejerøe-Olsen, T.P.J. Linsinger and H. Schimmel. 2007. Determination of tributyltin (TBT) in marine sediment using pressurised liquid extraction–gas chromatography–isotope dilution mass spectrometry (PLE–GC–IDMS) with a hexane–tropolone mixture. Anal. Bioanal. Chem. 388: 975–978.

Kösters, J., R.A. Diaz-Bone, B. Planer-Friedrich, B. Rothweiler and A.V. Hirner. 2003. Identification of organic arsenic, tin, antimony and tellurium compounds in environmental samples by GC-MS. J. Mol. Struct. 661–662: 347–356.

Kresage, C.T., M.E. Leonowicz, W.J. Roth, J.C. Vartuli and J.S. Beck. 1992. Ordered mesoporous molecular sieves synthesized by a liquid-crystal template mechanism. Nature. 359: 710–712.

Kristenson, E.M., L. Ramos and U.A.T. Brinkman. 2006. Recent advances in matrix solid phase dispersion. Trends Anal. Chem. 25: 96–111.

Kronholm, J., K. Hartonen and M.-L. Riekkola. 2007. Analytical extractions with water at elevated temperatures and pressures. Trends Anal. Chem. 26: 396–412.

Kuballa, J., R.D. Wilken, E. Jantzen, K.K. Kwan and Y.K. Chau. 1995. Speciation and genotoxicity of butyltin compounds. Analyst. 120: 667–673.

Kumar, U.T., N.P. Vela, J.G. Dorsey and J.A. Caruso. 1993. Supercritical fluid extraction of organotins from biological samples and speciation by liquid chromatography and inductively coupled plasma mass spectrometry. J. Chromatogr. A 655: 340–345.

Kun-Chih, H., C. Bing-Hung, E.Y. Liya and L.E. Yu. 2007. Cloud-point extraction of selected polycyclic aromatic hydrocarbons by nonionic surfactants. Sep. Purif. Tech. 57: 1–10.

Kuo, C.-Y., S.-J. Jiang and A.C. Sahayam. 2007. Speciation of chromium and vanadium in environmental samples using HPLC-DRC-ICP-MS. J. Anal. At. Spectrom. 22: 636–641.

Lambertsson, L. and E. Björn. 2004. Validation of a simplified field-adapted procedure for routine determinations of methyl mercury at trace levels in natural water samples using species-specific isotope dilution mass spectrometry. Anal. Bioanal. Chem. 380: 871–875.

Lambertsson, L., E. Lundberg, M. Nilsson and W. Frech. 2001. Applications of enriched stable isotope tracers in combination with isotope dilution GC-ICP-MS to study mercury species transformation in sea sediments during *in situ* ethylation and determination. J. Anal. At. Spectrom. 16: 1296–1301.

Lancas, F.M., M.E.C. Queiroz, P. Grossi and I.R.B. Olivares. 2009. Recent developments and applications of stir bar sorptive extraction. J. Sep. Sci. 32: 813–824.

Larsen, E.H., J. Sloth, M. Hansen and S. Moesgaard. 2003. Selenium speciation and isotope composition in ^{77}Se-enriched yeast using gradient elution HPLC separation and ICP-dynamic reaction cell-MS. J. Anal. At. Spectrom. 18: 310–316.

Latva, S., S. Peräniemi and M. Ahlgrén. 1999. Separation of microgram quantities of As(V), As(III) and organoarsenic species in aqueous solutions and determination by energy dispersive X-ray fluorescence spectrometry. Analyst. 124: 1105–1108.

Latva, S., M. Hurtta, S. Peräniemi and M. Ahlgrén. 2000. Separation of arsenic species in aqueous solutions and optimization of determination by graphite furnace atomic absorption spectrometry. Anal. Chim. Acta. 418: 11–17.

Le Gac, M., G. Lespes and M. Potin-Gautier. 2003. Rapid determination of organotin compounds by headspace solid-phase microextraction. J. Chromatog. A 999: 123–134.

Lee, T.-H. and S.-J. Jiang. 2005. Speciation of lead compounds in fish by capillary electrophoresis-inductively coupled plasma mass spectrometry. J. Anal. At. Spectrom. 20: 1270–1274.

Leenaers, J., W. Van Mol, H. Goenaga-Infante and F.C. Adams. 2002. Gas chromatography-inductively coupled plasma-time-of-flight mass spectrometry as a tool for speciation analysis of organomercury compounds in environmental and biological samples. J. Anal. At. Spectrom. 17: 1492–1497.

Lemes, M. and F. Wang. 2009. Methylmercury speciation in fish muscle by HPLC-ICP-MS following enzymatic hydrolysis. J. Anal. At. Spectrom. 24: 663–668.

Lemos Batista, B., J.L. Rodrigues, S.S. de Souza, V.C. Oliveira Souza and F. Barbosa Jr. 2011. Mercury speciation in seafood samples by LC–ICP-MS with a rapid ultrasound-assisted extraction procedure: Application to the determination of mercury in Brazilian seafood samples. Food Chem. 126: 2000–2004.

Lespes, G., C. Carlier-Pinasseau, M. Potin-Gautier and M. Astruc. 1996. Direct determination of butyl- and phenyltin compounds as chlorides using gas chromatography and flame photometric detection. Analyst. 121: 1969–1973.

Lespes, G., V. Desauziers, C. Montigny and M. Potin-Gautier. 1998. Optimization of solid-phase microextraction for the speciation of 1 butyl- and phenyltins using experimental designs. J. Chromatog. A 826: 67–76.

Leufroy, A., L. Noël, V. Dufailly, D. Beauchemin and T. Guérin. 2011. Determination of seven arsenic species in seafood by ion exchange chromatography coupled to inductively coupled plasma-mass spectrometry following microwave assisted extraction: Method validation and occurrence data. Talanta. 83: 770–779.

Li, K., S. Li and H.K. Lee. 1995. Determination of organolead and organotin compounds in water samples by micellar electrokinetic chromatography. J. Liq. Chromatgr. 18: 1325–1347.

Li, Y. and B. Hu. 2007. Sequential cloud point extraction for the speciation of mercury in seafood by inductively coupled plasma optical emission spectrometry. Spectrochim. Acta Part B. 62: 1153–1160.

Li, Y., B. Hu and Z. Jiang. 2006. On-line cloud point extraction combined with electrothermal vaporization inductively coupled plasma atomic emission spectrometry for the speciation of inorganic antimony in environmental and biological samples. Anal. Chim. Acta. 576: 207–214.

Li, Y., B. Hu, M. He and G. Xiang. 2008a. Simultaneous speciation of inorganic selenium and antimony in water samples by electrothermal vaporization inductively coupled plasma mass spectrometry following selective cloud point extraction. Water Res. 42: 1195–1203.

Li, P., J. Duan and B. Hu. 2008b. High-sensitivity capillary electrophoresis for speciation of organomercury in biological samples using hollow fiber-based liquid liquid-liquid microextraction combined with on-line preconcentration by large-volume sample stacking. Electrophoresis. 29: 3081–3089.

Li, H.-F., E. Lombi, J.L. Stroud, S.P. McGrath and F.-J. Zhao. 2010a. Selenium speciation in soil and rice: influence of water management and Se fertilization. J. Agric. Food Chem. 58: 11837–11843.

Li, L.F. Wang, B. Meng, M. Lemes, X. Feng and G. Jiang. 2010b. Speciation of methylmercury in rice grown from a mercury mining area. Environ. Pollut. 158: 3103–3107.

Li, P., X. Zhang and B. Hu. 2011. Phase transfer membrane supported liquid–liquid–liquid microextraction combined with large volume sample injection capillary electrophoresis–ultraviolet detection for the speciation of inorganic and organic mercury. J. Chromatog. A 1218: 9414–9421.

Liang, P. and R. Liu. 2007. Speciation analysis of inorganic arsenic in water samples by immobilized nanometer titanium dioxide separation and graphite furnace atomic absorption spectrometric determination. Anal. Chim. Acta. 602: 32–36.

Li-Ping, Y. 2005. Cloud point extraction preconcentration prior to high-performance liquid chromatography coupled with cold vapor generation atomic fluorescence spectrometry for speciation analysis of mercury in fish samples. J. Agric. Food Chem. 53: 9656–9662.

Liu, Q.Y. 2010. Determination of ultra-trace amounts of inorganic selenium species in natural water by ion chromatography-inductively coupled plasma-mass spectrometry coupled with nano-Al_2O_3 solid phase extraction. Central Eu. J. Chem. 8: 326–330.

Liu, T., S. Li, S. Liu and G. Lv. 2011. Optimization of supercritical fluid extraction/headspace solid-phase microextraction and gas chromatography–mass spectrometry method for determinating organotin compounds in clam samples. J. Food Process Eng. 34: 1125–1143.

Liu, W. and H.K. Lee. 2000. Continuous-flow microextraction exceeding 1000-fold concentration of dilute analytes. Anal. Chem. 72: 4462–4467.

Liu, Y., V. López-Ávila, M. Alcaraz and W.F. Beckert. 1994. Off-line complexation/supercritical fluid extraction and gas chromatography with atomic emission detection for the determination and speciation of organotin compounds in soils and sediments. Anal. Chem. 66: 3788–3796.

Liu, Y., E.C. Zhao, W.T. Zhu, H.X. Gao and Z.Q. Zhou. 2009. Determination of four heterocyclic insecticides by ionic liquid dispersive liquid–liquid microextraction in water samples. J. Chromatogr. A 1216: 885–891.

Llorente-Mirandes, T., M.J. Ruiz-Chancho, M. Barbero, R. Rubio and J.F. López-Sánchez. 2011. Determination of water-soluble arsenic compounds in commercial edible seaweed by LC-ICPMS. J. Agric. Food Chem. 59: 12963–12968.

Looser, P.W., M. Berg, K. Fent, J. Mühlemann and R. Schwarzenbach. 2000. Phenyl- and butyltin analysis in small biological samples by cold methanolic digestion and GC/MS. Anal. Chem. 72: 5136–5141.

López, I., S. Cuello, C. Cámara and Y. Madrid. 2010. Approach for rapid extraction and speciation of mercury using a microtip ultrasonic probe followed by LC–ICP-MS. Talanta. 82: 594–599.

Lopez-Avila, V., Y. Liu and W.F. Beckert. 1997. Interlaboratory evaluation of an off-line supercritical fluid extraction and gas chromatography with atomic emission detection method for the determination of organotin compounds in soil and sediments. J. Chromatogr. A 785: 279–288.

López-García, I., R.E. Rivas and M. Hernández-Córdoba. 2011. Use of carbon nanotubes and electrothermal atomic absorption spectrometry for the speciation of very low amounts of arsenic and antimony in waters. Talanta. 86: 52–57.

Lorenzo, R.A., M.J. Vázquez, A.M. Carro and R. Cela. 1999. Methylmercury extraction from aquatic sediments. A comparison between manual, supercritical fluid and microwave-assisted techniques. Trends Anal. Chem. 18: 410–416.

Louppis, A.P., D. Georgantelis, E.K. Paleologos and M.G. Kontominas. 2010. Determination of tributyltin through ultrasonic assisted micelle mediated extraction and GFAAS:

Application to the monitoring of tributyltin levels in Greek marine species. Food Chem. 121: 907–911.

Luque-García, J.L. and M.D. Luque de Castro. 2003. Ultrasound: A powerful tool for leaching. Trends Anal. Chem. 22: 41–46.

Ma, M. and F.F. Cantwell. 1998. Solvent microextraction with simultaneous back-extraction for sample cleanup and preconcentration: Quantitative extraction. Anal. Chem. 70: 3912–3919.

Ma, M. and F.F. Cantwell. 1999. Solvent microextraction with simultaneous back-extraction for sample cleanup and preconcentration: Preconcentration into a single microdrop. Anal. Chem. 71: 388–393.

Majors, R.E. 1996. Liquid extraction techniques for sample preparation. LC-GC. 14: 936–938.

Maldonado Santoyo, M., J.A. Landero Figueroa, K. Wrobel and K. Wrobel. 2009. Analytical speciation of mercury in fish tissues by reversed phase liquid chromatography–inductively coupled plasma mass spectrometry with Bi^{3+} as internal standard. Talanta. 79: 706–711.

Mallah, M.H., F. Shhmirani and M.G. Maragheh. 2009. Ionic liquids for simultaneous preconcentration of some lanthanoids using dispersive liquid–liquid microextraction technique in uranium dioxide powder. Environ. Sci. Technol. 43: 1947–1951.

Mao, X., B. Chen, C. Huang, M. He and B. Hu. 2011. Titania immobilized polypropylene hollow fiber as a disposable coating for stir bar sorptive extraction–high performance liquid chromatography–inductively coupled plasma mass spectrometry speciation of arsenic in chicken tissues. J. Chromatog. A 1218: 1–9.

Mao, Y., G. Liu, G. Meichel, Y. Cai and G. Jiang. 2008. Simultaneous speciation of monomethylmercury and monoethylmercury by aqueous phenylation and purge-and-trap preconcentration followed by atomic spectrometry detection. Anal. Chem. 80: 7163–7168.

Mao, Y., Y. Yin, Y. Li, G. Liu, X. Feng, G. Jiang and Y. Cai. 2010. Occurrence of monoethylmercury in the Florida Everglades: identification and verification. Environ. Pollut. 158: 3378–3384.

Marcic, C., G. Lespes and M. Potin-Gautier. 2005. Pressurised solvent extraction for organotin speciation in vegetable matrices. Anal. Bioanal. Chem. 382: 1574–1583.

Margetínová, J., P. Houserová-Pelcová and V. Kubáň. 2008. Speciation analysis of mercury in sediments, zoobenthos and river water samples by high-performance liquid chromatography hyphenated to atomic fluorescence spectrometry following preconcentration by solid phase extraction. Anal. Chim. Acta. 615: 115–123.

Martinis, E.M. and R.G. Wuilloud. 2010. Cold vapor ionic liquid-assisted headspace single-drop microextraction: A novel preconcentration technique for mercury species determination in complex matrix samples. J. Anal. At. Spectrom. 25: 1432–1439.

Martinis, E.M., L.B. Escudero, P. Berton, R.P. Monasterio, M.F. Filippini and R.G. Wuilloud. 2011. Determination of inorganic selenium species in water and garlic samples with on-line ionic liquid dispersive microextraction and electrothermal atomic absorption spectrometry. Talanta. 85: 2182–2188.

Mason, T.J. 1999. Sonochemistry. Oxford University Press, Oxford, UK.

Mason, T.J. and J.P. Lorimer. 2002. Applied sonochemistry: Uses of power ultrasounds in chemistry and processing. Wiley-VCH, Weinheim, Germany.

Masqué, N., R.M. Marcé and F. Borrull. 2001. Molecularly imprinted polymers: new tailor-made materials for selective solid-phase extraction. Trends Anal. Chem. 20: 477–486.

Masscheleyn, P.H., R.D. Delaune and W.H. Patrick. 1991. Selenium speciation in aqueous solutions using a hydride generation atomic absorption spectrophotometry technique. Spectrosc. Lett. 24: 307–322.

Mato-Fernández, M.J., J.R. Otero-Rey, J. Moreda-Piñeiro, E. Alonso-Rodríguez, P. López-Mahía, S. Muniategui-Lorenzo and D. Prada-Rodríguez. 2007. Arsenic extraction in marine biological materials using pressurised liquid extraction. Talanta. 71: 515–520.

Mazej, D., I. Falnoga, M. Veber and V. Stibilj. 2006. Determination of selenium species in plant leaves by HPLC–UV–HG-AFS. Talanta. 68: 558–568.

Mazej, D., J. Osvald and V. Stibilj. 2008. Selenium species in leaves of chicory, dandelion, lamb's lettuce and parsley. Food Chem. 107: 75–83.

McKiernan, J.W., J.T. Creed, C.A. Brockhoff, J.A. Caruso and R.M. Lorenzana. 1999. A comparison of automated and traditional methods for the extraction of arsenicals from fish. J. Anal. At. Spectrom. 14: 607–613.

Menéndez-García, A., M.C. Pérez-Rodríguez, J.E. Sánchez-Uria and A. Sanz-Medel. 1995. Sb(III) and Sb(V) separation and analytical speciation by a continuous tandem on-line separation device in connection with inductively coupled plasma atomic emission spectrometry. Fresenius' J. Anal. Chem. 353: 128–132.

Menéndez-Sánchez, W., B. Zwicker and A. Chatt. 2009. Determination of As(III), As(V), MMA and DMA in drinking water by solid phase extraction and neutron activation. J. Radioanal. Nucl. Chem. 282: 133–138.

Meng, W., F. Weiyue, S. Junwen, Z. Fang, W. Bing, Z. Motao, L. Bai, Z. Yuliang and C. Zhifang. 2007. Development of a mild mercaptoethanol extraction method for determination of mercury species in biological samples by HPLC–ICP-MS. Talanta. 71: 2034–2039.

Mester, Z. and J. Pawliszyn. 1999. Electrospray mass spectrometry of trimethyllead and triethyllead with in-tube solid phase microextraction sample introduction. Rapid Commun. Mass Spectrom. 13: 1999–2003.

Mester, Z. and J. Pawliszyn. 2000. Speciation of dimethylarsinic acid and monomethylarsonic acid by solid-phase microextraction-gas chromatography-ion trap mass spectrometry. J. Chromatogr. A 873: 129–135.

Mester, Z., J. Lam, R. Sturgeon and J. Pawliszyn. 2000a. Determination of methylmercury by solid-phase microextraction inductively coupled plasma mass spectrometry: a new sample introduction method for volatile metal species. J. Anal. At. Spectrom. 15: 837–842.

Mester, Z., H. Lord and J. Pawliszyn. 2000b. Speciation of trimethyllead and triethyllead by in-tube solid phase microextraction high-performance liquid chromatography electrospray ionization mass spectrometry. J. Anal. At. Spectrom. 15: 595–600.

Mester, Z., R.E. Sturgeon, J.W. Lam, P.S. Maxwell and L. Péter. 2001. Speciation without chromatography. Part I. Determination of tributyltin in aqueous samples by chloride generation, headspace solid-phase microextraction and inductively coupled plasma time of flight mass spectrometry. J. Anal. At. Spectrom. 16: 1313–1316.

Michel, P. and B. Averty. 1991. Tributyltin analysis in seawater by GC-FPD after direct aqueous-phase ethylation using sodium tetraethylborate. Appl. Organometal. Chem. 5: 393–397.

Millán, E. and J. Pawliszyn. 2000. Determination of butyltin species in water and sediment by solid-phase microextraction–gas chromatography–flame ionization detection. J. Chromatog. A 873: 63–71.

Miravet, R., E. Bonilla, J.F. López-Sánchez and R. Rubio. 2005. Antimony speciation in terrestrial plants. Comparative studies on extraction methods. J. Environ. Monit. 7: 1207–1213.

Mishra, S., R.M. Tripathi, S. Bhalke, V.K. Shukla and V.D. Puranik. 2005. Determination of methylmercury and mercury(II) in a marine ecosystem using solid-phase microextraction gas chromatography-mass spectrometry. Anal. Chim. Acta. 551: 192–198.

Mittal, K.L. and E.J. Fendler. 1982. Solution behavior of surfactants: theoretical and applied aspects. Plenum Press. New York.

Miura, J., H. Ishii and H. Watanabe. 1976. Extraction and separation of nickel chelate of 1-(2-thiazolylazo)-2-naphthol in nonionic surfactant solution. Bunseki Kagaku. 25: 808–809.

Mizuishi, K., M. Takeuchi and T. Hobo. 1998. Trace analysis of tributyltin and triphenyltin compounds in sea water by gas chromatography–negative ion chemical ionization ass spectrometry. J. Chrom. A 800: 267–273.

Moens, L., T. De Smaele and R. Dams. 1997. Sensitive, simultaneous determination of organomercury, -lead, and -tin compounds with headspace solid phase microextraction capillary gas chromatography combined with inductively coupled plasma mass spectrometry. Anal. Chem. 69: 1604–1611.

Monasterio, R.P. and R.G. Wuilloud. 2010. Ionic liquid as ion-pairing reagent for liquid–liquid microextraction and preconcentration of arsenic species in natural waters followed by ETAAS. J. Anal. At. Spectrom. 25: 1485–1490.

Mondal, B.C. and A.K. Das 2003. Determination of mercury species with a resin functionalized with a 1,2-bis(o-aminophenylthio)ethane moiety. Anal. Chim. Acta. 477: 73–80.

Monperrus, M., R.C. Rodríguez Martín-Doimeadios, J. Scancar, D. Amouroux and O.F.X. Donard. 2003a. Simultaneous sample preparation and species-specific isotope dilution mass spectrometry analysis of monomethylmercury and tributyltin in a certified oyster tissue. Anal. Chem. 75: 4095–4102.

Monperrus, M., O. Zuloaga, E. Krupp, D. Amouroux, R. Wahlen, B. Fairman and O.F.X. Donard. 2003b. Rapid, accurate and precise determination of tributyltin in sediments and biological samples by species specific isotope dilution-microwave extraction-gas chromatography-ICP mass spectrometry. J. Anal. At. Spectrom. 18: 247–253.

Montes, R., P. Canosa, J.P. Lamas, A. Piñeiro, I. Orriols, R. Cela and I. Rodríguez. 2009. Matrix solid-phase dispersion and solid-phase microextraction applied to study the distribution of fenbutatin oxide in grapes and white wine. Anal. Bioanal. Chem. 395: 2601–2610.

Montes-Bayón, M., E.G. Yanes, C. Ponce de León, K. Jayasimhulu, A. Stalcup, J. Shann and J.A. Caruso. 2002. Initial studies of selenium speciation in *Brassica juncea* by LC with ICPMS and ES-MS detection: an approach for phytoremediation studies. Anal. Chem. 74: 107–113.

Montes-Bayón, M., M.J. Díaz Molet, E. Blanco González and A. Sanz-Medel. 2006. Evaluation of different sample extraction strategies for selenium determination in selenium-enriched plants (*Allium sativum* and *Brassica juncea*) and Se speciation by HPLC-ICP-MS. Talanta. 68: 1287–1293.

Montperrus, M., Y. Bohari, M. Bueno, A. Astruc and M. Astruc. 2002. Comparison of extraction procedures for arsenic speciation in environmental solid reference materials by high-performance liquid chromatography-hydride generation-atomic fluorescence spectrometry. Appl. Organometal. Chem. 16: 347–354.

Morabito, R., H. Muntau, W. Cofino and Ph. Quevauviller. 1999. A new mussel certified reference material (CRM 477) for the quality control of butyltin determination in the marine environment. J. Environ. Monitor. 1: 75–82.

Morcillo, Y., Y. Cai and J.M. Bayona. 1995. Determination of methyltin compounds in aqueous samples using solid phase microextraction and capillary gas chromatography following *in situ* derivatization with sodium tetraethylborate. J. High Res. Chromatog. 18: 767–770.

Moreda-Piñeiro, A., E. Peña-Vázquez, P. Hermelo-Herbello, P. Bermejo-Barrera, J. Moreda-Piñeiro, E. Alonso-Rodríguez, S. Muniategui-Lorenzo, P. López-Mahía and D. Prada-Rodríguez. 2008. Matrix solid-phase dispersion as a sample pretreatment for the speciation of arsenic in seafood products. Anal. Chem. 80: 9272–9278.

Moreda-Piñeiro, A., J. Moreda-Piñeiro, P. Herbello-Hermelo, P. Bermejo-Barrera, S. Muniategui-Lorenzo, P. López-Mahía and D. Prada-Rodríguez. 2011. Application of fast ultrasound water-bath assisted enzymatic hydrolysis—high performance liquid chromatography–inductively coupled plasma-mass spectrometry procedures for arsenic speciation in seafood materials. J. Chromatogr. A 1218: 6970–6980.

Moreda-Piñeiro, A., M.C. Barciela-Alonso, R. Domínguez-González, E. Peña-Vázquez, P. Herbello-Hermelo and P. Bermejo-Barrera. 2009. Alternative solid sample pre-treatment methods in green analytical atomic spectrometry. Spectrosc. Lett. 42: 394–417.

Moreda-Piñeiro, J., E. Alonso-Rodríguez, A. Moreda-Piñeiro, C. Moscoso-Pérez, S. Muniategui-Lorenzo, P. López-Mahía, D. Prada-Rodríguez and P. Bermejo-Barrera. 2010. Simultaneous pressurized enzymatic hydrolysis extraction and clean up for arsenic speciation in seafood samples before high performance liquid chromatography–inductively coupled plasma-mass spectrometry determination. Anal. Chim. Acta. 679: 63–73.

Moreno, P., M.A. Quijano, A.M. Gutiérrez, M.C. Pérez-Conde and C. Cámara. 2002. Stability of total selenium and selenium species in lyophilised oysters and in their enzymatic extracts. Anal. Bioanal. Chem. 374: 466–476.

Moscoso-Pérez, C., J. Moreda-Piñeiro, P. López-Mahía, S. Muniategui-Lorenzo, E. Fernández-Fernández and D. Prada-Rodríguez. 2008. Pressurized liquid extraction followed by high performance liquid chromatography coupled to hydride generation atomic fluorescence spectrometry for arsenic and selenium speciation in atmospheric particulate matter. J. Chromatogr. A 1215: 15–20.

Mothes, S. and R. Wennrich. 1999. Solid phase microextraction and GC-MIP-AED for the speciation analysis of organomercury compounds. J. High Resol. Chromatogr. 22: 181–182.

Mothes, S. and R. Wennrich. 2000. Coupling of SPME and GC-AED for the determination of organometallic compounds. Mikrochim. Acta. 135: 91–95.

Mulugeta, M., G. Wibetoe, C.J. Engelsen and W. Lund. 2010. Speciation analysis of As, Sb and Se in leachates of cementitious construction materials using selective solid phase extraction and ICP-MS. J. Anal. At. Spectrom. 25: 169–177.

Muñoz, J., M. Gallego and M. Valcárcel. 2005a. Speciation analysis of mercury and tin compounds in water and sediments by gas chromatography–mass spectrometry following preconcentration on C_{60} fullerene. Anal. Chim. Acta. 548: 66–72.

Muñoz, J., M. Gallego and M. Valcárcel. 2005b. Speciation of organometallic compounds in environmetal samples by gas chromatography after flow preconcentration on fullerenes and nanotubes. Anal. Chem. 77: 5389–5395.

Murray, D.A. 1979. Rapid micro extraction procedure for analyses of trace amounts of organic compounds in water by gas choromatography and comparisons with macro extraction methods. J. Chromatogr. A. 177: 135–140.

Najafi, N.M., H. Tavakoli, R. Alizadeh and S. Seidi. 2010. Speciation and determination of ultra trace amounts of inorganic tellurium in environmental water samples by dispersive liquid–liquid microextraction and electrothermal atomic absorption spectrometry. Anal. Chim. Acta. 670: 18–23.

Nam, S.-H., H.-J. Oh, H.-S. Min and J.-H. Lee. 2010. A study on the extraction and quantitation of total arsenic and arsenic species in seafood by HPLC–ICP-MS. Microchem. J. 95: 20–24.

Narukawa, T., A. Takatsu, K. Chiba, K.W. Riley and D.H. French. 2005. Investigation on chemical species of arsenic, selenium and antimony in fly ash from coal fuel thermal power stations. J. Environ. Monit. 7: 1342–1348.

Ndungu, K. and L. Mathiasson. 2000. Microporous membrane liquid–liquid extraction technique combined with gas chromatography mass spectrometry for the determination of organotin compounds. Anal. Chim. Acta. 404: 319–328.

Nikolaos, M. and T.G.A. Zachariadis. 2010. Speciation of inorganic and tetramethyltin by headspace solid-phase microextraction and gas chromatography–mass spectrometry. J. Sep. Sci. 33: 1610–1616.

Nischwitz, V., B. Michalke and A. Kettrup. 2004. Investigations on extraction procedures for Pt species from spiked road dust samples using HPLC–ICP–MS detection. Anal. Chim. Acta. 521: 87–98.

Nord, L. and B. Karlberg. 1981. Automated extraction system for flame atomic-absorption spectrometry. Anal. Chim. Acta. 125: 199–202.

Nuhu, A.A., C. Basheer and B. Saad. 2011. Liquid-phase and dispersive liquid–liquid microextraction techniques with derivatization: Recent applications in bioanalysis. J. Chromatog. B 879: 1180–1188.

Okamoto, K., Y. Seike and M. Okumura. 2010. A Simple Speciation analysis using tristimulus colorimetry for arsenic (III) and Arsenic (V) in environmental water after selective coprecipitation with barium sulfate. Bunseki Kagaku. 59: 653–658.

Oliveira, V., J.L. Gómez-Ariza and D. Sánchez-Rodas. 2005. Extraction procedures for chemical speciation of arsenic in atmospheric total suspended particles. Anal. Bioanal. Chem. 382: 335–340.

Oliveira-Souza, J.M. and C.R. Teixeira-Tarley. 2008. Preconcentration and speciation of Sb(III) and Sb(V) in water samples and blood serum after cloud point extraction using chemometric tools for optimization. Anal. Lett. 41: 2465–2486.

Olsen, S., L.C.R. Pessenda, J. Růžička and E.H. Hansen. 1983. Combination of flow injection analysis with flame atomic-absorption spectrophotometry: determination of trace amounts of heavy metals in polluted seawater. Analyst. 108: 905–917.

Oudsema, J.W. and C.F. Poole. 1992. On-line supercritical fluid extraction and chromatography of organotins with packed microbore columns and formic acid modified carbon dioxide. Fresenius' J. Anal. Chem. 344: 426–434.

Paleólogos, E.K., D.L. Giokas and M.I. Karayannis. 2005. Micelle-mediated separation and cloud-point extraction. Trend Anal. Chem. 24: 426–436.

Pannier, F., A. Astruc and M. Astruc. 1996. Determination of butyltin compounds in marine biological samples by enzymatic hydrolysis and HG-GC-QFAAS detection. Anal. Chim. Acta. 327: 287–293.

Pardo-Martínez, M., P. Viñas, A. Fisher and S.J. Hill. 2001. Comparison of enzymatic extraction procedures for use with directly coupled high performance liquid chromatography-inductively coupled plasma mass spectrometry for the speciation of arsenic in baby foods. Anal. Chim. Acta. 441: 29–36.

Park, M., H. Yoon, C. Yoon and J.-Y. Yu. 2011. Estimation of mercury speciation in soil standard reference materials with different extraction methods by ion chromatography coupled with ICP-MS. Environ. Geochem. Health. 33: 49–56.

Pawliszyn, J. 1997. Solid Phase Microextraction: Theory and Practice, Wiley-VCH. New York.

Peachey, E., N. McCarthy and H. Goenaga-Infante. 2008. Acceleration of enzymatic hydrolysis of protein-bound selenium by focused microwave energy. J. Anal. At. Spectrom. 23: 487–492.

Peachey, E., K. Cook, A. Castles, C. Hopley and H. Goenaga-Infante. 2009. Capabilities of mixed-mode liquid chromatography coupled to inductively coupled plasma mass spectrometry for the simultaneous speciation analysis of inorganic and organically-bound selenium. J. Chromatogr. A 1216: 7001–7006.

Pécheyran, C., D. Amouroux and O.F.X. Donard. 1998. Field determination of volatile selenium species at ultra trace levels in environmental waters by on-line purging, cryofocusing and detection by atomic fluorescence spectroscopy. J. Anal. At. Spectrom. 13: 615–621.

Pedersen-Bjergaard, S. and K.E. Rasmussen. 1999. Liquid-liquid-liquid microextraction for sample preparation of biological fluids prior to capillary electrophoresis. Anal. Chem. 71: 2650–2656.

Pedrero, Z., J. Ruiz Encinar, Y. Madrid and C. Cámara. 2007. Application of species-specific isotope dilution analysis to the correction for selenomethionine oxidation in Se-enriched yeast sample extracts during storage. J. Anal. At. Spectrom. 22: 1061–1066.

Pedrero, Z., S. Murillo, C. Cámara, E. Schram, J.B. Luten, I. Feldmann, N. Jakubowski and Y. Madrid. 2011. Selenium speciation in different organs of African catfish (*Clarias gariepinus*) enriched through a selenium-enriched garlic based diet. J. Anal. At. Spectrom. 26: 116–125.

Pelizzetti, E. and E. Pramauro. 1985. Analytical applications of organized molecular assemblies. Anal. Chim. Acta. 169: 1–29.

Peña-Farfal, C., A. Moreda-Piñeiro, A. Bermejo-Barrera, P. Bermejo-Barrera, H. Pinochet-Cancino and I. de Gregori-Henríquez. 2004a. Use of enzymatic hydrolysis for the multi-element determination in mussel soft tissue by inductively coupled plasma-atomic absorption spectrometry. Talanta. 64: 671–681.

Peña-Farfal, C., A. Moreda-Piñeiro, A. Bermejo-Barrera, P. Bermejo-Barrera, H. Pinochet-Cancino and I. de Gregori-Henríquez. 2004b. Ultrasound bath-assisted enzymatic hydrolysis procedures as sample pretreatment for the multielement determination in mussels by inductively coupled plasma atomic absorption spectrometry. Anal. Chem. 76: 3541–3547.

Peñalver, R., N. Campillo and M. Hernández-Córdoba. 2011. Comparison of two derivatization reagents for the simultaneous determination of organolead and organomanganese compounds using solid-phase microextraction followed by gas chromatography with atomic emission detection. Talanta. 87: 268–275.

Pena-Pereira, F., I. Lavilla and C. Bendicho. 2009a. Miniaturized preconcentration methods based on liquid–liquid extraction and their application in inorganic ultratrace analysis and speciation: A review. Spectrochim. Acta B 64: 1–15.

Pena-Pereira, F., I. Lavilla and C. Bendicho. 2009b. Headspace single-drop microextraction with *in situ* stibine generation for the determination of antimony (III) and total antimony by electrothermal-atomic absorption spectrometry. Microchim. Acta. 164: 77–83.

Pena-Pereira, F., I. Lavilla, C. Bendicho, L. Vidal and A. Canals. 2009c. Speciation of mercury by ionic liquid-based single-drop microextraction combined with high-performance liquid chromatography-photodiode array detection. Talanta. 78: 537–541.

Pena-Pereira, F., I. Lavilla and C. Bendicho. 2010. Liquid-phase microextraction techniques within the framework of green chemistry. Trends Anal. Chem. 29: 617–628.

Perna, L., A. LaCroix-Fralish and S.J. Stürup. 2005. Determination of inorganic mercury and methylmercury in zooplankton and fish samples by speciated isotopic dilution GC-ICP-MS after alkaline digestion. J. Anal. At. Spectrom. 20: 236–238.

Pongratz, R. and K.G. Heumann. 1998. Determination of concentration profiles of methyl mercury compounds in surface waters of polar and other remote oceans by GC-AFD. Int. J. Environ. Anal. Chem. 71: 41–56.

Poperechna, N. and K.G. Heumann. 2005. Simultaneous multi-species determination of trimethyllead, monomethylmercury and three butyltin compounds by species-specific isotope dilution GC–ICP–MS in biological samples. Anal. Bioanal. Chem. 383: 153–159.

Priego-Capote, F. and M.D. Luque de Castro. 2006. Speciation of chromium by in-capillary derivatization and electrophoretically mediated microanalysis. J. Chromatogr. A 1113: 244–250.

Prieto, A., O. Zuloaga, A. Usobiaga, N. Etxebarria, L.A. Fernández, C. Marcic and A. de Diego. 2008. Simultaneous speciation of methylmercury and butyltin species in environmental samples by headspace-stir bar sorptive extraction–thermal desorption–gas chromatography–mass spectrometry. J. Chromatog. A 1185: 130–138.

Prieto, A., O. Basauri, R. Rodil, A. Usobiaga, L.A. Fernandez, N. Etxebarria and O. Zuloaga. 2010. Stir-bar sorptive extraction: A view on method optimisation, novel applications, limitations and potential solutions. J. Chromatogr. A 1217: 2642–2666.

Psillakis, E. and N. Kalogerakis. 2001. Application of solvent microextraction to the analysis of nitroaromatic explosives in water samples. J. Chromatogr. A 907: 211–219.

Psillakis, E. and N. Kalogerakis. 2003. Developments in liquid-phase microextraction. Trend Anal. Chem. 22: 565–574.

Pu, X., B. Chen and B. Hu. 2009. Solvent bar microextraction combined with electrothermal vaporization inductively coupled plasma mass spectrometry for the speciation of inorganic arsenic in water samples. Spectrochim. Acta Part B. 64: 679–684.

Puk, R. and J.H. Weber. 1994. Determination of mercury(II), monomethylmercury cation, dimethylmercury and diethylmercury by hydride generation, cryogenic trapping and atomic absorption spectrometric detection. Anal. Chim. Acta. 292: 175–183.

Pyrzynska, K., P. Drzewicz and M. Trojanowicz. 1998. Preconcentration and separation of inorganic selenium species on activated alumina. Anal. Chim. Acta. 363: 141–146.

Quevauviller, P., M. Astruc, L. Ebdon, V. Desauziers, P.M. Sarradin, A. Astruc, G.N. Kramer and B. Griepink. 1994a. Certified reference material (CRM 462) for the quality control of dibutyl- and tributyl-tin determinations in coastal sediment. Appl. Organomet. Chem. 8: 629–637.

Quevauviller, P., M. Astruc, L. Ebdon, G.N. Kramer and B. Griepink. 1994b. Interlaboratory study for the improvement of tributyltin determination in harbour sediment (rm 424). Appl. Organomet. Chem. 8: 639–644.

Quijano, M.A., P. Moreno, A.M. Gutiérrez, M.C. Pérez-Conde and C. Cámara. 2000. Selenium speciation in animal tissues after enzymatic digestion by high-performance liquid chromatography coupled to inductively coupled plasma mass spectrometry. J. Mass Spectrom. 35: 878–884.

Rahman, G.M.M., H.M. Skip Kingston, T.G. Towns, R.J. Vitale and K.R. Clay. 2005. Determination of hexavalent chromium by using speciated isotope-dilution mass spectrometry after microwave speciated extraction of environmental and other solid materials. Anal. Bioanal. Chem. 382: 1111–1120.

Rahman, F., Z.L. Chen and R. Naidu. 2009a. A comparative study of the extractability of arsenic species from silverbeet and amaranth vegetables. Environ. Geochem. Health. 31: 103–113.

Rahman, M.M., Z.L. Chen and R. Naidu. 2009b. Extraction of arsenic species in soils using microwave-assisted extraction detected by ion chromatography coupled to inductively coupled plasma mass spectrometry. Environ. Geochem. Health. 31: 93–102.

Rai, R., W. Maher and F. Kirkowa. 2002. Measurement of inorganic and methylmercury in fish tissues by enzymatic hydrolysis and HPLC-ICP-MS. J. Anal. At. Spectrom. 12: 1560–1563.

Rao, T.P., S. Daniel and J.M. Gladis. 2004. Tailored materials for preconcentration or separation of metals by ion-imprinted polymers for solid-phase extraction (IIP-SPE). Trends Anal. Chem. 23: 28–35.

Rao, T.P., R. Kala and S. Daniel. 2006. Metal ion-imprinted polymers-Novel materials for selective recognition of inorganics. Anal. Chim. Acta. 578: 105–116.

Rasmussen, K.E. and S. Pedersen-Bjergaard. 2004. Developments in hollow fibre-based, liquid-phase microextraction. Trend Anal. Chem. 23: 1–10.

Rattanachongkiat, S., G.E. Millward and M.E. Foulkes. 2004. Determination of arsenic species in fish, crustacean and sediment samples from Thailand using high performance liquid chromatography (HPLC) coupled with inductively coupled plasma mass spectrometry (ICP-MS). J. Environ. Monit. 6: 254–261.

Reuther, R., L. Jaeger and B. Allard. 1999. Determination of organometallic forms of mercury, tin and lead by *in situ* derivatization, trapping and gas chromatography—atomic emission detection. Anal. Chim. Acta. 394: 259–269.

Reyes, L.H., G.M.M. Rahman, T. Fahrenholz and H.M. Skip Kingston. 2008. Comparison of methods with respect to efficiencies, recoveries, and quantitation of mercury species interconversions in food demonstrated using tuna fish. Anal. Bioanal. Chem. 390: 2123–2132.

Rezaee, M., Y. Assadi, M.R.M. Hosseini, E. Aghaee, F. Ahmadi and S. Berijani. 2006. Determination of organic compounds in water using dispersive liquid–liquid microextraction. J. Chromatogr. A 1116: 1–9.

Río-Segade, S. and J.F. Tyson. 2007. Determination of methylmercury and inorganic mercury in water samples by slurry sampling cold vapor atomic absorption spectrometry in a flow injection system after preconcentration on silica C_{18} modified. Talanta. 71: 1696–1702.

Rivas, R.E., I. López-García and M. Hernández-Córdoba. 2009. Speciation of very low amounts of arsenic and antimony in waters using dispersive liquid–liquid microextraction and electrothermal atomic absorption spectrometry. Spectrochim. Acta Part B. 64: 329–333.

Rodil, R., A.M. Carro, R.A. Lorenzo, M. Abuín and R. Cela. 2002. Methylmercury determination in biological samples by derivatization, solid-phase microextraction and gas chromatography with microwave-induced plasma atomic emission spectrometry. J. Chromatog. A 963: 313–323.

Rodríguez Martín-Doimeadios, R.C., M. Monperrus, E. Krupp, D. Amouroux and O.F.X. Donard. 2003. Using speciated isotope dilution with GC-inductively coupled plasma MS to determine and unravel the artificial formation of monomethylmercury in certified reference sediments. Anal. Chem. 75: 3202–3211.

Rodríguez Pereiro, I. and A. Carro Díaz. 2002. Speciation of mercury, tin, and lead compounds by gas chromatography with microwave-induced plasma and atomic-emission detection (GC–MIP–AED). Anal. Bioanal. Chem. 372: 74–90.

Rodríguez-Pereiro, I., A. Wasik and R. Łobiński. 1998. Purge-and-trap isothermal multicapillary gas chromatographic sample introduction accessory for speciation of mercury by microwave-induced plasma atomic emission spectrometry. Anal. Chem. 70: 4063–4069.

Rodríguez-González, P., J.I. García Alonso and A. Sanz-Medel. 2004. Development of a triple spike methodology for validation of butyltin compounds speciation analysis by isotope dilution mass spectrometry. Part 2. Study of different extraction procedures for the determination of butyltin compounds in mussel tissue CRM 477. J. Anal. At. Spectrom. 19: 767–772.

Rodríguez-González, P., J.I. García Alonso and A. Sanz-Medel. 2005. Single and multiple spike procedures for the determination of butyltin compounds in sediments using isotope dilution GC-ICP-MS. J. Anal. At. Spectrom. 20: 1076–1084.

Roerdink, A. and J.H. Aldstadt III. 2004. Sensitive method for the determination of roxarsone using solid-phase microextraction with multi-detector gas chromatography. J. Chromatog. A. 1057: 177–183.

Ruiz Encinar, J., P. Rodríguez González, J.I. García Alonso and A. Sanz-Medel. 2002a. Evaluation of extraction techniques for the determination of butyltin compounds in sediments using isotope dilution-GC/ICPMS with ^{118}Sn and ^{119}Sn-enriched species. Anal. Chem. 74: 270–281.

Ruiz Encinar, J., P. Rodríguez-González, J. Rodríguez Fernández, J.I. García Alonso, S. Díez, J.M. Bayona and A. Sanz-Medel. 2002b. Evaluation of accelerated solvent extraction for butyltin speciation in PACS-2 CRM using double-spike isotope dilution-GC/ICPMS. Anal. Chem. 74: 5237–5242.

Ruiz Encinar, J., D. Schaumlöffel, Y. Ogra and R. Łobinski. 2004. Determination of selenomethionine and selenocysteine in human serum using speciated isotope dilution-capillary HPLC-inductively coupled plasma collision cell mass spectrometry. Anal. Chem. 76: 6635–6642.

Ruiz-Chancho, M.J., R. Sabé, J.F. López-Sánchez, R. Rubio and P. Thomas. 2005. New approaches to the extraction of arsenic species from soils. Microchim. Acta. 151: 241–248.

Saitoh, T. and W.L. Hinze. 1991. Concentration of hydrophobic organic compounds and extraction of protein using alkylammoniosulfate zwitterionic surfactant mediated phase separations (cloud point extractions). Anal. Chem. 63: 2520–2526.

Sakakibara, M., D. Wang, R. Takahasshi and S. Mori. 1996. Influence of ultrasound irradiation on hydrolysis of sucrose catalyzed by invertase. Enzyme Microb. Technol. 18: 444–448.

Saleh, A., Y. Yamini, M. Faraji, M. Rezaee and M. Ghambarian. 2009a. Ultrasound-assisted emulsification microextraction method based on applying low density organic solvents followed by gas chromatography analysis for the determination of polycyclic aromatic hydrocarbons in water samples. J. Chromatog. A 1216: 6673–6679.

Saleh, A., Y. Yamini, M. Faraji, S. Shariati and M. Rezaee. 2009b. Hollow fiber liquid phase microextraction followed by high performance liquid chromatography for determination of ultra-trace levels of Se(IV) after derivatization in urine, plasma and natural water samples. J. Chromatog. B 877: 1758–1764.

Sánchez, D.M., R. Martín, R. Morante, J. Marín and M.L. Munuera. 2000. Preconcentration speciation method for mercury compounds in water samples using solid phase extraction followed by reversed phase high performance liquid chromatography. Talanta. 52: 671–679.

Sánchez-Rojas, F., C. Bosch-Ojeda and J.M. Cano-Pavón. 2009. A Review of Stir Bar Sorptive Extraction Chromatographia. 69: S79–S94.

Santiago-Rivas, S., A. Moreda-Piñeiro, P. Bermejo-Barrera, J. Moreda-Piñeiro, E. Alonso-Rodríguez, S. Muniategui-Lorenzo, P. López-Mahía and D. Prada-Rodríguez. 2007. Pressurized liquid extraction-assisted mussel cytosol preparation for the determination of metals bound to metallothionein-like proteins. Anal. Chim. Acta. 603: 36–43.

Santos, H.M. and J.L. Capelo. 2007. Trends in ultrasonic-based equipment for analytical sample treatment. Talanta. 73: 795–802.

Santos, Jr. D., F.J. Krug, M. de Godoy Pereira and M. Korn. 2006. Currents on ultrasound-assisted extraction for sample preparation and spectroscopic analytes determination. Appl. Spectrosc. Rev. 41: 305–321.

Sanz, E., R. Muñoz-Olivas and C. Cámara. 2005a. Evaluation of a focused sonication probe for arsenic speciation in environmental and biological samples. J. Chromatogr. A 1097: 1–8.

Sanz, E., R. Muñoz-Olivas and C. Cámara. 2005b. A rapid and novel alternative to conventional sample treatment for arsenic speciation in rice using enzymatic ultrasonic probe. Anal. Chim. Acta. 535: 227–235.

Sanz, E., R. Muñoz-Olivas, C. Cámara, M.K. Sengupta and S. Ahamed. 2007. Arsenic speciation in rice, straw, soil, hair and nails samples from the arsenic-affected areas of Middle and Lower Ganga plain. J. Environ. Sci. Health A 42: 1695–1705.

Sanz-Landaluze, J., C. Dietz, Y. Madrid and C. Cámara. 2004. Volatile organoselenium monitoring in production and gastric digestion processes of selenized yeast by solid-phase microextraction-multicapillary gas chromatography coupled microwave-induced plasma atomic emission spectrometry. Appl. Organometal. Chem. 18: 606–613.

Sargar, B.M. and M.A. Anuse. 2001. Liquid–liquid extraction study of tellurium(IV) with N-n-octylaniline in halide medium and its separation from real samples. Talanta. 55: 469–478.

Sato, K., M. Kohri and H. Okochi. 1996. The speciation of organotin compounds in seawater by hydride purge-and-trap/ICP-MS. Bunseki Kagaku. 45: 575–582.

Saygi, K.O., E. Melek, M. Tuzen and M. Soylak. 2007. Speciation of selenium(IV) and selenium(VI) in environmental samples by the combination of graphite furnace atomic absorption spectrometric determination and solid phase extraction on Diaion HP-2MG. Talanta. 71: 1375–1381.

Schmidt, A.C., W. Reisser, J. Mattusch, P. Popp and R. Wennrich. 2000. Evaluation of extraction procedures for the ion chromatographic determination of arsenic species in plant materials. J. Chromatogr. A 889: 83–91.

Schmidt, A.C., W. Reisser, J. Mattusch, R. Wennrich and K. Jung. 2004. Analysis of arsenic species accumulation of plants and the influence on their nitrogen uptake. J. Anal. At. Spectrom. 19: 172–177.

Schmidt, A.-C., K. Kutschera, J. Mattusch and M. Otto. 2008. Analysis of accumulation, extractability, and metabolization of five different phenylarsenic compounds in plants by ion chromatography with mass spectrometric detection and by atomic emission spectroscopy. Chemosphere. 73: 1781–1787.

Schramm, L.L. 2005. Emulsions, Foams, and Suspensions. Fundamentals and Applications; Wiley-VCH. Weinheim.

Schwarz, J., G. Henze and F.G. Thomas. 1995. Voltammetric speciation of butyltin compounds. Fresenius' J. Anal. Chem. 352: 479–482.

Sellergren, B. 1994. Direct drug determination by selective sample enrichment on an imprinted polymer. Anal. Chem. 66: 1578–1582.

Seppänen, M.M., J. Kontturi, I. Lopez Heras, Y. Madrid, C. Cámara and H. Hartikainen. 2010. Agronomic biofortification of *Brassica* with selenium—enrichment of SeMet and its identification in *Brassica* seeds and meal. Plant Soil. 337: 273–283.

Serafimovska, J.M., S. Arpadjan and T. Stafilov. 2011. Speciation of dissolved inorganic antimony in natural waters using liquid phase semi-microextraction combined with electrothermal atomic absorption spectrometry. Microchem. J. 99: 46–50.

Serafimovski, I., I. Karadjova, T. Stafilov and J. Cvetković. 2008. Determination of inorganic and methylmercury in fish by cold vapour atomic absorption spectrometry and inductively coupled plasma atomic emission spectrometry. Microchem. J. 89: 42–47.

Serra, A.M., J.M. Estela and V. Cerdà. 2009. An MSFIA system for mercury speciation based on an anion-exchange membrane. Talanta. 78: 790–794.

Shah, A.Q. and G.A. Kandhro. 2012. Arsenic speciation and other parameters of surface and ground water samples of Jamshoro, Pakistan. Int. J. Environ. Anal. Chem. 92: 28–42.

Shemirani, F., M. Baghdadi and M. Ramezani. 2005. Preconcentration and determination of ultra trace amounts of arsenic(III) and arsenic(V) in tap water and total arsenic in biological samples by cloud point extraction and electrothermal atomic absorption spectrometry. Talanta. 65: 882–887.

Shen, Y., J.A. Jönsson and L. Mathiasson. 1998. On-line microporous membrane liquid–liquid extraction for sample pretreatment combined with capillary gas chromatography applied to local anaesthetics in blood plasma. Anal. Chem. 70: 946–953.

Shi, Y., R. Acharya and A. Chatt. 2001. Speciation of arsenic in natural waters by HPLC-NAA. J. Radioanal. Nucl. Chem. 262: 277–286.

Shioji, H., S. Tsunoi, H. Harino and M. Tanaka. 2004. Liquid-phase microextraction of tributyltin and triphenyltin coupled with gas chromatography–tandem mass spectrometry Comparison between 4-fluorophenyl and ethyl derivatizations. J. Chromatog. A 1048: 81–88.

Shuai, Z. and F. Zhefeng. 2011. Determination of antimony(III) and total antimony by dispersive liquid-liquid microextraction combined with electrothermal atomic absorption spectrometry. At. Spectrosc. 32: 75–79.

Sigrist, M., A. Albertengo, H. Beldoménico and M. Tudino. 2011. Determination of As(III) and total inorganic As in water samples using an on-line solid phase extraction and flow injection hydride generation atomic absorption spectrometry. J. Hazard. Mat. 188: 311–318.

Silva, M.F., E.S. Cerutti and L.D. Martinez. 2006. Coupling cloud point extraction to instrumental detection systems for Metal Analysis. Microchim. Acta. 155: 349–364.

Simon, R., J.E. Tietge, B. Michalke, S. Degitz and K. Schramm. 2002. Iodine species and the endocrine system: thyroid hormone levels in adult *Danio rerio* and developing *Xenopus laevis*. Anal. Bioanal. Chem. 372: 481–485.

Siwek, M., B. Galunsky and B. Niemeyer. 2005. Isolation of selenium organic species from antarctic krill after enzymatic hydrolysis. Anal. Bioanal. Chem. 381: 737–741.

Slaets, S. and F.C. Adams. 2000. Determination of organomercury compounds with a miniaturised automated speciation analyser. Anal. Chim. Acta. 414: 141–149.

Slaets, S., F. Adams, I. Rodríguez-Pereiro and R. Łobiński. 1999. Optimization of the coupling of multicapillary GC with ICP-MS for mercury speciation analysis in biological materials J. Anal. At. Spectrom. 14: 851–857.

Šlejkovec, Z., J.T. van Elteren and A.R. Byrne. 1998. A dual arsenic speciation system combining liquid chromatographic and purge and trap-gas chromatographic separation with atomic fluorescence spectrometric detection. Anal. Chim. Acta. 358: 51–60.

Sloth, J.J., E.H. Larsen and K. Julshamn. 2003. Determination of organoarsenic species in marine samples using gradient elution cation exchange HPLC-ICP-MS. J. Anal. At. Spectrom. 18: 452–459.

Sloth, J.J., E.H. Larsen and K. Julshamn. 2005. Survey of inorganic arsenic in marine animals and marine certified reference materials by anion exchange high-performance liquid chromatography-inductively coupled plasma mass spectrometry. J. Agric. Food Chem. 53: 6011–6018.

Smichowski, P., L. Valiente and A. Ledesma. 2002. Simple method for the selective determination of As(III) and As(V) by ETAAS after separation with anion exchange mini-column. At. Spectrosc. 23: 92–97.

Smith, L.M., W.A. Maher, P.J. Craig and R.O. Jenkins. 2002. Speciation of volatile antimony compounds in culture headspace gases of *Cryptococcus humicolus* using solid phase microextraction and gas chromatography-mass spectrometry. Appl. Organometal Chem. 16: 287–293.

Smrkolj, P., V. Stibilj, I. Kreft and E. Kapolna. 2005. Selenium species determination in selenium-enriched pumpkin (*Cucurbita pepo* L.) seeds by HPLC-UV-HG-AFS. Anal. Sci. 21: 1501–1504.

Smrkolj, P., V. Stibilj, I. Kreft and M. Germ. 2006. Selenium species in buckwheat cultivated with foliar addition of Se(VI) and various levels of UV-B radiation. Food Chem. 96: 675–681.

Snell, J.P., W. Frech and Y. Thomassen. 1996. Performance improvements in the determination of mercury species in natural gas condensate using an on-line amalgamation trap or solid-phase micro-extraction with capillary gas chromatography–microwave-induced plasma atomic emission spectrometry. Analyst. 121: 1055–1060.

Soko, L., L. Chimuka, E. Cukrowska and S. Pole. 2003. Extraction and preconcentration of manganese(II) from biological fluids (water, milk and blood serum) using supported liquid membrane and membrane probe methods. Anal. Chim. Acta. 485: 25–35.

Sosa-Ferrera, Z., C. Padrón-Sanz, C. Mahugo-Santana and J.J. Santana-Rodríguez. 2004. The use of micellar systems in the extraction and pre-concentration of organic pollutants in environmental samples. Trend Anal. Chem. 23: 469–479.

Sounderajan, S., G.K. Kumar, S.A. Kumar, A.C. Udas and G. Venkateswaran. 2009. Characterization of As(V), As (III) by selective reduction/adsorption on palladium nanoparticles in environmental water samples. Talanta. 78: 1122–1128.

Stäb, J.A., W.P. Cofino, B. Hattum and U.A.T. Brinkman. 1993. Comparison of GC/MSD and GC/AED for the determination of organotin compounds in the environment. Fresenius´ J. Anal. Chem. 347: 247–255.

Stadlober, M., M. Sager and K.J. Irgolic. 2001. Effects of selenate supplemented fertilisation on the selenium level of cereals—identification and quantification of selenium compounds by HPLC–ICP–MS. Food Chem. 73: 357–366.

Stalikas, C.D. 2002. Micelle-mediated extraction as a tool for separation and preconcentration in metal analysis. Trend Anal. Chem. 21: 343–355.

Stalikas, C.D. and Y.C. Fiamegos. 2008. Microextraction combined with derivatization. Trends Anal. Chem. 27: 533–542.

Stefánka, Z., I. Ipolyi, M. Dernovics and P. Fodor. 2001. Comparison of sample preparation methods based on proteolytic enzymatic processes for Se-speciation of edible mushroom (*Agaricus bisporus*) samples. Talanta. 55: 437–447.

Stiboller, M., M. Damm, A.M. Barbera, D. Kuehnelt, K.A. Francesconi and C.O.A Kappe. 2011. A miniaturized microtiter plate protocol for the determination of selenomethionine in selenized yeast via enzymatic hydrolysis of protein-bound selenium. Anal. Methods. 3: 738–741.

Stripeikis, J., J. Pedro, A. Bonivardi and M. Tudino. 2004. Determination of selenite and selenate in drinking water: a fully automatic on-line separation/pre-concentration system coupled to electrothermal atomic spectrometry with permanent chemical modifiers. Anal. Chim. Acta. 502: 99–105.

Sturgeon, R.E., R. Wahlen, T. Brandsch, B. Fairman, C. Wolff-Briche, J.I. Garcia Alonso, P. Rodríguez González, J. Ruiz Encinar, A. Sanz-Medel, K. Inagaki, A. Takatsu, B. Lalere, M. Monperrus, O. Zuloaga, E. Krupp, D. Amouroux, O.F.X. Donard, H. Schimmel, B. Sejeroe-Olsen, P. Konieczka, P. Schultze, P. Taylor, R. Hearn Mackay, R. Myors, T. Win, A. Liebich, R. Philipp, L. Yang and S. Willie. 2003. Determination of tributyltin in marine sediment: Comité Consultatif pour la Quantité de Matière (CCQM) pilot study P-18 international intercomparison. Anal. Bioanal. Chem. 376: 780–787.

Sun, Y.C. and J.Y. Yang. 1999. Simultaneous determination of arsenic(III,V), selenium(IV,VI), and antimony(III,V) in natural water by coprecipitation and neutron activation analysis. Anal. Chim. Acta. 395: 293–300.

Suvarna-Sounderajan, G., K. Kumar and A.C. Udas. 2010. Cloud point extraction and electrothermal atomic absorption spectrometry of Se (IV)-3,3-Diaminobenzidine for the estimation of trace amounts of Se(IV) and Se(VI) in environmental water samples and total selenium in animal blood and fish tissue samples. J. Hazard. Mat. 175: 666–672.

Swearingen. J.W., A. Manuel, A.M.F. Plishker, C.P. Saavedra, C.C. Vásquez and T.G. Chasteen. 2004. Identification of biogenic organotellurides in Escherichia coli K-12 headspace gases using solid-phase microextraction and gas chromatography. Anal. Biochem. 331: 106–114.

Swinnerton. J.W., C.H. Cheek and V.J. Linnenbom. 1962. Determination of dissolved gas in aqueous solutions by gas chromatography. Anal. Chem. 34: 483–485.

Tajes-Martínez, P., E. Beceiro-González, S. Muniategui-Lorenzo and D. Prada-Rodríguez. 2006. Micro-columns packed with Chlorella vulgaris immobilized on silica gel for mercury speciation. Talanta. 68: 1489–1496.

Takahiko, B., K. Fumio, N. Susumu and T. Katsuroku. 2000. Study of drop coalescence behavior for liquid–liquid extraction operation. Chem. Eng. Sci. 55: 5385–5391.

Talukder, M.M.R., M.M. Zaman, Y. Hayashi, J.C. Wu and T. Kawanishi. 2006. Ultrasonication enhanced hydrolytic activity of lipase in water/isooctane two-phase systems. Biocatal. Biotransform. 24: 189–194.

Tamer, U., B. Yates, A. Galal, T. Gbatu, R. LaRue, C., Schmiesing, K. Temsamani, O. Ceylan and H.B. Mark. 2003. Electrochemically aided control of solid phase micro-extraction (EASPME) using conducting polymer-coated solid substrates applicable to neutral analytes. Microchim. Acta. 143: 205–215.

Tanzer, D. and K.G. Heumann. 1990. GC determination of dimethyl selenide and trimethyl selenonium ions in aquatic systems using element specific detection. Atmos. Environ. 24: 3099–3102.

Tao, H., R. Babu-Rajendran, C.R. Quetel, T. Nakazato, M. Tominaga and A. Miyazaki. 1999. Tin speciation in the femtogram range in open ocean seawater by gas chromatography/ inductively coupled plasma mass spectrometry using a shield torch at normal plasma conditions. Anal. Chem. 71: 4208–4215.

Taylor, V.F., B.P. Jackson and C.Y. Chen. 2008. Mercury speciation and total trace element determination of low-biomass biological samples. Anal. Bioanal. Chem. 392: 1283–1290.

Taylor, V.F., A. Carter, C. Davies and B.P. Jackson. 2011. Trace-level automated mercury speciation analysis. Anal. Methods. 3: 1143–1148.

Telford, K., W. Maher, F. Krikowa and S. Foster. 2008. Measurement of total antimony and antimony species in mine contaminated soils by ICPMS and HPLC-ICPMS. J. Environ. Monit. 10: 136–140.

Tessier, E., D. Amouroux, G. Abril, E. Lemaire and O.F.X. Donard. 2002. Formation and volatilisation of alkyl-iodides and -selenides in macrotidal estuaries. Biogeochem. 59: 183–206.

Theis, A.L., A.J. Waldack, S.M. Hansen and M.A. Jeannot. 2001. Headspace solvent microextraction. Anal. Chem. 73: 5651–5654.

Tian, Y., C. Ming-Li, C. Xu-Wei, W. Jian-Hua, Y. Hirano, H. Sakamoto and T. Shirasaki. 2011. Arsenic preconcentration via solid phase extraction and speciation by HPLC-gradient hydride generation atomic absorption spectrometry. J. Anal. At. Spectrom. 26: 133–140.

Tolosa, I., J.M. Tolosa, J. Bayona, J. Albaigés, L.F. Alencastro and J. Tarradellas. 1991. Organotin speciation in aquatic matrices by CGC/FPD, ECD and MS, and LC/MS. Fresenius' J. Anal. Chem. 339: 646–653.

Tolosa, I., J.W. Readman, A. Blaevoet, S. Ghilini, J. Bartocci and M. Hovart. 1996. Contamination of Mediterranean (Cote d'Azur) Coastal Waters by Organotins and Irgarol 1051 used in Antifouling Paints. Mar. Pollut. Bull. 32: 335–341.

Tor, A. and M.E. Aydin. 2006. Application of liquid-phase microextraction to the analysis of trihalomethanes in water. Anal. Chim. Acta. 575: 138–143.

Trivelin, L.A., J.J. Rodrigues-Rohwedder and S. Rath. 2006. Determination of pentavalent antimony in antileishmaniotic drugs using an automated system for liquid–liquid extraction with on-line detection. Talanta. 68: 1536–1543.

Tsunoi, S., H. Shioji and M. Tanaka. 2004. Derivatization of tributyltin with sodium tetrakis(4-fluorophenyl)borate for sensitivity improvement of tandem mass spectrometry. Anal. Sci. 20: 101–105.

Tu, Q., J. Qian and W. Frech. 2000. Rapid determination of methylmercury in biological materials by GC-MIP-AES or GC-ICP-MS following simultaneous ultrasonic assisted *in situ* ethylation and solvent extraction. J. Anal. At. Spectrom. 15: 1583–1588.

Tukai, R., W.A. Maher, I.J. McNaught and M.J. Ellwood. 2002. Measurement of arsenic species in marine macroalgae by microwave-assisted extraction and high performance liquid chromatography–inductively coupled plasma mass spectrometry. Anal. Chim. Acta. 457: 173–185.

Tutschku, S., S. Mothes and R. Wennrich. 1996. Preconcentration and determination of Sn- and Pb-organic species in environmental samples by SPME and GC-AED. Fresenius' J. Anal. Chem. 354: 587–591.

Tutschku, S., M.M. Schantz and S.A. Wise. 2002. Determination of methylmercury and butyltin compounds in marine biota and sediments using microwave-assisted acid extraction, solid-phase microextraction, and gas chromatography with microwave-induced plasma atomic emission spectrometric detection. Anal. Chem. 74: 4694–4701.

Tuzen M., K.O. Saygi and M. Soylak. 2007. Separation and speciation of selenium in food and water samples by the combination of magnesium hydroxide coprecipitation-graphite furnace atomic absorption spectrometric determination. Talanta. 71: 424–429.

Ugarte, A., N. Unceta, M.C. Sampedro, M. Aránzazu-Goicolea, A. Gómez-Caballero and R.J. Barrio. 2009. Solid phase microextraction coupled to liquid chromatography-inductively coupled plasma mass spectrometry for the speciation of organotin compounds in water samples. J. Anal. At. Spectrom. 24: 347–351.

Uluozlu, O.D., M. Tuzen, D. Mendil and M. Soylak. 2010. Determination of As(III) and As(V) species in some natural water and food samples by solid-phase extraction on Streptococcus pyogenes immobilized on Sepabeads SP 70 and hydride generation atomic absorption spectrometry. Food Chem. Toxicol. 48: 1393–1398.

Ulusoy, H.I., M. Akçay, S. Ulusoy and R. Gürkan. 2011a. Determination of ultra trace arsenic species in water samples by hydride generation atomic absorption spectrometry after cloud point extraction. Anal. Chim. Acta. 703: 137–144.

Ulusoy, H.I., M. Akçay and R. Gürkan. 2011b. Development of an inexpensive and sensitive method for the determination of low quantity of arsenic species in water samples by CPE–FAAS. Talanta. 85: 1585–1591.

Vale, G., R. Rial-Otero, A. Mota, L. Fonseca and J.L. Capelo. 2008. Ultrasonic-assisted enzymatic digestion (USAED) for total elemental determination and elemental speciation: A tutorial. Talanta 75: 872–884.

Vale, G., A. Rodrigues, A. Rocha, R. Rial, A.M. Mota, M.L. Gonçalves, L.P. Fonseca and J.L. Capelo. 2010. Ultrasonic assisted enzymatic digestion (USAED) coupled with high performance liquid chromatography and electrothermal atomic absorption spectrometry as a powerful tool for total selenium and selenium species control in Se-enriched food supplements. Food Chem. 121: 268–274.

Välimäki, I. and P. Perämäki. 2001. Determination of mercury species by capillary column GC-QTAAS with purge and trap preconcentration technique. Mikrochim. Acta. 137: 191–201.

Van Elteren, J.T., N.G. Haselager, H.A. Das, C.L. de Ligny and J. Agterdenbos. 1991. Determination of arsenate in aqueous samples by precipitation of the arsenic(V)-molybdate complex with tetraphenylphosphonium chloride and neutron activation analysis or hydride generation atomic absorption spectrometry. Anal. Chim. Acta. 252: 89–95.

Van, D.N., T.T.X. Bui and S. Tesfalidet. 2008. The transformation of phenyltin species during sample preparation of biological tissues using multi-isotope spike SSID-GC-ICPMS. Anal. Bioanal. Chem. 392: 737–747.

Vela, N.P. and D.T. Heitkemper. 2004. Total arsenic determination and speciation in infant food products by ion chromatography-inductively coupled plasma-mass spectrometry. J. AOAC Int. 87: 244–252.

Vela, N.P., D.T. Heitkemper and K.R. Stewart. 2001. Arsenic extraction and speciation in carrots using accelerated solvent extraction, liquid chromatography and plasma mass spectrometry. Analyst. 126: 1011–1017.

Vercauteren, J., A. De Meester, T. De Smaele, F. Vanhaecke, L. Moens, R. Dams and P. Sandra. 2000. Headspace solid-phase microextraction-capillary gas chromatography-ICP mass spectrometry for the determination of the organotin pesticide fentin in environmental samples. J. Anal. At. Spectrom. 15: 651–656.

Vercauteren, J., C. Pérès, C. Devos, P. Sandra, F. Vanhaecke and L. Moens. 2001. Stir bar sorptive extraction for the determination of ppq-Level traces of organotin compounds in environmental samples with thermal desorption-capillary gas chromatography-ICP mass spectrometry. Anal. Chem. 73: 1509–1514.

Vereda-Alonso, E., M.T. Siles-Cordero, A. García de Torres, P. Cañada-Rudner and J.M. Cano-Pavón. 2008. Mercury speciation in sea food by flow injection cold vapor atomic absorption spectrometry using selective solid phase extraction. Talanta. 77: 53–59.

Vidler, D.S., R.O. Jenkins, J.F. Hall and C.F. Harrington. 2007. The determination of methylmercury in biological samples by HPLC coupled to ICP-MS detection. Appl. Organometal. Chem. 21: 303–310.

Vilanó, M. and R. Rubio. 2001. Determination of arsenic species in oyster tissue by microwave-assisted extraction and liquid chromatography-atomic fluorescence detection. Appl. Organometal. Chem. 15: 658–666.

Vogrinčič, M., P. Cuderman, I. Kreft and V. Stibilj. 2009. Selenium and its species distribution in above-ground plant parts of selenium enriched buckwheat (*Fagopyrum esculentum* Moench). Anal. Sci. 25: 1357–1363.

Voice, T.C., L.V. Flores del Pino, I. Havezov and D.T. Long. 2011. Field deployable method for arsenic speciation in water. Phys. Chem. Earth. 36: 436–441.

Vonderheide, A.P., M. Montes-Bayon and J.A. Caruso. 2002. Solid-phase microextraction as a sample preparation strategy for the analysis of seleno amino acids by gas chromatography-inductively coupled plasma mass spectrometry. Analyst. 127: 49–53.

Wahlen, R. 2004. Fast and accurate determination of arsenobetaine in fish tissues using accelerated solvent extraction and HPLC-ICP-MS determination. J. Chromatogr. Sci. 42: 217–222.

Wahlen, R. and C. Wolff-Briche. 2003. Comparison of GC–ICP–MS and HPLC–ICP–MS for species-specific isotope dilution analysis of tributyltin in sediment after accelerated solvent extraction. Anal. Bioanal. Chem. 377: 140–148.

Wahlen, R. and T. Catterick. 2003. Comparison of different liquid chromatography conditions for the separation and analysis of organotin compounds in mussel and oyster tissue by liquid chromatography–inductively coupled plasma mass spectrometry. J. Chromatogr. B 783: 221–229.

Wahlen, R. and T. Catterick. 2004. Simultaneous co-extraction of organometallic species of different elements by accelerated solvent extraction and analysis by inductively coupled plasma mass spectrometry coupled to liquid and gas chromatography. Rapid Commun. Mass. Spectrom. 18: 211–217.

Wahlen, R., S. McSheehy, C. Scrivera and Z. Mester. 2004. Arsenic speciation in marine certified reference materials Part 2. The quantification of water-soluble arsenic species by high-performance liquid chromatography-inductively coupled plasma mass spectrometry. J. Anal. At. Spectrom. 19: 876–882.

Wai, C.M., Y. Lin, R. Brauer, S. Wand and W. Beckert. 1993. Supercritical fluid extraction of organic and inorganic mercury from solid materials. Talanta. 40: 1325–1330.

Wang, J.S. and K.-H. Chiu. 2007. Extraction of Chromium(III) and Chromium(VI) species from solid matrices using green solvent supercritical carbon dioxide. Anal. Sci. 23: 1337–1341.

Wang, R.-Y., Y.-L. Hsu, L.-F. Chang and S.-J. Jiang. 2007. Speciation analysis of arsenic and selenium compounds in environmental and biological samples by ion chromatography–inductively coupled plasma dynamic reaction cell mass spectrometer. Anal. Chim. Acta. 590: 239–244.

Wang, Y., X. Luo, J. Tang and X. Hu. 2011. Determination of Se(IV) using solidified floating organic drop microextraction coupled to ultrasound-assisted back-extraction and hydride generation atomic fluorescence spectrometry. Microchim. Acta. 173: 267–273.

Wasik, A. and T. Ciesielski. 2004. Determination of organotin compounds in biological samples using accelerated solvent extraction, sodium tetraethylborate ethylation, and multicapillary gas chromatography-flame photometric detection. Anal. Bioanal. Chem. 378: 1357–1363.

Wasik, A., I. Rodríguez-Pereiro and R. Łobiński. 1998. Interface for time-resolved introduction of gaseous analytes for atomic spectrometry by purge-and-trap multicapillary gas chromatography (PTMGC). Spectrochim. Acta Part B. 53: 867–879.

Wasik, A., B. Radke, J. Bolałek and J. Namieśnik. 2007. Optimisation of pressurised liquid extraction for elimination of sulphur interferences during determination of organotin compounds in sulphur-rich sediments by gas chromatography with flame photometric detection. Chemosphere. 68: 1–9.

Watts, M.J., J. O'Reilly, A.L. Marcilla, R.A. Shaw and N.I. Ward. 2010. Field based speciation of arsenic in UK and Argentinean water samples. Environ. Geochem. Health. 32: 479–490.

Wei-Xun, W., Y. Tzung-Jie, L. Zu-Guang, J. Ting-Ting and L. Maw-Rong. 2011. A novel method of ultrasound-assisted dispersive liquid–liquid microextraction coupled to liquid chromatography–mass spectrometry for the determination of trace organoarsenic compounds in edible oil. Anal. Chim. Acta. 690: 221–227.

Welz, B. and M. Sperling. 1999. Atomic Absorption Spectrometry. Wiley–VCH Verlag. Weinheim, Germany.

Wercinski, S.A.S. 1999. Solid Phase Micro-extraction: a Practical Guide, Marcel Dekker, New York.

Wickenheiser, E.B., K. Michalke, C. Drescher, A.V. Hirner and R. Hensel. 1998. Development and application of liquid and gas-chromatographic speciation techniques with element specific (ICP-MS) detection to the study of anaerobic arsenic metabolism. Fresenius' J. Anal. Chem. 362: 498–501.

Williams, G., J.M. West, I. Koch, K.J. Reimer and E.T. Snow. 2009. Arsenic speciation in the freshwater crayfish, *Cherax destructor* Clark. Sci. Total Environ. 407: 2650–2658.

Wójcik, P., K. Pyrzynska and M. Biesaga. 2003. Ion-chromatography of inorganic selenium species with a preliminary preconcentration step. Chromatog. 57: 67–71.

Wrobel, K., K. Wrobel, S.S. Kannamkumarath and J.A. Caruso. 2003a. Identification of selenium species in urine by ion-pairing HPLC–ICP–MS using laboratory-synthesized standards. Anal. Bioanal. Chem. 377: 670–674.

Wrobel, K., S.S. Kannamkumarath, K. Wrobel and J.A. Caruso. 2003b. Hydrolysis of proteins with methanesulfonic acid for improved HPLC-ICP-MS determination of seleno-methionine in yeast and nuts. Anal. Bioanal. Chem. 375: 133–138.

Wu, H., Y. Jin, W. Han, Q. Miao and S. Bi. 2006. Non-chromatographic speciation analysis of mercury by flow injection on-line preconcentration in combination with chemical vapor generation atomic fluorescence spectrometry. Spectrochim. Acta Part B. 61: 831–840.

Wu, H., Y. Jin, Y. Shi and S. Bi. 2007. On-line organoselenium interference removal for inorganic selenium species by flow injection coprecipitation preconcentration coupled with hydride generation atomic fluorescence spectrometry. Talanta. 71: 1762–1768.

Wu, H., X. Wang, B. Liu, Y. Liu, S. Li, J. Lu, J. Tian, W. Zhao and Z. Yang. 2011. Simultaneous speciation of inorganic arsenic and antimony in water samples by hydride generation -double channel atomic fluorescence spectrometry with on-line solid-phase extraction using single-walled carbon nanotubes micro-column. Spectrochim. Acta Part B. 66: 74–80.

Wu, J., Z. Mester and J. Pawliszyn. 2000. Speciation of organoarsenic compounds by polypyrrole-coated capillary in-tube solid phase microextraction coupled with liquid chromatography/electrospray ionization mass spectrometry. Anal. Chim. Acta. 424: 211–222.

Wu, J., Z. Mester and J. Pawliszyn. 2001. Determination of tributyltin by automated in-tube solid-phase microextraction coupled with HPLC-ES-MS. J. Anal. At. Spectrom. 16: 159–165.

Xia, L., B. Hu, Z. Jiang, Y. Wu and Y. Liang. 2004. Single-drop microextraction combined with low-temperature electrothermal vaporization ICPMS for the determination of trace Be, Co, Pd, and Cd in biological samples. Anal. Chem. 76: 2910–2915.

Xia, L., B. Hu, Z. Jiang, Y. Wu, R. Chen and L. Li. 2006. Hollow fiber liquid phase microextraction combined with electrothermal vaporization ICP-MS for the speciation of inorganic selenium in natural waters. J. Anal. At. Spectrom. 21: 362–365.

Xia, L., B. Hu and Y. Wu. 2007. Hollow fiber-based liquid–liquid–liquid microextraction combined with high-performance liquid chromatography for the speciation of organomercury. J. Chromatog. A 1173: 44–51.

Xiao, Y.B., J. Ling, S.H. Qian, A.Q. Lin, W.J. Zheng, W.Y. Xu, Y.X. Luo and M. Zhang. 2007. Preconcentration of trace arsenite and arsenate with titanium dioxide nanoparticles and subsequent determination by silver diethyldithiocarbamate spectrophotometric method. Water Environ. Res. 79: 1015–1022.

Xiong, C., M. He and B. Hu. 2008. On-line separation and preconcentration of inorganic arsenic and selenium species in natural water samples with CTAB-modified alkyl silica microcolumn and determination by inductively coupled plasma-optical emission spectrometry. Talanta. 76: 772–779.

Xu, L., C. Basheer and H.K. Lee. 2007. Developments in single-drop microextraction. J. Chromatogr. A 1152: 184–192.

Xu, S.Q., Z.-Q Xie, W. Liu, H.-X. Yang and B. Li. 2010. Extraction and determination of total bromine, iodine, and their species in atmospheric aerosol. Chin. J. Anal. Chem. 38: 219–224.

Xue-Bo, Y. 2007. Dual-cloud point extraction as a preconcentration and clean-up technique for capillary electrophoresis speciation analysis of mercury. J. Chromatog. A 1154: 437–443.

Yalei, C., Q. Wenqi, C. Jieshan and C. Mou-Sen. 1993. Determination of arsenic(V) and arsenic(III) species in environmental samples by coprecipitation with zirconium hydroxide and pre-atomization atomic absorption spectrometry. J. Anal. At. Spectrom. 8: 379–381.

Yang, G.D., J.H. Xu, L.J. Xu, G.N. Chen and F.F. Fu. 2010. Analysis of ultratrace triorganotin compounds in aquatic organisms by using capillary electrophoresis–inductively coupled plasma mass spectrometry. Talanta. 80: 1913–1918.

Yang, J., E. Sargent, S. Kelley and J.Y. Ying. 2009. A general phase-transfer protocol for metal ions and its application in nanocrystal synthesis. Nat. Mater. 8: 683–689.

Yang, L. and J.W.H. Lam. 2001. Microwave-assisted extraction of butyltin compounds from PACS-2 Sediment for quantitation by high-performance liquid chromatography inductively coupled plasma mass spectrometry. J. Anal. At. Spectrom. 16: 724–731.

Yang, L., V. Colombini, P. Maxwell, Z. Mester and R.E. Sturgeon. 2003a. Application of isotope dilution to the determination of methylmercury in fish tissue by solid-phase microextraction gas chromatography—mass spectrometry. J. Chromatogr. A 1011: 135–142.

Yang, L., Z. Mester and R.E. Sturgeon. 2003b. Determination of methylmercury in fish tissues by isotope dilution SPME-GC-ICP-MS. J. Anal. At. Spectrom. 18: 431–436.

Yang, L., R.E. Sturgeon, S. McSheehy and Z. Mester. 2004. Comparison of extraction methods for quantitation of methionine and selenomethionine in yeast by species specific isotope dilution gas chromatography–mass spectrometry. J. Chromatogr. A 1055: 177–184.

Yates, B.J., K.R. Temsamani, O. Ceylan, S. Oztemiz, T.P. Gbatu, R.A. LaRue, U. Tamer and H.B. Mark. 2002. Electrochemical control of solid phase micro-extraction: conducting polymer coated film material applicable for preconcentration/analysis of neutral species. Talanta. 58: 739–745.

Yeh, C.-F. and S.-J. Jiang. 2005. Speciation of arsenic compounds in fish and oyster tissues by capillary electrophoresis-inductively coupled plasma-mass spectrometry. Electrophoresis. 26: 1615–1621.

Yeuk-Ki, T., S. Tam and L.K. Sze-Yin. 2010. Rapid speciation of methylated and ethylated mercury in urine using headspace solid phase microextraction coupled to LC-ICP-MS. J. Anal. At. Spectrom. 25: 1758–1762.

Yin, X.-B. 2007. Dual-cloud point extraction as a preconcentration and clean-up technique for capillary electrophoresis speciation analysis of mercury. J. Chromatogr. A 1154: 437–443.

Yin, X., W. Frech, E. Hoffmann, C. Lüdke and J. Skole. 1998. Mercury speciation by coupling cold vapour atomic absorption spectrometry with flow injection on-line preconcentration and liquid chromatographic separation. Fresenius´ J. Anal. Chem. 361: 761–766.

Yong-Guang, Y., M. Chen, P. Jin-Feng, L. Jing-Fu and J. Gui-Bin. 2010. Dithizone-functionalized solid phase extraction–displacement elution-high performance liquid chromatography–inductively coupled plasma mass spectrometry for mercury speciation in water samples. Talanta. 81: 1788–1792.

Yousefi, S.R., F. Shemirani and M.R. Jamali. 2010. Determination of antimony(III) and total antimony in aqueous samples by electrothermal atomic absorption spectrometry after dispersive liquid-liquid microextraction (DLLME). Anal. Sci. 43: 2563–2571.

Yu, C., Q. Cai, Z.X. Guo, Z. Yang and S.B. Khoo. 2002. Antimony speciation by inductively coupled plasma mass spectrometry using solid phase extraction cartridges. Analyst. 127: 1380–1385.

Yu, C., Q. Cai, Z.X. Guo, Z. Yang and S.B. Khoo. 2003a. Speciation analysis of tellurium by solid-phase extraction in the presence of ammonium pyrrolidine dithiocarbamate and inductively coupled plasma mass spectrometry. Anal. Bioanal. Chem. 376: 236–242.

Yu, C., Q. Cai, Z.X. Guo, Z. Yang and S.B. Khoo. 2003b. Inductively coupled plasma mass spectrometry study of the retention behavior of arsenic species on various solid phase extraction cartridges and its application in arsenic speciation. Spectrochim. Acta Part B. 58: 1335–1349.

Yu, C., Q. Cai, Z.X. Guo, Z. Yang and S.B. Khoo 2004. Simultaneous speciation of inorganic selenium and tellurium by inductively coupled plasma mass spectrometry following selective solid-phase extraction separation. J. Anal. At. Spectrom. 19: 410–413.

Yu, X. and J. Pawliszyn. 2000. Speciation of alkyllead and inorganic lead by derivatization with deuterium-labeled sodium tetraethylborate and SPME-GC/MS. Anal. Chem. 72: 1788–1792.

Yu, Z.H., M. Jing, X.R. Wang, D.Y. Chen and Y.L. Huang. 2009. Simultaneous determination of multi-organotin compounds in seawater by liquid-liquid extraction-high performance liquid chromatography-inductively coupled plasma mass spectrometry. Spectrosc. Spectral Anal. 29: 2855–2859.

Yu, Z.-H., J.-Q. Sun, M. Jing, X. Cao, F. Lee and X.-R. Wang. 2010. Determination of total tin and organotin compounds in shellfish by ICP-MS. Food Chem. 119: 364–367.

Yuan, C.-G., G.-B. Jiang and B. He. 2005. Evaluation of the extraction methods for arsenic speciation in rice straw, *Oryza sativa* L., and analysis by HPLC-HG-AFS. J. Anal. At. Spectrom. 20: 103–110.

Yuan, C.-G., B. He, E.-L. Gao, J.-X. Lü and G.-B. Jiang. 2007. Evaluation of extraction methods for arsenic speciation in polluted soil and rotten ore by HPLC-HG-AFS analysis. Microchim. Acta. 159: 175–182.

Yuan, C., X. Lu, J. Qin, B.P. Rosen and X.C. Le. 2008. Volatile arsenic species released from *Escherichia coli* expressing the AsIII S-adenosylmethionine methyltransferase gene. Environ. Sci. Technol. 42: 3201–3206.

Yusof, A.M., S. Salleh and A.K.H. Wood. 1999. Speciation of inorganic arsenic and selenium in leachates from landfills in relation to water quality assessment. Biol. Trace Elemen. Res. 71: 139–148.

Zabaljauregui, M., A. Delgado, A. Usobiaga, O. Zuloaga, A. de Diego and J.M. Madariaga. 2007. Fast method for routine simultaneous analysis of methylmercury and butyltins in seafood. J. Chromatogr. A 1148: 78–85.

Zachariadis, G.A. and D.C. Kapsimali. 2008. Effect of sample matrix on sensitivity of mercury and methylmercury quantitation in human urine, saliva, and serum using GC-MS. J. Sep. Sci. 31: 3884–3893.

Zachariadis, G.A. and E. Rosenberg. 2009a. Speciation of organotin compounds in urine by GC–MIP-AED and GC–MS after ethylation and liquid–liquid extraction. J. Chrom. B 877: 1140–1144.

Zachariadis, G.A. and E. Rosenberg. 2009b. Determination of butyl- and phenyltin compounds in human urine by HS-SPME after derivatization with tetraethylborate and subsequent determination by capillary GC with microwave-induced plasma atomic emission and mass spectrometric detection. Talanta. 78: 570–576.

Zachariadis, G.A., A.N. Anthemidis, E.I. Daftsis and J.A. Stratis. 2005. On-line speciation of mercury and methylmercury by cold vapour atomic absorption spectrometry using selective solid phase extraction. J. Anal. At. Spectrom. 20: 63–65.

Zanjani, M.R.K., Y. Yamini, S. Shariati and J.A. Jonsson. 2007. A new liquid-phase micro-extraction method based on solidification of floating organic drop. Anal. Chim. Acta. 585: 286–293.

Zapf, A., R. Heyer and H. Stan. 1995. Rapid micro liquid-liquid extraction method for trace analysis of organic contaminants in drinking water. J. Chromatogr. A 694: 453–461.

Zeng, C., F. Yang and N. Zhou. 2011. Hollow fiber supported liquid membrane extraction coupled with thermospray flame furnace atomic absorption spectrometry for the speciation of Sb(III) and Sb(V) in environmental and biological samples. Microchem. J. 98: 307–311.

Zhai, Y., S. Duan, Q. He, X. Yang and Q. Han. 2010. Solid phase extraction and preconcentration of trace mercury(II) from aqueous solution using magnetic nanoparticles doped with 1,5-diphenylcarbazide. Microchim. Acta. 169: 353–360.

Zhang, L., Y. Morita, A. Sakuragawa and A. Isozaki. 2007. Inorganic speciation of As(III,V), Se(IV,VI) and Sb(III,V) in natural water with GF-AAS using solid phase extraction technology. Talanta. 72: 723–729.

Zhang, Y., T. William and J.R. Frankenberger. 2000. Formation of dimethylselenonium compounds in soil. Environ. Sci. Technol. 34: 776–783.

Zhao, R., W. Lao and X. Xu. 2004a. Headspace liquid-phase microextraction of trihalomethanes in drinking water and their gas chromatographic determination. Talanta. 62: 751–756.

Zhao, R.S., S. Chu and X.B. Xu. 2004b. Optimization of nonequilibrium liquid-phase microextraction for the determination of nitrobenzenes in aqueous samples by gas chromatography-electron capture detection. Anal. Sci. 20: 663–666.

Zheng, J. and H. Hintelmann. 2009. HPLC-ICP-MS for a comparative study on the extraction approaches for arsenic speciation in terrestrial plant, *Ceratophyllum demersum*. J. Radioanal. Nuclear Chem. 280: 171–179.

Zheng, J., A. Iijima and N. Furuta. 2001. Complexation effect of antimony compounds with citric acid and its application to the speciation of antimony(III) and antimony(V) using HPLC-ICP-MS. J. Anal. At. Spectrom. 16: 812–818.

Zhi-Perényi, K., P. Jankovics, E. Sugár and A. Lásztity. 2008. Solid phase chelating extraction and separation of inorganic antimony species in pharmaceutical and water samples for graphite furnace atomic absorption spectrometry. Spectrochim. Acta Part B. 63: 445–449.

Zhou, Q. X., X.G. Zhang and J.P. Xiao. 2009. Ultrasound-assisted ionic liquid dispersive liquid-phase micro-extraction: A novel approach for the sensitive determination of aromatic amines in water samples. J. Chromatogr. A 1216: 4361–4365.

Zhu, L., S.Z. Chen, D.B. Lu and X.L. Cheng. 2009. Single-wall Carbon Nanotubes for Speciation of Arsenic in Environmental Samples by Inductively Coupled Plasma Mass Spectrometry. At. Spectrom. 30: 218–222.

Zhu, X., X. Zhu and B. Wang. 2006. Cloud point extraction for speciation analysis of inorganic tin in water samples by graphite furnace atomic absorption spectrometry. J. Anal. At. Spectrom. 21: 69–73.

Zuliani, T., G. Lespes, R. Milačič, J. Ščančar and M. Potin-Gautier. 2006. Influence of the soil matrices on the analytical performance of headspace solid-phase microextraction for organotin analysis by gas chromatography-pulsed flame photometric detection. J. Chromatog. A 1132: 234–240.

Zuliani, T., G. Lespes, R. Milačič and J. Ščančar. 2010. Development of the extraction method for the simultaneous determination of butyl-, phenyl- and octyltin compounds in sewage sludge. Talanta. 80: 1945–1951.

Zúñiga, M.C., E. Jover, V. Arancibia and J.M. Bayona. 2008. Development of a methodology for the simultaneous determination of inorganic and organolead compounds using supercritical fluid extraction followed by gas chromatography–mass spectrometry and its application to environmental matrices. Talanta. 80: 504–510.

Separation Techniques for Elemental Speciation in Soil, Sediments, and Environmental Samples

Márcia F. Mesko,[1,a],* Carla A. Hartwig,[1,b] Cezar A. Bizzi,[2,c] Edson I. Müller,[2,d] Fábio A. Duarte[2,e] and Paola A. Mello[2,f]

Introduction

Knowledge of trace elements and their role in the life-cycle is of great importance. Almost a hundred species of alkylated metals and metal(loid)s have been found in different ecosystems. The level of contaminants can be very low (at ng range per dm^3 or kg) or very high and can come from

[1]Centro de Ciências Químicas, Farmacêuticas e de Alimentos, Universidade Federal de Pelotas, 96010-610, Pelotas, RS, Brazil.
[a]Email: marcia.mesko@pq.cnpq.br
[b]Email: carlahartwig@yahoo.com.br
[2]Departamento de Química, Universidade Federal de Santa Maria, 97105-900, Santa Maria, RS, Brazil.
[c]Email: c_bizzi@yahoo.com.br
[d]Email: edson_muller@yahoo.com.br
[e]Email: fabioand@gmail.com
[f]Email: paoladeazevedomello@gmail.com
*Corresponding author

natural sources or be a result of an enrichment. Whether through natural or anthropogenic activities, when metal ions enter the environment or living systems, only a fraction will remain as free ions. In general, the major portion will be complexed by ligands and can result in different species with specific behavior in the system, resulting in several bio(geo)chemical processes (Lobinski 1997, Hirner 2006). For example, natural methylation, which is generally prevalent in metal ions under specific conditions, can result in much more or less toxic species, depending on the element. Another example is the possibility of metal ions being incorporated into large molecular structures such as humic substances. On the other hand, due to different anthropogenic activities, some species can enter the environment, for example, lead and tin species (Cornelis et al. 2003, Hirner 2006).

Knowledge of different systems (air, natural water, soils, sediments, waste, sludge, biota and related matrices) has been the main focus of speciation analyses for environmental samples. Among the main analytes that have been studied, As, Hg, Pb and Sn species can be considered the most investigated and less frequently the species of Al, Bi, Cr, I, Sb, Se and V. The following aspects can be highlighted considering speciation analysis in environmental media: (i) the management and remediation of soils and waste are dependent on the biogeochemistry of a contaminant; (ii) the homeostatic control, metabolism, and detoxification of essential and toxic elements are related to their chemical forms; (iii) the impact of human exposure to environmental samples depends on the species of contaminants in the sample; and (iv) the processes that will be applied in an environmental area or sample as well as the methods for its analysis are dependent on the species which are contaminating that area or sample (as, e.g., if the contaminant is organic or inorganic and its oxidation state) (Lobinski 1997, Hirner 2006). Therefore, due to the different toxicities and the specific impact of a given species, speciation analysis is required (Ellis and Roberts 1997, Feldmann et al. 2009, Batley et al. 2009).

In this case, species-selective techniques which could be used for speciation, such as Mössbauer spectroscopy, X-ray photoelectron spectroscopy (XPS), electron spin resonance spectroscopy (ESR), or mass or tandem mass spectrometry (MS or MS/MS) are usually not sensitive enough to reach trace levels for the most investigated species. On the other hand, non-specific detectors such as ultraviolet (UV) and flame ionization detectors (FID) suffer from large background noise and poor sensitivity. In particular, an element selective detector is able to differentiate the analyte signal from complex matrix signals and detect trace levels (Lobinski 1997). Although many sensitive element-specific detectors are available for analysis purposes, they can only provide total element information. Therefore, speciation can be attained by coupling a separation technique with an element-specific detector (Alonso and Encinar 2003, Feldmann

2005). Considering speciation studies, methods frequently make use of a chromatographic technique (mainly gas or liquid chromatography) coupled to sensitive and selective (or specific) detection technique such as atomic absorption spectrometry (AAS), atomic emission spectrometry (AES), inductively coupled plasma optical emission spectrometry (ICP-OES), inductively coupled plasma mass spectrometry (ICP-MS), electron impact ionization mass spectrometry (EI-MS), and electrospray ionization mass spectrometry (ESI-MS) (Ackley and Caruso 2003, Alonso and Encinar 2003, Caruso et al. 2006, Hirner 2006).

Speciation analysis is a complex task that requires several analytical steps, which may include extraction, clean up, derivatization, preconcentration, separation and final measurement procedures. The choice of the correct matrix treatment and procedure for analysis (sample handling, preparation, species separation and detection) used to identify or measure the quantities of individual species are important factors to be considered (Needham et al. 2005). In addition, the stability of species through the entire analytical process (mainly in the sample preparation step) and the lack of suitable standards and certified reference materials are the main challenges in speciation analysis (Ackley and Caruso 2003, Alonso and Encinar 2003, Hirner 2006, Ellis and Roberts 1997).

Species separation and detection techniques are considered as a crucial step and most of the scientific publications in the area of speciation analyses are devoted to one or both of these processes. The most used instrumental separation techniques in speciation analysis are liquid chromatography (LC), gas chromatography (GC), and capillary electrophoresis (CE). It is important to mention that LC has the broadest applicability range in terms of type of samples and matrices in several applications (Ackley and Caruso 2003, Alonso and Encinar 2003).

The physicochemical and physical properties of the analyte, for example, volatility, charge, polarity, and molecular mass, must be considered when choosing the species separation method (de Leon et al. 2002, Lobinski 1997). In previous studies, LC with ultraviolet (UV) detection and GC with electron capture detection (ECD), with their inherent simplicity and drawbacks, were used (Lobinski 1997). With the growing interest in species-selective instrumental analysis, instrumentation was developed to the currently available equipment (Lobinski 1997). The instrumental set-up must attain the species separation at analyte(s) level(s) in the sample.

When analytes are volatile species presenting thermal stability or can be converted into volatile species using chemical modification (derivatization), the method of choice should be gas chromatography and various related methods. Nevertheless, if these requirements are not fulfilled, liquid chromatography and its techniques can be used, which are versatile considering the different mechanisms available, can be used [reversed-phase chromatography (RPC), ion exchange chromatography (IEC) and size exclusion chromatography (SEC)]. In addition, electrophoresis presents high resolution, combined with its ability to separate highly polar compounds [polyacrylamide gel electrophoresis (PAGE) and capillary zone electrophoresis (CZE)], making it a reliable tool for separation (Lobinski 1997).

In view of the inherent necessity to separate species using element-specific detectors and the complexity of samples, separation methods are extremely important in speciation analysis. In this chapter, a detailed discussion will be presented for the most commonly used instrumental techniques for separation, namely liquid chromatography, gas chromatography and electrophoresis.

Separation Techniques for Elemental Speciation

The development of coupled techniques allows the separation and detection of many analytes in different matrixes that focus on environmental speciation. An overview of coupled techniques usually used for speciation in environmental studies is presented in Fig. 3.1. The physicochemical properties of the analyte, such as polarity, thermal stability and volatility, determine the choice of the separation technique. On the other hand, for choosing the detection system to be coupled to the separation technique the analyte and its level are the basic requirements. In the following sections, the most frequently used instrumental techniques for separation will be presented.

Sample preparation is usually a mandatory step in meeting the basic requirements imposed by both the separation and the detection techniques. In general, sample preparation must be developed in accordance with the sample matrix, analytes, and their levels in the sample (Mester and Sturgeon 2003, Arruda 2007). A well planned sample preparation methodology is

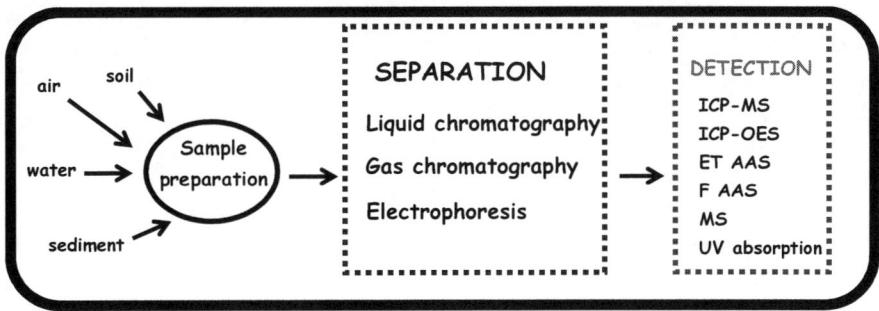

Figure 3.1 A simplified overview of hyphenated techniques used for speciation analyses in environmental samples.

important to assure the quality of the results and is also related to the time and costs of the overall procedure. The main objectives of this step can be extraction of analytes from the matrix, preconcentration or clean-up, and adaptation of species for analysis (Alonso and Encinar 2003). Different approaches have been used for sample preparation (Bouyssiere et al. 2003b) including liquid-liquid extraction (LLE) (Pena-Pereira et al. 2009), microwave-assisted extraction (MAE) (Namiesnik and Gorecki 2000, Reyes et al. 2011), ultrasound-assisted extraction (UAE) (Huerga et al. 2005, Najafi et al. 2012, Priego-Capote and de Castro 2007), solid-phase extraction (SPE), solid-phase microextraction (SPME) (Mester et al. 2001, Wuilloud et al. 2004, Kaur et al. 2006), accelerated solvent extraction (ASE) (Encinar et al. 2002, Marcic et al. 2005) and stir bar sorptive extraction (SBSE) (Wuilloud et al. 2004).

Liquid chromatography

Most methods of elemental speciation analysis include separation by LC coupled with element-specific detectors (Harrington et al. 2011). Interest in LC for element speciation in soil, sediments and environmental samples has increased in recent decades. A number of reviews have been published describing the benefits associated with using element-specific detectors for LC (Guerin et al. 1999, Harrington 2000, Szpunar et al. 2000, González-Toledo et al. 2003, Montes-Bayon et al. 2003, B'Hymer and Caruso 2004, 2006, Miravet et al. 2010, Popp et al. 2010, Chainet et al. 2011). It is important to mention that LC could be easily coupled to some element-specific detectors without significant changes to the original instruments (González-Toledo et al. 2003).

In general, separations by LC are performed with the sample introduced to a chromatographic column containing a solid stationary phase while the liquid mobile phase is pumped through to the column. The degree of interactions between the analytes and the stationary phases change their partition into the two phases (mobile and stationary), which results in different elution times. Unlike GC, LC has the capacity to separate non-volatile analytes and also the versatility of stationary and mobile phases allows several applications for LC.

The different types of chromatography which are commonly used for LC separations are RPC, IEC, SEC, normal phase chromatography (NPC), reversed phase ion pair chromatography (IPC), micellar chromatography, chiral liquid chromatography, and micro LC. For speciation purposes in environmental samples, NPC is less frequently used because it is suitable for analytes with low polarity which are not the case for species usually investigated in this field. In addition, the majority of element-specific detectors are incompatible with nonpolar mobile phases used in NPC. The use of bonded stationary phases for NPC (with chemically bonded polar functional groups), instead of unmodified silica or alumina, improves analyte separation and reproducibility in retention times. Size exclusion chromatography is conventionally applied for separation of analytes with high molecular weight. The separation phenomena are based on the ability of analytes to penetrate the pores of stationary phase. The composition of mobile phase is not a key parameter because in this type of chromatography analyte must be only solubilized into this phase. Due to its particular characteristic for separation of high molecular weight compounds, SEC is widely used for elemental speciation in biological samples and is less frequently used for species separation from soil, sediments and environmental samples (Montes-Bayon et al. 2003, Ackley and Caruso 2003, B'Hymer and Caruso 2006).

Among the types of chromatography used in LC, RPC and IEC are often used the most in elemental speciation analyses. The most common stationary phase in RPC is silica modified by silanization, usually replacing the polar silanol groups by C8 or C18 chains. These alkyl chains are able to separate nonpolar or slightly polar species, using polar eluents as water or mixtures with polar organic solvents (e.g., methanol or acetonitrile) as mobile phases, acting as modifiers mainly in terms of polarity of the mobile phase. Separation is also governed by changing the variables as the stationary phase functional group, pH, ionic strength, organic modifiers used in mobile phase, and the gradient program used for separation. However, particular attention should be provided when organic solvents are used in the mobile phase and when detection is performed by plasma-based techniques. In addition, detection with plasma-based techniques needs a careful choice of the eluent, and gradient profiles should be avoided (Tomlinson et al. 1994, González-Toledo et al. 2003). Reversed phase chromatography is frequently used to separate Hg species. Moreover, as a complement to RPC, the addition of an ion pair reagent (tetraalkyl ammonium salts, triethylalkyl ammonium salts and sodium alkyl sulfonates) could improve the simultaneous separation of anionic, cationic and neutral species. Similar to IPC, micellar chromatography is performed by adding a surfactant in the mobile phase (Ackley and Caruso 2003, Montes-Bayon et al. 2003).

Unlike RPC, IEC is applied for separation of ionized (or ionizable) species. Anion or cation exchange is performed most often using quaternary ammonium or sulfonate groups supported in polymeric or silica materials. The separation occurs via interaction between the stationary phase and an analyte with an opposite charge. Aqueous solutions of inorganic salts are the most common mobile phases, which make IEC separations compatible with element-specific detectors (Ackley and Caruso 2003, B'Hymer and Caruso 2006, Kotaś and Stasicka 2000). This type of chromatography is commonly applied to separate Al, As, Cr, Sb, Se, and Sn species. However, the separation of mono-substituted organotin compounds (OTC) is difficult due to the strong interaction with the stationary phase, requiring the use of complex mobile phases or pH-gradient elution (González-Toledo et al. 2003). Independent of elements, the pH and ionic strength of the eluent are important parameters in RPC and IEC. Depending on the column, the tolerable pH for RPC and IEC is from 2 to 10 and 1 to 14, respectively. These parameters can affect the (de)protonation and retention times of analytes (Ackley and Caruso 2003, Kotaś and Stasicka 2000). The use of IEC is widely used in most speciation studies, as the separation process is more reproducible and less prone to sample matrix interferences than with RPC (Sanchez-Rodas et al. 2010, B'Hymer and Caruso 2006).

Nowadays, the reduction of particle size and internal diameters of columns are suitable choices in order to improve chromatographic separation. The most common diameters of particles in commercial LC columns ranged from 5 to 10 μm. However, the availability to produce uniform particles with sizes at values lower than 3 μm allowed a reduction of internal diameters and lengths of columns. Separations using small particles and columns could be classified as micro LC (0.5–1.0 mm i.d.), capillary LC (100–500 μm) and nanoscale LC (10–100 μm) (Montes-Bayon et al. 2003, Ackley and Caruso 2003, Castillo et al. 2008). Small columns offer many advantages, for example, the low consumption of solvents and samples, reduction of peak broadening, improvement of resolution and a low mobile phase flow-rate (about μL min^{-1}), which results in higher tolerance to saline solutions and organic solvents by the detector. In addition, the low flow-rate allows direct sample introduction (without nebulization) into ICP instruments (Tangen et al. 1997, Shum et al. 1992, Powell et al. 1995, Pergantis et al. 1997).

Coupling LC to the detection techniques

The multi-elemental capability and compatibility with some LC mobile phases makes ICP-MS and ICP-OES the most attractive detectors for elemental speciation (Szpunar et al. 2000, Montes-Bayon et al. 2003, Al-Assaf et al. 2009, Gettar et al. 2000). In particular, ICP-MS offers excellent sensitivity, a wide linear dynamic range, the ability to be used for isotope dilution analyses, and offers high-speed analyses (Sutton and Caruso 1999). Atomic fluorescence spectrometry (AFS) is an interesting alternative, especially for As, Hg, Sb, and Se, because it provides low limits of detection (LOD) (Sanchez-Rodas et al. 2010, Yuan et al. 2007). The use of AAS (Grotti et al. 2001, Frankowski et al. 2010) and diode-array detector (DAD) (Cathum et al. 2002, Gao and Ma 2011) techniques have been less frequently applied. When flame atomic absorption spectrometry (F AAS) is used as detector, the short signal integration time allowed by most software available could be a limitation, because it is not long enough to perform a full analysis. In some cases, a typical Al separation by LC provides retention times of up to 60 min. However, with special software, the limitations related to integration time can be resolved (Frankowski et al. 2010).

Especially for hydride-forming analytes (As, Bi, Cd, Ge, Pb, Sb, Se, Sn and Te), the sensitivity of element-specific detectors (e.g., ICP techniques, AAS and AFS) could be significantly improved using post column hydride generation (HG) or cold vapor (CV, for Hg). In this approach, analytes in its inorganic forms can be converted into their hydrides (e.g., AsH_3, SbH_3, except for Hg which forms $Hg°$) by $NaBH_4$ (Cai 2003). It is important to point out that this alternative allows the separation of matrix and analyte. On the other hand, it is important to consider the inherent drawbacks associated with HG or CV as matrix separation technique, as long reaction times, the incomplete reaction for hydride generation that can occurs due to some particularity in the matrix, the thermodynamic inability for hydride formation by some species, or a considerable kinetic limitation to hydride formation, the relatively low sample throughput, and the increased use of chemicals that can lead to contamination (Balcerzak and Raynor 1998, Liu and Lee 1999, Cai 2003, Kumar and Riyazuddin 2007).

The composition of the mobile phase is a key issue in the hyphenation of LC with detection techniques, especially ICP-MS. The problem related to introducing organic solvents in the plasma, plasma extinguishing and carbon deposits in the interface, is well known. However, the use of a cooled spray chambers (0–5°C) can reduce the solvent loading on the plasma (González-Toledo et al. 2003, B'Hymer and Caruso 2006). The addition of low amounts of oxygen in the introduction system is an alternative to reduce carbon deposition. However, the disadvantage of using oxygen is the reduction of the cones lifetime (Sutton et al. 1997, González-Toledo et al. 2003). In addition, in order to increase the sample volume introduced to the plasma, ultrasonic nebulizers (USN) have been successfully used (increasing the efficiency 1–3 percent to 10–30 percent) (González-Toledo et al. 2003, B'Hymer and Caruso 2006). Another possibility is the use of special nebulizers for plasma techniques which are designed to operate specifically at low flow rates (0.01–0.1 mL min^{-1}). For example, the direct injection nebulizer (DIN) and oscillating capillary nebulizer (OCN) may provide almost 100 percent nebulization efficiency (González-Toledo et al. 2003, B'Hymer and Caruso 2006). It is important to mention that independent of the sample, the choice of mobile phase is strongly dependent on analytes/ species and the detection system used.

Although separations using LC provide much information related to species in different samples, knowledge of the molecular structure of the original compound is essential. However, it can be sometimes lost due to the high degree of atomization and/or ionization achieved in the detection systems, which can result in the loss of molecular information due to the breakage of bonds. In these cases, less severe ionization techniques should be chosen, as atmospheric pressure ionization techniques, based on ESI or atmospheric pressure chemical ionization (APCI) are recommended, mainly in combination with mass spectrometry (Amayo et al. 2011, González-Toledo et al. 2003).

Gas chromatography

Although separations performed by liquid chromatography have been used in most speciation studies, GC techniques present interesting alternatives (de Leon et al. 2002). In addition, one advantage of GC in comparison with other separation techniques is its inherent suitability to be used for naturally occurring volatile species such as those in gas samples. Gas chromatography is a high resolution chromatographic separation technique (Popp et al. 2010) that, is suitable for the separation of a given compound governed by its volatility and thermal stability (Alonso and Encinar 2003, Popp et al. 2010, Lobinski 1997). Separation follows the basic concept in which the components of a sample are partitioned between two phases; the stationary phase (liquid or solid) and the mobile phase (carrier gas), which percolates through the stationary bed. Initially, the sample is vaporized and then carried by the mobile gas phase through the column where the stationary phase is contained. The equilibrium (the partitioning/adsorption analyte-stationary phase) and also the interaction with the mobile phase is based on their solubility at the given temperature. Then, sample components are separated based on their vapor pressures and affinities to the stationary phase. Instrumentation for GC has continuously evolved since the introduction of the first chromatograph, but the basic apparatus is composed of a carrier gas, a flow controller, a sample inlet (injector), the oven for the column (where column is inserted to control the temperature program), the detector, and the data system (McNair and Miller 2008). Separations by GC for environmental speciation are mainly attained using three GC-based techniques: (i) cryogenic trapping, (ii) packed columns, and (iii) capillary columns.

Separations using cryogenic trapping are suitable for volatile analytes and for species that can be easily converted to volatile species. This method is based on the purge of analytes using an inert gas through a U-shaped tube filled with a chromatographic material, with this system immersed in liquid nitrogen (−192°C). In this step, analytes are trapped in the column and can be desorbed sequentially to the detector by removing the liquid nitrogen and heating the column (Alonso and Encinar 2003, Lobinski 1997). Apolar sorbents are normally used, and separations using cryogenic trap are considered more dependent on the differences in the analytes' boiling points than in the characteristics of the packing material. Even though the set-up is not commercially available, its components can be obtained from several suppliers and many laboratory operational systems have published studies using this principle (Lobinski 1997). In order to minimize losses, dead volumes, or peak broadening, connections must be as short as possible. In addition, it is recommended to heat the connection between the end of column and the detector to prevent condensation. Air and gas samples can be also analyzed, for which the use of a cold trap (e.g., dry ice-acetone) in the column before retaining water vapor is recommended (Alonso and Encinar 2003, Lobinski 1997). The main advantages that can be pointed out regarding the cryogenic trap are its simplicity and low cost, non-selectivity, the ability to preserve unstable species before desorption, easy coupling to the detector and large preconcentration factors. Resolution is relatively poor in comparison to the columns, but it is enough for many environmental applications (Lobinski and Adams 1997). Cryogenic trapping followed by thermal desorption has been used for speciation analysis of As, Hg, Pb, Sb, Se and Sn as well as volatile organic compounds, mainly in water and gas samples (Geng et al. 2009, Ye et al. 2010, Kumar and Riyazuddin 2007, Amouroux et al. 1998, Tseng et al. 1999, Wang et al. 1999, Cai et al. 1993).

The use of packed columns represents the first separations using GC for elemental speciation. The materials used were primarily composed of 5 to 10 percent polyethylene glycol (Carbowax®) supported in a polyaromatic cross-linked material (Chromosorb®) (Alonso and Encinar 2003). The more effective interaction of the analyte with the packing material contributes to the separation of analytes combined with differences in volatility. Usually, packed columns are filled with an inert support loaded with a nonpolar phase (about 3 to 10 percent). Chromatographic resolution and separation of analytes can be improved by an optimized temperature program. The main problem in using packed columns is that peaks are normally broad in comparison to capillary columns; thus, compounds of similar volatilities can be not resolved. In some cases, packed columns cannot allow the necessary resolution for complex matrices requiring a selective and specific detector. On the other hand, large sample amounts can be introduced (usually 10 to 50 µL), making it possible to obtain low limits of detection. In addition, packed columns are more stable and can be used many times (Lobinski 1997, Alonso and Encinar 2003).

Open-tubular or capillary columns offer higher resolution, providing sharper bands. At lower carrier gas flow-rates, sensitivity is also improved in comparison to packed columns (Lobinski 1997, Alonso and Encinar 2003). In contrast with packed columns, where analytes can strongly interact with the stationary phases, capillary columns are preferred in particular for separation of organometallic species containing very polar metal-halogen bonds (Lobinski 1997). There is a diversity of coatings available, from almost non-polar phases up to high polar types, with different film thickness and internal diameters (Dorman et al. 2010). Separation efficiency (number of theoretical plates per unit length) is increased using capillary columns by decreasing its internal diameter. However, the loading capacity will also be reduced. This loading limitation can be considered the main disadvantage of capillary GC. The typical injection volume is about 1 to 2 μL, which can be problematic particularly in speciation analysis where very low concentrations must be determined (Alonso and Encinar 2003). To overcome this limitation, sample preconcentration and the use of large volume and temperature programmed injectors can be chosen (Teske and Engewald 2002, Engewald et al. 1999, Lobinski and Adams 1992). Attention must be paid when using larger volumes to avoid problems in the detector by the presence of high solvent volumes. The use of methods for on-line preconcentration solvent removal (as programmed temperature vaporization) is an alternative that avoids the excessive injection of solvent (Lobinski and Adams 1992). Another alternative to overcome the limitation of low-volume injection is the use of multicapillary columns, which can improve resolution and reduce the time of analysis (Patrushev et al. 2010, Jitaru et al. 2005). Most applications of GC with capillary columns have been published for Hg, Pb, and Sn speciation (Stoichev et al. 2006, Takeuchi et al. 2000, Zufiaurre et al. 1997, CarlierPinasseau et al. 1996, Lalere et al. 1995, Munoz et al. 2005, Nevado et al. 2011, Centineo et al. 2004, Vahcic et al. 2011).

Derivatization

Concerning speciation analysis, some compounds cannot directly fulfill the basic requirements for GC, and chemical reactions that transform nonvolatile compounds (usually ionic) into volatile, thermally-stable compounds have to be performed (Liu and Lee 1999). A number of organometallic compounds are volatile enough to be separated by GC, for example, tetraalkyllead species, methylselenium compounds and some organomercury compounds (Lobinski and Adams 1997). On the other hand, it is important to consider that the majority of organometallic species exist in quasi-ionic polar forms, which have relatively high boiling points and poor thermal stability (Lobinski and Adams 1997). In order to solve this limitation, many derivatization reactions have been used for GC that enable the speciation of a large variety of compounds (Alonso and Encinar 2003). These reactions may include hydride generation, aqueous ethylation, and alkylation using Grignard reagents (Alonso and Encinar 2003, Liu and Lee 1999, Lobinski and Adams 1997).

Species derivatization, as previously presented for LC in combination with HG, can be also used for GC as a chemical modification approach.

Hydride generation is easier to handle than aqueous ethylation, propylation or alkylation (Liu and Lee 1999). A well-known derivatization reaction that has been applied to small inorganic and organometalic ions in order to form volatile covalent compounds of As, Bi, Cd, Ge, Hg, Pb, Sb, Se, Sn, and Te has been considered. The main advantages of HG versus alkylation are the slightly higher sensitivity of some species and the shorter time required for the reaction and purge in the reactor (Liu and Lee 1999). Nevertheless, taking into account thermodynamic and/or kinetic behavior or because of the low stability of the produced hydrides, this procedure cannot be applied under the same conditions to all species of the same element, as for some alkylated lead species (Weber 1997) and some compounds at the high oxidation states (e.g., for some organoarsenicals) (Anawar 2012, Gong et al. 2002). It was reported that the use of cysteine allows generation of volatile forms for different As species using the same condition (Le et al. 1994). Other works have shown that some of these species have to be reduced beforehand, and the fractionation information could be obtained by measuring the difference in the total element determination (after reduction) against the previously-determined low oxidation state species (Alonso and Encinar 2003, Gong et al. 2002).

Derivatization via alkylation (*in situ* alkylation) involves the addition of one or more ethyl groups to inorganic or alkylated metal species to form the di- (Hg, Se), tri- (Bi), or tetraalkylated species (Sn, Pb), which are hydrophobic, volatile and thermally stable and thus suitable for GC separations (Alonso and Encinar 2003). This modification has been mainly performed by aqueous ethylation using sodium tetraethylborate (NaBEt$_4$), which combines the advantages of working in the aqueous phase of hydride generation with the low matrix interference of Grignard reagents (Liu and Lee 1999, Honeycutt and Riddle 1961). It is considered a simple and quantitative derivatization procedure that can be performed in the aqueous phase and keep the metal-carbon bond already present in the species (Alonso and Encinar 2003, Liu and Lee 1999). Alkylation provides more reproducible results in comparison with hydride generation because it is not affected by inorganic interferents from extract and also by borohydride competition and it occurs in a foam-free medium. Additionally, the limits of detection of some organometallic compounds can be significantly improved when compared to hydride generation (Liu and Lee 1999, Cai et al. 1993, Zufiaurre et al. 1997). When compared to the modification using Grignard reagents, NaBEt$_4$ is stable in water, allowing derivatization in aqueous media, which can be particularly useful for speciation analysis of water samples (Lobinski 1997).

An alkylation reaction can be also performed using Grignard reagents (e.g., ethyl magnesium bromide). In this case, modification must be carried out in an anhydrous organic phase after the separation of the analytes from the matrix by liquid-liquid extraction. This is the main reason that the procedure is less prone to matrix interference when compared to the previously presented aqueous alkylation or hydride generation (Alonso and Encinar 2003, Fent and Muller 1991). Alkylation of organometallic compounds using Grignard reagents, which use alkyl-magnesium halides, can only be performed in a water-free organic phase and under inert atmospheric conditions. In this sense, a more complex experimental set-up is required to perform chemical modification (Alonso and Encinar 2003). However, if these conditions are provided, it can provide quick derivatization for OTC (Attar 1996), Hg (Leopold et al. 2010), Pb (Zufiaurre et al. 1997), and other elements (Liu and Lee 1999). Depending on the characteristics of the target species, alkylating reagents with different alkyl-groups are available, such as methyl-, ethyl-, propyl-, butyl-, pentyl-, hexyl-, and phenyl-. Alkylation by Grignard reagents with longer alkyl-groups (pentyl- and hexyl-) result in compounds with relatively low volatility, which contributes to the preconcentration step. On the other hand, alkylation using shorter alkyl-groups produce highly volatile compounds, and partial losses during the preconcentration step may occur (Liu and Lee 1999).

In general, the choice of a derivatization reaction for a given application is dependent on the type of analyte and matrix. It is important to point out that the derivatization reaction is strongly dependent on the matrix behavior (Alonso and Encinar 2003, Popp et al. 2010). Hence, in order to assure the quality of results, it is necessary to perform recovery tests, especially when different matrices have to be evaluated (Liu and Lee 1999). Independent of the method used for chemical modification, a great variety of higher boiling points and more polar compounds become accessible to GC if a suitable derivatization protocol is carried out, and the use of a derivatization reaction has broadened the field of GC applications for speciation.

Coupling GC to detection techniques

Flame ionization detection (FID), ECD, and flame photometric detection (FPD) are well know detection systems that are normally coupled to GC, and some options with this coupling are commercially available (de Leon et al. 2002). On the other hand, very powerful instruments for performing speciation analyses can be also obtained by coupling GC with atomic detectors and mass spectrometry (Ellis and Roberts 1997), mainly because the gas stream emerging from the GC is itself a heated gas (and not a liquid such in LC, which leads to dilution in the sample flow). In addition, GC instrumentation is compatible with ICP-MS and offers some advantages compared to LC-ICP-MS. Gas chromatography in conjunction with ICP-MS (GC-ICP-MS) provides higher resolution power and 100 percent introduction efficiency into the ICP (Popp et al. 2010). Stable plasmas and less spectral interference are obtained and the calibration could be simplified (Popp et al. 2010, de Leon et al. 2002, Bouyssiere et al. 2002, Lobinskii and Adams 1997). For coupling, it is necessary to heat the transfer line linking the two instruments (GC to the detector) to prevent condensation. It is also necessary to ensure that no dead volume and cold areas are present (Alonso and Encinar 2003).

Combining GC with F AAS and electrothermal atomic absorption spectrometry (ET AAS) is an attractive approach and can be a useful alternative due to the relative low cost of atomic absorption spectrometers compared to other equipment. The gas emerging from GC can be introduced through the nebulizer of a F AAS instrument. As the nebulizer presents only about 10 percent efficiency, the problems associated with the intensity of the flame are less pronounced (Ellis and Roberts 1997). Despite the higher sensitivity of ET AAS, the combination between a GC and this equipment as detector is more problematic due to the non-continuous characteristic of the signal. This problem can be solved using the graphite cuvette, which is maintained at the atomization temperature for the entire time of chromatogram. Many applications couple GC with atomic detectors, especially those using cryogenic trapping as a separation method (Lobinski and Adams 1992, Ye et al. 2010, Tseng et al. 1999).

The coupling of GC with plasma sources, such as ICP-MS and microwave-induced plasma atomic emission spectrometry (MIP-AES) has been extensively reported and presents many advantages for speciation analysis. The high sensitivity of ICP-MS (LODs in the sub-ng L^{-1} range), the wide linear dynamic range, and the multi-element capabilities combined with the high resolution of GC make this technique a powerful tool for speciation (Bouyssiere et al. 2003a, Feldmann 2005, Lobinski and Adams 1997). In this coupling, the gas effluent from the GC column must be transported to the torch inlet. The main prerequisite of any interface coupling a GC technique and the plasma torch is that the volatilized analytes remain in the gas phase during transport from the column to the plasma. This explains the necessity of heating the entire transfer line to avoid condensation maintaining peak sharpness, good resolution and low LODs (Wuilloud et al. 2004, Bouyssiere et al. 2002). Another aspect that must be pointed out is the use of a make-up carrier gas flow through the central channel of the torch to allow introduction of the analytes into the plasma once the normal flows used in GC systems (in the order of a few ml min^{-1}) (Wuilloud et al. 2004) do not reach the typical flow-rate in the ICP torch (typically in the range of 1 L min^{-1} for most manufacturers). Common applications of GC-ICP-MS include analysis of organometallic compounds of Hg, Pb and Sn. These compounds may contain a variety of alkyl groups (e.g., methyl, ethyl, propyl, and butyl) but also species containing aromatic groups (e.g., phenyl). In addition, the use of GC-ICP-MS for organic compounds containing hetero-atoms (e.g., As, Br, I, P and S) has also been demonstrated (Popp et al. 2010).

Electrophoresis

The use of the electrophoresis principle as a separation technique is considered a non-chromatographic alternative with a separation mechanism considered complementary to LC. The separation mechanism combined with its high resolution makes electrophoresis an important separation technique, especially for the separation of highly polar compounds. One significant difference that must be pointed out is that in electrophoresis, there is no stationary phase, thus avoiding the contact with surfaces that make undesired interactions possible. Consequently, species integrity in electrophoresis can be, in principle, less easily affected than in LC (Michalke 2003). This separation method is widely used as a bioanalytical tool for fundamental research and diagnostics settings that have been used mainly for the isolation and identification of high molecular weight biomolecules. The separation in electrophoresis is based upon the mobility of charged molecules under the influence of an electric field. Mobility is a fundamental property of a compound and it depends on the magnitude of its charge, its molecular weight and its structural shape (Mikkelsen and Cortón 2004, Lobinski 1997).

Separation and detection techniques already established with other applications have to be combined in new ways and modified according to a specific speciation problem (Michalke 2003). In this sense, the combination of separation technologies such as capillary and gel electrophoresis (GE) with element or molecule selective detection systems have been studied for speciation analyses.

Capillary electrophoresis

In the simplest instrumentation, CE separates the analytes using the electroosmotic flow phenomenon, which consists of the application of a constant direct-current potential through a capillary filled with a conducting aqueous buffer solution. The use of CE in speciation analysis is directly associated with the analytical activity of identifying and measuring species, including identification of the binding partners of elements. Today, interest in the CE technique as a powerful tool in environmental, biomedical and clinical, forensic and industrial analyses is growing. However, despite this increasing interest and several applications, many improvements in basic principles and methodology, selectivity, detection sensitivity, and sample pre-concentration strategies have been developed in order to consolidate the method as an alternative for speciation (Michalke 2003, Timerbaev 2004, Kuban and Timerbaev 2012). Some applications have shown that the CE method is sometimes superior to conventional LC methods for ionic multispecies analyses. In comparison to LC, the main advantages related to CE include high separation efficiency, low material and sample consumption, relatively short analysis times, low investment and operational costs, and greater tolerance to complex matrices that can be processed, thus avoiding laborious pretreatment (Timerbaev 2002, Michalke 2003). In addition, CE can be used as a primary separation technique and may also be used for quality control and species identification as a second dimension separation technique (in this case, LC is commonly used as the primary separation technique).

A basic CE set-up is operated by applying an electric field along an open tube column with a low internal diameter at a high voltage (typically about 30 kV). Each compound moves in the electric field at a different velocity. Thus, the obtained electropherograms can be similarly processed like LC chromatograms (Michalke 2003). The endoosmotic flow (a laminar flow in pH values higher than 3) is induced by the inner negative capillary surface where the silanoic groups of the fused silica capillary attract positive ions. The produced electrical double layer promotes the flow of cations to the cathode via the electric field, while the anions move in the opposite direction. In particular, a single CE instrument allows several different separation modes, mainly (i) CZE, (ii) micellar electrokinetic chromatography (MECC), (iii) capillary isoelectric focusing (cIEF), and (iv) capillary isotachophoresis (cITP). Considering the speciation analysis, these different separation mechanisms provide different possibilities of separation of species (Michalke 2003).

The separation mode can be modified by changing the electrolyte system, mainly by replacing a few milliliters of a buffer system with another. Then, one of the above mentioned separation principles can be selected to separate species according to their physicochemical properties. The charge-to-size ratio is the basic principle involved in separation by CZE, and target analytes are all charged molecules, amino acids and proteins. On the other hand, neutral molecules with different abilities to enter charged hydrophobic micelles are separated by MECC (for charged analytes, both electrophoretic migration in the aqueous electrolyte and solubilization into micelles are responsible for separation). With cIEF, which is mainly used for proteins and peptides, the isoeletrical point dictates the separation. Many metal cations, however, are of nearly identical charge and hydrated ionic radius, and their differences in mobility are not sufficient to provide suitable separations. In this case, cITP separates analytes by their specific conductivity (using discontinuous buffer systems with different conductivities), which is useful for different dissociated molecules (Kuhn and Hofstetter-Kuhn 1993).

It is important to point out that CE suffers from severe matrix effects that affect the migration time of species requiring suitable pretreatment (Liu and Lee 1999, Wolf et al. 2003). Ion mobility is dependent on the electric field and on the endoosmotic flow, which is pH-dependent. As the sample matrix can influence the endoosmotic flow, a high buffering capacity is necessary in order to avoid interferences. In addition, differences in analyte conductivity can change the electric field and consequently alter the migration time of a specific compound. In this way, the addition of an internal standard is used to correct species migration time according to standard electropherograms. Another possibility is the standard addition procedure of the compounds under investigation (Michalke 2003).

To improve separation or enhance detection, a suitable chemical modification method can be used. In this context, derivatization, complexation, ion dissociation and ion pairing are mainly employed. One of the main objectives of derivatization for CE is to increase the charge-to-size ratio among analytes, thus improving separation. Complexation has been the most used alternative, and several reagents are available for this purpose. Examples are S-containing derivatizing agents (e.g., diethylditiocarbamate—DDTC), polyaminocarboxilic acids (e.g., ethylenediaminetetracetic acid —EDTA), hydroxocarboxylates (e.g., citrate), and α-amino acids (e.g., cysteine). Complexation using aminocarboxylic acids or polyhydrocarboxoylates is mainly applied due to the good solubility of the produced salts in water. The formation of complexes as well as the charge of the complexes formed can be controlled by the pH (Liu and Lee 1999, Liu et al. 1999).

The main limitation of CE is its very small sample volume, typically only a few nanoliters. Sample homogeneity and choosing a representative amount are related drawbacks. Consequently, the detection capability is a crucial point, requiring a sensitive detector to obtain low LODs to reach the natural low levels of species in environmental samples. In general, LODs are orders of magnitude worse than those for LC separations using the same detector (Michalke 2003).

Gel electrophoresis

Gel electrophoresis is a separation technique used to complement CE and is useful for separating complex mixtures. This method allows detection with radiotracers. When it is used in two dimensional and then, the amount of material is larger, allowing off-line identification. In addition gels can be stored for years for further analysis (Chéry 2003). Although it allows the detection of metals (Sussulini et al. 2007), quantitative analysis is still difficult, and GE has primarily remained a semiquantitative technique.

As the main requirement for speciation analysis is to keep the analyte unchanged during the analytical steps, the choice of buffers, electrodes and pH is crucial in GE (Chéry 2003). One important aspect of GE that must be mentioned is the widespread availability of ready-to-use gels and buffers.

The principle of separation is based on the same mechanism of CE. An electrophoretic separation occurs in an intervening medium that separates two electrodes (the positively charged anode and the negatively charged cathode) under high voltage (up to 2000 V). The intervening medium consists of a liquid, usually a buffer that is supported by an inert solid material such as paper or a semisolid gel. The entire set-up can be maintained in a temperature-controlled chamber to minimize the Joule effect during the migration process. The liquid allows the movement of ions, while the solid support provides frictional drag. When voltage is applied across the electrodes, a current is generated from the movement of ions in the electric field. The electric field strength determines the rates of species migration and can be varied experimentally. The intervening (support) medium may be as short as 10 cm or as long as 1 m. Throughout this medium, positively charged species will migrate toward the cathode and negatively charged species will move toward the anode (Mikkelsen and Cortón 2004, Chéry 2003).

Applications of GE are mainly devoted to macromolecules (metallo-enzymes and proteins) in the field of bioanalytical separations (Mesko et al. 2011). The main applications include separations of DNA/RNA, humic acids and proteins, and the main set-up used has been two-dimensional gel electrophoresis (2DE).

Coupling CE and GE to detection techniques

The coupling of CE to other techniques is a crucial point to allow speciation purposes. This is a key aspect due to the low sample volume usually injected in CE. This is the main limitation of using an UV detector, despite the feasibility of its on-line coupling and widespread use as detector for electrophoresis. This sensitivity problem can be overcome by the use of ICP-MS equipment coupled with CE, but it is considered not so simple as the coupling of LC to ICP-MS (Feldmann 2005). Basically, some aspects must be observed for this coupling: (i) it is necessary to perform the grounding of the electrical current of the CE high-voltage supply; (ii) CE has a low flow-rate to match the gas flow in the ICP instrument; (iii) it is important to assure maximum efficiency in the transport of analytes from CE to the mass spectrometer (Sonke and Salters 2007); (iv) a low dead volume interface should be used to preserve high resolution and separation power; and (v) it is necessary to overcome the back pressure produced by the nebulizer gas when it exits the nebulizer or electric connections (de Leon et al. 2002). Applications of coupling between CE and ICP-MS include separation and quantification of lanthanides and actinides, metalloporphyrins, and different metal-species (Zhao et al. 2012, Janos 2003, Pyrzynska 2001, Sutton et al. 1997, Trojanowicz et al. 2003, Michalke and Schramel 1999).

Electrospray ionization mass spectrometry is also a suitable detector for CE. The soft ionization promoted by the electrospray interface allows the preservation of whole molecules which is an advantage for structure determination. Another important aspect is the low flow-rates used in CE and ESI enabling the coupling. One of the main drawbacks is that volatile buffers are preferentially used for ESI, thus limiting its capabilities for separation in CE, without losing sensitivity. In addition, the ionization process itself can lead to the formation of ion-solvent clusters and counter ions are substituted by solvent molecules. These facts can result in the splitting of one species into multiple signals, decreasing sensitivity and increasing spectral complexity (Schramel et al. 1999b, Michalke 2003). The coupling CE-ESI-MS has been proposed for speciation of As and Se (Schramel et al. 1999a, b, Michalke et al. 1999). Additionally, the high resolution power of CE can be improved with the advantages of ESI-MS (mainly molecular information) and the advantages of ICP-MS (lower LODs and direct element information and quantification) in combined protocols.

When using GE, one option is to choose a method that allows the measurement of the analytes in the entire gel or a method that allows the gel to be cut in subsamples. To detect trace elements in the gel subsamples, they must be decomposed for further metal analysis by ICP-MS, AAS, or ICP-OES (Sussulini et al. 2007, Michalke 2003). The method can be time consuming and laborious, but it allows low LODs for most elements. Another alternative with gel subsamples is the use of solid sample analysis. In this case, the gel subsamples are not decomposed, and solids are directly analyzed, mainly by electrothermal vaporization (ETV) coupled to ICP-MS (ETV-ICP-MS), direct solid sampling ET AAS (DSS-ET-AAS) or neutron activation analysis (NAA) (Sidenius and Gammelgaard 2000, Michalke 2003). Laser ablation inductively coupled plasma mass spectrometry (LA-ICP-MS) can be used as a detection technique to map the gels (Chassaigne et al. 2004, Michalke 2003).

Speciation Analysis in Environmental Samples

Quality control testing and mass balance in environmental analysis

Even though most of method developments in speciation analysis have incorporated quality assurance and quality control, these topics remain a challenge in this area (Sturgeon and Francesconi 2009). The most common way to ensure that generated data are reliable when developing a method is to use a certified reference material (CRM), to perform a comparison of results from different techniques or to analyze a proficiency test sample (Sturgeon and Francesconi 2009). Based on the definition of speciation analysis it is important to consider that the total concentration of an element in a given matrix should correspond to the sum of different individual chemical species. When the sum of the species does not correspond to the total concentration, it can be impossible to perform the speciation analysis rigorously. In addition, the knowledge of total concentration is also useful when speciation is highly dynamic and there is a need to know an upper limit for a certain species or when more specific information is unavailable. Mainly, speciation is directly related to the knowledge of the total concentration which is necessary as a quality check for speciation data, allowing the mass balance (Sturgeon and Francesconi 2009). This is an important point to be considered when performing speciation analysis and even with the reasonable availability of calibration standards for most analytes investigated in speciation studies and the small array of CRMs to support the validation of data, it is frequently absent (Sturgeon and Francesconi 2009).

The critical aspects in speciation analysis methods are related to all steps, from sample collecting and preservation, sample preparation, analysis (separation and detection) and interpretation of the results. Some simple quality criteria have been proposed (Emons 2002, Batley et al. 2009, Sturgeon and Francesconi 2009) and include the specification of the requirements and determination of method characteristics. However, the quality of data in speciation studies is not always reported and some papers also do not stipulate quality assurance and quality control requirements for works published in speciation area (Batley et al. 2009). In many works, the recovery of all species from the column can be considered one of the main problems. As compounds can vary from small ions (being inorganic or even organic forms) up to large molecules (such as complexed to organic ligands) it is difficult to have a column able to separate all compounds and elute all of them under the conditions applied in a given method. The optimization of conditions for separation, independent of the method of choice, must be carefully developed in order to assure the quality of results in speciation analysis. In this aspect, care must be taken with the mass balance and chromatographic recovery that helps to know if some species are being lost, retained or converted as well as any other problem that can occur during the development of the method.

Applications

Even considering the coupling separation techniques with element-specific detectors, it is important to mention that detection of element species presents some limitations related to the speciation analysis when it is performed on an environmental sample, which could affect the LOD and instrument sensitivity (Cornelis et al. 2003, Ellis and Roberts 1997, Wuilloud et al. 2004). Additionally, the stability of these species also requires some special care once transformations can occur during the pretreatment step that contribute to increasing the uncertainty related to speciation analysis (Mester and Sturgeon 2005, Wuilloud et al. 2004). However, if element species are not appropriately separated before detection, the entire analytical sequence could be unusable. In this sense, it is important to mention that each element could be present in many different forms in the same matrix; thus it must be considered when a specific detection is required (Craig 2003). On the other hand, the lack of standards could represent another limitation related to the levels of the separation selectivity and signal identification (Szpunar et al. 2000, de Leon et al. 2002).

As previously discussed, the essentiality and toxicity of some elements are strongly dependent on the chemical form of these elements in the environment. In this regards, finding information about total trace element concentration could be not enough, and speciation analysis has become an important tool to obtain information related to an environment (Cornelis et al. 2003, Cornelis et al. 2005). Taking into account the above mentioned discussion, it is intended here to present the examples used to perform speciation analysis of elements (Al, As, Cr, Hg, I, Pb, Sb, Se, Sn, and V) that are normally determined in environmental samples such as soil, water, landfill leachate, sediment and road dust. Some examples selected to cover most investigated analytes and samples in recent applications are summarized in Table 3.1. The selected applications are an example of all the speciation methods discussed in this chapter and were chosen to cover the last 15 years related to speciation analysis in environmental samples. Applications are organized by separation technique—LC, GC and electrophoresis—which were the focus of this chapter. Specific developments in each work are described as comments in Table 3.1.

Final Considerations

The determination of species concentration, and not the total element concentration, is essential for some applications. The speciation analysis field has received growing interest once it was proved that different species of the same element can present different behavior, from toxic to beneficial

Table 3.1 Summary of some of the chromatographic and electrophoretic techniques applied for speciation studies in environmental samples.

Sample	Species determined	Separation method	Detection technique	Comments	Reference
Water, soil, and leachate	Cr(III) and Cr(VI) as CrO_4^{2-} and $Cr_2O_7^{2-}$	LC	DAD	Species separations were performed using a C18 reversed phase carbamate column. Sample solution was previously reacted with 1-pyrrolidinecarbodithioic acid, ammonium salt and Cr species were separated without pH or buffering adjustment.	(Cathum et al. 2002)
Effluent water	As(III), As(V), MMA, DMA, AsC, TMAO, TETRA	LC	ICP-MS	Inorganic arsenic species of suspended and soluble forms were determined using a weak anion-exchange column and an eluent containing 5 mmol L^{-1} sodium carbonate, 40 mmol L^{-1} sodium hydroxide, and 4% (v/v) methanol. To separate anionic, neutral, and cationic arsenic species, nitric acid gradient elution was applied to anion-exchange column. During the chromatographic run, the nitric acid concentration was changed from 0.5 to 50 mmol L^{-1} (pH 3.3 to 1.3).	(Mattusch and Wennrich 1998)
Estuarine water	As(III), As(V), MMA Cr(III), Cr(VI) Se(IV), Se(VI)	LC	DRC-ICP-MS	The species were baseline-separated in an 11 min elution on an anion exchange column using a gradient of $NH_4NO_3/NH_4H_2PO_4$ as mobile phase at pH 6.0. EDTA was added to the sample solution before LC separation in order to allow the formation of EDTA-Cr(III).	(Tsoi and Sze-Yin Leung 2010)
Groundwater	IO_3^-, I^-	LC	ICP-MS	0.01% (m/v) KOH was used as the sample medium for both total iodine analysis and iodine speciation analysis by ICP-MS. An ion exchange column was used to promote species separation. 0.03 mol L^{-1} ammonium carbonate solution (pH 9.4) was selected for sample analysis. The LC flow rate was set as 1 mL min^{-1}.	(Yang et al. 2007)

Table 3.1 contd....

Table 3.1 contd.

Sample	Species determined	Separation method	Detection technique	Comments	Reference
Landfill leachates	As(V), As(III), AsB, TMAO, AsC, MMA, DMA, TETRA, DMDTA, DMMTA	LC	DRC-ICP-MS	An anion exchange column with 20 mmol L^{-1} $NH_4H_2PO_4$ (pH 5.6, flow rate 1.5 mL min^{-1}) was used to separate As(III), As(V), MMA and DMA. A C18 column with 5 mmol L^{-1} formic acid (pH 2.9, flow rate 1.3 mL min^{-1}) was used to separate thiol-organoarsenic compounds. A cation-exchange column with 20 mmol L^{-1} pyridine (pH 2.6 with formic acid, flow rate 1.5 mL min^{-1}) to separate AsB, TMAO, AsC, and TETRA.	(Li et al. 2010)
		LC	ESI-MS/MS	A C18 column with a pH of 2.9 and 5 mmol L^{-1} formic acid as mobile phase with a flow rate of 1.0 mL min^{-1} was used for arsenic species identification; both positive and negative ionization modes; cone voltages at 15, 30 and 45 V; desolvation gas (nitrogen) flow rate of 600 L h^{-1}; desolvation temperature at 350ºC; source temperature at 150ºC; and the potential applied on the capillary at 3.0 kV were used to obtain molecular (low cone voltage) and structural information (high cone voltage).	
River sediment, agricultural soil, sewage sludge	As(III), As(V), MMA, DMA	LC	AFS	Determination of As species was performed using an ion-exchange LC coupled on-line to atomic fluorescence detector through hydride generation performed after orthophosphoric acid extraction of the solids.	(Gallardo et al. 2001)

Seawater	$^{129}IO_3^-$, $^{129}I^-$	LC	NAA	Iodide and iodate were first separated from seawater matrix using the anion exchange method. The sample flow rate was 5 mL min^{-1}. The iodide retained on the resin column was then eluted using 500 mL of 2.0 mol L^{-1} KNO$_3$ solution at a flow rate of 2 mL min^{-1}. The solution was passed through a new ion exchange column in order to separate iodate, washing the column with 50 mL of 0.5 mol L^{-1} KNO$_3$. The iodate in seawater was eluted from the column with 500 mL of a 2.0 mol L^{-1} KNO$_3$ solution.	(Hou et al. 2001)
Seawater	Organotin species: MBT, MPhT, DBT, DPhT, TBT, TPhT	LC	Spectrofluorimetry	Chromatographic separations were carried out at room temperature using a 1.0 mL min^{-1} mobile phase flow rate and C18 column. The initial mobile phase starts from 30 mg L^{-1} oxalic acid and 0.03% (v/v) triethylamine in methanol-acetic acid-water (57.5:2.5:40) to 0.03% (v/v) triethylamine in methanol-acetic acid-water (84:1:15).	(González-Toledo et al. 2000)
Sediments, zoobenthos and river water	Hg^{2+}, $MeHg^+$, $EtHg^+$, $PhHg^+$	LC	CV-AFS	Microwave-assisted extraction of mercury species from sediments and zoobenthos samples were performed using a mixture containing 3 mol L^{-1} HCl, 50% aqueous methanol and 0.2 mol L^{-1} citric acid (for masking co-extracted Fe^{3+}). An isocratic elution profile of aqueous methanol (65%/35%, v/v) was used for chromatographic separation of mercury species at a flow rate 0.8 mL min^{-1} with a C18 column.	(Margetínová et al. 2008)
Soil	AlF^{2+}, AlF_2^+, AlF_3, AlF_4^-, and Al^{3+}	LC	F AAS	An ion-exchange column that was used contained mixed anion and cation beds with sulfonic acid and alkanol quaternary ammonium functional groups. The chromatographic run was at a gradient of 2 mL min^{-1} (from 100% of water to 100% of NH$_4$Cl 1.8 mol L^{-1}, pH 3 and return to initial) with an injection volume of 500 mL.	(Frankowski et al. 2010)

Table 3.1 contd....

Table 3.1 contd.

Sample	Species determined	Separation method	Detection technique	Comments	Reference
Water	Sb(V), Sb(III), TMSb	LC	ICP-MS	An anion exchange column was used to separate species using a flow rate of 1.5 mL min^{-1} of 3 mmol L^{-1} tetramethylammonium hydroxide as mobile phase.	(Lintschinger et al. 1998)
			ESI-MS	Electrospray was performed using 4.5 kV, 100 °C capillary temperature and 0.68 L min^{-1} N$_2$.	
Landfill gas	Mo(CO)$_6$, W(CO)$_6$	GC	ICP-MS	Gas samples were cryogenically preconcentrated by trapping the gases at –78°C (dry ice/acetone slush). A combination of thermodesorption of the cryotrapped sample and separation by using a non polar chromatographic column (–196 to 150°C within 3 min) was performed and gases were separated by using a He flow of 133 mL min^{-1}. A heated Teflon transfer line (120°C) was used to introduce the separated samples to the torch of the ICP-MS.	(Feldmann and Cullen 1997)
Landfill leachates	Organotin compounds: MMeT, DMeT, TMeT, TEtT, MBT, DBT, TBT, MPhT, DPhT, TPhT, MOcT, DOcT, TOcT	GC	ICP-MS	Sample injection (2 μL) was performed in a splitless mode. A 15 m capillary column (5% phenyl/methylpolysiloxane) with the following temperature program was used: 50°C (0.8 min) followed by a ramp up to 200°C (20°C min^{-1}), held for 2 min then raised to 220°C (40°C min^{-1}), held for 0.5 min and raised to 280°C (50°C min^{-1}). Separated analytes were carried by He flow (1 mL min^{-1}) and directly introduced to the ICP torch by a heated transfer line.	(Vahcic et al. 2011)

Road dust	Organolead: TML	GC	ICP-TOFMS	The sample was weighed (0.4 g) into a centrifugation vessel and then 10 mL HAc-NaAc buffer solution and 0.5 mL EDTA (masking agent) were added. Multicapillary-GC separation was performed after the previous derivatization step (0.0125% (m/v) NaBEt$_4$ and of 0.00625% (m/v) NaBPr$_4$). Helium was used as the carrier gas (155 mL min^{-1}), and a separation temperature program was 100°C to 150°C (3 min) at 50°C min^{-1}. A fused silica capillary heated at 150°C was used as the transfer line to ICP-TOFMS.	(Jitaru et al. 2004)
Sewage sludge	Organotin compounds: MBT, DBT, TBT, MOcT, DOcT, TOcT	GC	FPD	The derivatization of sample aliquots (1–4 mL) of the extracted sample (glacial acetic acid) was performed with NaBEt$_4$. Sample injection was performed in splitless mode (290°C). Nitrogen was used as the carrier gas (2 mL min^{-1}). A capillary column coated with poly(dimethyl-siloxane) was used for separation via the following temperature program: 80°C (1 min) followed by a ramp up to 180°C (30°C min^{-1}), a subsequent temperature raise up to 270°C (10°C min^{-1}), and held at the final temperature for 7 min.	(Zuliani et al. 2008)
Water (natural)	Inorganic and organomercury: Hg^{2+}, MeHg$^+$ Organolead: TML, TEL Organotin: MBT, DBT, TBT	GC	EI-MS	The sample was ethylated with NaBEt$_4$ (derivatization step) with simultaneous headspace solid phase microextraction for further determination by GC-MS. Splitless injection mode was used (260°C, 1 min). A fused-silica column (5% phenil/methylsiloxane) with an initial temperature of 50°C up to 250°C (30°C min^{-1}) was used. Helium was used as the carrier gas (1.2 mL min^{-1}). The transfer line temperature was set at 280°C.	(Centineo et al. 2004)

Table 3.1 contd....

Table 3.1 contd.

Sample	Species determined	Separation method	Detection technique	Comments	Reference
Water and marine sediments	Hg^{2+}, MeHg, EtHg, Sn^{4+}, MTB, DBT, TBT	GC	EI-MS	The analytes were complexed with sodium diethyldithiocarbamate, retained on a C_{60} fullerene column and eluted with ethyl acetate containing $NaBPr_4$ as a derivatizing reagent. Chromatographic separation was performed using a capillary column (5% phenyl/methylpolysiloxane). Helium was used as the carrier gas (1 mL min^{-1}). The initial temperature was set at 40°C (2 min) followed by a 15°C min^{-1} ramp up to 115°C (held for 0 min), and a new 20°C min^{-1} ramp to 250°C (held for 5 min). The injection port, transfer line, and ion source temperatures were maintained at 200, 250, and 200°C, respectively.	(Muñoz et al. 2005)
Water (natural) soil, sludge water	As(V), As(III), MMA, DMA, TMAO	GC	QF-AAS	A pH-selective reduction was used for speciation of As(III) and As(V). As (III+V), MMA, DMA, and TMAO were reduced with $NaBH_4$ in 1% oxalic acid solution (pH 1.5). As(III) was quantified in a 0.05 mol L^{-1} phosphate buffer (pH 6.8). Arsines and alkylarsines formed were cryo-trapped in the U-column maintained in liquid nitrogen. Subsequently, the column was removed from liquid nitrogen, and the volatilized hydrides were carried to an electrically heated quartz cell where they were atomized in a H_2–O_2 flame.	(Guerin et al. 2000)
		LC	ICP-MS	An anion-exchange column was used with $(NH_{4/2})HPO_4$ in a water mobile phase. pH was controlled to 8.5 with NH_4OH. The chromatographic system was directly interfaced to the ICP-MS detector and $^{75}As^+$ was monitored.	

Sample	Analytes		Technique	Description	Reference
Water	Se(IV), Se(VI)	GC	FID	Se(IV) was determined in the form of peazselenol (complex with 4-nitro-o-phenylenediamine) with subsequent extraction using an organic solvent, which was performed using ultrasound-assisted emulsification microextraction. Total inorganic Se was determined after reduction of Se(VI) to Se(IV). The sample was injected in a splitless mode (260°C). Helium was used as the carrier gas (4 mL min⁻¹). A 30 m fused-silica capillary column was used. The oven temperature program was set as follows: 100°C for 3 min, increase to 175°C (8°C min⁻¹), hold for 5 min, increase to 260°C (15°C min⁻¹), and then hold at 260°C for 2 min.	(Najafi et al. 2012)
Water	Hg²⁺, MeHg⁺	GC	pyro-AFS	Sample (1 μL) was injected in splitless mode (300°C) and He was used as the carrier gas (3 mL min⁻¹). A non-polar capillary column was used. The temperature program was: 40°C (2 min) followed by a ramp up to 200°C (40°C min⁻¹).	(Berzas Nevado et al. 2011)
			MS	Sample (1 μL) was injected in splitless mode (250°C) and He was used as the carrier gas (1 mL min⁻¹). The capillary column was 5% phenyl/95% dimethyl polysiloxane. The temperature program was 40°C (1 min) followed by a ramp up to 90°C (15°C min⁻¹) with subsequent ramp up to 200°C (50°C min⁻¹). The interface temperature was set at 280°C.	
			ICP-MS	The sample (1 μL) was injected in splitless mode (200°C), and He was used as the carrier gas (16 mL min⁻¹). The capillary column was 100% dimethyl polysiloxane. The temperature program was 50°C (1 min) followed by a ramp up to 200°C (40°C min⁻¹). A heated transfer line (170°C) was directly coupled to the ICP torch.	

Table 3.1 contd....

Table 3.1 contd.

Sample	Species determined	Separation method	Detection technique	Comments	Reference
Coal	Hg^{2+}, $MeHg^+$, as cysteine complex	CE	Ciclic voltametry UV	A fused-silica capillary (65 cm x 50 μm i. d.) was used to separate species (25°C). Hydrodynamic injection of the sample into the capillary was performed (5 s). The buffer electrolyte used was a 25 mmol L^{-1} sodium borate adjusted to pH 9.3 with NaOH.	(Martin et al. 2010)
Dried sediment	As(III), As(VI), MMA, DMA, Se(IV), Se(VI), SeCyst, SeMet, SeCystamine	CE	UV	A hydrodynamic injection was performed. These anionic and cationic species were separated using a negative separation voltage polarity (−25 kV) in a fused-silica capillary (64.5 cm x 50 μm i. d., 20°C) coated with poly(diallyldimethylammonium chloride) (PDDAC). A 15 mmol L^{-1} phosphate buffer at pH 10.6 was used.	(Sun et al. 2004)
Fouling sludge, and sewage sludge	Sb(V), Sb(III), TMSb	CE	ICP-MS	Two different CE methods were used for Sb speciation based on the use of different buffers and stacking electrolytes. One of them, Na_2CO_3 (pH 11.6), used as an electrolyte, makes it possible to perform a well resolved electropherogram, but it destroyed Sb species. The other one, NaH_2PO_4/Na_2HPO_4 (pH 5.6), was used as an electrolyte, which keeps the Sb species stable during analysis. A capillary electrophoretic system at a temperature of 20°C was used.	(Michalke and Schramel 1999)
Natural groundwater	La(III), Pu(III), Pu(IV), Pu(V), Pu(VI), Th(IV), Np(IV), Np(V), U(VI)	CE	ICP-MS UV-Vis	A homemade CE system was used with a fused-silica capillary (62.6 cm x 50 μm i. i.) under + 30 kV. A makeup flow rate (560 μL min^{-1}) was necessary to couple CE capillary to the ICP torch. The sample injection was performed by hydrodynamic mode (20–1000 mbar Ar) during 10 s with 100 mbar. An electrolyte buffer of AcOH (1 mol L^{-1}) at pH 2.47 was used.	(Kuczewski et al. 2003)

Spiked mineral water	As(III), As(V), MMA, DMA, AsB, AsC,	CE	ICP-MS UV	A sample injection was performed under kPa pressure (40 s plus 3 to 12 post-injection). A borate buffer (20 mmol L^{-1} for UV detection and 50 mmol L^{-1} for ICP-MS detection and 50 mmol L^{-1} for UV detection- pH 9.4, 2% osmotic flow modifier) was used. Fused-silica capillaries (54 cm x 50 μm i. d. for UV detection; or 88 cm x 75 μm i. d. for ICP-MS detection) were used as electrophoretic capillaries (–25 kV voltage, 25°C).	(Van Holderbeke et al. 1999)
Water	As(III), As(V), MMA, DMA	CE-HG	ICP-MS GC	Electrokinetic in conjunction with hydrodynamic modified electroosmotic flow (pressure increments of 0.1 psi) was used for sample injection. A fused-silica capillary (85 cm x 75 μm i.d.) coated with polyimide was used. Separation was performed under pH 9.03 (20 mmol L^{-1} potassium hydrogen phthalate/20 mmol L^{-1} boric acid buffer).	(Magnuson et al. 1997)
Electroplating bath and waste water	Bi(III), Co(II), Co(III), Cr(III), Cu(II), Fe(III), Hg(II), Ni(II), Pb(II), V(IV)	CZE	UV	Fused-silica capillaries (57 cm x 75 μm i.d.) were used. A hydrodynamic mode was used to inject the sample. Better resolution was achieved using counter-electroosmotic conditions (25 kV) with 5 mmol L^{-1} DTPA electrolyte at pH 8.5	(Padarauskas and Schwedt 1997)
Leached from waste-catalyst	V(IV), V(V)	CZE	UV	Vanadium species were chelated with EDTA to form anionic complexes. A fused-silica capillary column (100 cm x 75 μm i.d.) modified with hexadecyltrimethylammonium bromide was used to separate the V(IV)-EDTA and V(V)-EDTA complexes. A 50 mmol sodium acetate buffer (pH 4.6) containing 0.5 mmol EDTA was used as the electrolyte. Separation was performed at an applied voltage of –18 kV.	(Jen et al. 1997)

Table 3.1 contd....

Table 3.1 contd.

Sample	Species determined	Separation method	Detection technique	Comments	Reference
Soil leachate	As(III), As(V), MMA, DMA, Se(IV), Se(VI), SeCyst, SeMet, Sb(V), Te(IV), Te(VI)	CZE	ICP-MS	A homemade CE instrument was used to separate species with fused-silica capillary (80 cm x 75 µm i. d.). The separation voltage was kept at –20 kV. The sample solution was injected at the cathodic end of the capillary by hydrostatic or electromigrative mode (EML). The last injection mode resulted in a very low detection limit for the species with high electrophoretic mobility, but the sensitivity was drastically lower for species with low electrophoretic mobility, which was considered inappropriate for speciation study.	(Casiot et al. 2002)
Spiked groundwater	V(IV), V(V)	CZE	UV	On-column complexation of V species was performed with EDTA, DTPA, NTA and HEDTA. The optimum separation of anionic V forms was obtained using an electrolyte containing 5 mmol L⁻¹ EDTA (pH 4.0). Electrophoretic separation was performed using a fused-silica capillary (40 cm x 50 µm i. d.).	(Chen and Naidu 2002)

AsB, arsenobetaine; AsC, arsenocholine; AFS: atomic fluorescence spectrometry; DBT, dibutyltin; DEL, diethyllead; DMA, dimethylarsinic acid; DMDTA, dimethyldithioarsinic acid; DMeT, dimethyltin; DML, dimethyllead; DMMTA, dimethylmonothioarsinic acid; DOcT, dioctyltin; DPhT, diphenyltin; DRC-ICP-MS, dynamic reaction cell inductively coupled plasma mass spectrometry; EtHg⁺, ethylmercury; ICP-TOFMS, inductively coupled plasma time-of-flight mass spectrometry; ID-ICP-MS, isotope dilution inductively coupled plasma mass spectrometry; MBT, monobutyltin; MeHg⁺, methylmercury; MMA, monomethylarsonic acid; MMeT, monomethyltin; MOcT, monooctyltin; MPhT, monophenyltin; PhHg⁺, phenylmercury; OFM, pyro-AFS, atomic fluorescence detector via pyrolysis; QF AAS, quartz atomizing cell flame atomic absorption spectrometry; SFC, supercritical fluid chromatography; SeCyst, selenocysteine; Se-Cystamine, selenocystamine; SeMet, selenomethionine; TBT, tributyltin; TEL, triethyllead; TETRA, tetramethylarsonium ion; TEtT, triethyltin; TMAO, trimethylarsine oxide; TMeT, trimethyltin; TML, trimethyllead; TMSb, trimethylantimony; TOcT, trioctyltin; TPhT, triphenyltin; TrPhT, tetraphenyltin.

effects. To accomplish the objectives of speciation analysis it is necessary to develop a suitable combination between a separation technique and a detection technique. Speciation analysis can be performed by using these so-called hyphenated techniques. Whereas separation depends mainly of the analyte and the matrix, and detection is mainly dependent on the analyte concentration. However, as presented, the quality of the results is directly related to separation, which must assure the complete separation of different species, allowing them to reach the detector, preferentially in the absence of the matrix. The main instrumental methods that have been used for separation include LC, GC and CE. Each one possesses specific characteristics, advantages, drawbacks, and is dedicated to some types of analytes. Accordingly, volatility is the main aspect to be considered for GC, and polarity dictates the choice for a LC method. Electrophoresis can be used for those cases where separation cannot be achieved more easily than with other methods or when concentration is sufficiently high to be determined by the most common detectors. All of these separation techniques have been studied and proposed for speciation analysis in environmental samples (e.g., air, gas, soils and water) with a special focus on As, Hg, Pb, Se, and Sn.

Acknowledgements

The authors are grateful to CNPq, CAPES and FAPERGS for their financial support.

References

Ackley, K.L. and J.A. Caruso. 2003. Separation techniques—liquid chromatography. *In*: R. Cornelis, H. Crews. J. Caruso and K. Heumann [eds.]. Handbook of elemental speciation: techniques and methodology. John Wiley & Sons, Ltd., Chichester, England. pp. 147–239.

Al-Assaf, K.H., J.F. Tyson and P.C. Uden. 2009. Determination of four arsenic species in soil by sequential extraction and high performance liquid chromatography with post-column hydride generation and inductively coupled plasma optical emission spectrometry detection. Journal of Analytical Atomic Spectrometry. 24(4): 376–384.

Alonso, J.I.G. and J.R. Encinar. 2003. Separation techniques—gas chromatography and other based methods. *In*: R. Cornelis, H. Crews, J. Caruso and K. Heumann [eds.]. Handbook of elemental speciation: techniques and methodology. John Wiley & Sons, Ltd., Chichester, England. pp. 147–239.

Amayo, K.O., A. Petursdottir, C. Newcombe, H. Gunnlaugsdottir, A. Raab, E.M. Krupp and J.R. Feldmann. 2011. Identification and quantification of arsenolipids using reversed-phase HPLC coupled simultaneously to high-resolution ICPMS and high-resolution electrospray MS without species-specific standards. Analytical Chemistry. 83(9): 3589–3595.

Amouroux, D., E. Tessier, C. Pecheyran and O.F.X. Donard. 1998. Sampling and probing volatile metal(loid) species in natural waters by *in situ* purge and cryogenic trapping followed by gas chromatography and inductively coupled plasma mass spectrometry (P-CT-GC-ICP/MS). Analytica Chimica Acta. 377(2-3): 241–254.

Anawar, H.M. 2012. Arsenic speciation in environmental samples by hydride generation and electrothermal atomic absorption spectrometry. Talanta 88: 30–42.

Arruda, M.A. [ed.]. 2007. Trends in sample preparation. Nova Science Publishers, Inc., New York.

Attar, K.M. 1996. Analytical methods for speciation of organotins in the environment. Applied Organometallic Chemistry. 10(5): 317–337.

B'Hymer, C. and J.A. Caruso. 2004. Arsenic and its speciation analysis using high-performance liquid chromatography and inductively coupled plasma mass spectrometry. Journal of Chromatography A 1045(1-2): 1–13.

B'Hymer, C. and J.A. Caruso. 2006. Selenium speciation analysis using inductively coupled plasma-mass spectrometry. Journal of Chromatography A 1114(1): 1–20.

Balcerzak, M. and M.W. Raynor. 1998. Sample preparation for gas chromatographic separation methods in speciation analysis of metals and metalloids. Chemia Analityczna. 43(3): 287–299.

Batley, G.E., K.A. Francesconi and W.A. Maher. 2009. The role of speciation in environmental chemistry and the case for quality criteria. Environmental Chemistry. 6(4): 273–274.

Berzas Nevado, J.J., R.C. Rodríguez Martín-Doimeadios, E.M. Krupp, F.J. Guzmán Bernardo, N. Rodríguez Fariñas, M. Jiménez Moreno, D. Wallace and M.J. Patiño Ropero. 2011. Comparison of gas chromatographic hyphenated techniques for mercury speciation analysis. Journal of Chromatography A 1218(28): 4545–4551.

Borai, E.H., E.A. El-Sofany, A.S. Abdel-Halim and A.A. Soliman. 2002. Speciation of hexavalent chromium in atmospheric particulate samples by selective extraction and ion chromatographic determination. Trac-Trends in Analytical Chemistry. 21(11): 741–745.

Bouyssiere, B., J. Szpunar and R. Lobinski. 2002. Gas chromatography with inductively coupled plasma mass spectrometric detection in speciation analysis. Spectrochimica Acta Part B-Atomic Spectroscopy. 57(5): 805–828.

Bouyssiere, B., J. Szpunar, G. Lespes and R. Lobinski. 2003a. Gas chromatography with inductively coupled plasma mass spectrometric detection (GC-ICP MS). Advances in Chromatography. 42: 107–137.

Bouyssiere, B., J. Szpunar, M. Potin-Gautier and R. Lobinski. 2003b. Sample preparation techniques for elemental speciation studies. In: Handbook of elemental speciation: techniques and methodology. John Wiley & Sons, Ltd., Chichester, England. pp. 95–118.

Bueno, M.T. and M. Potin-Gautier. 2002. Solid-phase extraction for the simultaneous preconcentration of organic (selenocystine) and inorganic [Se(IV), Se(VI)] selenium in natural waters. Journal of Chromatography A 963(1–2): 185–193.

Cai, Y. 2003. Derivatization and vapor generation methods for trace element analysis and speciation. In: Z. Mester and R. Sturgeon [eds.]. Sample preparation for trace element analysis, vol XLI. Elsevier, Amsterdam. pp. 577–610.

Cai, Y., S. Rapsomanikis and M.O. Andreae. 1993. Determination of butylin compounds in sediment using gas chromatography-atomic absorption spectrometry: comparison of sodium tetrahydroborate and sodium tetraethylborate derivatization methods. Analytica Chimica Acta. 274(2): 243–251.

Carey, J.M., N.P. Vela and J.A. Caruso. 1994. Chromium determination by supercritical fluid chromatography with inductively coupled plasma mass spectrometric and flame ionization detection. Journal of Chromatography A 662(2): 329–340.

CarlierPinasseau, C., G. Lespes and M. Astruc. 1996. Determination of butyltin and phenyltin by GC-FPD following ethylation by NaBEt(4). Applied Organometallic Chemistry. 10(7): 505–512.

Caruso, J.A., R.G. Wuilloud, J.C. Altamirano and W.R. Harris. 2006. Modeling and separation-detection methods to evaluate the speciation of metals for toxicity assessment. Journal of Toxicology and Environmental Health-Part B-Critical Reviews 9(1): 41–61.

Casiot, C., O.F.X. Donard and M. Potin-Gautier. 2002. Optimization of the hyphenation between capillary zone electrophoresis and inductively coupled plasma mass spectrometry for the measurement of As-, Sb-, Se- and Te-species, applicable to soil extracts. Spectrochimica Acta Part B: Atomic Spectroscopy. 57(1): 173–187.

Castillo, A., A.F. Roig-Navarro and O.J. Pozo. 2008. Capabilities of microbore columns coupled to inductively coupled plasma mass spectrometry in speciation of arsenic and selenium. Journal of Chromatography A 1202(2): 132–137.

Cathum, S.C., C.B. Brown and W.W. Wong. 2002. Determination of Cr^{3+}, CrO_4^{2-} and $Cr_2O_7^{2-}$ in environmental matrixes by high-performance liquid chromatography with diode-array detection (HPLC-DAD). Analytical and Bioanalytical Chemistry. 373(1): 103–110.

Centineo, G., E.B. Gonzalez and A. Sanz-Medel. 2004. Multielemental speciation analysis of organometallic compounds of mercury, lead and tin in natural water samples by headspace-solid phase microextraction followed by gas chromatography-mass spectrometry. Journal of Chromatography A 1034(1-2): 191–197.

Chainet, F., C.P. Lienemann, M. Courtiade, J. Ponthus and O.F.X. Donard. 2011. Silicon speciation by hyphenated techniques for environmental, biological and industrial issues: A review. Journal of Analytical Atomic Spectrometry. 26(1): 30–51.

Chassaigne, H., C.C. Chery, G. Bordin, F. Vanhaecke and A.R. Rodriguez. 2004. 2-Dimensional gel electrophoresis technique for yeast selenium-containing proteins—sample preparation and MS approaches for processing 2-D gel protein spots. Journal of Analytical Atomic Spectrometry. 19(1): 85–95.

Chen, Z.and R. Naidu. 2002. On-column complexation and simultaneous separation of vanadium(IV) and vanadium(V) by capillary electrophoresis with direct UV detection. Analytical and Bioanalytical Chemistry. 374(3): 520–525.

Chéry, C.C. 2003. Gel electrophoresis for speciation purposes. *In*: R. Cornelis, H. Crews, J. Caruso and K. Heumann [eds.]. Handbook of elemental speciation: techniques and methodology John Wiley & Sons, Ltd., Chichester, England. pp. 224–239.

Cornelis, R., H. Crews, J. Caruso and K. Heumann. 2003. Handbook of elemental speciation: techniques and methodology. John Wiley & Sons, Ltd., Chichester, England.

Cornelis, R., J. Caruso, H. Crews and K. Heumann. 2005. Handbook of Elemental Speciation II—Species in the Environment, Food, Medicine and Occupational Health, vol. II. John Wiley & Sons Ltd., Chichester.

Craig, P.J. 2003. Organometallic Compounds in the Environment. 2nd edn. John Wiley & Sons Ltd., Chichester.

de Leon, C.A.P., M. Montes-Bayon and J.A. Caruso. 2002. Elemental speciation by chromatographic separation with inductively coupled plasma mass spectrometry detection. Journal of Chromatography A 974(1-2): 1–21.

Demuth, N. and K.G. Heumann. 2001. Validation of methylmercury determinations in aquatic systems by alkyl derivatization methods for GC analysis using ICP-IDMS. Analytical Chemistry. 73(16): 4020–4027.

Dorman, F.L., J.J. Whiting, J.W. Cochran and J. Gardea-Torresdey. 2010. Gas Chromatography. Analytical Chemistry. 82(12): 4775–4785.

dos Santos, J.S., M. de la Guárdia, A. Pastor and M.L.P. dos Santos. 2009. Determination of organic and inorganic mercury species in water and sediment samples by HPLC on-line coupled with ICP-MS. Talanta. 80(1): 207–211.

Ellis, L.A. and D.J. Roberts. 1997. Chromatographic and hyphenated methods for elemental speciation analysis in environmental media. Journal of Chromatography A 774(1-2): 3–19.

Emons, H. 2002. Artefacts and facts about metal(loid)s and their species from analytical procedures in environmental biomonitoring. Trac-Trends in Analytical Chemistry. 21(6-7): 401–411.

Encinar, J.R., P. Rodriguez-Gonzalez, J.R. Fernandez, J.I.G. Alonso, S. Diez, J.M. Bayona and A. Sanz-Medel. 2002. Evaluation of accelerated solvent extraction for butyltin speciation in PACS-2 CRM using double-spike isotope dilution-GC/ICPMS. Analytical Chemistry. 74(20): 5237–5242.

Engewald, W., J. Teske and J. Efer. 1999. Programmed temperature vaporiser-based injection in capillary gas chromatography. Journal of Chromatography A 856(1-2): 259–278.

Feldmann, J. 2005. What can the different current-detection methods offer for element speciation? Trends in Analytical Chemistry. 24(3): 228–242.

Feldmann, J. and W.R. Cullen. 1997. Occurrence of Volatile Transition Metal Compounds in Landfill Gas: Synthesis of Molybdenum and Tungsten Carbonyls in the Environment. Environmental Science & Technology. 31(7): 2125–2129.

Feldmann, J., P. Salaün and E. Lombi. 2009. Critical review perspective: elemental speciation analysis methods in environmental chemistry—moving towards methodological integration. Environmental Chemistry. 6(4): 275–289.

Fent, K. and M.D. Muller. 1991. Occurence of organotins in municipal wast-water and sewage-sludge and behavior in a treatment-plant. Environmental Science & Technology. 25(3): 489–493.

Frankowski, M., A. Zioła-Frankowska and J. Siepak. 2010. New method for speciation analysis of aluminium fluoride complexes by HPLC–FAAS hyphenated technique. Talanta. 80(5): 2120–2126.

Gallardo, M.V., Y. Bohari, A. Astruc, M. Potin-Gautier and M. Astruc. 2001. Speciation analysis of arsenic in environmental solids Reference Materials by high-performance liquid chromatography-hydride generation-atomic fluorescence spectrometry following orthophosphoric acid extraction. Analytica Chimica Acta. 441(2): 257–268.

Gao, Z. and X. Ma. 2011. Speciation analysis of mercury in water samples using dispersive liquid–liquid microextraction combined with high-performance liquid chromatography. Analytica Chimica Acta. 702(1): 50–55.

Geng, W.H., R. Komine, T. Ohta, T. Nakajima, H. Takanashi and A. Ohki. 2009. Arsenic speciation in marine product samples: Comparison of extraction-HPLC method and digestion-cryogenic trap method. Talanta. 79(2): 369–375.

Gettar, R.T., R.N. Garavaglia, E.A. Gautier and D.A. Batistoni. 2000. Determination of inorganic and organic anionic arsenic species in water by ion chromatography coupled to hydride generation–inductively coupled plasma atomic emission spectrometry. Journal of Chromatography A 884(1–2): 211–221.

Gong, Z.L., X.F. Lu, M.S. Ma, C. Watt and X.C. Le. 2002. Arsenic speciation analysis. Talanta. 58(1): 77–96.

González-Toledo, E., R. Compañó, M. Granados and M.D. Prat. 2000. Determination of butyltin and phenyltin species by reversed-phase liquid chromatography and fluorimetric detection. Journal of Chromatography A 878(1): 69–76.

González-Toledo, E., R. Compañó, M. Granados and M. Dolors Prat. 2003. Detection techniques in speciation analysis of organotin compounds by liquid chromatography. TrAC Trends in Analytical Chemistry. 22(1): 26–33.

Grotti, M., P. Rivaro and R. Frache. 2001. Determination of butyltin compounds by high-performance liquid chromatography-hydride generation-electrothermal atomization atomic absorption spectrometry. Journal of Analytical Atomic Spectrometry. 16(3): 270–274.

Guerin, T., A. Astruc and M. Astruc. 1999. Speciation of arsenic and selenium compounds by HPLC hyphenated to specific detectors: a review of the main separation techniques. Talanta. 50(1): 1–24.

Guerin, T., N. Molenat, A. Astruc and R. Pinel. 2000. Arsenic speciation in some environmental samples: a comparative study of HG-GC-QFAAS and HPLC-ICP-MS methods. Applied Organometallic Chemistry. 14(8): 401–410.

Harrington, C.F. 2000. The speciation of mercury and organomercury compounds by using high-performance liquid chromatography. TrAC Trends in Analytical Chemistry. 19(2–3): 167–179.

Harrington, C.F., R. Clough, L.R. Drennan-Harris, S.J. Hill and J.F. Tyson. 2011. Atomic spectrometry update. Elemental speciation. Journal of Analytical Atomic Spectrometry 26(8): 1561–1595.

Heisterkamp, M. and F.C. Adams. 2001. Gas chromatography inductively coupled plasma time-of-flight mass spectrometry for the speciation analysis of organolead compounds in environmental water samples. Fresenius Journal of Analytical Chemistry. 370(5): 597–605.

Hirner, A.V. 2006. Speciation of alkylated metals and metalloids in the environment. Analytical and Bioanalytical Chemistry. 385(3): 555–567.

Honeycutt, J. and J.M. Riddle. 1961. Preparation and Reactions of Sodium Tetraethylboron and Related Compounds. Journal of the American Chemical Society. 83(2): 369–&.

Hou, X., H. Dahlgaard and S.P. Nielsen. 2001. Chemical speciation analysis of 129I in seawater and a preliminary investigation to use it as a tracer for geochemical cycle study of stable iodine. Marine Chemistry, 74(2-3): 145–155.

Huerga, A., I. Lavilla and C. Bendicho. 2005. Speciation of the immediately mobilisable As(III), As(V), MMA 14 and DMA in river sediments by high performance liquid chromatography-hydride generation-atomic fluorescence spectrometry following ultrasonic extraction. Analytica Chimica Acta. 534(1): 121–128.

Janos, P. 2003. Analytical separations of lanthanides and actinides by capillary electrophoresis. Electrophoresis. 24(12-13): 1982–1992.

Jen, J.-F., M.-H. Wu and T.C. Yang. 1997. Simultaneous determination of vanadium(IV) and vanadium(V) as EDTA complexes by capillary zone electrophoresis. Analytica Chimica Acta. 339(3): 251–257.

Jitaru, P., H.G. Infante and F.C. Adams. 2004. Simultaneous multi-elemental speciation analysis of organometallic compounds by solid-phase microextraction and multicapillary gas chromatography hyphenated to inductively coupled plasma-time-of-flight-mass spectrometry. Journal of Analytical Atomic Spectrometry. 19(7): 867–875.

Jitaru, P., A. Birzu, R. Mocanu and F.C. Adams. 2005. Effect of the interface on separation in multicapillary gas chromatography-based hyphenated techniques for speciation analysis of organometallic compounds. Analytical and Bioanalytical Chemistry. 382(8): 1993–1998.

Kaur, V., A.K. Malik and N. Verma. 2006. Applications of solid phase microextraction for the determination of metallic and organometallic species. Journal of Separation Science 29(3): 333–345.

Kotaś, J. and Z. Stasicka. 2000. Chromium occurrence in the environment and methods of its speciation. Environmental Pollution. 107(3): 263–283.

Kuban, P. and A.R. Timerbaev. 2012. CE of inorganic species—A review of methodological advancements over 2009–2010. Electrophoresis. 33(1): 196–210.

Kuczewski, B., C.M. Marquardt, A. Seibert, H. Geckeis, J.V. Kratz and N. Trautmann. 2003. Separation of Plutonium and Neptunium Species by Capillary Electrophoresis–Inductively Coupled Plasma-Mass Spectrometry and Application to Natural Groundwater Samples. Analytical Chemistry. 75(24): 6769–6774.

Kuhn, R. and S. Hofstetter-Kuhn. 1993. Capillary electrophoresis: principles and practice. Spinger-Verlag, Berlin.

Kumar, A.R. and P. Riyazuddin. 2007. Non-chromatographic hydride generation atomic spectrometric techniques for the speciation analysis of arsenic, antimony, selenium, and tellurium in water samples—a review. International Journal of Environmental Analytical Chemistry. 87(7): 469–500.

Lalere, B., J. Szpunar, H. Budzinski, P. Garrigues and O.F.X. Donard. 1995. Speciation analysis for organotin compounds in sediments by capillary gas-chromatography with flame photometric detection after microwave-assisted acid leaching. Analyst. 120(11): 2665–2673.

Le, X.-C., W.R. Cullen and K.J. Reimer. 1994. Effect of cysteine on the speciation of arsenic by using hydride generation atomic absorption spectrometry. Analytica Chimica Acta. 285(3): 277–285.

Leopold, K., M. Foulkes and P. Worsfold. 2010. Methods for the determination and speciation of mercury in natural waters—A review. Analytica Chimica Acta. 663(2): 127–138.

Li, Y., G.K.C. Low, J.A. Scott and R. Amal. 2010. Arsenic speciation in municipal landfill leachate. Chemosphere. 79(8): 794–801.

Lintschinger, J., O. Schramel and A. Kettrup. 1998. The analysis of antimony species by using ESI-MS and HPLC-ICP-MS. Fresenius' Journal of Analytical Chemistry. 361(2): 96–102.

Liu, B.F., L.B. Liu and J.K. Cheng. 1999. Analysis of inorganic cations as their complexes by capillary electrophoresis. Journal of Chromatography A 834(1-2): 277–308.

Liu, W.P. and K. Lee. 1999. Chemical modification of analytes in speciation analysis by capillary electrophoresis, liquid chromatography and gas chromatography. Journal of Chromatography A 834(1-2): 45–63.

Lobinski, R. 1997. Elemental speciation and coupled techniques. Applied Spectroscopy. 51(7): A260–A278.

Lobinski, R. and F.C. Adams. 1992. Ultratrace speciation analysis of organolead in water by gas-chromatography atomic emission-spectrometry after in-liner preconcentration. Journal of Analytical Atomic Spectrometry. 7(6): 987–992.

Lobinski, R. and F.C. Adams. 1997. Speciation analysis by gas chromatography with plasma source spectrometric detection. Spectrochimica Acta Part B-Atomic Spectroscopy. 52(13): 1865–1903.

Magnuson, M.L., J.T. Creed and C.A. Brockhoff. 1997. Speciation of Arsenic Compounds in Drinking Water by Capillary Electrophoresis with Hydrodynamically Modified Electroosmotic Flow Detected Through Hydride Generation Inductively Coupled Plasma Mass Spectrometry With a Membrane Gas-Liquid Separator. Journal of Analytical Atomic Spectrometry. 12(7): 689–695.

Marcic, C., G. Lespes and M. Potin-Gautier. 2005. Pressurised solvent extraction for organotin speciation in vegetable matrices. Analytical and Bioanalytical Chemistry. 382(7): 1574–1583.

Margetínová, J., P. Houserová-Pelcová and V. Kubáň. 2008. Speciation analysis of mercury in sediments, zoobenthos and river water samples by high-performance liquid chromatography hyphenated to atomic fluorescence spectrometry following preconcentration by solid phase extraction. Analytica Chimica Acta. 615(2): 115–123.

Martin, L.G., L.T. Jongwana and A.M. Crouch. 2010. Capillary electrophoretic separation and post-column electrochemical detection of mercury and methyl mercury and applications to coal samples. Electrochimica Acta. 55(14): 4303–4308.

Mattusch, J. and R. Wennrich. 1998. Determination of Anionic, Neutral, and Cationic Species of Arsenic by Ion Chromatography with ICPMS Detection in Environmental Samples. Analytical Chemistry. 70(17): 3649–3655.

McNair, H.M. and J.M. Miller. 2008. Basic Gas Chromatography. John Wiley & Sons, Inc.

Mesko, M.F., C.A. Hartwig, C.A. Bizzi, J.S.F. Pereira, P.A. Mello and E.M.M. Flores. 2011. Sample preparation strategies for bioinorganic analysis by inductively coupled plasma mass spectrometry. International Journal of Mass Spectrometry. 307(1-3): 123–136.

Mester, Z. and R. Sturgeon [eds.]. 2003. Sample preparation for trace element analysis, vol XLI. Comprehensive Analytical Chemistry. Elsevier, Amsterdam.

Mester, Z. and R. Sturgeon. 2005. Trace element speciation using solid phase microextraction. Spectrochimica Acta Part B-Atomic Spectroscopy. 60(9-10): 1243–1269.

Mester, Z., R. Sturgeon and J. Pawliszyn. 2001. Solid phase microextraction as a tool for trace element speciation. Spectrochimica Acta Part B-Atomic Spectroscopy. 56(3): 233–260.

Michalke, B. 2003. Capillary electrophoresis in speciation analysis. *In*: R. Cornelis, H. Crews, J. Caruso and K. Heumann [eds.]. Handbook of elemental speciation: techniques and methodology John Wiley & Sons, Ltd., Chichester, England. pp. 201–223.

Michalke, B. and P. Schramel. 1999. Antimony speciation in environmental samples by interfacing capillary electrophoresis on-line to an inductively coupled plasma mass spectrometer. Journal of Chromatography A 834(1-2): 341–348.

Michalke, B., O. Schramel and A. Kettrup. 1999. Capillary electrophoresis coupled to inductively coupled plasma mass spectrometry (CE/ICP-MS) and to electrospray ionization mass spectrometry (CE/ESI-MS): An approach for maximum species information in speciation of selenium. Fresenius Journal of Analytical Chemistry 363(5-6): 456–459.

Mikkelsen, S.R. and E. Cortón. 2004. Principles of electrophoresis. *In:* Bioanalytical Chemistry. John Wiley & Sons, Inc. pp. 167–190.

Miravet, R., E. Hernández-Nataren, A. Sahuquillo, R. Rubio and J.F. López-Sánchez. 2010. Speciation of antimony in environmental matrices by coupled techniques. TrAC Trends in Analytical Chemistry. 29(1): 28–39.

Montes-Bayon, M., K. DeNicola and J.A. Caruso. 2003. Liquid chromatography-inductively coupled plasma mass spectrometry. Journal of Chromatography A 1000(1-2): 457–476.

Munoz, J., M. Gallego and M. Valcarcel. 2005. Speciation analysis of mercury and tin compounds in water and sediments by gas chromatography-mass spectrometry following preconcentration on C-60 fullerene. Analytica Chimica Acta. 548(1-2): 66–72.

Najafi, N.M., H. Tavakoli, Y. Abdollahzadeh and R. Alizadeh. 2012. Comparison of ultrasound-assisted emulsification and dispersive liquid–liquid microextraction methods for the speciation of inorganic selenium in environmental water samples using low density extraction solvents. Analytica Chimica Acta. 714(0): 82–88.

Namiesnik, J. and T. Gorecki. 2000. Sample preparation for chromatographic analysis of plant material. Jpc-Journal of Planar Chromatography-Modern Tlc. 13(6): 404–413.

Needham, L.L., D.G. Patterson, D.B. Barr, J. Grainger and A.M. Calafat. 2005. Uses of speciation techniques in biomonitoring for assessing human exposure to organic environmental chemicals. Analytical and Bioanalytical Chemistry. 381(2): 397–404.

Nevado, J.J.B., R.C.R. Martin-Doimeadios, E.M. Krupp, F.J.G. Bernardo, N.R. Farinas, M.J. Moreno, D. Wallace and M.J.P. Roper. 2011. Comparison of gas chromatographic hyphenated techniques for mercury speciation analysis. Journal of Chromatography A 1218(28): 4545–4551.

Padarauskas, A. and G. Schwedt. 1997. Capillary electrophoresis in metal analysis: Investigations of multi-elemental separation of metal chelates with aminopolycarboxylic acids. Journal of Chromatography A 773(1-2): 351–360.

Patrushev, Y.V., O.A. Nikolaeva and V.N. Sidelnikov. 2010. Investigation of the Loading Capacity of Multicapillary Chromatographic Columns. Journal of Analytical Chemistry. 65(11): 1129–1131.

Pena-Pereira, F., I. Lavilla and C. Bendicho. 2009. Miniaturized preconcentration methods based on liquid-liquid extraction and their application in inorganic ultratrace analysis and speciation: A review. Spectrochimica Acta Part B-Atomic Spectroscopy. 64(1): 1–15.

Pergantis, S.A., W.R. Cullen, D.T. Chow and G.K. Eigendorf. 1997. Liquid chromatography and mass spectrometry for the speciation of arsenic animal feed additives. Journal of Chromatography A 764(2): 211–222.

Popp, M., S. Hann and G. Koellensperger. 2010. Environmental application of elemental speciation analysis based on liquid or gas chromatography hyphenated to inductively coupled plasma mass spectrometry—A review. Analytica Chimica Acta. 668(2): 114–129.

Powell, M.J., D.W. Boomer and D.R. Wiederin. 1995. Determination of chromium species in environmental samples using high-pressure liquid chromatography direct injection nebulization and inductively coupled plasma mass spectrometry. Analytical Chemistry. 67(14): 2474–2478.

Priego-Capote, F. and L. de Castro. 2007. Ultrasound-assisted digestion: A useful alternative in sample preparation. Journal of Biochemical and Biophysical Methods. 70(2): 299–310.

Pyrzynska, K. 2001. Analysis of selenium species by capillary electrophoresis. Talanta. 55(4): 657–667.

Reyes, L., J. Mar, A. Hernández-Ramírez, J. Peralta-Hernández, J. Barbosa and H. Kingston. 2011. Microwave assisted extraction for mercury speciation analysis. Microchimica Acta. 172(1): 3–14.

Sanchez-Rodas, D., W.T. Corns, B. Chen and P.B. Stockwell. 2010. Atomic Fluorescence Spectrometry: a suitable detection technique in speciation studies for arsenic, selenium, antimony and mercury. Journal of Analytical Atomic Spectrometry. 25(7): 933–946.

Schramel, O., B. Michalke and A. Kettrup. 1999a. Application of capillary electrophoresis-electrospray ionisation mass spectrometry to arsenic speciation. Journal of Analytical Atomic Spectrometry. 14(9): 1339–1342.

Schramel, O., B. Michalke and A. Kettrup. 1999b. Capillary electrophoresis/electrospray ionization mass spectrometry (CE/ESI-MS) as a powerful tool for trace element speciation. Fresenius Journal of Analytical Chemistry. 363(5-6): 452–455.

Shum, S.C.K., H.M. Pang and R.S. Houk. 1992. Speciation of mercury and lead compounds by microbore column liquid chromatography-inductively coupled plasma mass spectrometry with direct injection nebulization. Analytical Chemistry. 64(20): 2444–2450.

Sidenius, U. and B. Gammelgaard. 2000. Direct determination of selenoproteins in polyvinylidene difluoride membranes by electrothermal atomic absorption spectrometry. Fresenius Journal of Analytical Chemistry. 367(1): 96–98.

Sonke, J.E. and V.J.M. Salters. 2007. Capillary electrophoresis-high resolution sector field inductively coupled plasma mass spectrometry. Journal of Chromatography A 1159(1-2): 63–74.

Stoichev, T., D. Amouroux, R.C.R. Martin-Doimeadios, M. Monperrus, O.F.X. Donard and D.L. Tsalev. 2006. Speciation analysis of mercury in aquatic environment. Applied Spectroscopy Reviews. 41(6): 591–619.

Sturgeon, R.E. and K.A. Francesconi. 2009. Enhancing reliability of elemental speciation results —quo vadis? Environmental Chemistry. 6(4): 294–297.

Sun, B., M. Macka and P.R. Haddad. 2004. Speciation of arsenic and selenium by capillary electrophoresis. Journal of Chromatography A 1039(1-2): 201–208.

Sussulini, A., J.S. Garcia, M.F. Mesko, D.P. Moraes, E.M.M. Flores, C.A. Perez and M.A.Z. Arruda. 2007. Evaluation of soybean seed protein extraction focusing on metalloprotein analysis. Microchimica Acta. 158(1-2): 173–180.

Sutton, K.L. and J.A. Caruso. 1999. Liquid chromatography–inductively coupled plasma mass spectrometry. Journal of Chromatography A 856(1-2): 243–258.

Sutton, K., R.M.C. Sutton and J.A. Caruso. 1997. Inductively coupled plasma mass spectrometric detection for chromatography and capillary electrophoresis. Journal of Chromatography A 789(1-2): 85–126.

Szpunar, J., S. McSheehy, K. Polec, V. Vacchina, S. Mounicou, I. Rodriguez and R. Lobinski. 2000. Gas and liquid chromatography with inductively coupled plasma mass spectrometry detection for environmental speciation analysis—advances and limitations. Spectrochimica Acta Part B-Atomic Spectroscopy. 55(7): 779–793.

Takeuchi, M., K. Mizuishi and T. Hobo. 2000. Determination of organotin compounds in environmental samples. Analytical Sciences. 16(4): 349–359.

Tangen, A., R. Trones, T. Greibrokk and W. Lund. 1997. Microconcentric nebulizer for the coupling of micro liquid chromatography and capillary zone electrophoresis with inductively coupled plasma mass spectrometry. Journal of Analytical Atomic Spectrometry. 12(6): 667–670.

Teske, J. and W. Engewald. 2002. Methods for, and applications of, large-volume injection in capillary gas chromatography. Trac-Trends in Analytical Chemistry. 21(9-10): 584–593.

Timerbaev, A.R. 2002. Recent advances and trends in capillary electrophoresis of inorganic ions. Electrophoresis. 23(22-23): 3884–3906.

Timerbaev, A.R. 2004. Capillary electrophoresis of inorganic ions: An update. Electrophoresis 25(23-24): 4008–4031.

Tomlinson, M.J., J. Wang and J.A. Caruso. 1994. Speciation of toxicologically important transition metals using ion chromatography with inductively coupled plasma mass spectrometric detection. Journal of Analytical Atomic Spectrometry. 9(9): 957–964.

Trojanowicz, M., E. Pobozy and G. Gubitz. 2003. Speciation of oxidation states of elements by capillary electrophoresis. Journal of Separation Science. 26(11): 983–995.

Tseng, C.M., A. De Diego, J.C. Wasserman, D. Amouroux and O.F.X. Donard. 1999. Potential interferences generated during mercury species determination using acid leaching, aqueous ethylation, cryogenic gas chromatography and atomic spectrometry detection techniques. Chemosphere. 39(7): 1119–1136.

Tsoi, Y.-K. and K. Sze-Yin Leung. 2010. Simultaneous determination of seven elemental species in estuarine waters by LC-ICP-DRC-MS. Journal of Analytical Atomic Spectrometry. 25(6): 880–885.

Vahcic, M., R. Milacic and J. Scancar. 2011. Development of analytical procedure for the determination of methyltin, butyltin, phenyltin and octyltin compounds in landfill leachates by gas chromatography-inductively coupled plasma mass spectrometry. Analytica Chimica Acta. 694(1-2): 21–30.

Van Holderbeke, M., Y. Zhao, F. Vanhaecke, L. Moens, R. Dams and P. Sandra. 1999. Speciation of six arsenic compounds using capillary electrophoresis-inductively coupled plasma mass spectrometry. Journal of Analytical Atomic Spectrometry. 14(2): 229–234.

Vela, N.P. and J.A. Caruso. 1993. Comparison of flame ionization and inductively-coupled plasma-mass spectrometry for the detection of organometallics separated by capillary supercritical-fluid chromatography. Journal of Chromatography. 641(2): 337–345.

Wang, J.L., S.W. Chen and C. Chew. 1999. Automated gas chromatography with cryogenic/ sorbent trap for the measurement of volatile organic compounds in the atmosphere. Journal of Chromatography A 863(2): 183–193.

Weber, J.H. 1997. Speciation of methylarsenic, methyl- and butyltin, and methylmercury compounds and their inorganic analogues by hydride derivatization. TrAC Trends in Analytical Chemistry. 16(2): 73–78.

Wolf, C., D. Schaumloffel, A.-N. Richarz, A. Prange and P. Bratter. 2003. CZE-ICP-MS separation of metallothioneins in human brain cytosols: comparability of electropherograms obtained from different sample matrices. Analyst. 128(6): 576–580.

Wuilloud, J.C.A., R.G. Wuilloud, A.P. Vonderheide and J.A. Caruso. 2004. Gas chromatography/ plasma spectrometry—an important analytical tool for elemental speciation studies. Spectrochimica Acta Part B-Atomic Spectroscopy. 59(6): 755–792.

Yang, H., W. Liu, B. Li, H. Zhang, X. Liu and D. Chen. 2007. Speciation Analysis for Iodine in Groundwater Using High Performance Liquid Chromatography-Inductively Coupled Plasma-Mass Spectrometry (HPLC-ICP-MS). Geostandards and Geoanalytical Research. 31(4): 345–351.

Ye, Y.S., J.C. Sang, H.B. Ma and G.H. Tao. 2010. Gas-phase chemiluminescence with ozone oxidation for the determination of total tin in environmental samples using flow injection hydride generation and cryotrapping. Analytica Chimica Acta. 677(2): 149–155.

Yuan, C.-G., B. He, E.-L. Gao, J.-X. Lü and G.-B. Jiang. 2007. Evaluation of extraction methods for arsenic speciation in polluted soil and rotten ore by HPLC-HG-AFS analysis. Microchimica Acta. 159(1-2): 175–182.

Zhao, Y.Q., J.P. Zheng, L. Fang, Q. Lin, Y.N. Wu, Z.M. Xue and F.F. Fu. 2012. Speciation analysis of mercury in natural water and fish samples by using capillary electrophoresis-inductively coupled plasma mass spectrometry. Talanta. 89: 280–285.

Zufiaurre, R., B. Pons and C. Nerin. 1997. Speciation of trimethyllead in rainwater by gas chromatography mass spectrometry after ethylation with sodium tetraethylborate— Comparison with other alkylation methods. Journal of Chromatography A 779(1-2): 299–306.

Zuliani, T., G. Lespes, R. Milačič, J. Ščančar and M. Potin-Gautier. 2008. Comprehensive study of the parameters influencing the detection of organotin compounds by a pulsed flame photometric detector in sewage sludge. Journal of Chromatography A 1188(2): 281–285.

Role and Importance of Hyphenated Techniques in Speciation Analysis

Rajmund Michalski,[a], Magdalena Jabłońska[b]* and *Sebastian Szopa[c]*

Speciation and Speciation Analysis

The word *speciation* is borrowed from biological terminology. It derives from the Latin term *species*, which means species or species evolution (Florence et al. 1980). Speciation defines the occurrence of an element in various, clearly defined chemical forms. It has a qualitative sense. The term "speciation analysis" first appeared in literature in 1993 and was described as "movement and transformation of element forms in the environment" (Forstner 1993). Even though it is comparatively expensive, speciation analysis has been gaining increasing importance. It is vital for both determining total content of elements and considering the forms in which those elements occur. It plays a significant role in processes such as

Institute of Environmental Engineering of Polish Academy of Sciences, 34 Skłodowska-Curie Street, 41-819 Zabrze, Poland.
[a]Email: michalski@ipis.zabrze.pl
[b]Email: mjablonska@ipis.zabrze.pl
[c]Email: sszopa@ipis.zabrze.pl
*Corresponding author

studying biochemical cycles of specific chemical compounds; determining toxicity and ecotoxicity of elements; food and pharmaceuticals quality control; technological process control; health risk evaluation; and clinical analysis (Kot and Namieśnik 2000).

The results of toxicological tests show that in many cases concentrations of different element forms, rather than the total element content, influence living organisms. For this reason, understanding the occurrence of various element forms is more crucial than knowing its total content. It must be emphasized that, for many elements, properties depend on the species or chemical form of the element present in the sample: toxicity or nutritional value; environmental mobility and persistence; bioavailability; volatility and chemical reactivity.

The quality of the analytical and toxicological input significantly determines the assessment of the elemental species influence in medicine, biology and food science. Proteomics and metallomics, which deal with the determination of trace metals in biomolecules, are throwing new light on the subject. These new scientific techniques will hopefully provide more profound understanding of the essential role that trace metals and metalloids play in life processes. They will also improve our comprehension of the physiological influence that metals and metalloids exert on different organisms. Undoubtedly, the most important practical application of elemental speciation is found in toxicology. In turn, better understanding of toxicology will lead to better and more specific regulatory regime governing hazardous substances.

Within speciation analysis, it is possible to differentiate between substances of anthropogenic origin and natural compounds that are formed as a result of biochemical transformations occurring in living organisms or environment. The former group is interesting for environmental analysts while the latter calls for the attention of biochemists and ecotoxicologists.

Literature definitions of speciation are equivocal. Various speciation analysis types, both physical and chemical, are enumerated. Physical speciation takes into account the existence of free analytes and analytes in bound forms. During surface water analysis, the fraction bound with suspension is separated from the solved fraction with a 0.45-μm membrane filter. Consequently, it is possible to demonstrate that approximately 80 percent of the determined analyte is adsorbed on the suspension. On the other hand, only 20 percent of the analyte is solved and passes into water—a natural solvent, according to which leaching of contaminations from the environmental samples is determined. The obtained filtrate and insoluble suspension constitute materials for speciation analysis (filtrate) and fractionation (suspension).

The notion of fractionation was separated from speciation analysis. It concerns solid matrices and suspensions. Substances such as dust, wastewater and municipal sludge, bottom sediments or compost and industrial samples can be fractionated thoroughly, first in the granulometric and then in the chemical way. Consequently, fractionation is understood as a classification process of the analyte or analyte group in a given sample in accordance with their physical (e.g., different molecule size or diversified solubility), or chemical properties (e.g., reactivity). It is claimed that the lower granulation is, the higher accumulation of organic and inorganic compounds is obtained. It results from a larger surface area of grains.

Fractionation distinguishes between diverse forms of metal occurrence. Sequential extraction procedures converse trace metals into chemical forms that can be released into the solution under different environmental conditions.

The most common practice is using Tessier's procedure of metal fractionation from bottom sediments, with its division into five fractions (Tessier et al. 1979). Fraction 1 is connected with the process of metal adsorption on the surface of solids and is the most easily available. Passing of a metal from its solid phase into water may occur when water ionic content is changed because of sorption-desorption balance shift. Fraction 2 involves carbonate-bound metals. They can be released because of pH level reduction. Fraction 3, i.e., metals bound with hydrated iron and manganese oxides, is sensitive to redox potential alterations. It is thermodynamically unstable under anoxic conditions. Fraction 4 comprises of metals bound with organic matter. They are temporarily unavailable. With time, they pass into pelagic zone or one of the other fractions as a consequence of anaerobic and aerobic digestion of the organic matter. Fraction 5, i.e., metals bound with the remaining fraction, mainly includes primary and secondary minerals containing metal atoms built into their crystal lattice. These metals are not practically available for living organisms under natural conditions and are perceived as permanently immobilized.

Chemical speciation encompasses four analysis types, i.e., screening, group, distribution and individual speciation (Caroli 1995). Screening speciation determines only one analyte, i.e., the most dangerous one in a sample. A good example is determination of tributyltin in seawater and tissues or methylmercury in tissues. Group speciation attempts to determine concentration levels of a given compound group or determined element at different oxidation states. It copes with problems such as simultaneous determination of Cr(III) and Cr(VI); Fe(II) and Fe(III); Mn(II), Mn(IV) and Mn(VII); BOD (Biochemical Oxygen Demand) and COD (Chemical Oxygen Demand); or organic and inorganic mercury. Distribution speciation examines physiological fluids, e.g., blood, serum, urine and saliva. Finally, individual speciation is the most difficult as it identifies and determines all individual chemicals containing a given element in their composition.

As the afore-mentioned definitions describe speciation in numerous ways, it became necessary to standardize this notion. The official documents of IUPAC (International Union of Pure and Applied Chemistry) describing speciation were evolving since 1992 (Muntau et al. 1992). The final recommendation was made in 2000 (IUPAC 2000). The definition states "speciation is the occurrence of an element in the forms of various chemical individuals that are defined by isotopic composition, electron structure, oxidation state and complex or molecule structure."

Analytical procedure is composed of sampling, sample fixation, transportation, storage and preparation for analysis. Analyte determination constitutes its final stage. Sample preparation is usually the most time-consuming process, and is also the most frequent source of errors. Even routine procedures used in speciation analysis (e.g., dilution, pH changes caused by sample fixation, temperature and pressure alternations) can cause irreversible changes in the primary analyte form.

Particular difficulties arise when sampling and analyzing take place under different conditions. Such a situation occurs when sampling takes place at lower strata in bodies of water. The pressure drop causes the emission of gaseous constituents. For CO_2, the sample pH rises, acid dissociation constants shift, complex stability increases and sparingly soluble sediments precipitate. Sample instability and its changeability are of vital importance, particularly when biological material is analyzed. Such samples are exposed to microbiological, enzymatic, photochemical and other processes that can still occur after sampling is completed. Moreover, their character is often unclear and unexpected. Apart from classical sample preparation methods (e.g., mineralization, digestion, extraction, purification or analyte enrichment), speciation analysis uses:

- barbotage (percolation of gas bubbles through a liquid);
- analyte derivatization techniques;
- recovery capillary cryocapture and analyte pre-enrichment techniques (Wasik et al. 1998);
- SPME (Solid-Phase MicroExtraction);
- HS-SPME (Headspace Solid-Phase MicroExtraction) (Mester et al. 2001);
- SBSE (Stir Bar Sorptive Extraction) (Vercauteren et al. 2001).

Lowering analyte detection limits to extremely low concentration levels is essential. As a result, some of the applied analytical methods do not meet the necessary requirements. For that reason, various separation techniques and detection methods are combined in the form of hyphenated techniques. An appropriate hyphenated technique should be selective towards analytes, sensitive within a wide range of concentrations and enable possibly the most unambiguous identification of determined substances.

Chromatographic methods are used in separation procedures (Ellis and Roberts 1997), while spectroscopic techniques are applied in detection processes. Nevertheless, the application of other methods is also possible (Viera et al. 2009). Development and application of hyphenated techniques combined with flow-injection and atomic spectrometry for environmental analysis were described by Wang et al. (Wang et al. 1995).

A suitable analytical technique for speciation analysis should address three issues (Szpunar and Łobiński 2003):

1. Selectivity of the separation technique must allow the target analyte species to enter the detector well separated from potential matrix interferents and from one other.
2. The detector sensitivity should allow determining particular speciation forms of analytes at adequately low concentration levels.
3. Species identification. When appropriate standards are not available, the use of a molecule-specific detection technique is mandatory.

These requirements can be met by means of hyphenated techniques. They are schematically presented in Fig. 4.1.

Chromatography as a separation method has been known since the beginning of the 20th century (Tswett 1906), however its rapid development occurred 50 yr later. Nowadays, chromatographic techniques belong to the most common instrumental methods in the analytical chemistry. They enable quick separation and determination of substances in samples with complex matrices.

Electrophoresis methods make up another group of popular separation techniques. The phenomenon of electrophoresis (the migration of charged molecules towards opposite charge electrode in the electric field) was discovered at the end of the 19th century. Arne Wilhelm Kaurin Tiselius was the first to use electrophoresis as a separation method in 1937. He was awarded the Nobel Prize in 1948 for his research concerning separation of proteins (Jorgenson and Lukacs 1981).

Combining electrophoresis and electro-osmotic flows facilitates separating a wide range of positive and negative ions as well as neutral compounds during one analysis. The matrix complexity of the analyzed sample may require two or more separation mechanisms connected in series.

At present, the term CE (Capillary Electrophoresis) is used to describe electromigration techniques such as CZE (Capillary Zone Electrophoresis), CIEF (Capillary Isoelectric Focusing), CGE (Capillary Gel Electrophoresis), and CITP (Capillary Isotachophoresis). Sometimes, MEKC (Micellar Electrokinetic Capillary Chromatography) is also mentioned.

MS (Mass Spectrometry), ICP-MS (Inductively Coupled Plasma—Mass Spectrometry), AFS (Atomic Fluorescence Spectrometry), ICP-OES

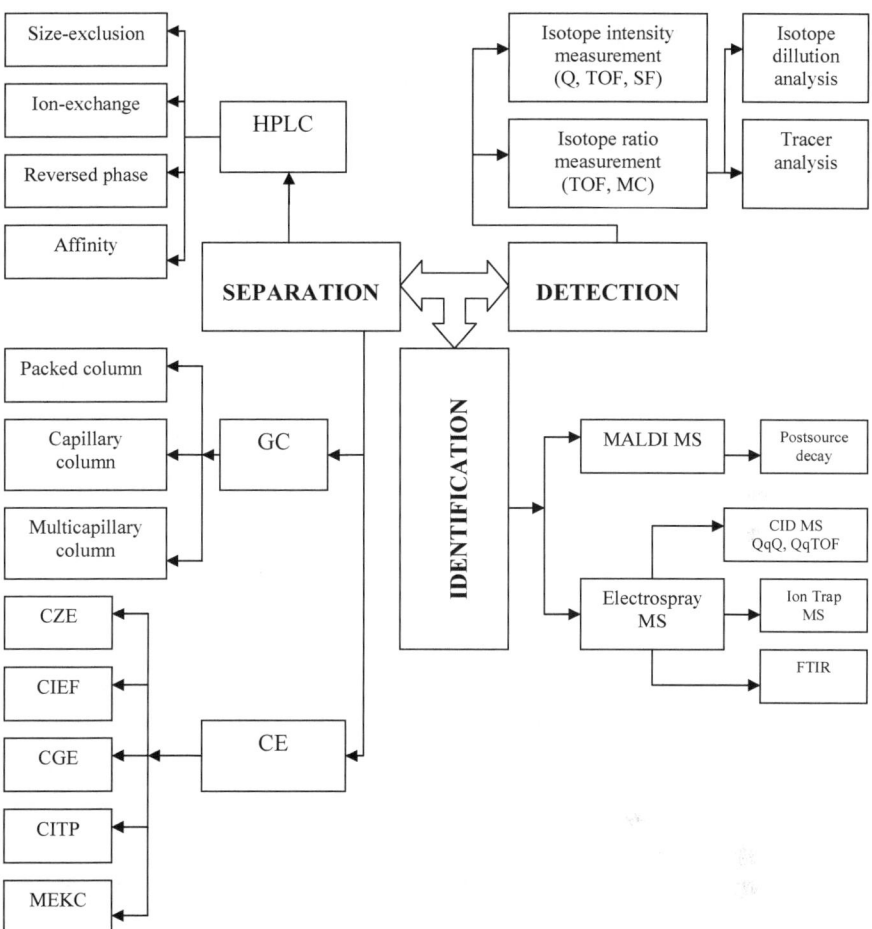

Figure 4.1 Examples of hyphenated techniques applied in species analysis.

(Inductively Coupled Plasma—Optical Emission Spectrometry), AAS (Atomic Absorption Spectrometry), AES (Atomic Emission Spectrometry), UV (Ultraviolet Spectrometry), FTIR (Fourier Transform Infrared Radiation), NMR (Nuclear Magnetic Resonance) are the most frequently used detection methods in speciation analysis.

Hyphenated techniques were first introduced by means of coupling GC (Gas Chromatography) with different detectors. The following systems were created: GC-AAS (Gas Chromatography—Atomic Absorption Spectrometry), GC-OES (Gas Chromatography—Optical Emission Spectrometry), GC-MS (Gas Chromatography—Mass Spectrometry) or GC-ICP-TOF-MS (Gas Chromatography–Inductively Coupled Plasma—Time of Flight Mass Spectrometry—Mass Spectrometry). Due to technological reasons, couplings using liquid chromatographic methods for separation of the analytes, such as HPLC-ICP-MS (High Performance Liquid Chromatography—Inductively Coupled Plasma—Mass Spectrometry), appeared later (Shalliker 2011).

The minimum requirements for all hyphenated methods coupled with ICP-MS detection include:

- The connecting of the interface (transfer line) must enable quantitative transfer of the analytes between the separation system and ICP-MS plasma.
- Separated components of the sample must not undergo degradation when they are transferred into the plasma.
- The separation system should be synchronized with the ICP-MS detector to allow simultaneous separation and detection of the analytes.
- The ICP-MS detector must have a sufficient dynamic range of measurement.

Other methods encompass the technique of hydride generation for selected volatile elements; and couplings of different high-resolution separation methods, including multidimensional techniques with sensitive detectors. Multidimensional and multimode chromatography involves two different mechanisms (e.g., ion exchange coupled with ion exclusion) and two or more columns connected in series. Bruno (Bruno 2011) reviewed hyphenated chromatographic instrumentation in species analysis.

High separation performance characterizes GC. Therefore, GC is the main speciation method used for gas analytes. It also allows reach very low detection limits. However, many element forms do not occur in the gas form and it is impossible to transform them into it with derivatization reactions. In fact, almost all coordination complexes of trace metals belong to this category. It also includes numerous organometallic compounds (containing covalent-bound metal or metalloid). The most common choice is the application of column separation methods in the liquid phase, like variants of HPLC (High-Performance Liquid Chromatography) or CE. They are popular because they are easily coupled in an on-line system. What is more, they provide a number of different separation mechanisms and mobile phase availability and thus help to preserve the determined substance in an unchanged form.

MS (Mass Spectrometry) is the most prevalent detection method used in speciation analysis. It provides information on the qualitative and quantitative composition of the sample. It also helps to define structure and molar masses of the examined analytes. The access to structural data is essential for identification of the already known and newly discovered compounds. Nevertheless, it presents a challenge for speciation analysis, particularly when a higher sensitivity of the applied methods contributes to the increase in the number of detected metal and metalloid forms.

The necessity to maintain a very low pressure in the spectrometer is one of the main difficulties related to using mass spectrometry detectors with chromatographic methods. It arises from the fact that separated analyte ions leave the chromatographic column under relatively high pressure.

Whereas coupling a gas chromatograph with mass spectrometry detector was relatively easy, larger volumes of the eluate proved to be a major obstacle in the introduction of HPLC-MS (High-Performance Liquid Chromatography-Mass Spectrometry) in the laboratory practice. HPLC-MS apparatus uses different sources of ionization, such as ESI (Electrospray Ionization), APCI (Atmospheric Pressure Chemical Ionization) or APPI (Atmospheric Pressure Photochemical Ionization).

The selection of the ionization method depends on the analyte characteristics and the required type of analytical information. "Hard ionization" methods, such as electron ionization or chemical ionization, make use of their fragmentation capabilities to gain structural information, typically concerning small molecules. On the other hand, "soft ionization" techniques, such as ESI or LD (Laser Desorption), are used to obtain mass spectra of molecules with little or no fragmentation. In laser desorption methods, a pulsed laser is used to desorb species from a target surface. Therefore, a mass analyzer compatible with pulsed ionization methods has to be used. Typically, TOF (Time of Flight) analyzers are employed. It is also possible to use several hybrid systems (Q-TOF). Moreover, high-resolution FT-ICR (Fourier Transform-Ion Cyclotron Resonance) analyzers have been successfully adapted recently.

The range of applications depends on the analyte polarity and mass as well as the eluent flow rate. MS detection can be performed in the SIM (Selected Ion Monitoring) or SM (Scan Mode) modes. The former provides information about the analyte mass while the latter focuses on mass spectra and mass distribution. In the case of large molecules, identification difficulties arise from the fact that the obtained spectra frequently originate from signals with the same mass-to-charge ratios.

Chromatographic or electrophoresis methods are often combined with an ICP-MS detector. The coupling is performed directly with a nebulizer for column techniques of liquid separation or with LA (Laser Ablation) for planar techniques. Gas chromatography with packed columns used at first in GC-ICP-MS (Gas Chromatography Coupled with Inductively Coupled Plasma Mass Spectrometry) system (Van Loon et al. 1986) was replaced with capillary gas chromatography. The coupling of capillary gas chromatography with ICP-MS techniques were first described in publications (Kim et al. 1992a, b).

Packed columns are adapted to work under high flow rate of the carrier gas stream and high sample volumes. On the other hand, their performance and separation are low due to large dispersion of the analyte after passing through the column. Moreover, large column volume negatively affects sensitivity and limits of detection, whereas the content itself can chemically react with many organometallic forms. Such a situation forces silanization and undermines the credibility of the obtained results (Amouroux et al. 1998).

GC-ICP-MS with capillary columns offers high resolution, which is important when separating mixtures of organometallic compounds present in complex matrix samples. Capillary gas chromatography counteracts the co-elution of solvents and volatile compounds (e.g., Me_4Sn or Me_2Hg). Therefore, it helps either to avoid or to minimalize unfavorable plasma cooling process.

Many studies concerning Flash GC were published in the late 1990s. This technique uses columns composed of a few thousand capillaries with a very short internal diameter (20–40 μm). They are called multicapillary columns (Lobinski et al. 1999). A large number of capillaries allow eliminating drawbacks related to the application of singular capillary or packed columns. It also retains advantages of both column types. MC-GC (Multicapillary Gas Chromatography) is characterized by a high flow rate of the carrier gas, which decreases the dilution factor and facilitates transportation of analytes into the plasma. MC-GC with ICP-MS used with an unheated switch enables reaching limits of detection (for Hg speciation examinations) at the level of 0.08 pg (Slaets et al. 1999).

Application of hyphenated techniques in speciation analysis of organic and inorganic forms of metals and metalloids

The range of hyphenated techniques applications is large and still expanding. The most important ones involve speciation of metals and metalloids bound with organic and inorganic compounds. Water is probably the most studied environmental sample and the majority of speciation studies concern various types of water (Das et al. 2001). Popp et al. (Popp et al. 2010) presented an up-to-date literature review regarding applications of hyphenated methods in speciation analysis of environmental samples. It concerns techniques based on gas or liquid chromatography coupled with ICP-MS detection. Sarzanini (Sarzanini 1999) described applications of liquid chromatography in speciation analysis of metals.

Cases in which a given element containing a covalent bond between carbon and metal (metalloid) falls into the scope of the environmental speciation analysis are given below:

1. Products of the environmental methylation of mercury, selenium, arsenic, tin, bismuth, molybdenum and tungsten;
2. Anthropogenic organometallic pollutants and products of their degradation or transformation in the environment, e.g., tetraethyl lead (antiknock gasoline additive), butyltin, octyltin or phenyltin (components of antifouling paints that penetrate water environment as well as products of their degradation or biomethylation);
3. Arsenic metabolism products enabling the formation of carbon-arsenic bond (e.g., arsenobetaine, arsenosugars);
4. Selenoamino acids (peptides and proteins biosynthesized by bacteria, fungi and plants);
5. Metal-binding peptides synthesized enzymatically in living organisms exposed to heavy metals.

Standard recovery methods, internal standard addition, inter-laboratory comparisons and comparisons of results obtained with other methods are used for quality control during quantitative determination of various element species. Although more and more certified reference materials containing defined analyte speciation forms (e.g., tribulytin or methylmercury standards) are available in the market, new materials are still in great demand.

A covalent bond between carbon and metal (metalloid) assures sufficient analyte stability during sample separation. Gas chromatography, which offers high separation performance and lack of condensed mobile phase, is used for the analyses of volatile forms. It ensures sensitive and specific detection of the element by means of atomic spectrometry (Bouyssiere et al. 2002).

The way in which metalloids like arsenic and selenium are metabolized by living organisms leads to the formation of a covalent bond between a heteroatom and carbon bound to a large molecule (e.g., arsenosugars, selenoproteins). Microorganisms and plants create many internal mechanisms allowing them to control concentrations of elements essential for them and to cope with stress caused by toxic elements (Prasad and Hagemeyer 1999).

Only hyphenated techniques enable understanding mechanisms that regulate detoxification. Some plants develop particularly effective mechanisms of metal (metalloid) homeostasis sustenance that allow them to function and propagate in the environment highly polluted with those metals. Such defensive tools include induction and activation of antioxidant enzymes and using organic acids (e.g., phytic, malonic, citric and oxalic acids). Plants also produce metal-binding peptides and proteins in order to improve the accumulation and tolerance of heavy metals. These reactions result in the formation of complexes whose characteristics have not been understood so far.

The obtained compounds are difficult to transform into gas forms and hence gas chromatography cannot be used to separate them. Liquid phase separation techniques such as HPLC and CZE with detection based on ICP-MS as well ESI-MS-MS (Electrospray Ionization MS MS) are applied for this type of determination.

ESI-MS or MALDI-MS (Matrix Assisted Laser Desorption Ionization MS) can be used for either column or planar separation techniques, respectively. MALDI differs from ESI because it works in a sequential way. Consequently, it is seldom used as a detector for high-performance separation methods. The ESI-MS interface is an interesting issue. The ESI technique is based on the introduction of a liquid stream, whose flow rate is low, into a strong electric field under atmospheric pressure. ESI chamber can be coupled with any available mass analyzer, i.e., Q (Quadrupole), IT (Ion Trap), TOF (Time of Flight), SF (Sector Field) and FT-ICR (Fourier Transform Ion Cyclotron Resonance). Double Q-Q or hybrid (e.g., Q-TOF) couplings are also in use.

ESI-MS is also applied to determine compounds in biological materials, such as nucleic acids, amino acids, peptides, proteins and their complexes with metals and metalloids.

It is recommended to use TOF-MS (Time of Flight-Mass Spectrometry), as its precision is high. This mass spectrometer measures the mass-dependent time it takes ions in different masses to move from the ion source to the detector. Structural data can be obtained with CID (Collision Induced Dissociation) of the ion selected with a quadrupole mass filter. The subsequent ionic scanning of products is performed with quadrupole mass or TOF analyzers. TOF-MS enables obtaining complete mass spectra at high frequency. Therefore, it constitutes a valuable tool for detection of transition signals produced as a result of using fast chromatographic techniques (Baena et al. 2001).

In practice, speciation analysis of volatile organometallic forms is dominated by three techniques, i.e., GC-MIP-AED (Gas Chromatography—Microwave Induced Plasma—Atomic Emission Detector), GC-ICP-MS and GC-EI-MS (Gas Chromatography with Electron Impact Mass Spectrometry). Methylmercury determination in environmental samples makes an exception in which GC-AAS and GC-AFS systems are still widely used.

As it is universal and provides detection limits below pg level, GC-MIP-AED is prevalent in speciation analyses of anthropogenic environmental pollutants and their degradation products. It is also popular because the appropriate apparatus, which enables such analyses, is available in the market. GC-MIP-AED system is used in speciation of organic compounds of lead and tin in environmental samples and methylmercury in biological samples.

AFS (Atomic Fluorescence Spectrometry) coupled with a gas chromatograph is another system available in the market that supports mercury speciation analyses (Armstrong et al. 1999).

A similar situation occurs in organotin determination. It can be performed with GC-FPD (Gas Chromatography-Flame Photometric Detector). Despite problems related to the complete elimination of background and hydrocarbon interferences, FDP (Flame Photometric Detector) is common as a selective detector for tin. PFPD (Pulse FPD) is advancement in FPD use, which guarantees sensitivity in organotin detection that is 10 times better (Tzanani and Amirav 1995).

Hyphenated techniques using ICP-MS are among the most rapidly developing methods and applications gaining interest in atomic spectrometry. Although an ICP-MS spectrometer does not provide information on the chemical or structural analyte forms (all analyte forms are transformed into positively charged ions in plasma), it is an excellent elemental analyzer. ICP-MS offers low detection limits for many analytes; high sensitivity; a wide range of linearity; good precision and accuracy.

The progress and state-of-the-art knowledge of this technique were summarized in publications (Becker 2003, 2005, Alvarez-Llamas et al. 2005, Nageswara and Kumar Talluri 2007).

Plasma is a highly ionized gas in which the number of free electrons amounts to the number of positive ions. ICP (Inductively Coupled Plasma) is a distinct kind of plasma. It is generated under atmospheric pressure in a gas (typically argon) that flows through induction coils that produce a high-energy magnetic field of high frequency. The coils usually work at approximately 30-MHz frequency and energy levels between 1,000 and 2,000 W. The temperature obtained in such plasma ranges between 6,000 and 10,000 K. It suffices to excite or ionize the majority of examined analytes.

The quartz torch in which the plasma is formed consists of three concentric glass tubes. Argon flows in each tube with different intensity. Three argon streams form three main plasma constituents. Cooling and auxiliary gases flow through the outer channel; a gas stabilizing plasma flame passes through the middle channel; and the central channel serves to introduce the sample in the gas or aerosol form.

A suitable switch is the key required for the appropriate coupling of a liquid chromatograph with ICP-MS. The easiest way is to link a chromatographic column outlet to a conventional nebulizer. In such a case, the main requirement is its compatibility with the applied flow rates and eluent composition. Water-based eluents can be harmful to nebulizer parts as they contain large amounts of salt. Organic liquid eluents or those containing organic constituents can cause plasma instability because their vapors have large thermal expansion. The pneumatic system is the most popular solution used for introducing samples by means of nebulization. Its application results in the fact that ICP becomes a detector largely dependent on flow rates. Pneumatic nebulizers are characterized by both constant nebulization efficiency and significantly reduced efficiency at high flow rates. Consequently, they generate linear dependence between mass analyte flow in the nebulizer and the observed number of counts for a given analyte in the mass spectrometer for low flow rates and a constant value for high flow rates.

A practical consequence of this phenomenon is a strong dependence between repeatability and stability of the flow rate and calibration precision. An additional problem is created by the existence of analytes whose determination with conventional nebulizers is difficult, particularly at low concentration levels. These analytes include polarized compounds of mercury, iodine, thallium and silver. Their mutual reactions with polymer nebulizer parts are unpredictable. Another good example is boron, which reacts intensely with the glass surface of the nebulizer chamber.

Ions originating from ICP are subsequently sucked inside the mass spectrometer passing through two cones (usually nickel ones). Later, they are focused with ion lenses and flow into the mass detector. The analyzer is made of four parallel rods placed symmetrically. Their profile is typically hyperbolic. The opposite rods are electrically connected and the voltage (the sum of AC and DC voltage) is applied between two pairs of rods.

Only ions with an adequate mass-to-charge ratio can move in the central part of such an electromagnetic field. Others become scattered and do not pass through the analyzer. It is possible to set the analyzer to let in ions with an established mass-to-charge ratio and defined accuracy. This is done through changing AC or DC voltage and altering the frequency of the voltage applied to the rods. Afterwards, ions separated in this way are counted in the electron multiplier.

Quadrupole analyzers are the most common analyzers in ICP-MS due to their comparatively low cost of manufacture and simplicity. They also enable quick separation of ions based on the mass-to-signal ratio. The most widespread is the application of a quadrupole or octapole mass analyzers in the ICP-MS spectrometer. The latest generation of those devices offers low detection levels (ppb or ppt levels) for many elements. It is possible to use a mass analyzer with double focusing or a DRC (Dynamic Reaction Cell) to obtain better resolution and thus reduce isobaric interferences.

A DRC is an additional quadrupole placed in the space between ion lenses system and the analyzing quadrupole (D'Ilio et al. 2011, Pick et al. 2010). The space is filled with a reactive gas, such as NH_3. The gas reacts with the ion beam and changes interferents into forms that do not interfere in the determination of specific analytes. The analytes are stable under chamber conditions and pass into the mass analyzer in unchanged forms.

The overlapping of isobaric ions does not actually pose a problem due to the separation of interfering ions, such as $^{40}Ar^{35}Cl$, $^{40}Ar^{12}C$, $^{40}Ar^{13}C$ and others, in the on-line system. It is advisable to use an ICP-MS spectrometer with a reaction or collision cell to eliminate isobaric interferences (Kuo et al. 2007, Salazar et al. 2011, Tanner et al. 2002).

Some spectrometers use the mechanism of polyatomic molecules collision. Others employ reactions in the gaseous phase in order to cause dissociation or creation of secondary molecules that may be rejected by the mass analyzer. Some of the available ICP-MS spectrometers are equipped with RC-ICP-MS (Reaction Cell Inductively Coupled Plasma Mass Spectrometry) reaction cells. The elimination of potential isobaric interferences, such as ArC+, is crucial, especially in the determination of chromium and its speciation forms. During the determination of chromium speciation forms with ICP-MS, a 2 percent addition of methanol can be used. Consequently, isobaric interferences increase and the chromium detection level rises. The discussed interferences can be excluded with the application of a reaction gas (such as ammonia or methane) in the reaction chamber.

Isobaric interferences do not pose a serious problem when a double focusing sector field instrument is used. It offers high resolution required to eliminate interferences during arsenic or chromium determination. On the other hand, increasing resolution inevitably leads to the substantial drop in the method sensitivity. Interestingly, the latest generation of quadrupole spectrometers reaches resolution level that is only two-three times lower than the one provided by high-resolution spectrometers (Szpunar and Łobiński 2003).

An additional problem is the loss of a particular analyte signal due to the effects of ion suppression and transport. Internal standardization is a strategy employed to solve this problem by using elements such as In, Rh, Hf, Ge.

Speciation analysis significantly benefited from using enriched isotopes in ICP-MS detectors. Isotopic specificity of the ICP-MS technique paves the way for the application of stable isotopes or forms enriched with stable isotopes for the research on transformations and artifact formation in extraction and derivatization processes. It also helps to introduce quantitative determination with ID-MS (Isotope Dilution-Mass Spectrometry) on a larger scale. ID-MS is a technique of high accuracy. Sources of systematic errors are well known and can be eliminated. Thus, ID-MS is considered to be a definitive analysis method. The principles of GC–ICP-ID-MS techniques in the specific analysis of a given chemical compound form were described thoroughly in the study (Gallus and Heumann 1996).

Another function that hyphenated techniques serve in speciation analysis is that of determining inorganic forms of metals and metalloids. Elements occurring in ionic forms are believed to exhibit biological activity and toxicity towards living organisms. Group speciation concerns elements occurring at different oxidation states. It consists in defining the form and concentration of each element. It is crucial for elements that demonstrate diversified toxic properties depending on their oxidation states. These include $Cr(III)/Cr(VI)$; $As(III)/As(V)$; or $Sb(III)/Sb(V)$.

The main HPLC varieties used in metal determinations are:

- ion chromatography (ion-exchange, ion-exclusion);
- reversed-phase ion interaction chromatography ;
- chelation chromatography;
- multi dimensional and multimode chromatography.

IC (Ion Chromatography) is the most popular method used in separation and determination of organic and inorganic ionic substances (Weiss 2004). It is used in hyphenated techniques and speciation analysis to determine selected water disinfection by-products (Michalski and Lyko 2010) and metal and metalloid ions (Sarzanini and Mentasti 1997).

Reversed-phase chromatography has been widely used for separating neutral or weakly charged metal complexes, but the more extensive applications are based on the ion-pairing mechanism (Gennaro and Angelino 1997). Chelation ion chromatography involves both an ion-exchange process and the formation of coordinate bonds (Jones and Nesterenko 1997). Multi dimensional methods (also known as "heart cut column switching") encompass all techniques in which the eluate flows from the primary column and into the secondary one within a precisely defined time.

Couplings of various liquid chromatography methods, such as HPLC, IC, I-EC (Ion-Exchange Chromatography) or GPC (Gel Permeation Chromatography), with ICP-MS or ESI-MS detectors are the most popular techniques used to determine different metal and metalloid ion forms. The most widespread hyphenated techniques employing IC are IC-ICP-MS (Ion Chromatography—Inductively Coupled Plasma—Mass Spectrometry); IC-ICP-OES (Ion Chromatography—Inductively Coupled Plasma—Optical Emission Spectrometry); and IC-MS (Ion Chromatography—Mass Spectrometry) (Wille et al. 2007).

Even though CE is less popular for such determinations, it is excellent for speciation analysis due to its high separation capacity; small amount of the sample required for determination; and lack of stationary phase subject to interactions with metals and able to influence complexation constant reaction (Kajiwara 1991). Coupling CE with ICP-MS entails four principles:

1. Providing electrical connection for the apparatus and constant current intensity (to ensure reproducible separation conditions in CE);
2. Adjusting liquid flow rate in the capillary (0.1 ÷ 0.9 µL per min, depending on the capillary internal diameter) to working conditions of the commercially available nebulizers (100 ÷ 1000 µL per min) as such a large difference necessitates introducing supplementary liquid stream;
3. Providing electro-osmotic flow in the capillary that eliminates the influence of the under-pressure formed by the nebulizing gas in the nebulizer;
4. Providing the minimal dead volume of the system (spray chamber and the part connecting the capillary with the nebulizer).

HPLC-CE-ICP-MS (High-Performance Liquid Chromatography-Capillary Electrophoresis-Inductively Coupled Plasma-Mass Spectrometry) is also possible to use but requires a suitable switch. The simplest way is to connect an HPLC column outlet (4.6–10 mm) to a conventional pneumatic or cross-flow nebulizer. Employing HPLC systems with traditional capillary columns or columns whose internal diameter varies between 0.32 and 1.0 mm is becoming more and more popular, particularly due to the RPC (Reversed Phase Chromatography) application. It requires using micro-nebulizers with DIN (Direct Injection Nebulizer), DIHEN (Direct Injection High Efficiency Micro-Nebulizer), or micro-nebulizers (e.g., Micromist) with a low-volume spray chamber.

Initially, IC applications in speciation analysis concerned determinations of $NO_2^-/NO_3^-/NH_4^+$ and $S^{2-}/SO_3^{2-}/SO_4^{2-}$ ions. They were most often determined with a conductivity detector. At present, IC is coupled with other detectors to determine inorganic drinking water disinfection by-products ($ClO_2^-/ClO_3^-/BrO_3^-$) and metal and metalloid ions at different oxidation states.

Protecting people against health-hazard microorganisms present in drinking water requires various methods of disinfection. Water chlorination, used for years, is a well-known and effective technology. On the other hand, health-hazard by-products such as trihalomethanes are formed during the chlorination process. As a result, other water disinfection methods have been looked for. Among them, ozonation has become the most popular one. Nevertheless, new water disinfection methods also have disadvantages and limitations. They mainly include the formation of inorganic halide derivatives such as bromates, chlorites and chlorates.

Bromates can form in raw water that contains bromides and are subjected to ozonation. IARC (International Agency for Research on Cancer) qualified it as potentially carcinogenic (B2 group). Initially, WHO (World Health Organization) and U.S. EPA (United States Environmental

Protection Agency) determined the safe bromate dose at the level of 0.8 µg per dm³. Nevertheless, due to the lack of a simple analytical method allowing determinations of such low concentrations, the provisional admissible content of bromates in drinking water was established at the level of 25 µg per dm³ (WHO 1996).

In highly industrialized countries, the present permissible bromate concentration in drinking water amounts to merely 10 µg per dm³. Keeping this in mind, IC becomes a useful tool. The methods of bromate determination with IC are grouped into (Michalski 2009b):

1. Direct methods (conductivity detection).
2. Indirect methods (UV/Vis detection).
3. Hyphenated techniques (mass spectrometry).

Direct methods are based on the selective separation of BrO_3^- ions in the presence of other anions in the sample and their detection in the conductivity detector after suppression. These methods are relatively simple and inexpensive. The main difficulty is the separation of BrO_3^- and Cl^- ions as their concentrations in real samples differ significantly. Indirect methods include derivatization methods that transform the determined substance into its derivatives. These can be detected with a UV/Vis detector. Finally, the third group (i.e., hyphenated techniques) encompasses IC-ICP-MS and IC-MS (Creed et al. 1996, Charles et al. 1996). These systems offer very high detection and precision levels for determinations. Nevertheless, they are not used routinely in laboratories due to their high cost. IC is also applied in speciation analysis for simultaneous separations and determinations of metal and metalloid ions at different oxidation states.

The emergence of new high-selective stationary phases for analytical columns and elaboration of new sample preparation methods constituted a breakthrough in the application of IC for speciation analysis of metal and metalloid ions (Michalski et al. 2011). It enables both simultaneous separation of different ions of the same element and their sensitive detection with suitable methods.

Summary and Conclusions

Speciation has evolved into an important sub-discipline of the analytical chemistry and it has a significant impact on the environmental monitoring and life sciences.

The progress in many sciences depends both on reliable analyses performed in laboratories equipped with the state-of-the-art analytical apparatus and on using sophisticated analytical systems, such as hyphenated techniques. They create new possibilities. Their advantages include extremely low limits of detection and quantification, high accuracy and repeatability of determinations. As other methods, hyphenated techniques have certain drawbacks, e.g., high price and complexity of the apparatus. As a result, they are not used frequently. Moreover, their application entails perfect understanding of analytical methodologies and apparatus. Thus, presently they are used in scientific research rather than in routine analyses.

The current situation in speciation analysis is complex. On the one hand, there are elemental species that have been thoroughly investigated and described. On the other hand, there are compounds whose scientific understanding is only preliminary. Nonetheless, elemental speciation has established itself as a well-respected link between organic and inorganic analytical chemistry. It uses the best aspects of both fields for its development, i.e., specific methodology and fundamental paradigms. The development of analytical methods in biochemical and inorganic speciation analysis is based on the combination of many scientific disciplines. The increasing number of applications and works describing hyphenated techniques proves that they are gaining more and more importance.

References

Alvarez-Llamas, G.M., F. del Rosario, A. de la Campa and A. Sanz-Medel. 2005. ICP-MS for specific detection in capillary electrophoresis. Trends Anal. Chem. 24: 28–36.

Amouroux, D., E. Tessier, C. Pecheyran and O.F.X. Donard. 1998. Sampling and probing volatile metal(loid) species in natural waters by *in situ* purge and cryogenic trapping followed by gas chromatography and inductively coupled plasma mass spectrometry (P-CT-GC-ICP/MS). Anal. Chim. Acta. 377: 241–254.

Armstrong, H.L., W.T. Corns, P.B. Stockwell, G. O'Connor, L. Ebdon and E.H. Evans. 1999. Comparision of AFS and ICP-MS detection coupled white gas chromatography for the determination of methylmercury in marine samples. Anal. Chim. Acta. 390: 245–253.

Baena, J.R., M. Gallego, M. Valcarcel, J. Leenaers and F.C. Adams. 2001. Comparision of three coupled gas chromatographic detectors (MS, MIP-AES, ICP-TOFMS) for organolead speciation analysis. Anal. Chem. 73: 3927–3934.

Becker, J.S. 2003. Mass spectrometry of long-lived radionuclides. Spectrochim. Acta B 58: 1757–1784.

Becker, J.S. 2005. Trace and ultratrace analysis in liquids by atomic spectrometry. Trends Anal. Chem. 24: 3, 243–254.

Bruno, T.J. 2000. A review of hyphenated chromatographic instrumentation. Separation & Purification Reviews. 29: 63–89.

Caroli, S. 1995. Element Speciation: Challenges and Prospects. Microchem J. 51: 64–70.

Charles, L., D. Pepin and B. Casetta. 1996. Electrospray Ion Chromatography—Tandem mass Spectrometry of Bromate at sub-ppb levels in water. Anal. Chem. 68: 2554–2558.

Creed, J.T., M.L. Magnuon, J.D. Pfaff and C. Brockhoff. 1996. Determination of bromate in drinking waters by ion chromatography with inductively coupled plasma mass spectrometric detection. J. Chromatogr. A 753: 261–267.

Das, A.K., M. Guardia and M.L.Cervera. 2001. Literature survey of on-line elemental speciation in aqueous solutions (Review). Talanta. 55: 1–28.

D'Ilio, S., N. Violante, C. Majorani and F. Petrucci. 2011. Dynamic reaction cell ICP-MS for determination of total As, Cr, Se and V in complex matrices: Still a challenge? A review. Anal. Chim. Acta. 698: 6– 13.

Ellis, L.A. and D.J. Roberts. 1997. Chromatographic and Hyphenated Methods for Elemental Speciation Analysis in Environmental Media. J. Chromatogr. A 774: 3–19.

Florence, T.M., G.E. Batley and P. Benes. 1980. Chemical Speciation in Natural Waters, Crit. Rev. Anal. Chem. 9: 219–296.

Forstner, U. 1993. Metal speciation—general concepts and applications, Intern. J. Environ. Anal Chem. 51: 5–23.

Gallus, S.M. and K.G. Heumann. 1996. Development of a gas chromatography inductively coupled plasma isotope dilution mass spectrometry system for accurate determination of volatile element species, part 1. Selenium speciation J. Anal. At. Spectrom. 11: 887–892.

Gennaro, M.C. and S. Angelino. 1997. Separation and determination of inorganic anions by reversed-phase high-performance liquid chromatography (Review). J. Chromatogr. A 789: 181–194.

Jones, P. and P.N. Nesterenko. 1997. High-performance chelation ion chromatography. A new dimension in the separation and determination of trace metals. J. Chromatogr. A 789: 413–435.

Jorgenson, J.W. and K.D. Lukacs. 1981. Zone Electrophoresis in Open-Tubular Glass Capillaries. Anal. Chem. 53: 1298–1301.

Kajiwara, H. 1991. Application of high-performance capillary electrophoresis to the analysis of conformation and intertion of metal-binding proteins. J. Chromatogr. A 559: 345–356.

Kim, A., S. Hill, L. Ebdon and S. Rowland. 1992a. Determination of organometallic compounds by capillary gas chromatography-inductively coupled plasma mass spectrometry, J. High Res. Chromatogr. 15: 665–668.

Kim, A.M.E. Foulkes, L. Ebdon, S. Hill, R. Patience, A.G. Barwise and S.J. Rowland. 1992b. Construction of a capillary gas chromatography inductively coupled plasma mass spectrometry transfer line and application of the technique to the analysis of alkyllead species in fuel. J. Anal. At. Spectrom. 7: 1147–1149.

Kot, A. and J. Namiesnik. 2000. The role of speciation in analytical chemistry. Trends Anal. Chem. 19: 69–79.

Kuo, C.Y. and S.J. Jiang. 2008. Determination of selenium and tellurium compounds in biological samples by ion chromatography dynamic reaction cell inductively coupled plasma mass spectrometry. J. Chromatogr. A 1181: 60–66.

Lobinski, R., V. Sidelnikov, Y. Patrushev, I. Rodriguez and A. Wasik. 1999. Multicapillary column gas chromatography with element-selective detection. TrAC, Trends Anal. Chem. 18: 449–460.

Mester, Z., R. Sturgeon and J. Pawliszyn. 2001. Solid Phase Microextraction as a Tool for Trace Element Speciation. Spectrochim. Acta, Part B. 56: 233–260.

Michalski, R. 2009a. Application of ion chromatography for the determination of inorganic cations, Crit. Rev. Anal. Chem. 39: 230–250.

Michalski, R. 2009b. Ion Chromatography Determination of Bromate—State of the Art. Trends in Chromatography. 5: 27–46.

Michalski, R. 2010. Inorganic Oxyhalide By-Products in Drinking Water: Ion Chromatographic Methods. *In:* J. Cazes [ed.]. Encyclopedia of Chromatography.3rd edn. Taylor & Francis, New York, USA. pp. 1212–1217.

Michalski, R. and A. Łyko. 2010. Determination of Bromate in Water Samples Using Post Column Derivatization Method with Triiodide. J. Environ. Sci. Health. Part A. 45: 1275–1280.

Michalski, R., M. Jablonska, S. Szopa and A. Łyko. 2011. Application of Ion Chromatography with ICP-MS or MS Detection to the Determination of Selected Halides and Metal/Metalloids Species. Crit. Rev. Anal. Chem. 41: 133–150.

Muntau, H., M. Bianchi, R. Cenci, K. Fytianos, R. Baudo and A. Lattanzio. 1992. Sequential metal extraction from sediments and soils: Applications and justfication for the use of a common scheme. Commission of the European Communities Joint Research Centre, Environment Institute, Ispra Italy.

Nageswara, R.R. and M.V.N Kumar Talluri. 2007. An overview of recent applications of inductively coupled plasma-mass spectrometry (ICP-MS) in determination of inorganic impurities in drugs and pharmaceuticals. J. Pharm. Biomed. Anal. 43: 1–13.

Popp, M., S. Hann and G. Koellensperger. 2010. Environmental application of elemental speciation analysis based on liquid or gas chromatography hyphenated to inductively coupled plasma mass spectrometry—A review. Anal. Chim. Acta. 668: 114–129.

Prasad, M.N.V. and J. Hagemeyer, 1999. Heavy Metal Stress in Plants—From Molecules to Ecosystem. Springer. Heidelberg, Germany.

Raab, A. and J. Feldman. 2005. Arsenic speciation in hair extracts. Anal. Bioanal. Chem. 381: 332–338.

Salazar, R.F.S., M.B.B. Guerra, E.R. Pereira-Filho and J.A. Nobrega. 2011. Performance evaluation of collision–reaction interface and internal standardization in quadrupole ICP-MS measurements. Talanta. 86: 241–247.

Sarzanini, C. 1999. Liquid chromatography: a tool for the analysis of metal species. J. Chromatogr. A 850: 213–228.

Sarzanini, C. and E. Mentasti. 1997. Determination and speciation of metals by liquid chromatography. J.Chromatogr. A 789: 301–321.

Shalliker, R.A. 2011. Hyphenated and alternative methods of detection in chromatography. Chromatographic Science series. CRC Press. Boca Raton, Florida. USA.

Slaets, S., F. Adams, I.R. Pereiro and R. Lobinski. 1999. Optimization of the coupling of multicapillary GC with ICP-MS for mercury speciation analysis in biological materials. J. Anal. At. Spectrom. 14: 851–857.

Szpunar, J. and R. Łobiński. 2003. Hyphenated techniques in speciation analysis. The Royal Society of Chemistry. R.S. Smith [ed.]. RSC Chromatography Monographs.

Tanner, S.D., V.I. Baranov and D.R. Bandura. 2002. Reaction cells and collision cells for ICP-MS: a tutorial review. Spectrochim. Acta, Part B. 57: 1361–1452.

Tessiere, A., P.G. Campbell and M. Kisson. 1979. Sequential Extraction Procedure for the Speciation of Particulate Trace Metals. Anal. Chem. 51: 844–851.

Tswett, M.S. 1906. Physikalisch-Chemische Studien Uber das Chlorophyll. Die Adsorptionen. Ber. Bot. Ges. 24: 316–332.

Tzanani, N. and A. Amirav. 1995. The Combined Pulsed Flame Photometric Ionization Detector. Anal. Chem. 67: 167–173.

Van Loon, J.C., L.R. Alcock, W.H. Pinchin and J.B. French. 1986. Inductively coupled plasma source mass spectrometry—a new element/isotope specific mass spectrometry detector for chromatography, Spectros. Lett. 19: 1125–1135.

Vercauteren, J., C. Peres, C. Devos, P. Sandra, F. Vanhaecke and L. Moens. 2001. Stir bar sorptive extraction for the determination op ppq-level traces of organotin compounds in environmental samples with thermal desorption-capillary gas chromatography-ICP mass spectrometry. Anal. Chem. 73: 1509–1514.

Viera, M.A., P. Grinberg, C.R.R. Bobeda, M.N.M. Reyes and R.C. Campos. 2009. Non-chromatographic atomic spectrometric methods in speciation analysis. A review, Spectrochim. Acta, Part B. 64: 459–476.

Wang, X., Z. Zhuang, P. Yang and B. Huang. 1995. Hyphenated Techniques Combined with Atomic Spectrometry for Environmental Studies. Microchim. J. 51: 88–98.

Wasik, A., I.R. Pereiro and R. Lobinski. 1998. Interface for time-resolved introduction of gaseous analytes for atomic spectrometry by purge-and-trap multicapillary gas chromatography (PTMGC). Spectrochim. Acta, Part B. 53: 867–879.

Weiss, J. 2004. Handbook of Ion Chromatography, Wiley-VCH, Heilderberg.

Wille, A., S. Czyborra and A. Steinbach. 2007. Hyphenated Techniques in Ion Chromatography. LCGC Asia Pacific. 12(3): 12–18.

Selenium Speciation in the Environment

Rodolfo G. Wuilloud[1,2], and *Paula Berton[1,2]*

Introduction

Selenium (Se) and its several species have been demonstrated to be essential for living organisms, including animals and humans (Pinset 1954). This element is required to obtain the biologically active selenol group (-SeH), which is a precursor of glutathione peroxidase, thioredoxin reductase and other seleno-containing enzymes involved in oxidative stress response (Holben and Smith 1999). Even though plants represent the main source of this element for animal diets, it is not essential for them. Since confirmation of the essentiality of Se, evaluation of its nutritional status is a crucial task to establish the right doses avoiding deficiency-linked diseases and promoting good health conditions (Lei et al. 2011). However, Se has a narrow interval between toxicity and essentiality, which is puzzling toxicologists and alarming nutritionists and legislators. Different diseases can be caused by its deficiency, but it can be toxic from levels one order of magnitude

[1]Laboratory of Analytical Chemistry for Research and Development (QUIANID), Instituto de Ciencias Básicas, Universidad Nacional de Cuyo, Padre Jorge Contreras 1300, Parque Gral. San Martín, C.P. M5502JMA Mendoza, Argentina.
[2]Consejo Nacional de Investigaciones Científicas y Técnicas (CONICET), Argentina.
*Corresponding author: rwuilloud@mendoza-conicet.gob.ar

higher than that required for good health (Wada et al. 1995). Thus, daily consumption of food containing less than 0.1 mg/kg of body weight will result in Se deficiency, whereas dietary levels above 1 mg/kg may lead to toxicity (Wada et al. 1995).

Selenium occurs in the environment and biological materials under inorganic and organic chemical forms or species, so the role of analytical speciation is essential to fully understand how bioavailability and toxicity depend on its current species. As other living organisms, the human body accumulates Se: its level in human blood is around 200 ng/mL, which is many times higher concentration than that found in surface water (Al-Kunani et al. 2001, Coudray et al. 1996, Shamberger 1981). It is evident that Se can be accumulated in biological systems, with specific tissues as ending fate. Furthermore, Se levels in marine fish have been observed to be about 2 µg/g, which is around 50,000-fold higher concentration than that found in the surrounding environment of these animals (Thiry et al. 2012, Yoshida et al. 2011, Hamilton 2004). Most soils contain between 0.02 and 2.5 µg Se/g (Cornelis et al. 2005). The average crustal abundance of Se is 0.09 µg/g. Selenium concentrations range from 0.47 to 8.1 µg/g in coal and from 2.4 to 7.5 µg/g in fuel oil (Shamberger 1981). Selenium dioxide (SeO_2) is of high concern for industries due to potential emissions to the atmosphere. The dioxide forms selenious acid with water or liquid media, and this acid reacts irritantly. Moreover, Se compounds released during coal or petroleum combustion might be a significant source of exposure (Wen and Carignan 2007, Frankenberger and Engberg 1998). However, transformations and associations of Se introduced into the environment by these sources are strictly dependent on the physicochemical properties of Se species under which Se occurs.

In this chapter, fundamental discussion regarding speciation studies of Se in environmental compartment is presented. Several aspects showing the crucial importance that Se speciation is having for different fields, such as toxicology and nutrition, are critically commented and discussed. A comprehensive summary of mostly-spread and useful state-of-the-art instrumental techniques and sample preparation methods to face the challenge raised by Se speciation analysis in environmental samples, mainly soil and water, is arbitrary organized and discussed.

Environmental Relevance of Selenium Speciation

The modification of nature's elemental life cycle by anthropogenic activities includes (i) the extraction, smelting, and processing of metal ores into products, (ii) the distribution and application of derived-products by consumers and industry, and (iii) the comeback of metals under the form of concentrated wastes to source environment. Thus, generally-accepted essential elements such as Se could become contaminants in the receiving environments at high concentrations. However, the complexity of metal and metalloids contaminated sites is usually simplified to a measure of the total elemental content. Despite total elemental content is a fundamental measure in assessing risks, it does not provide by itself a complete picture on the bioavailability, mobility, and fate of elements (Cornelis et al. 2003, Cornelis et al. 2005). Mobility is clearly dependent on chemical form or species, and it is fundamental to be considered for fully evaluating ecological risks and ultimately potential effects on human health.

Bioavailability and toxicity are not always obvious concepts and depend on biological species and their genetic characteristics. Bioavailability relates concentration of a particular element to its availability to organisms. In fact, Se bioavailability, governed by nature of chemical species and its uptake by animals and plants, determines biochemical functions in biological systems. Furthermore, Se bioavailability could be modified by different physicochemical processes in the environment, depending on the chemical properties of a specific Se species. Table 5.1 shows some of the most common Se species that could be found in the environment and biological systems. In addition, an extensive list of Se species can be easily drawn up from several bibliographic sources (Cornelis et al. 2005, Cornelis et al. 2003, Uden 2002).

The occurrence of Se in waters, sediments, and soils results mainly from anthropogenic activities. On the other hand, Se is emitted into the atmosphere from different sources, including natural and anthropogenic ones (Wen and Carignan 2007). Selenium speciation in surface water, groundwater and waste water involves not only the inorganic species selenite (SeO_3^{2-}) and selenate (SeO_4^{2-}), but also non-volatile organic species, such as selenoamino acids and volatile methylated species, mainly dimethylselenide (DMSe) and dimethyldiselenide (DMDSe) (Cutter 1992, Lenz et al. 2006). Frankenberger et al. have provided a comprehensive treatment of Se in environment and have already highlighted the importance of chemical speciation of this element (Frankenberger and Engberg 1998). Moreover, there is a thorough investigation demonstrating that toxicity of oxidized inorganic species is due to its high biological activity (Holben and Smith 1999).

Another concern related with Se speciation in the environment and its ecotoxicity arises from the ever-increasing field of nanotechnology. Although this is a major innovative scientific growing area around the world, nanoparticles (NPs) and other nanomaterials may have negative effects on human health and could raise potential risks for environmental biota, as well (Ema et al. 2010, Li et al. 2008). The size of nanomaterials is in the range of 1–100 nm and can be composed of many different based materials (carbon, silicon, gold, cadmium and selenium) with particular shapes. Nanoparticles could occur both naturally or as a consequence

Table 5.1 Selenium species occurring in environmental and biological compartments.

Species	Chemical formula	Abbreviation	Origin
Inorganic nature			
Selenate	$[H_2SeO_4, HSeO_4^-, SeO_4^{2-}]$	SeO_4^{2-}	Soil, sediment and water.
Selenite	$[H_2SeO_3, HSeO_3^-, SeO_3^{2-}]$	SeO_3^{2-}	Acidic environments, volcanic eruptions and combustion processes.
Selenium Dioxide	$[SeO_2]$	SeO_2	
Elemental Selenium	$[Se^0]$	Se^0	Microbial and inorganic processes.
Selenide	$[H_2Se]$	H_2Se	Microbial processes.
Organic nature			
Methylselenol	$[CH_3SeH]$	MSeH	Product of microbial methylation, in vitro experiments for mimicking biochemical processes, organic tissues, urine metabolite, plants metabolites.
Dimethylselenide	$[(CH_3)_2Se]$	DMSe	
Dimethyldiselenide	$[(CH_3)_2Se_2]$	DMDSe	
Trimethylselenonium cation	$[(CH_3)_3Se^+]$		
Dimethylselenone	$[(CH_3)SeO_2]$		
Dimethylselenoxide	$[(CH_3)_2SeO]$		
Methylselenic acid anion	$[(CH_3)Se(O)O^-]$		
Dimethylseleniumsulfide	$[(CH_3)_2SeS]$	DMSeS	
Dimethylseleniumdisulfide	$[(CH_3)_2SeS_2]$	DMSeDS	

Name	Formula	Abbreviation
Selenomethionine	$[H_3N^+CH(COO^-)CH_2CH_2Se\ CH_3]$	SeMet
Selenocysteine	$[H_3N^+CH(COO^-)CH_2SeH]$	SeCys
Selenocystine	$[H_3N^+CH(COO^-)CH_2Se_2CH_2CH(COO^-)NH_3^+]$	$SeCys_2$
Se-methylselenocysteine	$[H_3N^+CH(COO^-)CH_2SeCH_3]$	SeMCys
Gamma-glutamyl-Se-methylselenocysteine	$[H_3N^+CH_2CH_2CONHCH(COO^-)CH_2SeCH_3]$	γ-glutamyl-SeMCys
Selenocystathionine	$[H_3N^+CH(COO^-)CH_2CH_2SeCH_2CH(COO^-)NH_3^+]$	
Selenohomocysteine	$[H_3N^+CH(COO^-)CH_2CH_2SeH]$	
Se-adenosylselenohomocysteine	$[NH_2CH(COOH)CH_2CH_2SeCH_2C_4H_5O_3C_5N_4NH_2]$	

Adapted from (Fernández-Martínez and Charlet 2009) and (Cornelis et al. 2005).

of specific industrial processes. Furthermore, nanomaterial properties significantly differ compared with those of the source compounds since 50 percent of the atoms in NPs are on the surface, resulting in higher reactivity than raw materials (Fernández-Martínez and Charlet 2009). Therefore, NPs could cause substantially different biological effects than source compounds (Peng et al. 2007). Recent developments in the field of drug delivery have called attention regarding bioavailability of elemental Se-NPs. It was observed that inorganic Se species, i.e., oxyanions, would be reduced by bacteria to form Se^0 as NPs of different sizes and chemical structures (Oremland et al. 2004). Physicochemical properties of these elemental Se-NPs are still vastly unknown and different bioavailability and toxicity is expected. Moreover, it has been reported that solubility of some elements could increase as a consequence of size reduction of materials to nanometer scale, which raise further questions regarding bioavailability and the environmental fate of elemental Se-NPs (Sadeghian et al. 2011). Thus, the concept of bioavailability would need to be revised in light of colloidal systems, such as those obtained with NPs, as small-size particles are able to carry other adsorbed toxic compounds through membranes. Therefore, the concept of "bioaccessibility", meaning "the fraction of a substance that becomes soluble within the gut or lungs and hence available for absorption through a membrane" should be considered (Reeder et al. 2006). Both bioavailability and bioaccessibility strongly depend on different physicochemical properties (e.g., speciation, pH, ionic strength, redox potential, etc.) affecting solubility of particles, and eventually of compounds. For example, the release of manufactured elemental Se-NPs into the aquatic environment is very unknown (Farré et al. 2009). The small size of NPs and nanotubes provide surfaces that may allow the binding and transport of toxic pollutants, as well as possibly being toxic by themselves. Thus, a full picture of Se speciation should include the possible introduction of man-made elemental Se-NPs into the environment, leading to consider the possibility of changing our current knowledge on the speciation of this element and the need for further studies on this specific matter.

The question regarding why Se speciation should be studied in the environment has a simple answer, and is that it is part of an overall strategy to understand the complex chemistry and behavior of Se in the environment and organisms. The information about speciation is needed for several fields of environmental importance, including, elucidation of biogeochemical transformations of Se species, identification of natural and anthropogenic Se species with beneficial or detrimental effects, evaluation of plant species that better fulfill the concept of phytoremediation (Barillas et al. 2011, Feng and Wei 2012), and development of certain strategies to avoid or limit toxic effects in plants, animals and humans.

The study of Se ecotoxicity can be a difficult task as Se is usually present in many species, each of which showing distinguished bioavailability and toxic effects (Lemly 1998). Toxicity of Se in the environment changes, as it is captured by living organisms, metabolized, and transferred through the food chain (Lemly 1996, Weech et al. 2012). Selenium is metabolized to different chemical species than those to which they were exposed, particularly organoselenium (Org-Se) compounds. A schematic on the metabolic pathways of Se in plants is shown in Fig. 5.1. Selenomethionine (SeMet) and selenocysteine (SeCys) may represent an ecotoxic threat. Selenocysteine is a highly cytotoxic Se species that catalyze formation of free radicals from very low levels. On the other hand, elemental Se is an insoluble form of this element that has been generally thought to not pose great threat. However, this assumption may need some future revision from the scientific community on new findings related to surface properties of elemental Se-NPs and its potential biological action in organisms (Fernández-Martínez and Charlet 2009, Sieber 2003). Depending on its size, Se-NPs showed a Se accumulation and higher glutathione S-transferase (GST) activity. Since both GST activity and small molecular weight selenocompounds accumulated *in vivo* are important intermediates for chemoprevention by Se, these results suggest that Se-NPs should be more effective as a chemopreventive agent at smaller particle size (Peng et al. 2007). On the other hand, in comparison to SeO_3^{2-}, SeMet and methylselenocysteine (SeMCys), recent studies have shown different results regarding bioavailability and toxicity of Se-NPs: while in mice and rats bioavalibility is comparable and toxicity is lower (Wang et al. 2007); in aquatic animals Se-NPs were found to be more hyper-accumulated, and with strong toxicity for medaka with an approximately

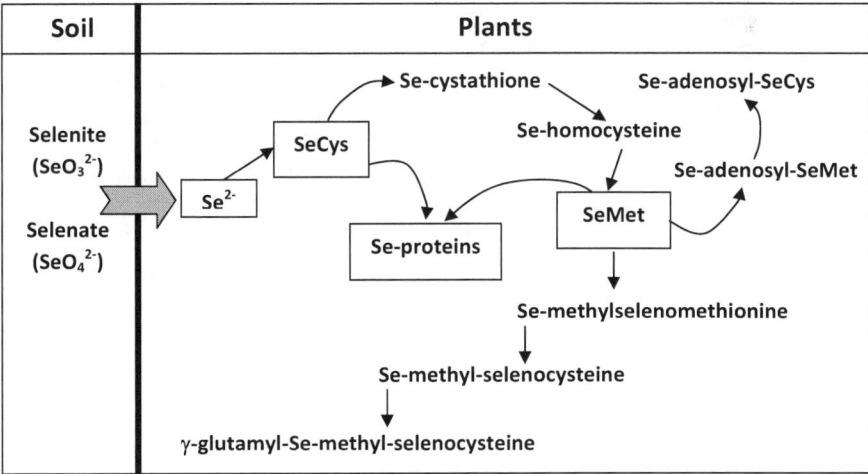

Figure 5.1 Current knowledge on metabolic pathways of Se in plants.

five-fold difference in terms of LC_{50} compared to SeO_3^{2-} (Li et al. 2008). These results suggest that nanotoxicological studies should be carried out on a case-by-case basis.

The chemical forms of Se may change while they are transferred from organisms placed at low trophic level towards others at higher levels. In order to understand mechanistic biochemistry of Se ecotoxicity, knowledge of the processes of biotransformation and accumulation of Se in the food chain must be achieved (Lemly 1996). Thus, to evaluate Se ecotoxicity in a given ecosystem, Se speciation in the different environmental compartments must be completely evaluated and understood. As described later, biogeochemical cycling of Se in aquatic systems involves geological erosion and anthropogenic influence to contribute to Se in seawater, mainly as SeO_3^{2-} and SeO_4^{2-} (Lemly 1999). These inorganic species are also predominant in soils, and levels of Se in this compartment range from low ng/g to hundred µg/g. Contribution of volatile Se species is estimated in 6×10^6 kg Se per year to the atmosphere (Wen and Carignan 2007). Organisms of low trophic level such as phytoplankton and zooplankton are considered to accumulate Org-Se species through a biomethylation process of Se. In most plants, Se levels are below 1 µg/g dry weight. However, as described later, the levels in certain family plants could be of thousands of µg/g dry weight.

The chemical and biochemical transformations involved in the environment require analytical identification and quantification of Se species at low concentrations, typically ng/g or ng/ml if solid or liquid samples are studied, respectively. Among the several Se species involved in the speciation of this element, Org-Se species have been studied to a lesser extent. Thus, coarse discrimination of Se species into total inorganic and organic Se has been made in different environmental samples. Non-destructive instrumental techniques such as X-ray absorption spectroscopy (XAS) have been employed in these cases (Wen et al. 2006). Furthermore, fractionation techniques involving non-chromatographic approaches have been used to aim this initial speciation analysis of Se (Gonzálvez et al. 2010, Gonzálvez et al. 2009). However, despite these ongoing investigations on Se speciation in ecosystems, Org-Se speciation still needs further development to clarify the complete scenario of Se speciation. Probably, the main difficulty is related to major complexity of Org-Se speciation as compared to inorganic chemical forms. Mass spectrometry (MS) is among the pack of analytical techniques for qualitative identification of unknown Org-Se species (Ogra and Anan 2009). Several Se species have been identified in plant tissues using this tool, but many others remain undiscovered (Ogra and Anan 2009, Caruso and Montes-Bayon 2003). In fact, a full description of Org-Se speciation in all parts of the ecosystem (i.e., soils, sediments, plants, insects, birds, etc.) is a matter of continuous research in analytical chemistry focusing on elemental speciation. Selenoamino acids such as SeCys and SeMet have been the determined Se species in most cases (Ogra and Anan 2009). However, selenoamino acids comprise an assorted family of compounds than just protein-related amino acids (e.g., SeMCys), for which information on toxicity is limited. As discussed earlier, the real toxicological impact of Se on animals and plants of a certain environment would be estimated not only by having fundamental information on total Se, but also on Se speciation existing throughout the different trophic levels in which organisms may be placed.

Selenium Biogeochemistry

Environmental cycle of selenium

Due to its complex chemical behavior, Se combines with a variety of elements in nature, thus making Se compounds widespread in atmospheric, marine and terrestrial environments (Wen and Carignan 2007). Historically, the main sources of the element were the natural ones, introducing 50–65 percent of total Se emissions. However, since the onset of industrialization, anthropogenic emissions have greatly increased compared to natural sources (Lemly 2004). As a result, the scientific community is paying increasing attention to the geochemistry of Se, in an effort to evaluate its relevance in the environment. The natural sources of Se include crustal weathering (wind-blown soil dust), volcanoes, sea salt and the continental and marine biospheres (Lemly 2004).

The main source of the element in the terrestrial system (40 percent of the total in the Earth's crust) is rocks. The transport of Se from rocks to other compartments is primarily due to biogeochemical processes such as weathering, rock–water interactions and biological activity. These processes distribute Se in an inhomogeneous way over the Earth (Fernández-Martínez and Charlet 2009). As a result, soil Se content is mostly influenced by Se content in the parent materials, being seleniferous soils those which are derived from Precambrian organic-rich carbonates, or from Cretaceous shales (Burau 1985). Concentrations and chemical forms of Se in soils or drainage water are ruled by several physical and chemical factors such as pH, chemical and mineralogical composition, adsorbing surface, and oxidation–reduction status (Dhillon and Dhillon 1999). The oxyanions SeO_3^{2-} and SeO_4^{2-} predominate in alkaline soils and oxidizing conditions, maintain the availability of biologically active forms, and cause plant uptake of the metal to be increased. On the other hand, insoluble selenides and elemental forms of Se are mostly present in poorly aerated, acid, organic-rich soils under strong reducing conditions, and the amount of biologically available Se should steadily decrease. Both in aquatic and terrestrial environments, Se is frequently associated with organic matter (OM) (Zhang and Moore

1996). OM plays a fundamental role among the soil biogeochemical variables governing the mobility of trace elements by complexation processes, including weathering of Se^0 or metallic selenides (Se-bearing FeS_2, $FeSe_2$), and oxidation of selenide (Se^{2-}) to SeO_3^{2-} under acidic conditions, or to SeO_4^{2-} under alkaline conditions (Fernández-Martínez and Charlet 2009). Either Se can form Se-metal-humic ternary complexes or associate with natural polysaccharides of different molecular weight. The bioavailability of Se depends directly on these associations (Zhang and Moore 1996). Additionally, Se speciation, mobility and bioavailability are highly affected by the presence of microorganisms in the environment. Their main influence on the bioavailability is through the control of Se oxidation state, which directly relates to the solubility of different Se compounds. Irrigation of seleniferous soils for crop production in arid and semi-arid regions is a major source of Se introduction into the environment, causing high levels of Se to be released into local aquatic systems (Lemly 2004).

Industrial processes and, in particular, coal and fossil fuel combustion are important sources of atmospheric Se emissions (67–79 percent of total emissions) (Wen and Carignan 2007). The majority of anthropogenic activity (70–80 percent of total emissions) occurs in the northern hemisphere. Selenium is present in coal, coal conversion products, oil, oil shale and their waste by-products. It is also found at high levels in the mineral fraction of fly-ash and bottom-ash from the combustion of coal, in nonferrous metal melting, and in manufacturing and utilization of agriculture products (Lemly 2004). In some coal deposits, Se is present as Org-Se compounds, chelated species or as the adsorbed element. Although marine biogenic Se is the dominant natural source of emission to the atmosphere (60–80 percent of total natural emissions) (Wen and Carignan 2007), volcanism also contributes with Se emissions to the atmosphere and hydrosphere (Fernández-Martínez and Charlet 2009). Due to the high chemical similarity of Se and S, as S forms sulfur dioxide, Se forms a gaseous flux of $SeO_2(g)$. Selenium containing dusts derived from volcanoes and wind erosion of the Earth's surface contribute to the transfer from land-to-atmosphere Se with 180 tons per year, while suspended sea salts contribute to an additional amount of 550 tons per year more (Floor and Román-Ross 2012). Both geologic and anthropogenic sources often release mostly SeO_4^{2-}, which is not reactive with particle surfaces, although some types of bacteria convert SeO_4^{2-} to Se^0 in sediments (Luoma and Presser 2009).

Before deposition, atmospheric Se may undergo various physical and chemical transformations. Because of their significantly different atmospheric behavior, at least three species of Se should be examined explicitly for emission accounts: (1) volatile organic Se [DMSe, DMDSe, MSeH, DMSeS, etc. (Table 5.1)], mainly emitted by living organisms;

(2) volatile inorganic Se (Se^0, H_2Se, SeO_2), occurred from volcanic and anthropogenic emissions; and (3) particulate Se, mostly originated from naturally released sea salts, wind-blown dust and volcanic ash, as well as man-derived fly-ash from coal combustion and refuse incineration dust from mining and metal melting. An extensive overview of atmospheric emission, transport, transformation, and removal of Se was recently published by Wen and Carignan (Wen and Carignan 2007).

Selenium returns to the Earth's surface through wet and dry depositional processes (Velinsky and Cutter 1991), the first type being responsible for about 80 percent of the total deposition. The importance of atmospheric deposition as a significant source of contamination is evidenced by the elevated levels of Se in the aquatic environment and in terrestrial plants such as lichens and mosses distant from anthropogenic emission sources (Wen and Carignan 2007).

In water column profiles of well-stratified aqueous environments, the speciation of Se changes from oxidized forms (e.g., SeO_4^{2-} and SeO_3^{2-}) to more reduced ones (Org-Se and H_2Se) with transit from the oxic surface waters to the anoxic bottom waters. The oxidized forms are easily leached from soils through geological erosion and anthropogenic influences, mainly from oil refineries, which release significant amounts into coastal waters and electric utilities, and produce contaminated aqueous discharges from the storage of coal, coal ash and landfill. Selenate is very soluble and less strongly adsorbed by particles than SeO_3^{2-} and Org-Se species, which are much more reactive (Luoma and Presser 2009). In fact, SeO_4^{2-} is easily leached from soils, transported to groundwaters , and taken up by plants (Terry et al. 2000, Zayed and Terry 1992, Zayed et al. 1998). SeO_4^{2-} is thermodynamically stable in alkaline and well-oxidized environments. On the other hand, SeO_3^{2-} occurs in mildly oxidizing neutral pH environments, is less soluble than SeO_4^{2-}, and may be reduced chemically or biochemically to Se^0 (Zayed et al. 1998). Selenides and Se-enriched sulfides exist in reducing acidic environments but are very insoluble, resistant to oxidation, and show low bioavailability to plant and animals (Stolz et al. 2006). Selenium is incorporated as Org-Se compound sequentially through phytoplankton and zooplankton, lower and higher vertebrates, or can be recycled through microorganisms to SeO_3^{2-} and SeO_4^{2-}, and also to Se^{2-} as HSe^- and colloidal Se^0. Org-Se is released back to the water column as these cells die or are consumed, where some SeO_3^{2-} is formed. But neither SeO_3^{2-} nor Org-Se is reconverted to SeO_4^{2-} because the back reaction has a half time of hundreds of years. The result is a build-up of proportionately more Org-Se and SeO_3^{2-} as Se is recycled through the base of food webs, and proportionately less SeO_4^{2-} (Luoma and Presser 2009).

Bioremediation of selenium

Immobilization and removal of Se from soils and waters represent complex and expensive chemical processes (Sheoran et al. 2011). Different technologies based on physical, chemical and biological methods have been developed to remove Se. An extensive review on the advantages and disadvantages of each treatment technology for Se removal was published by Frankenberger et al. (Frankenberger et al. 2004). They concluded that biological treatment holds the most promise because of its cost-effectiveness and permanent removal of Se. Accordingly, man-made bioremediation, that is, the removal of Se using biological processes, is a very important area of study today (Frankenberger et al. 2004). On the one hand, bacteria have been studied and described for use in removing or immobilizing Se species in contaminated water for many years. On the other hand, phytoremediation, the process by which certain plants can be used to remove Se *in situ* from contaminated soils and/or waters is actively being pursued (Bañuelos et al. 2002).

The ability of plants to tolerate and accumulate Se differs strikingly among species. Selenium accumulating plants belonging to the families *Allium*, *Brassica*, *Composite*, *Cruciferae*, *Asteraceae* and *Leguminosae* may accumulate Se up to concentrations of several thousand mg/kg dry mass when grown in Se-rich soil (Montes-Bayón et al. 2002a, Montes-Bayón et al. 2006, Bañuelos et al. 2002, Raskin and Ensley 1999). Selenium accumulation by plants is not only strongly influenced by the plant species but also by Se speciation and soil properties, such as sulfate concentrations and soil salinity, among others (Terry et al. 2000).

Different mechanisms have been proposed for Se uptake and distribution in these plants (shown in Fig. 5.1). Selenate is incorporated in plants via a sulfate transporter in the root, reduced to SeO_3^{2-}, and then assimilated into the selenoamino acids SeCys and SeMet, following the S metabolic pathway, which would eventually lead to Se protein incorporation (Terry et al. 2000, Pilon-Smits and LeDuc 2009). Plants can also take up organic forms of Se, such as SeMet actively. Some plants can even absorb volatile Se from the atmosphere via the leaf surface (Terry et al. 2000). The major mechanism whereby high Se accumulation in plant induces Se toxicity is associated with the incorporation of SeCys and SeMet into proteins. In fact, the main difference between Se accumulators from non-accumulators is that the former metabolize the SeCys and SeMet through methylation into various non-proteinogenic or nonessential selenoamino acids (Terry et al. 2000, Pilon-Smits and LeDuc 2009). Some of these amino acids are known to be effective inhibitors of tumor formation, such as SeMCys and derivatives such as γ-glutamyl-methyl-selenocysteine (γ-glutamyl-SeMCys), which have been found in *Allium* group vegetables (Pyrzynska 2009).

Many species have been evaluated for their efficacy in phytoremediation. The ideal plant species for Se phytoremediation is one that can accumulate and volatilize large amounts of Se, grow rapidly and produce large biomass on Se-contaminated soil. Most of these attributes are covered by the Indian mustard plant, *Brassica juncea*, which has been evaluated as a phytoextractor or phytoaccumulator of Se for removal from contaminated soils (Zayed et al. 1998, Yawata et al. 2010). After plants have accumulated and sequestered Se, they can be harvested and disposed in an environmentally safe manner. The concept of using a crop which reduces levels of Se in soil and also has economic value is an attractive approach for managing seleniferous soils in regions highly dependent on agriculture (Dhillon and Dhillon 2009). Frequently, Se-deficient and -excess areas are located close together. Such a scenario seems ideal for phytotechnological solutions where phytoextraction using Se-hyperaccumulating plants may be used to lower Se-burdens of Se-rich soils and the Se-rich plant material could be used for agronomic biofortification of the Se-deficient soil (Barceló and Poschenrieder 2011). Moreover, it may be feasible to use Se containing plant material as animal feed or animal food—after proper dilution to non-toxic levels—in regions with Se deficiency (Bañuelos 2001). Furthermore, a third potential use of the Se-rich plant material such as *Brassica* plants like canola and broccoli for the remediation of Se under field conditions could result in phyto-products enriched in an essential trace element in broccoli, feed meal, organic fertilizer and oil that can be used as a biofuel additive (Bañuelos 2006, Zhu et al. 2009).

An interesting option to increase the efficiency of Se removal and to decrease Se ecotoxicity is the volatilization of Se by plants and microbes. During biological volatilization, the Se removed from a contaminated site is transformed into forms with lower toxicity, such as DMSe, which is 500–700 times less toxic than SeO_4^{2-} or SeO_3^{2-} (LeDuc and Terry 2005, de Souza et al. 2000). This process, called phytovolatilization, is based on naturally-occurring or genetically-modified plants capable of absorbing elemental forms of Se from the soil and releasing Se into the atmosphere after biological conversion to gaseous species. It has been demonstrated that both Se non-accumulator and accumulator species volatilize Se. However, while the volatile Se compound released from the Se accumulator *Astragalus racemosus* was identified as DMDSe, Se released from alfalfa, a Se nonaccumulator, was identified as DMSe (Pilon-Smits and LeDuc 2009). *Chlorella* sp., a single-celled freshwater microalga, was presented as another example due to its ability to efficiently reduce SeO_4^{2-} (converts 87 percent of the SeO_4^{2-} accumulated in 24 hr) to DMSe. This capacity to efficiently reduce Se may have evolved in microalgae because their large surface to volume ratio means that their Se uptake rates can be relatively high while space available for storage of toxic Se compounds is small. Phytovolatilization has the added benefits of minimal site disturbance, less erosion and no need to dispose of contaminated plant material. It is suggested that the addition to atmospheric levels through this practice would not contribute significantly to the atmospheric pool, since the contaminants are likely to be subject to more effective or rapid natural degradation processes, such as photodegradation (Padmavathiamma and Li 2007). Therefore, this technique appears to be a promising tool for remediating Se-contaminated soils, as a permanent site solution. However, a major limitation of its use for bioremediation is the fact that uptake of SeO_4^{2-} is strongly inhibited by the presence of sulfate in the medium (LeDuc and Terry 2005).

The capacity of certain rhizobacteria to precipitate Se oxyanions, reducing their toxicity in contaminated matrices, could be seen as an alternative option to Se phytoextraction or phytovolatilization, so far emphasized for the removal of toxic metalloids from soil (Di Gregorio et al. 2006). Data suggest that bacteria in the rhizosphere of accumulator plants are required for Se volatilization from SeO_4^{2-} and SeO_3^{2-} but not from SeMet (Terry et al. 2000). In fact, *Brassica juncea* cultivation on the tested soil resulted in a higher volatilization rate when compared with non-vegetated soil. However, the presence of *Brassica juncea* in soil amended with Se as SeO_3^{2-} or SeO_4^{2-} has revealed to promote a significant Se precipitation by eliciting rhizobacteria capable of reducing the metalloid oxyanions to Se^0 (Di Gregorio et al. 2006). On the other hand, a reduction in plant biomass was also observed, evidencing the improved capability of *Brassica juncea* to accumulate Se at concentrations that are actually toxic for plants (Lampis et al. 2009).

The Se-tolerant microbes that are found in association with Se hyper-accumulators may perhaps be used for bio- or phytoremediation, by themselves or in concert with plants (El Mehdawi and Pilon-Smits 2011). The relevance of Se tolerant microorganisms has mainly been investigated in relation to Se volatilization, Se phytoextraction, and geochemical Se-reduction for remediation technology (Gadd 2010, Zayed et al. 1999). The metal-reducing bacteria *Geobacter sulfurreducens*, *Shewanella oneidensis* and *Veillonella atypica*, use different mechanisms to transform SeO_3^{2-} to less toxic, non-mobile Se^0 and then to selenide in anaerobic environments (Pearce et al. 2009). Some bacteria can use SeO_4^{2-} as a terminal electron acceptor in dissimilatory reduction and also reduce and incorporate Se into organic components, e.g., selenoproteins. The inorganic species SeO_4^{2-} and SeO_3^{2-} can also be reduced to Se^0. It has been demonstrated that *Bacillus subtilis* is a substantive Se bioaccumulator (Combs Jr. et al. 1996). Methylation of Se is a ubiquitous microbial property and can occur in soils, sediments and water. Bacteria and fungi are the most important Se-methylaters in soil, with DMSe being volatile and most frequently produced. Furthermore, the volatile species DMDSe is produced in smaller amounts (Gadd 2010).

Some limitations in the use of plant-based technologies result from the fact that autotrophs are not ideally suited for the metabolism and breakdown of organic compounds, the often slow time-scale for remediation to acceptable levels, and also due to toxicity to the plants themselves. These restrictions can be overcome through interactions with both endophytic bacteria and rhizosphere bacteria associated with plants, which have shown to have the potential to degrade organic compounds in association with plants (Dowling and Doty 2009).

Another approach to improve Se accumulation and/or Se tolerance was proposed by the genetic engineering of transgenic plants (White et al. 2007). Using existing knowledge about rate-limiting steps in plant Se metabolism and Se detoxification mechanisms in tolerant species, different genetic engineering strategies have been designed and used successfully to further enhance plant Se uptake, accumulation, volatilization and tolerance (Pilon-Smits and LeDuc 2009, Zhao and McGrath 2009). Moreover, through recombinant DNA technology, transgenic plants express specific genes from heterotrophic organisms such as bacteria and mammals to increase its tolerance and metabolism of organic chemicals or heavy metals (Dowling and Doty 2009). Extensive reviews on Se accumulation in plants, their mechanisms of Se metabolism, and application of genetic engineering to enhance plant Se accumulation were recently published (Kotrba et al. 2009, Barillas et al. 2011, Pilon-Smits and LeDuc 2009, Zhu et al. 2009).

Environmental Legislation Related to Selenium Speciation

Toxic effects of Se in the environment are associated with bioaccumulation and biotransformation of the element through food chain, which is definitively related to speciation as mentioned early in this chapter. Therefore, there is an imperative need to develop regulatory policy controlling its level in the environment. However, the great difficulty of Se speciation raises some problems to regulators for establishing regulatory policy. The knowledge of total Se levels in the environment does not provide information about Se species. This is particularly problematic in the case of Se, which exists in a great number of chemical forms in the environment, each of which has a distinguished bioavailability, bioaccessibility, toxicity and ecotoxic potential. The importance of bioavailability and its great relation with speciation was demonstrated by Amweg et al. (Amweg et al. 2003). In this study, the effects of an algae-bacterial Se-reduction system applied to irrigation waters of the San Joaquin Valley (California, USA) were evaluated. An 80 percent reduction of Se levels in the influent waters was obtained, although SeO_3^{2-} and Org-Se levels were increased by a factor of 8. Since these two species show a significantly higher bioavailability to biota than the original ones, the levels of Se in the inhabitant organisms of the ecosystem suffered a 2 to 4-fold increase. Speciation is thus a key factor to establish appropriate regulatory policy regarding Se in the environment, but more importantly about possible implementation of technologies used for environmental remediation as well.

An insoluble form of Se, i.e., Se^0, was initially thought to pose limited toxicological threat to organisms because of their low bioavailability, although biological activity has been reported for Se^0-NPs (Zhang et al. 2005). Highly cytotoxic Se forms, such as SeCys, catalyze the generation of free radicals and may be harmful at very low levels. Furthermore, SeMet, may represent a toxic threat in the way that is not directly cytotoxic, but highly stable instead, persisting in food chain, and could eventually be metabolized to other more toxic chemical forms. Implications for human health are large, and even more if the irrigated croplands are not appropriately remediated. For these reasons, measurements of total Se do not accurately represent the real risk of Se in the environment. Therefore, in order to develop a correct regulatory policy, both the chemical forms of Se occurring in a particular environment and the extent of toxicity and ecotoxicity of those species must be determined (Cornelis et al. 2005).

Analytical Methods for Se Speciation Studies

In environmental studies, the identification of the different chemical forms of an element has been a challenge. Generally, for species determination two complementary techniques are required in order to achieve an efficient and reliable separation of the species and an adequate quantification. Thus, for speciation analysis, the coupling of chromatographic techniques such as gas chromatography (GC), high performance liquid chromatography (HPLC), and capillary electrophoresis (CE) with highly sensitive and selective atomic spectrometry (AS) detector has been widely used. Even though GC has enjoyed particular attention because of its high efficiency and simplicity of coupling, HPLC has the ability of dealing with non-volatile compounds, extending the range of application and avoiding derivatization in some cases. Nowadays, the detector of choice is inductively coupled plasma mass spectrometry (ICP-MS) due to its high sensitivity, large linear range, relatively low limit of detection (LOD) and simplicity of coupling.

The hyphenation of common techniques such as chromatography and atomic absorption or fluorescence spectrometry (AAS or AFS) is a substitute of MS techniques of great interest (Capelo et al. 2006). In fact, from the sensitivity viewpoint, the MS detector is not better than AFS for some elements (Chen and Belzile 2010). Thus, determination of Se species can be carried out using AAS or AFS detectors in connection with hydride generation (HG) techniques, through the initial on-line reduction of SeO_4^{2-} forms (Sánchez-Rodas et al. 2010). Additionally, the application of on-line mineralization for Org-Se compounds dissociation allows to determine Org-Se species in ground water by HG-AAS (Niedzielski 2005). While HPLC-HG-AFS is an inexpensive and affordable technique; the literature associated to HPLC-HG/Cold vapor generation (CVG)-AFS technique is relatively limited (Chen and Belzile 2010). The major reason why HPLC-ICP-MS is gaining popularity is likely due to the simplicity of its interface. Unlike HG/CVG-AFS, no oxidation and prereduction steps are required with ICP-MS, unless HG is introduced in the system (Chen and Belzile 2010). Extensive and complete reviews on the state of the art of this technique have been recently published, focusing on sample preparation, post-column treatments and on the applications of this technique to various liquid and solid samples (Chen and Belzile 2010, Sánchez-Rodas et al. 2010).

Although there has been significant progress in the chromatography-AS hyphenation, it is important to remember that chromatographic techniques represent only a minor part of the separation procedures available and, in certain cases, efforts must be made to apply basic chemistry for sample treatment, which could provide quantitative information about specific chemical forms. Thus, the development of speciation analysis based on non-chromatographic separation approaches is a promising research area.

Non-chromatographic techniques

In order to guarantee safety to operators and the environment, the green analytical chemistry principles are focused on simplification, reagent selection, major sample throughput, minimization of consumption and detoxification of wastes (de la Guardia and Garrigues 2011). These principles are represented on the next developed techniques for Se species extraction and preconcentration from environmental samples.

Solid phase extraction (SPE) has been extensively used for preconcentration of inorganic and some organic Se species. Although most of the proposed SPE techniques for Se species determination in environmental samples were focused on the inorganic forms, SeO_3^{2-} and SeO_4^{2-}, there are also some reports for SPE of the main seleno-aminoacids (SeMet and SeCys$_2$), total Se and Org-Se. Most of the methods reported have been validated for real matrix applications in fresh or mineral waters, while few methods are being applied to more complex samples such as soil, sediments, saline waters or air particulate material. Besides the typical SPE phases, extensively reviewed by Wake et al. (Wake et al. 2004), other sorbents have been used. Nanometer-sized TiO$_2$ particles have been used for preconcentration and separation of inorganic Se species in natural water and sludge samples (Li and Deng 2002). A major improvement related to the use of these nanoparticles is that once they adsorbed Se inorganic species (from natural water), the solid phase can be prepared to form a slurry for determination by electrothermal atomic absorption spectrometry (ETAAS) (Zhang et al. 2007) or the colloid can be determined directly by HG-AFS (Fu et al. 2012). Compared with the traditional methods, which require desorption process, the proposed methods appear to be more advantageous with short-time separation and preconcentration, simple operation, fast analysis, and lower detection limits. A direct speciation of dissolved inorganic and organic Se species in environmental samples by flow-injection (FI) dual-column preconcentration/separation on-line coupled to ICP-MS was proposed (Huang et al. 2008). The authors used two columns: one packed with nanometer-sized Al_2O_3 and the other packed with mesoporous TiO_2 chemically modified by dimercaptosuccinic acid (DMSA). Due to the different absorption behaviors of inorganic Se and Org-Se on the columns, a selective elution of SeO_3^{2-}, SeO_4^{2-}, SeMet and $SeCys_2$ was obtained. The proposed method was successfully applied for the speciation of dissolved inorganic and organic Se species in environmental and biological samples (Huang et al. 2008).

Other materials were also proposed for selective retention of Se species from environmental samples, such as cross-linked chitosan with diethylene triamine (selective for SeO_4^{2-}) (Dai et al. 2011); activated charcoals (AC) (selective for SeO_3^{2-} after reduction to Se^0 by ascorbic acid) (Bertolino et al. 2006); aminoacid L-methionine (selective for SeO_4^{2-}) covalently immobilized on controlled pore glass (Pacheco et al. 2007); or cetyltrimethylammonium bromide (CTAB)-modified alkyl silica sorbent (selective for SeO_4^{2-}) (Xiong et al. 2008). These materials were mainly used for Se species determination from river, pool, tap, lake, rain and well waters. Since no chelating reagents were needed when these materials were employed, the possibilities of contamination risks were reduced. SPE resins have been also used for the removal of matrix effects and storage of the analyte. For example, XAD-7 and XAD-8 resins were proposed as adsorbents of fulvic acids and will remove organic material, without retaining inorganic Se (Johansson et al. 1993).

Solid phase microextraction (SPME) is a relatively new form of SPE, in which a silica fiber, coated with a polymeric substance of choice, is placed into the sample or the headspace (HS) of sample vial to adsorb the analytes. HS-SPME with a polypyrrole-coated fiber was applied as a sample preparation method for determination of SeO_3^{2-} following derivatization with 1,2-diaminobenzene to convert into the piaselenol form and analysis by ion mobility spectrometry. The method was applied satisfactorily for determination of Se in human serum and environmental surface (spring and well) water samples (Shahdousti and Alizadeh 2011). Dietz et al. proposed a method for speciation of volatile Org-Se compounds (DMSe and DMDSe) using SPME, thermal desorption by an in-house developed desorption unit, and detection by different detectors such as AAS and ICP-MS, without an additional chromatographic system for species separation. The method was applied to quantification of Org-Se in garlic samples (Dietz et al. 2003).

Coprecipitation is suitable for the preconcentration of inorganic Se species, based on the formation of an insoluble metal hydroxide that acts as a collector, forming a colloid with the analyte, and thereby coprecipitation. Most regularly, lanthanum(III) is used, which precipitates as $La(OH)_3$ and coprecipitates SeO_3^{2-} (Tang et al. 2005). The use of knotted reactors (KR) has allowed collection of the precipitate in continuous flow and FI systems (Tao and Hansen 1994). An easy and simple approach was proposed by Wu et al. to eliminate the interference of DMSe and DMDSe occurred in some environmental samples during SeO_3^{2-} determination by HG-AFS due to their contribution on HG-response. Lanthanum nitrate was added to the sample, promoting the coprecipitation of SeO_3^{2-} and quantitative collection by a PTFE KR. DMSe and DMDSe, however, were unretained and expelled from the KR (Wu et al. 2007). Magnesium hydroxide and ferric hydroxide were also proposed for coprecipitation of SeO_3^{2-} prior to its determination by ETAAS in natural waters and microwave digested food samples with satisfactory results (Tuzen et al. 2007, Rao et al. 1996).

Regarding liquid phase microextraction (LPME) methods, single drop microextraction (SDME), hollow fiber LPME (HF-LPME), dispersive liquid-liquid microextraction (DLLME), and ultrasound-assisted emulsification microextraction (USAEME) were used for Se species separation/preconcentration. Mainly, the LPME proposed were evaluated for Se inorganic species microextraction from water (tap, lake, river and sea) and soil samples (Ghasemi et al. 2010, Martinis et al. 2011, Najafi et al. 2012, Xia et al. 2006, Bidari et al. 2007).

Several methods based on catalytic kinetic spectrophotometry have been also proposed. Based on catalytic effects of SeO_3^{2-}, lower-cost and FI catalytic kinetic spectrophotometric methods were suggested mainly to determine Se inorganic species from water samples (Nakano et al. 2004, Zhengjun et al. 2005, Gürkan and Ulusoy 2010). Moreover, Zhengjun et al. proposed a model ZJ-la automatic home-made metallic elements analyzer to monitor inorganic Se species *in situ* in seas and oceans (Zhengjun et al. 2005).

Voltammetry has been used for Se species determination due to its high sensitivity and selectivity. Additionally, low capital investment is required, and the equipment can be portable to be used in onboard-ship analysis. Most of the voltammetric techniques reported for Se refer to inorganic Se species (SeO_3^{2-} and SeO_4^{2-}) in water (tap and highly saline media) and soil samples (Korolczuk and Grabarczyk 2003, Ferri and Frasconi 2006, Rúriková and Kunáková 2000, de Carvalho et al. 1999, Profumo et al. 2001). Since SeO_3^{2-} is the Se species determined by voltammetry, selective methods are required to convert SeO_4^{2-} into SeO_3^{2-} before determination. The reduction of SeO_4^{2-} to SeO_3^{2-} was mainly performed using either concentrated HCl at high temperature or UV photolysis in alkaline solution (Bertolino et al. 2006, de Carvalho et al. 1999). Furthermore, the UV irradiation pretreatment step was also employed for diluted OM photo-oxidation, which may interfere not only in voltammetric determinations (do Nascimento et al. 2009) but also in HG-AAS determinations (Bujdos et al. 2000).

Ochsenkühn-Petropoulou and Tsopelas have proposed two different methods; the first one was developed for the separation and determination of SeO_3^{2-}, SeO_4^{2-}, DMDSe, DMSe, SeMet, and selenocystine ($SeCys_2$) based on voltammetric techniques. In the other method, a flow sheet was used for the separation and electrochemical determination of inorganic (SeO_3^{2-} and SeO_4^{2-}) and Org-Se species (selenourea (SeU), selenocystamine (SeCM), DMDSe, and DMSe). Since SeU and SeCM have very close half-wave potential when they are simultaneously analyzed by voltammetric techniques, an ion-exchange separation—based on a cation exchanger Purolite C 100 H—was developed. Both procedures were tested on a coal fly ash reference material and on a soil sample of Thermopyles Spa (Greece) (Ochsenkühn-Petropoulou and Tsopelas 2002, Ochsenkühn-Petropoulou and Tsopelas 2004).

Hyphenated instrumental techniques for determination and structural elucidation of selenium species

LC techniques

HPLC is a powerful separation technique which can be directly applied to non-volatile compounds of high and low molecular weight, provides a great versatility derived from different separation modes and can be easily "on-line" interfaced to ICP-MS (Pedrero and Madrid 2009). Separation of Se species is mainly developed using ion exchange, ion-pairing, reverse phase (RP) or size-exclusion (SEC). Nevertheless, speciation of Se using LC–ICP-MS is still limited to the availability of standards of Se species. As a consequence, techniques such as electrospray ionization mass spectrometry (ESI-MS) and LC-electrospray ionization tandem mass spectrometry (LC–ESI-MS-MS) are needed to identify the unknown Se species (Vonderheide et al. 2006).

Determination of Se by ICP-MS has, however, problems of moderate ionization efficiency and isobaric interferences. Significant isotopic overlap from $^{40}Ar^{40}Ar$ on the most abundant isotope ^{80}Se (49.6 percent) often requires monitoring the less abundant isotopes ^{82}Se (8.6 percent) or ^{77}Se (7.6 percent) (Bednar et al. 2009). Elements such as Br and Cl, present in environmental matrices, can also produce the interfering species $^{81}BrH^+$ and $^{37}Cl^{40}Ar^+$ that have the nominal mass of ^{82}Se and ^{77}Se, respectively (Simon et al. 2005,

Wallschläger and London 2004). Currently, the collision/reaction cell (C/RC) technology is used to overcome the most common interferences caused by polyatomic ions on Se quantification by ICP-MS in a variety of environmental and geological samples, including waters, simulated biological leachates from soils and wildfire ashes (Wolf et al. 2011, Bednar et al. 2009, Darrouzes et al. 2005), crude oil, aqueous process stream samples and wastewaters samples from petroleum refinery (Stivanin de Almeida et al. 2009).

Since Se chemistry often parallels S chemistry, it raises considerable analytical challenges as Se species are regularly present at three orders of magnitude below S (Uden 2002). SPE was used for samples pretreatment in order to improve LOD of Se and overcome matrix interference (Wrobel et al. 2005, Liu 2010). Wrobel et al. discussed general aspects related to collection, processing and storage strategies for Se speciation analysis in several types of samples, including environmental applications (Wrobel et al. 2005). Ultrasonic or pneumatic nebulizers with a desolvation system can also be used in order to improve LOD (Gettar et al. 2005, Castillo et al. 2008). Gettar et al. published an extensive evaluation of different nebulizer/expansion chamber combinations to assess their performance for sample introduction in the direct coupling with an axial view ICP multi element spectrometer for on-line determination of As and Se inorganic species from natural water samples without derivatization (Gettar et al. 2005). In another study, Castillo et al. evaluated the performance of microHPLC-microconcentric nebulizer-ICP-MS coupled for the simultaneous determination of As and Se species in water samples. Microbore columns revealed several advantages such as a significant diminution of sample and solvent consumption without sacrificing sensitivity and the overall resolution in faster analysis time (less than 5 min). Good accuracy and repeatability were obtained for spiked influent and effluent water treatment plant (Castillo et al. 2008).

Due to the fact that Se is a hydride-forming element, the LOD can be further improved by coupling vapor generation (VG) to LC–ICP-MS. In this case, the volatile species generated from the analyte species present in the eluent can be introduced directly into the plasma (Sun et al. 2006). Some novel devices were proposed as interfaces between HPLC and ICP for CVG. Vilano et al. used UV irradiation for SeO_3^{2-}, SeO_4^{2-}, SeCys, and SeMet determination, previous separation by an anionic exchange column. The method was validated by analyzing two water certified reference materials, in which only SeO_3^{2-} was detected (Vilano and Rubio 2000). An improvement was made by Sun et al., who developed an UV/nano-TiO_2 CVG device, coupled between a chromatographic column and ICP-MS for the determination of SeO_3^{2-} and SeO_4^{2-} in water samples without the need to use the conventional borohydride reduction reaction. As a result, the stability of the plasma was improved because no molecular hydrogen is delivered to the ICP-MS unit (Sun et al. 2006).

Regarding ion-pair RP chromatographic systems for Se speciation analysis, many chromatographic RP systems used trifluoroacetic acid (TFA) or pentanesulfonate anion as ionic pairing reagents for Se species separation (B'Hymer and Caruso 2000, Bird et al. 1997, Olivas et al. 1996). Pentafluoropropanoic acid (PFPA) and heptafluorobutanoic acid (HFBA) were used in order to gain improvement in resolution of many Org-Se species (Kotrebai et al. 2000, Montes-Bayón et al. 2002b). Methods using perfluoroaliphatic acid as pairing agent and ICP-MS Se-specific detection for the separation of selenoamino acids and anionic, cationic, and neutral Se compounds in Se-enriched yeast, garlic, ramps and mushrooms, etc., have been proven effective (Uden 2002). McSheehy et al. used two-dimensional HPLC with semi-preparative RP columns to isolate fractions of selenized yeast extracts for on-line and off-line investigations by ICP-MS (McSheehy et al. 2001). An interesting work used separation by ion-pairing HPLC–ICP-MS was published by Vonderheide et al., who investigated Se-containing root exudates in *Brassica juncea* plants. Chromatographic peaks unable to be identified by retention-time matching were collected for analysis by ESI-MS. Several Se-containing compounds were identified in the exudate-containing solution and two were identified as SeCys and the selenosulfate ($SeSO_3^{2-}$) ion. Even though DMSe was detected, its presence cannot be attributed exclusively to plant exudation since plants were not grown in sterile conditions (Vonderheide et al. 2006).

Ion-exchange HPLC may be used in either cation or anion-exchange modes, both of which have been used in Se speciation analysis. Anion exchange chromatography (AEC) has been widely applied for separation of inorganic Se species (SeO_3^{2-} and SeO_4^{2-}) in different environmental samples, such as soil, sediment and water samples (Petrov et al. 2012, Stroud et al. 2012, Orero Iserte et al. 2004, Jitmanee et al. 2007). Besides inorganic Se species, AEC-HG-ICP-C/RC-MS was proposed for the determination of SeO_3^{2-}, SeO_4^{2-}, and SeCN⁻ in rain and sea waters (Wallschläger and London 2004), and in effluents from a petroleum refinery (Miekeley et al. 2005), where SeCN⁻ was by far the most abundant Se species (Stivanin de Almeida et al. 2009). An AEC-ICP-MS technique was also proposed to monitor the microbial Se reduction for application in the removal of Se from industrial waste water. The column separation performance for the determination of Se was investigated for both inorganic (SeO_3^{2-} and SeO_4^{2-}) and some organic Se species. The microbial removal of Se from industrial waste water was at first hampered by the presence of nitrate (denitrification) and the partial reduction of SeO_4^{2-} was observed after the complete removal of SeO_3^{2-} to the metallic form (Tirez et al. 2000). In another approach using AEC–ICP-MS with an anion self-regenerating suppressor (ASRS), 13 Se species in flue gas desulfurization (FGD) waters from coal-fired power plants were separated and quantified. Both $SeSO_3^{2-}$ and $SeCN^-$ were conclusively identified in such samples for the first time, using retention time matching and confirmation by ESI-MS (Petrov et al. 2012). As shown in Fig. 5.2, the Se chemistry in FGD waters is much more complex than commonly assumed, i.e., not just an interconversion between SeO_3^{2-} and SeO_4^{2-}. Besides SeO_3^{2-} and SeO_4^{2-}, $SeSO_3^{2-}$ was the most frequently occurring Se species (up to 63 percent) of the total Se concentration in waters obtained from plants using inhibited oxidation scrubbers. $SeCN^-$ occurred in about half of the tested samples, but was only a minor species (up to 6 percent of the total Se concentration). Nine additional unidentified Se-containing compounds were found in FGD waters (Petrov et al. 2012).

Regarding cation exchange, it was used to separate 10 standard Org-Se compounds and was capable of separating 30 Se species occurring in the hydrolysates of Se enriched yeast (Larsen et al. 2003) and to investigate cationic Se species present in leaf extract of wild-type *Brassica juncea* supplemented with SeO_3^{2-}. Major cationic Se species identified were methylselenomethionine (MSeMet) and MSeCys, while SeMet was found in minor quantities (Yathavakilla et al. 2005).

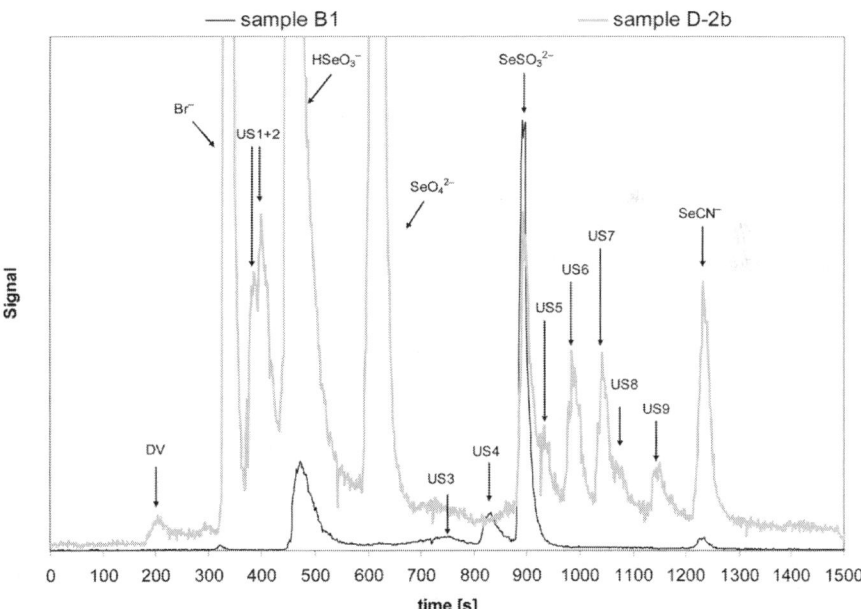

Figure 5.2 Analysis of 13 Se species in FGD waters by AEC-ASRS-ICP-DRC-MS. Chromatograms for two different samples are overlaid to show all Se species encountered. US = unidentified species; DV = unidentified species (eluting in dead volume). Peaks for SeO_3^{2-}, SeO_4^{2-} and bromide are off-scale, but within linear range, only trace for the main quantification isotope [80]Se is shown. Reprinted with permission from American Chemical Society.

By coupling two chromatographic methods, Ochsenkühn-Petropoulou et al. were able to determine SeU, selenoethionine (SeEt), SeMet, SeO_3^{2-}, SeO_4^{2-}, DMSe, and DMDSe. An AEC-ICP-MS was employed to separate and quantify SeU, SeEt, SeMet, SeO_3^{2-}, and SeO_4^{2-}; while an RP chromatography was used to separate MSe and DMDSe. This procedure was applied to a soil sample from the warm spring area of Thermopyles (Greece), and several extraction methods for leaching the Se species were evaluated. It was found that HCl method was the best extraction procedure for leaching. The investigated sediment sample was shown to contain SeU, SeEt, SeO_3^{2-}, and SeO_4^{2-}, while the concentrations of SeMet, DMSe, and DMDSe were below the LODs of the method (Ochsenkühn-Petropoulou et al. 2003).

Tolu et al. also proposed a mixed-mode chromatography (RP and AEC) coupled with ICP-MS to determine Se species present in soils at trace levels based on parallel single extractions (Tolu et al. 2011). Ultrapure water, phosphate buffer (pH 7) and NaOH were chosen as extractants owing to their efficiency in extracting Se and good compatibility with Se species stability. These extractants allowed assessing water-soluble Se (i.e., the most mobile Se fraction), exchangeable Se (i.e., sorbed onto soil component surface), and Se bound to soil organic matter. This on-line speciation analysis also highlighted the presence of dissolved and/or colloidal unknown compounds in some ultrapure water and NaOH soil extracts (Tolu et al. 2011).

Peachey and co-workers also proposed the mixed-mode (anion-exchange and RP) for the simultaneous retention and selective separation of a range of inorganic (SeO_3^{2-} and SeO_4^{2-}) and organically-bound Se species [SeCys$_2$, SeMCys, SeMet, methylseleninic acid (MSA), γ-glutamyl-SeMCys] in watercress extracts after enzymatic hydrolysis or leaching in water by accelerated solvent extraction (ASE). Both extracts seem to contain SeMCys, SeMet, and inorganic Se (Fig. 5.3). Three unknown Se peaks were also detected in the extracts (Peachey et al. 2009).

Size-exclusion HPLC is often used in Se speciation to characterize the various selenoproteins encountered from different sample matrices such as *Brassica juncea* (Indian mustard), used for phytoremediation (Mounicou et al. 2004). Besides ICP-MS or inductively coupled plasma optical emission spectrometry (ICP-OES), interest in couplings based on AFS has recently increased, since it is particularly efficient for trace analysis of As and Se. Simon et al. proposed the determination of SeO_3^{2-}, SeO_4^{2-}, SeCys$_2$, and SeMet from water and oyster samples by on-line coupling of HPLC-(UV decomposition)-HG-AFS (Simon et al. 2005). Moscoso-Pérez et al. proposed a pressurized liquid extraction (PLE) procedure using EDTA as extracting solvent as automated method for extracting As and Se inorganic species from atmospheric particulate matter. After extraction, concentrations of As and Se species were determined by HPLC–HG-AFS (Moscoso-Pérez et

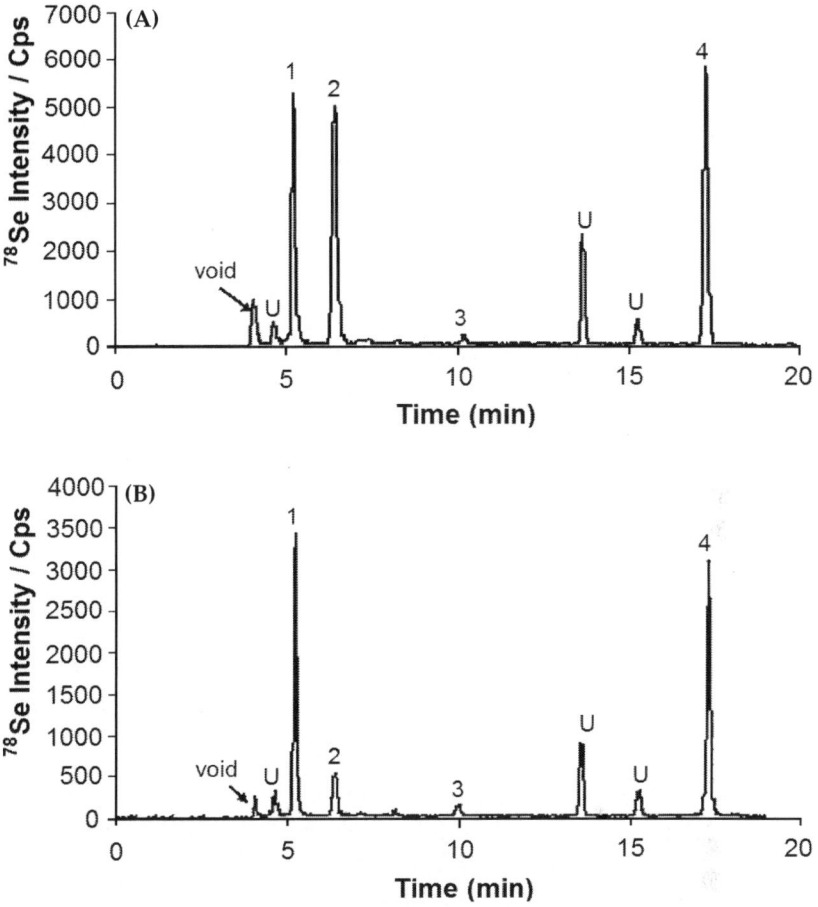

Figure 5.3 Reversed-phase in-line with mixed-mode HPLC-ICP-MS profile of ^{78}Se obtained for (A) an enzymatic extract and (B) an ASE water extract from watercress. Peak identity: SeMCys (1), SeMet (2), SeO$_3^{2-}$ (3), SeO$_4^{2-}$ (4), U = unknown. Reprinted with permission from Elsevier.

al. 2008). The same extraction, separation and determination procedures were used previously for the extraction of SeCys, SeMet, SeEt, SeO$_3^{2-}$, and SeO$_4^{2-}$ from spiked and native yeast (Gómez-Ariza et al. 2004). Wallschläger and Bloom used AEC–HG-AFS for SeCN$^-$, SeO$_3^{2-}$, and SeO$_4^{2-}$ determination. The method was applied to determine Se speciation in petroleum refinery waste water, where SeCN$^-$ was the only detected species, and in gold mine waste water, where smaller fractions of SeO$_3^{2-}$ and SeO$_4^{2-}$ were encountered, besides the major species SeCN$^-$ (Wallschläger et al. 1998).

GC techniques

Gas chromatography is highly suited for coupling with mass spectrometry, since compounds are already in the gas phase. The ionization of the gaseous molecules can be done either by electron impact or chemical ionization. In contrast to LC-MS, GC-MS did not find widespread use in Se speciation until now. GC-MS is normally used in combination with GC-ICP-MS for molecular identification and confirmation of Org-Se species. However, the application of microextraction and preconcentration techniques such as SPME and LPME allows both sensitive qualitative and quantitative analysis of volatile Se speciation. Thus, although sensitive analytical instrumentation might be available, Se speciation studies in environmental samples require the application of a preconcentration step. Besides SPME, different preconcentration techniques including, cryogenic trapping (CT) followed by thermal desorption (Tessier et al. 2002, de la Calle-Guntiñas et al. 1995, Pecheyran et al. 1998a, de la Calle-Guntiñas et al. 1999) or purge-and-trap (PT) without the cryogenic module (Campillo et al. 2005) have been used in combination with GC-based separation methods. SPME has proved to be a very useful choice for preconcentrating both inorganic (Guidotti 2000, Guidotti et al. 1999) and Org-Se compounds in environmental studies (Dietz et al. 2004b, 2004a, Lenz et al. 2008, Lenz et al. 2011, Meija et al. 2002). Recently, Ghasemi et al. have developed a simple and novel speciation method for the determination of DMSe and DMDSe combining HS-HF-LPME with capillary GC-MS. The effect of different variables on the extraction efficiency was studied simultaneously using an experimental design. The variables of interest in the HS-HF-LPME were sample volume, extraction time, temperature of sample solution, ionic strength, stirring rate and dwelling time. Under optimum conditions, preconcentration factors up to 1250 and 1170 were achieved for DMSe and DMDSe, respectively. The LOD and RSD for DMSe were 65 ng/L and 4.8 percent, respectively. They were also obtained for DMDSe as 57 ng/L and 3.9 percent, respectively. Despite the good LODs obtained with this method, this found sole application to spiked environmental (tap, river, lake and waste water) and several biological samples.

The occurrence of volatile Org-Se species in the atmosphere is mostly due to microbial activity. Bacterial activity might play a significant role in the biogeochemical cycling of Se given that an estimated 90 percent of all volatile Se emissions to the atmosphere are biogenic (Frankenberger and Engberg 1998). The current list of identified organoselenides originated from biological sources has been extended as toxicological interest and analytical methods are improved. Thus, several species such as hydrogen selenide, methaneselenol and DMDSe, were identified before the end of the last century (Doran and Alexander 1977, Chasteen et al. 1990). The generation of volatile, mixed S/Se species has been detected in the headspace of bacterial cultures. Moreover, dimethyl selenenyl sulfide (CH_3SeSCH_3, DMSeS) and dimethyl selenenyl disulfide ($CH_3SeSSCH_3$, DMSeDS) have been reported as well (Chasteen et al. 1990, Swearingen Jr et al. 2006). In an exhaustive GC-MS analysis, Burra et. al. have studied the production of volatile Org-Se species in the headspace of a metalloid-resistant bacterium identified as Bacillus sp. LHVE (Burra et al. 2010). Two novel compounds so far undetected in bacterial culture headspace, DMDSe and $CH_3SeSeSeCH_3$ (DMTSe) (Fig. 5.4) were identified by SPME-GC with either fluorine-induced chemiluminescence or MS. Differences in the EI[+] fragmentation pattern of the mixed S/Se compounds allowed differentiation between the symmetric and asymmetric isomers and $CH_3SeSeSCH_3$ (DMDSeS) isomer found in the headspace.

Interestingly, the possible formation of these novel species, i.e. DMDSeS and DMTSe, in biological systems was previously hypothesized (Meija and Caruso 2004). The formation of all possible Se- and S-containing trichalcogenide isomers (-SeSS-, -SSeS-, -SeSeS-, -SeSSe- and -SeSeSe-) was just observed at room temperature in synthetic solutions containing diselenide and trisulfide. Methyl and ethyl derivatives of these species were characterized using GC-TOFMS with electron impact, chemical and field ionization.

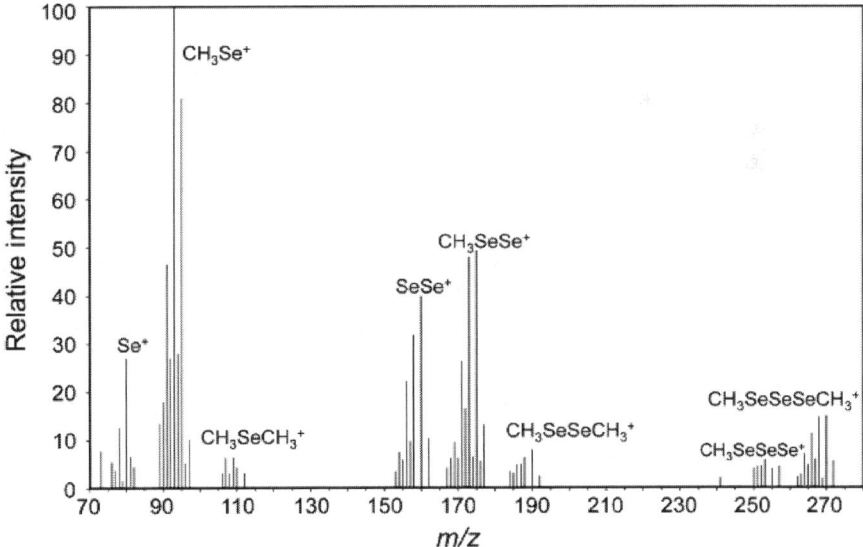

Figure 5.4 Mass spectrum of dimethyltriselenide (DMTSe) produced by *Bacillus* sp. LHVE amended with 1 mM SeO_3^{2-} from a GC/MS chromatographic peak eluting at 11.17 min. Reprinted with permission from Elsevier.

Gas chromatography with specific-elemental detection has been widely applied for analytical studies of volatile Se species. Among these are methylselenol (MSeH), DMSe and DMDSe by using different detectors (ICP-MS, MIP-AES and AFS). The combination of hyphenated instrumental techniques with SPME for extraction and preconcentration of volatile Se species has served as a useful tool for developing several studies. Most works using GC have been oriented to study the metabolism of Se in plants, which is of great interest due to generation of volatile species emitted to the environment and the potential application for phytoremediation of contaminated soils. For example, Carusos's group (Meija et al. 2002) performed determination of several volatile alkylselenides and their sulfur analogues by SPME-GC-ICP-MS. DMSe and DMDSe were the primary volatile species found in the headspace of a closed vial containing *Brassica juncea* seedlings. More recently, the complementary use of atomic and molecular mass spectrometric techniques was demonstrated for identification of yet unreported minor headspace Se-containing volatiles such as DMTSe, $CH_3SeSSeCH_3$, and $CH_3SeCH_2CH_3$ (Kubachka et al. 2007). Likewise, Dietz et al. use SPME prior to GC-ICP-MS or GC-MIP-AES to determine volatile Se compounds in lupine, yeast, Indian mustard, and garlic formed during growing in Se-enriched media (Dietz et al. 2004b, Dietz et al. 2003). GC has also been applied to non-volatile selenocompounds after derivatization. Derivatization with methyl chloroformiate and with cyanogen bromide (CNBr) has been employed to determine selenoamino acids (Yang et al. 2004). Vonderheide et al. used SPME to extract volatile compounds after derivatization by isopropylchloroformate obtaining sub-ppb detection limits (Vonderheide et al. 2002). The derivatized selenoamino acids were chromatographed on a 5% phenyl 95% methyl silicone (HP-5) column using ICP-MS detection. However, derivatization of non-volatile Se species for GC analytical separation is a time-consuming method and chemical stability of species could be compromised (B'Hymer and Caruso 2006). Likewise, the derivatization approach is not amenable to the analyses of peptides and proteins due to their thermal lability. Therefore, selenoamino acids and selenopeptides are preferably determined with LC as the separation method coupled to one of the elemental detectors already mentioned before, but mostly ICP-MS.

Sediment, air and water samples have often been analyzed. Donard and his co-workers (Amouroux et al. 1998, Amouroux et al. 2001, Tessier et al. 2002) have often utilized a purge and CT system previously published by Pechyran et al. (Pecheyran et al. 1998b) for the analysis of the various volatile dimethyl Se compounds. *In situ* sampling of the metalloid species was performed by collection of compounds in CT stored at −196°C until analysis in the laboratory. Cryogenic traps were then introduced into a flash desorption, cryofocusing GC-ICP-MS. More recently, CT-GC-ICP-MS instrumental set-up has been used to study volatile species of Se and Te produced by fermentation during compost of duck manure (Pinel-Raffaitin et al. 2008). In this case, the authors explored the use of quantitative structure-activity relationship (QSAR) model to identify or semi-identify the volatile species through the correlation between boiling point and retention time of each compound. The model however showed limitations for identification of the heaviest alkylated species of Se and Te within the series of compounds under study, which was assigned to reduced chromatographic resolution or high complexity of the compost gas mixture.

Gas chromatography with atomic plasma emission detection (GC-AED) and GC–microwave-induced plasma (MIP)-AED are well recognized and suitable analytical techniques to develop speciation studies related to volatile Se compounds in environmental and biological samples (Campillo et al. 2005, Campillo et al. 2007). These techniques have found widespread application for determination of Org-Se species present in or produced by plants (Arnault and Auger 2006, Uden et al. 2004, Uden et al. 1998, Cornelis et al. 2005). A very useful and less expensive alternative to ICP-based detectors for Se speciation in environmental samples has been MIP-AED (Dietz et al. 2004b, 2004a, Dimitrakakis et al. 2004, Lobinski and Adams 1993). Other detection techniques have included AFS, ETAAS, MS and photoionization (PID) (de la Calle-Guntiñas et al. 1999, Moreno et al. 2003, de la Calle-Guntiñas et al. 1997). Thus, selectivity and higher sensitivity are notorious advantages shown by MIP-AED with respect to ICP-MS and AFS detection for Se speciation analysis. This point has been illustrated in a work done by Cámara's group using SPME-MC-MIP-AED for evaluation of Se speciation in the headspace of growing plants (lupine and Indian mustard) (Dietz et al. 2004b). Detection limits obtained were 0.57, 0.47, and 0.19 ng/mL for DMSe, DEtSe, and DMDSe, respectively. In a different study, involving the determination of SeO_3^{2-}, DMSe, and DMDSe in several natural and tap water samples, the application of two derivatizing agents (sodium tetraethylborate and 4,5-dichloro-1,2-phenylenediamine) used to complex the inorganic species was evaluated by SPME-GC-AED (Campillo et al. 2007). Direct immersion (DI) mode and a relatively long extraction time (20 min) were selected for the method, with better sensitivity achieved for the three species (Fig. 5.5). Furthermore, DMSe and SeO_3^{2-} were found in several of the water samples analyzed at concentrations of 0.07–1.0 ng/mL.

CE techniques

Capillary electrophoresis represents a valuable alternative or complementary technique to HPLC due to short analysis time, low sample consumption and high separation efficiency. Due to the lack of interactions between the analytes and a stationary phase, the equilibrium between different species of

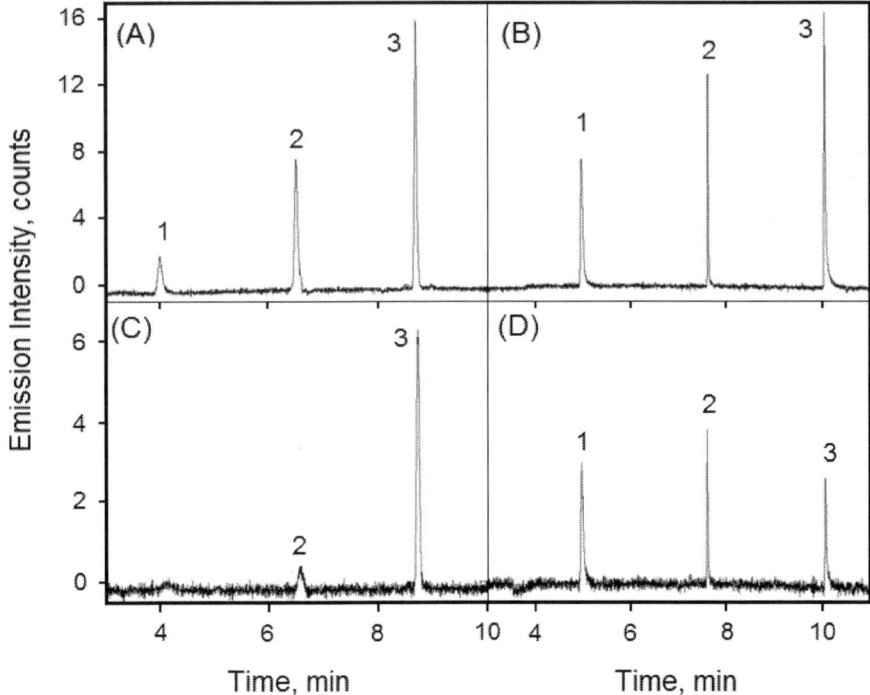

Figure 5.5 SPME-GC-AED chromatograms obtained from (A and B) a standard mixture (0.9, 0.15 and 0.6 ng/mL of DMSe, DMDSe and SeO_3^{2-}, respectively), (C) mining water 1 and (D) a seawater sample (containing 0.3, 0.05 and 0.07 ng/mL of DMSe, DMDSe and SeO_3^{2-}, respectively), when using sodium tetraborate (A and C) and 4,5-dichloro-1,2-phenylenediamine (B and D). 1: DMSe; 2:DMDSe; 3: SeO_3^{2-}. Reprinted with permission from Elsevier.

a given sample is undisturbed (Dzierzgowska et al. 2003). These advantages of CE must be considered, given the need for ultrasensitive detection, such as high resolution ICP-MS, because of the small sample amount injected (Pyrzynska 2001). However, one of the main problems to consider is the design of an interface due to the small volumes of injected sample and low flow-rate through the capillary (Michalke 2003). Commercial and home-made interfaces have been proposed for capillary zone electrophoresis (CZE)-ICP-MS in order to optimize electrophoretic and nebulizer flows with minimal dilution and sample consumption (Casiot et al. 2002, Gine et al. 2002, Schaumlöffel and Prange 1999, Yanes and Miller-Ihli 2003). Casiot et al. tested different nebulizers for Se species (SeO_3^{2-}, SeO_4^{2-}, SeMet, SeCys₂), As species, antimonate and Te species simultaneously determined by CZE-ICP-MS. The optimization of CE-ICP-MS interface operating parameters was discussed for each nebulizer-interface combination. Different nebulizer gases and liquid sheath flow rates were studied in detail and it was observed that they hardly affect electrophoretic resolution and peak width. The best analytical performance characteristics were obtained with the MicroMist nebulizer. Prior CE determination, soil samples were extracted using ultrapure water in order to obtain the "water soluble" fraction of the metalloids and, therefore, an indirect evaluation of their bioavailability. Analysis of soil extracts showed that it was possible to apply this technique to real samples (Casiot et al. 2002).

An extensive report on Se speciation studies based on CE techniques was published by Morales et al. They reviewed the different Se species determined, the samples analyzed and the detectors employed (Morales et al. 2008). Zhang et al. proposed a pressure-assisted electrokinetic injection (PAEKI) in CE couple with ESI-MS/MS for on-line preconcentration of As, Se and Br inorganic species in water samples. PAEKI is a sample preconcentration method which utilizes principles of both countercurrent electroconcentration and stacking. On-line enrichment of the target analytes was achieved with 1–3 ng/mL detection limits, which was below the maximum contaminant levels in drinking water for all five anions studied (Zhang et al. 2011).

Nutritional tablets based on selenized yeast were analyzed by CZE using capillaries coated dynamically with poly(vinyl sulfonate) and detected by ICP-MS. More than twenty different Se compounds were separated in the extract, with LODs better than 15 µg/L (Bendahl and Gammelgaard 2004). Mounicou et al. used a two-dimensional separation approach for selenized yeast speciation, based on size exclusion followed by CZE-ICP-MS, coupling to ICP via a self-aspirating total consumption nebulizer. Detection limits of low molecular mass Se species were in the range 7–18 µg/L, but problems were encountered with recovery of high molecular mass Se species from the CZE capillary (Mounicou et al. 2002). A C/RC ICP-MS was used as a CE detector for the speciation analysis of As and Se. Samples containing As species, SeO_3^{2-}, SeO_4^{2-}, SeCys, SeMet, and SeMCys were subjected to electrophoretic separation before being introduced into the microconcentric nebulizer (CEI-100) for their determination by ICP-MS. The LOD for As and Se was in the range of 0.6–1.8 ng/mL, and 0.5–1.4 ng/mL, respectively, based on peak height. This method has been applied in coal fly ash and dog fish liver reference materials and a Se dietary supplement (Hsieh et al. 2010).

Conclusions

Selenium is a toxic or essential element depending on its concentration, but more important on its chemical forms or species. Likewise, only bioaccessible species are taken up by organisms of a specific ecosystem. The great variety of Se species and their possible occurrence as organic and inorganic compounds gives Se a very complex behavior, and several species could be simultaneously present in an environment. Also, new issues and concerns are rising from growing development of nanotechnology and the presence and final destinations of elemental Se-NPs in the environmental compartments could be affected by its possible association to other toxic elements or compounds or even the modification of Se bioaccessibility and bioavailability by its association to others NPs. Therefore, speciation analysis of Se in environmental samples is not usually an easy task, and future analytical developments need to be considered in order to face the environmental study of coming issues.

Although there has been significant evolution of analytical chemistry to detect several Se species in complex environmental or biological matrices, both qualitative and quantitative analysis of speciation at very low Se levels remain challenging. High sensitive and interference-free analysis of speciation can be achieved by preconcentration, sample clean-up and separation of Se species. Moreover, derivatization with a variety of chemical reagents has found practical application along with GC-based methods. However, preservation of primitive Se speciation in the samples must be assured by these procedures. Finally, hyphenated instrumental techniques such as LC-ICP-MS and GC-ICP-MS, in combination with molecular characterization and identification of species by MS, allows high throughput and comprehensive data acquisition to perform Se speciation studies in the environment.

Acknowledgements

The authors would like to thank Atiya Jordan, from the Warner Research Group of Louisiana State University, Baton Rouge, Louisiana; for her valuable revision of the English language.

References

Al-Kunani, A.S., R. Knight, S.J. Haswell, J.W. Thompson and S.W. Lindow. 2001. The selenium status of women with a history of recurrent miscarriage. British J. Obst. Gyn. 108: 1094–1097.

Amouroux, D., E. Tessier, C. Pecheyran and O.F.X. Donard. 1998. Sampling and probing volatile metal(loid) species in natural waters by *in situ* purge and cryogenic trapping followed by gas chromatography and inductively coupled plasma mass spectrometry (P-CT-GC-ICP/MS). Anal. Chim. Acta. 377: 241–254.

Amouroux, D., P.S. Liss, E. Tessier, M. Hamren-Larsson and O.F.X. Donard. 2001. Role of oceans as biogenic sources of selenium. Earth Planet. Sci. Lett. 189: 277–283.

Amweg, E.L., D.L. Stuart and D.P. Weston. 2003. Comparative bioavailability of selenium to aquatic organisms after biological treatment of agricultural drainage water. Aquat. Toxicol. 63: 13–25.

Arnault, I. and J. Auger. 2006. Seleno-compounds in garlic and onion. J. Chromatog. A 1112: 23–30.

B'Hymer, C. and J.A. Caruso. 2000. Evaluation of yeast-based selenium food supplements using high performance liquid chromatography and inductively coupled plasma mass spectrometry. J. Anal. Atom. Spectrom. 15: 1531–1539.

B'Hymer, C. and J.A. Caruso. 2006. Selenium speciation analysis using inductively coupled plasma-mass spectrometry. J. Chromatogr. A 1114: 1–20.

Bañuelos, G.S. 2001. The green technology of selenium phytoremediation. BioFactors 14: 255–260.

Bañuelos, G.S. 2006. Phyto-products may be essential for sustainability and implementation of phytoremediation. Environ. Pollut. 144: 19–23.

Bañuelos, G.S., Z.Q. Lin, L. Wu and N. Terry. 2002. Phytoremediation of selenium-contaminated soils and waters: fundamentals and future prospects. Rev. Environ. Health 17: 291.

Barceló, J. and C. Poschenrieder. 2011. Hyperaccumulation of trace elements: from uptake and tolerance mechanisms to litter decomposition; selenium as an example. Plant Soil. 341: 31–35.

Barillas, J.R.V., C.F. Quinn and E.A. H. Pilon-Smits. 2011. Selenium accumulation in plants-phytotechnological applications and ecological implications. Int. J. Phytoremediat. 13: 166–178.

Bednar, A.J., R.A. Kirgan and W.T. Jones. 2009. Comparison of standard and reaction cell inductively coupled plasma mass spectrometry in the determination of chromium and selenium species by HPLC-ICP-MS. Anal. Chim. Acta. 632: 27–34.

Bendahl, L. and B. Gammelgaard. 2004. Separation of selenium compounds by CE-ICP-MS in dynamically coated capillaries applied to selenized yeast samples. J. Anal. At. Spectrom. 19: 143–148.

Bertolino, F.A., A.A.J. Torriero, E. Salinas, R. Olsina, L.D. Martinez and J. Raba. 2006. Speciation analysis of selenium in natural water using square-wave voltammetry after preconcentration on activated carbon. Anal. Chim. Acta. 572: 32–38.

Bidari, A., E. Zeini Jahromi, Y. Assadi and M.R. Milani Hosseini. 2007. Monitoring of selenium in water samples using dispersive liquid-liquid microextraction followed by iridium-modified tube graphite furnace atomic absorption spectrometry. Microchem. J. 87: 6–12.

Bird, S.M., H. Ge, P.C. Uden, J.F. Tyson, E. Block and E. Denoyer. 1997. High-performance liquid chromatography of selenoamino acids and organo selenium compounds. Speciation by inductively coupled plasma mass spectrometry. J. Chromatogr. A 789: 349–359.

Bujdos, M., J. Kubová and V. Stresko. 2000. Problems of selenium fractionation in soils rich in organic matter. Anal. Chim. Acta. 408: 103–109.

Burau, R.G. 1985. Environmental chemistry of selenium. Calif. Agr. 39: 16–18.

Burra, R., G.A. Pradenas, R.A. Montes, C.C. Vasquez and T.G. Chasteen. 2010. Production of dimethyl triselenide and dimethyl diselenenyl sulfide in the headspace of metalloid-resistant Bacillus species grown in the presence of selenium oxyanions. Anal. Biochem. 396: 217–222.

Campillo, N., N. Aguinaga, P. Viñas, I. López-García and M. Hernández-Córdoba. 2005. Gas chromatography with atomic emission detection for dimethylselenide and dimethyldiselenide determination in waters and plant materials using a purge-and-trap preconcentration system. J. Chromatog. A 1095: 138–144.

Campillo, N., R. Peñalver, M. Hernández-Córdoba, C. Pérez-Sirvent and M.J. Martíınez-Sánchez. 2007. Comparison of two derivatizing agents for the simultaneous determination of selenite and organoselenium species by gas chromatography and atomic emission detection after preconcentration using solid-phase microextraction. J. Chromatog. A 1165: 191–199.

Capelo, J.L., C. Fernandez, B. Pedras, P. Santos, P. Gonzalez and C. Vaz. 2006. Trends in selenium determination/speciation by hyphenated techniques based on AAS or AFS. Talanta. 68: 1442–1447.

Caruso, J.A. and M. Montes-Bayon. 2003. Elemental speciation studies—New directions for trace metal analysis. Ecotox. Environ. Safe. 56: 148–163.

Casiot, C., O.F.X. Donard and M. Potin-Gautier. 2002. Optimization of the hyphenation between capillary zone electrophoresis and inductively coupled plasma mass spectrometry for the measurement of As-, Sb-, Se- and Te-species, applicable to soil extracts. Spectrochim. Acta B 57: 173–187.

Castillo, A., A.F. Roig-Navarro and O.J. Pozo. 2008. Capabilities of microbore columns coupled to inductively coupled plasma mass spectrometry in speciation of arsenic and selenium. J. Chromatogr. A 1202: 132–137.

Chasteen, T.G., G.M. Silver, J.W. Birks and R. Fall. 1990. Fluorine-induced chemiluminescence detection of biologically methylated tellurium, selenium, and sulfur compounds. Chromatographia. 30: 181–185.

Chen, Y.-W. and N. Belzile. 2010. High performance liquid chromatography coupled to atomic fluorescence spectrometry for the speciation of the hydride and chemical vapour-forming elements As, Se, Sb and Hg: A critical review. Anal. Chim. Acta. 671: 9–26.

Combs, Jr., G.F., C. Garbisu, B.C. Yee, A. Yee, D.E. Carlson, N.R. Smith, A.C. Magyarosy, T. Leighton and B.B. Buchanan. 1996. Bioavailability of selenium accumulated by selenite-reducing bacteria. Biol. Trace Elem. Res. 52: 209–225.

Cornelis, R., J.A. Caruso, H. Crews and K.G. Heumann. 2003. Handbook of Elemental Speciation: Techniques and Methodology,. vol. 1. John Wiley & Sons Ltd., Chichester.

Cornelis, R., J.A. Caruso, H. Crews and K.G. Heumann. 2005. Handbook of Elemental Speciation II—Species in the Environment, Food, Medicine and Occupational Health, vol. 2. John Wiley & Sons Ltd., Chichester.

Coudray, C., H. Hida, F. Boucher, V. Tirard, J. De Leiris and A. Favier. 1996. Effect of selenium supplementation on biological constants and antioxidant status in rats. J. Trace Elem. Med. Biol. 10: 12–19.

Cutter, G.A. 1992. Kinetic controls on metalloid speciation in seawater. Mar. Chem. 40: 65–80.

Dai, J., F.L. Ren, C.Y. Tao and Y. Bai. 2011. Synthesis of cross-linked chitosan and application to adsorption and speciation of Se(VI) and Se(IV) in environmental water samples by inductively coupled plasma optical emission spectrometry. Int. J. Mol. Sci. 12: 4009–4020.

Darrouzes, J., M. Bueno, G. Lespes and M. Potin-Gautier. 2005. Operational optimization of ICP-octopole collision/reaction cell-MS for applications to ultratrace selenium total and speciation determination. J. Anal. Atom. Spectrom. 20: 88–94.

de Carvalho, M.L., G. Schwedt, G. Henze and S. Sander. 1999. Redox speciation of selenium in water samples by cathodic stripping voltammetry using an automated flow system. Analyst. 124: 1803–1809.

de la Calle-Guntiñas, M., R. Lobinski and F.C. Adams. 1995. Interference-free determination of selenium(IV) by capillary gas chromatography-microwave-induced plasma-atomic emission- spectrometry after volatilization with sodium tetraethylborate. J. Anal. Atom. Spectrom. 10: 111–115.

de la Calle-Guntiñas, M., C. Brunori, R. Scerbo, S. Chiavarini, P. Quevauviller, F. Adams and R. Morabito. 1997. Determination of selenomethionine in wheat samples: Comparison of gas chromatography-microwave-induced plasma atomic emission spectrometry, gas chromatography-flame photometric detection and gas chromatography-mass spectrometry. J. Anal. Atom. Spectrom. 12: 1041–1046.

de la Calle-Guntiñas, M., F. Laturnus and F.C. Adams. 1999. Purge and trap/thermal desorption device for the determination of dimethylselenide and dimethyldiselenide. Fresenius J. Anal. Chem. 364: 147–153.

de la Guardia, M. and S. Garrigues. 2011. An Ethical Commitment and an Economic Opportunity. *In:* M.D.L. Guardia and S. Garrigues [eds.]. Challenges in Green Analytical Chemistry. RSC Publishing, Cambridge, UK. pp. 1–12.

de Souza, M.P., C.M. Lytle, M.M. Mulholland, M.L. Otte and N. Terry. 2000. Selenium assimilation and volatilization from dimethylselenoniopropionate by Indian Mustard. Plant Physiol. 122: 1281–1288.

Dhillon, K.S. and S.K. Dhillon. 1999. Adsorption-desorption reactions of selenium in some soils of India. Geoderma. 93: 19–31.

Dhillon, S.K. and K.S. Dhillon. 2009. Phytoremediation of selenium-contaminated soils: the efficiency of different cropping systems. Soil Use Manage. 25: 441–453.

Di Gregorio, S., S. Lampis, F. Malorgio, G. Petruzzelli, B. Pezzarossa and G. Vallini. 2006. *Brassica juncea* can improve selenite and selenate abatement in selenium contaminated soils through the aid of its rhizospheric bacterial population. Plant Soil. 285: 233–244.

Dietz, C., T. Perez-Corona, Y. Madrid-Albarran and C. Camara. 2003. SPME for on-line volatile organo-selenium speciation. J. Anal. Atom. Spectrom. 18: 467–473.

Dietz, C., J.S. Landaluze, P. Ximenez-Embun, Y. Madrid-Albarran and C. Camara. 2004a. SPME-multicapillary GC coupled to different detection systems and applied to volatile organo-selenium speciation in yeast. J. Anal. Atom. Spectrom. 19: 260–266.

Dietz, C., J.S. Landaluze, P. Ximenez-Embun, Y. Madrid-Albarran and C. Camara. 2004b. Volatile organo-selenium speciation in biological matter by solid phase microextraction-moderate temperature multicapillary gas chromatography with microwave induced plasma atomic emission spectrometry detection. Anal. Chim. Acta. 501: 157–167.

Dimitrakakis, E., C. Haberhauer-Troyer, Y. Abe, M. Ochsenkühn-Petropoulou and E. Rosenberg. 2004. Solid-phase microextraction-capillary gas chromatography combined with microwave-induced plasma atomic-emission spectrometry for selenite determination. Anal. Bioanal. Chem. 379: 842–848.

do Nascimento, P.C., C.L. Jost, L.M. de Carvalho, D. Bohrer and A. Koschinsky. 2009. Voltammetric determination of Se(IV) and Se(VI) in saline samples: Studies with seawater, hydrothermal and hemodialysis fluids. Anal. Chim. Acta. 648: 162–166.

Doran, J. and M. Alexander. 1977. Microbial transformations of selenium. Appl. Environ. Microbiol. 33: 31–37.

Dowling, D.N. and S.L. Doty. 2009. Improving phytoremediation through biotechnology. Curr. Opin. Biotech. 20: 204–206.

Dzierzgowska, M., K. Pyrzyñska and E. Pobozy. 2003. Capillary electrophoretic determination of inorganic selenium species. J. Chromatogr. A 984: 291–295.

El Mehdawi, A.F. and E.A.H. Pilon-Smits. 2011. Ecological aspects of plant selenium hyperaccumulation. Plant Biol. 14: 1–10.

Ema, M., N. Kobayashi, M. Naya, S. Hanai and J. Nakanishi. 2010. Reproductive and developmental toxicity studies of manufactured nanomaterials. Reprod. Toxicol. 30: 343–352.

Farré, M., K. Gajda-Schrantz, L. Kantiani and D. Barceló. 2009. Ecotoxicity and analysis of nanomaterials in the aquatic environment. Anal. Bioanal. Chem. 393: 81–95.

Feng, R.W. and C.Y. Wei. 2012. Antioxidative mechanisms on selenium accumulation in Pteris vittata L., a potential selenium phytoremediation plant. Plant Soil Environ. 58: 105–110.

Fernández-Martínez, A. and L. Charlet. 2009. Selenium environmental cycling and bioavailability: a structural chemist point of view. Rev. Environ. Sci. Biotechnol. 8: 81–110.

Ferri, T. and M. Frasconi. 2006. Determination of Se(IV) and Se(VI) in Italian mineral waters. Ann. Chim. 96: 647–656.

Floor, G.H. and G. Román-Ross. 2012. Selenium in volcanic environments: A review. Appl. Geochem. 27: 517–531.

Frankenberger, J.W.T. and R.A. Engberg. 1998. Environmental Chemistry of Selenium. Marcel Dekker, New York.

Frankenberger, W.T., C. Amrhein, T.W.M. Fan, D. Flaschi, J. Glater, E. Kartinen, K. Kovac, E. Lee, H.M. Ohlendorf, L. Owens, N. Terry and A. Toto. 2004. Advanced treatment technologies in the remediation of seleniferous drainage waters and sediments. Irrigat. Drain Syst. 18: 19–42.

Fu, J., X. Zhang, S. Qian and L. Zhang. 2012. Preconcentration and speciation of ultra-trace Se (IV) and Se (VI) in environmental water samples with nano-sized TiO$_2$ colloid and determination by HG-AFS. Talanta. 94: 167–171.

Gadd, G.M. 2010. Metals, minerals and microbes: geomicrobiology and bioremediation. Microbiology. 156: 609–643.

Gettar, R.T., P. Smichowski, R.N. Garavaglia, S. Farías and D.A. Batistoni. 2005. Effect of nebulizer/spray chamber interfaces on simultaneous, axial view inductively coupled plasma optical emission spectrometry for the direct determination of As and Se species separated by ion exchange high-performance liquid chromatography. Spectrochim. Acta B 60: 567–573.

Ghasemi, E., N.M. Najafi, F. Raofie and A. Ghassempour. 2010. Simultaneous speciation and preconcentration of ultra traces of inorganic tellurium and selenium in environmental samples by hollow fiber liquid phase microextraction prior to electrothermal atomic absorption spectroscopy determination. J. Hazard. Mater. 181: 491–496.

Gine, M.F., A.P.G. Gervasio, A.F. Lavorante, C.E.S. Miranda and E. Carrilho. 2002. Interfacing flow injection with capillary electrophoresis and inductively coupled plasma mass spectrometry for Cr speciation in water samples. J. Anal. Atom. Spectrom. 17: 736–738.

Gómez-Ariza, J.L., M.A. Caro de la Torre, I. Giráldez and E. Morales. 2004. Speciation analysis of selenium compounds in yeasts using pressurised liquid extraction and liquid chromatography-microwave-assisted digestion-hydride generation-atomic fluorescence spectrometry. Anal. Chim. Acta. 524: 305–314.

Gonzálvez, A., M.L. Cervera, S. Armenta and M. de la Guardia. 2009. A review of non-chromatographic methods for speciation analysis. Anal. Chim. Acta. 636: 129–157.

Gonzálvez, A., S. Armenta, M.L. Cervera and M. de la Guardia. 2010. Non-chromatographic speciation. Trac-Trends Anal. Chem. 29: 260–268.

Guidotti, M. 2000. Determination of Se4+ in Drinkable Water by Solid-Phase Microextraction and Gas Chromatography/Mass Spectrometry. J. AOAC Int. 83: 1082–1085.

Guidotti, M., G. Ravaioli and M. Vitali. 1999. Selective determination of Se4+ and Se6+ using SPME and GC/MS. J. High Resolut. Chromatogr. 22: 414–416.

Gürkan, R. and H.İ. Ulusoy. 2010. The Investigation of a Novel Indicator System for Trace Determination and Speciation of Selenium in Natural Water Samples by Kinetic Spectrophotometric Detection. Bull. Korean Chem. Soc. 31: 1907–1914.

Hamilton, S.J. 2004. Review of selenium toxicity in the aquatic food chain. Sci. Total Environ. 326: 1–31.

Holben, D.H. and A.M. Smith. 1999. The diverse role of selenium within selenoproteins: A review. J. Am. Diet. Assoc. 99: 836–843.

Hsieh, M.-W., C.-L. Liu, J.-H. Chen and S.-J. Jiang. 2010. Speciation analysis of arsenic and selenium compounds by CE-dynamic reaction cell-ICP-MS. Electrophoresis. 31: 2272–2278.

Huang, C., B. Hu, M. He and J. Duan. 2008. Organic and inorganic selenium speciation in environmental and biological samples by nanometer-sized materials packed dual-column separation/preconcentration on-line coupled with ICP-MS. J. Mass Spectrom. 43: 336–345.

Jitmanee, K., N. Teshima T. Sakai, and K. Grudpan. 2007. DRC™ ICP-MS coupled with automated flow injection system with anion exchange minicolumns for determination of selenium compounds in water samples. Talanta. 73: 352–357.

Johansson, K., U. Ornemark and A. Olin. 1993. Solid-phase extraction procedure for the determination of selenium by capillary gas chromatography. Anal. Chim. Acta. 274: 129–140.

Korolczuk, M. and M. Grabarczyk. 2003. Determination of Se(IV) in on-line system by cathodic stripping voltammetry. Electroanal. 15: 821–826.

Kotrba, P., J. Najmanova, T. Macek, T. Ruml and M. Mackova. 2009. Genetically modified plants in phytoremediation of heavy metal and metalloid soil and sediment pollution. Biotechnol. Adv. 27: 799–810.

Kotrebai, M., J.F. Tyson, E. Block and P.C. Uden. 2000. High-performance liquid chromatography of selenium compounds utilizing perfluorinated carboxylic acid ion-pairing agents and inductively coupled plasma and electrospray ionization mass spectrometric detection. J. Chromatogr. A 866: 51–63.

Kubachka, K.M., J. Meija, D.L. Leduc, N. Terry and J.A. Caruso. 2007. Selenium volatiles as proxy to the metabolic pathways of selenium in genetically modified Brassica juncea. Environ. Sci. Technol. 41: 1863–1869.

Lampis, S., A. Ferrari, A. Cristina, A.C.F. Cunha-Queda, P. Alvarenga, S. Di Gregorio and G. Vallini. 2009. Selenite resistant rhizobacteria stimulate SeO$_3^{2-}$ phytoextraction by Brassica juncea in bioaugmented water-filtering artificial beds. Environ. Sci. Pollut. R. 16: 663–670.

Larsen, E.H., J. Sloth, M. Hansen and S. Moesgaard. 2003. Selenium speciation and isotope composition in ^{77}Se-enriched yeast using gradient elution HPLC separation and ICP-dynamic reaction cell-MS. J. Anal. Atom. Spectrom. 18: 310–316.

LeDuc, D. and N. Terry. 2005. Phytoremediation of toxic trace elements in soil and water. J. Ind. Microbiol. Biot. 32: 514–520.

Lei, C., X. Niu, X. Ma and J. Wei. 2011. Is selenium deficiency really the cause of Keshan disease? Environ. Geochem. Health. 33: 183–188.

Lemly, A.D. 1996. Assessing the toxic threat of selenium to fish and aquatic birds. Environ. Monit. Assess. 43: 19–35.

Lemly, A.D. 1998. A position paper on selenium in ecotoxicology: A procedure for deriving site-specific water quality criteria. Ecotox. Environ. Safe. 39: 1–9.

Lemly, A.D. 1999. Selenium transport and bioaccumulation in aquatic ecosystems: A proposal for water quality criteria based on hydrological units. Ecotox. Environ. Safe. 42: 150–156.

Lemly, A.D. 2004. Aquatic selenium pollution is a global environmental safety issue. Ecotox. Environ. Safe. 59: 44–56.

Lenz, M., A. Gmerek and P.N.L. Lens. 2006. Selenium speciation in anaerobic granular sludge. Int. J. Environ. Anal. Chem. 86: 615–627.

Lenz, M., M. Smit, P. Binder, A.C. Van Aelst and P.N.L. Lens. 2008. Biological alkylation and colloid formation of selenium in methanogenic UASB reactors. J. Environ. Qual. 37: 1691–1700.

Lenz, M., E.D. Van Hullebusch, F. Farges, S. Nikitenko, P.F.X. Corvini and P.N.L. Lens. 2011. Combined speciation analysis by X-ray absorption near-edge structure spectroscopy, ion chromatography, and solid-phase microextraction gas chromatography—mass spectrometry to evaluate biotreatment of concentrated selenium wastewaters. Environ. Sci. Technol. 45: 1067–1073.

Li, H., J. Zhang, T. Wang, W. Luo, Q. Zhou and G. Jiang. 2008. Elemental selenium particles at nano-size (Nano-Se) are more toxic to Medaka (Oryzias latipes) as a consequence of hyper-accumulation of selenium: A comparison with sodium selenite. Aquat. Toxicol. 89: 251–256.

Li, S. and N. Deng. 2002. Separation and preconcentration of Se(IV)/Se(VI) species by selective adsorption onto nanometer-sized titanium dioxide and determination by graphite furnace atomic absorption spectrometry. Anal. Bioanal. Chem. 374: 1341–1345.

Liu, Q. 2010. Determination of ultra-trace amounts of inorganic selenium species in natural water by ion chromatography-inductively coupled plasma—mass spectrometry coupled with nano-Al$_2$O$_3$ solid phase extraction. Cent. Eur. J. Chem. 8: 326–330.

Lobinski, R. and F.C. Adams. 1993. Recent advances in speciation analysis by capillary gas chromatography—microwave induced plasma atomic emission spectrometry. Trac-Trends Anal. Chem. 12: 41–49.

Luoma, S.N. and T.S. Presser. 2009. Emerging opportunities in management of selenium contamination. Environ. Sci. Technol. 43: 8483–8487.

Martinis, E.M., L.B. Escudero, P. Berton, R.P. Monasterio, M.F. Filippini and R.G. Wuilloud. 2011. Determination of inorganic selenium species in water and garlic samples with on-line ionic liquid dispersive microextraction and electrothermal atomic absorption spectrometry. Talanta. 85: 2182–2188.

McSheehy, S., P. Pohl, J. Szpunar, M. Potin-Gautier and R. Lobinski. 2001. Analysis for selenium speciation in selenized yeast extracts by two-dimensional liquid chromatography with ICP-MS and electrospray MS-MS detection. J. Anal. Atom. Spectrom. 16: 68–73.

Meija, J. and J.A. Caruso. 2004. Selenium and sulfur trichalcogenides from the chalcogenide exchange reaction. Inorg. Chem. 43: 7486–7492.

Meija, J., M. Montes-Bayon, D.L. Le Duc, N. Terry and J.A. Caruso. 2002. Simultaneous monitoring of volatile selenium and sulfur species from se accumulating plants (wild type and genetically modified) by GC/MS and GC/ICPMS using solid-phase microextraction for sample introduction. Anal. Chem. 74: 5837–5844.

Michalke, B. 2003. Element speciation definitions, analytical methodology, and some examples. Ecotox. Environ. Safe. 56: 122–139.

Miekeley, N., R.C. Pereira, E.A. Casartelli, A.C. Almeida and M. de F.B. Carvalho. 2005. Inorganic speciation analysis of selenium by ion chromatography-inductively coupled plasma-mass spectrometry and its application to effluents from a petroleum refinery. Spectrochim. Acta B 60: 633–641.

Montes-Bayón, M., T.D. Grant, J. Meija and J.A. Caruso. 2002a. Selenium in plants by mass spectrometric techniques: developments in bio-analytical methods—Plenary Lecture. J. Anal. Atom. Spectrom. 17: 1015–1023.

Montes-Bayón, M., E.G. Yanes, C.P. De León, K. Jayasimhulu, A. Stalcup, J. Shann and J.A. Caruso. 2002b. Initial studies of selenium speciation in *Brassica juncea* by LC with ICPMS and ES-MS detection: An approach for phytoremediation studies. Anal. Chem. 74: 107–113.

Montes-Bayón, M., M.J.D. Molet, E.B. Gonzalez and A. Sanz-Medel. 2006. Evaluation of different sample extraction strategies for selenium determination in selenium-enriched plants (*Allium sativum* and *Brassica juncea*) and Se speciation by HPLC-ICP-MS. Talanta. 68: 1287–1293.

Morales, R., J.F. López-Sánchez and R. Rubio. 2008. Selenium speciation by capillary electrophoresis. Trac-Trends Anal. Chem. 27: 183–189.

Moreno, M.E., C. Perez-Conde and C. Camara. 2003. The effect of the presence of volatile organoselenium compounds on the determination of inorganic selenium by hydride generation. Anal. Bioanal. Chem. 375: 666–672.

Moscoso-Pérez, C., J. Moreda-Piñeiro, P. López-Mahía, S. Muniategui-Lorenzo, E. Fernández-Fernández and D. Prada-Rodríguez. 2008. Pressurized liquid extraction followed by high performance liquid chromatography coupled to hydride generation atomic fluorescence spectrometry for arsenic and selenium speciation in atmospheric particulate matter. J. Chromatogr. A 1215: 15–20.

Mounicou, S., S. McSheehy, J. Szpunar, M. Potin-Gautier and R. Lobinski. 2002. Analysis of selenized yeast for selenium speciation by size-exclusion chromatography and capillary zone electrophoresis with inductively coupled plasma mass spectrometric detection (SEC-CZE-ICP-MS). J. Anal. Atom. Spectrom. 17: 15–20.

Mounicou, S., J. Meija and J. Caruso. 2004. Preliminary studies on selenium-containing proteins in *Brassica juncea* by size exclusion chromatography and fast protein liquid chromatography coupled to ICP-MS. Analyst. 129: 116–123.

Najafi, N.M., H. Tavakoli, Y. Abdollahzadeh and R. Alizadeh. 2012. Comparison of ultrasound-assisted emulsification and dispersive liquid-liquid microextraction methods for the speciation of inorganic selenium in environmental water samples using low density extraction solvents. Anal. Chim. Acta. 714: 82–88.

Nakano, S., M. Yoshii and T. Kawashima. 2004. Flow-injection simultaneous determination of selenium(IV) and selenium(IV + VI) using photooxidative coupling of p-hydrazinobensenesulfonic acid with N-(1-naphthyl)ethylenediamine. Talanta. 64: 1266–1272.

Niedzielski, P. 2005. The new concept of hyphenated analytical system: Simultaneous determination of inorganic arsenic(III), arsenic(V), selenium(IV) and selenium(VI) by high performance liquid chromatography-hydride generation-(fast sequential) atomic absorption spectrometry during single analysis. Anal. Chim. Acta. 551: 199–206.

Ochsenkühn-Petropoulou, M., B. Michalke, D. Kavouras and P. Schramel. 2003. Selenium speciation analysis in a sediment using strong anion exchange and reversed phase chromatography coupled with inductively coupled plasma-mass spectrometry. Anal. Chim. Acta. 478: 219–227.

Ochsenkühn-Petropoulou, M. and F. Tsopelas. 2002. Speciation analysis of selenium using voltammetric techniques. Anal. Chim. Acta. 467: 167–178.

Ochsenkühn-Petropoulou, M.T. and F.N. Tsopelas. 2004. Separation of organoselenium compounds and their electrochemical detection. Anal. Bioanal. Chem. 379: 770–776.

Ogra, Y. and Y. Anan. 2009. Selenometabolomics: Identification of selenometabolites and specification of their biological significance by complementary use of elemental and molecular mass spectrometry. J. Anal. Atom. Spectrom. 24: 1477–1488.

Olivas, R.M., O.F.X. Donard, N. Gilon and M. Potin-Gautier. 1996. Speciation of organic selenium compounds by high-performance liquid chromatography—inductively coupled plasma mass spectrometry in natural samples. J. Anal. Atom. Spectrom. 11: 1171–1176.

Oremland, R.S., M.J. Herbel, J.S. Blum, S. Langley, T.J. Beveridge, P.M. Ajayan, T. Sutto, A.V. Ellis and S. Curran. 2004. Structural and spectral features of selenium nanospheres produced by Se-respiring bacteria. Appl. Environ. Microbiol. 70: 52–60.

Orero Iserte, L., A.F. Roig-Navarro and F. Hernández. 2004. Simultaneous determination of arsenic and selenium species in phosphoric acid extracts of sediment samples by HPLC-ICP-MS. Anal. Chim. Acta. 527: 97–104.

Pacheco, P.H., R.A. Gil, P. Smichowski, G. Polla and L.D. Martinez. 2007. On-line preconcentration and speciation analysis of Se(IV) and Se(VI) using l-methionine immobilised on controlled pore glass. J. Anal. Atom. Spectrom. 22: 305–309.

Padmavathiamma, P. and L. Li. 2007. Phytoremediation technology: Hyper-accumulation metals in plants. Water Air Soil Poll. 184: 105–126.

Peachey, E., K. Cook, A. Castles, C. Hopley and H. Goenaga-Infante. 2009. Capabilities of mixed-mode liquid chromatography coupled to inductively coupled plasma mass spectrometry for the simultaneous speciation analysis of inorganic and organically-bound selenium. J. Chromatogr. A 1216: 7001–7006.

Pearce, C.I., R.A.D. Pattrick, N. Law, J.M. Charnock, V.S. Coker, J.W. Fellowes, R.S. Oremland and J.R. Lloyd. 2009. Investigating different mechanisms for biogenic selenite transformations: *Geobacter sulfurreducens*, *Shewanella oneidensis* and *Veillonella atypica*. Environ. Technol. 30: 1313–1326.

Pecheyran, C., D. Amouroux and O.F.X. Donard. 1998a. Field determination of volatile selenium species at ultra trace levels in environmental waters by on-line purging, cryofocusing and detection by atomic fluorescence spectroscopy. J. Anal. Atom. Spectrom. 13: 615–621.

Pecheyran, C., C.R. Quetel, F.M.M. Lecuyer and O.F.X. Donard. 1998b. Simultaneous determination of volatile metal (Pb, No, Sn, In, Ga) and nonmetal species (Se, P, As) in different atmospheres by cryofocusing and detection by ICPMS. Anal. Chem. 70: 2639–2645.

Pedrero, Z. and Y. Madrid. 2009. Novel approaches for selenium speciation in foodstuffs and biological specimens: A review. Anal. Chim. Acta. 634: 135–152.

Peng, D., J. Zhang, Q. Liu and E.W. Taylor. 2007. Size effect of elemental selenium nanoparticles (Nano-Se) at supranutritional levels on selenium accumulation and glutathione S-transferase activity. J. Inorg. Biochem. 101: 1457–1463.

Petrov, P.K., J.W. Charters and D. Wallschläger. 2012. Identification and determination of selenosulfate and selenocyanate in flue gas desulfurization waters. Environ. Sci. Technol. 46: 1716–1723.

Pilon-Smits, E.A.H. and D.L. LeDuc. 2009. Phytoremediation of selenium using transgenic plants. Curr. Opin. Biotech. 20: 207–212.

Pinel-Raffaitin, P., C. Pécheyran and D. Amouroux. 2008. New volatile selenium and tellurium species in fermentation gases produced by composting duck manure. Atmos. Environ. 42: 7786–7794.

Pinset, J. 1954. Need for selenite and molybdate in the formation of formic dehydogenase by members of the *Eschesichia Coli-Aerp-Bacte Aerpgemes* group of bacteria. Biochem. J. 57: 10.

Profumo, A., G. Spini, L. Cucca and B. Mannucci. 2001. Sequential extraction procedure for speciation of inorganic selenium in emissions and working areas. Talanta. 55: 155–161.

Pyrzynska, K. 2001. Analysis of selenium species by capillary electrophoresis. Talanta 55: 657–667.

Pyrzynska, K. 2009. Selenium speciation in enriched vegetables. Food Chem. 114: 1183-1191.

Rao, T.P., M. Anbu, M.L.P. Reddy, C.S.P. Iyer and A.D. Damodaran. 1996. Coprecipitative preconcentration and differential pulse cathodic stripping voltammetric determination of Selenium (IV). Anal. Lett. 29: 2563–2571.

Raskin, I. and B.D. Ensley [eds.]. 1999. Phytoremediation of Toxic Metals: Using Plants to Clean Up the Environment. 1st edn. Wiley-Interscience. New York, USA.

Reeder, R.J., M.A. Schoonen and A. Lanzirotti. 2006. Metal speciation and its role in bioaccessibility and bioavailability. *In:* J.J. Ross [ed.]. The emergent field of medical mineralogy and geochemistry. Mineralogical Society of America and the Geochemical Society, New York. pp. 59–113.

Rúriková, D. and I. Kunáková. 2000. Determination of selenium in soils by cathodic stripping voltammetry. J. Trace Microprobe T. 18: 193–199.

Sadeghian, S., G.A. Kojouri and A. Mohebbi. 2011. Nanoparticles of Selenium as species with stronger physiological effects in sheep in comparison with sodium selenite. Biol. Trace Elem. Res. 1–7.

Sánchez-Rodas, D., W.T. Corns, B. Chen and P.B. Stockwell. 2010. Atomic Fluorescence Spectrometry: a suitable detection technique in speciation studies for arsenic, selenium, antimony and mercury. J. Anal. Atom. Spectrom. 25: 933–946.

Schaumlöffel, D. and A. Prange. 1999. A new interface for combining capillary electrophoresis with inductively coupled plasma-mass spectrometry. Fresen. J. Anal. Chem. 364: 452–456.

Shahdousti, P. and N. Alizadeh. 2011. Headspace-solid phase microextraction of selenium(IV) from human blood and water samples using polypyrrole film and analysis with ion mobility spectrometry. Anal. Chim. Acta. 684: 67–71.

Shamberger, R.J. 1981. Selenium in the environment. Sci. Total Environ. 17: 59–74.

Sheoran, V., A.S. Sheoran and P. Poonia. 2011. Role of hyperaccumulators in phytoextraction of metals from contaminated mining sites: A review. Crit. Rev. Env. Sci. Tec. 41: 168–214.

Sieber, F. 2003. Selenium in oxidation state zero is a potent and selective anti-leukemia/lymphoma agent. Exp. Hematol. 31: 119–120.

Simon, S., A. Barats, F. Pannier and M. Potin-Gautier. 2005. Development of an on-line UV decomposition system for direct coupling of liquid chromatography to atomic-fluorescence spectrometry for selenium speciation analysis. Anal. Bioanal. Chem. 383: 562–569.

Stivanin de Almeida, C.M., A.S. Ribeiro, T.D. Saint'Pierre and N. Miekeley. 2009. Studies on the origin and transformation of selenium and its chemical species along the process of petroleum refining. Spectrochim. Acta B 64: 491–499.

Stolz, J.F., P. Basu, J.M. Santini and R.S. Oremland. 2006. Arsenic and selenium in microbial metabolism. Annu. Rev. Microbiol. 60: 107–130.

Stroud, J.L., S.P. McGrath and F.J. Zhao. 2012. Selenium speciation in soil extracts using LC-ICP-MS. Int. J. Environ. An. Ch. 92: 222–236.

Sun, Y.C., Y.C. Chang and C.K. Su. 2006. On-Line HPLC-UV/Nano-TiO$_2$-ICPMS system for the determination of inorganic selenium species. Anal. Chem. 78: 2640–2645.

Swearingen, Jr., J.W., D.P. Frankel, D.E. Fuentes, C.P. Saavedra, C.C. Vasquez and T.G. Chasteen. 2006. Identification of biogenic dimethyl selenodisulfide in the headspace gases above genetically modified Escherichia coli. Anal. Biochem. 348: 115–122.

Tang, X., Z. Xu and J. Wang. 2005. A hydride generation atomic fluorescence spectrometric procedure for selenium determination after flow injection on-line co-precipitate preconcentration. Spectrochim. Acta B 60: 1580–1585.

Tao, G. and E.H. Hansen. 1994. Determination of ultra-trace amounts of selenium(IV) by flow injection hydride generation atomic absorption spectrometry with on-line preconcetntration by coprecipitation with lantanum hydroxide. Analyst. 119: 333–337.

Terry, N., A.M. Zayed, M.P.d. Souza and A.S. Tarun. 2000. Selenium in higher plants. Annu. Rev. Plant Physiol. Plant Mol. Biol. 51: 401–432.

Tessier, E., D. Amouroux, G. Abril, E. Lemaire and O.F.X. Donard. 2002. Formation and volatilisation of alkyl-iodides and -selenides in macrotidal estuaries. Biogeochemistry. 59: 183–206.

Thiry, C., A. Ruttens, L. De Temmerman, Y.J. Schneider and L. Pussemier. 2012. Current knowledge in species-related bioavailability of selenium in food. Food Chem. 130: 767–784.

Tirez, K., W. Brusten, S. Van Roy, N. De Brucker and L. Diels. 2000. Characterization of inorganic selenium species by ion chromatography with ICP-MS detection in microbial-treated industrial waste water. J. Anal. Atom. Spectrom. 15: 1087–1092.

Tolu, J., I. Le Hécho, M. Bueno, Y. Thiry and M. Potin-Gautier. 2011. Selenium speciation analysis at trace level in soils. Anal. Chim. Acta. 684: 126–133.

Tuzen, M., K.O. Saygi and M. Soylak. 2007. Separation and speciation of selenium in food and water samples by the combination of magnesium hydroxide coprecipitation-graphite furnace atomic absorption spectrometric determination. Talanta. 71: 424–429.

Uden, P.C. 2002. Modern trends in the speciation of selenium by hyphenated techniques. Anal. Bioanal. Chem. 373: 422–431.

Uden, P.C., S.M. Bird, M. Kotrebai, P. Nolibos, J.F. Tyson, E. Block and E. Denoyer. 1998. Analytical selenoamino acid studies by chromatography with interfaced atomic mass spectrometry and atomic emission spectral detection. Fresenius J. Anal. Chem. 362: 447–456.

Uden, P.C., H.T. Boakye, C. Kahakachchi and J.F. Tyson. 2004. Selective detection and identification of Se containing compounds-review and recent developments. J. Chromatogr. A 1050: 85–93.

Velinsky, D.J. and G.A. Cutter. 1991. Geochemistry of selenium in a coastal salt marsh. Geochim. Cosmochim. Acta. 55: 179–191.

Vilano, M. and R. Rubio. 2000. Liquid chromatography—UV irradiation-hydride generation-atomic fluorescence spectrometry for selenium speciation. J. Anal. Atom. Spectrom. 15: 177–180.

Vonderheide, A.P., M. Montes-Bayon and J.A. Caruso. 2002. Solid-phase microextraction as a sample preparation strategy for the analysis of seleno amino acids by gas chromatography-inductively coupled plasma mass spectrometry. Analyst. 127: 49–53.

Vonderheide, A.P., S. Mounicou, J. Meija, H.F. Henry, J.A. Caruso and J.R. Shann. 2006. Investigation of selenium-containing root exudates of Brassica juncea using HPLC-ICP-MS and ESI-qTOF-MS. Analyst. 131: 33–40.

Wada, O., N. Kurihara and N. Yamazaki. 1995. Toxicity of Essential Trace Elements. Jpn. J. Nutr. Assess. 11: 48–54.

Wake, B.D., A.R. Bowie, E.C.V. Butler and P.R. Haddad. 2004. Modern preconcentration methods for the determination of selenium species in environmental water samples. TrAC-Trend Anal. Chem. 23: 491–500.

Wallschläger, D. and J. London. 2004. Determination of inorganic selenium species in rain and sea waters by anion exchange chromatography-hydride generation-inductively-coupled plasma-dynamic reaction cell-mass spectrometry (AEC-HG-ICP-DRC-MS). J. Anal. Atom. Spectrom. 19: 1119–1127.

Wallschläger, D., M.V.M. Desai, M. Spengler, C.C. Windmoller and R.D. Wilken. 1998. How humic substances dominate mercury geochemistry in contaminated floodplain soils and sediments. J. Environ. Qual. 27: 1044–1054.

Wang, H., J. Zhang and H. Yu. 2007. Elemental selenium at nano size possesses lower toxicity without compromising the fundamental effect on selenoenzymes: Comparison with selenomethionine in mice. Free Radical Bio. Med. 42: 1524–1533.

Weech, S.A., A.M. Scheuhammer and M.E. Wayland. 2012. Selenium accumulation and reproduction in birds breeding downstream of a uranium mill in northern Saskatchewan, Canada. Ecotoxicology. 21: 280–288.

Wen, H. and J. Carignan. 2007. Reviews on atmospheric selenium: Emissions, speciation and fate. Atmos. Environ. 41: 7151–7165.

Wen, H., J. Carignan, Y. Qiu and S. Liu. 2006. Selenium speciation in kerogen from two Chinese selenium deposits: Environmental implications. Int. J. Environ. Sci. Technol. 40: 1126–1132.

White, P.J., M.R. Broadley, H.C. Bowen and S.E. Johnson. 2007. Selenium and its relationship with sulfur. *In:* M.J. Hawkesford and L.J.D. Kok [eds.]. Sulfur in Plants—an Ecological Perspective. Springer, Dordrecht, The Netherlands. pp. 225–252.

Wolf, R.E., S.A. Morman, P.L. Hageman, T.M. Hoefen and G.S. Plumlee. 2011. Simultaneous speciation of arsenic, selenium, and chromium: Species stability, sample preservation, and analysis of ash and soil leachates. Anal. Bioanal. Chem. 401: 2733–2745.

Wrobel, K., K. Wrobel and J.A. Caruso. 2005. Pretreatment procedures for characterization of arsenic and selenium species in complex samples utilizing coupled techniques with mass spectrometric detection. Anal. Bioanal. Chem. 381: 317–331.

Wu, H., Y. Jin, Y. Shi and S. Bi. 2007. On-line organoselenium interference removal for inorganic selenium species by flow injection coprecipitation preconcentration coupled with hydride generation atomic fluorescence spectrometry. Talanta. 71: 1762–1768.

Xia, L., B. Hu, Z. Jiang, Y. Wu, R. Chen and L. Li. 2006. Hollow fiber liquid phase microextraction combined with electrothermal vaporization ICP-MS for the speciation of inorganic selenium in natural waters. J . Anal. Atom. Spectrom. 21: 362–365.

Xiong, C., M. He and B. Hu. 2008. On-line separation and preconcentration of inorganic arsenic and selenium species in natural water samples with CTAB-modified alkyl silica microcolumn and determination by inductively coupled plasma-optical emission spectrometry. Talanta. 76: 772–779.

Yanes, E.G. and N.J. Miller-Ihli. 2003. Characterization of microconcentric nebulizer uptake rates for capillary electrophoresis inductively coupled plasma mass spectrometry. Spectrochim. Acta. B 58: 949–955.

Yang, L., R.E. Sturgeon, W.R. Wolf, R.J. Goldschmidt and Z. Mester. 2004. Determination of selenomethionine in yeast using CNBr derivatization and species specific isotope dilution GC ICP-MS and GC-MS. J. Anal. Atom. Spectrom. 19: 1448–1453.

Yathavakilla, S.V.K., M. Shah, S. Mounicou and J.A. Caruso. 2005. Speciation of cationic selenium compounds in Brassica juncea leaves by strong cation-exchange chromatography with inductively coupled plasma mass spectrometry. J. Chromatogr. A 1100: 153–159.

Yawata, A., Y. Oishi, Y. Anan and Y. Ogra. 2010. Comparison of selenium metabolism in three brassicaceae plants. J. Health Sci. 56: 699–704.

Yoshida, S., M. Haratake, T. Fuchigami and M. Nakayama. 2011. Selenium in seafood materials. J. Health Sci. 57: 215–224.

Zayed, A.M. and N. Terry. 1992. Selenium Volatilization in Broccoli as Influenced by Sulfate Supply. J. Plant Physiol. 140: 646–652.

Zayed, A., C.M. Lytle and N. Terry. 1998. Accumulation and volatilization of different chemical species of selenium by plants. Planta 206: 284–292.

Zayed, A., E. Pilon-Smits, M. deSouza, Z.-Q. Lin and N. Terry. 1999. Remediation of Selenium-Polluted Soils and Waters by Phytovolatilization. *In:* N. Terry and G. Bañuelos [eds.]. Phytoremediation of Contaminated Soil and Water. CRC Press, Boca Ratón, USA. pp.

Zhang, H., J. Gavina and Y.-L. Feng. 2011. Understanding mechanisms of pressure-assisted electrokinetic injection: Application to analysis of bromate, arsenic and selenium species in drinking water by capillary electrophoresis-mass spectrometry. J. Chromatogr. A 1218: 3095–3104.

Zhang, J., H. Wang, X. Yan and L. Zhang. 2005. Comparison of short-term toxicity between Nano-Se and selenite in mice. Life Sci. 76: 1099–1109.

Zhang, L., Y. Morita, A. Sakuragawa and A. Isozaki. 2007. Inorganic speciation of As(III, V), Se(IV, VI) and Sb(III, V) in natural water with GF-AAS using solid phase extraction technology. Talanta. 72: 723–729.

Zhang, Y. and J. N. Moore. 1996. Selenium fractionation and speciation in a wetland system. Environ. Sci. Technol. 30: 2613–2619.

Zhao, F.-J. and S.P. McGrath. 2009. Biofortification and phytoremediation. Curr. Opin. Biotech. 12: 373–380.

Zhengjun, G., Z. Xinshen, C. Guohe and X. Xinfeng. 2005. Flow injection kinetic spectrophotometric determination of trace amounts of Se(IV) in seawater. Talanta. 66: 1012–1017.

Zhu, Y.-G., E.A.H. Pilon-Smits, F.-J. Zhao, P.N. Williams and A.A. Meharg. 2009. Selenium in higher plants: understanding mechanisms for biofortification and phytoremediation. Trends Plant Sci. 14: 436–442.

Speciation of Chromium and Vanadium in Soil Matrices

Khakhathi L. Mandiwana[a],* and *Nikolay Panichev*[b]

Speciation as a Tool in Analytical Chemistry

Analytical chemists have realized that the determination of the total amount (concentration) of specific chemical elements cannot provide the required information about the mobility, bioavailability and the impact of the specific chemical element on the ecological systems or biological organisms. Only knowledge about the exact concentration of the specific compounds can lead to an understanding of chemical and biochemical behavior involving these compounds. These will provide correct information about dose-linked "toxicity" or "essentiality" of different "species" of the same element in biological systems. The growing awareness of the strong dependence of the toxicity of elements on their specific chemical forms has led to an increasing interest in the qualitative and quantitative methods of determination of specific species.

In 2000 in a comprehensive document, the IUPAC (International Union of Pure and Applied Chemistry) gave guidelines for terms related to chemical speciation and fractionation of elements. It defines speciation

Department of Chemistry, Tshwane University of Technology, P. O. Box 56208, Arcadia, 0007, Pretoria, South Africa.
[a]Email: MandiwanaKL@tut.ac.za
[b]Email: Panichevn@tut.ac.za
*Corresponding author

of an element as its distribution amongst defined chemical species and uses the terms speciation analysis and fractionation to refer to analytical activities (Templeton et al. 2000).

"Chemical species" denotes an element present in a specific molecular or complex structure, or oxidation state, this means that an element existing in different forms must be regarded as representing different species. Speciation analysis is the analytical activity of identifying and measuring one or more chemical species of an element present in a sample. It aims at the differentiation of oxidation states or between simple and coordinated ions, cationic, neutral and anionic forms, monomeric and polymeric species.

Element species are responsible for the mobility, bioavailability and the ecological or toxicological impact of the element rather than the total element concentration (Florence 1983, Turner 1984). Therefore, the determination of chemical species of elements, represent the key to the understanding of biogeochemical cycles of contaminants (inorganic and organometallic) in the environment and their harmful effects to biota and humans (Ebdon et al. 2001). Legislation governing the maximum permissible levels of polluting element in environmental samples, for example, chromium in water, up to now refers only to the total concentrations (50 µg L^{-1}) rather than Cr(VI) concentration, mainly due to analytical problems associated with species determination. However, the toxicity of metals, their environmental mobility, their interaction with solid and liquid phases, the tendency to be accumulated in living systems and their resultant toxicity are strictly correlated with their chemical forms. That is why knowledge of the total concentration does not give sufficient information about the potential risk or benefits (Morabito 1995, Caruso et al. 2000). Speciation analysis promises to become an essential tool for useful risk assessment of elements present in the environment. The need for speciation analysis arises from the necessity to determine the concentration of those species characterized by the highest toxicity and mobility with respect to other species of the same element. Present day knowledge on solid matter speciation of trace elements, in particular Cr(VI) and V(V) compounds, is still somewhat unsatisfactory because of the absence of appropriate analytical methods.

Chromium and Vanadium Speciation: Analytical and Environmental Problems

The increase in demand for chromium (Cr) and vanadium (V) products has prompted a surge in the development of increasing Cr and V ore production in South Africa. During mining and processing, some amounts of Cr and V compounds are discharged into the environment and due to the extremely harmful properties of Cr(VI) and V(V) compounds, they could have undesirable ecological effects. Cr(VI) is highly toxic while Cr(III) is an essential micronutrient. If Cr(VI) is absorbed into the body, it converts to semi-stable Cr(V) and then to Cr(III) by giving off free radicals which can cause DNA damage and subsequently cancer (Henline et al. 2011). Vanadium(IV) and V(V) species are toxic but V(V) is more toxic (Taylor and van Staden 1994).

In a review devoted to Cr speciation (Marqués et al. 1998, Marqués et al. 2000), it was shown that out of 451 articles published in 1983–1997, most of them (76 percent) were devoted to water analysis. The remaining papers were connected with the analysis of industrial materials (12 percent), soils, sediments, geological materials and other solid samples (7 percent), chemical and biological materials (5 percent). From these publications, it can be concluded that in spite of the presence of well established procedures suitable for the correct measurement of Cr(III) and Cr(VI) species in natural waters, additional efforts must be made to improve analytical procedures for Cr speciation in solid samples and especially in soils, sediments and plants. The speciation of Cr and V is an important facet of the chemistry of pollutants specific for countries that mine these minerals with respect to biological uptake of metals (Otabbong 1989).

Currently, there is not much information on the chemical forms of Cr and V species present in complex samples such as soils, sediments and plants. The knowledge about the actual distribution of these species will significantly advance our understanding of the role of Cr and V in the natural and human environment. Therefore, speciation has become necessary to understand the potential toxicity of elements as well as their biological activity as many elements may exist in various (chemical) forms with contrasting chemical properties. Biochemical and toxicological investigations have shown that, for living organisms, the chemical form of a specific element or the oxidation state in which that element exists in the environment, is crucial (Ebdon et al. 2001). One of the most difficult problems encountered in speciation analysis is the development of an analytical procedure that does not disturb the equilibria between the different forms of the element that exist in a given solid matrix. In normal routine analytical laboratories, the most practical procedures for the analytical determination of specific species in a matrix is dominated by the use of different separation operations, chemical conversion into detectable species and different detection methods (Templeton et al. 2000).

The contrasting chemical properties of the same element in different oxidation states, like Cr(III) and Cr(VI) or V(IV) and V(V), require separate determination of both chemical species of the same element in soil and agricultural products, because of their high toxicity.

The most common procedure for sample preparation for the determination of the mobility of an element in soil, is based on the five-stage methodology of Tessier (Tessier et al. 1979), in which a sequential multiple extraction technique using different extractants is applied to the soil sample. Each successive treatment is chemically more aggressive than the previous one, so as to dissolve as selectively as possible the fraction of heavy metals associated with well-defined phases.

It may, however, be interesting to come closer to the species as they exist in the original samples, by making them available for the measurements in the same way as they are liberated into the environment and enter the bio-cycle. "This implies that no aqua regia, strong acids or strong ligands are used to leach the "maximum amount" of the metals out of soils, sediments and sludges, but only that natural circumstances are mimicked" (Cornelis et al. 2001, Cornelis 2002).

In spite of the fact that V(V) is toxic, little attention has been paid to the determination of V(V) species in solid matrices like soil, sediments and plants. The U.S. Environmental Protection Agency has not listed vanadium as a pollutant requiring urgent research and legislation. Consequently, there are few international standards and regulations relating to environmental pollution by vanadium (Moskalyk and Alfantazi 2003).

Literature Methods of Chromium Speciation in Solids

Chromium speciation analysis has been performed by a large number of researchers and are summarized in the paragraphs that follow. The most stable species of chromium in solid matrices are Cr(III) and Cr(VI) that coexist and differ enormously with regards to their biological and chemical properties. It is therefore essential to discriminate quantitatively between the two, given that Cr(III) forms compounds that are essential trace elements in the human body playing a vital role in the metabolism of glucose and certain lipids (mainly cholesterol) while Cr(VI) compounds are very toxic and carcinogenic (Florence and Batley 1980, De Flora and Wetterhahn 1989).

Speciation of Cr in solids has been carried out since late 70s. Since then, many studies and new analytical protocols for analysis for Cr(VI) speciation in solid matrices have been proposed (Gómez and Callao 2006). The sample pre-treatment procedures and risks of Cr(III)-Cr(VI) interconversions during Cr(VI) determination in solid environmental samples were reported by Pettine and Capri (2005).

The USEPA proposed Method 3060 consisted of digesting a sample in alkaline solution of 0.28 M Na_2CO_3 and 0.5 M NaOH (EPA 1984). Nitric acid was added to the filtrate to adjust the pH to approximately 7.5 prior determination of Cr(VI) with DPC method under standard pH conditions. This extraction protocol was unsuccessful with the extraction of Cr(VI) in reducing sediments and was therefore removed from the third edition of USEPA methods (EPA 1986). The USEPA Method 3060 was revised in

1996 (to become USEPA method 3060A) for extracting Cr(VI) from soil, sludges, sediments and solid wastes. This was achieved by digesting the solid sample in 0.28 M Na_2CO_3/0.5 M NaOH at 90–95°C followed by the addition of $MgCl_2$ (The addition of Mg^{2+} in the presence of a phosphate buffer prevents Cr(III)-Cr(VI) inter-conversions) and 1.0 M phosphate buffer (0.5 M K_2HPO_4/0.5 M KH_2PO_4). The exchangeable Cr(VI) in soil was determined after shaking soil sample at pH 7.2 with K-phosphate buffer and followed by standard DPC method (James and Bartlett 1983).

The leaching efficiency of five different extractants for Cr(VI) from soil were tested, viz. distilled water (pH 5.7), phosphate buffer (5 mM K_2HPO_4/5 mM KH_2PO_4, pH 7.0), carbonate-hydroxide solutions (0.28 M Na_2CO_3/0.5 M NaOH, pH 11.8), with and without heating and hydroxide solutions (0.1 M NaOH, pH 13) with sonication (James et al. 1995). Cr(VI) was leached completely in soil with carbonate-hydroxide solution by heating the mixture at 85°C. The fractions of Cr(VI) were also leached by less aggressive reagent solutions, viz., 1 M NH_4NO_3, 0.1 M K_2HPO_4 (DIN 19730, James and Bartlett 1983). Cr(VI) was also determined in soil after extraction into phosphate buffer (DIN 19734, Rüdel and Terytze 1999).

The extraction Cr(VI) as soluble or sparingly soluble $SrCrO_4$ was achieved with ammonium sulfate-ammonium hydroxide buffer (Luque-Garcia and Luque de Castro 2002, Morales-Munoz et al. 2004). Cr(VI) was leached completely from plants samples (Panichev et al. 2005, Mandiwana and Panichev 2011), cement (Potgieter et al. 2003), soil (Panichev et al. 2003, Mandiwana et al. 2007a, Mandiwana et al. 2007b), sediment (Elci et al. 2010) and dust (Mandiwana et al. 2006b, Mandiwana et al. 2007) with 0.1 M Na_2CO_3. This was achieved by boiling the soil-reagent solution mixture for a period of at least 10 min. The leached Cr(VI) was then quantified by electrothermal atomic absorption spectrometry (ET-AAS) after filtration of the sample solutions through Hydrophilic Millipore PVDF 0.45 µm filter. The application of 0.01 M Na_3PO_4 (Mandiwana 2008) as a leaching reagent solution reduced the leaching time to 5 min.

Tirez et al. determined Cr(VI) in soil and waste materials by applying alkaline digestion described in EPA method 3060A followed by ion chromatography but the method was not that successful because of poor extraction of Cr(VI) insoluble compounds (Tirez et al. 2007). Grabarczyk et al. determined Cr(VI) in soil after extraction with ammonia buffer containing diethylenetriaminepentaacetic acid, DTPA (Grabarczyk et al. 2006) or S,S-ethylenediamine-N,N'-trisodium salt solution, EDDS (Grabarczyk 2008) based on complexation of insoluble chromate with a chelating agent to form a soluble complex.

A number of studies applied water or phosphate buffer as extractants of Cr(VI) in soil that could leach to groundwater or absorbed by plants and microorganisms as these compounds extract only soluble and exchangeable forms of Cr(VI) (Milacic and Scancar 2000, Long et al. 2006, Scancar et al. 2007, Mandiwana 2008).

Scancar et al. (2005) determined Cr(VI) in cement by using both high performance liquid chromatography inductively coupled plasma mass spectrometry (HPLC-ICP-MS) and fast protein liquid chromatography (FPLC) with AAS detection (Scancar et al. 2005). A cement sample of 10.0 g was mixed with 40 mL of water and filtered through a 0.45 mm filter. For the HPLC-ICP-MS procedure, aliquots of the diluted sample were injected onto the column and eluted with 0.35 mol L^{-1} nitric acid. Under these chromatographic conditions, positively charged Cr(III) species and hydroxo Cr(III) species that were retained by the sulfonic acid functional groups were not eluted from the column resin. For FPLC-ETAAS procedure, the diluted sample was mixed with a sawdust with pH of 8.0 and 5 percent of quicklime was added to raise the pH to 12. The sample was injected onto the column and eluted with TRIS–HCl buffer (0.005 mol L $^{-1}$, pH 8.0) and the same buffer with NaCl (0.5 mol L^{-1})(Milacic and Scancar 2000). The separated Cr species were determined "off line" by ETAAS in 0.5 mL fractions. The results of HPLC-ICP-MS and FPLC-ETAAS were in good agreement for cement samples containing Cr(VI) concentration higher than 6 mg kg^{-1}.

A dynamic reaction cell inductively coupled plasma mass spectrometry (HPLC-DRC-ICPMS) was used in the determination of Cr(VI) in soil and plant samples (Kuo et al. 2007). Five millilitres (5 mL) solution of 2% v/v HF in mobile phase (0.5 mmol L^{-1} TBAP and 5 mmol L^{-1} EDTA in 4% v/v CH_3OH at pH = 6.85) was added to approximately 0.1000 g sample in a polyethylene centrifuge tubes. After centrifuged, the supernatant was diluted 1.25 to 200-fold and filtered through a PVDF filter (Millipore) of 0.22 μm porosity before HPLC separation and detection by DRC-ICP-MS.

Chromium (VI) was determined in soil (Malherbe et al. 2011), oxy-fuel ash (Jiao et al. 2011), fly and bottom ash (Huggins et al. 1999) without pre-treatment by X-ray absorption fine-structure (XAFS). Huggins et al. determined Cr(VI) in coal combustion by-products (fly ash and bottom ash) ground to particle size of approximately 0.075 mm top size (Huggins et al. 1999). The determination was based on the relative heights of the pre-edge peaks for the different Cr oxidation state in X-ray absorption near-edge structure (XANES) spectra. This technique is restricted by high limit of detection (LOD) since at least 50 ppm could be quantified by this technique.

Chromium(III) species in solids can be determined by subtracting Cr(VI) concentration from total Cr concentration. Lu et al. determined Cr(III) in soil after extracting Cr(VI) and total Cr separately by KCl extracting agent and alkali fusion, respectively (Lu et al. 2009). Preconcentration of Cr(VI) and total Cr was achieved after adjusting the pH of the extractant to an appropriate pH followed by cloud point extraction (CPE) with ammonium pyrrolidine dithiocarbamate (APDC) as the chelating agent and Triton X-114 as the cloud point extractant. Chromium(VI) and total Cr concentrations were then determined with double-slotted quartz tube atom trap-flame atomic absorption spectrometry (STAT-FAAS). The concentration of Cr(III) in soil was then equivalent to the difference between total Cr and Cr(VI) concentration.

Literature Methods of Vanadium Speciation Determination in Solids

Several methods have been developed for the determination of V(IV) and V(V) using the two-step procedure of measuring one species and obtaining the other by difference from the total measurement (Murthy and Reddy 1989). Other techniques have also been applied after derivatization (Patel et al. 1990, Hirayama et al. 1992, Komarova et al. 1991, Jen and Jang 1994). Lewis et al. (1989) studied the speciation of trace metals by coupling ion chromatography with direct current plasma atomic emission spectrometry. These approaches were applied in the determination of V(V) in liquid samples and speciation of vanadium in solids is limited.

Ethylenediaminetetraacetic (EDTA) was investigated as the complexing agent for the speciation of vanadium (V/IV) ions in ash by capillary electrophoresis (Pozdniakova and Padarauskas 1998). The sample was mixed with EDTA before analysis in a fused-silica capillary (57 cm × 75 µm i.d.), at −30 kV, with 5×10^{-3} M EDTA of pH 4 as the electrolyte and detection was carried out at 214 nm with satisfactory results.

The sequential extraction analysis procedure of Tessier et al. for heavy metals was used by Poledniok and Hull for the vanadium separation in soil (Poledniok and Hull 2003). The method involved sequential leaching of the soil samples to separate five fractions of metals: (1) exchangeable, (2) bound to carbonates, (3) bound to Fe-Mn oxides, (4) bound to organic matter and (5) residual. The leaching solutions of Tessier were used for the vanadium leaching but for the residual fraction, $HClO_4$ was replaced with H_2SO_4. These procedures do not discriminate between V(IV) and V(V) species. Furthermore, a spectrophotometric method based on the ternary complex V(IV) with Chrome Azurol S and benzyldodecyldimethylammonium bromide was applied for the vanadium determination after separation of V(V) by solvent extraction using mesityl oxide and reduction of V(V) using ascorbic acid (Poledniok and Hull 2003). The results of the investigation showed that content of vanadium in the fractions of Upper Silesia soil were the highest in the organic fraction, indicating input by anthropogenic activities. The total content of vanadium in the five soil fractions was in good correlation with the total content of this element in soils found after HF-H$_2$SO$_4$ digestion.

Vanadium(V) was determined in fertilizers (phosphate rocks and nitrogen, phosphorus and potassium fertilizer) as vanadium(V) ternary complex formed with 4-(2-pyridylazo) resorcinol (PAR) and hydrogen peroxide using ion-interaction reversed-phase high-performance liquid chromatography (Vachirapatama et al 2002). The optimal mobile phase was a methanol–water solution (32:68, v/v) containing 3×10^{-3} M tetrabutylammonium bromide, 5×10^{-3} M acetic acid and 5×10^{-3} M citrate buffer at pH 7, with absorbance detection at 540 nm and optimum mole ratio of V(V):PAR:H$_2$O$_2$ was 1:1:1. The optimal conditions for precolumn formation of the ternary complex were 0.01 M acetate, 7×10^{-3} M H$_2$O$_2$, 3×10^{-4} M PAR, and pH 6.

The liquid chromatographic method was used for the separation of V(IV) and V(V) as ethylenediaminetetraacetic acid (EDTA) complexes in sediment, mussel and fish muscle samples collected from Lake Maracaibo, Venezuela (Colina et al. 2005). The method was carried out using reversed-phase ion-pair liquid chromatography with inductively coupled plasma-mass spectrometry detection. A C-8 reversed-phase column, 15 cm long, was used to separate the species. A solution containing 0.06 M ammonium acetate, 0.01 M tetrabutylammonium hydroxide, 0.01 M ammonium di-phosphate and 2.5×10^{-3} M EDTA at pH 6 was used as the mobile phase in order to avoid the use of organic solvents that reduce the sensitivity of the determination. To prevent changes in distribution of the vanadium species, samples were prepared freshly.

Speciation of vanadium was carried out using a seven-step sequential extraction procedure of the coal bottom ash each releasing species of vanadium: water soluble, exchangeable, carbonate, reducible, oxidizable, sulphide and residual fractions (Aydin et al. 2012). This approach does not identify between V(VI) and V(V) species in the solid matrix and therefore cannot be used to evaluate risk assessment of the environmental contamination with vanadium.

Similarly, as with Cr(VI), V(V) was leached completely from soil (Mandiwana and Panichev 2004) and plants (Mandiwana and Panichev 2006a) with 0.1 M Na$_2$CO$_3$. It was also found that 1 M (NH$_4$)$_2$HPO$_4$ or 0.01 M Na$_3$PO$_4$ could be applied to leach V(V) in soil (Mandiwana et al. 2005).

A dynamic reaction cell inductively coupled plasma mass spectrometry (HPLC-DRC-ICPMS) was also used in the determination of V(V) in soil and plant samples (Kuo et al. 2007). Five mililiters (5 mL) solution of 2% v/v HF in mobile phase (0.5 mmol L^{-1} TBAP and 5 mmol L^{-1} EDTA in 4% v/v CH$_3$OH at pH = 6.85) was added to approximately 0.1000 g sample in a polyethylene centrifuge tubes. After centrifuged, the supernatant was diluted 1.25 to 200-fold and filtered through a PVDF filter (Millipore) of 0.22 μm porosity before HPLC separation and detection by DRC-ICP-MS.

Natural Leaching of Cr(VI) and V(V) Species from Solids in the Environment

To leach metal species in the laboratory, one has to understand how these metal species are leached naturally in the environment and transform that knowledge and apply it in the laboratory environment. Therefore, the theory behind natural leaching of metals including Cr and V will be discussed. Metal species could be leached naturally from solids by forming complexes that are soluble in water at a specific pH of solution. There are a number of leaching agents that are able to covert insoluble Cr(VI) and V(V) into soluble forms, e.g., CO_2, CO_3^{2-} and PO_4^{3-}. The discussion that follows will only be restricted to CO_2, CO_3^{2-}, PO_4^{3-}.

Carbon dioxide, CO_2, is one of the main components of atmospheric air which results due to the decaying of vegetation, liberation of CO_2 by micro-organisms, plant respiration in the rooting zone and the microbial decomposition of organic matter. CO_2 dissolves in the soil water system to form carbonate ions (CO_3^{2-}) that are capable of changing the solubility of an element's species, which exist in anionic forms, e.g., Cr(VI), V(V) compounds.

$$CO_2(g) + H_2O(l) \leftrightarrow H_2CO_3(l) \leftrightarrow 2H^+(aq) + CO_3^{2-}(aq)$$

$$MA(s) + CO_3^{2-}(aq) \leftrightarrow MCO_3(s) + A^{2-}(aq)$$

The interaction of carbonate ions with mineral surfaces plays a significant role in the reactions leading to the leaching of some elements from the solid to the dissolved state (Ferguson 1982). The concentration of carbonic acid, H_2CO_3, is proportional to the amount of CO_2 dissolved and this, in turn, according to Henry's Law (Brandy and Holum 1996), is proportional to the partial pressure of CO_2, i.e.,

$$[CO_2(sol)] = K \times pCO_2$$

where $K = 3.34 \times 10^{-4}$ mol l^{-1} kPa

From the equations above, it follows that the amount of leached Cr(VI) should be proportional to the partial pressure of CO_2 in the soil under the given conditions:

$$[CrO_4^{2-}] = K \times pCO_2$$

While the amount of CO_2 in rainwater is in equilibrium with atmospheric concentration of CO_2, the CO_2 content can become quite high when in contact with the soil due to the presence of decaying vegetation, plant respiration in the rooting zone and microbial decomposition of organic matter (Bolt and Bruggenwert 1978). The partial pressure of CO_2 in soils ranges from the levels equivalent to that in atmospheric air (0.032 kPa) to 1–5 kPa.

Sometimes levels as high as 100–200 kPa have been reported (Sposito 1989). Additionally, areas contaminated with wastes having high organic carbon content can generate CO_2, and thus carbonates, at rates greatly exceeding natural sources.

To transform this natural reaction to laboratory environment, one has to leach metal anion with reagents MCO_3^{2-}, where M is the element that forms soluble complex with the metal species, Na_2CO_3. Therefore, the qualitative determination of anions is achieved in the laboratory after the leaching of anions from solid samples by the treatment of the sample with Na_2CO_3 solutions. During the treatment, PO_4^{3-}, CrO_4^{2-}, SO_4^{2-}, VO_3^- and some other anions, such as AsO_4^{3-} and AsO_3^{3-}, a total of 18 anions go into solution. When a solid sample is boiled with an aqueous solution of Na_2CO_3, anions go into solution by transposition. The driving force of the reaction is the difference in solubility products of the initial and final compound (Svehla 1996).

The treatment of samples with Na_2CO_3 transfer insoluble Cr(VI) ions into soluble sodium chromate. Both $PbCrO_4$ and $PbCO_3$ are non-soluble compounds and the solubility product (SP) of $PbCrO_4$ is less than that of $PbCO_3$, such that the equilibrium of the reaction would be shifted to the left. Therefore, to convert all non-soluble chromates to the soluble form, leaching should be carried out in an excess Na_2CO_3. Usually, a 0.1 M solution of Na_2CO_3 is sufficient as illustrated by the equations below:

At equilibrium,

$$PbCrO_4(s) + Na_2CO_3(aq) \leftrightarrow PbCO_3(s) + Na_2CrO_4(aq)$$

$$SP_{PbCrO_4} = [Pb^{2+}][CrO_4^{2-}] \text{ and } SP_{PbCO_3} = [Pb^{2+}][CO_3]$$

The value of K= 0.29 shows that the conversion of chromate from a solid into a solution with Na_2CO_3 is possible.

$$CuCrO_4 + CO_3^{2-} \rightarrow CuCO_3 + CrO_4^{2-} \qquad K = \frac{SP_{CuCrO_4}}{SP_{CuCO_3}} = \frac{3.6 \times 10^{-6}}{2.3 \times 10^{-10}} = 1.6 \times 10^4$$

The extreme high value of K indicates that $CuCrO_4$ is easily converted to its carbonate naturally at a rapid rate, so that it would not exist naturally in nature. For 0.1 M Na_2CO_3, the CO_3^{2-} concentration will also be about 0.1 M, therefore the concentration of CrO_4^{2-} ions is determined by the ratio of solubility products of both salts.

According to the literature data, the solubility products of all carbonates are less than those of corresponding chromates. Therefore, all chromate ions can be transferred to solution from their insoluble salts present in solid samples.

Because chemical reactions are reversible, the position of a chemical equilibrium is described in quantitative terms by means of equilibrium constants. The expression for an equilibrium constant of a reaction can be written in an exact form (thermodynamic expression) and in an approximate (conditional) form that is applicable to a limited set of conditions only.

In the case of leaching of Cr(VI) species from solid samples, according to the reaction:

$$BaCrO_4(s) + Na_2CO_3(aq) \leftrightarrow BaCO_3(s) + Na_2CrO_4(aq)$$

the equilibrium constant K for this transposition reaction can be expressed as follows:

$$K = \frac{[CrO_4^{2-}]}{[CO_3^{2-}]} = \frac{SP_{BaCrO_4}}{SP_{BaCO_3}}$$

From the equation above, it follows that K is a ratio of SPs of both insoluble salts.

$$[CrO_4^{2-}] = K [CO_3^{2-}]$$

When leaching of chromate ions takes place from solids samples with known or unknown chemical composition, the exact knowledge of K is not necessary, because the ratio of solubility products of chromate and carbonate compounds will be constant under specific experimental conditions, i.e., in 0.1 M Na_2CO_3 solution either equilibrium constant is expressed in exact thermodynamic form or concentration units. The amount of extracted Cr(VI) is determined only by the amount of CO_3^{2-} ions.

The expression of solubility product constant required the use of activities value for the calculations of chromate ions concentrations in saturated solutions of sparingly soluble chromate salts, like $PbCrO_4$ or $BaCrO_4$ in pure water. For this purpose, it is necessary to know activity coefficients (a_A), because activity and equilibrium concentrations are connected with each other via activity coefficients:

$$a_A = f_A [A]$$

where a_A is activity of ion A, f_A is the coefficient of activity and [A] equilibrium concentration of ion A.

In saturated aqueous solution of $PbCrO_4$, activity coefficient can be calculated using Debye-Hückel limiting low equation

$$-lg\, f_A = A \times Z^2 \times I^{1/2}$$

where A is a constant, which depends on temperature and dielectric properties of solvent;

Z is a charge on the species;

I is a ionic strength of the solution;

The thermodynamic value of SP $PbCrO_4 = 1.8 \times 10^{-14}$ (Lide 1999);

The value of A at 25°C and water as a solvent is 0.512 (Midgley and Torrance 1991).

In saturated solution of $PbCrO_4$, concentrations of Pb^{2+} and CrO_4^{2-} ions can be found from the value of SP constant: $[Pb^{2+}] = [CrO_4^{2-}] = \sqrt{SP} = 1.34 \times 10^{-7}$ M. The value of ionic strength is calculated according to the following equation:

$$I = \frac{1}{2}\left(M_{Pb^{2+}} \times Z^2_{Pb^{2+}} + M_{CrO_4^{2-}} \times Z^2_{CrO_4^{2-}}\right)$$

$I = 5.36 \times 10^{-7}$

Calculations, using Debye–Hückel equation give the following value for activity coefficient:

$$-lg\,(f_{CrO_4^{2-}}) = 0.512 \times 4 \times \sqrt{5.36} \times 10^{-7} = 0.001499$$

$f_{CrO_4^{2-}} = 0.9966.$

The same value of activity coefficient has been obtained for Pb^{2+}

$f_{Pb^{2+}} = 0.9966.$

From the results of calculations, it follows that deviation of activity coefficient ($f = 0.9966$) from the ideal system, in which $f=1$ is only 0.34 percent and that deviation can be considered as negligible. Therefore calculations confirm that a constant expressed in units of activity and in units of molar concentration is the same and therefore the difference between the influence of activity and concentration is negligible.

Similarly, Cr(VI) compounds are leached by PO_4^{3-} compounds. The leaching of samples with Na_3PO_4 transform insoluble Cr(VI) compounds, like $BaCrO_4$, to soluble form according to the equation

$$3BaCrO_4(s) + 2Na_3PO_4(aq) \rightarrow 3Na_2CrO_4(aq) + Ba_3(PO_4)_2(s)$$

The procedure removes all common metals, including Cr(III) as insoluble phosphates, oxides or hydroxides (Svehla 1996).

Because V(V) and Cr(VI) have similar chemical properties (Greenwood and Earnshaw 1990), they can exist under natural environmental conditions in corresponding anionic forms, as VO_3^- and CrO_4^{2-} ions. Therefore, V(V) is also separated from V(IV) compounds by treatment with a solution of Na_2CO_3. The leaching of V(V) in solids is illustrated by the following equations:

$$2Na_2CO_3(aq) + 4MVO_3(s) \leftrightarrow 2M_2CO_3(s) + 4NaVO_3(aq)$$

$$6(NH_4)_2HPO_4(aq) + 3MVO_3(s) \leftrightarrow M_3PO_4(s) + 3NH_4VO_3(aq) + 3H_2O(l)$$

$$Na_3PO_4(aq) + 3MVO_3(s) \leftrightarrow M_3PO_4(s) + 3NaVO_3(aq)$$

The amount of V(V) leached is also proportional to the pH of the leaching reagent solution wherein V(V) or Cr(VI) is leached completely in soil at pH ~ 12 (Fig. 6.1).

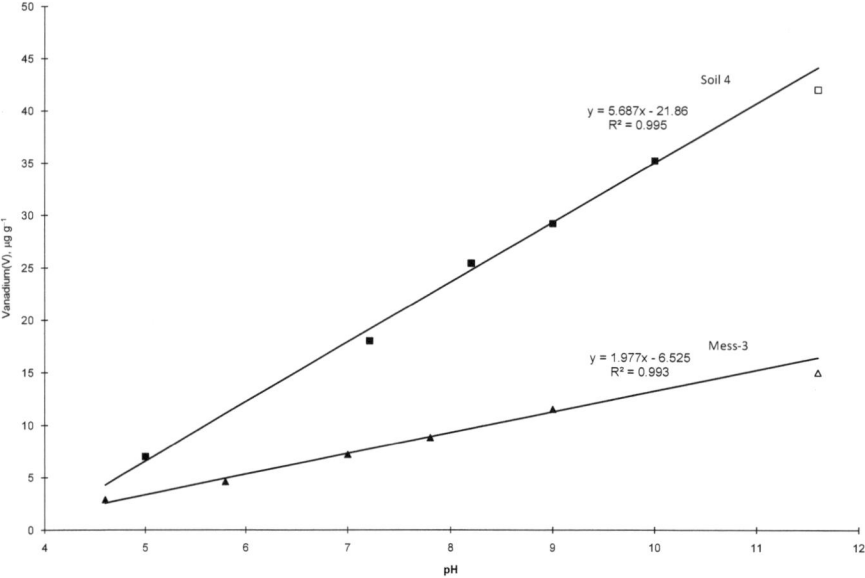

Figure 6.1 Influence of pH of the saturated CO_2 solution on the extraction of V(V) from soil. (■,▲), With continuous bubbling of carbon dioxide at different pH. (□,△), correspond to the pH of 0.1 M Na_2CO_3 (Mandiwana and Panichev 2004).

Leaching of Cr(VI) and V(V) in Solids in the Laboratory Environment

To leach a metal species in the laboratory, one needs to know the most stable metal species of the element in solid and apply conditions that enable their separations as generally species of the same element have normally contrasting properties. Cr(III) and Cr(VI) compounds are only stable species of Cr that co-exist in solids. Cr(VI) is relatively mobile whereas Cr(III) tends to form inert precipitates (Nriagu and Nieboer 1988, Ebdon et al. 2001). The treatment of samples with Na_2CO_3 solution transforms insoluble Cr(VI) compounds to soluble forms while common metals, including Cr(III) are removed as insoluble carbonates, oxides or hydroxides. After filtration the sample solution through 0.45 µm filter, the solution contains only soluble chromates, CrO_4^{2-} or Cr(VI) in solution. Similarly, V(V) is separated from

V(IV) compounds in the same way as all non-soluble V(V) compounds turn to the soluble sodium metavanadate, Na_2VO_3, while V(IV) compounds remained in the precipitates. The analysis of the residue would give the amount of V(IV) (or Cr(III) in case of Cr) in the sample. Even if solubility differences are moderately unfavorable, increased concentration of Na_2CO_3 assures adequate conversion (Bolt and Bruggenwert 1978). Leaching of Cr(VI) or V(V) by PO_4^{3-} follows the same trend.

Techniques Employed for Speciation of Metals in Soil, Sediments and Environmental Samples

Modern analytical techniques provide appropriate sensitivity for the determination of total metal concentration at mg L^{-1}, however, most of them are unsuitable for the speciation of Cr or/and V as they exist in solids at lower levels ($\mu g\ g^{-1}$ and $\mu g\ kg^{-1}$). This generally exempt instrumentation like neutron activation analysis, NAA (LOD~5000 ng mL^{-1}), X-ray fluorescence, XRF (LOD ~ 5000 ng mL^{-1}) and flame atomic absorption spectroscopy, FAAS (LOD ~ 20–200 ng mL^{-1}) thereby limiting their application to the determination of minority of metal species in real samples (de la Guardia and Morales-Rubio 2003).

Due to their intrinsic selectivity in the presence of matrix and inter elemental interferences, ET-AAS, ICP-MS, HPLC-ICP-MS, FPLC-ETAAS, etc., are the methods of choice to examine the presence and distribution of Cr and V species in samples leached with appropriate reagents. The determination of metal species by ICP-OES or ICP MS is a very tedious long process since it takes time to filter enough sample solution through 0.45 μm filter as this instrumentation consumes a lot of samples. On the other hand, the application of XANES technology in the speciation of metal species is unsuitable for routine analysis of environmental samples due to its high LOD, moreover, the instrument is very expensive and unaffordable by many institutions of the developing world.

Validations of methodologies available for Cr and V speciation are limited by the lack of solid certified reference materials, CRMs. There are only limited solid CRMs with known Cr(VI) content, viz. CRM 545, Cr(VI) in welding dust (Community Bureau of Reference, BCR) and CRM060-030, Cr(VI) in clay (RTC) whereas no solid CRMs with certified values of V(V) species are available (Kuo et al. 2007).

Conclusions

Although many methods have been developed, there is no uniform procedure recommended for Cr(III)-Cr(VI) speciation in solid samples. All the methods applied are based on the leaching of Cr(VI) compounds, but the composition of leaching solutions in many proposed methods is not justified. Some of the recommended methods for Cr(VI) speciation in solid samples contradict the chemical properties of hexavalent chromium. Extraction with $BaCl_2$, $MgCl_2$ or NaCl in the presence of buffers is also unlikely because the reactions of Cr(VI) with strong electrolytes are unknown. The leaching of Cr(VI) with Na_2SO_4 or $(NH_4)_2SO_4$ is possible but complete leaching is unlikely (Wang et al. 1997). The official US EPA method 3060A (SW 846 EPA 3060A 1992), which is based on the leaching of Cr(VI) with Na_2CO_3 in alkaline media with spectrophotometric determination of Cr(VI), is theoretically the correct method, but it cannot be used for determination of small amounts of Cr(VI). The limit of detection (LOD) is rather high, that is, 500 µg L^{-1} (SW 846 EPA7196A 1992).

There are limited number of methods for the speciation of vanadium as V(V) in solids and more effort is needed to develop better methods of leaching V(V) with minimum inter-conversion to V(IV). Validations of methodologies with the use of solids reference materials containing certified values of V(V) content is a necessity. Unfortunately, such solid reference materials with known V(V) content are not currently available thereby hampering the development of reliable methods of vanadium speciation in solids. To reduce this problem, V(V) concentrations were determined in several solid reference materials like CRMs of sediments, PACS-2 and MESS-3 (Mandiwana and Panichev 2004) and in soil CRMs (Mandiwana and Panichev 2010).

The application of synchron-based XANES speciation of chromium without pre-treatment is successful only to solid matrices with higher analyte concentration due to the high LOD (~ ppm) and therefore is unsuitable for analysis of environmental samples in which the analyte exist in ppb.

References

Aydin, F., A. Saydut, B. Gunduz, I. Aydin and C. Hamamci. 2012. Chemical speciation of vanadium in coal bottom ash, Clean-Soil, Air, Water 00: 1–5.

Bartlett, R.J. and B.R. James. 1996. Chromium. *In:* D.L. Sparks [ed.]. Methods of Soil Analysis, part 3-Chemical Methods. SSSA, WI: Madison, USA.

Bolt, G.H. and G.H. Bruggenwert. 1978. Soil chemistry: A basic elements. Developments in Soil Science: 5A. Elsevier, New York.

Brandy, J.E. and J.R. Holum. 1996. Chemistry. 2nd edn. John Willey & Sons, Inc., New York.

Caruso, J.A., K.L. Sutton and K.L. Ackley. 2000. Elemental Speciation: New Approaches for trace element analysis. *In:* J.A.Caruso, K.L. Sutton and K.L. Ackley [eds.]. Wilson & Wilson's Comprehensive Analytical Chemistry, vol. XXXIII, Elsevier Science B.V., Amsterdam.

Colina, M., P.H.E. Gardiner, Z. Rivas and F. Troncone. 2005. Determination of vanadium species in sediment, mussel and fish muscle tissue samples by liquid chromatography–inductively coupled plasma-mass spectrometry. Anal. Chim. Acta 538: 107–115.

Cornelis, R. 2002. Speciation of trace elements: a way to a safer world. Anal. Bioanal. Chem. 373: 123–124.

Cornelis, R., H. Crews, O.X.F. Donard, L. Ebdon and P. Quevauviller. 2001. Trends in certified reference materials for the speciation of trace elements. Fresenius J. Anal. Chem. 370: 120–125.

De Flora, S. and K.E. Wetterhahn. 1989. Mechanisms of chromium metabolism and genotoxicity. Life Chemistry Reports 7.

De Flora, S., M. Bagnasco, D. Serra and P. Zanacchi. 1990. Genotoxicity of chromium compounds. Mutat. Res. 238: 99–172.

De la Guardia, M. and A. Morales-Rubio. 2003. Sample preparation for chromium speciation. *In:* D. Barcelo [ed.]. Wilson & Wilson's Comprehensive Analytical Chemistry, vol. XLI, Elsevier B.V., Amsterdam. pp. 1115–1171.

DIN (Deutsches Institut Fur Normung Hrsg.) 19730, 1997. Extraction of trace elements in soils using ammonium nitrate solution. Beuth Verlag, Berlin.

Ebdon, L., L. Pitts, R. Cornelis, H. Crews, O.F.X. Donard and P. Quevauviller. 2001. Trace element speciation for environment and, food and health. The Royal Society of Chemistry. Cambridge.

Elci, L., U. Divrikli, A. Akdogan, A. Hol, A. Cetin and M. Soylak. 2010. Selective extraction of chromium(VI) using a leaching procedure with sodium carbonate from some plant leaves, soil and sediment samples, J. Hazard Matter. 173: 778–782.

EPA Method 3060, 1984. *In:* Test methods for evaluating solid wastes (SW), physical/chemical methods, SW–846. US Environment Protection Agency, Washington.

EPA Rep. 600/4–86/039.1986. Cincinnati.

Fairhurst, S. and C.A. Minty. 1989. Health and Safety Executive Toxicity Review. 21. HMSO. London.

Ferguson, J.E. 1982. Inorganic Chemistry and the Earth, vol. 6: Pergamon Series on Environmental Science. Pergamon Press, Oxford.

Florence, T.M. 1983. Trace element speciation and aquatic toxicology. Trend Anal. Chem. 2: 162–166.

Florence, T.M. and G.E. Batley. 1980. Chemical speciation in natural waters. CRC Crit. Rev. Anal. Chem., 9: 219–296.

Gómez, V. and M.P. Callao. 2006. Chromium determination and speciation since 2000. TrAC, Trends Anal. Chem. 25: 1006–1015.

Grabarczyk, M. 2008. Protocol for Extraction and Determination of Cr(VI) in Solid Materials with a High Cr(III)/Cr(VI) Ratio Using EDDS as a Leaching Agent for Cr(VI) and a Masking Agent for Cr(III). Electroanalysis. 20: 1857–1862.

Grabarczyk, M., M. Korolczuk and K. Tyszczuk. 2006. Extraction and determination of hexavalent chromium in soil samples. Anal. Bioanal. Chem. 386: 357–362.

Greenwood, N.N. and A. Earnshaw. 1984. Chemistry of the elements. Pergamon Press. Oxford. pp. 1187–1200.

Greenwood, N.N. and A. Earnshaw. 1990. Chemistry of elements. Pergamon Press, Oxford.

Henline, K., B. Hennessey, D. Herr, J. Janicki and B. Jean. 2011. Chromium(VI) versus chromium(III), Ethics Forum, Pittsburgh.

Hiriyama, K., S. Kageyama and N. Unohara. 1992. Mutual separation and preconcentration of vanadium(V) and vanadium(IV) in natural waters with chelating functional group immobilized silica gels followed by determination of vanadium by inductively coupled plasma atomic emission spectrometry. Analyst. 117: 13–17.

Huggins, F.E., M. Najih and G.P. Huffman. 1999. Direct speciation of chromium in coal combustion by-roducts by X-ray absorption fine-structure spectroscopy. Fuel. 78: 233–242.

James, B.R and R.J. Bartlett. 1983. Behavior of chromium in soils: VII. Adsorption and reduction of hexavalent forms. J. Environ. Qual. 12: 177–181.

James, B.R., J.C. Petura, R.J. Vitale and R. Mussoline. 1995. Hexavalent chromium extraction from soils: a comparison of five methods. Environ. Sci. Technol. 29: 2377–2381.

Jen, J.F. and S.M. Jang. 1994. Simultaneous speciation determination of vanadium(IV) and vanadium(V) as EDTA complexes by liquid chromatography with UV detection. Anal. Chim. Acta. 289: 97–104.

Jiao, F., N. Wijaya, L. Zhang, Y. Ninomiya and R. Hocking. 2011. Synchrotron-based Xanes speciation of chromium in the oxy-fuel fly ash collected from lab-scale drop-tube furnace. Environ. Sci. Technol. 45: 6640–6646.

Kawashima, T. 1992. Flow–injection analysis of trace elements by use of catalytic reactions. Anal.Chim. Acta. 261: 167–182.

Komarova, T.V., O.N. Obrezkov and O.A. Shpigun. 1991. Ion chromatographic behaviour of anionic EDTA complexes of vanadium(IV) and vanadium(V). Anal. Chim. Acta. 254: 61–63.

Koreman, Y.I., S. Kopach, Y. Kalembkievich, L. Filar and B. Papchak. 2000. Solvent extraction–atomic absorption determination of chromium(VI) in soils. J. Anal. Chem. 55: 25–28.

Kuo, C., S. Jiang and A.C. Sahayam. 2007. Speciation of chromium and vanadium in environmental samples using HPLC-DRC-ICO-MS. J. Anal. At. Spectrom. 22: 636–641.

Lewis, V.D., S.H. Nam and I.T. URASA. 1989. Direct determination of Chromium(III) and chromium(VI) with ion chromatography using direct current plasma emission as element–selective detectotor. J. Chromatogr. Sci. 27: 30–37.

Lide, D.R. 1999. CRC handbook of chemistry and physics: a ready-reference book of chemical and physics data. 80th edn. CRC Press. Florida.

Long, X., M. Miro and E.H. Hansen. 2006. On-line dynamic extraction and automated determination of readily bioavailable hexavalent chromium in solid substrates using micro-sequential injection bead-injection lab-on-valve hyphenated with electrothermal atomic absorption spectrometry. Analyst. 131:132–140.

Luque-Garcia, J.L. and M.D. Luque de Castro. 2002. Continuous ultrasound-assisted extraction of hexavalent chromium from soil with or without on-line preconcentration prior to photometric monitoring. Analyst. 127: 1115–1120.

Lu, J., J. Tian, H. Wu and C. Zhao. 2009. Speciation determination of chromium(VI) and chromium(III) in soil samples after cloud point extraction. Anal. Lett. 42: 1662–1677.

Malherbe, J., M. Isaure, F. Séby, R.P. Watson, P. Rodriquez-Gonzalez, P.E. Stutzman, C.W. Davis, C. Maurizio, N. Unceta, J.R. Sieber, S.E. Long and O.F.X. Donard. 2011. Evaluation of hexavalent chromium extraction method EPA method 3060A for soils using XANES spectroscopy. Environ. Sci. Technol. 45: 10492–10500.

Mandiwana, K.L. 2008. Rapid leaching of Cr(VI) in soil with Na_3PO_4 in the determination of hexavalent chromium by electrothermal atomic absorption spectrometry. Talanta. 74: 736–740.

Mandiwana, K.L. and N. Panichev. 2004. Electrothermal atomic absorption spectrometric determination of vanadium(V) in soil after leaching with Na_2CO_3. Anal. Chim. Acta. 517: 201–206.

Mandiwana, K.L. and N. Panichev. 2010. Analysis of soil reference materials for vanadium(+5) species by electrothermal atomic absorption spectrometry. J. Hazard. Matter. 178: 1106–1108.

Mandiwana, K.L. and N. Panichev. 2011. Determination of chromium(VI) in black, green and herbal teas. Food Chemistry. 129: 1839–1843.

Mandiwana, K.L., N. Panichev and R. Molatlhegi. 2005. The leaching of V(V) with PO_4^{3-} in the speciation analysis of soil. Anal. Chim. Acta. 545: 239–243.

Mandiwana, K.L. and N. Panichev. 2006a. Speciation analysis of plants in the determination of V(V) by ETAAS. Talanta. 70: 1153–1156.

Mandiwana, K.L., N. Panichev and T. Resane. 2006b. Electrothermal atomic absorption spectrometric determination of total and hexavalent chromium in atmospheric aerosols. J. Hazard. Mater. B123: 379–382.

Mandiwana, K.L., N. Panichev and P. Ngobeni. 2007a. Electrothermal atomic absorption spectrometric determination of Cr(VI) during ferrochrome production. J. Hazard. Mater. 145: 511–514.

Mandiwana, K.L., N. Panichev, M. Kataeva and S. Siebert. 2007b. The solubility of Cr(III) and Cr(VI) compounds in soil and their availability to plants. J. Hazard. Mater. 147: 540–545.

Marqués, M.J., A. Salvador, A.E. Morales–Rubio and M. De La Guardia. 1998. Analytical methodologies for chromium speciation in solid matrices: a survey of literature. Fresenius J. Anal. 362: 239–248.

Marqués, M.J, A. Salvador, A.E. Morales–Rubio and M. De La Guardia. 2000. Chromium speciation in liquid matrices: a survey of the literature. Fresenius J. Anal. 367: 601–613.

Midgley, D. and K. Torrance. 1991. Potentiometric water analysis. 2nd edn. John Willey and Sons, New York.

Milacic, R. and J. Scancar. 2000. Determination of hexavalent chromium in lime-treated sewage sludge by anion-exchange fast protein liquid chromatography with electrothermal atomic absorption spectrometry detection. Analyst. 125: 1938–1942.

Morabito, R. 1995. Extraction techniques in speciation analysis of environmental samples. Fresenius J Anal. Chem. 351: 378–385.

Morales-Munoz, S., J.L. Luque-Garcia and M.D. Castro. 2004. A continuous approach for the determination of Cr(VI) in sediment and soil based on the coupling of microwave-assisted water extraction, preconcentration, derivatization and photometric detection. Anal. Chim. Acta. 515: 343–348.

Moskalyk, R.R. and A.M. Alfantazi 2003. Processing of vanadium: A review. Mineral Engineering. 16: 793–805.

MSDS. 2002. Material safety data sheet for vanadium. Canadian Centre for Occupational Health and Safety. Toronto.

Murthy, G.V.R. and T.S. Reddy. 1989. Extraction and simultaneous spectrophotometric determination of vanadium(IV) and vanadium(V) in admixture with 2-hydroxyacetophenone oxime. Analyst. 114: 493–495.

Nriagu, J.O. and E. Nieboer. 1988. Chromium in Natural and Human Environments. Wiley Interscience. New York. pp. 189–215.

OSHA, United States Occupational Safety and Health Administration.1998. OSHA Analytical Methods Manual, 2nd edn. Method ID–215. Occupational Safety and Health Administration. Salt Lake City.

Otabbong, E. 1989. Chemistry of Cr in some Swedish soils. I. Chromium speciation in soil extracts: a comparison of different methods. Acta Agric. Scand. 39: 119–129.

Panichev, N., K. Mandiwana and G. Foukaridis. 2003. Electrothermal atomic absorption spectrometric determination of Cr(VI) in soil after leaching of Cr(VI) species with carbon dioxide. Anal. Chim. Acta. 491: 81–89.

Panichev, N., K. Mandiwana, M. Kataevab and S. Siebert. 2005. Determination of Cr(VI) in plants by electrothermal atomic absorption spectrometry after leaching with sodium carbonate, Spectrochim. Acta Part B. 60: 699–703.

Patel, B., G.E. Henderson, S.J. Haswell and R. Grzeskowiak. 1990. Speciation of vanadium present in a model yeast system. Analyst. 115: 1063–1066.

Pettine, M. and S. Capri. 2005. Digestion treatments and risks of Cr(III)-Cr(VI) interconversions during Cr(VI) determination in soils and sediments-a review. Anal. Chim. Acta. 540: 231–238.

Poledniok, J. and F. Hull. 2003. Speciation of vanadium in soil. Talanta. 59: 1–8.

Potgieter, S.S., N. Panichev, J.H. Potgieter and S. Panicheva. 2003. Determination of hexavalent chromium in South African cements. Cem. Concr. Res. 33: 1589–1593.

Pozdniakova, S. and A. Padarauskas. 1998. Speciation of metals in different oxidation states by capillary electrophoresis using pre-capillary complexation with complexones. Analyst. 123: 1497–1500.

Rüdel, H. and K. Terytze. 1999. Determination of extractable chromium(VI) in soils using a photometric method. Chemosphere 39: 697–708.

Scancar, J., R. Milacic, F. Seby and O.F.X. Donard. 2005. Determination of hexavalent chromium in cement by the use of HPLC-ICP-MS, FPLC-ETAAS, spectrophotometry and selective extraction techniques. J. Anal. At. Spectrom. 20: 871–875.

Scancar, J., M. Zupancic and R. Milacic. 2007. Development of analytical procedure for the determination of exchangeable Cr(VI) in soils by anion-exchange fast protein liquid chromatography with electrothermal atomic absorption spectrometry detection. Water, Air, Soil Pollut. 185: 121–129.

Sposito, G. 1989. The chemistry of soils. Oxford Univ. Press. Oxford.

Svehla, G. 1996. Vogel's qualitative inorganic analysis. 7th edn. Longman Singapore Publishers. Delhi.

SW 846 EPA. 1992. Method 3060A: Alkaline Digestion for Hexavalent Chromium, Test Methods for Evaluating Solid Waste. Physical/Chemical Methods. US Environment Protection Agency. Washington.

SW 846 EPA. 1992. Method 7196A: Chromium, Hexavalent (colorimetric), Test Methods for Evaluating Solid Waste. Physical/Chemical Methods. US Environment Protection Agency. Washington.

Taylor, M. and J. van Staden. 1994. Spectrophotometric determination of vanadium(IV) and vanadium(V) in each other's presence. Analyst. 119: 1263–1276.

Templeton, D.M., F. Ariese, R. Cornelis, G. Danielsson, H. Muntau, H.P. Van Leeuwen and R. Lobinski. 2000. Guidelines for terms related to chemical speciation and fractionation of elements: definitions, structural aspects and methodical approaches. Pur. Appl. Chem. 72: 1453–1470.

Tessier, A., P.G.C. Campbell and M. Bisson. 1979. Sequential extraction procedure for the speciation of particulate trace metals. Anal. Chem. 51: 844–851.

Tirez, K., H. Scharf, D. Calzolari, R. Cleven, M. Kisser and D. Luck. 2007. Validation of European standard for the determination of hexavalent chromium in solid material. J. Environ. Monit. 9: 749–759.

Turner, D.R. 1984. Relationships between biological availability and chemical measurement. Met. Ions Biol. Syst. 18: 137–144.

Underwood, E.J. 1977. Trace elements in human and animal nutrition. 4th edn. Academic Press. London.

USEPA, United States Environmental Protection Agency. 1996. Method 3060A. *In:* Test Methods for Evaluating Solid Wastes, Physical/Chemical Methods, SW–846, Update, Office of Solid Waste and Emergency Response. US Environment Protection Agency. Washington.

Vachirapatama, N., G. Dicinski, A.T. Towsend and P.R. Haddad. 2002. Determination of vanadium as 4-(2-pyridylazo)resorcinol–hydrogen peroxide ternary complexes by ion-interaction reversed-phase liquid chromatography. J. Chromatogr. A 956: 221–227.

Vitale, R.J., G.R. Mussoline, K.A. Rinehimer, J.C. Petura and B.R. James. 1997. Extraction of Sparingly Soluble Chromate from Soils: Evaluation of Methods and E_h–pH Effects. Environ. Sci. Technol. 31: 390–394.

Wang, J., K. Ashely, E.R. Kennedy and C. Neumeister. 1997. Determination of hexavalent chromium in industrial hygiene samples using ultrasonic extraction and flow injection analysis. Analyst. 122: 1307–1312.

Speciation and Solubility of Thallium in Low Temperature Systems

Additional Aqueous and Solid Thallium Species Potentially Important in Soil Environments

Yongliang Xiong

Introduction

Thallium has numerous applications in industry. It is also of great environmental concern because of its high toxicity. Therefore, stabilities of its aqueous and solid species under low temperature environments are fundamentally important to its impact on environments. In previous publications (Xiong 2007, 2009), a number of aqueous and solid thallium species and their stabilities were addressed. However, several thallium species that are potentially important to soil environments, especially saline soil environments, have not been covered.

Sandia National Laboratories, Carlsbad Programs, 4100 National Parks Highway, Carlsbad, NM 88220.
Email: yxiong@sandia.gov

In this chapter, additional aqueous and solid thallium species that are potentially important to solubility of thallium in soil environments are presented. These species include TlBr(cr), $TlIO_3$(cr), TlI(cr), TlI_3(cr), and TlI(aq). In addition, solubility product constants for TlCl(cr) at 0°C and 50°C are also evaluated in this study. The solubility product constants or stability constants for these species are evaluated by using the Specific Interaction Theory (SIT) model. Relevant solubility data in NH_4I, $MgSO_4$, KCl, KI, KNO_3, and Na_2SO_4 solutions from the literature compiled by Silcock (1979) are utilized for all evaluations. In combination with the relevant SIT coefficients obtained in previous works (Xiong 2007, 2009), the SIT coefficients evaluated in this study would enable us to model thallium chemistry/geochemistry in moderate to high ionic strength environments such as saline soil environments. Unless otherwise noted, uncertainties presented in this study are two standard deviations (2σ).

Solubility Product Constant of $TlIO_3$(cr) at 25°C

In previous studies (Xiong 2007, 2009), thermodynamic solubility product constants potentially important in soil environments at 25°C for some solid phases including lanmuchangite ($TlAl(SO_4)_2 \cdot 12H_2O$), dorallcharite ($TlFe_3(SO_4)_2(OH)_6$), lorandite ($TlAsS_2$), and TlCl (cr), were provided.

In this chapter, the thermodynamic solubility product constants of $TlIO_3$(cr) at 25°C are evaluated by using the SIT model. The evaluation procedure is similar to that described in Grenthe et al. (1992).

The solubility reaction of $TlIO_3$(cr) can be expressed as follows,

$$TlIO_3(cr) \leftrightarrow Tl^+ + IO_3^- \tag{1}$$

According to the SIT model, the solubility product constant of $TlIO_3$(cr) at infinite dilution is evaluated in the following form,

$$\log K_{sp} = \log Q_{sp} - 2D + \Delta\varepsilon(Eq.\ 1) \times I_m \tag{2}$$

In Eq. (2), $\log K_{sp}$ is a solubility product constant at infinite dilution, whereas $\log Q_{sp}$ is a solubility product quotient at a constant ionic strength (e.g., conditional solubility product constant). The D in Eq. (2) is the Debye-Hückel term, and $\Delta\varepsilon$ is the stoichiometric summation of specific interaction coefficients regarding Eq. (1) in a respective medium. The Debye-Hückel term is given by the following equation,

$$D = \frac{A_\gamma \sqrt{I}_m}{1 + 1.5\sqrt{I}_m} \qquad (3)$$

In Eq. (3), A_γ is Debye-Hückel slope for activity coefficient, which is 0.5092 at 25°C (Helgeson and Kirkham 1974); and I_m ionic strength on molal scale.

In Fig. 7.1, a plot showing [log Q_{SP} – 2D] as a function of molality of KNO_3 is presented. In this plot, the intercept is log K_{sp}, which is –5.48 ± 0.02 (Table 7.1), and the slope is –Δε(Eq. 1), which is –0.48 ± 0.06. The Δε(Eq. 1) in a KNO_3 medium is given by

$$\Delta\varepsilon(\text{Eq. 1}) = \varepsilon(Tl^+, NO_3^-) + \varepsilon(K^+, IO_3^-) \qquad (4)$$

Based on Δε(Eq. 1) = –0.48 ± 0.06, and $\varepsilon(Tl^+, NO_3^-)$ = –0.24 ± 0.02 (Xiong 2009), $\varepsilon(K^+, IO_3^-)$ is derived as –0.24 ± 0.06 (Table 7.2).

The log K_{sp} for $TlIO_3$(cr) is also evaluated from solubilities of $TlIO_3$(cr) in an $MgSO_4$ medium (Fig. 7.2). The log K_{sp} obtained is –5.49 ± 0.01, which is in good agreement with that evaluated from a KNO_3 medium (Table 7.1). The Δε(Eq. 1) in an $MgSO_4$ medium is given by

$$\Delta\varepsilon(\text{Eq. 1}) = \varepsilon(Tl^+, SO_4^{2-}) + \varepsilon(Mg^{2+}, IO_3^-) \qquad (5)$$

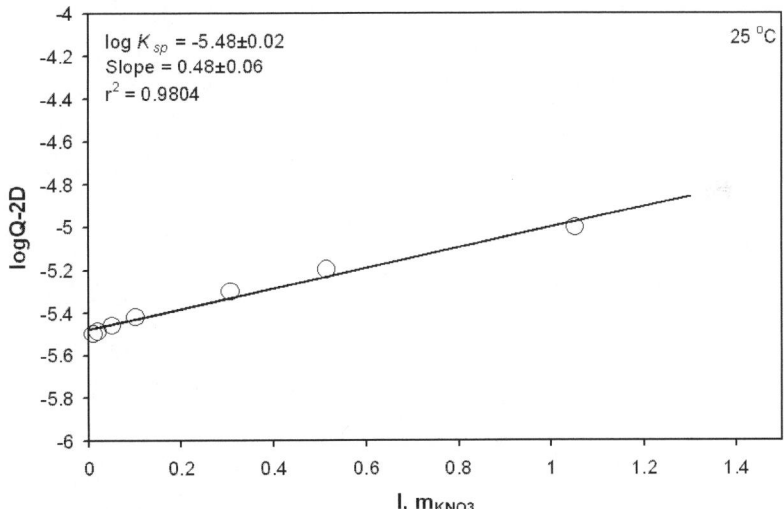

Figure 7.1 A plot showing [log Q_{SP} – 2D] as a function of molality of KNO_3 at 25°C. The log Q_{SP} is conditional solubility product of $TlIO_3$(cr) in a KNO_3 medium.

Table 7.1 Thermodynamic solubility product constants and complex formation constant obtained in this study.

Species	Reactions	T, °C	$\log K_{sp} \pm 2\sigma$ or $\log \beta_1 \pm 2\sigma$	Remarks*
$TlIO_3(cr)$	$TlIO_3(cr) = Tl^+ + IO_3^-$	25	-5.48 ± 0.02	Evaluated from solubility data of $TlIO_3(cr)$ in KNO_3 and $MgSO_4$ solutions.
$TlCl(cr)$	$TlCl(cr) = Tl^+ + Cl^-$	0	-4.37 ± 0.03	Evaluated from solubility data of $TlCl(cr)$ in KNO_3 solutions.
		25	-3.62 ± 0.03	Evaluated from solubility data of $TlCl(cr)$ in $MgSO_4$ and KCl solutions.
		50	-3.06 ± 0.01	Evaluated from solubility data of $TlCl(cr)$ in KNO_3 solutions.
$TlBr(cr)$	$TlBr(cr) = Tl^+ + Br^-$	20	-5.60 ± 0.01	Evaluated from solubility data of $TlBr(cr)$ in $MgSO_4$ solutions.
		20	-5.60 ± 0.01	Evaluated from solubility data of $TlBr(cr)$ in Na_2SO_4 solutions.
$TlI(cr)$	$TlI(cr) = Tl^+ + I^-$	20	-7.46 ± 0.01	Evaluated from solubility data of $TlI(cr)$ in $MgSO_4$ solutions.
		20	-7.46 ± 0.02	Evaluated from solubility data of $TlI(cr)$ in Na_2SO_4 solutions.
$TlI(aq)$	$Tl^+ + I^- = TlI(aq)$	20	1.86 ± 0.02	Evaluated from solubility data of $TlI(cr)$ in KI solutions.
$TlI_3(cr)$	$TlI_3(cr) = Tl^{3+} + 3I^-$	25	-24.40 ± 0.24	Evaluated from solubility data of $TlI_3(cr)$ in NH_4I solutions.

*Solubility data used for evaluation of $\log K_{sp}$ or $\log \beta_1$ were compiled by Silcock (1979).

Table 7.2 SIT interaction coefficients at 25°C derived in this study.

SIT Coefficient	Value ± 2σ	Reference
$\varepsilon(Tl^+, NO_3^-)$	-0.24 ± 0.02	Xiong (2009)
$\varepsilon(Tl^+, SO_4^{2-})$	-0.29 ± 0.03	This work
$\varepsilon(Tl^+, Cl^-)$	-0.63 ± 0.12	This work
$\varepsilon(Tl^+, I^-)$	-0.24 ± 0.14	This work
$\varepsilon(TlI(aq), K^+, I^-)$	-0.29 ± 0.20	This work
$\varepsilon(K^+, IO_3^-)$	-0.24 ± 0.06	This work
$\varepsilon(Mg^{2+}, IO_3^-)$	-0.05 ± 0.04	This work
$\varepsilon(Na^+, Br^-)$	-0.28 ± 0.14	This work
$\varepsilon(Mg^{2+}, Br^-)$	0.10 ± 0.03	This work
$\varepsilon(Na^+, I^-)$	-0.18 ± 0.14	This work
$\varepsilon(Mg^{2+}, I^-)$	0.12 ± 0.03	This work
$\varepsilon(Tl^{3+}, I^-)$	-0.77 ± 0.16	This work

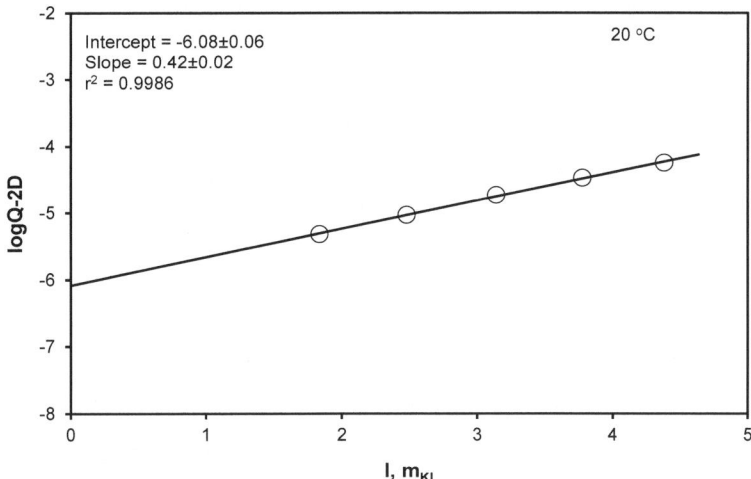

Figure 7.2 A plot showing [log Q_{SP} – 2D] as a function of molality of MgSO$_4$ at 25°C. The log Q_{SP} is conditional solubility product of TlIO$_3$(cr) in an MgSO$_4$ medium.

Solubility Product Constant of TlCl(cr) at 0°C–50°C

In the work of Xiong (2009), the log K_{sp} for TlCl(cr) at 25°C was evaluated from solubility data of TlCl(cr) in KNO$_3$ solutions. In order to obtain ε(Mg^{2+}, IO$_3^-$) in Eq. (5), ε(Tl$^+$, SO$_4^{2-}$) is evaluated from solubility data of TlCl(cr) at 25°C in MgSO$_4$ solutions.

In Fig. 7.3, a plot showing [log Q_{SP} – 2D] for solubility of TlCl(cr) as a function of ionic strengths in MgSO$_4$ solutions is displayed. The conditional solubility product constant, log Q_{SP}, refers to the following reaction,

$$TlCl(cr) = Tl^+ + Cl^- \tag{6}$$

According to the SIT model, the solubility product constant of TlCl(cr) at infinite dilution is evaluated in the following form,

$$\log K_{sp} = \log Q_{sp} - 2D + \Delta\varepsilon \text{ (Eq. 6)} \times I_m \tag{7}$$

The $\Delta\varepsilon$(Eq. 6) in an MgSO$_4$ medium is given by

$$\Delta\varepsilon(\text{Eq. 6}) = \varepsilon(Tl^+, SO_4^{2-}) + \varepsilon(Mg^{2+}, Cl^-) \tag{8}$$

According to $\Delta\varepsilon$(Eq. 6) = –0.08 ± 0.02, and ε(Mg^{2+}, Cl$^-$) = 0.209 ± 0.022 from Xiong (2006), ε(Tl$^+$, SO$_4^{2-}$) is derived as –0.29 ± 0.03 (Table 7.2). In addition, the log K_{sp} (–3.62 ± 0.03) of TlCl(cr) at 25°C evaluated from solubility data in MgSO$_4$ solutions are in excellent agreement with that evaluated from a KNO$_3$ medium (Xiong 2009), which is –3.65 ± 0.01.

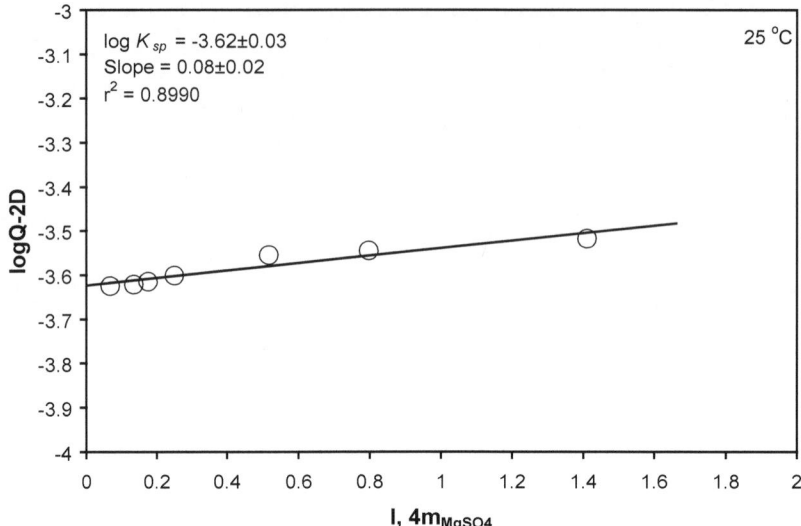

Figure 7.3 A plot showing [log Q_{SP} – 2D] as a function of molality of $MgSO_4$ at 25°C. The log Q_{SP} is conditional solubility product of TlCl(cr) in an $MgSO_4$ medium.

Substituting $\Delta\varepsilon$(Eq. 1) = –0.34 ± 0.02 in an $MgSO_4$ medium (see Fig. 7.2) and $\varepsilon(Tl^+, SO_4^{2-})$ = –0.29 ± 0.03 into Eq. (5), $\varepsilon(Mg^{2+}, IO_3^-)$ is derived as –0.05 ± 0.04 (Table 7.2).

As TlCl(cr) could be important in saline soil solutions, the solubility product constants of TlCl(cr) at 0°C and 50°C are also evaluated from solubilities of TlCl(cr) in a KNO_3 medium (Figs. 7.4 and 7.5, Table 7.1). The log K_{sp} for TlCl(cr) at 0°C and 50°C obtained in this study are –4.37 ± 0.03 and –3.07 ± 0.01, respectively.

Additionally, in order to enable researchers to model speciation and solubility of thallium in solutions with high ionic strengths such as saline soil solutions, the interaction coefficient, $\varepsilon(Tl^+, Cl^-)$ is required. This interaction coefficient is important because Cl^- is expected to be a major ion in such high ionic strength solutions. However, this interaction coefficient is not available in the existing literature and should be construed. To remedy this situation, solubility data of TlCl(cr) in KCl solutions are evaluated to obtain this interaction coefficient.

The $\Delta\varepsilon$(Eq. 6) in a KCl medium is given by

$$\Delta\varepsilon(Eq.\ 6) = \varepsilon(Tl^+, Cl^-) + \varepsilon(K^+, Cl^-) \qquad (9)$$

According to $\Delta\varepsilon$(Eq. 6) = –0.63 ± 0.12 (the negative slope) in Fig. 7.6 and $\varepsilon(K^+, Cl^-)$ = 0.00 ± 0.01 from Ciavatta (1980), $\varepsilon(Tl^+, Cl^-)$ is derived as –0.63 ± 0.12 (Table 7.2).

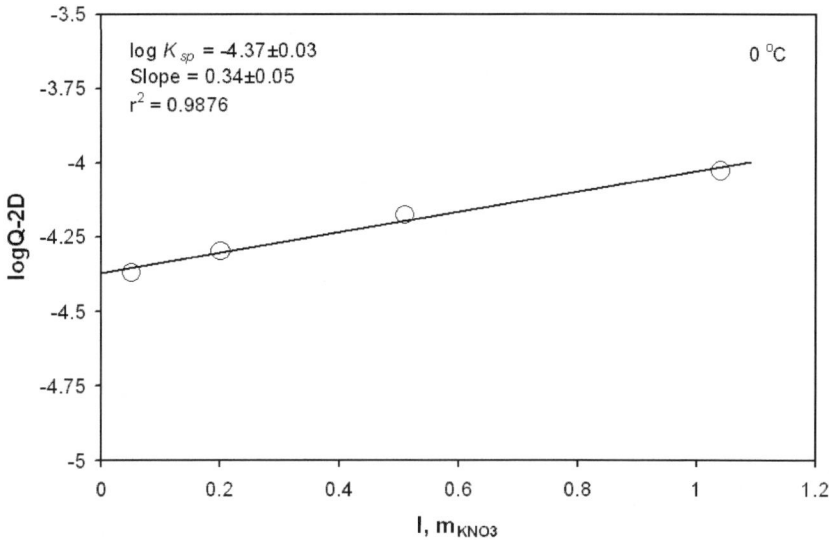

Figure 7.4 A plot showing [log Q_{SP} – 2D] as a function of molality of KNO_3 at 0°C. The log Q_{SP} is conditional solubility product of TlCl(cr) in a KNO_3 medium.

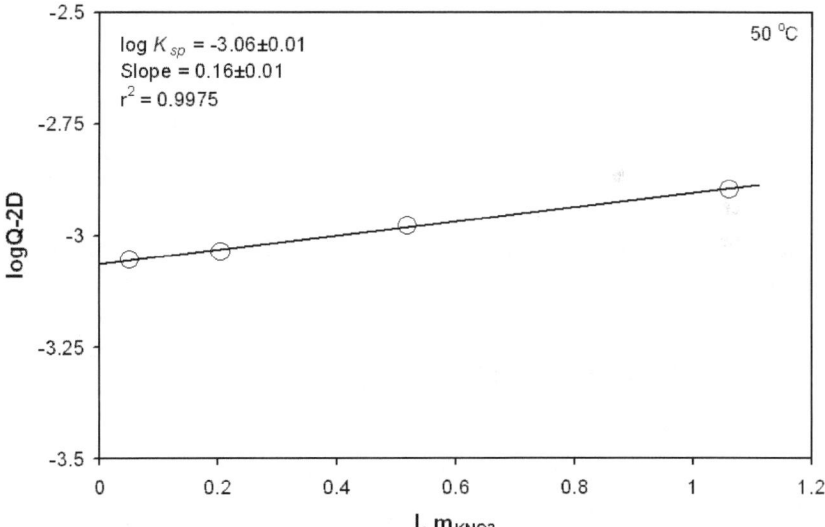

Figure 7.5 A plot showing [log Q_{SP} – 2D] as a function of molality of KNO_3 at 50°C. The log Q_{SP} is conditional solubility product of TlCl(cr) in a KNO_3 medium.

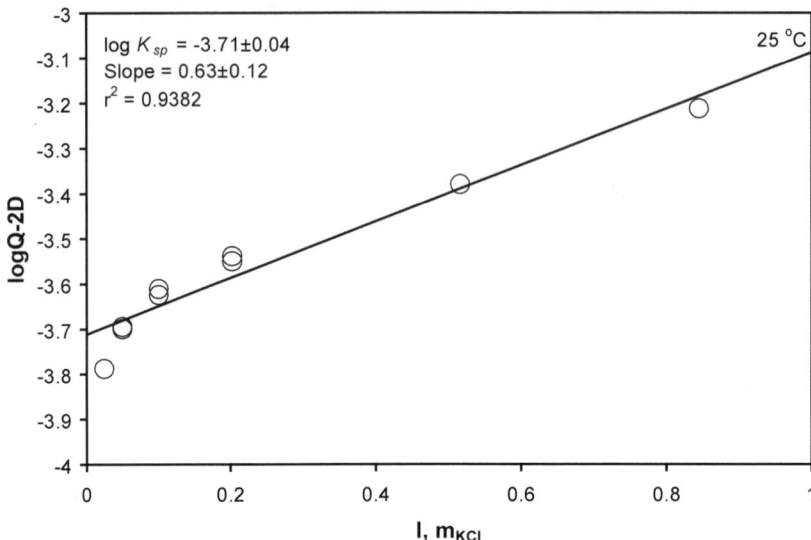

Figure 7.6 A plot showing [log Q_{sp} – 2D] as a function of molality of KCl at 25°C. The log Q_{sp} is conditional solubility product of TlCl(cr) in a KCl medium.

Solubility Product Constant of TlBr(cr) at 20°C

In Figs. 7.7 and 7.8, log K_{sp}'s for TlBr(cr) at 20°C are evaluated from solubility data of TlBr(cr) in MgSO$_4$ and Na$_2$SO$_4$ solutions, respectively. The dissolution reaction of TlBr(cr) is expressed as,

$$TlBr(cr) = Tl^+ + Br^- \tag{10}$$

Based on the SIT model, its solubility product constant at infinite dilution is evaluated in the following form,

$$\log K_{sp} = \log Q_{sp} - 2D + \Delta\varepsilon \text{ (Eq. 10)} \times I_m \tag{11}$$

The $\Delta\varepsilon$(Eq. 10) in an MgSO$_4$ medium is given by

$$\Delta\varepsilon(\text{Eq. 10}) = \varepsilon(Tl^+, SO_4^{2-}) + \varepsilon(Mg^{2+}, Br^-) \tag{12}$$

and the $\Delta\varepsilon$(Eq. 10) in an Na$_2$SO$_4$ medium is given by

$$\Delta\varepsilon(\text{Eq. 10}) = \varepsilon(Tl^+, SO_4^{2-}) + \varepsilon(Na^+, Br^-) \tag{13}$$

The log K_{sp} at 20°C evaluated from solubility data of TlBr(cr) in MgSO$_4$ and Na$_2$SO$_4$ media are –5.60 ± 0.01 and –5.60 ± 0.01 (Table 7.1), respectively, which are in excellent agreement. Based on $\Delta\varepsilon$(Eq. 10) = –0.19 ± 0.01 for Eq. (12) at 20°C and $\varepsilon(Tl^+, SO_4^{2-})$ = –0.29 ± 0.03 derived above, $\varepsilon(Mg^{2+}, Br^-)$

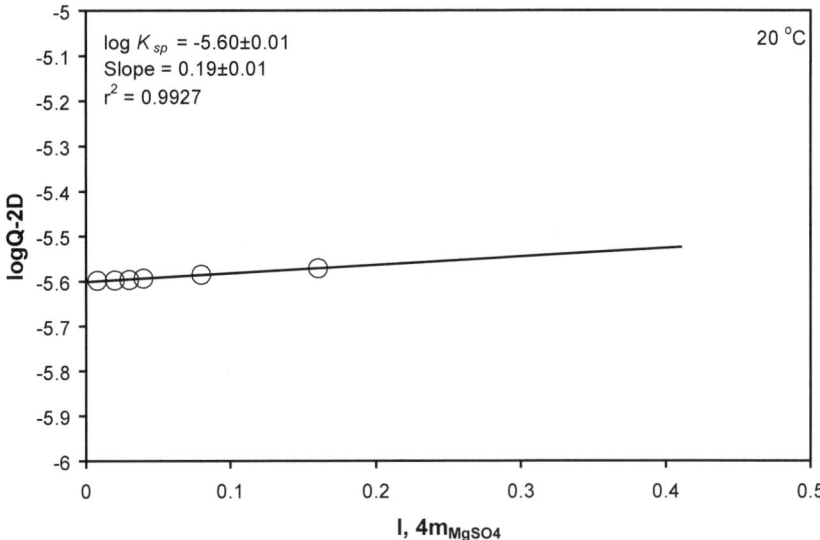

Figure 7.7 A plot showing [log $Q_{SP} - 2D$] as a function of ionic strengths in $MgSO_4$ solutions at 20°C. The log Q_{SP} is conditional solubility product of TlBr(cr) in an $MgSO_4$ medium.

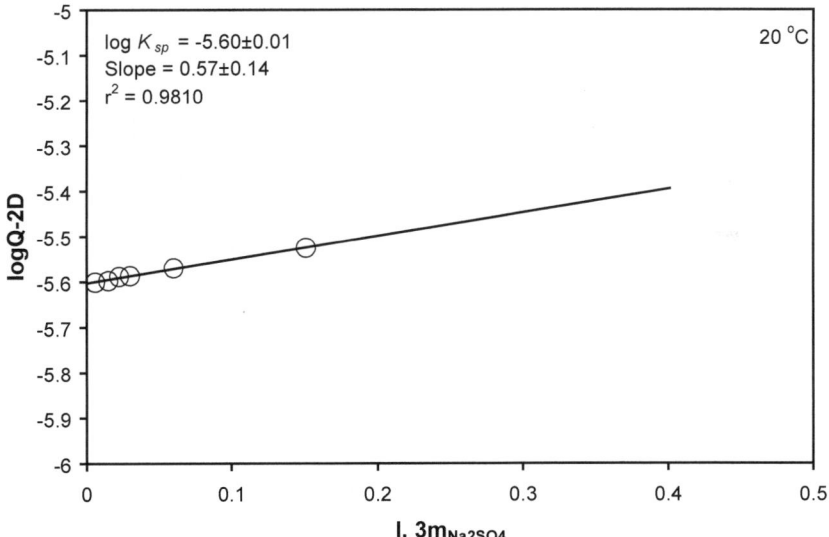

Figure 7.8 A plot showing [log $Q_{SP} - 2D$] as a function of ionic strengths in Na_2SO_4 solutions at 20°C. The log Q_{SP} is conditional solubility product of TlBr(cr) in an Na_2SO_4 medium.

is calculated to be 0.10 ± 0.03 (Table 7.2). Similarly, based on $\Delta\varepsilon$(Eq. 10) = -0.57 ± 0.14 for Eq. (13) at 20°C, $\varepsilon(Na^+, Br^-)$ is calculated to be -0.28 ± 0.14 (Table 7.2).

Solubility Product Constants of TlI(cr) at 20°C

In Figs. 7.9 and 7.10, log K_{sp}'s for TlI(cr) at 20°C are evaluated from solubilities of TlI(cr) in $MgSO_4$ and Na_2SO_4 solutions, respectively. The dissolution reaction for TlI(cr) can be cast as,

$$TlI(cr) = Tl^+ + I^- \tag{14}$$

According to the SIT model, the solubility product constant of TlI(cr) at infinite dilution can be evaluated in the following form,

$$\log K_{sp} = \log Q_{sp} - 2D + \Delta\varepsilon \text{ (Eq. 14)} \times I_m \tag{15}$$

The $\Delta\varepsilon$(Eq. 14) in an $MgSO_4$ medium is governed by the following equation,

$$\Delta\varepsilon(\text{Eq. 14}) = \varepsilon(Tl^+, SO_4^{2-}) + \varepsilon(Mg^{2+}, I^-) \tag{16}$$

Similarly, the $\Delta\varepsilon$(Eq. 14) in an Na_2SO_4 medium is related to the following equation,

$$\Delta\varepsilon(\text{Eq. 14}) = \varepsilon(Tl^+, SO_4^{2-}) + \varepsilon(Na^+, I^-) \tag{17}$$

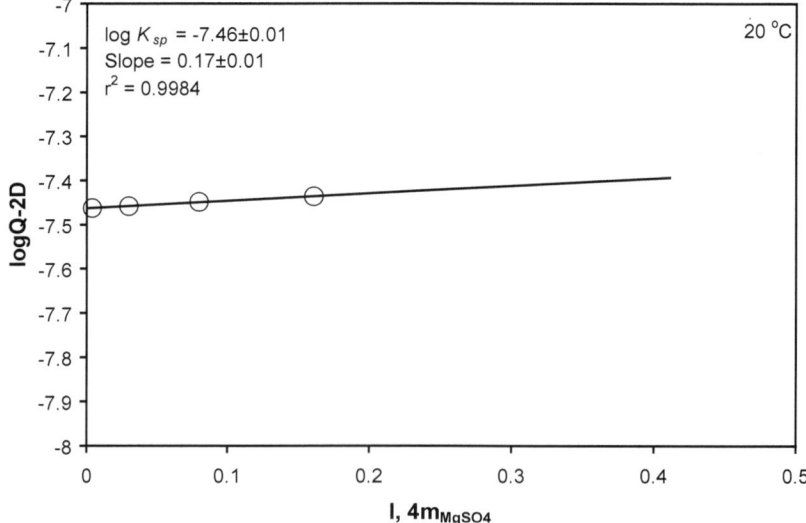

Figure 7.9 A plot showing [log Q_{sp} – 2D] as a function of ionic strengths in $MgSO_4$ solutions at 20°C. The log Q_{sp} is conditional solubility product of TlI(cr) in an $MgSO_4$ medium.

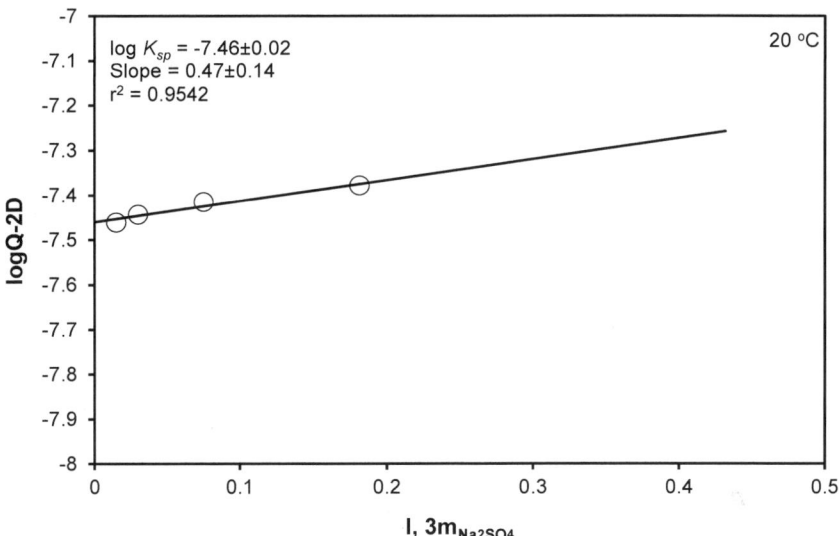

Figure 7.10 A plot showing [log Q_{sp} – 2D] as a function of ionic strengths in Na_2SO_4 solutions at 20°C. The log Q_{sp} is conditional solubility product of TlI(cr) in an Na_2SO_4 medium.

Figures 7.9 and 7.10 indicate that the log K_{sp} for TlI(cr) at 20°C evaluated from solubilities of TlI(cr) in $MgSO_4$ and Na_2SO_4 media are –7.46 ± 0.01 and –7.46 ± 0.02 (Table 7.1), respectively. These two values are in excellent agreement. According to $\Delta\varepsilon$(Eq. 14) = –0.17 ± 0.01 for Eq. (16) at 20°C and $\varepsilon(Tl^+, SO_4^{2-})$ = –0.29 ± 0.03 derived above, $\varepsilon(Mg^{2+}, I^-)$ is derived as 0.12 ± 0.03 (Table 7.2). Similarly, based on $\Delta\varepsilon$(Eq. 14) = –0.47 ± 0.14 for Eq. (17) at 20°C, $\varepsilon(Na^+, I^-)$ is computed as –0.18 ± 0.14 (Table 7.2).

Solubility Product Constants of TlI$_3$(cr) at 25°C

In Fig. 7.11, log K_{sp} for TlI$_3$(cr) at 25 °C are evaluated from solubilities of TlI$_3$(cr) in NH_4I solutions. The dissolution reaction for TlI$_3$(cr) can be written as,

$$TlI_3(cr) = Tl^{3+} + 3I^- \tag{18}$$

According to the SIT model, the solubility product constant of TlI$_3$(cr) at infinite dilution can be evaluated in the following form,

$$\log K_{sp} = \log Q_{sp} - 12D + \Delta\varepsilon \text{ (Eq. 18)} \times I_m \tag{19}$$

The $\Delta\varepsilon$(Eq. 18) in an NH_4I medium is governed by the following equation,

$$\Delta\varepsilon(\text{Eq. 18}) = \varepsilon(Tl^{3+}, I^-) + 3 \times \varepsilon(NH_4^+, I^-) \tag{20}$$

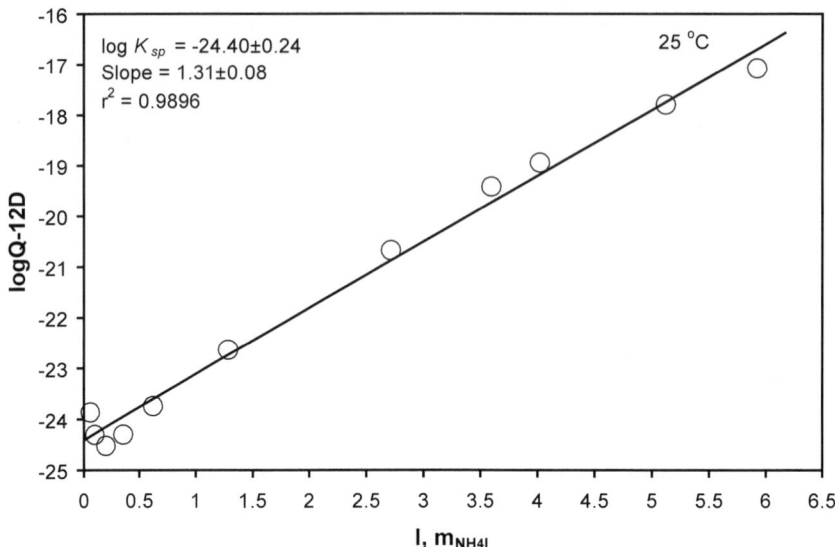

Figure 7.11 A plot showing [log Q_{SP} – 12D] as a function of ionic strengths in NH$_4$I solutions at 20°C. The log Q_{SP} is conditional solubility product of TlI$_3$(cr) in an NH$_4$I medium.

Figure 7.11 shows that the log K_{sp} for TlI$_3$(cr) at 25°C is –24.40 ± 0.24. According to Δε(Eq. 18) = –1.31 ± 0.08, and using ε(Na$^+$, I$^-$) = –0.18 ± 0.14 derived in this study, ε(Tl^{3+}, I$^-$) is derived as –0.77 ± 0.16 (Table 7.2).

Formation Constant of TlI(aq) at 25°C

Figure 7.12 demonstrates that if total thallium concentrations are not corrected for TlI(aq) regarding solubilities of TlI(cr) in KI solutions, the log K_{sp} (i.e., –6.08, the intercept in Fig. 7.12) evaluated would be orders of magnitude higher than that evaluated from MgSO$_4$ and Na$_2$SO$_4$ (log K_{sp} = –7.46 ± 0.02, Table 7.1). The log K_{sp} evaluated from MgSO$_4$ and Na$_2$SO$_4$ is much lower and reliable, as there is no contribution from TlI(aq) because of very low concentrations of I$^-$. Therefore, the solubility data of TlI(cr) in KI solutions imply that Tl$^+$ forms a relatively strong complex with I$^-$, and TlI(aq) substantially increases solubilities of TlI(cr) in KI solutions.

In this study, the conditional formation constant of TlI(aq) are calculated based on the log K_{sp} determined in MgSO$_4$ and Na$_2$SO$_4$ media. First, log Q_{sp}, conditional solubility product quotient in KI solutions, are calculated utilizing a rearranged version of Eq. (15),

$$\log Q_{sp} = \log K_{sp} + 2D - \Delta\varepsilon \text{ (Eq. 14)} \times I_m \qquad (21)$$

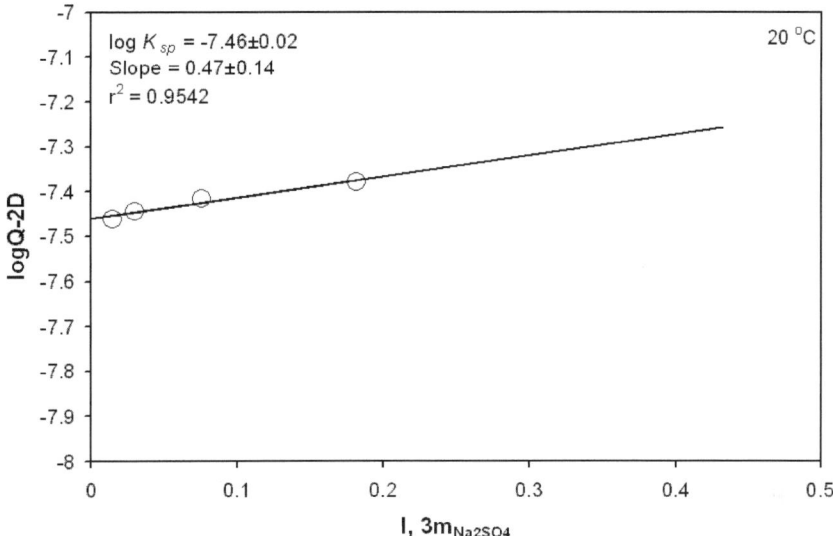

Figure 7.12 A plot showing [log Q_{SP} – 2D] as a function of ionic strengths in KI solutions at 20°C. The log Q_{SP} is conditional solubility product of TlI(cr) without corrections for TlI(aq) in a KI medium.

where $\Delta\varepsilon$ (Eq. 14)in a KI medium should be

$$\Delta\varepsilon(\text{Eq. 14}) = \varepsilon(\text{Tl}^+, \text{I}^-) + \varepsilon(\text{K}^+, \text{I}^-) \tag{22}$$

As the corrections for TlI(aq) are expected to be proportional to concentrations of I$^-$, the negative slope in Fig. 7.12 should be the same as $\Delta\varepsilon$(Eq. 14) for Eq. (22). Using $\Delta\varepsilon$(Eq. 14) = –0.42 ± 0.02, log Q_{sp} at different concentrations of KI are calculated (Table 7.3). Based on $\Delta\varepsilon$(Eq. 14) = –0.42 ± 0.02, and assuming $\varepsilon(\text{K}^+, \text{I}^-) \approx \varepsilon(\text{Na}^+, \text{I}^-)$, which is –0.18 ± 0.14 derived in this study, $\varepsilon(\text{Tl}^+, \text{I}^-)$ is calculated to be –0.24 ± 0.14 (Table 7.2).

Second, the total concentrations of thallium can be expressed as,

$$m_{\Sigma Tl} = m_{Tl^+} + m_{TlI(aq)} \tag{23}$$

Rearranging Eq. (23),

$$m_{TlI(aq)} = m_{\Sigma Tl} - m_{Tl^+} \tag{24}$$

As the equation for Q_{SP} is,

$$Q_{SP} = m_{Tl^+} \times m_{I^-} \tag{25}$$

Table 7.3 Conditional solubility product quotients of TlI(cr) and formation constants of TlI(aq) at 20°C obtained in this study.

KI, m*	Solubility of TlI(cr), m*	log Q_{sp}	log β_1^I
1.84	9.14E-06	−6.15 ± 0.04	1.09±0.04
2.48	1.35E-05	−5.86 ± 0.05	0.98±0.05
3.14	2.18E-05	−5.57 ± 0.05	0.90±0.05
3.78	3.38E-05	−5.30 ± 0.06	0.81±0.06
4.38	4.96E-05	−5.04 ± 0.06	0.72±0.06

*Solubility data were from compilations of Silcock (1979).

Substituting Eq. (25) into Eq. (24),

$$m_{TlI(aq)} = m_{\Sigma Tl} - \frac{Q_{SP}}{m_{I^-}} \tag{26}$$

Therefore, based on log Q_{sp} calculated above, and solubility data for total thallium concentrations and I⁻, $m_{TlI(aq)}$ can be calculated. According to known concentrations of m_{Tl^+}, m_{I^-} and $m_{TlI(aq)}$, conditional formation constant for TlI(aq) can be computed following the reaction below,

$$Tl^+ + I^- = TlI(aq) \tag{27}$$

$$\beta_1^I = \frac{m_{TlI(aq)}}{m_{Tl^+} \times m_{I^-}} \tag{28}$$

The conditional formation constants calculated in this way are listed in Table 7.3. It should be mentioned that the log β_1^I= 0.72 ± 0.06 at I = 4.38 m and 20°C evaluated by this study is consistent with the log β_1^I= 0.74 ± 0.02 at I = 4 M and 25°C in the literature (Kulba and Mironov 1960, cited in Smith et al. 2004). For these calculations, it is assumed that total iodide concentrations are equal to free iodide concentrations without corrections for concentrations of TlI(aq), as concentrations of the latter species are negligible in comparison with high concentrations of total iodide. Based on those conditional formation constants listed in Table 7.3, the thermodynamic formation constant is evaluated in accordance with the following equation (Fig. 7.13),

$$\log \beta_1 = \log \beta_1^I + 2D + \Delta\varepsilon \text{ (Eq. 27)} \times I_m \tag{29}$$

The log β_1 obtained is 1.86 ± 0.02, which is a relatively strong complex. According to $\Delta\varepsilon$ (Eq. 27) = 0.13 ± 0.01, $\varepsilon(Tl^+, I^-)$ = −0.24 ± 0.14, and $\varepsilon(K^+, I^-)$ ≈ $\varepsilon(Na^+, I^-)$ = −0.18 ± 0.14 derived in this study, $\varepsilon(TlI(aq), K^+, I^-)$ is calculated to be −0.29 ± 0.20 (Table 7.2).

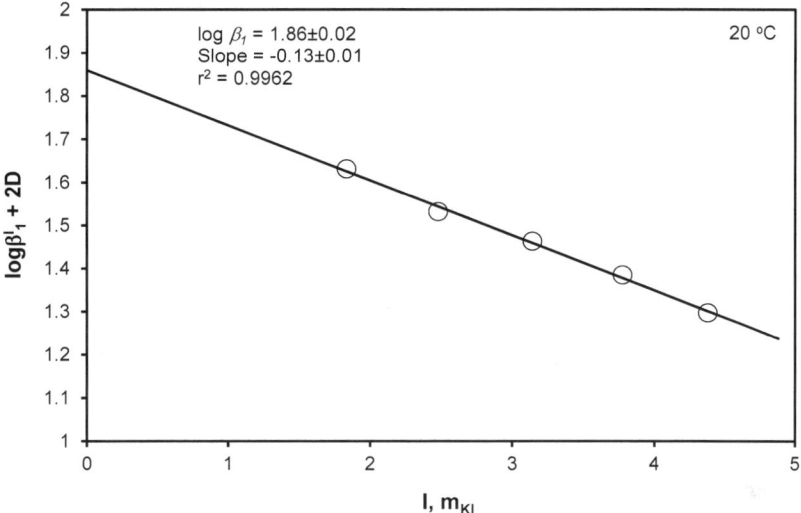

Figure 7.13 A plot showing $[\log \beta_1^I + 2D]$ as a function of ionic strengths in KI solutions at 20°C. The $\log \beta_1^I$ is conditional formation constant of TlI(aq) in a KI medium.

Discussions and Summary

The above results indicate that, in the order of increasing solubility for thallium(I), thallium iodide has the lowest solubilities, followed by thallium bromide, thallium iodate and thallium chloride. Under very oxidizing conditions, thallium could form $TlI_3(cr)$, which has very low solubilities. Therefore, in the soil environments where there are relatively high concentrations of iodide or bromide, TlI(cr) or TlBr(cr) could become solubility-limiting phase(s) for thallium. As the aqueous complex, TlI(aq), is relatively strong, it would become a dominant species in soil solutions where iodide concentrations are sufficiently high.

In conjunction with the SIT coefficients related to thallium aqueous species obtained before (Xiong 2007, 2009), the relevant SIT coefficients provided by this study would make it feasible to model the speciation and solubility of thallium in natural waters with moderate and high ionic strengths. For instance, seawater contains relatively high concentrations of Cl^-, Na^+, Mg^{2+}, SO_4^{2-}, Ca^{2+}, and K^+ with an ionic strength of about 0.7 m. In saline soil solutions, the major components are Na^+ and Cl^-. In strongly saline soil solutions, salinities may be higher than 16 ds/m in terms of electric conductivity, which would be higher than ~ 10 g/L in terms of total dissolved salts (or higher than ~ 0.2 m in terms of molality for NaCl equivalent). To model speciation and solubility of thallium under these conditions, the SIT model with the relevant SIT coefficients for thallium aqueous species would be appropriate.

Acknowledgements

Sandia National Laboratories is a multi-program laboratory managed and operated by Sandia Corporation, a wholly owned subsidiary of Lockheed Martin Corporation, for the U.S. Department of Energy's National Nuclear Security Administration under contract DE-AC04-94AL85000.

References

Ciavatta, L. 1980. The specific interaction theory in evaluating ionic equilibria. Annali di Chimica. 70: 551–567.
Grenthe, I., J. Fuger, R.J.M. Konings, R.J. Lemire, A.B. Muller, C. Nguyen-Trung and H. Wanner. 1992. Chemical Thermodynamics of Uranium. Elsevier Science Publishers, New York. USA.
Helgeson, H.C. and D.H. Kirkham. 1974. Theoretical prediction of the thermodynamic behavior of aqueous electrolytes at high pressures and temperatures. II. Debye-Hückel parameters for activity coefficients and relative partial molal properties. American Journal of Sciences. 274: 1199–1261.
Kulba, F.Y. and V.E. Mironov. 1960. Stability of $TlBr_n^{1-n}$ and TlI_n^{1-n} ions. Russian Journal of Inorganic Chemistry. 5: 922.
Silcock, H. 1979. Solubilities of Inorganic and Organic Compounds, vol. 3. Pergamon Press. New York. USA.
Smith, R.M., A.E. Martell and R.J. Motekaitis. 2004. NIST Critical Selected Stability Constants of Metal Complexes Database Version 8.0 for Windows. NIST Standard Reference Database 46, National Institute of Standards and Technology, US Department of Commerce, Gaithersburg, MD. USA.
Xiong, Y.-L. 2006. Estimation of medium effects on equilibrium constants in moderate and high ionic strength solutions at elevated temperatures by using specific interaction theory (SIT): Interaction coefficients involving Cl^-, OH^- and Ac^- up to 200 degrees C and 400 bars. Geochemical Transactions. 7: (4)1–19, doi: 10.118611467-4866-7-4.
Xiong, Y.-L. 2007. Hydrothermal thallium mineralization up to 300°C: A thermodynamic approach. Ore Geology Reviews. 32: 291–313.
Xiong, Y.-L. 2009. The aqueous geochemistry of thallium: speciation and solubility of thallium in low temperature systems. Environmental Chemistry. 6: 441–451.

Fractionation and Speciation Analysis of Antimony in Atmospheric Aerosols and Related Matrices

Patricia Smichowski

Introduction

Antimony is a fascinating element that, in spite of its widespread uses and potential toxic effects, has been much less studied in comparison with other elements such as As, Cr, Hg or Pb. Antimony may produce adverse effects on humans and the environment and has no known physiological functions. The toxic properties of Sb depend on the solubility of their compounds in biofluids, the presence of complexing agents and the antimony's oxidation state. Antimony interacts with –SH groups in cells, particularly with enzymes, inhibiting their enzymatic activity (Maeda 1994) and causes possible serious effects depending on the doses. Antimony(III) is more toxic

Comisión Nacional de Energía Atómica, Gerencia Química, Av. Gral Paz 1499, B1650KNA-San Martín, Provincia de Buenos Aires, Argentina.
Email: smichows@cnea.gov.ar

than Sb(V), and organic forms of Sb are less toxic than the inorganic forms. Maeda (1994) reported that another factor that could explain the higher toxicity of Sb(III) over Sb(V) is that Sb(III) tends to be retained for longer periods of time in the body, making its effects long lasting.

The United States Environmental Protection Agency (USEPA) and European Union (EU) deemed Sb to be a pollutant of priority interest and the USEPA has designated the maximum contaminant level (MCL) for Sb to be 6 µg L^{-1}, while the EU set their Sb MCL for drinking water at 5 µg L^{-1} (Fillela et al. 2002).

Antimony has been used by human cultures since the Early Bronze Age. The natural sulfide of Sb was known and used in Biblical times as medicine and as a cosmetic (http://minerals.usgs.gov/minerals/pubs/commodity/antimony/). Antimony (as Sb_2S_3) was used as a pigment to make mascara and is mentioned in an Egyptian papyrus dating from 1600 BC (Shotyk et al. 2005). Antimony has had, along the years, a remarkable history in alchemy and early medicine. More recently, Sb_2O_3 was used as a white pigment for paint. At present, Sb oxide combined with a halide (such as chlorine) is widely used as flame retardants in different products including papers, adhesives, textiles, tires, brake linings and plastics. Flame retardants account for approximately 70 percent of the demand for primary Sb and 90 percent of the demand for Sb_2O_3 (He et al. 2012).

Antimony and its compounds are natural components of the Earth's crust. Its abundance is about 0.3 µg g^{-1}, with no significant difference between the upper and the lower continental crust (Wedepohl 1995). Natural and anthropogenic sources are responsible for the presence of Sb in the environment. Volcanoes are the main source of natural Sb in the environment. Its contribution to the atmosphere was estimated in ~ 5 t per year (Shotyk et al. 2005). In spite of this, volcanic emissions only account for 3–5 percent of global Sb emission to the atmosphere.

The determination of Sb species is fundamental for environmental studies because the toxicity and biological behavior of this metalloid depends on the oxidation state. Most of the proposed methods dealing with Sb speciation have only been based on the determination of Sb(III) and Sb(V). This could be explained by the fact that in biological and environmental matrices, Sb is mainly found in two (III and V) oxidation states being, in addition, the most toxic ones. Therefore, the role played by inorganic and organic species of Sb in different environmental compartments merits further research.

The atmosphere is an important medium for global element transport to even faraway regions. As a consequence, people are exposed to metals and metalloids, including Sb, contained in airborne particles that often are well above natural background levels. In this regard, trace elements can be placed among major atmospheric pollutants and their determination in total airborne particulate matter (APM) as well as in the corresponding inhalable fractions represents an important parameter in assessing possible implications for public health (Rizzio et al. 2000).

The presence of Sb in the atmosphere is mainly due to anthropogenic inputs. Its contribution is 100 to 200 fold higher than that of atmospheric Sb emission from natural sources (Nriagu 1990). It is believed that when Sb is released into the atmosphere as an aerosol, it is oxidized to Sb_2O_3 by reaction with airborne oxidants.

The atmospheric Sb signal preserved in the ice core reflects contamination from industrialization, the economic boom which followed World War II, as well as the comparatively recent introduction of flue gas filter technologies and emission reduction efforts (Gómez et al. 2005). Today, Sb is still used in several industrial applications: as an alloying element to harden Pb and other metals, in casting alloys, bearings, cable sheathing, paint pigments, ceramics, enamels, plastics (it was found that Sb_2O_3 possesses certain properties which impart opacity to the plastics) and also in the chemical industry. Antimony trioxide (Sb_2O_3) is largely used in the production of glasses, ceramics and flame retardants in textiles, mattresses and different kind of fibers. It is applied as a pigment in dyes and paints and is also the catalyst used to manufacture polyethylene terephthalate (PET) used in the manufacture of plastic bottles. Shotyk and Krachler (2007) determined Sb in 132 brands of bottled water from 28 countries and observed that two of the brands were at or above the maximum allowable Sb concentration for drinking water in Japan (2 µg L^{-1}). The leaching of Sb from PET bottles showed variable reactivity and, in general increased with time of storage.

In recent years Sb was associated with traffic and identified as a traffic-related element (TRE). For a long time it was assumed that fuel combustion was the primary mechanism responsible for particles formation and also metal emission. Today, it is generally acknowledged that other important processes are involved and that metal emission can arise, among others, also from the different parts of a vehicle. The most important sources, other than fuel combustion, of potentially toxic elements include abrasion products from tires, brake lining, bearings, road surface, corrosion in general of vehicle components and resuspension of soil and road dust. Many parts of vehicles contain Sb alloys and other Sb compounds. Up to 7 percent of Sb as stibnite (Sb_2S_3) are used in brake linings as lubricant to reduce vibrations during braking and to improve friction stability. Brake pads are composed by a polymer matrix of more than 10 different compounds consisting of fibers, fillers and modifiers held together by a binder. During the braking process, high temperatures are accomplished due to friction effect. The surface contact temperature at the interface between brake pads and metallic counter faces can vary greatly, from 600 to 1500°C (Ingo et al.

2004). In these conditions, Sb_2S_3 is oxidized to Sb_2O_3 between 300 and 450°C (Jang and Kim 2000, Cho et al. 2006).

In the framework of a TRE, Sb_2O_3 is also used as an effective flame retardant in vulcanization of rubber. It is not by chance that due to its stabilizing properties, Sb is used in a number of alloys including those used in motor bearings. Brake pad manufactures are currently developing Sb-free brake pads that were first introduced into the market about three years ago. Their impact is yet unknown (von Uexküll et al. 2005). Road traffic emission of Sb has also been ascribed to the use of certain organic Sb-containing compounds such as Sb-dialkyldithiocarbamates in greases and motor oil (Huang et al. 1994). Smichowski reviewed investigations done over the last few years to quantify key traffic-related elements, namely, Pb, Pt, Rh and Sb in atmospheric aerosols (Smichowski et al. 2008a).

It would obviously be impracticable to cite, even briefly, all the contributions on speciation and/or fractionation of Sb in atmospheric aerosols that have enriched the relevant literature to date. Nonetheless, the author considers it important to mention some representative papers that have significantly contributed to shed light on the selective determination of Sb species in airborne particulate matter and related matrices in this chapter. These papers, classified according to the different matrices studied, are discussed briefly below.

Antimony in Atmospheric Aerosols and Related Matrices

As with many other trace elements, Sb travels through the atmosphere over long distances as part of its global biogeochemical cycling process. Pacyna and Pacyna (2001) have estimated that a total of 4.0 kt y^{-1} with 1.6 kt y^{-1} are released into the environment only from anthropogenic emissions.

The increasing concentrations and enrichment factors (EF) of Sb in atmospheric aerosols from urban areas observed in the last years (EF ≥ 10^5) has encouraged many research groups around the world to study this metalloid and its species in airborne particles, volcanic ashes and road dust, as well as to identify the origin (Cornelis et al. 2006, Miravet et al. 2006, Furuta et al. 2005, Miravet et al. 2007, Iijima et al. 2010, Araujo et al. 2010, Fujiwara et al. 2011). Perhaps, the first question to answer would be: why is it necessary to learn more about the species of Sb present in atmospheric aerosols? This question has been answered well by Filella et al. (2009) who stated that speciation will affect not only Sb transport, but also the form in which it will be deposited into an environmental compartment. In addition, speciation will greatly influence the subsequent behavior: the fate of Sb in waters and soils might be very different if the origin of the deposited aerosol is a combustion process rather than brake-wear.

The analytical determination of Sb in atmospheric aerosols and related matrices can be problematic because Sb concentrations are often very low. However, the wide range of analytical techniques and hyphenated systems available nowadays allows for the adequate quantification of this metalloid at trace and ultratrace levels. The problem of detecting very low concentrations of Sb is more severe when species or size fractionated samples should be analyzed. In this context, sample preparation is a critical step and greatest care should be devoted to the scrupulous preparation of blanks and the minimization of exogenous contamination. Sample preparation of atmospheric aerosols for elemental analysis and fractionation studies has been addressed by Smichowski et al. (2007) in a book chapter.

The on-line combination of separation techniques, mainly high performance liquid chromatography (HPLC) with suitable element-specific detectors, has been frequently used for the separation and subsequent determination of Sb(III) and Sb(V), and in some cases for organoantimony species in different kind of matrices (Smichowski et al. 1995, Zheng et al. 2000a, Krachler and Emons 2001a, Bellido-Martín et al. 2009, Iijima et al. 2010). In 1998, Smichowski and coworkers summarized and discussed the analytical methods, techniques and couplings most frequently used for the determination of Sb species at trace and ultratrace levels in waters and other matrices. In the same direction, some years later Miravet et al. (2010) reviewed different aspects of Sb speciation in environmental matrices using hyphenated techniques. Most revision papers paid little attention to the speciation analysis of Sb in atmospheric aerosols probably due to scarce information on this topic available in the literature.

In methods involving HPLC separation, the conditions of separation should be carefully investigated in order to achieve good efficiency and resolution in a reasonable analysis time. The use of complexing mobile phases has been a useful alternative to alleviate the problems of broad picks associated with the elution of Sb(III). However, Sb(III) is thought to present, in solution, rather complicated equilibrium partitions of agglomerations and non-defined species in the presence of these mobile phases. This is a controversial point that requires for further research. Mobile phases containing complexing ligands prevent obtaining any information about the original Sb species present in samples because the complexing ability of the mobile-phase components changes the original species during the chromatographic run (Filella et al. 2009).

Antimony in Airborne Particulate Matter

Research devoted to the selective determination of Sb species has been, in general in the framework of analytical chemistry studies. In these cases, only a limited number of real samples are analyzed and mainly with the objective to corroborate the success of the methodology developed. To a lesser extent, some work has more recently focused to elucidate environmental issues where Sb is merely included with several other trace elements. On the other hand, in environmental studies it is mandatory to collect and analyze a large number of samples preferably extending over long periods. Only in a limited number of environmental studies, that will be described below, Sb and its species were specifically and studied at length in atmospheric aerosols.

Airborne particulate matter is introduced into the atmosphere through a variety of processes, including natural (crustal weathering, sea-salt aerosol generation, volcanism) and anthropogenic (fossil fuel combustion, industrial activity, incineration, mining) sources. Fly ashes generated in combustion processes are also important carriers of hazardous substances (such as toxic metals) into the environment.

The environmental effects of aerosol particles depend upon their sizes and chemical compositions. Aerosol particles influence solar radiation transfer and cloud-aerosol interactions, and control the optical, electrical and radiative properties. Deposition, residence time, particle dispersion, transports and inhalation processes are predominantly influenced by the size of the particles. Particles larger than 1.0 μm occur as coarse material; these are generally produced during mechanical processes and are rapidly removed (in timescales of hours to a few days) near the source by gravitational sedimentation. Small particles have a considerably longer residence time in the atmosphere and are much more efficiently transported. To illustrate the importance of size distribution in APM, it is worth mentioning that the degree of respiratory penetration and retention is directly related to the aerodynamic particle size. Particles with aerodynamic diameters < 1.0 μm are deposited in the alveolar regions of the lungs, where the adsorption efficiency for trace elements is 60–80 percent (Infante and Acosta 1991), and this can affect lung physiology, especially if the particles contain biologically-available toxic metals (Fernández Espinosa et al. 2002).

To gain information on the potential toxicities of trace elements, it is also necessary to know the distributions of their chemical forms, because the bioavailability of elements depends on their characteristic surfaces, on the strength of their bonds and on the properties of solutions in contact with APM.

Most studies on the elemental determinations of trace elements in APM focus only on total metal concentration without distinguishing the various species that are present. This information is useful and necessary to evaluate overall pollution levels but is insufficient because the effect of a trace element in the environment strongly depends upon the associated form in the solid phase to which the element is bound (Ure 2001).

To gain information about the bioavailability, mobility, solubility, metal cycles, fate and toxicity of trace elements in APM samples has encouraged many researchers to design and investigate extraction schemes aimed at the sequential solubilization of metals bound to substrates such as soils, sediments and aerosols. Sequential chemical fractionation is therefore a way to determine the actual metal activity in the environment and provides a new perspective on analytical control (Fernández et al. 2000). At this point it is necessary to define the term "fractionation". The International Union for Pure and Applied Chemistry (IUPAC) defined fractionation as the "process of classifying an analyte or group of analytes from a certain sample according to physical (e.g., size, solubility) or chemical (e.g., bonding, reactivity) properties" (Templeton et al. 2000). It should be noted that this definition does not strictly correspond to determining the chemical forms of elements, but rather to measuring broad forms of elements. Indeed, the results obtained by sequential extraction procedures are operationally defined (in other words the forms of the trace elements are defined by the scheme used for their determination). Smichowski et al. (2005) recapitulated the different approaches followed, via sequential chemical extraction, for metal fractionation of atmospheric aerosols. The review focused mainly on sequential chemical fractionation schemes published over the last 15 yr which have been classified in five categories according to the main procedures described in the literature.

Little information is available about the fractionation of Sb by chemical bonding and published material only focuses to the determination of Sb in the aqueous fraction. Some of the studies on fractionation and/or speciation analysis of Sb in APM are described below.

The use of a powerful separation technique such as HPLC in combination with plasma-based techniques, in particular inductively coupled plasma-mass spectrometry (ICP-MS) resulted in being especially suited for Sb speciation studies. A number of papers reported the determination of Sb(III) and Sb(V) using an anionic polymer-based column (PRP-X100). Aside from this popular column, size-exclusion, cationic and reversed-phase columns have also been used for Sb speciation studies in APM.

A series of comprehensive research studies on the determination and identification of Sb species in APM was reported by Furuta and his group (Furuta et al. 2005, Zheng et al. 2000, 2001a, 2001b, 2001c). Different Sb species were determined in APM collected in Tokyo, Japan. Total and water-soluble Sb compounds were determined in a Japanese quality control APM sample, the Standard Reference Material (SRM) NIST 1648 (Urban Particulate Matter) and an APM sample collected in Tokyo (Zheng et al. 2000a). Speciation analysis was carried out using HPLC coupled to ICP-MS. With the objective to shorten the retention time of Sb(III) and peak tailing in the chromatographic column, a shorter silica-based anion-exchange column namely, Synchropak Q300 was tested and proved to be more adequate than the very well known PRP-X100. In the chromatographic separation of Sb(III) and Sb(V), 2 mmol L^{-1} phthalic acid was selected as a mobile phase in a pH range between 4.3 and 5.0. The authors reported that the addition of 5 mmol L^{-1} ethylenediaminetetraacetic acid (EDTA) improved the separation of the inorganic species. In spite of their efforts, a broad peak for Sb(III) was still observed and no significant improvements in the separation of inorganic species of Sb was obtained in comparison with previous reported studies (Smichowski et al. 1998). Total concentrations of Sb measured in APM and in the aqueous extract resulted to be 195 ± 13 and 37.9 ± 0.8 µg g^{-1}, respectively. A low recovery of Sb (19.4 percent) was reached in the water-soluble fraction. Detection limits were found to be 0.1 µg L^{-1} for Sb(V) and 0.3 µg L^{-1} for Sb(III). An interesting finding was that the chromatogram obtained evidenced the presence of organic Sb species in APM collected in Tokyo.

One aspect of Sb speciation that is severe and difficult to handle is associated with the identification of peaks in the HPLC chromatograms. It may be mainly attributed to the lack of appropriate Sb standard compounds for peak identification. Krachler et al. (2001b) discusses different aspects that have hampered the preparation of appropriate Sb standards. They report that attempts to synthesize soluble mono and dimethylated Sb compounds have failed so far because they tend to polymerize upon dissolution or cannot even be synthesized as a monomer, or are not stable under ambient air. Moreover, mono- and dimethylated Sb compounds can sometimes only be dissolved under drastic conditions, which is not appropriate for speciation in liquid extracts at all. Another important problem hampering the speciation of Sb compounds (and speciation analysis of other analytes) is the lack of stability of the Sb compounds throughout the entire analytical process. This scenario is probative of the difficulties that hamper speciation studies.

In an attempt to shed light on the identification of the unknown Sb species, a new HPLC-ICP-MS procedure for the speciation analysis of Sb in APM was developed by Zheng et al. (2000b). In a first step, APM samples were subjected to MW-assisted acid digestion and a total concentration of 56.0 ± 4.1 µg g^{-1} of Sb was quantified by ICP-MS in the solution. For speciation studies, Sb species were extracted from the filters loaded with APM using three solvents and their performance was compared. Each solvent (Milli-Q water, phosphate and EDTA) was selected according to a different affinity for trace metals. Milli-Q water was used to extract water-soluble Sb compounds, phosphate to remove metal fraction which is adsorbed by ion exchange mechanism, and EDTA to put in solution carbonate-bound and organically-bound fractions of Sb. In this attempt, anion-exchange and size-exclusion HPLC-ICP-MS approaches were investigated. In the aqueous extract, using the coupling HPLC-ICP-MS, the authors found that Sb(V) was the predominant species and three non-identified Sb species were detected. To gain information on these species, further experiments were carried out using hydride generation (HG). Derivatization by HG is a reliable, simple, fast and very used methodology for Sb determination at trace and ultratrace levels. The study showed that these species were active Sb species with different efficiency of generation. The aqueous extracts were injected in a size-exclusion chromatography-ICP-MS (SEC-ICP-MS) coupling that evidenced the presence of TMSb (trimethyl Sb). Sb(V) was the predominant species (~ 80%). The presence of TMSb and hydride active species in APM was indicative of the presence of organic species. This important information gave evidence that special attention should be paid when using hydride generation to the selective determination of Sb(III) and Sb(V) in these kind of samples. The authors proposed electrospray-mass spectrometry (ES-MS) as a possible alternative to identify the unknown species.

Another contribution from Furuta's group was the complexation of Sb compounds with citric acid and its application to the speciation analysis of Sb(III) and Sb(V) using HPLC-ICP-MS (Zheng et al. 2001c). As suggested previously, ES-MS was used to investigate the effects/advantages of the complexation of both Sb compounds with citric acid on speciation analysis. Both inorganic species of Sb formed very stable complexes with citric acid in the aqueous extracts of APM. This allowed the separation of Sb(III) and Sb(V) on the typical PRP X-100 anion exchange column with 10 mmol L^{-1} EDTA-1.0 mmol L^{-1} phthalic acid at pH 4.5 as mobile phase. The main advantage was that, both complexes were retained and no elution of any of them was observed in the solvent front. As a consequence, the stabilizing effect of the complexing agent allowed a significant improvement of the limits by one order of magnitude in comparison with those previously reported (Zheng et al. 2001a, b). The detection limits reported for Sb(III) and Sb(V) were 0.005 µg L^{-1} and 0.07 µg L^{-1}, respectively. The authors analyzed an APM sample containing 104 ± 3 µg g^{-1} of Sb. Both inorganic species of Sb were detected and the ratio Sb(V) to Sb(III) was found to be ~ 4.6:1.

Electrospray time-of-flight mass spectrometry (ES-TOF-MS) was used to investigate positive and negative ion electrospray mass spectra of frequently encountered organic Sb compounds namely, trimethlyantimony dichloride (TMSbCl$_2$) and trimethylantimony dihydroxide (TMSb(OH)$_2$), and inorganic Sb compounds, potassium hexahydroxyantimoniate (Sb(V)) and potassium antimonyl tartrate (Sb(III)) (Zheng et al. 2001c). A mixture of methanol-water (v/v, 50:50) was used for particles extraction and then the solutions were introduced into the ion source. Based on their findings, they proposed a mechanism for the solution chemistry of TMSb(OH)$_2$. With respect to TMSbCl$_2$, the authors speculated that in aqueous solution it might be hydrolyzed to form $(CH_3)Sb(OH)(Cl)$. Furuta and coworkers (Zheng et al. 2000a, 2001c) detected for the first time the presence of Sb(III), TMSb and several hydride active unknown Sb species in aqueous extracts of APM collected in Tokyo, Japan.

The studies described above are probative that the complexation of inorganic Sb species prevents the oxidation of Sb(III) to Sb(V) during the ultrasonic-assisted and MW-assisted extraction of APM samples. Recently, Hansen and Pergantis further investigated the formation of an Sb(V)-citrate complex by HPLC-ICP-MS and HPLC-ES-MS/MS due to their relevance and the role played in food chemistry (Hansen and Pergantis 2006).

Recently, Marconi et al. (2011) collected in Italy, (i) 15 pairs of PM-10 samples at a site heavily impacted by traffic in the urban area of Rome, (ii) dust deposited on the side of an urban road close to the APM sampling site, and (iii) dust samples from three kinds of brake pads before being analyzed were sieved to get particles < 63 µm prior to analysis. Antimony was extracted from all categories of samples using EDTA and potassium hydrogen phthalate (KHP). The extract obtained was separated in two aliquots. The first one was analyzed by ICP-MS to obtain total extractable Sb. The other one was subjected to ion chromatograph (IC)-ICP-MS for Sb(III) and Sb(V) determination. The analysis of coarse particles produced by brake pad abrasion and road dust showed that particles contained extractable Sb almost totally as inorganic forms being Sb(III) the predominant species in brake pad dust. These results indicate that the conversion of Sb(III) to Sb(V) is mainly driven by the temperature reached by the pad during the braking process. The analyses of PM-10 samples collected at an urban site showed that Sb(III) concentrations in APM were higher than those of Sb(V), in contrast with the findings in other environmental matrices. The results obtained from APM size-segregated samples indicated that a relevant fraction of total extracted Sb in particles < 1 µm (fine fraction) is in the form of solid nanoparticles suspended in the solution. The nanoparticles contribution may be easily estimated as the difference between ICP-MS and IC-ICP-MS analysis.

In another study, the same research group (Canepari et al. 2010) aimed to maximize the extraction efficiency and tested different extracting solutions (deionized water, citric acid 30 mM, EDTA 5 mM/KHP 0.5 mM) on the NIST SRM 1648 (Urban Particulate Matter). They observed that with respect to the mixture tartaric acid/NH_4HCO_3, the isocratic elution with ammonium tartrate allowed retention times reduction and better Sb(III) peak symmetry. EDTA reacted in a similar way to ammonium tartrate. The best separation was achieved by EDTA/KHP, as the addition of KHP to the particulate matter improved Sb(III) peak symmetry and reduced elution times. With respect to the relative distribution between Sb(III) and Sb(V), results confirmed that Sb(III) concentration in APM is comparable, or even higher, than that of Sb(V) and that brake pad abrasion is the main source of both inorganic soluble Sb species in coarse particles.

Antimony(III) oxide exists as two crystalline polymorphs: the cubic senarmontite and the orthorhombic valentinite. The oxidation of senarmontite takes place from 531°C and starts the oxidation product Sb_2O_4, which is a mixed oxide of Sb(III) and Sb(V) (Jenkins et al. 1998). No further oxidization takes place and, at temperatures higher than 531°C, Sb_2O_5 forms Sb_2O_4 (Jenkins et al. 1998, Zheng et al. 2000a). In conclusion, it is reported that the potential traffic related Sb compounds are: Sb_2S_3, Sb_2O_3 and Sb_2O_4.

Recently, an investigation was carried out to study the optimal leaching conditions for Sb_2O_3 or Sb_2O_4 that are formed during brakes operation (Zih-Perényi et al. 2010). Several steps were followed : (i) solubility test of the above mentioned compounds, and (ii) leaching of the compounds from dust spiked with them. In addition, real samples from two relevant busy junctions of Budapest were analyzed. As in other studies, citric acid resulted to be an efficient extractant for Sb. In this case, a solution of 0.5 mol L^{-1} citric acid was efficient to leach the whole Sb_2O_3 content while extracting less than 10 percent Sb_2S_3 and no Sb_2O_4 at all. Both extracted species resulted to be quantitatively and selectively soluble in a 6 mol L^{-1} HCl solution. Graphite furnace atomic absorption spectrometry (GFAAS) in conjunction with HG was used for Sb quantification in the extracts. The evolved SbH_3 was retained on the wall of a pyrolitic graphite tube previously coated with Pt/Zr by atomizing 20-µL Zr and 50-µL Pt solutions. The results of the leaching tests showed that the concentration of extracted Sb was 40 µg g^{-1} in the settled dust of Budapest, about half of which corresponded to Sb_2O_3. The Sb_2O_4 content calculated as the difference of total and leachable fraction was about 10 percent but according to the authors it was obtained with high uncertainty. The predominant Sb species was Sb_2S_3 with measured concentrations of 19.0 ± 1.9 and 24 ± 2.6 µg g^{-1}.

The determination of metal and metalloids in the soluble fraction of airborne particles has stimulated the interest of different research groups. Since the aqueous fraction is the most bioavailable, the researchers considered it more important to study only this fraction and APM samples were not subjected to further fractionation. Unfortunately, and probably due to the low levels at which Sb can be found in the soluble fraction, only few papers included Sb in their studies.

Furuta et al. (2005) reported that Sb had the highest enrichment factor (EF) among the elements determined in APM samples that were separated by size (d < 2 µm, 2–11 µm and > 11 µm). The EF is an index of anthropogenic source contribution, defined as the ratio (element/X) $|_{sample}$/ (elemenent/X) $|_{crustal\ rock}$; where X is an element taken as reference and the crustal rock chemical profile is based on the average crustal composition in the depth between 0–40 km (Wedepohl 1995). Furuta's experiments gave evidence that the source of Sb in squared-shape APM particles < 2 μm were from brake pad wear.

A study of concentration distribution of dissolved Sb(III) and Sb(V) in size-classified inhalable APM demonstrated that Sb(III) was the most abundant species in the coarse fraction whereas Sb(V) was distributed in both the fine and coarse fraction (Iijima et al. 2010). This information is useful to evaluate health risks by inhalation exposure to this metalloid.

Ferreira et al. (2011) reported the first application of slurry sampling in conjunction with HG-atomic absorption (HG-AAS) spectrometry quartz tube atomic absorption spectrometry (HG-QT-AAS) for the selective determination of inorganic Sb and Sb(III) in APM collected in the Bahia State, Brazil. The optimization step was performed by using full two-level factorial and Box–Behnken designs involving three factors: flow rate and concentration of reductant and hydrochloric acid concentration, having as chemometric response absorbance. This method allowed the determination of total Sb and Sb(III) with limits of quantification of 0.3 and 0.2 mg L^{-1}, respectively. The accuracy was confirmed by the analysis of the NIST SRM 1649a (Urban Dust). Total Sb concentrations in the four samples analyzed ranged from 4.32 to 4.60 ng m^{-3}, and Sb(III) concentrations varied from 0.33 to 0.67 ng m^{-3}.

The efficient extraction of particles from filters is of prime importance to reach reliable results. In spite of this, only few papers have specifically tackled this topic for the speciation analysis of Sb. Bellido-Martín et al. (2009) tested different extraction media (diammonium tartrate, hydroxylammonium chlorhydrate, citric acid + ascorbic acid, phosphoric acid and citrate solutions) with assistance of an ultrasonic probe for Sb(III) and Sb(V) speciation analysis in APM. Two key parameters were carefully optimized: the operation power and time of extraction. The higher extraction recoveries (> 90%) were obtained with a 100 mmol L^{-1} hydroxylammonium chlorhydrate aqueous solution assisted by the ultrasound probe operated at 50 W during 3 min. A mobile phase of 200 mmol L^{-1} diammonium resulted to be the best alternative for the efficient chromatographic separation of Sb(III) and Sb(V) and subsequent analysis of the extracts by HPLC-HG coupled to atomic fluorescence spectroscopy (AFS). AFS has received growing attention in the last years for Sb determination because it offers good analytical performance in terms of detection limits and selectivity. Analyzed samples were collected in heavy traffic streets from Buenos Aires, Argentina. The results showed the presence of both Sb species in all analyzed samples. Antimony concentrations ranged as follows: Sb(III) from 3.2 to 10.7 ng m^{-3} and Sb(V) from 4.0 to 7.3 ng m^{-3}.

Antimony in Volcanic Ashes

Potentially toxic trace metals, which are normally rare from a geochemical point of view, such as Sb, may be enriched in volcanic ashes with respect to their crustal abundance (Craig 2003). A significant amount of these trace elements has been found at inhalable sub-micrometer size, thereby representing a serious risk to the population. In addition, the effects of the air-transported volcanic ashes continue for months, even at distant locations (Miravet et al. 2006).

Samples of volcanic ashes were collected in Argentina during the latest eruptive activity of the Copahue volcano located along the Chile-Argentina border for the subsequent determination of trace elements by neuron activation analysis (NAA) and plasma-based techniques (Gómez et al. 2002, Smichowski et al. 2003). A simple fractionation procedure was applied to evaluate the distribution of different elements in the samples. Four particles size fractions (namely, A, B, C and D) of diameters, A < 36 µm, B = 36–45 µm, C = 45–150 µm and D = 150–300 µm were obtained. Among the elements investigated, As, Cd, Cu and Sb deserve particular attention because significant enrichment factors, in the smallest size fraction, were calculated as follows (fraction A): Sb, 6.26; Cd, 4.18; As, 373; Cu, 2.72. Antimony showed the highest EF and it is important to state that fraction A is the smallest one (A < 36 µm), i.e., the most important from the health point of view.

In an attempt to continue the previous study as well as to gain more information about not only the fractionation of Sb but also its speciation, Miravet and coworkers (2007) carried out an investigation to assess the presence of inorganic Sb species in ashes from the Copahue volcano. Antimony species were extracted from each fraction using a good extractant for Sb such as citrate (1 mol L^{-1} citrate buffer at pH 5). Both inorganic species of Sb were separated and determined in the extracts by HPLC combined on-line with ICP-MS (HPLC-ICP-MS). Total Sb ranged from 0.43 µg g^{-1} (fraction D) to 1.07 µg g^{-1} (fraction A). Antimony species concentration (expressed as µg g^{-1}) in the four fractions varied from 0.14 to 0.67 for Sb(III) and 0.02 to 0.03 for Sb(V). The results showed the occurrence of both inorganic Sb species in the extractable portion of volcanic ash samples being Sb(III) the predominant species in all cases. These results are consistent with the predominance of Sb species characteristic of reducing processes. The authors proposed as probable Sb species under reducing conditions the following volatile compounds: $SbCl_3$, Sb_2O_3, Sb_2S_3 as well as various antimonite and antimonate species, which form volcanic, sublimates. While Sb(V) is the most stable species under oxidizing conditions, minor amounts of Sb_2O_5 or other complex oxides or antimonates could also be present in the volcanic ash particulates. Although the oxidation of Sb(III) to Sb(V) by atmospheric O_2 may proceed slowly, it could also feasibly account for the presence of Sb(V) in the volcanic ashes analyzed.

Antimony in Fly Ashes

Coal combustion is recognized as one of the major anthropogenic sources of many metals and metalloids into the atmosphere. Fly ashes generated in combustion processes namely, coal-fired power stations and industrial waste incinerators are important carriers of hazardous substances into the environment. For this reason, the determination of toxic and potentially toxic elements in this matrix is an issue of permanent concern.

Particles of fly ashes are inhomogeneous, highly diverse in chemical composition, dispersed in a broad range of sizes and exhibits differing morphologies. The availability and mobility of elements present in fly ashes will depend on the physical-chemical forms of the elements. In this context, the knowledge of the chemical composition, physical characteristics and fractionation of the ashes are issues of prime importance to assess their environmental impact and health risks.

The chemical composition of fly ashes varies depending on the type of carbon used, but in general terms are enriched in many elements including the potentially toxic ones. As a chalcophilic element, Sb is enriched in coal to such an extent that up to 3000 µg Sb g^{-1} have been detected in the ashes of some German coals (Smichowski 2008b). Valkovik (1983) reported a worldwide average concentration of Sb in coals of 3 µg g^{-1}, which is evidence of its enrichment with respect to its abundance in the Earth's crust. Antimony vaporizes during combustion, resulting in the largest single source of anthropogenic Sb to the global atmosphere.

One area of particular interest in environmental chemistry is the impact of fine particles containing potentially toxic inorganic compounds when released from power stations and deposited downwind of the emission source. Hence, to investigate the leachability of elements from coal fly ashes is an important issue in the assessment of the environmental and health risks of coal uses. In extraction studies, special attention has to be paid in selecting the appropriate extraction conditions to avoid changes in the original form of Sb in the fly ashes by means of hydrolysis or complexed species formation, even when no oxidation/reduction reaction takes place.

The analytical speciation of Sb in this matrix is necessary to understand the mechanisms of transport and thus the subsequent environmental impact of the different species. In spite of this, only few studies based on chemical extraction are available in the literature. Chemical sequential extraction procedures provide useful information about solubility, origin, mode of occurrence, biological and physical-chemical availability, mobilization, fate and transport of trace metals in the environment. These procedures are operationally defined and their nature may be considered a weakness. In spite of this, sequential extraction methods provide knowledge that can improve our understanding of the risks for human health and the environment. Recently, Smichowski et al. (2005) reviewed metal fractionation of atmospheric aerosols via sequential chemical extraction procedures.

The presence of chemical species of As, Se and Sb in fly ashes from six coal fuel thermal power stations from different countries was investigated by Narukawa et al. (2005). To determine the mode of occurrence of this metalloid, a five-step chemical fractionation scheme was applied. Total Sb in fly ash samples was measured by ICP-MS following MW-assisted digestion using HCl, HF and HNO_3. In fly ash, total Sb concentration ranged from 1.0 to 3.9 µg g^{-1} while in the most bioavailable fraction, Sb levels varied from 0.05 to 1.14 µg g^{-1}. It appeared that the dominant chemical forms of Sb in the fly ashes were as extractable species. Inorganic Sb species were determined by HPLC-ICP-MS in aqueous extracts. Less than 3 percent of total Sb was as Sb(V) while Sb(III) was the dominant species.

The solubility of Sb inorganic compounds from five different size-segregated coal fly ash samples was also investigated (Seames et al. 2002). Antimony concentration in coal varied from 0.3 to 2.3 µg g^{-1}. In terms of Sb leachability, it was highly influenced by pH, being the metalloid partially soluble at pH 5 and exhibiting higher solubility at lower pH levels. Antimony resulted fairly soluble at pH 5.0 and very soluble at pH 2.9. The presence of following possible compounds was proposed: Sb, Sb_2O_4, and Sb_2O_5. The pH 5 leachability data constitutes useful information to assess the potential of Sb contained in small particles to migrate into the water supply after ground deposition downwind of the power plant.

A new three-step fractionation scheme was applied to study the distribution of Al, As, Cd, Cr, Cu, Fe, Mn, Mo, Ni, Pb, S, Sb, Ti, V and Zn in fly ashes collected in the electrostatic precipitator of a thermal power plant in the city of San Nicolás, Argentina (Smichowski et al. 2008c). This fractionation scheme was based on a previous one reported by Fernández Espinosa and coworkers (Fernández Espinosa et al. 2002) that was optimized for the analysis of fine atmospheric aerosols. The scheme applied consisted of extracting the elements in three fractions: (i) soluble and exchangeable elements; (ii) carbonates, oxides and reducible elements; and (iii) residual elements. Antimony was quantified at µg g^{-1} level in each fraction by ICP-MS. For validation purposes, the NIST SRM 2711 (Montana Soil) was subjected to the same chemical sequential extraction procedure adopted for the samples. The leachability of the 15 elements under study proved to be different. The distribution study among the four fractions indicated that 6.9 percent of Sb was present in the soluble fraction. Fortunately, Sb was mostly associated with the residual fraction (the most immobile one).

The potential toxicity of fly ashes is related to the leachable fraction of contaminants, since the total toxic amount of fly ashes is not extractable under natural environmental conditions. In a study of leachability and analytical speciation of Sb in coal fly ash, several single extraction procedures were tested while two of them were examined in depth: (i) using an aqueous solutions in the pH range 1–12, and (ii) using 1 mol L^{-1} citrate at pH 5 citrate (Miravet et al. 2006). Speciation analysis of the coal fly ash extracts by HPLC-ICP-MS and HPLC-HG-AFS was carried out in order to identify the presence of individual Sb species. Antimony(V) was the main Sb species in the leachates, although minor amounts of Sb(III) were also detected in some extracts. While citrate at pH 5 exhibited the best extraction efficiency, Sb species were also fairly soluble in aqueous solutions at acidic pHs. Fly ashes are a very complex matrix with different contents of many elements and ions that may influence Sb species distribution. The presence of high concentrations of ions such as Ca, Fe and Pb in the leachates reduced stibine generation in the HPLC-HG-AFS analysis.

Antimony in Road Dust

Road dust is particulate matter generated from the surface of streets and roads due to road surface wear. There are other sources that contribute to increase particles apportionment such as: tailpipes, abrasion products from tires, brake lining, bearings, road surface, corrosion of vehicle components and resuspension of soil and road dust (Fujiwara et al. 2011). These particles may be primary or secondary, originated by anthropogenic (demolition, construction, industrial stacks) and natural sources (short-and long-range transport of resuspended soils) and are deposited and accumulated in street dust (Amato et al. 2009). Emissions originating from different sources vary from location to location due to the impact of climate, road surface characteristics and traffic patterns.

The information on the content of metals and metalloids in this matrix can be considered a valuable indicator of environmental pollution. For this reason, many studies have been conducted to determine metallic elements in street dust in numerous cities (Sezgin et al. 2003, Banerjee 2003, Herngren et al. 2006) using different analytical techniques and sample treatment procedures. On the other hand, studies on the fractionation and/ or speciation of trace elements in street dust have been reported to a minor extent. Again, little attention has been paid to this kind of important study and most papers including fractionation only are focused on analyzing the aqueous fraction.

Traffic is the main source of several polluting elements (therefore called TREs) that can also be considered as emerging contaminants in road dust samples. Various studies have dealt with traffic-emitted elements such as

Cu, Mo, Pd, Pt, Rh, Sb and Zn in different environmental matrices (Gómez et al. 2003, Morcelli et al. 2005, Dietl et al. 1997, Ma et al. 2001, Rauch et al. 2000).

In addition to the well-known input of Sb into the environment through brake pads, road traffic emissions of this element have also been ascribed to the use of certain organic Sb-containing compounds such as Sb-dialkyldithiocarbamates in greases and motor oil (Huang et al. 1994).

Furuta et al. (2005) reported that the mean Sb released from vehicle brake pads in Japan is about 1.24 g per car per year. Amounts of Sb released during braking depend on: (i) concentration in specific brake pad brands, (ii) design, and (iii) braking patterns. After released, a considerable fraction of Sb reacts with oxygen in the air to form Sb_2O_3 which is considered as a potential carcinogen (von Uexküll et al. 2005).

Recent research from Amato et al. (2009) focused on road dust re-suspension as one of the major sources of atmospheric pollutants in the urban environment of Barcelona, Spain. This investigation reported average concentration of Sb in the PM 10 fraction of road dust varying from, 102 μg g^{-1} in the ring roads, to 200 μg g^{-1} in the city center.

In spite of the interest and the increasing number of studies on road dust reported in recent years, those including Sb are mainly focused on the determination of its total content. Little work has been reported on the determination of Sb species, or fractionation studies of Sb, in this matrix.

Fujiwara and coworkers (Fujiwara et al. 2011) carried out a study aimed to improve the knowledge on the distribution of Sb in size-classified road dust samples and reported for the first time concentrations of this metalloid in samples collected in the megacity of Buenos Aires, Argentina. Nineteen road dust samples were collected in the metropolitan area of Buenos Aires during two months covering an area of ~ 7 km x 12 km. Samples were collected from urban zones with different traffic patterns and urban characteristics. Once in the laboratory, samples were dried at 100°C for 24 hr and then sieved. After sieving, four fractions were obtained with the following particle diameter: F1 < 37 μm, F2: 37–55 μm, F3: 55–75 μm and F4: 55–100 μm. Larger particles (> 100 μm) were not considered because if resuspended have a short residence time in the urban atmosphere. Samples were digested using an acid mixture containing 2 mL HNO$_3$, 6 mL HCl and 1 mL HF. Antimony was determined in each fraction of digested samples by inductively coupled plasma optical emission spectrometry (ICP OES) and / or HG-AAS. The NIST SRM 2709 (San Joaquin Soil) was used for checking accuracy. Antimony was detected in 49 out of 73 road dust sub-samples. Total concentrations varied from < 0.5 μg g^{-1} to 41.1 μg $^{-1}$. Minimum and maximum concentrations (in μg g^{-1}) measured in individual samples in the four fractions were: Fraction 1, < 0.5–20.4; Fraction 2, < 0.5–18.4; Fraction 3, < 0.5–6.3; Fraction 4, < 0.5–7.7. The highest concentrations of Sb were detected in the two smaller fractions (F1 and F2). The fine fractions are potentially more dangerous from the health and environmental standpoints. On the other hand, the authors reported levels of Sb in the largest size fractions (coarse fraction) of road dust namely, F3 and F4 between those of natural matrices and soil.

In another study, Amato and coworkers (Amato et al. 2011) investigated the spatial and chemical properties of the strength of the emission source in road dust particles below 10 μm sampled in eight sites of three different European cities namely, Barcelona (Spain), Girona (Spain) and Zürich (Switzerland). With the aim to reduce losses during the sampling procedure and given the absence of a definitive sampling protocol, an innovative sampling device was proposed for road dust. Using the developed device only particles < 10 μm were collected. In other words, the authors used a field resuspension chamber directly vacuuming *in situ*, at an air flow rate of 25 L min^{-1}, the resuspended fraction < 10 μm of road dust onto filters. Then, the samples were digested using a mixture of 5 mL HF, 2.5 mL HNO$_3$, 2.5 mL HClO$_4$ and the digests were analyzed by ICP OES and ICP-MS. For method validation, the NIST SRM 1633b (Coal Fly Ash) was used. Mean concentrations of Sb for the three cities were (in μg g^{-1}): 64 (Girona) < 196 (Barcelona) < 324 (Zürich). Averaged levels of Sb and other elements (As, Cr, Co, Cu Mo, Nb, Ni, Zn, and Zr,) resulted to be in all cases over 40 percent higher in Zürich than in Barcelona (in relative concentrations). For demonstrating metal enrichment the authors calculated the enrichment factors, normalized with respect to Al. The highest enrichments were found in Zürich, with a distinctive signature attributed mainly to the influence of traffic contamination.

Conclusions

The increase in the last years in the number of investigations in the field of element speciation and fractionation provides clear evidence of the push forward that this area gives to many disciplines. In the field of atmospheric chemistry and in the particular case of Sb information is still scarce considering the environmental and toxicological importance of this element. In fact, all the information reported above clearly testifies that a gap exists with respect to other much more studied elements and that more systematic studies on total Sb determination and its speciation in atmospheric aerosols are necessary.

An important drawback to tackle is the need for ready availability of certified reference materials with known contents of Sb species in different solid matrices. Another topic that needs more research is the low extraction efficiency of Sb compounds in solid environmental matrices. This issue needs to be improved for a reliable assessment of the occurrence of Sb species in APM and related matrices.

The determination of total Sb concentrations in APM, fly ashes, and road dust is readily achievable by well-consolidated plasma-based and atomic spectroscopic techniques that are available in most laboratories. For its intrinsic characteristics, ICP-MS is to date the more used technique. In addition, a significant reduction of spectral interferences in these complex matrices can be achieved thanks to mass spectrometric detection.

With respect to fractionation studies, it is recommended to harmonize the different chemical sequential procedures reported in order to facilitate comparability of data as well as to optimize the operating conditions, and to introduce on-line procedures aimed at reducing reagents consumption and the time demand of the different steps. It is important to state that accuracy and precision of measurements should be guaranteed in all kind of speciation/fractionation studies. In this connection and for validation purposes, there is both a need and a demand for more specific reference materials and especially those with certified extractable content.

Furthermore, it goes without saying that the extent of work that remains to be done on the determination of Sb and its species in environmental matrices should be directed for completing reliable eco-toxicological studies to shed light on some little explored aspects such as: (i) the lithogenic contribution of this metalloid in a given environmental compartment to evaluate natural versus anthropogenic input; (ii) acquiring information on all the possible Sb species present in solid environmental matrices (soils, sediments, APM).

Acknowledgements

The author would like to thank Roberto Carballa and Darío Gómez for their kind cooperation.

References

Amato, F., M. Pandolfi, M. Viana, X. Querol, A. Alastuey and T. Moreno. 2009. Spatial and chemical patterns of PM 10 in road dust deposited in urban environment. Atmos. Environ. 43: 1650–1659.

Amato, F., M. Pandolfi, T. Moreno, M. Furger, J. Pey, A. Alastuey, N. Bukowiecki, A.S.H. Prevot, U. Baltensperger and X. Querol. 2011. Sources and variability of inhalable road dust particles in three European cities. Atmos. Environ. 45: 6777–6787.

Araujo, R.G.O., B. Weltz, I.N.B. Castillo, M. Goreti, R. Vale, P. Smichowski, S.L.C. Ferreira and H. Becker-Ross. 2010. Determination of antimony in airborne particulate matter collected on filters using direct solid sampling and high-resolution continuum source graphite furnace atomic absorption spectrometry. J. Anal. At. Spectrom. 25: 580–584.

Banerjee, A.D.K. 2003. Heavy metal levels and solid phase speciation in street dust of Delhi, India. Environ. Pollut. 123: 95–105.

Bellido-Martín, A., J.L. Gómez-Ariza, P. Smichowski and D. Sánchez-Rodas. 2009. Speciation of antimony in airborne particulate matter using ultrasound probe fast extraction and analysis by HPLC-HG-AFS. Anal. Chim. Acta. 549: 191–195.

Canepari, S., E. Marconi, M.L. Astolfi and C. Perrino. 2010. Relevance of Sb(III), and Sb-containing nano-particles in urban atmospheric particulate matter. Anal. Bioanal. Chem. 397: 2533–2442.

Cho, M., J. Ju, S. Kim and H. Jang. 2006. Tribiological properties of solid lubricants (graphite, Sb_2S_3, MoS_2) for automotive brake friction materials. Wear. 260: 855–860.

Cornelis, G., T. Van Gerven and C. Vandecasteele. 2006. Antimony leaching from uncarbonated and carbonated MSW1 bottom ash. J. Hazard. Mat. 137: 1284–1292.

Craig, P.J. 2003. Organometallic compounds in the environment. John Wiley & Sons. West Sussex, UK.

Dietl, C., W. Reifenhäuser and L. Peichl. 1997. Association of antimony with traffic- occurrence in airborne dust, deposition and accumulation in standardized grass cultures. Sci. Total Environ. 205: 235–244.

Fernández, A.J., M. Ternero, F.J. Barragán and J.C. Jiménez. 2000. An approach to characterization of sources of urban airborne particles through heavy metal speciation. Chemosphere. 2: 123–136.

Fernández Espinosa, A.J., M. Ternero Rodríguez, F.J. Barragán de la Rosa and J.C. Jiménez Sánchez. 2002. A chemical speciation of trace metals for fine urban particles. Atmos. Environ. 36: 773–789.

Ferreira, S.L.C., S.M. Macedo, D.C. dos Santos, R.M. de Jesús, W.N.I. dos Santos, A.F. de S. Queiroz and J.B. de Andrade. 2011. Speciation analysis of inorganic antimony in airborne particulate matter employing slurry sampling and HG QT AAS. J. Anal. At. Spectrom. 26: 1887–1891.

Filella, M., N. Belzile and Y.-W. Chen. 2002. Antimony in the environment: a review focused on natural waters. I. Occurrence. Earth-Sci. Rev. 57: 125–176.

Filella. M., P.A. Williams and N. Belzile. 2009. Antimony in the environment: knowns and unknowns. Environ. Chem. 6: 95–105.

Fujiwara, F., R. Jiménez Rebagliati, J. Marrero, D. Gómez and P. Smichowski. 2011. Antimony as a traffic-related element in size-fractionated road dust samples collected in Buenos Aires. Microchem. J. 97: 62–67.

Furuta, N., A. Iijima, A. Kambe, K. Sakai and K. Sato. 2005. Concentrations, enrichment and predominant sources of Sb and other trace elements in size-classified airborne particulate matter collected in Tokyo from 1995 to 2004. J. Environ. Monit. 7: 1162–1168.

Gómez, D., P. Smichowski, G. Polla, A. Ledesma, S. Resnizky and S. Rosa. 2002. Fractionation of elements by particle size of ashes ejected from Copahue Volcano, Argentina. J. Environment. Monit. 4: 972–977.

Gómez, D.R., M.F. Giné, A.C. Claudia Sánchez Bellato and P. Smichowski. 2005. Antimony: a traffic-related element in the atmosphere of Buenos Aires, Argentina. J. Environ. Monit. 7: 1155–1161.

Gómez, M.B., M.M. Gómez and M.A. Palacios. 2003. ICP-MS determination of Pt, Pd and Rh in airborne and road dust after tellurium coprecipitation. J. Anal. At. Spectrom. 18: 80–83.

Hansen, H.R. and S.A. Pergantis. 2006. Investigating the formation of an Sb(V)–citrate complex by HPLC-ICP-MS and HPLC-ES-MS(/MS). J. Anal. At. Spectrom. 21: 1240–1248.

He, M., X. Wang, F. Wu and Z. Fu. 2012. Antimony pollution in China. Sci. Total Environ. 421–422: 41–50.

Herngren, L., A. Goonetilleke and G.A. Ayoko. 2006. Analysis of heavy metals in road-deposited sediments. Anal. Chim. Acta. 571: 270–278.

Huang, X., I. Olmez, N.K. Aras and G.E. Gordon. 1994. Emissions of trace elements from motor vehicles: potential marker elements and source composition profile. Atmos. Environ. 26: 1385–1391.

Iijima, A., K. Sato, T. Ikeda, H. Sato, K. Kozawa and N. Furuta. 2010. Concentration distributions of dissolved Sb(III) and Sb(V) species in size-classified inhalable airborne particulate matter. J. Anal. At. Spectrom. 25: 356–363.

Infante, R. and I.L. Acosta. 1991. Size distribution of trace metals in Ponce, Puerto Rico air particulate matter. Atmos. Environ. 25: 121–131.

Ingo, G.M., M. D'Uffizi, G. Falso, G. Bultrini and G. Padeletti. 2004. Thermal and microchemical investigation of automotive brake pad wear residues. Thermochim. Acta. 418: 61–68.

Jang, H. and S.J. Kim. 2000. The effects of antimony trisulfide-(Sb_2S_3) and zirconium silicate ($ZrSiO_4$) in the automotive brake friction material on friction characteristics. Wear. 239: 229–236.

Jenkins, R.O., P.J. Craig, D.P. Miller, L.C.A.M. Stoop, N. Ostah and T.-A. Morris. 1998. Antimony biomethylation by mixed cultures of micro-organisms under anaerobic conditions. Appl. Organomet. Chem. 12: 449–455.

Krachler, M. and H. Emons. 2001a. Urinary antimony speciation by HPLC-ICP-MS. J. Anal. At. Spectrom. 16: 20–25.

Krachler, M., H. Emons and J. Zheng. 2001b. Speciation of antimony for the 21st century: promises and pitfalls. Trends Anal. Chem. 20: 79–90.

Lough, G.J.J. Schauer, J. Soopark, M. Shafer, J. Deminter and J.P. Weinstein. 2005. Emissions of metals associated with motor vehicle roadways, Environ. Sci. Technol. 39: 826–836.

Ma, R., L. Staton, C.W. McLeod, M.B. Gómez, M.M. Gómez and M.A. Palacios. 2001. Assessment of airborne platinum contamination via ICP-mass spectrometric analysis of tree bark. J. Anal. At. Spectrom. 16: 1070–1075.

Maeda, S. 1994. Safety and environmental effects. *In*: S. Patai [ed.]. The Chemistry of Organic Arsenic, Antimony, and Bismuth Compounds, John Wiley and Sons, New York, USA. pp. 725–759.

Marconi, E., S. Canepari, M.L. Astolfi and C. Perrino. 2011. Determination of Sb(III), Sb(V) and identification of Sb-containing nanoparticles in airborne particulate matter. Environ. Sci. 4: 209–217.

Miravet, R., J.F. López-Sánchez and R. Rubio. 2006. Leachability and analytical speciation of antimony in coal fly ash, Anal. Chim. Acta. 576: 200–206.

Miravet, R., J.F. López-Sánchez, R. Rubio, P. Smichowski and G. Polla. 2007. Speciation analysis of antimony in extracts of size-classified volcanic ash by HPLC-ICP-MS. Anal. Bioanal. Chem. 387: 1949–1954.

Miravet, R., E. Hernández-Nataren, A. Sahuquillo, R. Rubio and J.F. López-Sánchez. 2010. Trends in Ana. Chem. 29: 28–39.

Morcelli, C.P.R., A.M.G. Figueiredo, J.E.S. Sarkis, J. Enzweiler, M. Kakazi and J.B. Sigolo. 2005. PGEs and other traffic-related elements in roadside soils from São Paulo, Brazil. Sci. Total Environ. 345: 81–91.

Narukawa, T., A. Takatsu, K. Chiba, K.W. Riley and D.H. French. 2005. Investigation into the relationship between major and minor element contents and particle size and leachability of boron in fly ash from coal fuel thermal power plants. J. Environ. Monit. 7: 1342–1346.

Nriagu, J.O. 1990. Global metal pollution: poisoning the biosphere. Environment. 32: 7–11.

Pacyna, J.M and E.G. Pacyna. 2001. An assessment of global and regional emissions of trace metals to the atmosphere from anthropogenic sources worldwide. Environ. Rev. 9: 269–298.

Rauch, S., M. Motelica-Heino, G.M. Morrison and O.F.X. Donard. 2000. Critical assessment of platinum group element determination in road and urban river sediments using ultrasonic nebulization and high resolution ICP-MS. J. Anal. At. Spectrom. 15: 329–334.

Rizzio, E., G. Giaveri and M. Gallorini. 2000. Some analytical problems encountered for trace elements determination in the airborne particulate matter of urban and rural areas. Sci. Total Environ. 256: 11–22.

Seames, W.S., J. Sooroshian and J.O.L. Wendt. 2002. Assessing the solubility of inorganic compounds from size-segregated coal fly ash aerosol impactor samples J. Aerosol Sci. 33: 77–90.

Sezgin, N., H. Ozcan, G. Demir, S. Nemlioglu and C. Bayat. 2003. Determination of heavy metal concentrations in street dust in Istanbul E-5 highway. Environ. Int. 29: 979–985.

Shotyk, W. and M. Krachler. 2007. Contamination of bottled waters with antimony leaching from Polyethylene Terephthalate (PET) increases upon storage. Environ. Sci. Technol. 41: 1560–1563.

Shotyk, W., M. Krachler and B. Chen. 2005. Anthropogenic impacts on the biogeochemistry and cycling of antimony. *In*: A. Sigel, H. Sigel and R.K.O Sigel [eds.]. Metal ions in biological systems, M. Dekker, New York, USA. pp. 171–203.

Smichowski, P., Y. Madrid, M.B. de la Calle Guntiñas and C. Cámara. 1995. Separation and determination of antimony(III) and antimony(V) species by High-performance liquid chromatography with hydride generation atomic absorption spectrometric and inductively coupled plasma mass spectrometric detection. J. Anal. At. Spectrom. 16: 815–821.

Smichowski, P., Y. Madrid and C. Cámara. 1998. Analytical methods for antimony speciation in Waters at trace and ultratrace levels. A review. Fresenius J. Anal. Chem. 360: 623–629.

Smichowski, P., D. Gómez, S. Rosa and G. Polla. 2003. Trace elements content in size-classified volcanic ashes as determined by inductively coupled plasma-mass spectrometry. Microchem. J. 75: 109–117.

Smichowski, P., G. Polla and D. Gómez. 2005. Metal fractionation of atmospheric aerosols via sequential chemical extraction: a review. Anal. Bioanal. Chem. 381: 302–316.

Smichowski, P., D. Gómez and G. Polla. 2007. Sample preparation of atmospheric aerosols for elemental analysis and fractionation studies. *In*: M.A. Zezzi Arruda [ed.]. Sample preparation, Nova Science Publishers, Inc. New York, USA. pp. 83–136.

Smichowski, P., D. Gómez, C. Frazzoli and S. Caroli. 2008a. Traffic-related elements in airborne particulate matter. Appl. Spectrosc. Rev. 43: 22–48.

Smichowski, P. 2008b. Antimony in the environment as a global pollutant: A review on analytical methodologies for its determination in atmospheric aerosols. Talanta. 75: 2–14.

Smichowski, P., G. Polla, D. Gómez, A.J. Fernández Espinosa and A. Calleja López. 2008c. A three-step metal fractionation scheme for fly ash collected in an Argentine thermal power plant. Fuel 87: 1249–1258.

Templeton, D.M., F. Arise, R. Cornelis, L.-G. Danielson, H. Muntau, H.I.P. Van Leeuwen and R. Lobinski. 2000. Guidelines for terms related to chemical speciation and fractionation of elements. Definitions, structural aspects, and methodological approaches. IUPAC Recommendations. Pure Appl. Chem. 72: 1453–1470.

Ure, A.M. 2001. Chemical speciation in the environment. Blackie, Glasgow, UK.

Valkovic, V. 1983. Trace Elements in Coal, CRC Press. Inc. Boca Raton, Florida, US.

von Uexküll, O., S. Skerfving, R. Doyle and M. Braungart. 2005. Antimony in brake pads: a carcinogenic component? J. Clean. Prod. 13: 19–31.

Wedepohl, K.H. 1995. The composition of the continental crust. Geochim. Cosmochim. Acta. 59: 1217–1232.

Zheng, J., M. Ohata and N. Furuta. 2000a. Antimony speciation in environmental samples by using high-performance liquid chromatography coupled to inductively coupled plasma mass spectrometry. Anal. Sci. 16: 75–79.

Zheng, J., M. Ohata and N. Furuta. 2000b. Studies on the speciation of inorganic and organic antimony compounds in airborne particulate matter by HPLC-ICP-MS. Analyst. 125: 1025–1028.

Zheng, J., A. Akihiro and N. Furuta. 2001c. Complexation effect of antimony compounds with citric acid and its application to the speciation of antimony(III) and antimony(V) using HPLC-ICP-MS. J. Anal. At. Spectrom. 16: 812–818.

Zih-Perényi, K., K. Neuróhr, G. Nagy, M. Balla and A. Lásztity. 2010. Selective extraction of traffic-related antimony compounds for speciation analysis by graphite furnace atomic absorption spectrometry. Spectrochim. Acta Part B. 65: 847–851.

Speciation of Arsenic in Soil, Sediment and Environmental Samples

Selin Bora,[1,] Işıl Aydın,[2,a] Ersin Kılınç[2,b] and Fırat Aydın[2,c]*

Introduction

Arsenic is found on the Earth's crust and its concentration varies in the environment. It has 0.0001 percent of abundance and exists with the ores of lead, gold and copper in nature. Its concentration is relatively lower in different samples. For example, As concentration is < 4.0 μg/L in river and marine waters, < 200 μg/L in sediment pore waters, ~ 5.0 mg/kg in unconsolidated sediments and ~ 7.0 mg/kg in soils (Smedley and Kinniburgh 2002). Arsenic levels may increase in the environment by weathering and dissolution of arsenic-containing minerals in surface and waters (Chowdhury et al. 2000). In addition, it is known that arsenic is introduced into the environment by anthropogenic activities. The principle

[1]Department of Chemistry, Middle East Technical University, 06800 Ankara, Turkey.
Email: sbora@metu.edu.tr
[2]Department of Chemistry, Dicle University, 21280 Diyarbakir, Turkey.
[a]Email: iaydin@dicle.edu.tr
[b]Email: ekilinc@dicle.edu.tr
[c]Email: faydin@dicle.edu.tr
*Corresponding author

anthropogenic sources are the base metal smelters, mining of arsenic ores (Krysiak and Karczewska 2007, Palumbo-Roe and Klinck 2007) and wastes from arsenic-processing plants. Moreover, in the poultry and livestock industries, several phenyl arsenical compounds have been used as feed additives to prevent coccidiosis and enhance animal growth. Poultry or livestock excrete most of the arsenic in the feed and then arsenic is easily passed into the soil because of the use of poultry litter as a type of fertilizer (Stolz et al. 2007, Garbarino et al. 2003).

In addition, it is known that arsenic has been used in glass making to remove the small bubbles from glass (Aksentijevic et al. 2012). It is also added to some alloys in order to improve both the hardness and corrosion resistance. Additionally, this element is used in the semiconductor industry (Ga-As) and in the production of catalysts (Azcue and Nriagu 1994, Léonard 1991, Moore and Ramamoorthy 1984, Blasius 1990, Haase 1994). Arsenic has different chemical forms (inorganic/organic). So far, more than 40 different arsenic containing species have been identified in environmental and biological samples. Arsenic exists in four oxidation states: As(III), As(0), As(III) and As(V). Inorganic As(III) and As(V) are the predominant and most toxic forms. In an abiotic systems (minerals, soils and water), inorganic As(III) and As(V) are found to combine with oxygen or sulfur (Cullen and Reimer 1989). While arsenate, As(V), plays an important role in many biochemical reactions as a molecular analogue of phosphate, arsenite, As(III), can inhibit proteins activities by reacting with thiols. Apart from inorganic forms, arsenic has several organic forms including methylarsonite (MA^{III}), methylarsonate (MA^{V}), dimethylarsinite (DMA^{III}), dimethylarsinate (DMA^{V}), trimethylarsineoxide (TMAO), tetramethylarsonium (TETRA), arsenobetaine (AsB) [(CH$_3$)$_3$N$^+$CH$_2$COO$^-$], arsenocholines (AsC) and arsenosugars (AsS). Since As(III), As(V), MA^{III}, MA^{V}, DMA^{III} and DMA^{V} are found in ionic forms, when they take hydrogen(s) to their structures, new forms are referred to as acids. For example, hydrogenated form of MA^{V} is called monomethylarsonic acid while hydrogenated DMA^{V} is known as dimethylarsinic acid. Chemical structures of some arsenic species and acid forms of some of mentioned species are shown in Fig. 9.1. It is mentioned in literature that AsB, AsC and AsS are not harmful and plants and animals accumulate arsenic in their body mostly as non-toxic organoarsenic molecules. It was also found that, larger diversity of organoarsenical compounds have been identified in marine biota when compared to terrestrial environment. In addition, the proportion of these possible non-toxic species to the total arsenic content in terrestrial animals is lower than the proportion in marine organisms (Reimer et al. 2010).

Arsenic is known as a highly toxic heavy metal. International Agency for Research on Cancer (IARC) has classified arsenic as a carcinogen for humans. Arsenic toxicity depends on its chemical form. Organic forms of

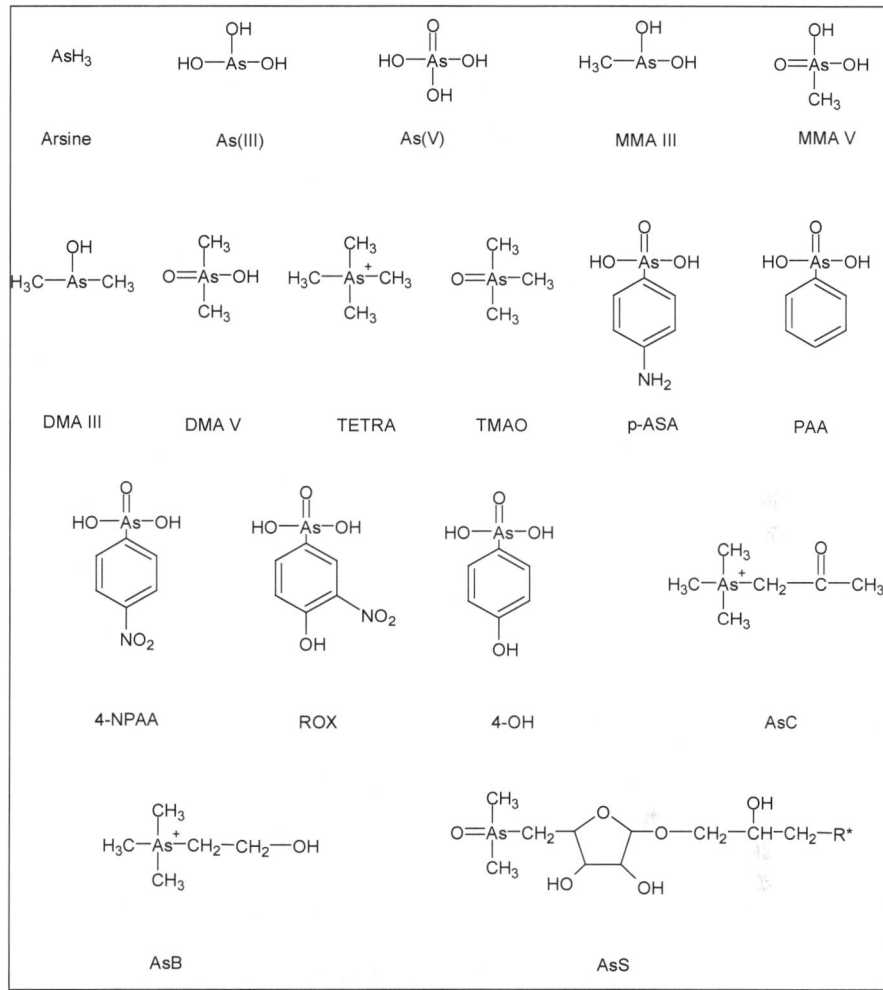

Figure 9.1. Chemical structures of some arsenic species; *p*-ASA: *p*-arsanilic acid, PAA: phenylarsonate, 4-NPAA: 4-nitrophenylarsonic acid, ROX: roxarsone or 3-nitro-4-hydrophenylarsonic acid, 4-OH: 4-hydroxyphenylarsonic acid (Sun et al. 2002, Terlecka 2005, Wang et al. 2010, Williams et al. 2009).

this element have lower toxicity than inorganic arsenic species. There is also a difference in the toxicity of inorganic arsenic species. It was stated that As(III) is more toxic than As(V) (Elci et al. 2008, Shah et al. 2009, Council 1999). Although methylated forms are considered to be moderately toxic, AsB, AsC and AsS are not harmful for human beings. On the other hand, arsine gas (AsH$_3$) is the most toxic compound causing serious health problems (Yamauchi and Fowler 1994). Toxicity of this element also depends on other factors such as physical state (gas, solution or powder), particle size of its compound, absorption rate into cells, elimination rate from the body, the nature of chemical substituents in the toxic compound and the pre-existing state of the patient (Hindmarsh and McCurdy 1986). Although it is generally accepted that methylation is the principal detoxification way of this element, recent literature findings have shown that methylated metabolites may be partly responsible for the adverse effects associated with arsenic exposure (Vega et al. 2001).

Arsenic mostly enters the human body by the consumption of foods and beverages containing arsenic species. Dietary arsenic species include mostly AsB, AsC and various AsS compounds. These species are found in fish, crustaceans and seaweeds (WHO 2001). It was reported that total As concentration in marine organisms ranges from 1.0 to 100 mg/kg due to bioaccumulation and bio-transformation processes (Léonard 1991). Therefore, seafood could be the main source of dietary arsenic. On the other hand, exposure to the toxic inorganic arsenic is through consumption of the foods and drinking water contaminated by this element. Exposure to inorganic arsenic from drinking water causes serious health problems worldwide (Kapaj et al. 2006). In addition to drinking water and seafood, some other foodstuffs can also be the source of arsenic. For example, arsenic concentration in grain is 80–200 µg/kg in the normal global range. If it is assumed that grain contains 50 percent inorganic arsenic, daily intake of this element may reach 8.0–20 µg. This is not good for the people living in Asia because of the high consumption of rice in this region (adults consumes about 200 g rice per day). Acceptable WHO limit is 10 µg/L for total arsenic in drinking water (Zavala and Duxbury 2008). However, there is still no definite value established for foodstuffs. Food and Agriculture Organization/World Health Organization (FDA/WHO) has recommended a tolerable weekly intake of max. 15 µg inorganic arsenic/kg body weight (Kohlmeyer et al. 2003). From this point of view, speciation studies are crucial to know about the amount of toxic and non-toxic forms for the evaluation of dietary risks (Bissen and Frimmel 2003).

Once arsenic contaminated foods are ingested, this element is absorbed and distributed through the body and finally excreted via urine rapidly. In the human body, absorption of inorganic arsenic species is more than organic forms, and among inorganic forms, As(V) is more readily absorbed than As(III) (Léonard 1991). It was found that some of inorganic arsenic species absorbed are metabolized to methylated arsenic species (mono-, di- and tri- methyl arsenic species) in the liver. Due to high reactivity, methylated compounds of trivalent arsenic are formed as intermediates. After the methylation process, no further conversion of DMA^V is observed while small portion of MA^V is converted to DMA^V. Methylated arsenic forms are excreted via urine from the human body (Roy and Saha 2002, Cohen et al. 2006, Baowei et al. 2009). Furthermore, AsS can be metabolized to several compounds including DMA^V. Like DMA^V, AsB remains in the same structure in the body (Francesconi et al. 2002). Although most of the ingested arsenic is excreted via urine, this element can partly accumulate in different tissues such as hair and nails because of the high content of keratin (Ma and Le 1998). In biological fluids, arsenic is most likely to be in the form of AsB, As(III), As(V) and mono- and dimethylated arsenic compounds (Ito et al. 2010). In the aerobic environment, arsenic is found in the oxidation state of +5. As(V) appears as H_3AsO_4 at pH values below 2. In the pH range between 2–11, H_3AsO_4 dissociates to $H_2AsO_4^-$ and $HAsO_4^{2-}$. On the other hand, at lower potentials, arsenic is found in the oxidation state of +3 as in the form of H_3AsO_3. Up to pH 9, H_3AsO_3 does not dissociate. At higher pH values, it appears as $H_2AsO_3^-$, $HAsO_3^{2-}$ and AsO_3^{3-}. At potentials below –250 mV, arsenic compounds, such as As_2S_3 can form in the presence of sulfur or hydrogen sulfide. Arsine and elemental arsenic can form under very strong reducing conditions (Ferguson and Gavis 1972, Ferguson and Anderson 1974, Heinrichs and Udluft 1993).

Speciation of Arsenic

For speciation studies of any aqueous sample for their arsenic species contents, hyphenation of chromatography with atomic and/or molecular spectroscopy has been used. Gas Chromatography Inductively Coupled Plasma Mass Spectrometry (GC-ICP-MS) (Leermakers et al. 2006), Liquid Chromatography-Inductively Coupled Plasma Mass Spectrometry (LC-ICP-MS) (Baba et al. 2008), Ion Chromatography Inductively Coupled Plasma Mass Spectrometry (IC-ICP-MS) (Karthikeyan and Hirata 2004), Liquid Chromatography Atomic Fluorescence Spectroscopy (LC-AFS) (Yang et al. 2012) and Liquid Chromatography Inductively Coupled Plasma Optical Emission Spectrometry (LC-ICP-OES) (Al-Assaf et al. 2009) are most commonly used chromatographic techniques for the separation and detection of different arsenic species (Ciardullo et al. 2010, Cornelis et al. 2003). Apart from chromatographic techniques, non-chromatographic techniques also provide economical and efficient tools for the species determination of toxic elements including arsenic (Gonzalvez et al. 2010, Kumar and Riyazuddin 2007).

Hydride Generation (HG) is one of the non-chromatographic speciation techniques (Chen et al. 2009, Reyes et al. 2008). In this technique, volatile hydride form of arsenic, AsH_3, can be detected by AAS. With suitable reducing agent, the oxidized form can be converted to As(III), and concentration of total arsenic can be calculated in terms of As(III). Without a reducing agent, only As(III) content of the sample can be determined. Therefore, subtraction of As(III) concentration from total concentration gives that of As(V). Electrothermal Atomic Absorption Spectrometry (ETAAS) is the other non-chromatographic speciation technique. Like HG, ETAAS can only determine the inorganic arsenic species. In addition, Capillary Electrophoresis (CE) (Akter et al. 2005) and X-ray Absorption Near Edge Structure Spectroscopy (XANES) (Caumette et al. 2011) have been applied in literature for the speciation of arsenic species in different matrices.

In the speciation of this element, the first step is the extraction of different arsenic species from the matrices without any conversion. Then, the extracted arsenic species should be separated from each other. After the separation step, arsenic species should be sensitively determined by using different detection systems.

Extraction techniques for arsenic speciation

In order to obtain accurate results from the speciation technique, the extraction step is required for solid samples. The extraction is quite a critical step since high recoveries should be obtained in addition to preservation of species after this step. In literature, different methods have been proposed for the extraction of arsenic species from solid samples. Some of these are microwave-assisted treatments (Heitkemper et al. 2009), ultrasound-assisted extraction (Hirata and Toshimitsu 2007), enzymatic extraction (Reyes et al. 2009) and Soxhlet extraction (Lin et al. 2008). In addition, solid-phase extraction (SPE) (Anthemidis and Martavaltzoglou 2006), liquid-liquid extraction (LLE) (Munoz et al. 1999), cloud-point extraction (CPE) (Shemirani et al. 2005), liquid phase microextraction (LPME) (Jiang et al. 2009) and coprecipitation methods have been used for the separation and preconcentration of arsenic species. Among the extraction methods, SPE and LLE are known as time consuming, and use large amounts of organic solvents which limit their applications. Extraction efficiencies depend on the sample matrix and the extractants used. To get maximum extraction efficiency, many different extractants have been tried for arsenic speciation. In addition to providing high recoveries, the extractant must not cause interconversion of species to each other in the extraction process (Reyes et al. 2008, Milstein et al. 2003).

For the extraction of arsenic from soil and sediments, phosphoric acid and/or hydroxylamine hydrochloride can be used. These chemicals preserve the two redox states of arsenic (Ellwood and Maher 2003). Ammonium oxalate (Vassileva et al. 2001), phosphoric acid and ascorbic acid (Garcia-Manyes et al. 2002) have been also used for these purposes. In general, low recovery and oxidation of As(III) to As(V) can be the drawbacks of the last two extractants. Detailed extraction procedures with the chemicals used for different environmental samples are given in the later.

Speciation of arsenic by HPLC-ICP-MS

High Performance Liquid Chromatography coupled to Inductively Coupled Plasma Mass Spectrometry (HPLC-ICP-MS) has been the most frequently used technique for arsenic speciation due to its lower detection limits and separation capability for non-volatile arsenic species. Most of the researchers use ion interaction reversed-phase chromatography. In addition, anion and cation exchange columns have been applied for the separation of most of the arsenic species (Hirata and Toshimitsu 2007). Many researchers have used this combination due to very low detection limits for many arsenic species.

Actual separation depends on pH of the medium. At neutral pH, As(V), (pKa_1 = 2.3), MA^V (pKa = 3.6) and DMA^V (pKa = 6.2) are present as anions, AsC, TMAO (pKa = 3.6) and TETRA ions as cations; AsB (pKa = 2.18) is as zwitterion and As(III) (pKa_1 = 9.3) is as an uncharged species. Therefore, both anion exchange (As(V), As(III), MA^V and DMA^V) and cation exchange (AsC, TMAO, TETRA and AsB) have been often used in literature (Leermakers et al. 2006).

Ion chromatography hyphenated with ICP-MS is one of the powerful analytical techniques due to its several advantages such as isotope ratio analysis, wide dynamic range and low detection limits for a variety of samples and simultaneous measurements (Chen et al. 2007, Nam et al. 2003). Unfortunately, arsenic has only one isotope, ^{75}As, to be monitored. Interference of chloride in matrix due to formation of $^{40}Ar^{35}Cl$ and $^{38}Ar^{37}Cl$ is reported a limitation when ^{75}As is measured in the arsenic speciation (Chen et al. 2007). Separation of chloride by ion chromatography (Quaghebeur et al. 2003, Gallagher et al. 2001), dynamic reaction cell with a flow of CH_4 (Wang et al. 2007) and octopole collision/reaction cell with a flow of He or H_2 are presented to overcome this trouble. Chen et al. reported that Cl signal could be completely eliminated in the case that H_2 was used as reaction gas (Chen et al. 2007).

Anion exchange and cation exchange columns have been used to separate arsenic species prior to their determination at ultra trace level by ICP-MS (Gong et al. 2002, Ponthieu et al. 2007). Cationic arsenical species except for As(III) and As(V) could be separated by cation exchange whereas anion exchange columns can be used for the separation of As(III), As(V) and other ionizable arsenicals (Larsen 1998). Special attention should be focused on the composition of the mobile phase in these studies for arsenic speciation. The mobile phase contains phosphate species that should be avoided due to polymeric depositions on ICP-MS sampler and skimmer cones (Saverwyns et al. 1997, Ammann 2010). Nitric acid as eluent combination with ion pairing reagent could be used for separation of different arsenic species. It was noted in literature that ion pairing reagents were not gradient compatible. Additionally, As(III) could oxidize to As(V) in the case of HNO_3 (Londesborough et al. 1999, Ammann 2010).

There are many studies in literature where HPLC-ICP-MS has been used for the speciation of different arsenic species at trace levels. Pizarro et al. used phosphoric acid for the extraction of arsenic species from the soil (Pizarro et al. 2003). For one extraction run, they obtained extraction efficiency of 82 percent. Both anionic and cationic columns were used in this study. For anionic column, ammonium dihydrogen phosphate at pH 6.0 and for cationic column, pyridine formate at pH 2.8 were used as mobile phases. The authors also checked chloride interference, $^{40}Ar^{35}Cl$ and $^{40}Ar^{37}Cl$ peaks (m/z 75 and 77) and no interference was reported due to low concentration of chloride in soil extracts. In chromatographic conditions where anionic column was used, As(III) and AsB coelute. Therefore, cationic one was used to separate them (Pizarro et al. 2003). According to this study, four types of arsenic species were found in soil; As(V), As(III), MA^V and DMA^V. In addition, As(V) was the most abundant species found in the soil samples.

Sanz et al. (Sanz et al. 2005) applied sonication system for the extraction of arsenic species from soil and sediment samples. With this technique, they aimed to reduce the treatment period, get maximum recovery and avoid inter-conversion and/or degradation of species. The authors tried water, ammonium oxalate and H_3PO_4 to find the relatively best extractant. Extraction period was optimized to obtain maximum recovery. The best recovery was obtained when H_3PO_4 was used as extraction reagent for As(III), As(V), MA^V and DMA^V species. In this study, anion-exchange column was used with $HPO_4^{2-}/H_2PO_4^-$ buffer at pH 8.5 as the mobile phase. Detection limits they obtained were 19.6, 12.7, 14.3 and 19.4 pg/mL for As(III), DMA^V, MA^V and As(V), respectively. The authors observed that As(V) was the dominant species in soil samples.

Arsenic speciation in river water was performed with HPLC-ICP-MS. Norshidah et al. determined As(III) and As(V) in their study (Norshidah et al. 2011). No interference was observed at m/z 75. The detection limits found in this study were 1.0 and 0.5 ng/mL for As(III) and As(V), respectively.

Zheng and Hintelmann performed As speciation in a terrestrial plant sample by HPLC-ICP-MS. They applied different extraction procedures. Among them, water extraction gave the best result with an extraction yield of approximately 82 percent. While Sector-Field-MS was used for the speciation of arsenic, TOF-MS was used for the determination of total arsenic. Germanium was used as an internal standard. Both cation and anion exchange columns were used for the determination of As(III), As(V), MA^V, DMA^V, TETRA and TMAO (Zheng and Hintelmann 2009).

Speciation of arsenic in terrestrial plants from arsenic contaminated areas was performed by other researchers (Jedynak et al. 2009). For extraction, water, tris-HCl and SDS (Sodium Dodecyl Sulphate) were tried as extractants. The highest extraction efficiency was obtained in this study by using SDS. The authors applied two step extractions before speciation analysis, and species found in each fraction were determined. Predominant species for all plant samples was found to be As(V).

Speciation analysis of thioarsenates in aquatic samples was studied by using anion-exchange chromatography coupled to ICP-DRC-MS (Planer-Friedrich et al. 2007, Wallschläger and Stadey 2007). As a novelty, As/S ratios were monitored. Concentrations of mono-, di-, tri- and tetrathioarsenate at trace levels were investigated in a variety of sulfidic, natural waters (Wallschläger and Stadey 2007). They obtained instrumental detection limits of about 0.01 ng/mL for arsenic species. Analytical characteristics of different LC-ICP-MS methods for arsenic speciation in a variety of samples are presented in Table 9.1 with experimental details.

Speciation of arsenic by ET-AAS

ET-AAS (or GF-AAS) is a common technique which has been applied for the speciation of arsenic in food, water, environmental and geological samples. GF-AAS, which is a highly sensitive and simple technique, is available in most of the laboratories (Serafimovski et al. 2006). In order to increase the sensitivity of this technique, preconcentration is needed.

It is clear that speciation of arsenic in surface and ground water is important to understand arsenic exposure for humans and animals. Shirkhanloo et al. developed a new speciation and preconcentration method for As(III) and As(V) determination in water samples (Shirkhanloo et al. 2011). This method based on dispersive liquid-liquid microextraction (DLLME) and then determination of arsenic species by ET-AAS. In the extraction process, As(III) in tap water samples was complexed with ammonium pyrrolidine dithiocarbamate (APDC). Then, As(III) was extracted into 1-hexyl-3-metylimidazolium hexafluorophosphate as an ionic liquid and determined by ETAAS. For total arsenic determination, As(V) in tap water was reduced to As(III) by using KI and ascorbic acid in HCl solution (Shirkhanloo et al. 2011). Similarly, Liang et al. performed a speciation study for inorganic arsenic in water samples by ET-AAS (Liang et al. 2009). The detection limit they obtained was 36 pg/mL for As (III) based on S/N ratio of 3. The enrichment factor was found to be 45 for 5.0 mL sample solution with this extraction procedure. Finally, the recoveries for As(III) and As(V) ranged between 96–103 percent.

Elci et al. developed a simple, sensitive and economical method for As(III) and As(V) determination in water samples by ET-AAS (Elci et al. 2008). In this method, As (III) was coprecipitated with Ce (IV) hydroxide in ammonium/ammonia buffer at pH 9.0. After collection on filter paper, the coprecipitate was dissolved in HNO_3. Then it was diluted to required

Table 9.1 Comparison of analytical characteristics of LC-ICP-MS techniques for As speciation.

Technique	Experimental conditions	Detection limits, ng/mL							Ref.
		AsB	As(III)	DMA	MMA	As(V)	p-ASA	ROX	
IC-DRC-ICP-MS	Mobile phase: gradient elution of 10 and 50 mmol/L (NH4)$_2$CO$_3$ in 2% v/v methanol (pH 9.0), separation on Hamilton PRP-X100, monitored ion ^{75}As^{12}CHH$^+$ *m/z* 89, DRC conditions: 0.6 mL/min CH$_4$ as reactive cell gas	0.003	0.010	0.004	0.003	0.002	-	-	Wang et al. 2007
IC-DRC-ICP-MS	Mobile phase: gradient elution of 0.5 mmol/L ammonium citrate in 1% methanol (pH 4.5) and 15 mmol/L ammonium citrate in 1% methanol (pH 8.0), separation on Hamilton PRP-X100, monitored ion ^{75}As^{12}CH$_2^+$ at *m/z* 89, DRC conditions: 1.0 mL/min CH$_4$ as reactive cell gas	-	0.009	0.009	0.006	0.006	-	-	Tsai and Jiang 2011
ORC-IC-ICP-MS[a]	Mobile phase: 20 mM (NH$_4$)$_2$HPO$_4$ at pH 8.90, separation on Hamilton PRP-100, monitored ion ^{75}As, Octopole collision/reaction cell with a flow rate of 3.0 mL/min for H$_2$	-	0.3	0.4	0.4	0.6	-	-	Chen et al. 2007
IC-ICP-MS[b]	Mobile phase: 10 mmol/L NH$_4$H$_2$PO$_4$, 10 mmol/L NH$_4$NO$_3$, pH 6.3 adjusted with NH$_3$(aq). separation on Hamilton PRP-100, monitored ion ^{75}As,	-	0.13	0.16	0.15	0.24	-	-	Kannamkumarath et al. 2004
IC-HR-ICP-MS[c]	Mobile phase: gradient elution of 0.5 and 100 mM NH$_4$NO$_3$, both at the same pH, separation on Dionex AG11, AG7-AS11, monitored ion ^{75}As	10	5	5	5	5	-	-	Ammann 2010

Method	Mobile phase								Reference
IC-ICP-MS[d]	Mobile phase: gradient elution of deionized water and 10 mM PO_4^{3-}, pH 7.2, separation on AS14, monitored ion ^{75}As	-	0.112	0.044	0.061	0.079	0.076	0.254	Jackson and Bertsch 2001
IC-ICP-MS[d]	Mobile phase: gradient elution of deionized water and 50 mM NaOH, separation on AS16, monitored ion ^{75}As	-	0.015	0.011	0.014	0.029	0.018	0.061	Jackson and Bertsch 2001
IC-ICP-MS[d]	Mobile phase: gradient elution of 2.5 mM HNO_3 and 50 mM HNO_3, separation on AS7, monitored ion ^{75}As	-	0.024	0.019	0.006	0.008	0.053	0.027	Jackson and Bertsch 2001
IC-ICP-MS[d]	Mobile phase: isocratic 20 mmol/L NH_4NO_3, pH 10.0, separation on Anion (IC-Pak A) HC, monitored ion ^{75}As	-	0.4	0.4	0.5	0.5	-	-	Pantstar-Kallio and Manninen 1999
IC-ICP-MS	Mobile phase: 30 mM $NH_4H_2PO_4$, at pH 5.6, separation on Hamilton PRP-100, monitored ion ^{75}As	-	0.1	0.1	0.2	0.3	-	-	Chen et al. 2008

[a]ORC Octopole reaction cell
[b]Detection limits were given as ng/L
[c]Ion chromatography high resolution inductively coupled plasma mass spectrometry
[d]Detection limits were given as µg/L

volume prior to measurement. For the reduction of As(V), NaI was used. They found preconcentration factor as 75 with a recovery of $\geq$ 95% and LOD as 0.05 ng/mL. Another study for inorganic arsenic speciation was performed by Baig et al. in natural water samples (Baig et al. 2009). Different extraction procedures, such as CPE and SPE, were applied for the determination of As(III) and total inorganic arsenic. As(III) was complexed with APDC and extraction was performed by Triton X-114. LOD values were 0.03 and 0.02 ng/mL for As (III) and total arsenic, respectively. Recoveries were calculated as $\geq$ 98% and preconcentration factor was found to be 40 (Baig et al. 2009). Inorganic arsenic speciation was also performed for marine sediments as an environmental sample besides water analysis. Bermejobarrera et al. determined As(III) and total arsenic by using ET-AAS with different modifiers. In this study, As(III) was chelated with sodium diethyldithiocarbamate in water and then extracted with isobutylmethyl ketone. Palladium was selected as a chemical modifier (Bermejobarrera et al. 1995).

Speciation of arsenic by HG

One of the most popular speciation techniques for some of the elements including arsenic is based on the separation of the element from the sample matrix to form a volatile hydride. The hydride generation technique is based on the reaction of arsenic with hydrogen which is formed during the reaction. The volatile hydride occurs at room temperature. Rapid generation of volatile hydrides takes place by the addition of an acidified aqueous solution of the sample to a small volume of a 1 percent aqueous solution of sodium borohydride ($NaBH_4$). A typical reaction is shown by the equation given below (Skoog et al. 1998, Radke et al. 2012).

$$3BH_4^- + 3H^+ + 4H_3AsO_3 \rightarrow 3H_3BO_3 + 4AsH_3 \uparrow + 3H_2O$$

HG coupled to AAS and AFS are sensitive analytical tools for speciation of inorganic arsenic at trace levels. The major drawback of this technique is the interfering reactions caused by other hydride forming elements that influence the formation of related hydride. Interfering reactions can be eliminated by developing an effective separation methods (Ulusoy et al. 2011).

There are many studies in literature to perform arsenic speciation at trace levels using the hydride generation method. Tuzen et al. performed arsenic speciation study in natural water and soil samples by HG-AAS (Tuzen et al. 2009). They extracted the As(V) first in their study. As(V) was co-precipitated with $Al(OH)_3$ in the pH range of 8.0–10.0. After dissolving the collected precipitate in concentrated HCl, reduction of As(V) to As(III) was performed. By this way, concentration of As(V) was determined. In order to determine the total arsenic concentration, oxidation of As(III) to As(V) was achieved by using $KMnO_4$ solution. Then, coprecipitation was followed by the reduction using potassium iodide and ascorbic acid. Concentration of As(III) was calculated by the difference of total arsenic and As(V) concentration in the sample. Optimized parameters are given in Table 9.2.

Preconcentration factor was calculated as 25, and LOD was 12 pg/mL. Another procedure was reported for the arsenic speciation in phosphate rocks by slurry sampling with HG-AAS (Macedo et al. 2009). As(III) was determined directly, and total arsenic was detected after the reduction reaction. For slurry preparation of rock samples, dry samples with the optimized volume and concentration of HCl were put into the ultrasonic bath. Then, slurries were treated with citric acid/sodium citrate buffer of pH 7.1 and diluted with de-ionized water for As(III) determination. Determination of total As was performed after the reduction process. In order to control the accuracy of slurry sampling method, temperature controlled acid digestion method was applied to rock samples by using HNO_3 and H_2O_2 mixture. The optimized values for the flow rate of $NaBH_4$, concentrations of HCl, KI, ascorbic acid and $NaBH_4$ are given in Table 9.2. LOD value for As(III) was calculated as 0.1 ng/mL (Macedo et al. 2009).

Another study for the speciation of arsenic species in water samples was performed by using HG-AAS (Ulusoy et al. 2011). The applied preconcentration method was CPE. In this study, As(III) was separated by forming an ion pairing complex with Pyronine B-sodium dodecyl sulfate at pH 10. Although As(V) forms complex with Pyronine B at the same pH value like As(III); 1:2 metal-ligand ratio of As(V) causes formation of weaker ion-pair complex. Therefore, researchers optimized the parameters

Table 9.2 Optimized parameters for extraction and hydride forming chemicals.

	Tuzen et al. 2009	Macedo et al. 2009	Ulusoy et al. 2011
For extraction			
KI, % (w/v)	0.75	10	-
Ascorbic acid, % (w/v)	1.25	2.0	-
Pyronine B, M	-	-	0.72×10^{-5}
SDS, M	-	-	0.9×10^{-5}
For hydride formation			
HCl as a carrier solution, M	-	-	5
$NaBH_4$, % (w/v)	0.3	2.0	4.0
NaOH, % (w/v)	0.1	0.5	1.0
$NaBH_4$ flow rate, mL/min	-	2.7	2.0
HCl flow rate, mL/min	-	-	2.0
Sample flow rate, mL/min	-	-	5.0
Argon flow rate, mL/min	70	100	70

for As(III) by considering the fact that As(III) forms a more stable complex with Pyronine B. In the reduction process, sodium thiosulfate was used at room temperature. Optimized values of the system are given in Table 9.2. The preconcentration factor was calculated as 60, and LOD value was found to be 8 pg/mL. It is clear that the applied method is a very versatile tool to perform arsenic speciation in water samples at ultra trace levels.

Concentration and type of arsenic species in the sample which is analyzed are the most important parameters to select the technique. Moreover, extraction has to be done efficiently to get accurate results. Therefore, the preconcentration factor and recovery values are also important. Analytical performances and extraction efficiencies of the common techniques which have been used in literature are compared in Table 9.3.

Table 9.3 Analytical performances and extraction efficiencies of some studies using common techniques for arsenic speciation.

	Sample	Extraction Technique	Preconc. factor	Recovery %	LOD, ng/mL As(III)
HPLC-ICP-MS					
Pizarro et al. 2003	soil	Liquid-liquid extraction	-	> 90	1.2
Sanz et al. 2005	Soil & sediment	Sonication	-	85.6	0.02
Norshidah et al. 2011	water		-		1.0
Zheng and Hintelmann 2009	plant	Liquid-liquid extraction	-	82	-
ET-AAS					
Shirkhanloo et al. 2011	water	DLLME	-	> 90	0.01
Liang et al. 2009	water	DLLME	45	96–103	0.04
Elci et al. 2008	water	Coprecipitation	75	≥ 95	0.05
Baig et al. 2009	water	CPE & SPE	40	≥ 98	0.03
HG-AAS					
Tuzen et al. 2009	Soil & water	Coprecipitation	25	> 96	0.01 As(V)
Macedo et al. 2009	rock	SS	-	-	0.1
Ulusoy et al. 2011	water	CPE	60	97–103.3	0.008

Speciation of arsenic by CE

Recently, capillary electrophoresis has been employed as a separation technique for arsenic speciation. This technique requires much smaller sample volume with respect to other techniques. Advantages of CE over the other methods are ease of operation, rapid separation, low cost, sensitivity and separation efficiency (Yang et al. 2009, Wu and Ho 2004).

After separation by capillary electrophoresis, arsenic species can be detected by different techniques. These are UV detection (Broeck and Vandecasteele 1998), Mass Spectrometry (Debusschere et al. 2000, Yang et al. 2012), Capacitively Coupled Contactless Conductivity Detection (Nguyen et al. 2007), Hydride Generation-Atomic Fluorescence Spectrometry (HG-AFS) (Yin 2004), ICP-MS (Yang et al. 2009), AFS (Wu and Ho 2004) and Laser-Induced Fluorescence (LIF) (Zhang et al. 2002). Among these detectors, lower detection limits have been achieved by using ICP-MS. As an alternative to element specific detectors such as ICP-MS or ICP-OES, molecular detection techniques such as ESI-MS could give molecular information about the species of interest. Advantages and possible limitations of CE-ICP-MS in addition to its application in speciation studies were discussed in recent reviews with details (Alvarez-Llamas et al. 2005, Kannamkumarath et al. 2002). An interface is required to introduce the analyte in mobile phase to ICP-MS. Different kinds of interface such as no-sheath-flow interface, sheath-flow interface and hydride generation interface were discussed in literature together with advantages and disadvantages (Yang et al. 2009, Richardson et al. 2004, Lu et al. 1995, Yang et al. 2008).

In order to reduce the back and laminar flow, the type of nebulizer in the hyphenation of CE to ICP-MS is very important. Casiot et al. tried different nebulizers and reported that the best analytical performance characteristics were obtained with the MicroMist nebulizer. When electromigrative injection was used, LOD was calculated as 0.08 ng/mL and linearity range reached up to 100 ng/mL. They used this system to determine the concentrations of arsenic species in soil (Casiot et al. 2002).

Sun et al. separated some organic and inorganic arsenic species by using $NaHCO_3$-Na_2CO_3 solution buffered to pH 10 by CE, and species were detected by UV at 192 nm. LOD values were found as 0.30, 1.83, 1.62, 0.34, 0.40, 0.18, 1.45, 0.40 and 6.22 µg/mL As, respectively for PAO, DMA, As(III), *p*-ASA, NPAA, PAA, MMA, Roxarsone and As(V) in case of counter electrophoretic flow (Sun et al. 2002). The stacking technique has been used in literature to improve the sensitivities for arsenic species. This technique is based on the differences between conductivities of sample solution and CE background electrolyte (BGE). By using this technique, the sample plug length can be increased without loss of resolution and sensitivity can be improved considerably. It was noted that mismatch in electro-osmotic flow (EOF) between the sample solution and the BGE is a basic limitation possibly causing peak broadening and loss of resolution in case too much sample solution is injected (Sun et al. 2003). CE-UV was employed for the determination of the most common inorganic and methylated arsenic species and some phenylarsenic compounds. Based on the separation method for anions using hydrodynamic sample injection, LOD values were 0.52, 0.25, 0.27, 0.12, 0.37, 0.6, 0.6, 1.2 and 1.0 µg As/mL for phenylarsine oxide, *p*-aminophenylarsonic acid, *o*-aminophenylarsonic, phenylarsonic acid, 4-hydroxy-3-nitrobenzenearsonic acid (roxarsone), MMA, DMA, As(III) and As(V), respectively. LOD values were improved by large-volume sample stacking in this study (Kutschera et al. 2007). A capillary zone electrophoresis-ultraviolet detection (CZE-UV) technique was developed by Gui-Di et al. for the separation and analysis of five arsenic species, As(III), DMA, *p*-AsA, MMA, and As(V) in sea foods. Optimization studies for the detection wavelength, ingredient of buffer solution, pH value, buffer concentration, separation voltages and injection times were performed. Under the optimum conditions, they achieved to separate five arsenic species within 11 min. LODs were in the range of 0.004–0.30 ng/ mL for all species (Gui-Di et al. 2009). The technique was applied to dried shrimps samples for arsenic speciation.

The application of CE based separation techniques for a variety of samples are listed in Table 9.4 along with their analytical performances.

Speciation of arsenic by XANES

X-ray absorption spectroscopy (XAS) is a fast way of sample analysis minimizing sample manipulation. It can be applied to both solution and solid-phase samples. Furthermore, XAS is an element and oxidation state specific technique and detection limits obtained using this technique are in parts per million range. XAS is divided into two regions: X-ray absorption near-edge structure (XANES) and extended X-ray absorption fine structure (EXAFS). Analysis performed by XANES focuses on the absorption-edge region of the XAS spectrum from a few electronvolts below an element's K-absorption edge to approximately 30 eV above the edge. The most probable chemical environment around the arsenic can be identified with XANES by the identification of some features such as oxidation state, atom type and arrangement (Foster et al. 1998). The EXAFS part of the spectrum has a typical range of 30 eV up to 1000 eV above the K-edge and provides direct structural information such as near-neighbor type, distance and coordination number (Smith et al. 2005). XANES can be used to identify the inorganic forms of arsenic in solid samples. Nonextractable arsenic species could be determined by using XANES as a complementary technique. The technique is based on a shift in the white line energy which depends on the oxidation state of arsenic. For As(III), absorption occurs at 11872.2 eV while this value is 11875.7 eV for As(V) (Niazi et al. 2011). The main disadvantage of this technique reported in literature is the poor sensitivity at parts per million levels (Smith et al. 2005).

XANES has been applied to a variety of samples including plankton organisms from contaminated lakes (Caumette et al. 2011), treated (Manning 2005) and contaminated soil (Niazi et al. 2011, Arcon et al. 2005,

Table 9.4 Comparison of CE separation hyphenated with different detection techniques for As speciation.

Sample	Analytical technique	Conditions for CE separation of As species	LOD values for As species	Ref.
Tap and lake water	CE-UV	2 s for 50 mbar, separation voltage was +18 kV, fused-silica capillary of 75 µm i.d., 48.5 cm, effective length to the detector of 40 cm	DMA:1.83 µg/mL, As(III): 1.62 µg/mL, p-ASA:0.34 µg/mL, NPAA:0.40 µg/mL, PAA: 0.18 µg/mL, MMA: 1.45 µg/mL, As(V): 6.22 µg/mL, Roxarsone: 0.40 µg/mL	Sun et al. 2003
Water	CE-UV	20 mM of sodium phosphate and 0.75 mM TTAB (tetradecyltrimethylammonium bromide) at pH 9.0, −20 kV of injection, fused silica capillary with 40 cm (34.5 cm × 50 µm i.d.)	As(V):0.3 nmol/mL, As(III):0.5 nmol/mL, DMA:1.0 nmol/mL	Chen et al. 2003
Chicken litter and soil samples	CE-UV	50 mbar for 150 s pressure injection, 15 mM phosphate buffer at a pH of 10.6, separation, voltage was +25 kV, fused silica capillary, 50 µm, i.d. x 360 µm od x 66.5 cm effective length	As(III): 1.29 ng/mL, DMA:1.13 ng/mL, PAA: 0.34 ng/mL, MMA: 0.89 ng/mL, Roxarsone: 1.06 ng/mL, As(V): 1.33 ng/mL	Jaafar et al. 2007
Natural and contaminated waters	CE-UV	10 mM Na$_2$MoO$_4$ 10 mM NaClO$_4$ at pH 3.0, pneumatic injection: 30 mbar, 100 s. voltage: −16 kV, fused-silica capillary of the total length of 70 cm, effective length of 60 cm and 75 mm i.d.	As(V): 0.005 µg/mL, MMA: 0.020 µg/mL, As(III): 0.005 µg/mL, DMA: 0.016 µg/mL	Koshcheeva et al. 2009
Tap and mineral water	CE-LIF	2.0 mM NaHCO$_3$ (pH 9.28) with 10^{-7} M fluorescein, separation voltage was +20 kV, 50 µm i.d. x 50 cm total length fused silica capillary (effective length 40 cm)	As(V): 0.54 µg/mL, MMA: 0.15 µg/mL, As(III): 0.4 µg/mL, DMA: 0.12 µg/mL	Zhang et al. 2002
NA[a]	CE-ESI-MS	5 mM ammonium acetate and the total voltage applied was: 0 to 6.5 min (30 + 4) kV, 6.5 to 20 min: (30–3.8) kV, fused silica capillary, 40 µm i.d., 78 cm of length	As(III): 0.5 µg/mL, DMA: 2.0 µg/mL, PAA: 1.4 µg/mL, MMA: 0.7 µg/mL, As(V): 0.2 µg/mL, AsC:3.0 µg/mL, AsB: 3.3 µg/mL for m/z:201, 7.5 µg/mL for m/z:179	Debusschere et al. 2000

Table 9.4 contd....

Table 9.4 contd....

Sample	Analytical technique	Conditions for CE separation of As species	LOD values for As species	Ref.
Fish and oyster tissues	CE-ICP-MS	5.5 μL/min flow rate of 15 mM Tris (pH 9.0) containing 15 mM sodium dodecyl sulfate (SDS) as electrolyte, +22 kV of separation voltage, 70 cm length x 75 μm i.d. fused-silica capillary	AsB:0.3 ng/mL, As(III): 0.5 ng/mL, AsC: 0.5 ng/mL, DMA: 0.3 ng/mL, MMA: 0.3 ng/mL, As(V): 0.4 ng/mL,	Yeh and Jiang 2005
Lavers collected from coastal waters	CE-ICP-MS	18 kV of separation voltage, 50 mM H_3BO_3-12.5 mM $Na_2B_4O_7$ at pH 9.10, fused capillary i.d. 75 μm; o.d. 365 μm; 70 cm long,	AsB, As(III), AsC, DMA, MMA, As(V)-detection limits were not given in related reference	Yang et al. 2012
Mya arenaria Linnaeus and Shrimp (sea foods)	CE-ICP-MS	20 mM NaH_2PO_4 –5 mM $Na_2B_4O_7$, pH 6.50, separation voltage was +12 kV, fused-silica capillary of i.d. 75μm; o.d. 365 μm; 60 cm long	As(V): 0.1 ng/mL, MMA: 0.19 ng/mL, As(III): 0.1 ng/mL, DMA: 0.18 ng/mL	Yang et al. 2009
Soil	CE-ICP-MS	20 μL/min flow rate of sodium chromate (5 mM) at pH 11.2, –20 kV of separation voltage, fused silica capillary 75 μm i.d., 150 mm o.d. with a total length of 80 cm	As(V): 0.08 ng/mL, MMA: 0.26 ng/mL, As(III): n.d., DMA:0.23 ng/mL	Casiot et al. 2002

[a]NA: Not applied to sample

Cancès et al. 2008), desert plant *Parkinsonia florida* (Castillo-Michel et al. 2011), rice root (Seyfferth et al. 2010) and rice tissues (Smith et al. 2009), hyper-accumulating fern (*Pteris vittata* L.) (Webb et al. 2003), blue mussels (*Mytilus edulis*) (Whaley-Martin et al. 2012), fine particulate matter derived from coal combustion (Shoji et al. 2002), corn seedlings (Parsons et al. 2008), cucumber (Meirer et al. 2007), waters and sediments of ephemeral floodplain pools (Hudson-Edwards et al. 2005) and pea (*Pisum sativum*) plants (Castillo-Michel et al. 2007). For XANES analysis, it was reported that soil samples were prepared for analysis by drying and sieving at 2 mm. Then, dried and sieved samples were grinded in a zirconium planetary mill to a submicrometer particle size (Arcon et al. 2005, Castillo-Michel et al. 2007). Homogenization by a mortar and pestle is recommended. In literature, subsamples of the wet homogenized tissue (Smith et al. 2009, Whaley-Martin et al. 2012), freshly harvested plant leaves (Webb et al. 2003) and lyophilized corn samples (Parsons et al. 2008) were placed between Kapton tape before XANES analysis as sample preparation technique.

In literature, there are many studies that apply XANES for arsenic speciation. In a study, XANES was used for the speciation of inorganic arsenic species, As(III) and As(V), in treated soils by fitting linear combinations of XANES spectra derived from several synthetic and well characterized As(III) and As(V) treated model compounds. In this study, the As(III) and As(V) are derived from the $NaAsO_2$ and Na_2HAsO_4. K-edge energies differ by approximately 2 eV per unit oxidation state change in arsenic and are used for absolute energy comparisons of As(III)/As(V) speciation in solids (Manning 2005). In another study, metallic arsenic, trivalent arsenic compounds including realgar (AsS), oripiment (As_2S_3), $NaAsO_2$ and arsenolite (As_2O_3), pentavalent organoarsenic compounds such as monomethylarsenic acid, MMAA and dimethylarsenic acid, DMAA, pentavalent arsenic minerals such as scorodite and pharmacosiderite and also soil samples were investigated by XANES and EXAFS (Arcon et al. 2005). In one of the XANES studies in literature, biologically important arsenic species such as MMA(III), DMA(III), DMPS: 2,3-dimercapto-1-propane sulfonic acid, As(Glu)$_3$: arsenic glutathione; MMA(V), DMA(V), TMAO, TETRA, AsB, C2-AB: arsenobetaine 2; C3-AB: arsenobetaine 3, AC and sugar A: a dimethylarsinyl riboside were determined in biological environmental samples by using XANES (Smith et al. 2005). Beam damage is one of the problems in this technique. At lower analyte concentrations, the damage observed could be attributed to total beam exposure for arsenic standards. In addition, beam damage was reported on the higher flux undulator beamline. Samples could be analyzed at −20°C to reduce the possible damages (Bacquart et al. 2010).

Conclusion

Speciation studies are very important to make sure that a sample containing analyte species are dangerous for human health or not. Arsenic is one of the toxic elements for human beings and has many sources to cause environmental contamination. Toxicity of arsenic species depends on its chemical form. Hence, detection of different arsenic species at trace levels in a variety of environmental samples is very crucial. Although there are many studies related to the determination of different species of arsenic in literature, problems such as low extraction efficiencies, accuracy checking are still waiting to be overcome. There are many extraction strategies to extract arsenic species from matrices to get higher extraction yields. In general, different chromatographic methods have been applied in the separation step while AAS, GF-AAS, ICP-OES, ICP-MS, XANES have been mostly used in the detection step. Researchers should make a choice among the speciation techniques for the determination of arsenic species in environmental samples by considering the detection limits, availability of the instrument, analysis time and extraction procedures.

In this chapter, some of the techniques which have been used for arsenic speciation in soil, sediments and water samples were discussed. The advantages and disadvantages of each method over other methods were evaluated by using proper literature values.

References

Akter, K.F., Z. Chen, L. Smith, D. Davey and R. Naidu. 2005. Speciation of arsenic in ground water samples: A comparative study of CE-UV, HG-AAS and LC-ICP-MS. Talanta. 68: 406–415.

Aksentijevic, S., J. Kiurski and M.V. Vasic. 2012. Arsenic distribution in water/sediment system of Sevojno. Environ. Monitor. Assess. 184: 335–341.

Al-Assaf, K.H., J.F. Tyson and P.C. Uden. 2009. Determination of four arsenic species in soil by sequential extraction and high performance liquid chromatography with post-column hydride generation andinductively coupled plasma optical emission spectrometry detection. J. Anal. Atom. Spectrom. 24: 376–384.

Alvarez-Llamas, G., M.R.F. laCampa and A. Sanz-Medel. 2005. ICP-MS for specific detection in capillary electrophoresis. TRAC-Trend Anal. Chem. 24: 28–36.

Ammann, A.A. 2010. Arsenic speciation by gradient anion exchange narrow bore ion chromatography and high resolution inductively coupled plasma mass spectrometry detection. J. Chromatogr. A 1217: 2111–2116.

Anthemidis, A.N. and E.K. Martavaltzoglou. 2006. Determination of arsenic(III) by flow injection solid phase extraction coupled with on-line hydride generation atomic absorption spectrometry using a PTFE turnings-packed micro-column. Anal. Chim. Acta. 573: 413–418.

Arai, Y., A. Lanzirotti, S. Sutton, J.A. Davis and D.L. Sparks. 2003. Arsenic speciation and reactivity in poultry litter. Environ. Sci. Technol. 37: 4083–4090.

Arcon, I., J.T. van Elteren, H.J. Glass, A. Kodre and Z. Slejkovec. 2005. EXAFS and XANES study of arsenic in contaminated soil. X-Ray Spectrum. 34: 435–438.

Arroyo-Abad, U., J. Mattusch, M. Moder, M.P. Elizalde-Gonzalez, R. Wennrich and F.M. Matysik. 2011. Identification of roxarsone metabolites produced in the system: Soil-chlorinated water-light by using HPLC-ICP-MS/ESI-MS, HPLC-ESI-MS/MS and High Resolution Mass Spectrometry (ESI-TOF-MS). J. Anal. Atom. Spectrom. 26: 171–177.

Azcue, J.M. and J.O. Nriagu. 1994. Arsenic: Historical perspectives. *In:* J.O. Nriagu [ed.]. Arsenic in the Environment, part I: Cycling and Characterization. Wiley, New York.

Baba, K., T. Arao, Y. Maejima, E. Watanabe, H. Eun and M. Ishizaka. 2008. Arsenic speciation in rice and soil containing related compounds of chemical warfare agents. Anal. Chem. 80: 5768–5775.

Bacquart, T., G. Deves and R. Ortega. 2010. Direct speciation analysis of arsenic in sub-cellular compartments using micro-X-ray absorption spectroscopy. Environ. Res. 110: 413–416.

Baig, J.A., T.G. Kazi, A.Q. Shah, M.B. Arain, H.I. Afridi, G.A. Kandhro and S. Khan. 2009. Optimization of cloud point extraction and solid phase extraction methods for speciation of arsenic in natural water using multivariate technique. Anal. Chim. Acta. 651: 57–63.

Baowei, C., H. Naramandura, L. Meiling and X.C. Le. 2009. Metabolism, toxicity and biomonitoring of arsenic species. Prog. Chem. 21: 474–482.

Bermejobarrera, P., M.C. Barcielaalonso, M. Ferronnovais and A. Bermejobarrera. 1995. Speciation of arsenic by the determination of total arsenic and arsenic(III) in marine sediment samples by electrothermal atomic absorption spectrometry. J. Anal. Atom. Spectrom. 10: 247–252.

Bissen, M. and F.H. Frimmel. 2003. Arsenic—a review. Part 1: Occurrence, toxicity, speciation, mobility. Acta Hydroch. Hydrob. 31: 9–18.

Blasius, J. 1990. Lehrbuch der analytischen und präparativen anorganischen Chemie. S. Hirzel, Stuttgart.

Broeck, K.V. den and C. Vandecasteele. 1998. Capillary electrophoresis for the speciation of arsenic. Mikrochim. Acta. 128: 79–85.

Cancès, B., F. Juillot, G. Morin, V. Laperche, D. Polya, D.J. Vaughan, J.-L. Hazemann, O. Proux, G.E. Brown Jr. and G. Calas. 2008. Changes in arsenic speciation through a contaminated soil profile: A XAS based study. Sci. Total Environ. 397: 178–189.

Casiot, C., O.F.X. Donard and M. Potin-Gautier. 2002. Optimization of the hyphenation between capillary zone electrophoresis and inductively coupled plasma mass spectrometry for the measurement of As-, Sb-, Se- and Te-species, applicable to soil extracts. Spectrochim. Acta B 57: 173–187.

Castillo-Michel, H., J.G. Parsons, J.R. Peralta-Videa, A. Martinez-Martinez, K.M. Dokken and J.L. Gardea-Torresdey. 2007. Use of X-ray absorption spectroscopy and biochemical techniques to characterize arsenic uptake and reduction in pea (Pisum sativum) plants. Plant Physiol. Bioch. 45: 457–463.

Castillo-Michel, H., J. Hernandez-Viezcas, K.M. Dokken, M.A. Marcus, J.R. Peralta-Videa and J.L. Gardea-Torresdey. 2011. Localization and speciation of arsenic in soil and desert plant Parkinsonia florida using μXRF and μXANES. Environ. Sci. Technol. 45: 7848–7854.

Caumette, G., I. Koch, E. Estrada and K.J. Reimer. 2011. Arsenic speciation in plankton organisms from contaminated lakes: transformations at the base of the freshwater food chain. Environ. Sci. Technol. 45: 9917–9923.

Chen, M.L., Y.M. Huo and J.H. Wang. 2009. Speciation of inorganic arsenic in a sequential injection dual mini-column system coupled with hydride generation atomic fluorescence spectrometry. Talanta. 78: 88–93.

Chen, Z.L., J. Lin and R. Naidu. 2003. Separation of arsenic species by capillary electrophoresis with sample-stacking techniques. Anal. Bioanal. Chem. 375: 679–684.

Chen, Z., N.I. Khan, G. Owens and R. Naidu. 2007. Elimination of chloride interference on arsenic speciation in ion chromatography inductively coupled mass spectrometry using an octopole collision/reaction system. Microchem. J. 87: 87–90.

Chen, Z., K.F. Akter, M.M. Rahman and R. Naidu. 2008. The separation of arsenic species in soils and plant tissues by anion-exchange chromatography with inductively coupled mass spectrometry using various mobile phases. Microchem. J. 89: 20–28.

Chowdhury, U.K., B.K. Biswas, T.R. Chowdhury, G. Samanta, B.K. Mandal, G.C. Basu, C.R. Chanda, D. Lodh, K.C. Saha, S.K. Mukherjee, S. Roy, S. Kabir, Q. Quamruzzaman and D. Chakraborti. 2000. Groundwater arsenic contamination in Bangladesh and West Bengal, India. Environ. Health Persp. 108: 393–397.

Ciardullo, S., F. Aureli, A. Raggi and F. Cubadda. 2010. Arsenic speciation in freshwater fish: Focus on extraction and mass balance. Talanta. 81: 213–221.

Cohen, S.M., L.L. Arnold, M. Eldan, A.S. Lewis and B.D. Beck. 2006. Methylated arsenicals: The implications of metabolism and carcinogenicity studies in rodents to human risk assessment. Crit. Rev. Toxicol. 36: 99–133.

Cornelis, R., J. Caruso, H. Crews and K.G. Heumann. 2003. Handbook of Elemental Speciation: techniques and methodology. Wiley. New York, USA.

Council, N.R. 1999. National Academy Press. Washington, DC. USA.

Cullen, W.R. and K.J. Reimer. 1989. Arsenic Speciation in the Environment. Chem. Rev. 89: 713–764.

Debusschere, L., C. Demesmay and J. L. Rocca. 2000. Arsenic speciation by coupling capillary zone electrophoresis with mass spectrometry. Chromatographia. 51: 262–268.

Elci, L., U. Divrikli and M. Soylak. 2008. Inorganic arsenic speciation in various water samples with GFAAS using coprecipitation. Int. J. Environ. Anal. Chem. 88: 711–723.

Ellwood, M.J. and W.A. Maher. 2003. Measurement of arsenic species in marine sediments by high-performance liquid chromatography–inductively coupled plasma mass spectrometry. Anal. Chim. Acta. 477: 279–291.

Ferguson, J.F. and J. Gavis. 1972. A review of the arsenic cycle in natural waters. Water Res. 6: 1259–&.

Ferguson, J.F. and M.A. Anderson. 1974. Chemical forms of arsenic in water supplies and their removal. *In:* A.J. Rubin [ed.]. Chemistry of Water Supply, Treatment, and Distribution. Ann Arbor Science Publishers, Michigan.

Foster, A.L., G.E. Brown, T.N. Tingle and G.A. Parks. 1998. Quantitative arsenic speciation in mine tailings using X-ray absorption spectroscopy. Am. Mineral. 83: 553–568.

Francesconi, K.A. and D. Kuehnelt. 2004. Determination of arsenic species: A critical review of methods and applications, 2000–2003. Analyst. 129: 373–395.

Francesconi, K.A., R. Tanggaard, C.J. McKenzie and W. Goessler. 2002. Arsenic metabolites in human urine after ingestion of an arsenosugar. Clin. Chem. 48: 92–101.

Gallagher, P.A., J.A. Shoemaker, X. Wei, C.A. Brockhoff-Schwegel and J.T. Creed. 2001. Extraction and detection of arsenicals in seaweed via accelerated solvent extraction with ion chromatographic separation and ICP-MS detection. Fresenius' J. Anal. Chem. 369: 71–80

Garbarino, J.R., A.J. Bednar, D.W. Rutherford, R.S. Beyer and R.L. Wershaw. 2003. Environmental fate of roxarsone in poultry litter. I. Degradation of roxarsone during composting. Environ. Sci. Technol. 37: 1509–1514.

Garcia-Manyes, S., G. Jimenez, A. Padro, R. Rubio and G. Rauret. 2002. Arsenic speciation in contaminated soils. Talanta. 58: 97–109.

Gong, Z., X. Lu, M. Ma, C. Watt and X.C. Le. 2002. Arsenic speciation analysis. Talanta. 58: 77–96.

Gonzalvez, A., S. Armenta, M.L. Cervera and M. de la Guardia. 2010. Non-chromatographic speciation. Trac-Trend. Anal. Chem. 29: 260–268.

Gui-Di, Y., Z. Jin-Ping, H. Hong-Xia, Q. Guo-Min, X. Jin-Hua and F. Feng-Fu. 2009. Speciation analysis of arsenic in seafood with capillary electrophoresis—UV detection. Chinese J. Anal. Chem. 37: 532–536.

Haase, H.J. 1994. Abwasser in der Halbleiterfertigung. Metalloberfläche. 48: 571–575.

Heinrichs, G. and P. Udluft. 1993. Die geochemische Herkunft von Arsen. *In:* M. Jekel [ed.]. Arsen in der Trinkwasserversorgung. DVGW-Schriftenreihe Wasser Nr. 82, Eschborn.

Heitkemper, D.T., K.M. Kubachka, P.R. Halpin, M.N. Allen and N.V. Shockey. 2009. Survey of total arsenic and arsenic speciation in US-produced rice as a reference point for evaluating change and future trends. Food Addit. Contam. B 2: 112–120.

Hindmarsh, J.T. and R.F. McCurdy. 1986. Clinical and environmental aspects of arsenic toxicity. Crit. Rev. Clin. Lab. Sci. 23: 315–347.

Hirata, S. and H. Toshimitsu. 2007. Determination of arsenic species and arsenosugars in marine samples by HPLC-ICP-MS. Appl. Organomet. Chem. 21: 447–454.

Hudson-Edwards, K.A., H.E. Jamieson, J.M. Charnock and M.G. Macklin. 2005. Arsenic speciation in waters and sediment of ephemeral floodplain pools, Rios Agrio–Guadiamar, Aznalco llar, Spain. Chem. Geol. 219: 175–192.

Ito, K., C.D. Palmer, A.J. Steuerwald and P.J. Parsons. 2010. Determination of five arsenic species in whole blood by liquid chromatographycoupled with inductively coupled plasma mass spectrometry. J. Anal. Atom. Spectrom. 25: 1334–1342.

Jaafar, J., Z. Irwan, R. Ahamad, S. Terabe, T. Ikegami and N. Tanaka. 2007. Online preconcentration of arsenic compounds by dynamic pH junction-capillary electrophoresis. J. Sep. Sci. 30: 391–398.

Jackson, B.P. and P.M. Bertsch. 2001. Determination of arsenic speciation in poultry wastes by IC-ICP-MS. Environ. Sci. Technol. 35: 4868–4873.

Jedynak, L., J. Kowalska, J. Harasimowicz and J. Golimowski. 2009. Speciation analysis of arsenic in terrestrial plants from arsenic contaminated area. Sci. Total Environ. 407: 945–952.

Jiang, H.M., B. Hu, B.B. Chen and L.B. Xia. 2009. Hollow fiber liquid phase microextraction combined with electrothermal atomic absorption spectrometry for the speciation of arsenic (III) and arsenic (V) in fresh waters and human hair extracts. Anal. Chim. Acta. 634: 15–21.

Kannamkumarath, S.S., K. Wrobel, K. Wobel, C. B'Hymer and J.A. Caruso. 2002. Capillary electrophoresis—inductively coupled plasma-mass spectrometry: an attractive complementary technique for elemental speciation analysis. J. Chromatogr. A 975: 245–266.

Kannammumarath, S.S., K. Wroabel, K. Wroabel and J.A. Carusa. 2004. Speciation of arsenic in different types of nuts by ion chromatography-inductively coupled plasma mass spectrometry. Journal of Agriculture and Food Chem. 52: 1458–1463.

Kapaj, S., H. Peterson, K. Liber and P. Bhattacharya. 2006. Human Health Effects From Chronic Arsenic Poisoning—A Review. J. Environ. Sci. Heal. A 41: 2399–2428.

Karthikeyan, S. and S. Hirata. 2004. Ion chromatography–inductively coupled plasma mass spectrometry determination of arsenic species in marine samples. App. Organomet. Chem. 18: 323–330.

Katsoyiannis, I.A. and A.A. Katsoyiannis. 2006. Arsenic and other metal contamination of groundwaters in the industrial area of Thessaloniki, Northern Greece. Environ. Monitor. Assess. 123: 393–406.

Kohlmeyer, U., E. Jantzen, J. Kuballa and S. Jakubik. 2003. Benefits of high resolution IC–ICP–MS for the routine analysis of inorganic and organic arsenic species in food products of marine and terrestrial origin. Anal. Bioanal. Chem. 377: 6–13.

Koshcheeva, O.S., O.V. Shuvaeva and L.I. Kuznetzova. 2009. Arsenic speciation in natural and contaminated waters using CZE with in situ derivatization by molybdate and direct UV-detection. Electrophoresis. 30: 1088–1093.

Krysiak, A. and A. Karczewska. 2007. Arsenic extractability in soils in the areas of former arsenic mining and smelting, SW Poland. Sci. Total Environ. 379: 190–200.

Kumar, A.R. and P. Riyazuddin. 2007. Non-chromatographic hydride generation atomic spectrometric techniques for the speciation analysis of arsenic, antimony, selenium, and tellurium in water samples—a review. Int. J. Environ. An. Ch. 87: 469–500.

Kutschera, K., A. Schmidt, S. Köhler and M. Otto. 2007. CZE for the speciation of arsenic in aqueous soil extracts. Electrophoresis. 28: 3466–3476.

Leermakers, M., W. Baeyens, M. De Gieter, B. Smedts, C. Meert, H.C. De Bisschop, R. Morabito and P. Quevauviller. 2006. Toxic arsenic compounds in environmental samples: Speciation and validation. TRAC-Trend. Anal. Chem. 25: 1–10.

Larsen, E.H. 1998. Method optimization and quality assurance in speciation analysis using high performance liquid chromatography with detection by inductively coupled plasma mass spectrometry. Spectrochim. Acta Part B. 53: 253–265.

Léonard, A. 1991. Arsenic. *In:* E. Merian [ed.]. Metals and their Compounds in the Environment – Occurrence, Analysis, and Biological Relevance. VCH, Weinheim.

Liang, P., L.L. Peng and P. Yan. 2009. Speciation of As(III) and As(V) in water samples by dispersive liquid-liquid microextraction separation and determination by graphite furnace atomic absorption spectrometry.Microchim. Acta. 166: 47–52.

Lin, H.T., S.W. Chen, C.J. Shen and C. Chu. 2008. Arsenic speciation in fish on the market. J. Food Drug Anal. 16: 70–75.

Londesborough, S., J. Mattusch and R. Wennrich. 1999. Separation of organic and inorganic arsenic species by HPLC-ICP-MS. Fresenius J. Anal. Chem. 363: 577–581.

Lu, Q., S.M. Bird and R.M. Barnes. 1995. Interface for capillary electrophoresis and inductively coupled plasma mass spectrometry. Anal. Chem. 67: 2949–2956.

Ma, M. and X.C. Le. 1998. Effect of arsenosugar ingestion on urinary arsenic speciation, Clin. Chem. 44: 539–550.

Macedo, S.M., R.M. de Jesus, K.S. Garcia, V. Hatje, A.F.D. Queiroz and S.L.C. Ferreira. 2009. Determination of total arsenic and arsenic (III) in phosphate fertilizers and phosphate rocks by HG-AAS after multivariate optimization based on Box-Behnken design. Talanta. 80: 974–979.

Manning, B. 2005. Arsenic speciation in As(III)- and As(V)-treated soil using XANES spectroscopy. Microchim. Acta. 151: 181–188.

Meirer, F., G. Pepponi, C. Streli, P. Wobrauschek, V.G. Mihucz, G. Zaray, V. Czech, J.A.C. Broekaert, U.E.A. Fittschen and G. Falkenberg. 2007. Application of synchrotron-radiation-induced TXRF-XANES for arsenic speciation in cucumber (Cucumis sativus L.) xylem sap. X-Ray Spectrom. 36: 408–412.

Milstein, L.S., A. Essader, C. Murrell, E.D. Pellizzari, R.A. Fernando, J.H. Raymer and O. Akinbo. 2003. Sample Preparation, Extraction Efficiency, and Determination of Six Arsenic Species Present in Food Composites. J. Agr. Food Chem. 51: 4180–4184.

Moore, J.W. and S. Ramamoorthy. 1984. Heavy Metals in Natural Waters. Springer, New York.

Munoz, O., D. Velez and R. Montoro. 1999. Optimization of the solubilization, extraction and determination of inorganic arsenic [As(III) + As(V)] in seafood products by acid digestion, solvent extraction and hydride generation atomic absorption spectrometry. Analyst. 124: 601–607.

Nam, S., J. Kim and S. Han. 2003. Direct determination of total arsenic and arsenic species by ion chromatography coupled with inductively coupled plasma mass spectrometry. Bull. Korean Chem. Soc. 24: 1805–1808.

Nguyen, H.T.A., P. Kubán, V.H. Pham and P.C. Hauser. 2007. Study of the determination of inorganic arsenic species by CE with capacitively coupled contactless conductivity detection. Electrophoresis. 28: 3500–3506.

Niazi, N.K., B. Singh and P. Shah. 2011. Arsenic speciation and phytoavailability in contaminated soils using a sequential extraction procedure and XANES spectroscopy. Environ. Sci. Technol. 45: 7135–7142.

Norshidah, B., S. Norashikin, O. Rozita, Z.S. Md, J. Hafizan and S.R. Saari. 2011. HPLC-ICP-MS Speciation Analysis of Arsenic in River Water of Sungai Kinta Malaysia. Res. J. Chem. Environ. 15: 45–48.

Quaghebeur, M., Z. Rengel and M. Smirk. 2003. Arsenic speciation in terrestrial plant material using microwave-assisted extraction, ion chromatography and inductively coupled plasma mass spectrometry. J. Anal. At. Spectrom. 18: 128–134.

Palumbo-Roe, B. and B. Klinck. 2007. Bioaccessibility of arsenic in mine waste-contaminated soils: A case study from an abandoned arsenic mine in SW England (UK). J. Environ. Sci. Heal. A 42: 1251–1261.

Pantsar-Kallio, M. and P.K.G. Manninen. 1999. Optimizing ion chromatography-inductively coupled plasma mass spectrometry for speciation analysis of arsenic, chromium and bromine in water samples, Int. J. Environ. Anal. Chem. 75: 43–55.

Parsons, J.G., A. Martinez-Martinez, J.R. Peralta-Videa and J.L. Gardea-Torresdey. 2008. Speciation and uptake of arsenic accumulated by corn seedlings using XAS and DRC-ICP-MS. Chemospere. 70: 2076–2083.

Pizarro, I., M. Gomez, C. Camara and M.A. Palacios. 2003. Arsenic speciation in environmental and biological samples Extraction and stability studies. Anal. Chim. Acta. 495: 85–93.

Planer-Friedrich, B., J. London, R.B. McCleskey, D.K. Nordstrom and D. Wallschläger. 2007. Thioarsenates in geothermal waters of yellowstone national park: determination, preservation, and geochemical importance. Environ. Sci. Technol. 41: 5245–5251.

Ponthieu, M., P. Pinel-Raffaitin, I. Le Hecho, L. Mazeas, D. Amouroux, O.F.X. Donard and M. Potin-Gautier. 2007. Speciation analysis of arsenic in landfill leachate. Water Res. 41: 3177–3185.

Radke, B., L. Jewell. and J. Namiesnik. 2012. Analysis of arsenic species in environmental samples. Crit. Rev. Anal. Chem. 42: 162–183.

Reimer, K.J., I. Koch and W.R. Cullen. 2010. Organoarsenicals. Distribution and Transformation in the Environment. Organometallics in Environ. Toxicol. 7: 165–229.

Reyes, L.H., J.L.G. Mar, G.M.M. Rahman, B. Seybert, T. Fahrenholz and H.M.S. Kingston. 2009. Simultaneous determination of arsenic and selenium species in fish tissues using microwave-assisted enzymatic extraction and ion chromatography–inductively coupled plasma mass spectrometry. Talanta. 78: 983–990.

Reyes, M.N.M., M.L. Cervera, R.C. Campos and M. de la Guardia. 2008. Non-chromatographic speciation of toxic arsenic in vegetables by hydride generation-atomic fluorescence spectrometry after ultrasound-assisted extraction. Talanta. 75: 811–816.

Richardson, D.D., S.S. Kannamkumarath, R.G. Wuilloud and J.A. Caruso. 2004. Hydride generation interface for speciation analysis coupling capillary electrophoresis to inductively coupled plasma mass spectrometry. Anal. Chem. 76: 7137–7142.

Roy, P. and A. Saha. 2002. Metabolism and toxicity of arsenic: A human carcinogen. Curr. Sci. India 82: 38–45.

Sanz, E., R. Munoz-Olivas and C. Camara. 2005. Evaluation of a focused sonication probe for arsenic speciation in environmental and biological samples. J. Chromatogr. A 1097: 1–8.

Saverwyns, S., X.R. Zhang, F. Vanhaecke, R. Cornelis, L. Moens and R. Dams. 1997. Speciation of six arsenic compounds using high-performance liquid chromatography-inductively coupled plasma mass spectrometry with sample introduction by thermospray nebulization J. Anal. At. Spectrom. 12: 1047–1052.

Serafimovski, I., I.B. Karadjova, T. Stafilov and D.L. Tsalev. 2006. Determination of total arsenic and toxicologically relevant arsenic species in fish by using electrothermal and hydride generation atomic absorption spectrometry. Microchem. J. 83: 55–60.

Seyfferth, A.L., S.M. Webb, J.C. Andrews and S. Fendorf. 2010. Arsenic localization, speciation, and co-occurrence with iron on rice (*Oryza sativa* L.) roots having variable Fe coatings. Environ. Sci. Technol. 44: 8108–8113.

Shah, A.Q., T.G. Kazi, M.B. Arain, M.K. Jamali, H.I. Afridi, N. Jalbani, G.A. Kandhro, J.A. Baig, R.A. Sarfraz and R. Ansari. 2009. Comparison of electrothermal and hydride generation atomic absorption spectrometry for the determination of total arsenic in broiler chicken. Food Chem. 113: 1351–1355.

Shemirani, F., M. Baghdadi and M. Ramezani. 2005. Preconcentration and determination of ultra trace amounts of arsenic(III) and arsenic(V) in tap water and total arsenic in biological samples by cloud point extraction and electrothermal atomic absorption spectrometry. Talanta. 65: 882–887.

Shirkhanloo, H., H.Z. Mousavi and A. Rouhollahi. 2011. Speciation and determination of trace amount of inorganic arsenic in water, environmental and biological samples. J. Chin. Chem. Soc.-Taip. 58: 623–628.

Shoji, T., F.E. Huggins, G.P. Huffman, W.P. Linak and C. Andrew Miller. 2002. XAFS Spectroscopy Analysis of Selected Elements in Fine Particulate Matter Derived from Coal Combustion. Energ. Fuel. 16: 325–329.

Skoog, D.A., F.J. Holler and T.A. Nieman. 1998. Principles of Instrumental Analysis. 5th edn. Saunders College Publishing. Philadelphia.

Smedley, P.L. and D.G. Kinniburgh. 2002. A review of the source, behaviour and distribution of arsenic in natural waters. Appl. Geochem. 17: 517–568.

Smith, E., I. Kempson, A.L. Juhasz, J. Weber, W.M. Skinner and M. Gräfe. 2009. Localization and speciation of arsenic and trace elements in rice tissues. Chemosphere. 76: 529–535.

Smith, P.G., I. Koch, RA. Gordan, D.F. Mandoli, B.D. Chapman and K.J. Remimer. 2005. X-ray absorption near-edge structure analysis of arsenic species for application to biological environmental samples. Environ. Sci. Technol. 39: 248–254.

Stolz, J.F., E. Perera, B. Kilonzo, B. Kail, B. Crable, E. Fisher, M. Ranganathan, L. Wormer and P. Basu. 2007. Biotransformation of 3-nitro-4-hydroxybenzene arsonic acid (roxarsone) and release of inorganic arsenic by Clostridium species. Environ. Sci. Technol. 41: 818–823.

Straif, K., L. Benbrahim-Tallaa, R. Baan, Y. Grosse, B. Secretan, F. El Ghissassi, V. Bouvard, N. Guha, C. Freeman, L. Galichet, V. Cogliano and W.H.O.I.A.R.C.M. Workin. 2009. A review of human carcinogens—Part C: metals, arsenic, dusts, and fibres. Lancet Oncol. 10: 453–454.

Sun, B., M. Macka and P.R. Haddad. 2002. Separation of organic and inorganic arsenic species by capillary electrophoresis using direct spectrophotometric detection. Electrophoresis. 23: 2430–2438.

Sun, B., M. Macka and P.R. Haddad. 2003. Trace determination of arsenic species by capillary electrophoresis with direct UV detection using sensitivity enhancement by counter- or co-electroosmotic flow stacking and a high-sensitivity cell. Electrophoresis. 24: 2045–2053.

Terlecka, E. 2005. Arsenic speciation analysis in water samples: A review of the hyphenated techniques. Environ. Monitor. Assess. 107: 259–284.

Tsai, C. and S. Jiang. 2011. Microwave-assisted extraction and ion chromatography dynamic reaction cell inductively coupled plasma mass spectrometry for the speciation analysis of arsenic and selenium in cereals. Anal. Sci. 27: 271–276.

Tuzen, M., D. Citak, D. Mendil and M. Soylak. 2009. Arsenic speciation in natural water samples by coprecipitation-hydride generation atomic absorption spectrometry combination. Talanta. 78: 52–56.

Ulusoy, H.I., M. Akcay, S. Ulusoy and R. Gurkan. 2011. Determination of ultra trace arsenic species in water samples by hydride generation atomic absorption spectrometry after cloud point extraction. Anal. Chim. Acta. 703: 137–144.

Vassileva, E., A. Becker and J.A.C. Broekaert. 2001. Determination of arsenic and selenium species in groundwater and soil extracts by ion chromatography coupled to inductively coupled plasma mass spectrometry. Anal. Chim. Acta. 441: 135–146.

Vega, L., M. Styblo, R. Patterson, W. Cullen, C. Wang and D. Germolec. 2001. Differential effects of trivalent and pentavalent arsenicals on cell proliferation and cytokine secretion in normal human epidermal keratinocytes. Toxicol. Appl. Pharm. 172: 225–232.

Wallschläger, D. and C.J. Stadey. 2007. Determination of (Oxy)thioarsenates in Sulfidic Waters. Anal. Chem. 79: 3873–3880.

Wang, R., Y. Hsu, L. Chang and S. Jiang. 2007. Speciation analysis of arsenic and selenium compounds in environmental and biological samples by ion chromatography—inductively coupled plasma dynamic reaction cell mass spectrometer. Anal. Chim. Acta. 590: 239–244.

Wang, Y., G. Morin, G. Ona-nguema, F. Juillot, F. Guyot, G. Calas and G.E. Brown. 2010. Evidence for Different Surface Speciation of Arsenite and Arsenate on Green Rust: An EXAFS and XANES Study Environ. Sci. Technol. 44: 109–115.

Webb, S.M., J. Gaillard, L.Q. Ma and C. Tu. 2003. XAS Speciation of arsenic in a hyper-accumulating fern. Environ. Sci. Technol. 37: 754–760.

Whaley-Martin, K.J., I. Koch, M. Moriarty and K.J. Reimer. 2012. Arsenic speciation in Blue Mussels (Mytilus edulis) along a highly contaminated arsenic gradient. Environ. Sci. Technol. 46: 3110–3118.

WHO. 2001. Environmental Health Criteria 224: Arsenic and arsenic compounds second edition. World Health Organization, Geneva.

Williams, G., J.M. West, I. Koch, K.J. Reimer and E.T. Snow. 2009. Arsenic speciation in the freshwater crayfish, Cherax destructor Clark. Sci. Total Environ. 407: 2650–2658.

Wu, J. and P.C. Ho. 2004. Speciation of inorganic and methylated arsenic compounds by capillary zone electrophoresis with indirect UV detection Application to the analysis of alkali extracts of As_2S_2 (realgar) and As_2S_3 (orpiment). J. Chromatogr. A 1026: 261–270.

Yamauchi, H. and B.A. Fowler. 1994. Toxicity and metabolism of inorganic and methylated arsenicals. In: J.O. Nriagu [eds.]. Arsenic in the Environment, part II. Human Health and Ecosystem Effects. Wiley, New York.

Yang, G.D., X.-Q. Xu, W. Wang, L.J. Xu, G.-N. Chen and F.-F. Fu. 2008. A new interface used to couple capillary electrophoresis with an inductively coupled plasma mass spectrometry for speciation analysis. Electrophoresis. 29: 2862–2868.

Yang, G., J. Xu, J. Zheng, X. Xu, W. Wang, L. Xu, G. Chen and F. Fum. 2009. Speciation analysis of arsenic in *Mya arenaria* Linnaeus and *Shrimp* with capillary electrophoresis-inductively coupled plasma mass spectrometry. Talanta. 78: 471–476.

Yang, G., J. Zheng, L. Chen, Q. Lin, Y. Zhao, Y. Wu and F. Fu. 2012. Speciation analysis and characterisation of arsenic in lavers collected from coastal waters of Fujian, south-eastern China. Food Chem. 132: 1480–1485.

Yeh, C. and S. Jiang. 2005. Speciation of arsenic compounds in fish and oyster tissues by capillary electrophoresis-inductively coupled plasma-mass spectrometry. Electrophoresis. 26: 1615–1621.

Yin, X. 2004. On-line preconcentration for capillary electrophoresis-atomic fluorescence spectrometric determination of arsenic compounds. Electrophoresis. 25: 1837–1842.

Zavala, Y.J. and J.M. Duxbury. 2008. Arsenic in rice: I. Estimating normal levels of total arsenic in rice grain. Environ. Sci. Technol. 42: 3856–3860.

Zhang, P., G. Xu, J. Xiong, Y. Zheng, Q. Yang and F. Wei. 2002. Capillary electrophoretic analysis of arsenic species with indirect laser induced fluorescence detection. J. Sep. Sci. 25: 155–159.

Zheng, J. and H. Hintelmann. 2009. HPLC-ICP-MS for a comparative study on the extraction approaches for arsenic speciation in terrestrial plant, Ceratophyllum demersum. J. Radioanal. Nucl. Ch. 280: 171–179.

Methods for Mercury Speciation in Environmental Samples

Zhenli Zhu,[1,a,]* *Qian He*[1,2] and *Zhifu Liu*[1,2]

1. Introduction

Mercury is one of the most toxic elements impacting on human and ecosystem health and therefore is one of the most intensively studied environmental pollutants. Mercury exists in a large number of physical and chemical forms with a large variety of properties that affect its complex distribution, biogeochemical transformation and toxicity. Table 10.1 lists common mercury species in environmental and biological samples. All mercury species are toxic, and organic mercury compounds generally are more toxic than inorganic species. In addition, the toxic effects of methylmercury (MeHg) can be significant due to its tendency to bioaccumulate or biomagnify within the aquatic food chain. Moreover, inorganic Hg^{2+} can undergo biomethylation, resulting in the formation of MeHg and dimethylmercury (DMeHg). These reactions are reversible

[1]State Key Laboratory of Biogeology and Environmental Geology, China University of Geosciences, Wuhan, China, 430074.
[a]Email: zhuzl03@gmail.com; zlzhu@cug.edu.cn
[2]Faculty of Earth Sciences, China University of Geosciences, Wuhan, China, 430074.
*Corresponding author

Table 10.1 Typical chemical forms of Hg and its abbreviations.

Mercury species	Molecular formula	Abbreviations	Source
Elemental mercury	Hg^0	Hg(0)	Natural waters, air samples, etc.
Mercurous	Hg_2^{2+}	Hg(I)	Soils and sediments, etc.
Mercuric(inorganic Hg)	Hg^{2+}	IHg/Hg(II)	Natural waters, sediments, sea food, fish, etc.
Methylmercury	CH_3Hg^+	MeHg/MMHg/MMeHg	Natural waters, sediment, sea food, fish, human hair, etc.
Ethylmercury	$CH_3CH_2Hg^+$	EtHg	Natural waters, sediment, fish, sea food, etc.
Dimethylmercury	CH_3HgCH_3	DMeHg	Natural waters, fish, sediment, etc.
Phenylmercury	$C_6H_5Hg^+$	PhHg	Natural waters, biological samples, etc.
Diethylmercury	$(CH_3CH_2)_2Hg$	DEtHg/Et$_2$Hg/DIEM	Organic synthesis
Diphenylmercury	$(C_6H_5)_2Hg$	DPhHg/Hg^{II}Ph$_2$	Organic synthesis
Thimerosal	$C_9H_9HgNaO_2S$	—	Medicine preservative
Dibenzylmercury	$(C_6H_5CH_2)_2Hg$	DBM	Cosmetics
4-(Hydroxymercury) benzoic acid sodium salt	$C_7H_5HgNaO_3$	PHMB	Chemical industry
Phenyl mercury benzoate	$C_{13}H_{10}HgO_2$	PMB	Pesticide and veterinary medicine

with demethylation facilitated by microorganisms and/or photolytic decomposition. Figure 10.1 illustrates the potential biotic and abiotic reactions, transformations, exchanges within and among reservoirs, and biological uptake of the primary Hg species in marine systems (Fitzgerald et al. 2007).

In order to better understand mercury geochemical cycling in the environment, it is necessary to study various processes: how and in what forms does mercury enter the ecosystem, what transformations does it undergo in a given environmental condition. In all these transformation processes, it is essential to determine the exact chemical form, or speciation of mercury, which includes detecting its oxidation state and ligands (Leopold et al. 2010, Stoichev et al. 2006, Mousavi et al. 2011). Therefore, more and more scientists are paying attention to perform mercury speciation in environmental and biological samples. Figure 10.2 shows the evolution of the number of publications on mercury speciation since 2000 (the data

Biogeochemical Cycling of Mercury

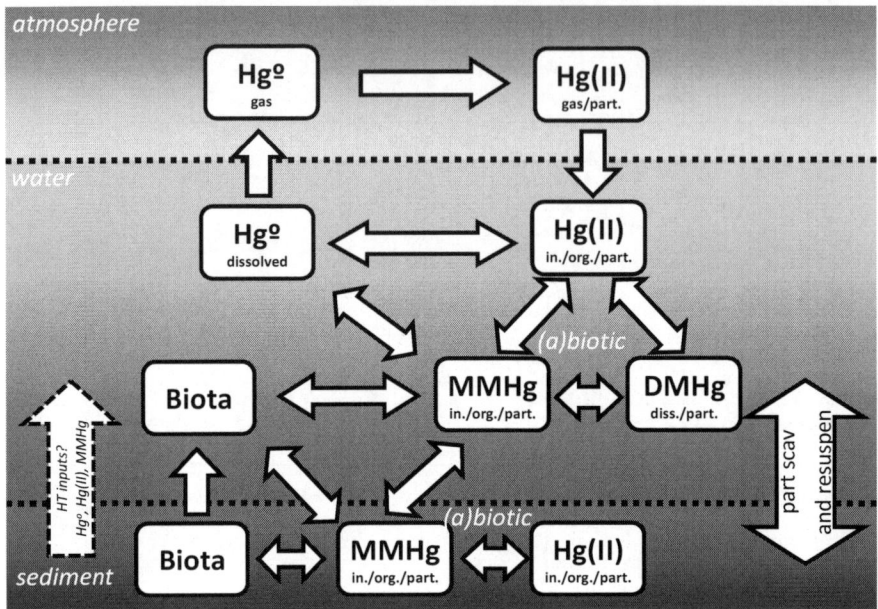

Figure 10.1 Biogeochemical cycling of Hg in the ocean (Fitzgerald et al. 2007).

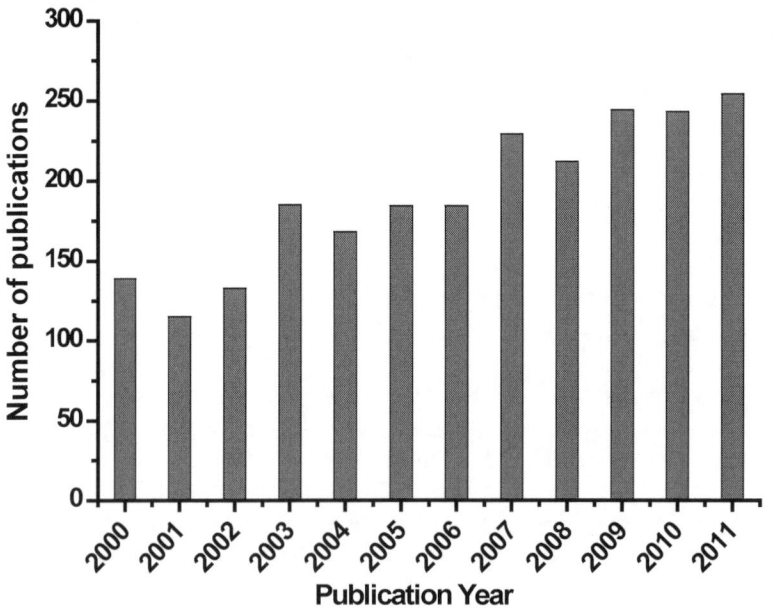

Figure 10.2 Evolution of the number of publications on mercury speciation.

was obtained from search of Web of Science using "mercury speciation" as keywords). As can be observed, there is a steady increase of the number of publications on mercury speciation.

Though various analytical methods (Pereiro and Díaz 2002) including atomic and molecular spectrometry have been reported for mercury speciation, this chapter will focus on the evolvement of atomic spectrometry-based mercury speciation methods, and their application in mercury speciation for environmental samples. Methods for mercury speciation can be classified into two general approaches (Leopold et al. 2010): non-chromatographic methods based on the chemical and physical properties of different Hg species and chromatographic methods (including gas chromatography, liquid chromatography and capillary electrophoresis, etc.). Non-chromatographic methods for Hg speciation mostly determine only one species or one fraction rather than the simultaneous determination of all Hg species present. These methods are based on different chemical and/or physical behavior of Hg species, i.e., solubility, volatility, or redox potential. Extraction of the analytes is a necessary step for mercury speciation in environmental and biological samples, especially for solid samples (sediment or soil), in order to separate the analytes from the matrix. It generally involves liquid-liquid extraction, acid extraction, base extraction, ultrasound-assisted extraction and microwave-assisted extraction (Leermakers et al. 2005, Ramalhosa et al. 2001). Special attention should be given so as not to alter the speciation (species decomposition, artifact methylation) during the extraction/digestion procedures. The extraction step in mercury speciation analysis is not the focus of this chapter, but it should be noted that some non-chromatographic methods discussed below are also currently used as extraction and clean-up techniques prior to chromatographic separations.

2. Non-chromatographic Methods

The most common tools for trace element speciation are the combination of separation techniques with a highly sensitive detector. These hyphenated techniques can provide the most complete information on the species' distribution and even structure. However, they are usually time-consuming, relatively costly and require complicated sample preparation processes. On the other hand, it is noticeable that, in some cases, the application of basic chemistry in sample treatment before detection can provide quantitative information about specific chemical forms of some elements. In this sense, non-chromatographic procedures can offer sufficient information about the elemental speciation for a range of situations. Moreover, these non-chromatographic strategies can be less time consuming, more cost-effective, and present competitive limits of detection. The main principles of various non-chromatographic method for trace elements and their advantages and drawbacks have been summarized by several reviews (Gonzalvez et al. 2010, Vieira et al. 2009, Gonzalvez et al. 2009).

2.1 Extraction methods

Extraction is frequently used to collect desired species from the sample for further studies. In this sense, it is considered as a pre-treatment step. However, extraction can also separate only one or a group of species from the sample for further identification, contributing to inter-separation. Different non-chromatographic speciation methods using extraction have been developed and some of them are listed in Table 10.2.

2.1.1 Liquid-Liquid Extraction (LLE)

Liquid-liquid extraction of metal species is the most popular method because of its simplicity and low cost. Other advantages are the reduction of matrix effects, high sensitivity and the potential of automation. Frequently, a selective complex of one of the species considered is extracted to an organic solvent, and thus is separated from the rest of the elemental species. In some cases, an additional back extraction step is necessary.

Different methods of non-chromatographic mercury speciation using liquid-liquid extraction have been reported. For example, total mercury and MeHg were determined in human permanent healthy teeth samples by electrothermal atomic absorption spectrometry (ET-AAS) (Saber-Tehrani et al. 2007). In this study, MeHg was transferred to the toluene phase by acid leaching method for determination, and total Hg was extracted to toluene phase using diethyldithiocarbamate (DDC) as a chelating agent after the sample was dissolved. The formation of a complex between MeHg and DDC also enhanced its signal and increased its reproducibility.

For Hg speciation in sediments, Sakamoto et al. (Sakamoto et al. 1992) and Tomiyasu et al. (Tomiyasu et al. 1996) developed similar extraction methods. Mercury species were first extracted by shaking a sample with 1 mol L^{-1} hydrochloric acid containing 3 percent sodium chloride in the presence of copper(I) chloride. In order to separate organic mercury from inorganic mercury, the extract was shaken with chloroform to extract organic mercury only. Inorganic mercury was left in the hydrochloric acid phase. The Hg in each extraction was determined by cold vapor atomic absorption spectrometry (CV-AAS). The procedure was also used for differential determination of Hg species in soil and aquatic organisms (Tomiyasu et al. 1996).

Speciation of mercury in fish samples by solvent extraction was described by Rezende et al. (Rezende et al. 1993). MeHg was extracted as bromide

Table 10.2 Selected applications of extraction method for non-chromatographic mercury speciation.

Matrix	Species	Extraction conditions	Analytical technique	Analytical features	Ref.
Liquid–liquid extraction					
Human teeth	MeHg, total mercury	Methylmercury was determined in organic phase after extraction with toluene.	ET-AAS/ CV-AAS	LOD (μg g^{-1}): 0.15 (MeHg); 0.12 (THg).	Saber-Tehrani et al. 2007
Sediments	Hg(II), organic Hg, Hg(II) oxide, Hg(II) sulfide	Successive extraction of Hg compounds with CHCl$_3$, a 0.05 mol L^{-1} H$_2$SO$_4$ and a 1 mol L^{-1} HCl containing 3% NaCl, respectively.	CV-AAS	LOD: 0.1 ng Hg.	Sakamoto et al. 1992
Sediment, soil and aquatic organisms	Hg(II) and organic Hg	Hg(II) was extracted from the materials with 1 mol L^{-1} HCl containing 3% NaCl in the presence of CuCl, and organic Hg was extracted with CHCl$_3$ from the HCl solution.	CV-AAS	LOD (mg kg^{-1}): 0.001 (Hg(II)); 0.008 (OHg).	Tomiyasu et al. 1996
Fish	Hg(II) and MeHg	Methylmercury as bromide was extracted from fish samples by CHCl$_3$. Inorganic Hg was determined in the residual aqueous phase by reduction with NaBH$_4$.	CV-AAS	LOD: 25 ng g^{-1} Hg (as MeHg)	Rezende et al. 1993
Fish	organic Hg, Hg(II),and total Hg	Organomercury compounds were selectively extracted from KBr solution by CHCl$_3$. Inorganic Hg remained in KBr solution.	CV-ETAAS	LOD (ng g^{-1}): 1(Hg(II)); 5(organic Hg) ; 1(THg).	Duarte et al. 2009
Water and urine	Hg(II) and MeHg	Selective continuous liquid–liquid extraction of methylmercury, into xylene, as bromide.	ICP-AES	LOD: 4 ng mL^{-1}	Garcia et al. 1996
Seafood	Hg(II) and MeHg	Triton X-114 was used as the extractant for selective extraction of MG–Hg(II)–I ion-associated complex; APDC was used as the chelating agent for CH$_3$Hg$^+$ determination.	ICP-OES	LOD (ng L^{-1}): 56.3 (Hg(II)); 94.6(MeHg).	Li and Hu 2007

Table 10.2 contd....

Table 10.2 contd.

Matrix	Species	Extraction conditions	Analytical technique	Analytical features	Ref.
Solid phase extraction					
Water	Hg(II) and organic Hg	Inorganic Hg species was retained on a column charged with a Dowex 1X-8 resin as the anionic complex formed with MTB at pH 6.3.	FI-CV AAS	LOD (ng L^{-1}): 0.8. RSD (%): 4.4.	Wuilloud et al. 2002
River water	Hg(II) and organic Hg	Retention on a column containing 2-mercaptobenzimidazol loaded on silica gel and then quantitatively eluted with 0.05 M KCN solution and 2.0 M HCl to desorp inorganic and methylmercury species, respectively.	CF-AFS	LOD (ng L^{-1}): 0.07 (THg); 0.05 (Hg(II)).	Bagheri and Gholami 2001
biological and environmental water samples	Hg(II) and MeHg	Selective retention inorganic mercury and methylmercury on the inner wall of a knotted reactor by using ammonium diethyl dithiophosphate and dithizone as complexing agents respectively.	CV-AFS	LOD (ng L^{-1}): 3.6 (Hg(II)); 2.0 (MeHg).	Wu et al. 2006
Seafood	MeHg	Direct coupling between headspace-SPME and quartz tube-AAS detection for speciation of	CV-AAS	LOD (ng mL^{-1}): 0.4 using silica fiber and 0.06 using the PDMS/DVB fiber.	Fragueiro et al. 2004

and determined directly in the organic phase by CV-AAS with sodium borohydride (NaBH$_4$) dissolved in dimethylformamide as a reducing agent; inorganic Hg was determined in the residual aqueous phase by reduction with aqueous NaBH$_4$. Duarte et al. developed another solvent extraction method for organic, inorganic and total mercury determination in fish tissue (Duarte et al. 2009). After drying and cryogenic grinding, potassium bromide and hydrochloric acid solution (1 mol L^{-1} KBr in 6 mol L^{-1} HCl) was added to samples. After centrifugation, total mercury was determined in the supernatant. Organic mercury compounds were selectively extracted from potassium bromide (KBr) solution using chloroform and the resulted solution was back extracted with 1% m/v L-cysteine. This back-extracted solution was used for organic Hg determination. Inorganic Hg remaining in KBr solution was directly determined by chemical vapor generation electrothermal atomic absorption spectrometry (CV-ET-AAS).

A selective continuous liquid-liquid extraction method was proposed for MeHg speciation in spiked seawater and spiked urine samples (Garcia et al. 1996). In a sample containing Hg(II) and MeHg, only the latter one is extracted into xylene as methylmercury bromide in a continuous mode, which was determined on-line by inductively coupled plasma atomic emission spectroscopy (ICP-AES) after mercury cold vapor generation from the organic phase. Sequential and selective extraction procedures were also used for mercury speciation in mine waste (Sladek and Gustin 2003). Pyrolytic methods were applied to determine volatile phases; leaching with Cl$^-$ solutions was utilized to determine soluble or mobile species; and acid digestions were used to extract strongly bound Hg species and for total Hg analyses.

Miniaturization of extraction procedures is a recent trend in sample pre-treatment techniques which greatly reduce the usage of a solvent. Liquid phase microextraction (LPME) is a novel miniaturized sample preparation technique that has gained extensive attention in analytical chemistry. It is an inexpensive, easy to operate, and nearly solvent free extraction technique that incorporates sampling, extraction, concentration and sample introduction into a single step. Hollow fiber liquid phase microextraction (HF-LPME) has been developed for the determination of MeHg content in human hair and sludge by graphite furnace atomic absorption spectrometry (GFAAS) (Jiang et al. 2008). MeHg was extracted into the organic phase (toluene) prior to its determination by GFAAS, while inorganic mercury remained as a free species in the sample solution. Total mercury was determined by inductively coupled plasma-mass spectrometry (ICP-MS), and the concentration of inorganic mercury was obtained by subtracting MeHg from total mercury.

The above extraction approaches can effectively decrease the detection limit and eliminate matrix interference. However, toxic organic solvents were generally used as extractants in these approaches, which may cause environmental and safety problems due to their high volatilization. Recently, room temperature ionic liquid(RTIL) began to take the place of organic solvents as extractants, due to their desirable properties, including low volatility, good chemical and thermal stability, and good solubility with most organic solvents (Jia et al. 2011). RTIL has been widely used for pre-concentration and separation of mercury (Jia et al. 2011, Martinis and Wuilloud 2010, Martinis et al. 2009, Pena-Pereira et al. 2009, Li et al. 2007). Martinis et al. (Martinis and Wuilloud 2010) developed a cold vapor ionic liquid-assisted head-space single drop microextraction (CVILAHS-SDME) method for Hg species determination at trace levels. A low cost RTIL, i.e., tetradecyl(trihexyl)phosphonium chloride (CYPHOS® IL 101), was used for Hg atomic vapor capture. Hg^{2+} was reduced to volatile Hg^0 by $SnCl_2$ for inorganic Hg determination. In order to evaluate total Hg concentration, the pre-treated samples were irradiated for 3 hr with a 15 W UV lamp in order to photo-oxidize organic Hg species. Organic Hg species concentration was evaluated based on the difference between total Hg and IHg concentration.

Another type of LLE is based on the use of surfactants. The separation of metal species is achieved by incorporating a selective complex of metal species in the surfactant structures followed by cloud-point extraction (CPE). The small volume of the surfactant-rich phase obtained with this methodology permits the design of extraction techniques that are simple, cheap, highly efficient, rapid and of lower toxicity to the environment than those conventional LLE that use organic solvents. Li et al. (Li and Hu 2007) developed a sequential cloud point extraction for the speciation of mercury in seafood. The method was based on that, Hg^{2+} was complexed with I^- to form HgI_4^{2-}, which reacted with the methyl green (MG) cation to form hydrophobic ion-associated complex, and this ion-associated complex was extracted into the surfactant-rich phase of the non-ionic surfactant octylphenoxypolyethoxyethanol (Triton X-114), which are subsequently separated from MeHg in the initial solution by centrifugation. The supernatant is also subjected to the similar CPE procedure for the pre-concentration of MeHg by adding a chelating agent, ammonium pyrrolidine dithiocarbamate (APDC) to form water-insolvable complex with MeHg. The extract solution was diluted with 0.5 mol L^{-1} HNO_3 for ICP-AES determination.

2.1.2 Solid Phase Extraction (SPE)

The use of SPE has been frequently proposed as a technique for speciation analysis, which is suitable for both in-batch and in-flow experiments. SPE has been proved to be a particularly useful method to analyze trace amounts with improved limits of detection (LODs). Usually, solid-phase extraction of species is applied in flow injection analysis (FIA). The extraction columns filled with anion or cation exchangers, complex forming resin, activated or modified alumina among others, are connected on-line to the detector. The analytes are extracted by sorption, eluted with a small amount of an organic solvent, and then derivatized or directly detected. Speciation analysis takes place by selective sorption or selective elution. The advantages of SPE over liquid–liquid extraction include a higher enrichment factor, lowering both solvent consumption and contamination risk, and the feasibility for application in field sampling and automation.

An on-line separation and pre-concentration of inorganic and organic mercury species system using a selective reagent and an anion exchange resin was described by Wuilloud et al. (Wuilloud et al. 2002). The inorganic Hg species were retained on a column loaded with a Dowex 1X-8 resin (particle size 50–100 mesh) as the anionic complex formed with methylthymol blue, at pH 6.3. Pre-oxidation of the organic mercury species permitted the determination of total mercury. The difference between total and inorganic mercury determined the organic mercury content. A similar procedure was described by Alonso et al. for mercury speciation in seafood (Alonso et al. 2008). The only difference is that, the column is packed with a chelating resin aminopropyl-controlled pore glass (550 Å) functionalized with [1,5-bis (2 pyridyl)-3-sulphophenyl methylene thiocarbonohydrazyde].

Bagheri et al. (Bagheri and Gholami 2001) reported another solid phase extraction method to determine very low levels of dissolved Hg(II) and MeHg in river waters. Inorganic and organic Hg were pre-concentrated simultaneously on a laboratory-made column containing 2-mercaptobenzimidazol loaded on silica gel and then quantitatively eluted with KCN solution and HCl to release Hg(II) and MeHg species, respectively. After irradiation with an intensive UV source, MeHg was decomposed and Hg vapor was generated from both species using $SnCl_2$ reductant in a continuous-flow system. Other modified silica gels with some chelating agents such as silica gel-immobilized-dithiocarbamate derivatives (Mahmoud 1999) and aminopropylbenzoylazo-2-mercaptobenzothiazole bonded to silica gel (Ma et al. 2000), have also been prepared and used in the selective determination of Hg(II) in waters or in other environmental samples.

Wu et al. (Wu et al. 2006) developed a mercury speciation method based on selective retention Hg(II) and MeHg on the inner wall of a knotted reactor by using ammonium diethyl dithiophosphate and dithizone as complexing agents respectively. With the sample pH kept at 2.0, the pre-concentration of Hg(II) on the inner walls of the knotted reactor was carried out based on the exclusive retention of Hg–DDP complex in the presence of MeHg via on-line merging the sample solution with ammonium diethyl dithiophosphate solution, and selective pre-concentration MeHg was achieved with dithizone instead of ammonium diethyl dithiophosphate. A 15% (v/v) HCl was introduced to elute the retained mercury species and merge with KBH_4 solution for atomic fluorescence spectrometry (AFS) detection.

In the last few years, some solid-phase microextraction (SPME) procedures have been applied to speciation analysis with analytical performance characteristics comparable to SPE procedures. In comparison with SPE, the advantages of SPME for elemental speciation arise from its simplicity, its capacity for *in situ* pre-concentration, ease of automation, selectivity and reduced breakthrough volume. Additionally, SPME offers the possibility of eliminating most interferences that are normally accompanied with continuous vapor generation (Mester et al. 2001). The selective determination of MeHg in food samples has been achieved by headspace solid-phase microextraction with quartz tube-atomic absorption spectrometry detection (Fragueiro et al. 2004). For MeHg determination in seafood, hydride or chloride generation and headspace SPME were performed after ultrasound-assisted extraction.

2.1.3 Other extraction methods

Supercritical fluid extraction (SFE) is an attractive alternative to conventional solvent extraction: it is relatively inexpensive; extraction is faster and more environmentally friendly than organic solvent based extraction. These procedures could be used for speciation in principle, because different oxidation states of the same metal often possess vastly different formation and extraction constants with the same reagent. In addition, when using SFE, the difference in solubility in supercritical carbon dioxide for the chelated metal species would also contribute to the potential of speciation analysis. For instance, organic and inorganic Hg species were extracted with the SeF_2, by varying the operating parameters and it was possible to selectively extract methyl, phenyl and inorganic Hg from different matrices (Foy and Pacey 2003).

Microwave-assisted extraction (MWAE) is also an expeditious, inexpensive and efficient extraction technique. MWAE provides a technique whereby intact organic and organometallic compounds can be selectively extracted. Berzas Nevado et al. (Berzas Nevado et al. 2006) developed a closed vessel microwave-assisted extraction to determine total, inorganic and organomercury in biological tissues. Total Hg was extracted using $HNO_3:H_2O_2$ (4:1) mixture, while only inorganic Hg species were extracted from a separate sample using tetramethylammoniumhydroxide (TMAH). The difference between total and inorganic mercury yielded the organic mercury content. Mercury in both extracts was determined using flow-injection cold vapor atomic fluorescence spectrometry (FI-CV-AFS).

2.2 Selective reduction

Derivatization by chemical vapor generation and subsequent detection of the analyte in the gas phase by different techniques has become a popular method for non-chromatographic trace-element speciation. Introducing analyte in the gas phase into the atomization system significantly improves the sensitivity and yields LODs several orders of magnitude lower than those from conventional nebulization. Selective reduction is based on the differences of the reduction potential of the different species. The reduction potential can be controlled by the concentration of the reductant, pH and by introducing catalysts or chelating agents. Table 10.3 summarizes some non-chromatographic mercury speciation utilizing selective reduction.

Probably the Magos method (Magos 1971, Magos and Clarkson 1972) is the best known speciation procedure by selective reduction. In this case, only inorganic mercury is determined by $SnCl_2$ reduction in alkaline medium. Nevertheless, total mercury is measured by $SnCl_2$ in alkaline medium in the presence of $CdCl_2$ as catalyst. The difference between inorganic mercury and total mercury yielded the MeHg content. However, the method can not discriminate all Hg species. Alkyl mercury species are not reduced by $SnCl_2$ in alkaline medium, but aryl mercury species are. Therefore, some organic species (phenyl mercury, for instance) can be mistaken as inorganic Hg. In the absence of the catalyst, it is also possible to discriminate inorganic and organic mercury. Inorganic mercury was determined after reduction with $SnCl_2$, while total mercury was determined after an oxidation step prior to the reduction step to elemental mercury with the same reducing agent. Organic mercury is calculated as the difference between them. The degradation of organic mercury is usually performed off-line or on-line with strong oxidant (e.g., $K_2S_2O_8$ (Burguera et al. 1999, Shao et al. 2006), KBr/ $KBrO_3$ (Cava-Montesinos et al. 2004), $KMnO_4$ (Capelo et al. 2004, Torres et al. 2009), sometimes with the assistance of UV (Aranda et al. 2009), microwave (Rivaro et al. 2007), ultrasound (Capelo et al. 2000, Fernandez et al. 2006).

There are several variations of the Magos method, where other reductants are proposed. Na/KBH_4 can reduce not only inorganic mercury but also alkyl mercury to element mercury. Therefore, inorganic mercury is reduced by $SnCl_2$, while total mercury is determined with Na/KBH_4 as reductant. This principle was used in mercury speciation in fish (Ubillus et al. 2000), mussel tissue (Rio-Segade and Bendicho 1999) and marine biological samples (Park and Do 2008). An alternative method for the determination of inorganic and total mercury in biological and environmental samples was developed, where the only reducing agent was $NaBH_4$ (Segade and Tyson 2003). In this case, the potential was adjusted by the reductant concentration: inorganic Hg was selectively determined after reduction with $10^{-4}\%$ m/v $NaBH_4$, while total Hg was determined after reduction with 0.75% w/v

Table 10.3 Selected applications of selective reduction method for non-chromatographic mercury speciation.

Matrix	Species	Conditions	Analytical technique	Analytical features	Ref.
Biological samples and blood	Hg(II), MeHg	Digestion with l-cysteine, NaCl and NaOH. Selective reduction and reaction with $SnCl_2$ or $SnCl_2$–$CdCl_2$.	CV-AAS	-	Magos 1971, Magos and Clarkson 1972
Fish, urine, biological samples, biodiesel and water	Hg(II), total mercury	Inorganic mercury was determined after reduction with $SnCl_2$, while organic mercury is determined by difference after determination of total Hg following an oxidative sample digestion.	CV-AAS/ CV-AFS/ CV-ICPMS	LOD: < 0.5µg L^{-1} for AAS; < 2.9 ng L^{-1} for AFS; < 0.8 µg L^{-1} for ICPMS	Burguera et al. 2006, Cava-Montesinos et al. 2004, Torres et al. 2009a, Aranda et al. 2009, Rivaro et al. 2007, Capelo et al. 2000, Fernandez et al. 2006
Fish, mussel tissue, marine biological samples	Hg(II), total mercury	Inorganic mercury was determined after reduction with $SnCl_2$, while organic mercury is determined by difference after determination of total Hg with Na/KBH_4 as reductant.	CV-AAS ICP-MS	LOD: 125 ng g^{-1} (inorganic Hg), 183 ng g^{-1} (total Hg) for CV-AAS; 0.018 µg g^{-1} for ICP-MS	Ubillus et al. 2000, Rio-Segade et al. 2008
Biological and environmental samples	Hg(II) and total mercury	The optimized flow-injection Hg system permitted the separate determination of Hg(II) and total Hg using $NaBH_4$ in different concentrations.	FI-CV-AAS	LOD (ng L^{-1}): 24 (total Hg); 3.9 (inorganic Hg).	Segade and Tyson 2003
Tap water	Hg(II), total mercury	Hg^{2+} and total mercury were detected under Vis light and UV light, respectively.	CVG-AFS	LOD (µg L^{-1}): 0.003(total Hg); 0.2 (IHg).	Zheng et al. 2005
Biological samples	Total mercury, MeHg	Total mercury was determined in formic acid or 2.5% TMAH, while methylmercury was selectively determined in 0.125%TMAH.	CV-AAS	LOD (µg L^{-1}): 0.03 (Hg^{2+}); 0.05 (methylmercury).	Vieira et al. 2007
River water	Hg(II), MeHg	Methylmercury was determined at high ultrasonic fields (> 10 W cm−3 of sample solution) and Hg^{2+} was detected conventional ultrasonic field.	CV-AAS	LOD (µg L^{-1}): 0.04 (Hg^{2+})	Ribeiro et al. 2007

NaBH$_4$. A similar procedure based on the sequential selective reduction of Hg(II) and MeHg was developed (Monteiro et al. 2002). Hg(II) is reduced by a 0.01% NaBH$_4$ solution, while the MeHg species is reduced by the same reductant but at a 0.3 percent concentration in the presence of iron (III) chloride. Except SnCl$_2$ and Na/KBH$_4$, TMAH was also proposed for mercury cold vapor generation (CVG) and used for speciation analysis of ultratrace inorganic mercury and total mercury in biological samples by inductively coupled plasma mass spectrometry (ICP-MS) (Wu et al. 2012). Inorganic mercury was directly determined by CVG-ICP-MS, while microwave-assisted oxidation with H$_2$O$_2$ was needed for total mercury detection. The concentration of organic mercury was obtained by the difference between total mercury and inorganic mercury.

Due to the selective reduction capability of inorganic mercury and organic mercury with a different light source, photochemical vapor generation (PVG) can be used for non-chromatographic mercury speciation. Inorganic and total Hg can be determined by PVG coupled to AFS (Zheng et al. 2005). In the presence of UV irradiation, both Hg^{2+} and CH$_3$Hg$^+$ can be converted to Hg^0 to determine total mercury; while only Hg^{2+} can be determined with the visible light: their difference yield the concentration of CH$_3$Hg$^+$. PVG has also been combined with AAS to determine total mercury and MeHg in biological samples. Tissues were digested in either formic acid or TMAH. Adjusting parameters such as acid or TMAH concentration, UV irradiation intensity and pH can generate Hg vapor from selected Hg species. Total Hg was determined in tissues digested in either formic acid or TMAH following reduction of both species by exposure of the solution to UV irradiation, while MeHg was selectively measured in tissues processed by TMAH and diluted to a final concentration of 0.125% m/v TMAH (Vieira et al. 2007). On-line preconcentration has also been combined with PVG for speciation analysis of ultratrace Hg(II) and MeHg by using AFS detection (Gao et al. 2011). Diethyldithiocarbamate (DDTC) served not only as a chelating reagent to form hydrophobic compound of mercury for on-line pre-concentration but also as a reductant for *in situ* photochemical vapor generation and desorption of mercury from the coiled reactor. Because of the different vapor generation efficiency with different concentration of DDTC, speciation analysis of Hg^{2+} and CH$_3$Hg$^+$ can be achieved.

An ultrasound-assisted vapor generation system was also investigated for selective reduction of mercury species (Ribeiro et al. 2007). As reduction of CH$_3$Hg$^+$ to Hg^0 occurs only at power density of more than 10 W cm^{-3} of sample solution, speciation was achieved by adjusting the power density during the measurement.

2.3 Atomization

Atomization is a promising technique for selective discrimination of mercury species. Table 10.4 summarizes some published papers with various atomization methods. It discriminates species based on the temperature of the atomization cell. The volatile Hg derivatives are generated from reaction with sodium borohydride ($< 0.05\%$ (m/v)): MeHg content is determined as the difference between total mercury measured at elevated temperature and inorganic Hg measured as elemental Hg vapor at room temperature.

Kaercher et al. (Kaercher et al. 2005) determined Hg (II) and MeHg by vapor generation AAS at different atomization temperatures. Inorganic mercury was determined by keeping the quartz cell furnace at room temperature and total mercury was determined by heating the quartz cell furnace to 650°C. Similarly, Torres et al. (Torres et al. 2005) determined inorganic and total mercury separately in biological samples by using different atom cell temperatures. Inorganic mercury was measured at room temperature and total mercury was measured by heating the quartz cell atomizer with an air-acetylene flame. The method is simple and easy to implement; however, the sensitivity of inorganic mercury at room temperature is higher than that at high temperatures, which means that, the sensitivity is limited by the involved temperature change in the method. In addition, the adjustment of temperature between measurements takes some time, which decreases the sample frequency. Alp et al. (Alp and Ertas 2009) reported a fast method to determine inorganic and total mercury using a tungsten coil (W-coil) atomizer, where the analysis can be completed in a very short time due to the fast heating and cooling rate of W-coil. Another interesting non-chromatographic approach for the speciation of Hg(II) and MeHg has been developed by AAS based on low temperature dielectric barrier discharge (DBD) atomizer (Zhu et al. 2010). Inorganic mercury is reduced to elemental mercury using 0.01% (m/v) $NaBH_4$, whereas the methylmercury forms an intermediate volatile methylmercury hydride (CH_3HgH). A low temperature DBD atomizer was used for the atomization of CH_3HgH. Only inorganic mercury is measured in the absence of the DBD plasma. However, in addition to inorganic mercury, CH_3HgH can be atomized and total mercury is determined in the presence of the plasma. The MeHg absorbance signal can then be obtained from the difference.

2.4 Miscellaneous methods

Due to the differential volatility of Hg and its compounds, thermal desorption/atomic absorption techniques can be used to distinguish their different species. This method was first applied for the identification of Hg compounds in rock samples (Windmoller et al. 1996). Analysis of

Table 10.4 Selected applications of atomization method for non-chromatographic mercury speciation.

Matrix	Species	Conditions	Analytical technique	Analytical features	Ref.
Fish	Hg(II), total mercury	Inorganic mercury was determined by keeping the quartz cell furnace at room temperature and total mercury was determined by heating the quartz cell furnace to 650°C.	CV-AAS	LOQ: 55 ng g^{-1}.	Kaercher et al. 2005
Biological samples	Hg(II), total mercury	Inorganic Hg is determined by keeping the quartz cell at room temperature, while total Hg is obtained by heating the quartz cell in an air–acetylene flame.	CV-AAS	0.13 µg g^{-1} (THg), 0.025 µg g^{-1} (Hg^{2+}).	Torres et al. 2005
Fish	Hg(II), total mercury	Inorganic mercury was measured when the W-coil is at room temperature and total mercury was measured when the W-coil temperature was set to 500°C.	CV-AAS/ ICP-MS	LOD (ng mL^{-1}): 0.60 (Hg(II)); 0.89 (MeHg).	Alp and Ertas 2009
Fish	Hg(II) and total mercury	Only inorganic mercury can be measured in the absence of the DBD plasma; total mercury is determined in the presence of the plasma	FI-CV-AAS	LOD (ng mL^{-1}): 0.35 (Hg(II)); 0.54 (MeHg).	Zhu et al. 2010

samples from different, Hg-contaminated sites shows that it is possible to discriminate different Hg-binding forms in soils and sediments by their thermal release characteristics (Biester and Scholz 1997). Recently, this technique was also used in Hg speciation analysis of contaminated soil or sediment samples (Bombach et al. 1994, Shuuaeua et al. 2008).

Mercury speciation by distillation was also proposed by Nagase et al. (Nagase et al. 1980) and Horvat et al. (Horvat et al.1988). Non-chromatographic separation of Hg(II) and MeHg were performed by vapor distillation, in a stream of air or nitrogen at 150°C, from a homogenate of the solid sample in diluted H_2SO_4 or HCl with excess of NaCl. The separated MeHg was then collected in a closed tube. This tube is water-cooled and stored in the dark in order to keep degradation of the extracted MeHg at a minimum before its final determination (Nagase et al. 1980).

The speciation of mercury in atmospheric samples has also been reported. A collecting method to prepare a fractional determination of ambient forms of mercury in the air was proposed (Takizawa et al. 1981). Particulate mercury (PM) is collected by a glass fiber filter. The assembly consisted of a train of four connected slender quartz tubes, in which Chromosorb W treated with HCl gas for Hg(II), Chromosorb W treated with 0.1 mol L^{-1} NaOH for MeHg, a silver-wire for metallic mercury and a gold plate for DMeHg were packed. The collection efficiency for these trapping tubes ranged from 85 to 100 percent at the μg or ng level. With this method, it was possible to indicate that in the proximity of volcanic and hot-spring regions, mercury was mostly observed as Hg(II), followed by metallic mercury, MeHg, DMeHg and PM in this order. Lu et al. (Lu and Schroeder 1999) compared the performance of a quartz filter for the collection of particulate phase Hg against a filtration system fitted with a front ended denuder (gold-coated) to remove gas phase Hg. In each case, PM is collected on a quartz fiber disk held in a miniaturized device and is analyzed using a pyrolysis/gold amalgamation/thermal desorption/ CVAFS) technique. The denuder-based system gave higher results in parallel sampling trials. They attributed this result to the Hg-coated gold particles flaking off the denuder and ending up on the filter. Feng et al. (Feng et al. 2000) described the determination of a Hg species referred to as "gaseous divalent mercury." The species was collected on KCl-coated denuder tubes and thermally desorbed at 450°C. The conversion to atomic vapor was achieved at 900°C and determination was performed by AFS after collection on a gold trap. The speciation of gaseous mercury, particulate mercury, and gaseous elemental mercury in the air over arid lands of south central New Mexico were also proposed using KCl-coated annular denuders coupled to an automated gas phase mercury analyzer (Caldwell et al. 2006).

Mercury speciation and transformation in the environment can also be determined using X-ray absorption spectrometry (XAS) (Andrews 2006), a method that is element-specific, and therefore can be used to study elements in their complex original matrices. Mercury can be studied in various types of complex samples, without chemical treatment that would change its speciation. Because speciation of mercury is quite dependent on the chemical environment, which includes factors such as pH, temperature, oxygen content, and presence of various species, including sulfate, sulfide, dissolved organic matter and dissolved oxygen, this method to study the chemistry of mercury can shed important light on the structure and bonding of mercury in various environments. XAS techniques such as Hg L3- and L2-edges, Hg L3 extended X-ray absorption fine structure, and sulfur and selenium K-edges are used to study mercury speciation and binding in minerals, coal flue gas emissions, chelation with organic matter, sulfide precipitation and in physiological forms. However, high mercury concentration is generally required for XAS analysis. In addition, speciation using XAS is not absolute, and it is possible to miss various components or assign them to the wrong species, because some spectra are similar.

3. Chromatographic Methods

In contrast to non-chromatographic approaches, chromatographic methods are able to separate "all" Hg species in a single step and obtain a complete picture of the species present in samples. The chromatographic separation technique for trace Hg speciation analysis is most commonly achieved by the coupling of gas or liquid chromatography with an element selective detector, e.g., atomic spectroscopy or mass spectrometry, with or without a previous pre-concentration step.

3.1 GC Separation methods

The principal advantage of gas chromatography (GC) is the quantitative transfer of analytes from the chromatographic column to the detector, which considerably improves the limits of detection with respect to liquid chromatography (LC). A drawback of the GC is the need for derivatization of the ionic mercury species to obtain volatile forms and a very widely accepted strategy is to form peralkylated volatile compounds with ethylation (Stoichev et al. 2006). GC separation technologies for mercury speciation were widely used in environmental samples—mainly fish and marine samples, water samples, hair, gaseous samples, soil samples and food samples, etc. Table 10.5 summarizes the selected applications of different gas-chromatographic methods for the determination of Hg species.

In GC separation, hydration with $NaBH_4$, aqueous phase ethylation with $NaBEt_4$ and derivatization with a Grignard reagent (e.g., ethylation, butylation and propylation) are the most commonly used derivatization

Table 10.5 Selected applications of different gas-chromatographic methods for the determination of Hg species.

Matrix	Hg species	Sample pre-treatment	Analytical technique	Detection limit	Ref.
Hair	Total Hg, MeHg	Total Hg: pre-concentration of Hg on a gold amalgamator after catalytic combustion of the sample. MeHg : Derivatization with sodium tetraethylborate	Total Hg : Combustion–AAS, MeHg : HS–GC–AFS	THg: 1.5 ng g^{-1}, MeHg:0.04 ng g^{-1}	Gao et al. 2010
Fish tissues	MeHg, Hg^{2+}	Derivatized with sodium tetraethylborate and extracted by SPME	GC–FAPES	MeHg:1.5 ng g^{-1}, Hg^{2+}: 0.7 ng g^{-1}	Grinberg et al. 2003a
Fish extracts	MeHg, Hg^{2+}	Sodium tetraethylborate derivatization and purge-and-trap injection	MCGC–ICP–TOFMS	MeHg: 16 fg g^{-1}, Hg^{2+}: 257 fg g^{-1}	Jitaru et al. 2003
Drinking water	MeHg, DiEtHg, EtHg, Hg^{2+}	Stir bar sorptive extraction with in situ propyl derivatization	TD–GC–MS	MeHg:0.02 ng mL^{-1}, DiEtHg:0.2 ng mL^{-1}, EtHg:0.01 ng mL^{-1}, Hg^{2+}: 0.2 ng mL^{-1}	Ito et al.2008
Sea water	MeHg, EtHg, Hg^{2+}	Sodium tetraphenylborate derivatization and headspace—solid phase microextraction	GC–MIP–AED	MeHg: 0.1 ng mL^{-1}, EtHg :0.1 ng mL^{-1}, Hg^{2+}:0.3 ng mL^{-1}	Carro et al. 2002
Refrozen fish	Total Hg, MeHg, Hg^{2+}	Total Hg: microwave digestion, MeHg /Hg^{2+}: derivatization with NaBEt$_4$	Total Hg: AFS, MeHg/Hg^{2+}: GC–ICPMS	—	Krystek and Ritsema 2005
Fresh waters	MeHg, Hg^{2+}	Derivatisation with NaBEt$_4$ and species-specific isotope dilution	GC–ICPMS	MeHg:0.005 ng L^{-1}, Hg^{2+}:0.074 ng L^{-1}	Jackson et al. 2009
Biological tissues	MeHg, Hg^{2+}	Ethylated with sodium tetraethylborate and purged to pre-concentrate on Tenax-TA	GC–FAPES	MeHg: 0.2 ng g^{-1}, Hg^{2+}: 1.4 ng g^{-1}	Jimenez and Sturgeon 1997
Fish tissues	MeHg, EtHg, Hg^{2+}	Extracted by sodium diethyldithiocarbamate and butylmagnesium chloride	GC–rf-HC-GD-AES	MeHg: 0.2 µg L^{-1}, EtHg: 0.2 µg L^{-1}, Hg^{2+}: 0.3 µg L^{-1}	Velado et al. 2000
Biological samples	MeHg, EtHg, Hg^{2+}	Derivatized with sodium tetraphenylborate or sodium tetrapropylborate and extracted by SPME	GC–FAPES	MeHg: 0.55 ng g^{-1}, EtHg: 0.34 ng g^{-1}, Hg^{2+}: 0.23 ng g^{-1}	Grinberg et al. 2003b
Seawater	MeHg, Hg^{2+}	Ethyl/propyl derivatization and extracted by SPME	GC–ICPMS	MeHg: 0.11/0.17 ng L^{-1}, Hg^{2+}: 1.6/0.35 ng L^{-1}	Bravo-Sanchez et al. 2004

methods. Solid-phase microextraction (SPME) can be used for the extraction of organometallic compounds after they have been derivatized to a sufficiently volatile form. SPME can be performed either in the aqueous phase or in the headspace. After SPME extraction, species were separated by GC and analyzed by furnace atomization plasma emission spectrometry (FAPES), microwave induced plasma-atomic emission spectrometry (MIP-AES) (Tutschku et al. 2002, Reuther et al. 1999, Pereiro et al. 1998), AFS, GC-mass spectrometry (GC–MS) or ICP-MS (Berzas Nevado et al. 2011) and so on.

Grinberg et al. (Grinberg et al. 2003a) evaluated the determination of MeHg and Hg(II) in fish tissue using SPME in conjunction with tandem gas chromatography–furnace atomization plasma emission spectrometry (SPME–GC–FAPES). Samples were digested with methanolic potassium hydroxide, derivatized with sodium tetraethylborate and extracted by SPME. After the SPME extraction, species were separated by GC and detected by FAPES. Detection limits of 1.5 ng g^{-1} for MeHg and 0.7 ng g^{-1} for Hg(II) in biological tissues were obtained. Speciation of Hg in human hair was carried out with combustion-atomic absorption spectrometry for total Hg (THg) and headspace–gas chromatography–atomic fluorescence spectrometry (HS–GC–AFS) for MeHg by Gao et al. (Gao et al. 2010). The ethylation step is without any doubt the most problematic one in the entire HS–GC–AFS analysis procedure. The detection limit of the MeHg in human hair, which amounts to 0.04 ng g^{-1} for a 20 mg sample, is far below the concentrations observed in natural samples. The most important advantage of this method is that conversely to all other methods reported in literature it does not require a clean-up procedure prior to injection into the GC.

Hollow cathode glow discharge (HCGD) atomic emission spectrometry (AES) with a radiofrequency (rf) source was also investigated as a detector in gas chromatography (GC) for elemental speciation of MeHg, EtHg and Hg(II) following a Grignard derivatization reaction (butylmagnesium chloride) (Velado et al. 2000). The absolute detection limits were found to be 0.2, 0.2 and 0.3 pg for MeHg, EtHg and Hg(II), respectively. Pereiro et al. (Pereiro et al. 1998) developed a compact device based on purge-and-trap multicapillary GC for sensitive analysis of CH$_3$Hg$^+$ and Hg^{2+} by MIP-AES. The operating mode includes *in situ* conversion of the analyte species to Methylethylmercury (MeEtHg) and Diethylmercury (Et$_2$Hg) and cryotrapping of the derivatives formed in a 0.53-mm i.d. capillary, followed by their flash (< 30 s) isothermal low-temperature separation on a minimulticapillary (22 cm) column. The device allowed speciation of CH$_3$Hg$^+$ and Hg^{2+} down to 5 pg g^{-1} in urine.

The usage of ICP-MS in Hg speciation analysis has also increased significantly due to its high sensitivity and selectivity. Hintelmann et al. (Hintelmann et al. 1995) described a novel technique for the calculation of mercury methylation rates in sediments by using GC-ICP-MS with stable isotope dilution. MeHg was isolated from sediments by distillation, converted to MeEtHg by $NaBEt_4$ and analyzed after purge-and-trap pre-collection on a Tenax adsorber and thermo-desorption onto the GC column. Detection limits were found to be approximate to 1 pg (as Hg) absolute or 0.02 ng g^{-1} dry sediment. Jackson et al. (Jackson et al. 2009) developed a method for aqueous MeHg quantification in lake water samples using GC-ICP-MS. Figure 10.3 depicts the purge and trap GC-ICP-MS chromatogram of 1 pg of CH_3Hg^+ and Hg^{2+} using this method. Jitaru et al. (Jitaru et al. 2003) developed a simple, rapid and accurate method on the basis of multicapillary gas chromatography (MCGC) combined with inductively coupled plasma–time-of-flight mass spectrometry (ICP–TOF-MS) for speciation analysis of MeHg and Hg(II). Using the purge-and-trap injection, after *in situ* derivatization of the ionic mercury species with $NaBEt_4$, a baseline separation was achieved within 35 sec. Detection limits for MeHg (as Hg) and Hg(II) were 16 and 257 fg g^{-1}, respectively. Yan et al. (Yan et al. 2008) designed and constructed an alternative thermodiffusion interface (TDI) for the effective on-line coupling of capillary gas chromatography (cGC) and ICPMS. Hg^{2+}, CH_3Hg^+ and $C_2H_5Hg^+$ were derived as $(C_4H_9)_2Hg$, $CH_3HgC_4H_9$, and $C_2H_5HgC_4H_9$ when butylmagnesium bromide was used as a derivatization reagent, avoiding the loss of their species specific

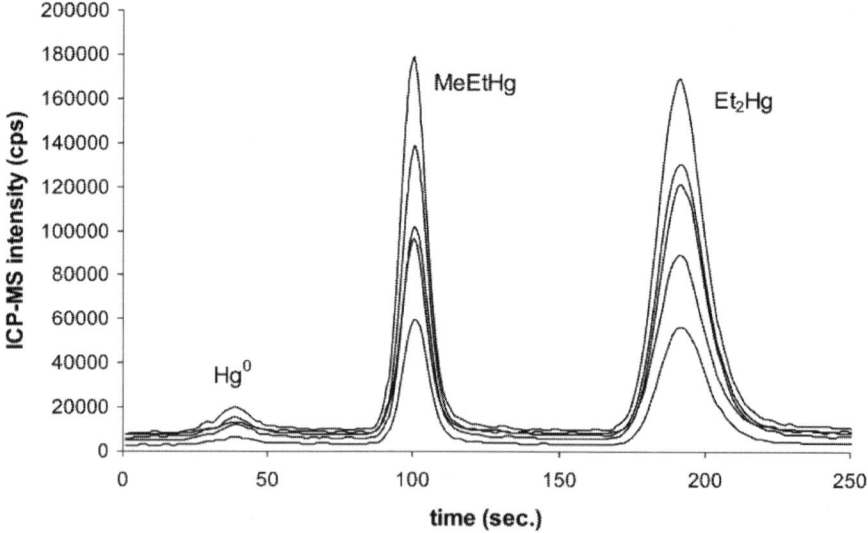

Figure 10.3 Purge and trap isotope dilution GC-ICP-MS chromatogram of 1 pg of MeHg (peak at 100 s) and Hg^{2+} (peak at 190 s). Peaks, in order of increasing signal intensity, are Hg isotopes 198, 199, 200, 201, and 202. Concentrations correspond to an aqueous concentration of 14 pg L^{-1} based on a sample volume of 70 mL (Jackson et al. 2009).

information. The method detection limits for MeHg, EtHg, and Hg(II) are 0.09, 0.1, and 0.2 pg g^{-1}, respectively. Mao et al. (Mao et al. 2008) developed a new method for the detection of trace levels of organo-mercury species by combining the high enrichment capacity of purge and trap with aqueous phenylation derivatization. Phenylation products of MeHg and EtHg were first separated by cGC and then detected by AFS or ICPMS. This new method has been validated for the direct detection of trace organomercury species in fresh-water samples and has the additional benefits of being free from interference by Cl$^-$ and dissolved organic matter. Chung et al. (Chung and Chan 2011) developed and validated a reliable and sensitive method to simultaneously determine MeHg and EtHg in various types of foods with GC-ICP-MS. Samples were digested with pancreatin and then hydrochloric acid. MeHg and EtHg in the extract were derivatized in an aqueous buffer with NaBH$_4$. After phase separation, the extract was directly used for analysis.

Emphasis should also be given to the three most common forms of mercury in atmospheric samples [i.e., gaseous elemental mercury (GEM, Hg^0), reactive gaseous mercury (RGM), and particle-bound mercury (Hg$_p$)]. GC-based methods have most frequently been used in the study of organic-bound mercury [e.g., mono-methylmercury (MeHg), dimethylmercury (DMeHg), and diethylmercury (DEtHg)]. Kato et al. (Kato et al. 1992) combined GC with axially viewed ICP-AES to analyze MeHg in air samples. These authors collected air samples on a column with Tenax sorbent and then thermally eluted MeHg with benzene. Using the ICP-AES detection method, they were able to determine MeHg at the level of no less than 3 pg level per sample. Lee et al. (Lee et al. 2003) used a high flow refluxing mist chamber for sampling. They determined atmospheric MeHg using a GC-CV-AFS method in a stepwise manner by: (1) collecting MeHg in an aerated water sample; (2) aqueous phase ethylation; and, (3) pre-collection onto Carbotrap column. The LOD for MeHg with this method was estimated as 1 pg m^{-3}. Thus, they were able to measure MeHg in ambient air in the concentration range of 3–22 pg m^{-3} during a sampling campaign in Gothenburg (Sweden).

3.2 HPLC separation methods

The main advantage of high performance liquid chromatographic (HPLC) would eliminate the need for derivatization of the mercury species, which in turn would reduce the possibilities for loss or conversion before analysis, reduce the number of sample preparation steps and decrease the potential for contamination. However, the main disadvantage of HPLC lies in the poor sensitivity of the detectors. Development of more sensitive detectors, such as AFS and ICP-MS, has resulted in wider applications in environmental studies. However, sensitivity of HPLC coupled to these detectors is even lower than GC due to the small amount of sample which is transferred to the column. For HPLC separation, a reversed phase column based on alkyl-silica and a mobile phase containing an organic modifier, together with a chelating or ion pair reagent (and in some cases a pH buffer) are usually used.

The interface to couple HPLC columns with the atomizer can be very simple, with the exit of the column directly connected to the nebulizer of the AAS or plasma detector. Unfortunately, nebulizer efficiency is very low (1–3 percent) and limits sensitivity, especially for flame AAS. Generally, a way out of this lack of sensitivity is post-column derivatization to form cold vapor of Hg. HPLC separation technologies for mercury species determination were widely used in sediment, fish and marine samples, water samples, hair, etc.

An automatic system, based on the on-line coupling of HPLC separation, post column microwave digestion, and CV-AFS detection, was proposed for the speciation analysis of four mercury compounds (Liang et al. 2003). Post column microwave digestion, in the presence of potassium persulfate (in HCl), was applied in the system to improve the conversion efficiency of three organic mercury compounds into inorganic mercury. Yin et al. (Yin et al. 2007b) developed a hyphenation of HPLC with AFS for the speciation of Hg(II), MeHg, EtHg and phenylmecury (PhHg) based on photo-induced chemical vapor generation (CVG) with formic acid instead of the conventional $K_2S_2O_8/KBH_4$ system. Under UV irradiation, the decomposition of organic mercury species and the reduction of Hg^{2+} could be completed in one step with this proposed photo-induced vapor generation (PVG) system. The novel PVG system used formic acid only, which simplified the flow system and avoided the possibility of contamination originating from additional chemicals. Afterwards, PVG with formic acid in mobile phase as reaction reagent was developed as interface to on-line couple HPLC with AFS for the separation and determination of inorganic mercury, MeHg, EtHg and PhHg (Yin et al. 2008). In their developed procedure, formic acid in mobile phase was used to decompose organomercuries and reduce Hg^{2+} to mercury cold vapor under UV irradiation. Therefore, no post-column reagent was used and the flow injection system in the traditional procedure was omitted. Figure 10.4 was the chromatogram of four mercury species using this developed HPLC–UV-CVG–AFS system. Similarly, there were other reports like using mercaptoethanol (Yin et al. 2009) and L-cysteine (Wang et al. 2010) in the mobile phase as photo chemical reagents to decompose organomercuries and reduce Hg^{2+} to mercury cold vapor under UV irradiation. Angeli et al. (Angeli et al. 2011) described a novel HPLC-MW/UV combined reactor coupled to a CVG-AFS detection system for the determination of p-hydroxymercurybenzoate (PHMB)-tagged thiols. The

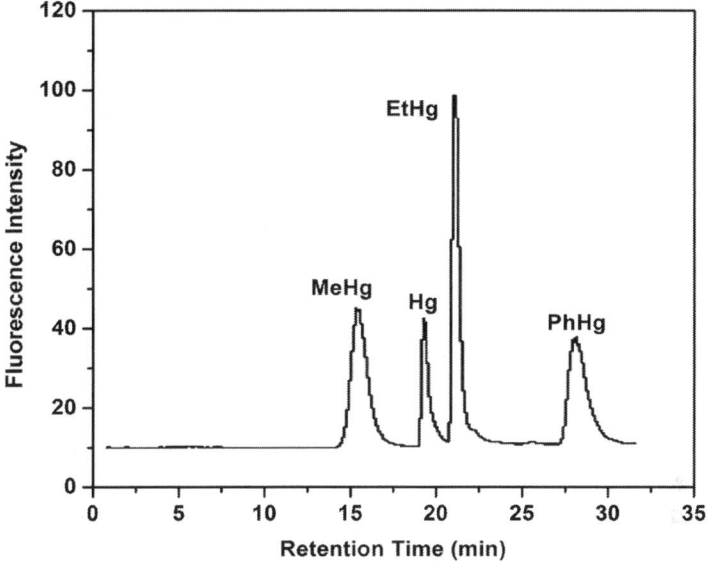

Figure 10.4 Chromatogram of four mercury species using HPLC–UV-CVG–AFS system. Concentration of mercury species: 10 µg L^{-1} (Yin et al. 2008).

use of a fully integrated MW/UV photochemical reactor allowed to obtain the on-line digestion of PHMB and thiol-PHMB complexes, to Hg(II) with a yield between 91 and 98 percent. Hg(II) was reduced to Hg^0 in a knitted reaction coil with NaBH$_4$ solution. He et al. (He et al. 2011) developed a novel solution cathode glow discharge (SCGD) that induced vapor generation as interface to the on-line couple HPLC with AFS for the speciation of Hg(II), MeHg and EtHg. The decomposition of organic mercury species and the reduction of Hg^{2+} could be completed in one step with SCGD.

Rai et al. (Rai et al. 2002) described a measurement of Hg(II) and MeHg in fish muscle tissues after protease type XIV extraction by HPLC-ICP-MS. A SphereClone 5 mum ODS2 80A PEEK column (100 mm x 4.6 mm, Phenomenex, USA) and a 5% v/v CH$_3$OH-water mobile phase containing 0.06 M CH$_3$COONH$_4$ and 0.1% w/v cysteine, pH 6.8 (flow rate, 1.0 ml min^{-1}) was used for the separation of mercury species. Chen et al. (Chen et al. 2009a) developed a sensitive method for speciation analysis of Hg(II) and MeHg by using HPLC combined with ICP-MS after cloud point extraction. The analytes were complexed with DDTC and pre-concentrated by a non-ionic surfactant Triton X-114. Mercury species were effectively separated by HPLC in less than 6 min. The enhancement factors for 25 mL sample solution were 42 and 21, and the limits of detection were 4 and 10 ng L^{-1} for Hg^{2+} and CH$_3$Hg$^+$, respectively. Ionic liquid based dispersive liquid–liquid microextraction (IL-DLLME) combined with HPLC-ICP-MS for the determination of mercury species in liquid cosmetic samples was described by Jia et al. (Jia et al. 2011). Firstly, Hg(II), MeHg and EtHg were complexed with ammonium pyrrolidinedithiocarbamate (APDC), and then the complexes were extracted into 1-hexyl-3-methylimidazolium hexafluorophosphate ([C_6MIM][PF$_6$]) using DLLME. Under the optimized conditions, the enhancement factors of 760, 115, 235 for Hg(II), MeHg and EtHg were obtained from only 5.00 mL sample solution. The detection limits of the analytes (as Hg) were 1.3 ng L^{-1} for Hg(II), 7.2 ng L^{-1} for MeHg and 5.4 ng L^{-1} for EtHg, respectively. Table 10.6 selected applications of different liquid chromatographic methods for the determination of Hg species.

Ion chromatography (IC) provides the possibility of separating more polar and ionic species directly, so that sample pre-treatment can be simplified. Chen et al. (Chen et al. 2009b) have combined short-column ion chromatographic separation and on-line photocatalyst-assisted techniques with ICPMS to develop a simple and sensitive hyphenated method for the determination of aqueous Hg(II) and MeHg species. This hyphenated method also provided detection limits 0.1 and 0.03 ng mL^{-1} for Hg(II) and MeHg, respectively. Liu (Liu 2010) developed the same technology of PVG as interface to on-line coupled Hg-cysteine IC with AFS with formic acid in mobile phase for rapid determination of MeHg in seafood. The LODs were found to be 0.1 ng mL^{-1} for Hg(II) and 0.08 ng mL^{-1} for MeHg.

3.3 Capillary electrophoresis separation methods

Capillary electrophoresis (CE) uses charge and frictional forces for the separation of Hg species (Kuban et al. 2009). Separation by CE is usually faster than that with LC and therefore is potentially a rapid and highly efficient separation technique for mercury speciation. Unfortunately, because of the very small sample volumes used in CE, the detection limit (µg L^{-1} range) is often insufficient for Hg speciation in natural waters. Current research is therefore focused on the development of suitable sample pre-concentration methods and/or coupling to element specific detection to achieve lower detection limits. CE separation technologies for mercury species determination can be used for biological samples and water samples.

Yin (Yin 2007) proposed a dual-cloud point extraction (dCPE) technique for the sample pre-treatment of CE speciation analysis of mercury. With CE separation and on-line UV detection, the detection limits were 45.2, 47.5, 4.1, and 10.0 µg L^{-1} (as Hg) for EtHg, MeHg, PhHg, and Hg(II), respectively. As an analysis method, the present dCPE–CE with UV detection obtained similar detection limits as of some CE–ICPMS hyphenation technique, but with simple instrumental setup and obviously low costs. Fan and Liu

Table 10.6 Selected applications of different liquid chromatographic methods for the determination of Hg species.

Matrix	Hg species	Sample pre-treatment	Mobile phase and column	Analytical technique	Detection limit	Ref.
Sediment	MeHg	Oxidized in the oxidation coil by UV irradiation and reduced in a reduction coil by tin(II) chloride	Methanol and water (5 + 95, v/v) containing 0.01% v/v 2-mercaptoethanol buffered at pH 5 with 0.06% v/v acetic acid and 0.15% m/v ammonium acetate with Nucleosil ODS (RP C_{18}, 25 cm × 4.6 mm, 5 mm)	UV-PCO-CV-AFS	12 ng g^{-1}	Ramalhosa, Segade et al. 2001b
Seafood	MeHg Hg^{2+} EtHg PhHg	Post column microwave digestion with potassium persulfate (in HCl) and reduced by KBH$_4$	CLC-ODS column, 150 × 6 mm i.d., 10 μm Mobile phase: 50% (v/v) CH$_3$OH: 10mmol L^{-1} TBA: 0.1 mol L^{-1} NaCl	CV-AFS	MeHg: 0.2 ng Hg^{2+}: 0.3 ng EtHg: 0.17 ng PhHg: 0.14 ng	Liang et al. 2003
Seafood	MeHg Hg^{2+} EtHg PhHg	Post-column oxidation with K$_2$S$_2$O$_8$ in HCl and reduced by KBH$_4$	RP-C$_{18}$ column with a mixture of methanol, acetonitrile and water (75 : 5 : 20) containing 0.01% m/v ammonium pyrrolidine dithiocarbamate (APDC)	CV-AFS	MeHg: 0.27 μg L^{-1} Hg^{2+}: 0.19 μg L^{-1} EtHg: 0.26 μg L^{-1} PhHg: 0.21 μg L^{-1}	Yin et al. 2008
Seafood	MeHg Hg^{2+} EtHg PhHg	Post-column UV irradiation with formic acid makes the decomposition of organic mercury species and the reduction of Hg^{2+} complete in one step	Shim-pack CLC-ODS (Shimadzu), 15 cm × 6.0 mm with mobile phase A: 5% (v/v) CH$_3$CN, 10 mmol L^{-1} NH$_4$Ac, 0.01% 2-mercaptoethanol; B: 30% (v/v) CH$_3$CN, 10 mmol L^{-1} NH$_4$Ac, 0.01% 2-mercaptoethanol. 0–11 min, 100% A; 11–12 min, 100% A – 100% B, 12 min–30 min: 100% B, 30–31 min: 100% B - 100% A	PVG-AFS	MeHg: 0.81 μg L^{-1} Hg^{2+}: 1.01 μg L^{-1} EtHg: 0.20 μg L^{-1} PhHg: 0.87 μg L^{-1}	Li and Hu 2007
Polluted lake sediment and seafood	MeHg Hg^{2+} EtHg PhHg	Post-column digestion by UV/TiO$_2$ with formic acid and sodium formate mixture as a hole scavenger	Shim-Pack VP-ODS column (150 × 4.6 mm) using a mixture of acetonitrile and water (65 : 35) containing 1.5 mmol L^{-1} ammonium pyrrolidine dithiocarbamate (APDC) (pH 5.5 buffered by acetic acid–ammonium acetate) as a mobile phase	PVG-AFS	MeHg: 20 pg mL^{-1} Hg^{2+}: 10 pg mL^{-1} EtHg: 30 pg mL^{-1} PhHg: 70 pg mL^{-1}	Yin et al. 2007a

Table 10.6 contd....

Table 10.6 contd.

Matrix	Hg species	Sample pre-treatment	Mobile phase and column	Analytical technique	Detection limit	Ref.
Seafood	MeHg Hg²⁺ EtHg PhHg	Formic acid in mobile phase was used to decompose organomercuries and reduce Hg²⁺ to mercury cold vapor under UV irradiation	Shim-pack CLC-ODS (Shimadzu), 15 cm length, 6.0mm I.D., 5µm ; Mobile phase A: 3% (v/v) CH_3CN, 240 mmol L⁻¹ $HCOONH_4$–HCOOH (pH 2.8), 0.01% 2-mercaptoethanol; B: 30% (v/v) CH_3CN, 240 mmol L⁻¹ $HCOONH_4$– HCOOH (pH 2.8), 0.01% 2-mercaptoethanol; 0–11 min: 100% A; 11–12 min: 100% A→100% B; 12–30 min: 100% B; 30–31 min: 100% A	PVG-AFS	MeHg: 0.033 µg L⁻¹ Hg²⁺: 0.085 µg L⁻¹ EtHg: 0.029 µg L⁻¹ PhHg: 0.038 µg L⁻¹	(Yin et al. 2008)
sediments, zoobenthos and river water samples	MeHg Hg²⁺ EtHg PhHg	C18 solid phase extraction (SPE) microcolumns for pre-concentration with Br⁻/BrO_3^- oxidation and $SnCl_2$ reduction	Zorbax SB-C₁₈ column (5µm particle size, 4.6mm×150mm, Agilent) with a mobile phase of aqueous methanol (65%/35%,v/v)	CV-AFS	MeHg: 4.3 µg L⁻¹ Hg²⁺: 0.8 µg L⁻¹ EtHg: 1.4 µg L⁻¹ PhHg: 0.8 µg L⁻¹	(Margetínová et al. 2008)
Fish and mollusks	MeHg Hg²⁺ EtHg PhHg	Acetic acid and mercaptoethanol in mobile phase as photochemical reagent to decompose organomercuries and reduce Hg²⁺ to mercury cold vapor under UV irradiation	Shim-pack CLC-ODS (Shimadzu, 15 cm × 6.0 mm ID × 5 µm) with mobile phase A: 3% (v/v) CH_3CN, 60 mmolL⁻¹ NH_4Ac (pH 6.9), 0.02% 2-mercaptoethanol; B: 30% (v/v) CH_3CN, 60mmolL⁻¹ NH_4Ac (pH 6.9), 0.02% 2-mercaptoethanol 0–11 min: 100% A; 11–12 min: 100%A→100% B; 12–30 min: 100%B; 30–31 min: 100% B→100% A	PVG-AFS	MeHg: 0.53 µg L⁻¹ Hg²⁺: 0.22 µg L⁻¹ EtHg: 0.18 µg L⁻¹ PhHg: 0.25 µg L⁻¹	(Yin et al. 2009)
Biological samples	MeHg Hg²⁺ EtHg	L-cysteine in the mobile phase to decompose organic mercury and Hg²⁺ reduced to Hg⁰by KBH_4	Shim-pack CLC-ODS column (15 cm × 6 mm I.D. × 5 µm) with mobile phase 1 g L⁻¹ L-cysteine, 0.06 mol L⁻¹ ammonium acetate in water	CV-AFS	MeHg: 0.05 µg L⁻¹ Hg²⁺: 0.1 µg L⁻¹ EtHg: 0.07 µg L⁻¹	(Wang et al. 2010)

Sample	Analytes	Preconcentration/method	Column/conditions	Detection	LOD/results	Reference
Biological samples	MeHg, Hg^{2+}, EtHg	Decomposition of organic mercury and reduction of Hg^{2+} by solution cathode glow discharge (SCGD) induced vapor generation	ZORBAX SB-C18 (2.1 mm × 50 mm × 5 μm) with mobile phase 0.06 mol L^{-1} ammonium acetate, 0.1% 2-mercaptothanol, pH 6.8	SCGD-AFS	MeHg: 0.55 μg L^{-1}; Hg^{2+}: 0.67 μg L^{-1}; EtHg: 1.19 μg L^{-1}	(He et al. 2011)
Fish muscle tissues	MeHg, Hg^{2+}	—	5% v/v CH_3OH–water mobile phase containing 0.06 M CH_3COONH_4 and 0.1% w/v cysteine, pH 6.8 with ODS2 80A PEEK column (100 mm × 4.6 mm)	ICP-MS	Lowest measurable mercury: 0.5 mg L^{-1}	(Rai et al. 2002)
Biological samples	MeHg, Hg^{2+}	—	5% (v/v) methanol, 0.1% v/v 2-mercaptoethanol and 0.06M CH_3COONH_4 with RPC_{18} column (150 mm × 3.9 mm)	ICP-MS	MeHg:0.2 μg L^{-1}; Hg^{2+}:0.2 μg L^{-1}	(Wang et al. 2007)
Seawater	MeHg, Hg^{2+}	On-line pre-concentration micro-column	C-18 3μm column (Alltech, Deerfield IL, USA) with a mobile phase of 0.5% lcysteine(m/v) and 0.05% 2-mercaptoethanol (v/v) dissolved in ultra-pure water.	ICP-MS	MeHg:0.02 ng L^{-1}; Hg^{2+}: 0.07 ng L^{-1}	(Cairns et al. 2008)
Water and human hair sample	MeHg, Hg^{2+}	Cloud point extraction (CPE) pre-concentration	Discovery C_{18} with a mobile phase of 90% (v/v) methanol–10% (v/v) water containing DDTC 1.0×10−4 mol L^{-1}	ICP-MS	MeHg: 10 ng L^{-1}; Hg^{2+}: 4 ng L^{-1}	(Chen et al. 2009a)
Liquid cosmetic samples	MeHg, Hg^{2+}, EtHg	Dispersive liquid–liquid microextraction (DLLME) preconcentration	C18 (5 μm,150 mm × 4.6 mm) with mobile phase 0.05 mol L^{-1} ammonium acetate, 4% v/v methanol and 10 mM L-cysteine	ICP-MS	MeHg: 7.2 ng L^{-1}; Hg^{2+}: 1.3 ng L^{-1}; EtHg: 5.4 ng L^{-1}	(Jia et al. 2011)

(Fan and Liu 2008) developed a novel method for determination of MeHg and PhHg by liquid–liquid–liquid microextraction (LLLME) coupled by CE with UV technique. The enhancement factors of 324 for MeHg and 210 for PhHg were obtained with 40 min LLLME and the detection limits (S/N = 3) of MeHg and PhHg were 0.94 and 0.43μg L^{-1} (as Hg), respectively. Li et al. (Li et al. 2011) improved the sample pre-treatment technique of LLLME for the simultaneous extraction of inorganic and organic mercury species. In LLLME, an intermediate solvent (acetonitrile) was added into the donor phase to improve the contact between target mercury species and complexing reagent. Following phase transfer/membrane supported (PT/MS)-LLLME, the acceptor solutions were directly analyzed by large volume sample stacking capillary electrophoresis/ultraviolet detection (LVSS-CE/UV). By combining PT/MS-LLLME with LVSS-CE/UV, enhancement factors were magnified up to 12138-fold and the limits of detection were at sub ppb level. Table 10.7 selected applications of capillary electrophoresis methods for the determination of Hg species.

In addition to developing efficient pre-concentration techniques, the design of the new interface between CE and the detectors, especially for atomic spectrometry determination, is also an active research area in CE-related speciation technique. CE directly interfaced to flame-heated furnace atomic absorption spectrometry (FHF-AAS) via a home-made thermospray interface for nanoliter trace mercury speciation was developed by Li et al. (Li et al. 2005). The CE-FHF-AAS interface integrated the superiority of stable CE separation, complete sample introduction, and continuous vaporization for AAS detection without the need of extra external heat sources and any post-column derivation steps. The detection limit of three mercury species (MeHg, PhHg and Hg(II)) was 3.0 ± 0.15 pg (as Hg) for 60 nL sample injection. Later, the interactions of Hg(II), MeHg, EtHg, and PhHg with DNA have been probed by CE with on-line electrothermal atomic absorption spectrometric detection (CE-ETAAS) in combination with circular dichroism and Fourier transform infrared spectroscopy by the same group (Li et al. 2006). Deng et al. (Deng et al. 2009) developed a novel on-line coupled CE cold vapor generation with electrothermal quartz tube furnace atomic absorption spectrometry system for mercury speciation. The detection limits of MeHg and Hg(II) were 0.035 and 0.027 μg mL^{-1}, respectively.

3.4 Other chromatographic methods

Among the chromatographic analytical techniques available, thin layer chromatography (TLC) is the simplest, reasonably powerful technique and most economic chromatographic method for detection and separation of small quantities of organic and inorganic compounds (Misra and Pachauree

Table 10.7 Selected applications of capillary electrophoresis methods for the determination of Hg species.

Matrix	species	Sample pre-treatment	Detection technique	Detection limit	Ref.
Biological sample	MeHg Hg^{2+} PhHg	Extraction of mercury species using a 3-mercptopropyltrimethoxysilane coated capillary	CE-UV	MeHg: 12 µg L^{-1} Hg^{2+}:7 µg L^{-1} PhHg: 3 µg L^{-1}	Bai and Fan 2010
Contact lenses solutions	MeHg Hg^{2+} EtHg	A microconcentric nebulizer (MCN) as the sample introduction device	CE-ICP-MS	MeHg: 80 µg L^{-1} Hg^{2+}: 170 µg L^{-1} EtHg: 100 µg L^{-1}	Lee and Jiang 2000
Natural water and tilapia muscle	MeHg Hg^{2+} EtHg PhHg	Dual-cloud point extraction (dCPE) pre-concentration	CE-UV	MeHg: 47.5 µg L^{-1} Hg^{2+}: 10.0 µg L^{-1} EtHg: 45.2 µg L^{-1} PhHg: 4.1 µg L^{-1}	Yin 2007
Water samples	MeHg PhHg	Liquid–liquid–liquid microextraction	CE-UV	MeHg:0.94 µg L^{-1} PhHg: 0.43 µg L^{-1}	Fan and Liu 2008
Dry goldfish muscle	MeHg Hg^{2+}	Hydride generation with NaBH$_4$ and HCl	CE-CVG-EQTFAAS	MeHg: 0.035 mg L^{-1} Hg^{2+}: 0.027 mg L^{-1}	Deng et al. 2009
Water samples	MeHg Hg^{2+} EtHg PhHg	Hydride generation with NaBH$_4$ and HCl and with hydrostatically modified electroosmotic flow (HSMEOF)	CE-VSG-AFS	MeHg: 16.5 µg L^{-1} Hg^{2+}: 6.8 µg L^{-1} EtHg: 15.9 µg L^{-1} PhHg: 13.3 µg L^{-1}	Yan et al. 2003
Biological samples	MeHg Hg^{2+} EtHg	Hydride generation with NaBH$_4$ and HCl	CE-VSG-ICPMS	MeHg: 30 µg L^{-1} Hg^{2+}: 1 µg L^{-1} —	Da Rocha, Soldado et al. 2001
Real water samples	MeHg Hg^{2+}	A MicroMist nebulizer was employed to increase the nebulization efficiency	SC-CE-ICP-MS	MeHg: 9.7 mg L^{-1} Hg^{2+}: 12.0 mg L^{-1}	Li 2011
Environmental water and biological samples	MeHg Hg^{2+} EtHg PhHg	Sample pretreatment technique termed phase transfer based liquid–liquid microextraction (PT-LLLME)	LVSS-CE/UV	MeHg: 0.087 µg L^{-1} Hg^{2+}: 0.37 µg L^{-1} EtHg: 0.12 µg L^{-1} PhHg:0.042 µg L^{-1}	Li et al. 2011

2001). The major advantage of TLC lies in the simplicity and the speed. However, a few works have reported separating mercury species with TLC. TLC separation technologies for mercury species determination may be a useful screening tool for clinical samples, cosmetics, pharmaceutical samples and industrial samples. Agarwal and Behari (Agarwal and Behari 2007) developed a method for the detection of mercury in aqueous samples and spiked human urine screening based on TLC comprising silica gel 'G' as a stationary phase and benzene as a mobile phase. Mercury was measured by complexation with dithizone followed by TLC, in the presence of other heavy metals, including arsenic, cadmium, lead, copper, iron, zinc and nickel. The method is simple, cheap, and does not involve any interference of matrix, prevailing in natural water or aqueous industrial effluent samples obtained from the field. A broad range for the detection of mercury, from 20 µg L^{-1} to 1000 mg L^{-1} was established.

Margler et al. (Margler and Mah 1981) presented a method for determination of MeHg by combining TLC and AAS. MeHg was extracted from biological samples by distilling and then extracting with dithizone–chloroform. The analytical procedure was as follows: separating MeHg on a TLC plate, scraping absorbent into an ignition tube, heating the tube and moving the mercury vapor under vacuum into an absorption cell where it absorbed light at 257.3 nm. The method had a detection limit of 0.7 ng Hg. A similar method was presented for inorganic and organic mercury analysis (Bruno et al. 1985). Densitometric detection was used to detect mercury in stead of AAS. Hg(II), MeHg, EtHg and PhHg have been separated as dithizonates by HPTLC and determined *in situ* at the subnanogram level by densitometric automatic scanning of the plate. Shekhovtsova, et al. (Shekhovtsova et al. 1997) proposed another mercury speciation technique by coupling TLC and enzymatic determination. Organomercury compounds and Hg(II) were preliminarily separated by TLC with further enzymatic determination using *o*-dianisidine oxidation catalyzed by horseradish peroxidase. In the course of enzymatic reaction the intermediate and final products of o-dianisidine oxidation (green and red colors, respectively) were observed visually. The rate of enzymatic reaction was characterized by the time of appearance of the red color of the final product. The existence of a proportional dependence of the time taken for the appearance of red color on the concentration of methyl-, ethyl- and phenyl-mercury cations allowed developing chromagenic spot tests for their visual determination on "Silufol". Inorganic mercury and some organo-mercury species have been separated as dithizonates by HPTLC and determined *in situ* at the subnanogram level by densitometric automatic scanning of the plate. Recently, Zhu's group developed a novel method for the speciation analysis of mercury by TLC coupled with AFS with low temperature plasma desorption/atomization source as the interface (Liu

Table 10.8 Thin-layer chromatography for the speciation of mercury.

Species	TLC plate	Developing solvent	R_f	Analytical technique	Samples	Detection limits (ng)	Ref.
Hg(II), MeHg	Silica gel 7	Petroleum ether - acetone (9:1)	0.43 (HgD); 0.83 (CH$_3$HgD)	AAS	Fish, sewage sludge	0.7 (total Hg); 2.3 (CH$_3$Hg-D).	Margler and Mah 1981
Hg(II), CH$_3$HgCl, C$_2$H$_5$HgCl, C$_6$H$_5$HgCl	Silica gel plates (Merck G60 HPTLC)	n-hexanel – acetone (85:15)	0.15 (HgD); 0.32 (CH$_3$HgD); 0.35 (C$_2$H$_5$HgD); 0.28 (C$_6$H$_5$HgD).	Densitometric detection	Water, urine	0.05 (Hg(II)); 0.11 (CH$_3$HgD); 0.23 (C$_2$H$_5$HgD); 0.07 (C$_6$H$_5$HgD).	Bruno et al. 1985
Hg(II), CH$_3$HgI, C$_2$H$_5$HgBr, C$_6$H$_5$HgCl	Silufol	Hexane – ethanol (5:1) for the fist step; Hexane – ethanol (9:1) for the second step.	0 (Hg(II)); 0.23 (CH$_3$HgI); 0.85 (C$_2$H$_5$HgBr); 0.41 (C$_6$H$_5$HgCl)	Enzymatic determination	-	LOD (M): 5.7 (CH$_3$HgI$^+$); 6.0 (C$_2$H$_5$HgBr$^+$); 6.7 (C$_6$H$_5$HgCl$^+$).	Shekhovtsova et al.1997

D: dithizone

LOD(M): the detection limit of analytical concentration, mol L^{-1}

et al. 2012). Mercury species were spotted on the TLC plate and separated simultaneously. After TLC speciation, the TLC plate was placed in an ablation cell where the analytes are desorbed and atomized by DBD plasma. The analytes were introduced to the AFS detector via a silicone tube. The limits of detection were 5.8 pg, 3.8 pg, and 8.9 pg for mercury dithizonates (HgD), methyl mercury dithizonates (CH_3HgD) and phenyl mercury dithizonates (C_6H_5HgD), respectively.

3.5 Isotope dilution analysis

Isotope dilution (ID) involves adding a known mass and concentration of an enriched stable isotope to a known mass of the sample. In addition to excellent precision and accuracy, another advantage of isotope dilution is that, the isotope spike could equilibrate with all the particular species in the sample given sufficient equilibration time. Hence, there is no need to know the pre-concentration or dilution factor of the sample or to take into account any non-quantitative separation or evaporation step. It provides highly accurate analytical results because analyte losses occurring after the isotope equilibration do not affect the final result. Thus, it has also been applied for elemental speciation (Rodriguez-Gonzalez and Garcia Alonso 2010, Meija and Mester 2008, Rodriguez-Gonzalez et al. 2005, Heumann 2004).

The application of isotope dilution analysis for elemental speciation can be performed under two different modes (the so-called species-specific and species-unspecific spiking mode) (Rodriguez-Gonzalez et al. 2005). In the species-unspecific spiking mode, the addition of the isotope tracer or spike is carried out after the complete separation of the naturally occurring species in the sample has taken place (post-column spiking). Conversely, the species-specific spiking mode requires the use of a spike solution containing the species to be analyzed in an isotopically labeled form, as in the classical organic isotope dilution technique.

The application of enriched stable isotopes in mercury speciation was first reported by Hintelmann et al. (Hintelmann et al. 1995). This pioneer work focused on the development of an alternative method to radiotracer methods to calculate mercury methylation rates in sediments. In 2001 the same group observed the above mentioned MeHg decomposition during ethylation in chloride containing solutions using GC coupled to ICP-IDMS (Demuth and Heumann 2001). Jackson et al. (Jackson et al. 2009) developed a method for aqueous MeHg quantification in lake samples using species-specific isotope dilution, purge and trap gas chromatography inductively coupled plasma mass spectrometry and provided instrument detection limits of about 0.3 fM (0.06 pg L^{-1}) and method detection limits of 15 fM (0.003 ng L^{-1}) for MeHg, which are among the lowest reported. Rahman et al. (Rahman and Kingston 2004) evaluated the inter-transformation between

Hg(II) and MeHg in extraction processes by speciated isotope dilution mass spectrometry (HPLC-ICP-MS). Two of the five tested methods were highly prone to form Hg(II) from MeHg. Some published methods converted MeHg to Hg(II) with a transformation ratio of nearly 100 percent. Epov et al. (Epov et al. 2008) presented the simultaneous on-line determination of the isotopic composition of different Hg species in a single sample by the hyphenation of GC with multicollector-inductively coupled plasma mass spectrometry (MC-ICPMS). With the use of commercially available instrumentation, precise and accurate species-specific Hg isotope δ values (per mil deviation of the Hg isotope ratio in the sample relative to a reference standard) have been obtained on-line from consecutive GC transient signals. Castillo et al. (Castillo et al. 2010) demonstrated the applicability of multiple spiking isotope dilution analysis to molecular mass spectrometry exemplified by the speciation analysis of mercury using GC-(EI)-MS instrumentation. A double spike isotope dilution approach using isotopically enriched mercury isotopes had been applied for the determination of Hg(II) and MeHg in fish reference materials. The estimated method detection limits were below 10 ng g^{-1} for both mercury species.

4. Conclusions and Future Trends

Obviously, mercury speciation analysis rather than total mercury determination offers more useful information to properly assess the toxicity and health risks of mercury and to further understand biogeochemical cycling of mercury compounds. The use of non-chromatographic methods for mercury speciation analysis is rapid, at a low cost, convenient, and can be used in screening or environmental monitoring, food security and clinical diagnostics application. To develop more reliable pre-treatment and pre-concentration techniques that are transportable, robust and easy to handle is essential for routine Hg speciation analysis. Non-chromatographic methods potentially meet these requirements, but at present they are usually limited to Hg(II) species and their detection limits are often insufficient. Therefore, to development more sensitive simple methods that are applicable to other Hg species, particularly MeHg, is desirable (Leopold et al. 2010).

To obtain a complete picture of species present in samples, chromatographic based methods are still required. The combination of GC with an element-specific detector, such as ICP-MS, is a primary approach for mercury speciation which offers the possibility to integrate sample derivatization, clean-up and pre-concentration into one single step by SPME or purge-and-trap. The LODs and sensitivity of GC-related techniques are comparable or even superior to those of other hyphenated approaches. However, with the fast development of metallomics and proteomics, the combination of HPLC with ICP-MS will play an important role in understanding metabolic pathways, bioavailabilities and toxicological effects of mercury species in organism as well as quantification of proteins. But some problems, such as insufficient sensitivity, the usage of large amounts of organic solvent, needs to be solved. ID analysis of speciation methods is the most powerful technique to optimize pre-treatment procedures and validation. However, it cannot overcome problems caused by contamination during sample pre-treatment. The prevention of contamination, minimization of trace analytes loss in sample preparation and determination process should be determined thoroughly.

Furthermore, the presence of mercury species with extremely low concentration demands more sensitive and robust analytical techniques. The development of automatic, reagent-free or reduced-reagent-consumption and environmentally friendly analytical methods is also desired. The miniaturized instrumentation is expected to have great application potential in the field analysis of mercury and mercury species.

References

Agarwal, R. and J.R. Behari. 2007. Screening for Mercury in Aqueous Environmental Samples and Urine Samples Using Thin Layer Chromatography. Water Environment Research. 79(12): 2457–2463.

Alonso, E.V., M.T.S. Cordero, A. G.de Torres, P.C. Rudner and J.M.C. Pavon. 2008. Mercury speciation in sea food by flow injection cold vapor atomic absorption spectrometry using selective solid phase extraction. Talanta. 77(1): 53–59.

Alp, O. and N. Ertas. 2009. Determination of inorganic and total mercury by flow injection vapor generation atomic absorption spectrometry using a W-coil atomizer. Journal of Analytical Atomic Spectrometry. 24(1): 93–96.

Andrews, J.C. 2006. Mercury speciation in the environment using X-ray absorption spectroscopy. Recent Developments in Mercury Science. 120: 1–35.

Angeli, V., C. Ferrari, I. Longo, M. Onor, A. D'Ulivo and E. Bramanti. 2011. Microwave-assisted photochemical reactor for the online oxidative decomposition and determination of p-hydroxymercurybenzoate and its thiolic complexes by cold vapor generation atomic fluorescence detection. Analytical Chemistry. 83: 338–343.

Aranda, P.R., P.H. Pacheco, R.A. Olsina, L.D. Martinez and R.A. Gil. 2009. Total and inorganic mercury determination in biodiesel by emulsion sample introduction and FI-CV-AFS after multivariate optimization. Journal of Analytical Atomic Spectrometry. 24(10): 1441–1445.

Bagheri, H. and A. Gholami. 2001. Determination of very low levels of dissolved mercury(II) and methylmercury in river waters by continuous flow with on-line UV decomposition and cold-vapor atomic fluorescence spectrometry after pre-concentration on a silica gel-2-mercaptobenzimidazol sorbent. Talanta. 55(6): 1141–1150.

Bai, X. and Z. Fan. 2010. 3-Mercaptopropyltrimethoxysilane coated capillary micro-extraction coupled to capillary electrophoresis for the determination of methylmercury, phenylmercury and mercury in biological sample. Microchimica Acta. 170(1-2): 107–112.

Berzas Nevado, J.J., R.C. Rodriguez Martin-Doimeadios, F.J. Guzman Bernardo and M. Jimenez Moreno. 2006. Indirect mercury speciation in biological tissues by closed-vessel microwave-assisted digestion and flow-injection cold-vapor atomic fluorescence detection. Analytical Letters. 39(14): 2657–2669.

Berzas Nevado, J.J., R.C. Rodriguez Martin-Doimeadios, E.M. Krupp, F.J. Guzman Bernardo, N. Rodriguez Farinas, M. Jimenez Moreno, D. Wallace and M.J. Patino Roper. 2011. Comparison of gas chromatographic hyphenated techniques for mercury speciation analysis. Journal of Chromatography. A 1218(28): 4545–4551.

Biester, H. and C. Scholz. 1997. Determination of mercury binding forms in contaminated soils: Mercury pyrolysis versus sequential extractions. Environmental Science & Technology. 31(1): 233–239.

Bombach, G., K. Bombach and W. Klemm. 1994. Speciation of mercury in soils and sediments by thermal evaporation and cold vapor atomic-absorption. Fresenius Journal of Analytical Chemistry. 350(1-2): 18–20.

Bravo-Sanchez, L.R., J. Ruiz Encinar, J.I. Fidalgo Martinez and A. Sanz-Medel. 2004. Mercury speciation analysis in sea water by solid phase microextraction-gas chromatography-inductively coupled plasma mass spectrometry using ethyl and propyl derivatization. Matrix effects evaluation. Spectrochimica Acta Part B: Atomic Spectroscopy. 59(1): 59–66.

Bruno, P., M. Caselli and A. Traini. 1985. Inorganic and organic mercury analysis by HPTLC and in situ densitometric detection. Application to real samples. Journal of High Resolution Chromatography. 8(3): 135–139.

Burguera, J.L., I.A. Quintana, J.L. Salager, M. Burguera, C. Rondon, P. Carrero R.A. de Salager and Y.P. de Pena. 1999. The use of emulsions for the determination of methylmercury and inorganic mercury in fish-eggs oil by cold vapor generation in a flow injection system with atomic absorption spectrometric detection. Analyst. 124(4): 593–599.

Cairns, W., M. Ranaldo, R. Hennebelle, C. Turetta, G. Capodaglio, C. Ferrari, A. Dommergue, P. Cescon and C. Barbante. 2008. Speciation analysis of mercury in seawater from the lagoon of Venice by on-line pre-concentration HPLC-ICP-MS. Analytica Chimica Acta. 622(1-2): 62–69.

Caldwell, C.A., P. Swartzendruber and E. Prestbo. 2006. Concentration and dry deposition of mercury species in arid south central New Mexico (2001–2002). Environmental Science & Technology. 40(24): 7535–7540.

Capelo, J.L., I. Lavilla and C. Bendicho. 2000. Room temperature sonolysis-based advanced oxidation process for degradation of organomercurials: Application to determination of inorganic and total mercury in waters by flow injection-cold vapor atomic absorption spectrometry. Analytical Chemistry. 72(20): 4979–4984

Capelo, J.L., C. Maduro and A.M. Mota. 2004. Advanced oxidation processes for degradation of organomercurials: determination of inorganic and total mercury in urine by FI-CV-AAS. Journal of Analytical Atomic Spectrometry. 19(3): 414–416.

Carro, A., I. Neira, R. Rodil and R. Lorenzo. 2002. Speciation of mercury compounds by gas chromatography with atomic emission detection. Simultaneous optimization of a headspace solid-phase microextraction and derivatization procedure by use of chemometric techniques. Chromatographia. 56(11): 733–738.

Castillo, A., P. Rodríguez-González, G. Centineo, A.F. Roig-Navarro and J.I.G. Alonso. 2010. Multiple spiking species-specific isotope dilution analysis by molecular mass spectrometry: simultaneous determination of inorganic mercury and methylmercury in fish tissues. Analytical Chemistry. 82(7): 2773–2783.

Cava-Montesinos, P., A. Domínguez-Vidal, M.L. Cervera, A. Pastor and M. de la Guardia. 2004. On-line speciation of mercury in fish by cold vapour atomic fluorescence through ultrasound-assisted extraction. J. Anal. At. Spectrom. 19(10): 1386–1390.

Chen, H., J. Chen, X. Jin and D. Wei. 2009a. Determination of trace mercury species by high performance liquid chromatography-inductively coupled plasma mass spectrometry after cloud point extraction. Journal of Hazardous Materials. 172(2-3): 1282–1287.

Chen, K., I. Hsu and Y. Sun. 2009b. Determination of methylmercury and inorganic mercury by coupling short-column ion chromatographic separation, on-line photocatalyst-assisted vapor generation, and inductively coupled plasma mass spectrometry. Journal of Chromatography. A 1216(51): 8933–8938.

Chung, S.W. and B.T. Chan. 2011. A reliable method to determine methylmercury and ethylmercury simultaneously in foods by gas chromatography with inductively coupled plasma mass spectrometry after enzymatic and acid digestion. Journal of Chromatography. A 1218: 1260–1265.

Da Rocha, M.S., A. Soldado, E. Blanco and A. Sanz-Medel. 2001. Speciation of mercury using capillary electrophoresis coupled to volatile species generation-inductively coupled plasma mass spectrometry. Journal of Analytical Atomic Spectrometry. 16(9): 951–956.

Demuth, N. and K.G. Heumann. 2001. Validation of methylmercury determinations in aquatic systems by alkyl derivatization methods for GC analysis using ICP-IDMS. Analytical Chemistry. 73(16): 4020–4027.

Deng, B., Y. Xiao, X. Xu, P. Zhu, S. Liang and W. Mo. 2009. Cold vapor generation interface for mercury speciation coupling capillary electrophoresis with electrothermal quartz tube furnace atomic absorption spectrometry: Determination of mercury and methylmercury. Talanta. 79(5): 1265–1269.

de Wuilloud, J.C.A., R.G. Wuilloud, R.A. Olsina and L.D. Martinez. 2002. Separation and preconcentration of inorganic and organomercury species in water samples using a selective reagent and an anion exchange resin and determination by flow injection-cold vapor atomic absorption spectrometry. Journal of Analytical Atomic Spectrometry. 17(4): 389–394.

Duarte, F.A., C.A. Bizzi, F.G. Antes, V.L. Dressler and E.M. de Moraes Flores. 2009. Organic, inorganic and total mercury determination in fish by chemical vapor generation with collection on a gold gauze and electrothermal atomic absorption spectrometry. Spectrochimica Acta Part B-Atomic Spectroscopy. 64(6): 513–519.

Epov, V.N., P. Rodriguez-Gonzalez, J.E. Sonke, E. Tessier, D.Amouroux, L.M. Bourgoin and O.F.X. Donard. 2008. Simultaneous determination of species-specific isotopic composition of Hg by gas chromatography coupled to multicollector ICPMS. Analytical Chemistry. 80(10): 3530–3538.

Fan, Z. and X. Liu. 2008. Determination of methylmercury and phenylmercury in water samples by liquid-liquid-liquid microextraction coupled with capillary electrophoresis. Journal of Chromatography. A 1180(1-2): 187–192.

Feng, X., J. Sommar, K. Gårdfeldt and O. Lindqvist. 2000. Improved determination of gaseous divalent mercury in ambient air using KCl coated denuders. Fresenius' Journal of Analytical Chemistry. 366(5): 423–428.

Fernandez, C., A.C.L. Conceicao, R. Rial-Otero, C. Vaz and J.L. Capelo. 2006. Sequential flow injection analysis system on-line coupled to high intensity focused ultrasound: Green methodology for trace analysis applications as demonstrated for the determination of inorganic and total mercury in waters and urine by cold vapor atomic absorption spectrometry. Analytical Chemistry. 78(8): 2494–2499.

Fitzgerald, W.F., C.H. Lamborg and C.R. Hammerschmidt. 2007. Marine biogeochemical cycling of mercury. Chemical Reviews. 107(2): 641–662.

Foy, G.P. and G.E. Pacey. 2003. Supercritical fluid extraction of mercury species. Talanta. 61(6): 849–853.

Fragueiro, S., I. Lavilla and C. Bendicho. 2004. Direct. coupling of solid phase microextraction and quartz tube-atomic absorption spectrometry for selective and sensitive determination of methylmercury in seafood: an assessment of chloride and hydride generation. Journal of Analytical Atomic Spectrometry. 19(2): 250–254.

Gao, Y., S. De Galan, A. De Brauwere, W. Baeyens and M. Leermakers. 2010. Mercury speciation in hair by headspace injection-gas chromatography-atomic fluorescence spectrometry (methylmercury) and combustion-atomic absorption spectrometry (total Hg). Talanta, 82(5): 1919–1923.

Gao, Y., W. Yang, C. Zheng, X. Hou and L. Wu. 2011. On-line preconcentration and in situ photochemical vapor generation in coiled reactor for speciation analysis of mercury and methylmercury by atomic fluorescence spectrometry. Journal of Analytical Atomic Spectrometry. 26(1): 126–132.

Garcia, A.M., M.L.F. Sanchez, J.E.S. Uria and A.S. Medel. 1996. Speciation of mercury by continuous flow liquid-liquid extraction and inductively coupled plasma atomic emission spectrometry detection. Mikrochimica Acta. 122(3-4): 157–166.

Gonzalvez, A., M. Cervera, S. Armenta and M. De la Guardia. 2009. A review of non-chromatographic methods for speciation analysis. Analytica Chimica Acta. 636(2): 129–157

Gonzalvez, A., S. Armenta, M. Cervera and M. De la Guardia. 2010. Non-chromatographic speciation. TrAC Trends in Analytical Chemistry. 29(3): 260–268.

Grinberg, P., R.C. Campos, Z. Mester and R.E. Sturgeon. 2003a. Solid phase microextraction capillary gas chromatography combined with furnace atomization plasma emission spectrometry for speciation of mercury in fish tissues. Spectrochimica Acta Part B: Atomic Spectroscopy. 58(3): 427–441.

Grinberg, P., R.C. Campos, Z. Mester and R.E. Sturgeon. 2003b. A comparison of alkyl derivatization methods for speciation of mercury based on solid phase microextraction gas chromatography with furnace atomization plasma emission spectrometry detection. Journal of Analytical Atomic Spectrometry. 18(8): 902–909.

He, Q., Z. Zhu, S. Hu and L. Jin. 2011. Solution cathode glow discharge induced vapor generation of mercury and its application to mercury speciation by high performance liquid chromatography-atomic fluorescence spectrometry. Journal of Chromatography. A 1218(28): 4462–4467.

Heumann, K.G. 2004. Isotope-dilution ICP-MS for trace element determination and speciation: from a reference method to a routine method? Analytical and Bioanalytical Chemistry. 378(2): 318–329.

Hintelmann, H., R.D. Evans and J.Y. Villeneuve. 1995. Measurement of Mercury Methylation in Sediments by Using Enriched Stable Mercury Isotopes Combined with Methylmercury Determination by Gas-Chromatography Inductively-Coupled Plasma-Mass Spectrometry. Journal of Analytical Atomic Spectrometry. 10(9): 619–624.

Horvat, M., K. May, M. Stoeppler and A.R. Byrne. 1988. Comparative studies of methylmercury determination in biological and environmental samples. Applied Organometallic Chemistry. 2(6): 515–524.

Ito, R., M. Kawaguchi, N. Sakui, H. Honda, N. Okanouchi, K. Saito and H. Nakazawa. 2008. Mercury speciation and analysis in drinking water by stir bar sorptive extraction with in situ propyl derivatization and thermal desorption-gas chromatography-mass spectrometry. Journal of Chromatography. A 1209(1-2): 267–270.

Jackson, B., V. Taylor, R.A. Baker and E. Miller. 2009. Low-level mercury speciation in freshwaters by isotope dilution GC-ICP-MS. Environmental Science & Technology. 43(7): 2463–2469.

Jia, X., Y. Han, C. Wei, T. Duan and H. Chen. 2011. Speciation of mercury in liquid cosmetic samples by ionic liquid based dispersive liquid-liquid microextraction combined with high-performance liquid chromatography-inductively coupled plasma mass spectrometry. Journal of Analytical Atomic Spectrometry. 26(7): 1380–1386.

Jiang, H., B. Hu, B. Chen and W. Zu. 2008. Hollow fiber liquid phase microextraction combined with graphite furnace atomic absorption spectrometry for the determination of methylmercury in human hair and sludge samples. Spectrochimica Acta Part B: Atomic Spectroscopy. 63(7): 770–776.

Jimenez, M.S. and R.E. Sturgeon. 1997. Speciation of methyl-and inorganic mercury in biological tissuesusing ethylation and gas chromatography with furnace atomization plasma emission spectrometric detection. Journal of Analytical Atomic Spectrometry. 12(5): 597–601

Jitaru, P., H. Goenaga Infante and F.C. Adams. 2003. Multicapillary gas chromatography coupled to inductively coupled plasma-time-of-flight mass spectrometry for rapid mercury speciation analysis. Analytica Chimica Acta. 489(1): 45–57.

428   *Speciation Studies in Soil, Sediment and Environmental Samples*

Kaercher, L.E., F. Goldschmidt, J.N.G. Paniz, E.M.D. Flores and V.L. Dressler. 2005. Determination of inorganic and total mercury by vapor generation atomic absorption spectrometry using different temperatures of the measurement cell. Spectrochimica Acta Part B-Atomic Spectroscopy. 60(5): 705–710.

Kato, T., T. Uehiro, A. Yasuhara and M. Morita. 1992. Determination of methylmercury species by capillary column gas chromatography with axially viewed inductively coupled plasma atomic emission spectrometric detection. Journal of Analytical Atomic Spectrometry. 7(1): 15–18

Krystek, P. and R. Ritsema. 2005. Mercury speciation in thawed out and refrozen fish samples by gas chromatography coupled to inductively coupled plasma mass spectrometry and atomic fluorescence spectroscopy. Analytical and Bioanalytical Chemistry. 381(2): 354–359.

Kuban, P., P. Pelcova, J. Margetinova and V. Kuban. 2009. Mercury speciation by CE: An update. Electrophoresis. 30(1): 92–99.

Lee, T.H. and S.J. Jiang. 2000. Determination of mercury compounds by capillary electrophoresis inductively coupled plasma mass spectrometry with microconcentric nebulization. Analytica Chimica Acta. 413(1-2): 197–205.

Lee, Y., I. Wängberg and J. Munthe. 2003. Sampling and analysis of gas-phase methylmercury in ambient air. The Science of the Total Environment. 304(1-3): 107–113.

Leermakers, M., W. Baeyens, P. Quevauviller and M. Horvat. 2005. Mercury in environmental samples: Speciation, artifacts and validation. Trac-Trends in Analytical Chemistry. 24(5): 383–393.

Leopold, K., M. Foulkes and P. Worsfold. 2010. Methods for the determination and speciation of mercury in natural waters-A review. Analytica Chimica Acta. 663(2): 127–138.

Liang, L.N., G.B. Jiang, J.F. Liu and J.T. Hu. 2003. Speciation analysis of mercury in seafood byusing high-performance liquid chromotogrphy on-line with cold-vapor atomic fluorescence spectrometry via a post column microwave digestion. Analytica Chimica Acta. 477(1): 131–137.

Li, B.H. 2011. Rapid speciation ananlysis of mercury by short column capillary electrophoresis on-line coupled with inductively coupled plasma mass spectrometry.Analytical Methods. 3(1): 116–121.

Liu, Q. 2010. Determination of mercury and methylmercury in seafood by ion chromotography using photo-induced chemical vapor generation atomic fluorescence spectrometric detection. Microchemical Journal. 95(2): 255–258.

Li,Y. and B. Hu. 2007. Sequential cloud point extraction for the the speciation of mercury in seafood by inductively coupled plasma optical emission spectrometry. Spectrochimica Acta, Part B-Atomic Spectroscopy. 62(10): 1153–1160.

Li,Y.,Y. Jiang and X.P. Yan. 2005. On-line hyphenation of capillary electrophoresis with flame-heated furnace atomic absorption spectrometry for trace mercury speciation. Electrophoresis. 26(3): 661–667.

Li,Y., Y. Jiang and X.P. Yan. 2006 Probing mercury species-DNA interactions by capillary electrophoresis with on-line electrothermal atomic absorption spectrometric detection. Analytical Chemistry. 78(17): 6115–6120.

Li, Z., Q. Wei, R. Yuan, X. Zhou, H. Liu, H. Shan and Q. Song. 2007. A new room temperature ionic liquid 1-butly1-3-trimethylsilylimidazolium hexafluorophosphate as a solvent for extraction and preconcentration of mercury with determination by cold vapor atomic absorption spectrometry. Talanta. 719 (1): 68–72.

Li, P., X. Zhang and B. Hu. 2011. Phase transfer membrane supported liquid-liquid-liquid microextraction combined with large volume sample injection capillary electrophoresis-ultraviolet detection for the speciation of inorganic and organic mercury. Journal of Chromotoghraphy. 26(3): 661–667.

Liu, Z.F., Z.L. Zhu, H.T. Zheng and S.H. Hu. 2012. Plasma Jet Desorption Atomization-Atomic Fluorescence Spectrometry and its Application to Mercury Speciation by Coupling with Thin Layer Chromotography. Analytical Chemistry. 84: 10170–10174

Lu, J.Y. and W.H. Schroeder. 1999. Comparision of conventional filtration and a denuder-based methodology for sampling of particulate-phase mercury in ambient air. Talanta. 49(1): 15–24.

Ma, W.X., F. Liu, K.A. Li, M. Chen and S. Tong. 2000. Preconcentration, separation and determination of trace Hg(II) in environmental samples with aminopropylbenzoylazo-2-mercaptobenzothiazole bonded to silica gel. Analytica Chimica Acta. 416(2): 191–196.

Magos, L. 1971. Selective atomic-absorption determination of inorganic mercury and methylmercury in undigested biological samples. Analyst. 96(1149): 847–653.

Magos,L. and T. Clarkson. 1972. Atomic absorption determination of total inorganic and organic mercury in blood.Journal-Association of Official Analytical Chemists. 55(5): 966–971.

Mahmoud, M.E. 1999. Selective solid phase extraction of mercury(II) by silica gel-immobilized-ditjiocarbamate derivatives. Analytica Chimica Acta. 398(2-3): 297–304.

Mao, Y., G. Liu, G. Meichel, Y. Cai and G. Jiang. 2008. Simultaneous speciation of monomethylmercury and monoethylmercury by acqueous phenylation and purge-and-trap preconcentration followed by atomic spectometry detection. Analytical Chemistry. 80(18): 7163–7168.

Margetinová, J., P. Houserová-Pelcová and V. Kuban. 2008. Speciation analysis of mercury in sediments, zoobenthos and river water samples by high-performance liquid chromatography hyphenated to atomic fluorescence spectrometry following preconcentration by solid phase extraction. Analytica Chimica Acta. 615(2): 115–123.

Martinis, E.M. and R.G. Wuilloud. 2010. Cold vapor ionic liquid-assisted-headspace single-drop microextraction: A novel preconcentration technique for mercury species determination in complex matrix samples. Journal of Analytical Atomic Spectrometry. 25(9): 1432–1439.

Martinis, E.M., P. Berton, R.A. Olsina, J.C. Altamirano and R.G. Wuilloud. 2009. Trace mercury determination in drinking and natural water samples by room temperature ionic liquid based-preconcentration and flow injection-cold vapor atomic spectrometry. Journal of Hazardous Materials. 167(1-3): 475–481.

Meija, J. and Z. Mester. 2008. Paradigms in istope dilution mass spectrometry for elemental speciation analysis. Analytica Chimica Acta. 607(2): 115–125.

Mester, Z., R. Sturgeon and J. Pawliszyn. 2001. Solid phase microextraction as a tool for trace element speciation. Spectrochimica Acta Part B-Atomic Spectroscopy. 56(3): 233–260.

Misra, A. and S. Pauchauree. 2001. Retention behaviour of synthetic dyes on zirconium (IV) phosphoantimonate papers. Acta Chromatographic. 139–153.

Monteiro, A.D.P., L.S.N. de Andrade, A.S. Luna and R.C. de Campos. 2002. Sequential quantification of methyl mercury in biological materials by selective reduction in the presence of mercury(II), using two gas-liquid separators. Spectrochimica Acta Part B-Atomic Spectroscopy. 57(12): 2103–2112.

Mousavi, A., R.D. Chavez, A.M.S. Ali and S.E. Cabaniss. 2011. Mercury in Natural Water: A Mini-Review. Environmental Forensics. 12(1): 14–18.

Nagase, H., Y. Ose, T. Sato, T. Ishikawa and K. Mitani. 1980. Differential determination of alklmercury and inorganic mercury in river sediment. International Journal of Environmental Analytical Chemistry. 7(4): 261–271.

Park, C.J. and H. Do. 2008. Determination of inorganic and total in marine biological samples by cold vapor generation inductively coupled plasma mass spectrometry after tetramethylammonium hydroxide digestion. Journal of Analytical Spectrometry. 23(7): 997–1002.

Pena-Pereiro, F., I. Lavilla, C. Bendicho, L. Vidal and A. Canals. 2009. Speciation of mercury by ionic liquid-based single-drop microextraction combined with high-performance liquid chromotography-photodiode array detection. Tanata. 78(2): 537–541.

Pereiro, I.R., A. Wasik and R. Lobinski. 1998. Purge-and-trap isothermal multicapillary gas chromotographic sample introduction accessory for speciation of mercury by microwave-induced plasma atomic emission spectrometry. Analytical Chemistry. 70(9): 4063–4069.

Pereiro, R.I. and C.A. Diaz. 2002. Speciation of mercury, tin and lead compounds by gas chromotography with microwave-induced plasma and atomic-emission detection (GC-MIP-AED). Analytical and Bioanalytical Chemistry. 373(1): 74–90

Rahman, G.M.M. and H.M. Kingston. 2004. Application of speciated isotope dilution mass spectrometry to evaluate extraction methods for determining mercury speciation in soils and sediments. Analytical Chemistry. 76(13): 3548–3555.

Rai, R., W. Maher and F. Kirowa. 2002. Measurement of inorganic and methylmercury in fish tissues by enzymatic hydrolysis and HPLC-ICP-MS. Journal of Analytical Atomic Spectrometry. 17(11): 1560–1563.

Ramalhosa, E., S.R. Segade, E. Pereira, C. Vale and A. Duarte. 2001. Microwave-assisted extraction for methylmercury determination in sediments by high performance liquid chromatography-cold vapor fluorescence spectrometry. Journal of Analytical Atomic Spectrometry. 16(6): 643–647.

Reuther, R., L. Jaeger and B. Allard. 1999. Determination of organometallic forms of mercury, tin and lead by in situ derivatization, trapping and gas chromatography- atomic emission detection. Analytica Chimica Acta. 39(2-3): 259–269.

Rezende, M.D.R., R.C. Campos and A.J. Curtius. 1993. Speciation of mercury in fish samples by solvent-extraction, methylmercury reduction directly in the organic medium and cold vapor atomic absorption spectrometry. Journal of Analytical Atomic Spectrometry. 8(2): 247–251.

Ribeiro, A.S., M.A. Vieira, S. Willie and R.E. Styrgeon. 2007.Ultrasound-assisted vapor generation of mercury. Analytical and Bioanalytical Chemistry. 388(4): 849–857.

Rio-Segade, S. and C. Bendicho. 1999. Selective reduction method for separate determination of inorganic and total mercury in mussel tissue by flow-injection cold vapor technique. Ecotoxicology and Environmental Safety. 42(3): 245–252.

Rivaro, P., C. Ianni, F. Soggia and R. Frache. 2007. Mercury speciation in environmental samples by cold vapur spectrometry wi9th in situ preconcentration on a gld trap. Michrchimica Acta. 158.

Rodriguez Farinas, N., M. Jimenez Moreno, D. Wallace and M.J. Patino Roper. 2011. Comparison of gas chroatographic hyphenated techniques for mercury speciation ananlysis. Journal of Chromatography. A 1218(28): 4545–4551.

Rodriguez-Gonzalez, P. and J.I. Garcia Alonso. 2010. Recent advances in isotope dilution ananlysis for elemental speciation. Journal of Ananlytical Atomic Spectrometry. 25(3): 239–259.

Rodriguez-Gonzalez, P., J.M. Marchante-Gayon J.I.G. Alonso and A. Sanz-Medel. 2005. Isotope dilution ananlysis for elemental speciation: A tutorial review. Spectrochimica Acta Part B-Atomic Spectroscopy. 60(2): 151–207.

Saber-Tehrani, M., M.H. Givianrad and H. Hashemi-Moghaddam. 2007. Determination of total and methyl mercury in human permanent healthy teeth by electrothermal atomic aborption spectrometry after extraction in organic phase. Talant. 71(3): 1319–1325.

Sakamoto, H., T. Tomiyasu and N. Yonehara. 1992. Differential determination of organic mercury, mercury(ii) oxide and mercury(ii) sulfide in sediments by cold vapor atomic-absorption spectrometry. Analytical Sciences. 8(1): 35–39.

Segade, S.R. and J.F. Tyson. 2003. Determination of inorganic mercury and total mercury in biological and environmental samples by flow injection-cold vapor-atomic absorption spectrometry using sodium borohydride as the sole reducing agent. Spectrochimica Acta Part B-Atomic Spectroscopy. 58(51): 797–807.

Shao, L.J., W.E. Gan and Q.D. Su. 2006. Determination of total and inorganic mercury in fish samples with on-line oxidation coupled to atomic fluorescence spectrometry. Analytica Acta. 562(1): 128–133.

Shekhovtsova, T.N., S.V. Muginova and N.A. Bagirova. 1997. Enzymatic determination of organomercury compounds and mercury(II) after their separation by thin-layer chromatography on Silufol". Mendeleev Communications. (3): 119–120.

Shuuaeua, O.V., M.A. Gustaytis and G.N. Anoshin. 2008. Mercury speciation in environmental solid samples using thermal release technique with atomic aborption detection. Analytica Chimica Acta. 621(2): 148–154.

Sladek, C. and M.S. Gustin. 2003. Evaluation of sequential and selective extraction methods for determination of mercury speciation and mobilty in mine waste. Applied Chemistry. 18(4): 567–576.

Stoichev, T., D. Amouroux, R.C.R. Martin-Doimeadios, M. Monperrus, O.F.X. Donard and D.L. Tsalev. 2006. Speciation analysis of mercury in aquatic environment. Applied Spectroscopy Review. 41(6): 591–619.

Takazawa, Y., K. Minagawa and M. Fujii. 1981. A practical and simple method in fractional determination of ambient forms of mercury in air. Chemosphere. 10(8), 801–809.

Tomiyasu, T., A. Nagano, H. Sakamoto and N. Yonehara. 1996. Differential determination of organic mercury and inorganic mercury in sediment, soil and aquatic organisms by cold-vapor atomic absorption spectrometry. Analytical Sciences. 12(3): 477–481.

Torres, D.P., M.A. Vieira, A.S. Ribeiro and A.J. Curtius. 2005. Detemination of inorganic and total mercury in biological samples treated with tetramethylammonium hydroxide by cold vapor atomic absorption spectrometry using different temperatues in the quartz cell.Journal of Analytical Atomic Spectrometry. 20(4): 289–294.

Torres, D.P., D.L.G. Borges, V.L.A. Frescura and A.J. Curtius. 2009. A simple and fast approach for the determination of inorganic and total mercury in aqueous slurries of biological samples using cold vapor atomic absorption spectrometry and in situ oxidation. Journal of Analytical Atomic Spectrometry. 24(8): 1118–1122.

Tutschku, S., M.M. Schantz and S.A. Wise. 2002. Determination of methylmercury and butyltin compounds in marine biota and sediments using microwave-assisted acid extraction, solid-phase microextraction, and gas chromatography with microwave-induced plasma atomic emission spectrometric detection. Analytical Chemistry. 74(18): 4694–4701.

Ubillus, F., A. Alegria, R. Barbera, R. Farre and M.J. Lagarda. 2000. Methylmercury and inorganic mercury determination in fish by cold vapor generation atomic abssorption spectrometry. Food Chemistry. 71(4): 529–533.

Velado, N.G.O., R. Pereiro and A. Sanz-Medel. 2000. Mercury speciation by capillary gas chromatography with radiofrquency hollow cathode discharge atomic detction. J. Anal. At. Spectrom. 15(1): 49–53.

Vieira, M.A., A.S. Ribeiro, A.J. Curtius and R.E. Sturgeon. 2007. Determination of total mercury and methylmercury in biological samples by photochemical vapor generation. 388(4): 837–847.

Vieira, M.A., P. Grinberg, C.R.R. Bobeda, M.N.M. Reyes and R.C. Campos. 2009. Non-chromatographic atomic spectrometric methods in speciation analysis: A review. Spectrochimica Acta Part B:Atomic Spectroscopy. 64(6): 459–476.

Wang, M., W. Feng, J. Shi, F. Zhang, B. Wang, M. Zhu, B. Li, Y. Zhao and Z. Chai. 2007. Development of a mid mercaptoethanol extraction method for determination of mercury species in biological samples by HPLC-ICP-MS. Talanta. 71(5): 2034–2039.

Wang, Z., Y. Yin, B. He, J. Shi, J. Liu and G. Jiang. 2011. L-cysteine-induced degradation of organic mercury as a novel interface in the HPLC-CV-AFS hyphenated system for speciation of Mercury. Journal of Analytical Atomic Spectrometry. 25(6): 810–814.

Windmoller, C.C., R.D. Wilken and W.D. Jardim. 1966. Mercury speciation in contaminated soils by thermal release ananlysis. Water, Air and Soil Pollution. 89(3-4): 399–416.

Wu, H., Y. Jin, W. Han, Q. Miao and S. Bi. 2006. Non-chromatographic speciation ananlysis of mercury by flow injection on-line preconcentration in combination with chemical vapor generation atomic fluiorescence spectrometry. Spectrochimca Acta Part B-Atomic Spectroscopy. 61(7): 831–840.

Wu, Y., Y.-I. Lee, L. Wu and X. Hou. 2012. Simple mercury speciation analysis by CVG-ICP-MS following TMAH pre-treatment and microwave-assisted digestion. Microchemical Journal. 103: 105–109.

Yan, D., L. Yang and Q. Wang. 2008. Thermodiffusion Interface for Simultaneous Speciation of Organic and Inorganic Lead and Mercury Species by Capillary GC-ICPMS Using Tri-n-propyl-lead Chloride as an Internal Standard. Analytical Chemistry. 80(15): 6104–6109.

Yan, X.P., X.B. Yin, D.Q. Jiang and X.W. He. 2003. Speciation of mercury by hydrostatically modified electroosmotic flow capillary electrophoresis coupled with volatile species generation atomic fluorescence spectrometry. Analytical Chemistry. 75(7): 1726–1732.

Yin, Y., J. Liang, L. Yang and Q. Wang. 2007a. Vapour generation at a UV/TiO_2, photocatalysis reaction device for determination and speciation of mercury by AFS and HPLC-AFS. Journal of Analytical Atomic Spectrometry. 22(3): 330–334.

Yin, Y., J. Liu, B. He, E. Gao and G. Jiang. 2007b. Photo-induced chemical vapour generation with formic acid: novel interface for high performance liquid chromatography-atomic fluorescence spectrometry hyphenated system and application in speciation of mercury, Journal of Analytical Atomic Spectrometry. 22(7): 822–826.

Yin, Y., J. Liu, B. He, J. Shi and G. Jiang. 2008. Simple interface of high-performance liquid chromatography-atomic fluorescence spectrometry hyphenated system for speciation of mercury based on photo-induced chemical vapour generation with formic acid in mobile phase as reaction reagent. Journal of Chromatography. A 1181(1-2): 77–82.

Yin, Y., J. Liu, B. He, and G. Jiang. 2009. Mercury speciation by a high performance liquid chromatography—atomic fluorescence spectrometry hyphenated system with photo-induced chemical vapour generation reagent in the mobile phase. Microchimica Acta. 169(3): 289–295.

Yin, X.B. 2007. Dual-cloud point extraction at a preconcentration and clean-up technique for capillary electrophoresis speciation analysis of mercury. Journal of Chromatography. A 1154(1-2): 437–443.

Zheng, C.B., Y. Li, Y.H. He, Q. Ma and X.D. Hou. 2005. Photo-induced chemical vapor generation with formic acid for ultrasensitive atomic fluorescence spectrometric determination of mercury: potential application to mercury speciation in water. Journal of Analytical Atomic Spectrometry. 20(8): 746–750.

Zhu, Z., Z. Liu, H. Zheng and S. Hu. 2010. Non-chromatographic determination of inorganic and total mercury by atomic absorption spectrometry based on a dielectric barrier discharge atomizer. Journal of Analytical Atomic Spectrometry. 25(5): 697–703.

Zinc Speciation Studies in Soil, Sediment and Environmental Samples

Todd P. Luxton,[1,a],* *Bradley W. Miller*[2] *and Kirk G. Scheckel*[1,b]

Introduction

Zinc (Zn) is ubiquitous in the environment, occurring on the Earth's crust at an average concentration of about 70 mg/kg (Lindsay and Norvell 1978, Lindsay 1979). Zinc metal is very rarely found in nature, requiring extreme reducing conditions to exist. Rather, Zn typically occurs in the +2 oxidation state, primarily as various geogenic minerals such as sphalerite (zinc sulfide), smithsonite (zinc carbonate), zincite (zinc oxide), hemimorphite (zinc silicate), and willemeite (zinc silicate) (Lindsay 1979, Van Damme 2010, Jacquat 2011). Sorption is the dominant reaction mechanism governing fate and transport of zinc in soil and sediment environments at or just below

---

[1]United States Environmental Protection Agency, 5995 Center Hill Avenue, Cincinnati, OH 45224 USA.
[a]Email: Luxton.Todd@epa.gov
[b]Email: Scheckel.Kirk@epa.gov
[2]Oak Ridge Institute for Science and Education, 5995 Center Hill Avenue, Cincinnati, OH 45224 USA.
Email: Miller.BradleyW@epa.gov
*Corresponding author

circumneutral pH (Puls and Bohn 1988, Hesterberg et al. 1997a, Elzinga et al. 1999, Smolders et al. 2004, Grafe and Sparks 2005). In higher pH environments, zinc can partition as solid-phase precipitates, independently or on the surfaces of minerals, and is less soluble than sorption complexes (Roberts et al. 2003, Nachtegaal et al. 2005, Panfili et al. 2005). Remediation efforts to Zn contaminated soils have demonstrated that *in situ* amendments and pH adjustments can alter the chemistry of zinc to less extractable and bioavailable forms (Agbenin 1998, Mench et al. 2000, Basta et al. 2001, Zwonitzer et al. 2003, Brown et al. 2004, Nachtegaal et al. 2005, Williams et al. 2011).

The primary anthropogenic sources of zinc in the environment are from metal smelters and mining activities (Manceau et al. 2000, Roberts et al. 2002, Scheinost et al. 2002, Juillot et al. 2003, Isaure et al. 2005, Nachtegaal et al. 2005, Jacquat et al. 2009c, Vespa et al. 2010). The production and use of zinc in brass, bronze, die castings metal, alloys, rubber, galvanized steel, nanoparticles and paints may also lead to its release into the environment through various waste streams, e.g., biosolids, material degradation/ decomposition or leaching of Zn from galvanized steel (Nriagu 1980, Jacquat et al. 2009c). In addition to anthropogenic inputs, geogenic sources from Zn enriched bedrock and sediments have been linked with elevated Zn soil concentrations (Manceau et al. 2003, Manceau et al. 2005, Jacquat et al. 2009a).

Because the primary anthropogenic source of Zn in soils is attributed to smelting activities, the largest body of research on Zn speciation in soils is associated with contamination from metal smelting operations (Manceau et al. 2000, Roberts et al. 2002, Scheinost et al. 2002, Juillot et al. 2003, Isaure et al. 2005, Nachtegaal et al. 2005, Jacquat et al. 2009c, Vespa et al. 2010).

Zinc has been identified at nearly 1000 Superfund sites across the United States (USEPA 2011a). Contamination of soils by smelter activities generally occurs through atmospheric deposition but contamination also occurs through land application of dredged sediments impacted by smelting operations and via leaching from metal tailings piles. There are a variety of zinc minerals commonly produced during smelter operations including: sphalerite and wurtzite (ZnS), franklinite ($ZnFe_2O_4$), willemite (Zn_2SiO_4), hemimorphite ($Zn_4Si_2O_7(OH)_2 \cdot H_2O$), Zn substituted magnetite (($Zn,Fe)Fe_2O_4$), gahnite ($ZnAl_2O_4$), zincite (ZnO), gunningite ($ZnSO_4 \cdot H_2O$), hydrozincite ($Zn_5(CO_3)_2(OH)_6$), and hardystonite ($CaZnSi_2O_7$) (Manceau et al. 2000, Roberts et al. 2002, Scheinost et al. 2002, Juillot et al. 2003, Isaure et al. 2005, Nachtegaal et al. 2005, Vespa et al. 2010). Historical smelting operations in Europe and the United States have resulted in soils with Zn concentrations exceeding 100 g kg^{-1} (Manceau et al. 2000, Roberts et al. 2002, Scheinost et al. 2002, Juillot et al. 2003, Isaure et al. 2005, Nachtegaal et al. 2005, Van Damme et al. 2010, Vespa et al. 2010). While zinc is common at hazardous waste sites, it is often not the primary risk driver since elements such as lead (Pb), arsenic (As), cadmium (Cd) and chromium (Cr) have more critical toxicological objectives to address. However, the impact of zinc in soil, sediment and water with respect to toxicity of sensitive human and ecological receptors must be explored in these mixed contaminant environments.

The impact on human health with respect to zinc in soils is manifested in the exposure pathways, defined as the path from sources of pollutants via, soil, water or food to man and other species or settings (McKone and Daniels 1991). The primary soil related exposure pathways for humans are:

1. soil–human (direct exposure pathway)
2. soil–crop plant–human
3. soil-crop plant–livestock-human
4. soil-groundwater (leachate)-drinking water-human
5. soil-groundwater (leachate)-surface water-human
6. soil–soil borne dust–translocation to indoor dust-human
7. soil-groundwater (leachate)–soil borne dust–translocation to indoor dust-human

Thus, potential exposure routes for humans from Zn contaminated soil include direct soil ingestion, dermal absorption, inhalation, water ingestion and food ingestion. Inhalation of dust (Pathways 6 and 7) is thought to be a minor exposure route for the majority of the population with the exception being industrial workers and those in close proximity to extremely contaminated sites (Rohrs 1957, Beckett et al. 2005). Dermal absorption of Zn contaminated soil (Pathways 1, 6 and 7) is rarely a risk (Hostynek et al. 1993) to humans, but recent evidence suggests zinc nanoparticles in sunscreens and cosmetics can penetrate the skin barrier (Nohynek et al. 2007, Zvyagin et al. 2008). Dermal exposure of Zn in soils is a much higher risk to soil ecological receptors, such as earthworms or aquatic organisms (Saxe et al. 2001, Lock and Janssen 2003). Water ingestion of Zn (Pathways 4 and 5) to impact human health is relatively localized and rare; and most often occurs in private residential wells or drinking water storage systems utilizing galvanized steel tanks. Health risks of Zn from crop plant ingestion (Pathways 2 and 3) is nearly impossible as most cultivated plants are highly sensitive to Zn phytotocixity that the plant would die long before appreciable amounts of Zn can be accumulated (Chaney 1993, Ibekwe et al. 1995, Brown et al. 1998, Long et al. 2003). Similarly, meat from livestock is rarely a risk (Pathway 3). Nonetheless, some meat products, such as oysters, can accumulate high levels of Zn (Solomons and Jacob 1981). While the most common exposure route for humans of Zn impacted soils is direct soil ingestion (Pathway 1) there is limited evidence that soil borne zinc is a real risk to humans (Wilhelm et al. 1994, Barceloux 1999, Granero and Domingo

2002, Hooda et al. 2002, Noonan et al. 2003, Pruvot et al. 2006, Roussel et al. 2010). Most often, excessive Zn in soil is present as a co-contaminant with other more harmful elements such as lead, cadmium or arsenic which become the primary element(s) of concern for risk assessors and regulatory agencies on a particular site.

Many regulatory agencies, states, and countries have imposed screening levels for Zn in soils as criteria for cleanup standards or ecoreceptor toxicity. The US EPA Regional Screening Level (RSL) for residential soils lists an ingestion screening level of 23,000 mg kg^{-1} for zinc (USEPA 2011b). Most states within the US have adopted this level. Worldwide soil-zinc screening levels range considerably from 100–23,000 mg kg^{-1} for residential areas and 360–100,000 mg kg^{-1} for industrial land use (Provoost et al. 2006). Most screening levels are carefully derived based on soil ingestion rates for human health, while some are based on ecoreceptor (plant or animal) toxicity. However, an agreed upon soil ingestion rate is hard to find in the literature (Binder et al. 1986, LaGoy 1987, Barnes 1990, Vanwijnen et al. 1990, Sedman and Mahmood 1994, Stanek and Calabrese 1995, Calabrese et al. 1997a,b). Literature derived values of soil ingestion rates vary from 25 to 500 mg day^{-1} with average values being suggested somewhere between 100 to 200 mg day^{-1}. For regulatory purposes, the US EPA uses a soil ingestion rate of 50 mg day^{-1} for adult health risk assessment (USEPA 1997) and a conservative estimate of 200 mg day^{-1} for children. A child with the pica eating disorder, characterized as persistent and compulsive ingestion of non-food items such as soil, can consume between 5,000 mg day^{-1} (LaGoy 1987) and 10,000 mg day^{-1} (USEPA 1997).

More often than not, human diets are deficient in Zn and require supplements. Zinc is an essential nutrient necessary for cellular metabolism with respect to growth and development, immune response, neurological function and reproduction (Cousins 2006). Many foods, such as oysters, crab, beef, pork, and chicken, can provide ample amounts of dietary Zn, but strict vegan diets can result in severe deficiencies without supplements (Food and Nutrition Board Institute of Medicine 2001). The recommended daily allowance for Zn varies from 2 to 15 mg day^{-1} depending on age and gender. However, the tolerable upper intake level of zinc for adults is 40 mg day^{-1}; beyond this level signs of copper deficiencies begin (Food and Nutrition Board Institute of Medicine 2001). Interestingly, a simple internet search demonstrates that Zn vitamin supplements can contain 12 to 100 mg of zinc per daily dose. These values, based on a soil ingestion rate of 200 mg day^{-1}, are equivalent to soil concentrations of 60,000 to 500,000 mg Zn kg^{-1} soil, respectively, suggesting that soil screening levels mentioned earlier are quite conservative to protect human health from soil ingestion of Zn. In fact, there are very few reported cases of soil Zn impacting human health (Fosmire 1990).

The biggest concern of excessive Zn in the environment is with respect to ecological receptors (organisms, plants and animals). Excessive Zn in agricultural soils can significantly reduce production via phytotoxic indicators with concentrations as low as 50 mg kg^{-1} in acidic environments (Smolders et al. 2009). Likewise, elevated concentrations of Zn in agricultural soils can hinder microbial activity with respect to nutrient cycling, e.g., nitrogen (Mertens et al. 2007, Ruyters et al. 2010). Soil organisms are especially sensitive to high levels of Zn. Earthworms (Spurgeon et al. 1994, Spurgeon and Hopkin 1996), nematodes (Korthals et al. 2000), microorganisms (van Beelen and Doelman 1997, Renella et al. 2003, Smolders et al. 2003, Smolders et al. 2004, Broos et al. 2007) and other soil receptors (Smit and Van Gestel 1996, Smit and Van Gestel 1998, Lock et al. 2000, Stevens et al. 2003, He et al. 2005) show significant response to Zn contamination. Likewise, Zn contaminated sediments can be detrimental to aquatic plants and organisms when high levels of free Zn are present in water (Ankley et al. 1993, Casas and Crecelius 1994, Ankley et al. 1996, Wang et al. 1999, Williams et al. 2011). Therefore, with ecological receptors being the truly sensitive population, risk assessment should focus on remedial actions that promote survival and growth of these organisms.

Remediation of Zn contaminated soils and sediments has been accomplished through simple means. A number of researchers have demonstrated significant increases in sorption and/or decreases in desorption of Zn by increasing the pH of the soil system just above circumneutral levels (McBride and Blasiak 1979, Maguire et al. 1981, Harter 1983, Msaky and Calvet 1990, Stahl and James 1991, Metwally et al. 1993, Rodda et al. 1993, Spurgeon and Hopkin 1996, Brown et al. 2004). Zinc also has high affinity for soil organic matter (Haghiri 1974, Marinsky et al. 1979, McGrath et al. 1988, Shuman 1988, Spurgeon and Hopkin 1996, Shuman 1999), consequently biosolids and organic composts inputs have been successful in reducing Zn phytotoxicity (Shuman 1999, Li et al. 2000, Basta et al. 2001, Brown et al. 2005, Brown et al. 2007). Further remediation efforts have examined other in-situ amendments, such as phosphate, to sequester Zn through mineralization (Barrow 1987, Agbenin 1998, Badora et al. 1998, Lothenbach et al. 1999, Mench et al. 2000, Pearson et al. 2000, Basta et al. 2001, Cao et al. 2004, Sneddon et al. 2008). While the options for remediation of Zn contaminated soils are numerous, the overall outcome of these remedial efforts results in a change in speciation leading to an end-product that is lower in mobility, toxicity and bioavailability.

It is well established that regulations on the fate and effects of metals in the environment based solely on total concentrations are no longer (perhaps never were) valid, state-of-the-art, or scientifically sound (Scheckel et al. 2009). The response of an at-risk population is not controlled by the total metal concentration, but instead is controlled by only the bioavailable portion, which is dependent on the route of exposure, the physiology of the organism, and the speciation of the contaminant. There is an immense body of literature illustrating the decisive role of metal speciation on predicting the fate, transport, and toxicity of elements in the environment (Sparks 1999). The speciation, or chemical form, of a metal, such as Zn, governs its fate, toxicity, mobility, and bioavailability in contaminated soils, sediments and water. Different chemical forms of a metal can vary greatly in the amounts taken up by organisms. The varying bioavailability values of different metal species are a large reason for the wide range of toxicological responses measured when using standardized methods. Other interactions between metals and soil components also govern speciation and affect bioavailability. The influence of the soil matrix on element availability is in constant dynamic equilibria with multiple independent variables such as solid mineral phases, exchangeable ions and surface adsorption, nutrient uptake by plants, soil air, organic matter, and microorganisms and water flux (Scheckel et al. 2009).

It is not a trivial task to determine element speciation in soil (D'Amore et al. 2005). Yet, the effort to determine zinc speciation in the environment can provide significant scientific advantages. In terms of remediation, speciation data can provide insights to the potential fate and transport of Zn, and, likewise, provide key information to allow manipulation of the soil chemistry to reduce toxicological effects. Remedial design strategies of dig and haul, in-situ amendments, monitored natural attenuation, and pH adjustments all benefit from the knowledge of metal speciation for a final decision. After remediation, speciation studies can evaluate the effectiveness of the effort. For bioavailability and toxicity to at-risk populations, speciation aids in understanding the variability of biologically available zinc uptake from different sites with similar concentrations. Further, a clearer comprehension of zinc speciation can assist in the development predictive models on the impact of Zn toxicity. However, to accomplish these tasks, researchers must employ a variety of appropriate spectroscopic and laboratory techniques to elucidate the forms of Zn present in the environment. There are a plethora of analytical instruments and chemical extractions that claim the ability of elemental speciation.

The objective of this chapter is to highlight a few true speciation techniques that are commonly used for environmental samples, identify potential Zn aqueous and solid phase species present in terrestrial environments, and examine how soil and sediment properties influence Zn pedogenic speciation. Particularly, solid phase Zn speciation in soils and sediments has been demonstrated by synchrotron-based X-ray absorption spectroscopy (XAS) in tandem with X-ray fluorescence (XRF) methods that spatially resolve the distribution of elements. X-ray diffraction has been successful in situations where excessive Zn concentrations and crystallinity are observed. For dissolved Zn speciation, researchers often utilize differential pulse cathodic stripping voltammetry (DPCSV) and differential pulse anodic stripping voltammetry (DPASV) for quantitative determination of ionic Zn species in solution.

Anodic Voltammetric Analyses for Solution Phase Speciation

The total concentration of Zn in soils, sediments or waters may be a misleading factor when evaluating the risk of exposure to contaminated materials. Different Zn species may be more or less labile in the environment, due in part to changes in pH, redox potential and the presence of chelating ligands affecting mobility. The risk to humans and the environment due to exposure to Zn is typically dependent upon dissolution and desorption of Zn from the solid phase or mineral surfaces, respectively. After entering the aqueous phase, the specific Zn species present will control the toxicity of to the biota present, e.g., fish (Vicente-Martorell et al. 2009).

Recent advances in voltammetric analyses have made it possible for scientists to accurately determine the presence and concentration of Zn species in aqueous environments. The greatest advantage of voltammetric analyses is rapid *in situ* quantification of toxic metals (Pb, Cd, in Zn) in surface waters (Magnier et al. 2011), during food manufacturing and processing (Sanna et al. 2000, Abbasi et al. 2011), and in soils and sediments (Olsen et al. 1994, Cigala et al. 2010). Anodic-stripping voltammetry (ASV) or adsorptive stripping voltammetry (AdSV) analyses are based upon measuring the flow of current between a working electrode (WE), typically hanging mercury-drop electrodes (HMDE), and an auxiliary electrode based on redox changes in the metal of interest as a function of the potential imposed on the WE and expressed with respect to the reference electrode (Buffle and Tercier-Waeber 2005). Advances in electronics and the technology that comprises voltammetric systems allows measurements of various ions and ligands in the range 10^{-6}–10^{-12} mol L^{-1} for AdSV and 10^{-6} – 10^{-11} mol L^{-1} for ASV both with a high degree of precision (Buffle and Tercier-Waeber, 2005, Cigala et al. 2010). Therefore, the electrodes can be placed into soils, sediments, wetlands, or attached to buoys in estuaries to collect data over time, as long as the WE does not become fouled.

Gel-protected Hg-plated iridium microelectrodes, a few hundred μm thick, have been used individually and interconnected forming an array to speciate Zn after equilibrating with the test soil solution by AVS interface (Fig. 11.1) (Buffle and Tercier-Waeber 2005). Typically, measurements from the reduction of Zn and other trace metals by ASV and AdSV are first made after a pre-concentration step. During ASV, the metal is reduced on a freshly

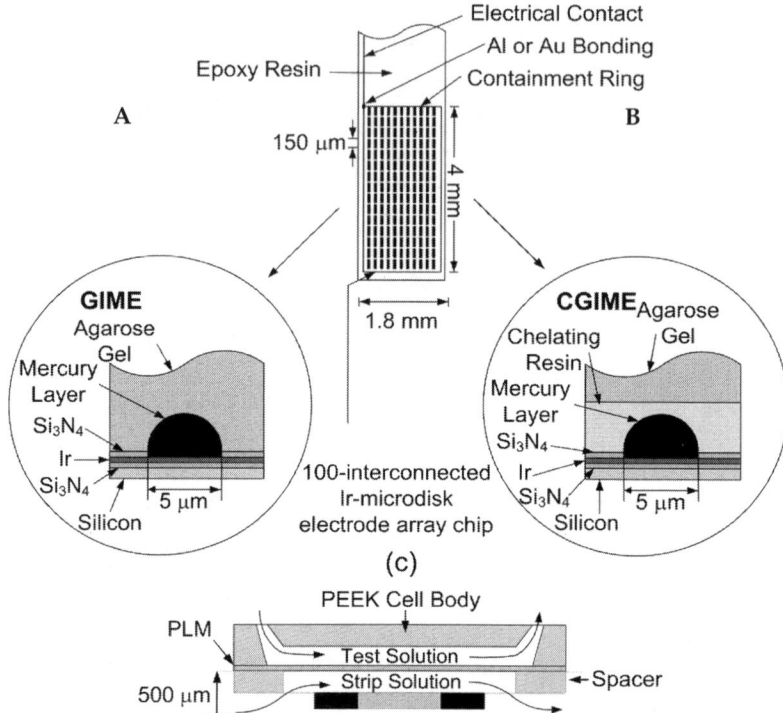

Figure 11.1 (A) Schematic drawing of the structure of an interconnected GIME array. The array of Ir microdiscs (r = 2.5 µm) is formed by depositing on Si wafers, successive thin layers of Si_3N_4, Ir, and Si3N4, and forming holes in this latter layer by plasma etching. The array is surrounded by a ring of SU-8 Epon resin (300 µm high) which contains the agarose gel layer. Hg semi-drops are deposited on Ir through the gel, by electrolysis. (B) Schematic drawing of a CGIME. The structure is the same as GIME except that an additional layer (5 µm) is deposited on top of the electrode array. (C) Schematic drawing of a PLM-GIME-ITAS. Figure reproduced with permission (Buffle and Tercier-Waeber 2005).

prepared HMDE at a constant potential approximately 1.2 V, followed by the re-oxidation of the metal at step intervals ranging from 0.1 µA to 1 mA resulting in a peak(s) in the current (Fig. 11.2) (Velasquez et al. 2002, Stephan et al. 2008). During AdSV, selective ligands are used to complex the analyte adjacent the WE during the pre-concentration step. This is followed by stripping the anode by re-oxidation of the ligand analyte to determine the concentration of the different analytes like ASV ions in water, soils and sediments (Buffle and Tercier-Waeber 2005, Cigala et al. 2010).

There have been a variety of papers published recently on Zn speciation in soils and sediments using ASV and AdSV. Stephan et al. (2008) used HMDE for differential pulse anodic-stripping voltammetry (DPASV) to speciate Zn from soils contaminated from smelters or sewage sludges, agricultural fields and vineyards, forests, and urban soils from North

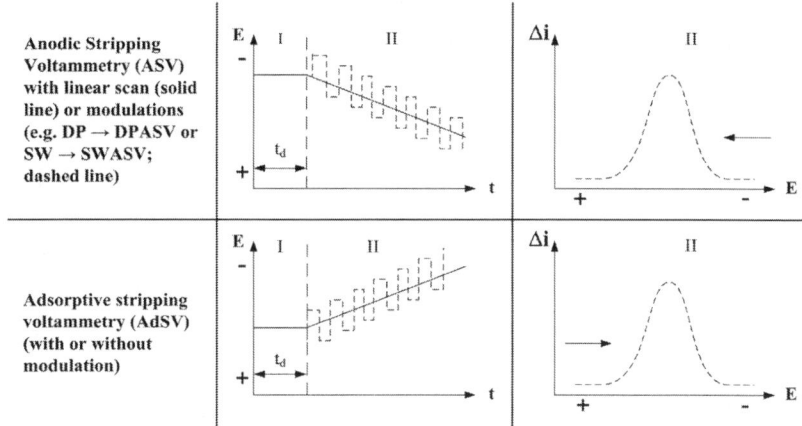

Figure 11.2 The major voltammetric techniques used for trace-metal analysis and their typical concentration ranges. E = electrode potential; t = time; td = preconcentration time; i = current. Figure reproduced with permission (Buffle and Tercier-Waeber 2005).

American and Europe (total of 66). The researchers extracted the "soil solution" with 20 ml 0.01 M KNO_3 solution by shaking 10 g of soil overnight, and filtering the solution at 0.45 µm. The results of the DPASV analyses delineated the Zn contamination into three pools 1) organically-complexed Zn, 2) inorganic Zn ion-pairs, and 3) free ionic Zn^{2+}. The results showed that most of the dissolved Zn (60–98 percent) in the 66 soils was organically-complexed Zn (Stephan et al. 2008). While 40–60 percent of the soil solution Zn was free ionic Zn^{2+} which is nearly as phytotoxic as Al^{3+} (Chaney 1993). One of the strengths in the Stephan et al. (2008) study is their ASV results were compared to and correlated with a variety of different Zn analytical techniques.

Luo et al. (2006) used a hanging mercury drop electrode for differential pulse anodic-stripping voltammetry (DPASV) to speciate Cd, Cu, and Zn in 40 contaminated agriculture soils in and around a copper mine of Nanjing City, China. Results showed that about half of the dissolved Zn was DPASV-labile at pH < 6 and when the pH > 6, 92% of the Zn was associated with soil organic matter. Additionally, the mean DPASV Zn^{2+} in the soil solutions was 36.9 µg L^{-1} (± 1.32 S.D.) in comparison to "dissolved Zn" 113 mg kg^{-1} soil (± 2.20 S.D.) regarded as extracted with 10 mmol L^{-1} KNO_3. The authors noted the elevated Zn concentrations in the soils due to the contamination and the need to determine the ecological relevance of the DPASV Zn in the soil solution.

Disadvantages to this technique during *in situ* analyses include fouling the WE by biofilms, dissolved organic matter, or inorganic (hydr) oxide colloids that have the potential to decrease or completely block the analytical signal (Buffle and Tercier-Waeber 2005, Nedeltcheva et al. 2005).

Technologies such as agarose-gel membranes have been developed to reduce interferences on the working electrodes (Buffle and Tercier-Waeber 2005). Fouling is also limited by generation of new electrodes with fresh mercury. Bostick et al. (2001) reported that Zn speciation in a wetland was dynamic and heterogeneous as a result of a recalcitrant oxidized fraction associated with (hydr)oxides. The Zn ions in the soil solution or adsorbed can be readily measured by DPASV but, if aggregates were protected against the strong reducing conditions at the beginning of the analyses, a potential pool of Zn may not be quantified. Lastly, it has been speculated that some portion of Zn that is weakly bound to natural organic ligands in the soil can be reduced and wrongly partitioned into either inorganic Zn ion-pairs or free ionic Zn^{2+} (Zhang 2004, Stephan et al. 2008).

Despite its limitations, the benefit of DPASV includes rapid *in situ* analyses. Voltammetric analyses may be more appealing compared to other common techniques such as chemical fractionation and chromatography. These *in vitro* analyses have a greater potential to introduce error by changing physiochemical state of samples after being collected from the field. Computational techniques such as GEOCHEM, WHAM, and MINTEQA generate data on Zn speciation based on chemical equilibrium models which make significant assumptions about the reactions of metal at the adsorption surface, concentrations of reactive sites and their distribution, the reactions of dissolved organic carbon, and these models require stability constants of humic and fulvic acids that are site-specific (Zachara and Westall 1999, Luo et al. 2006). The relatively low cost required for quantitative long-term Zn speciation makes ASV or AdSV an attractive *in situ* technology for use in waters, soils and sediments.

Zinc Solid Phase Speciation

As previously discussed, several spectroscopic techniques have been used to speciate Zn solid phases and Zn associated with solid phases in soils and sediments. In the most heavily contaminated soils (Zn > 10,000 mg kg⁻¹) traditional XRD maybe used to speciate the crystalline Zn phases present. However, XRD only accounts for the principal crystalline contaminants (source material) since XRD cannot resolve/detect mineral phases comprising less than 0.5 to 5 percent of the bulk material based on crystal properties. As a result, minor and non-crystalline phases (i.e., adsorption complexes) are missed. In heavily contaminated soils the Zn pedogenic fraction is generally a minor fraction and many of the Zn species present in soils are non-crystalline or have poor crystallinity and are not detectable by XRD. To identify and quantitatively determine non-crystalline minor Zn species in soils, with either high or low concentrations, a more robust element specific technique is required. The predominant technique used is synchrotron based X-ray absorption fine structure (XAFS) spectroscopy in tandem with additional micro analysis techniques and chemical/physical separation techniques.

X-ray absorption fine structure spectroscopy

X-ray absorption fine structure (XAFS) refers to the absorption of x-rays by an atom at energies near and above the core-level electron binding energies of that particular atom. XAFS is the modulation of an atom's X-ray absorption probability due to the chemical and physical state of the atom (Teo 1986, Bunker 2010). XAFS spectra are especially sensitive to the formal oxidation state, coordination chemistry/environment, interatomic distances, and coordination number of the atoms in the surrounding proximity of the selected element. As a result, XAFS provides a practical way to determine the chemical state and local atomic structure for a selected atomic species. A comprehensive explanation of the XAFS phenomenon and applications is available from Teo (1986) and Bunker (2010).

Since XAFS is an atomic probe, nearly all substances can be studied. Crystallinity is not a factor, so analysis of non-crystalline material, disordered compounds, solutions and gases is feasible. XAFS is capable of detecting and providing structural information on elements with concentrations as low as a few parts per million, making it ideal for evaluating trace quantities of Zn in environmental samples. An important aspect from an environmental perspective is that XAFS is an *in situ* spectroscopic technique allowing for the investigation of samples in their natural state with little to no pre-treatment of the sample.

The XAFS spectra is broken into two regimes: the X-ray Absorption Near-Edge Structure (XANES) and the Extended X-ray Absorption Fine-Structure (EXAFS) which contain related, but slightly different information about the element's chemical state and coordination environment (Fig. 11.3) (Bunker 2010). The XANES portion of the spectrum refers to the energy region from approximately −30 eV below to +220 eV above the adsorption edge, where the absorption edge is defined as the inflection point of the edge identified by the maximum of the first derivative (Fig. 11.3). The EXAFS portion of the spectrum occurs from just above the adsorption edge (+10 to 20 eV) to 1000 eV or more beyond the absorption edge (Kelly et al. 2008) (Fig. 11.3).

The XANES portion of the absorption spectrum is comprised of information regarding the atomic energy levels available from the excitation and production of a photoelectron (Kelly et al. 2008, Bunker 2010). As a result, the shape of the absorption spectrum in the XANES region is related to oxidation state and the atomic coordination environment for the element of interest. Different Zn minerals and adsorption complexes

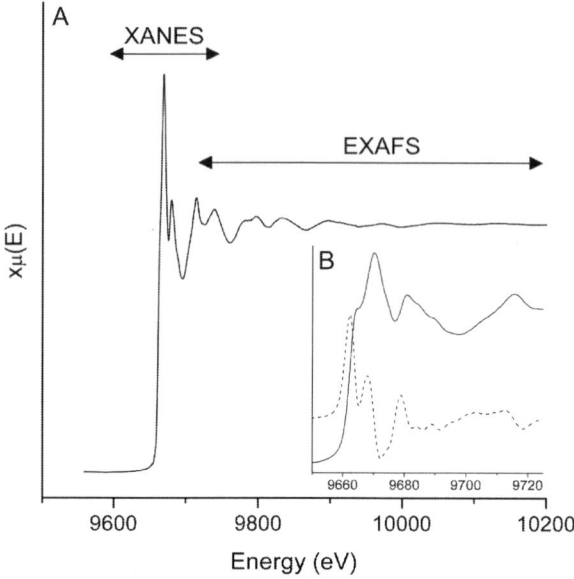

Figure 11.3 Standard XAFS spectra of ZnO. A) Illustration of the energy regions associated with the XANES and EXAFS portion of a spectra. B) XANES portion of the spectra with the first derivative (dashed line) of the ZnO spectra. The maximum of the first derivative is equal to the absorption energy (Authors data).

will have a unique XANES spectrum based on the shape and/or location of the absorption edge (Fig. 11.4). A change in the shape of the XANES spectrum between samples indicates a change in the Zn species present. In addition to identifying changes in Zn speciation, the XANES region can be used to quantify the number and relative percentage of each species present using principal component analysis (PCA) and subsequent linear combination fitting (LCF) of known standards. In Fig. 11.4B, LCF was used to determine the speciation of Zn in an agricultural soil using the Zn species in Fig. 11.4A.

The EXAFS region contains information on the local molecular bonding environment of the element of interest including type and number of atoms in coordination with the adsorber, the distance between atoms and the degree of bonding disorder (Kelly et al. 2008). The EXAFS portion of the spectra is best understood in terms of the wave behavior of the photoelectron, therefore it is customary to transform the energy data to k-space or the wave number of the photoelectron ($\chi(k)$), referred to as the chi-function (Fig. 11.5A). The different frequencies present in $\chi(k)$, correspond to the scattering of the photoelectron from different nearest neighbor atoms or coordination shells, which can be described and modeled. Applying a Fourier transform to the $\chi(k)$ data yields a pseudo radial

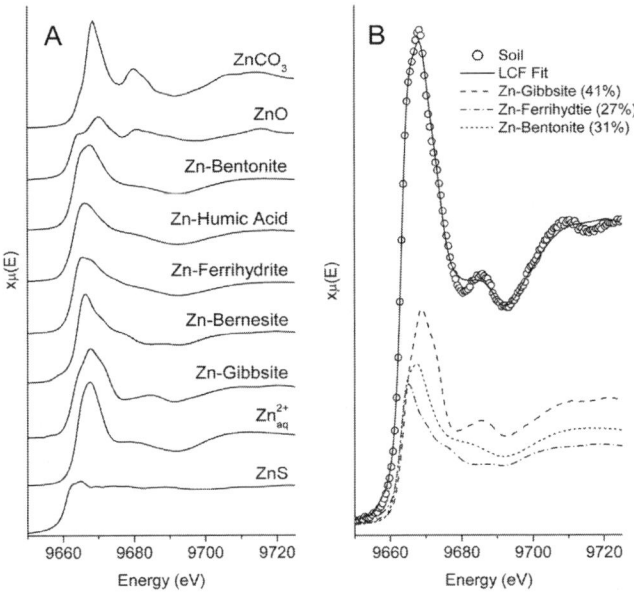

Figure 11.4 A) XANES spectra for a suite of Zn species commonly found in soils and sediments. The Zn- refers to Zn adsorbed to the mineral or organic phase identified. Changes in the shape and position of the adsorption edge are readily identifiable even for the adsorbed Zn species. B) Best linear combination fit for a soil (40 mg Zn kg^{-1}) using the Zn species in Panel A. Relative percentage of each phase is indicated in the figure (Authors Data). Linear combination fitting conducted using the IFEFFIT computer software package (Ravel and Newville 2005).

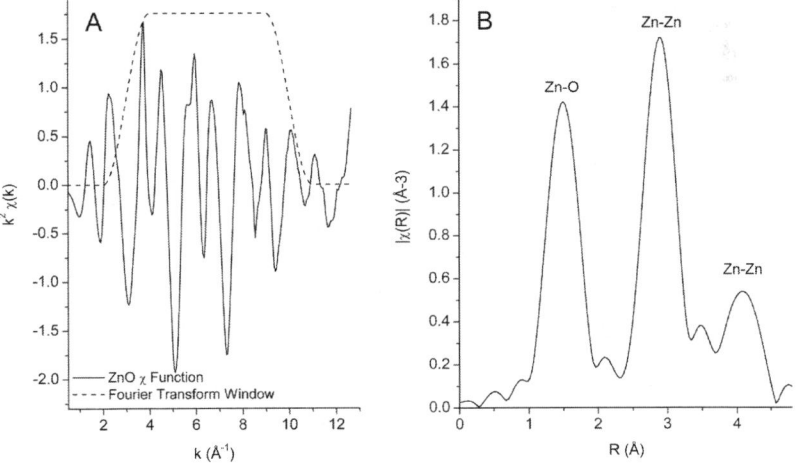

Figure 11.5 A) EXAFS χ function for ZnO spectra in Fig. 11.3. Energy values for the spectra have been transformed into k-space. B) Fourier transform of the ZnO k-space data to produce a radial distribution function. The transform was applied to the region of the χ function outlined by the dashed line (Authors data).

distribution function. Plotting the magnitude of the Fourier transform/ radial distribution function provides a visual illustration of the scattering distances between the adsorbing atom and its nearest neighbors (Fig. 11.5B). The peaks in the Fourier transform correspond to the different coordination shells in the ZnO crystal structure.

Synchrotron based micro X-ray fluorescence spectroscopy (μXRF) is often used in conjunction with XAFS analysis. The power of synchrotron based XRF analysis is the ability to focus the X-ray source down to micron and sub-micron dimensions. The X-ray source can then be rastered across the surface of a sample to provide maps for the spatial distribution and qualitative or quantitative concentration of elements (Brown and Parks 2002, Sutton et al. 2002 and references there in) (Fig. 11.6). This is particularly useful for determining elemental associations with other elements present in the sample. The technique can help identify the site and location of sorption and precipitation reactions (Manceau et al. 2000, Isaure et al. 2002, Roberts et al. 2002, Manceau et al. 2003).

As with any analytical technique, there are some disadvantages. However, minimal to moderate pretreatments of samples or additional physical and chemical analysis/information can usually overcome these issues. The first potential issue is the inability to differentiate between scattering atoms with similar atomic weights, e.g., differentiating between Zn adsorbed to Fe or Mn oxides, or differentiating between O and C (Newville 2004, Kelly et al. 2008, Bunker 2010). As a result additional information is required to effectively speciate Zn solid phases. Additional information may include X-ray diffraction patterns of the soils to identify the dominant mineralogy of the sample, bulk elemental composition, selective chemical extractions or leaching procedures (Manceau et al. 2000, Isaure et al. 2002, Roberts et al. 2002, Scheinost et al. 2002, Juillot et al. 2003, Schuwirth et al. 2007). The second commonly encountered issue is low elemental concentrations. This can normally be overcome by fractionating the soil by particle size. Research has shown that the largest pedogenic Zn fraction is associated with the clay fraction (< 2 μm) (Manceau et al. 2000, Isaure et al. 2002, Juillot et al. 2003, Vespa et al. 2010). Therefore, isolating this fraction will concentrate Zn in the soil thus allowing for improved data quality and quantification. Micro-XRF is useful for identifying areas of elevated elemental concentrations in soil, sediment and geological samples and subsequent μXAFS (XAFS spectra collected using a focused X-ray source) analysis of the spot can provide information on speciation. The third commonly encountered issue is identifying minor species present due to a large concentration of one or two species. This is a common issue when evaluating Zn species in smelter contaminated soils, where the concentration of the principal Zn component (smelter materials) is orders of magnitude higher than pedogenic species (Manceau et al. 2000, Isaure et al.

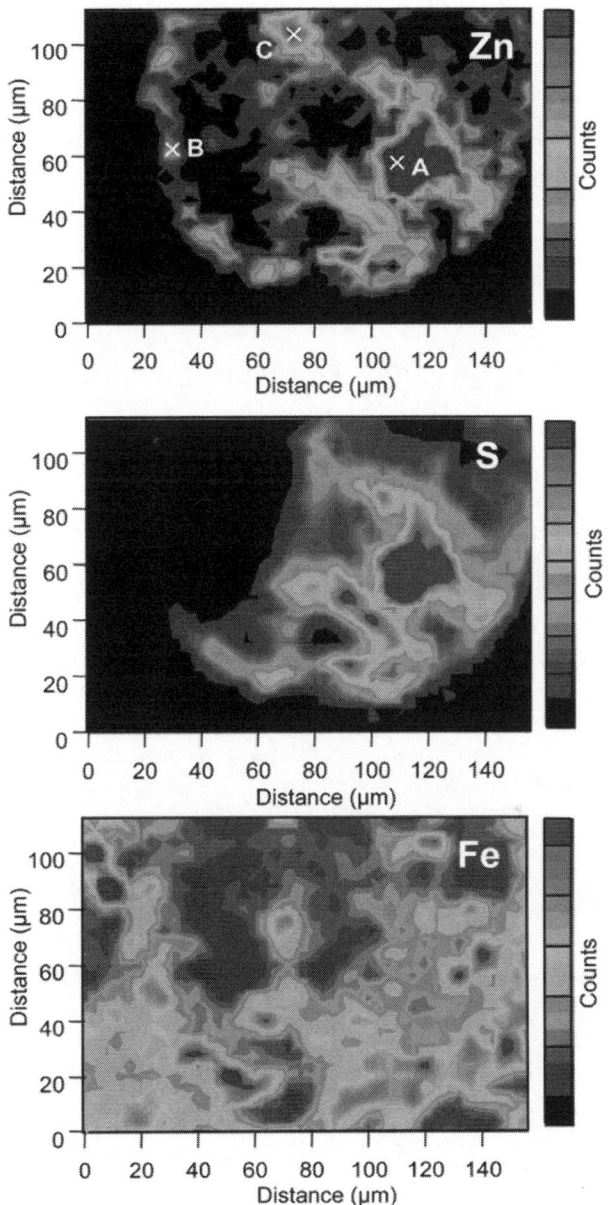

Figure 11.6 X-ray fluorescence maps for zinc, sulfur and iron from a porous black slag material. The scale bars to the right of the images denote intensity of the fluoresce signal which is related to abundance of the element. Maps show the co-localization of zinc with sulfur and iron. Figure reproduced with permission (Isaure et al. 2002).

Color image of this figure appears in the color plate section at the end of the book.

2002, Juillot et al. 2003, Manceau et al. 2003, Manceau et al. 2005, Jacquat et al. 2008, Jacquat et al. 2009a, Legros et al. 2010, Vespa et al. 2010). The most common solutions are fractionating the soil by size and selective chemical extractions. The final issue is differentiating between species in similar crystalline phases. XAFS is a local atomic probe, so determining mineral structures of an element in a similar immediate coordination environment is generally beyond the scope of the technique. However, (Manceau et al. 2000) developed a technique utilizing polarized XAFS measurements to differentiate between specific mineral structures, most notably di- and trioctahedral phyllosilicates.

Analytical and interpretation techniques for evaluating Zn XAFS data

Due to the relative abundance of specific Zn species present in soils, identifying all of the Zn solid phases can be difficult. To overcome the issue many researchers have utilized known reference compounds, μXRF coupled with μXAFS, and/or selective chemical extractions in conjunction with traditional XANES and EXAFS analysis techniques (Hesterberg et al. 1997b, Manceau et al. 2000, Bostick et al. 2001, Isaure et al. 2002, Roberts et al. 2002, Scheinost et al. 2002, Voegelin et al. 2002, Juillot et al. 2003, Peltier et al. 2003, Voegelin et al. 2003, Bang and Hesterberg 2004, Isaure et al. 2005, Nachtegaal et al. 2005, Sheinost et al. 2005, Martinez et al. 2006, Terzano et al. 2007, Jacquat et al. 2008, Jacquat et al. 2009a, c, Legros et al. 2010, Van Damme et al. 2010, Vespa et al. 2010, Donner et al. 2011, Williams et al. 2011). Generally, one or more of the approaches is utilized to aid in Zn speciation. The most common method for elucidating Zn solid phase species in soils is the comparison of the sample spectra with the spectra of known Zn standards. Previous research on the solid phase speciation of Zn, combined with soil chemical parameters can be used to narrow down the potential Zn species present in a given soil environment. Individual spectra of the potential Zn components can then be used to interpret the spectra or analyze the spectra utilizing principal component analysis, target transformation and/or linear combination fitting procedures (Isaure et al. 2002, Bang and Hesterberg 2004, Nachtegaal et al. 2005, Terzano et al. 2007, Jacquat et al. 2009c, Van Damme et al. 2010).

Micron resolved μXRF maps of soil samples coupled with μXAFS provide spatially resolved Zn speciation within the soil matrix (Fig. 11.6) (Manceau et al. 2000, Isaure et al. 2002, Martinez et al. 2002, Roberts et al. 2002, Juillot et al. 2003, Manceau et al. 2003, Isaure et al. 2005, Manceau et al. 2005, Nachtegaal et al. 2005, Voegelin et al. 2005, Martinez et al. 2006, Terzano et al. 2007, Jacquat et al. 2008, Jacquat et al. 2009a, Donner et al. 2011). X-ray fluorescence maps of Zn and other elements help to identify the collocation of Zn with other metals commonly present in soils and can provide insight on how specific Zn species are partitioned in the soil. Utilizing μXAFS or μXRD can provide further information (Manceau et al. 2003, Isaure et al. 2005, Manceau et al. 2005, Terzano et al. 2007, Vespa et al. 2010). The biggest draw backs to μXRF and μXAFS are the preparation of sample thin sections prior to analysis and the use of specialized beam lines capable of focusing synchrotron x-rays down to a micron sized beam.

Selective chemical extractions preferentially dissolve/remove specific Zn species based on the chemical extractant. Subsequent XAFS analysis of the sample can be used to identify the more recalcitrant Zn species or aid in the identification of the extracted/dissolved Zn species (Bostick et al. 2001, Scheinost et al. 2002, Juillot et al. 2003, Schuwirth et al. 2007, Terzano et al. 2007, Jacquat et al. 2008, Jacquat et al. 2009c). Selective extractions can increase sensitivity of minor Zn phases by removing a more abundant fraction. The biggest drawback of selective extraction is the unintentional removal of a specific species or the formation of a new species post extraction (Scheinost et al. 2002, Juillot et al. 2003). For example, Juillot et al. (2003) indicated extraction of the organic fraction of a soil, using 0.1 M Na_3PO_4, resulted in the removal/extraction of the exchangeable Zn fraction due to the high ionic strength. Further, the Na_3PO_4 may remove other non-organic fractions such as Mn-oxides, and promote dissolution of other Zn-species. One potential method for overcoming the shortcomings of chemical extractions is applying the extractions to reference materials in order to understand how chemical treatments will alter Zn species prior to the analysis of the soil or sediment sample (Juillot et al. 2003).

Zinc Adsorption/Sequestration Mechanisms

As previously discussed in heavily contaminated soils the dominant Zn species present is likely the source material. Over time these materials breakdown releasing Zn into the soil which is sequestered through a variety of different mechanisms based on soil chemical properties and mineralogy. Here we focus on Zn species derived from pedogenic processes, i.e. transformation and uptake of Zn released from either anthropogenic or geogenic sources.

Ultimately the fate and bioavailability of Zn in soils is controlled by adsorption/desorption and precipitation/dissolution reactions. Soil chemical and mineralogical properties will dictate the type and stability of Zn solid phase species. In order to identify specific Zn species in soil, it is important to understand the potential Zn species that may be present. X-ray absorption fine structure spectroscopy of Zn adsorption and precipitation on single mineral and organic phases, e.g., metal (hydr)oxides, phyllosilicates, organic matter (OM) is incredibly important in understanding the specific type of Zn species that may be present under different chemical conditions (pH, conc of Zn, soil solution composition) and mineralogical conditions (predominant mineral phase). Detailed analysis of the collected XANES and EXAFS data have revealed information on the immediate Zn coordination environment (octahedral v. tetrahedral), sorption mechanism (inner/outer-sphere, surface precipitate), number of surface bonds (mono v. multi dentate), adsorption locations on the mineral surface (face, edge, interlayer or corner sharing metal polyhedron), and precipitation or incorporation of Zn into secondary minerals (Sarret et al. 1998, Ford and Sparks 2000, Scheinost and Sparks 2000, Trainor et al. 2000, Schlegel et al. 2001, Elzinga and Reeder 2002, Manceau et al. 2002, Voegelin et al. 2002, Waychunas et al. 2002, Fomina et al. 2007, Jacquat et al. 2009b, He et al. 2011). Based on this work Zn adsorbs to mineral and biological surface through five mechanisms: outer-sphere surface complex (OSC), inner-sphere surface complex (ISC), surface precipitate, incorporation or substitution in the mineral structure, and precipitation of secondary mineral phase (Fig. 11.7). Examples of EXAFS spectra for different Zn sequestration mechanisms are present in Fig. 11.8, and Tables 11.1–11.3 provide lists specific Zn adsorption complexes and mechanism as a function of pH, Zn concentration, and mineralogy for laboratory based Zn adsorption studies on pristine minerals and organic matter (OM). These pristine mineral systems are compared with XAFS data collected from natural soils to compare and contrast Zn speciation in pristine and natural systems.

Outer-Sphere Adsorption Complex (OSC)

Typically OSC are labile in soils and are defined as ions or molecules retaining their complete aqueous solvation shell adsorbed to a mineral or organic surface (Stumm and Morgan 1996). Several XAFS studies involving pristine mineral surfaces have identified Zn-OSC, and Zn-OSC have been identified in soils (Table 11.1) (Sarret et al. 1997, Trivedi et al. 2001a, b, Roberts et al. 2002, Scheinost et al. 2002, Juillot et al. 2003).

In laboratory studies by Trivedi et al. (2001a, b) Zn OSC were identified with poorly crystalline and amorphous Fe and Mn hydroxides, respectively over a broad pH range (4–7) and Sarret et al. (1997) found evidence of OSC associated with humic substances only at the highest Zn concentrations (500 mg g^{-1}) investigated. In smelter contaminated soils OSC were identified by Juillot et al. (2003) and (Roberts et al. 2002) by leaching the soil with CaCl$_2$ and analyzing the quantity of Zn leached and changes in the XAFS spectra for Zn. Other studies identifying Zn-OSC complexes noted they were most commonly found in acidic pH environments (Roberts et al. 2002, Scheinost et al. 2002, Juillot et al. 2003, Jacquat et al. 2009a, c).

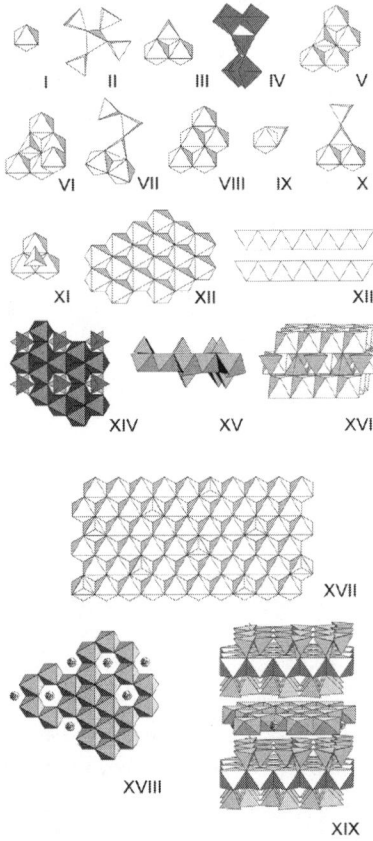

Figure 11.7 Types of Zn^{2+} sorption, coprecipitation, and model compound structures. (I) Isolated octahedral complex, form of equilibrium aqueous Zn^{2+}. (II) Tetrahedral Zn sharing vertices; the form of several $Zn(OH)_2$ polymorphs. (III) Bidentate tetrahedral Zn sharing vertices with edge sharing $Fe^{3+}O_6$ units. (IV) Polydentate version of structure III. (V) Planar polydentate cluster. (VI) Another type of polydentate cluster showing increased Zn-Fe next-nearest neighbor. (VII) Extended polymerization of $Zn(OH)_2$ on a bidentate cluster. (VIII) Octahedral Zn substituting into a planar $Fe^{3+}O_6$ structure. (IX) Edge sharing Zn tetrahedra on a $Fe^{3+}O_6$ polyhedron. (X) Bidentate cluster as in structure III with added polymerizing tetrahedral Zn^{2+} (XI) One type of outer-sphere tetrahedral Zn complex. The oxygens on the vertices of the tetrahedral complex reside just above the midpoint of triads of oxygens on $Fe^{3+}O_6$ faces. (XII) Layered double hydroxide brucite structure for $Zn(OH)_2$. (XIII) Structure XII seen from the side showing non-bonded layers. (XIV) Part of the structural motif of sclarite $([Zn,Mn,Mg]_4Zn_3[CO_3]_2[OH]_{10})$. (XV) Side view of sclarite structure. (XVI) View of one layer of the namuwite $([Zn,Cu]_4[SO_4]_6 \cdot 4H_2O)$ structure. (XVII) Part of the layer in the hydrozincite $(Zn_5[CO_3]_2[OH]_6)$ structure showing octahedral vacancies topped with tetrahedral units. All polyhedra filled with Zn^{2+} (XVIII) Side view of hydrozincite structure unit shown in XVII. (XIX) Franklinite $(ZnFe_2O_4)$ structure showing octahedral Fe^{3+} layers alternating with mixed Zn^{2+} tetrahedra and Fe^{3+} octahedra layers. (XX) Mixed Al-Zn hydroxide layer. (XXI) Mixed Al-Zn hydroxide layer in hydroxy-interlayered montmorillonite. Figure reproduced with permission (Scheinost et al. 2002 (XX and XXI) and Waychunas et al. 2002 (I-XIX).

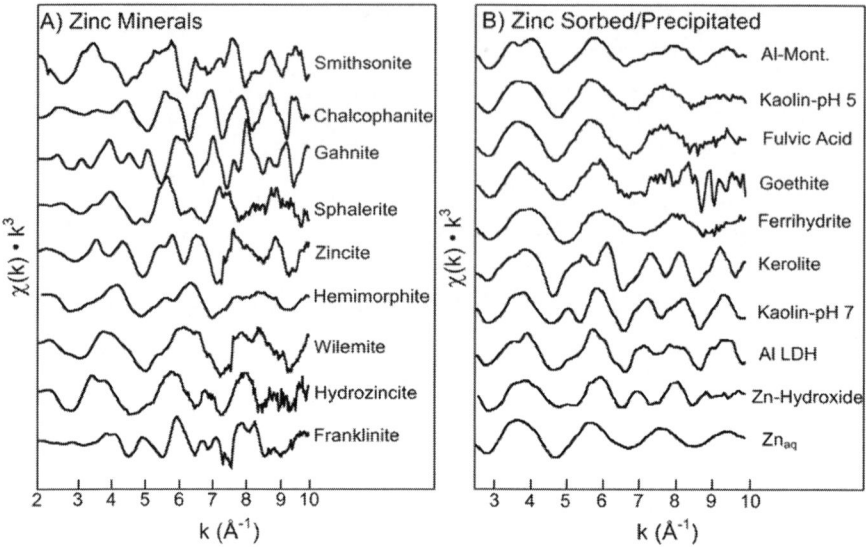

Figure 11.8 Zinc EXAFS spectra (k^3 weighted χ-spectra) for (A) Zn minerals and (B) adsorbed and co-precipitated Zn. Figure reproduced with permission (Nachtegaal et al. 2005).

Table 11.1 Review of previous XAFS studies documenting the formation of Zn outer-sphere adsorption complexes.

Material	pH	Zn Conc.	Zinc Species Identified	Reference
\multicolumn{5}{c}{Outer-Sphere Adsorption complex}				
Hydrous Manganese Oxide	3–7	10^{-4}–10^{-2} mol g^{-1}	• Octahedral outer-sphere complex over all pH values and Zn conc investigated	(Trivedi et al. 2001b)
Hydrous Ferric Oxide	6, 8	10^{-3}, 10^{-2} mol g^{-1}	• Octahedral outer-sphere complex for all conditions investigated	(Trivedi et al. 2001a)
Extracted Humic Substance	5	NL*	• Low Zn conc-mixture of tetrahedral and octahedral inner-sphere complexes with O ligands • High Zn conc-octahedral outer-sphere complexes	(Sarret et al. 1997)

*NL, Not Listed

Inner-Sphere Adsorption Complex (ISC)

Inner-sphere adsorption complexes are defined as ions or molecules that coordinate directly with the mineral/organic surface through a chemical bond (ionic, covalent or a combination there of) and are considered to be fairly stable under favorable chemical conditions (pH, competing molecules/ions, and/or oxidation reduction potential (ORP)) (Stumm and Morgan 1996). A large body of XAFS research exists for the formation and persistence of Zn-ISC (Table 11.2).

Zinc-ISC have been identified with most of the (hydr)oxides and phyllosilicates found in soils (Sarret et al. 1997, Nagy et al. 1998, Sarret et al. 1998, Bochatay and Persson 2000, Trainor et al. 2000, Trivedi et al. 2001a, Elzinga and Reeder 2002, Manceau et al. 2002, Waychunas et al. 2002, Roberts et al. 2003, Lee et al. 2004, Li et al. 2004, Nachtegaal and Sparks 2004, Pan et al. 2004, Cotter-Howells et al. 2005, Toner et al. 2005, Elzinga et al. 2006, Guine et al. 2006, Fomina et al. 2007, Karlsson and Skyllberg 2007, Ha et al. 2009, He et al. 2011). Zinc adsorption occurs through both tetrahedrally and octahedrally coordinated Zn and the specific coordination state is a function of total Zn, pH, and mineral phase. However, coordination geometry is not based solely on a single parameter (Ford and Sparks 2000, Manceau et al. 2000, Trainor et al. 2000, Schlegel et al. 2001).

Zinc/organic substrate complexes are predominantly inner-sphere. Studies evaluating Zn adsorption onto organic substrates have utilized organic materials typically found in soils: humic acids, bacteria, macro fauna and fungi in addition to isolated organic functional groups. The heterogeneity of organic functional groups associated with these materials have provided additional insight into the partitioning of Zn between different functional groups present (Sarret et al. 1997, Sarret et al. 1998, Cotter-Howells et al. 2005, Toner et al. 2005, Guine et al. 2006, Fomina et al. 2007, Karlsson and Skyllberg 2007).

Zinc-ISC have been identified in all the papers published on the speciation of Zn in soil. Based on the number of Zn species present in a specific soil, elucidating the specific adsorption mechanism can be difficult. In general terms Zn ISC are predominantly associated with OM and Fe and Mn (hydr)oxides in near neutral to basic pH soils. Further, Zn-ISC are generally the dominant Zn species in soils when the Zn to sorption site ratio in soils is low (Hesterberg et al. 1997b, Manceau et al. 2000, Isaure et al. 2002, Roberts et al. 2002, Manceau et al. 2003, Bang and Hesterberg 2004, Isaure et al. 2005, Jacquat et al. 2009c, Van Damme et al. 2010, Donner et al. 2011, Voegelin et al. 2011).

Table 11.2 Review of previous XAFS studies documenting the formation of Zn inner-sphere adsorption complexes.

Material	pH	Zn Concentration	Zinc Species Identified	Reference
Inner Sphere Surface Complex				
Calcite	8.3	1–10 µmol m^{-2}	• Adsorption of a tetrahedral tridentate	(Elzinga and Reeder 2002)
Birnessite	4.0	Zn/Mn = 0.2–12.8%	• Low surface coverage tetrahedral inner-sphere tridentate	(Manceau et al. 2002)
Manganite	6–9.8	0.9–9.7☐µmol m^{-2}	• Zn adsorption complex changed from tetrahedral to octahedral inner- sphere complex with increasing pH.	(Bochatay and Persson 2000)
Manganite	7.5	1–1.5 µmol m^{-2}	• Zn adsorbed as mixture of octahedral and tetrahedral inner-sphere bidentate complexes	(Pan et al. 2004)
Biogenic Birnessite in Hydrated Bacterial film	6.9	0.04–1.6 mmol kg^{-1}	• Low Zn-tetrahedral-inner-sphere complex • Increasing Zn resulted in a mix of tetrahedral and octahedral inner-sphere complexes.	(Toner et al. 2005)
δ-MnO$_2$	5.5	1.88–3.04 µmol m^{-2}	• Octahedral inner-sphere tridentate complex • Adsorption highly reversible	(Li et al. 2004)
Anatase	6.3, 6.8	2.0, 3.8 µmol m^{-2}	• Mixture of inner-sphere tetrahedral bidentate binuclear and five coordinated bidentate mononuclear	(He et al. 2011)
Ferrihydrite	6.5, 6.7	0.02–2.7 mmol g^{-1}	• Low surface coverage-inner-sphere tetrahedral bidentate binuclear at low surface coverage	(Waychunas et al. 2002)
Goethite	6, 8	12–20 µmol g^{-1}	• Tetrahedral inner-sphere bidentate complex for all conditions investigated	(Trivedi et al. 2001a)
Hematite	5.5	0.4–3.81 µmol m^{-2}	• Nano-Hematite (10.5 nm)-Mixture of tetrahedral and octahedral inner-sphere mononuclear bidentate complex • Micro Hematite (0.6 µm)-Octahedral inner-sphere mononuclear bidentate complex	(Ha et al. 2009)

Table 11.2 contd....

Table 11.2 contd.

Material	pH	Zn Concentration	Zinc Species Identified	Reference
Inner Sphere Surface Complex				
γ-Al$_2$O$_3$	7–8.2	1.1–3.5 μmol m^{-2}	• Tetrahedral inner-sphere bidentate complex low Zn conc	(Trainor et al. 2000)
Gibbsite	6–7.5	0.3–3.8 μmol m^{-2}	• High surface area gibbsite-mixture of octahedral and tetrahedral inner-sphere bidentate complexes	(Roberts et al. 2003)
Silica	5, 7	0.07–2.9 μmol m^{-2}	• Octahedral inner-sphere complex changing to tetrahedral with increasing pH and Zn loading/	(Roberts et al. 2003)
Kaolinite	5	~0.1–3 μmol m^{-2}	• Octahedral inner-sphere monodentate complex	(Nachtegaal and Sparks 2004)
Montmorillonite	7.0	1.5–2.7 μmol m^{-2}	• 1h–10d monomeric octahedral inner-sphere complex	(Lee et al. 2004)
Goethite coated Kaolinite	5	0.25–3.6 μmol m^{-2}	• Mix of octahedral inner-sphere complexes with goethite and kaolinite	(Nachtegaal and Sparks 2004)
Fungal and Ectomycorrhizal Biomass	NP	15 mM	• Octahedral inner sphere complexes with predominantly carboxylic functional groups.	(Fomina et al. 2007)
Biofilm from *P. putida*	6.9	50–650 μmol g^{-1}	• Octahedral inner-sphere complex with phosphoryl (85%) and carboxyl (15%) functional groups	(Toner et al. 2005)
Penicillium chrysogenum	6	3–970 μmol g^{-1}	• Tetrahedral inner-sphere complex with phosphoryl (~95%) and carboxyl (~5%) ligand	(Sarret et al. 1998)
Extracted Humic Substance	5	NL	• Low Zn conc-mixture of tetrahedral and octahedral inner-sphere complexes • High Zn-octahedral outer-sphere complexes	(Sarret et al. 1997)
Earthworm	NL	2200 mg kg^{-1}	• Zn present as a mixture of complexes resembling Zn-cysteine and Zn-phytate	(Cotter-Howells et al. 2005)

Precipitates

For the current discussion, Zn precipitates include: heterogeneous nucleation and precipitation on a mineral surface, neoformation of a secondary mineral phases, and incorporation into an existing or precipitating mineral phase.

These processes represent a continuum of sequestration mechanisms for Zn precipitates in soil. The point at which a surface precipitate becomes a separate mineral phase is not clear and absolute distinction between the two is difficult. The same may be said for Zn substitution into a mineral structure. The defining point may require a change in physical or chemical properties (diffraction pattern or dissolution rates), but detecting these changes in soils is often difficult even with the introduction of micro analysis techniques. However, Zn precipitates represent the largest proportion of Zn pedogenic species in soils and act as a long term sink for Zn (Ford and Sparks 2000, Manceau et al. 2000, Trainor et al. 2000, Gaillard et al. 2001, Isaure et al. 2002, Roberts et al. 2002, Scheinost et al. 2002, Juillot et al. 2003, Manceau et al. 2003, Isaure et al. 2005, Manceau et al. 2005, Nachtegaal et al. 2005, Martinez et al. 2006, Jacquat et al. 2008, Jacquat et al. 2009a, b, c, Van Damme et al. 2010, Vespa et al. 2010). Understanding the chemical and mineralogical conditions controlling their formation and persistence in soil is critical for addressing issues of risk assessment and bioavailability. Zn precipitates commonly found in soils include: mixed metal layered double hydroxides, Zn-phyllosilicates, and Zn substitution in metal (hydr)oxide and phyllosilicate minerals (Table 11.3).

Nucleation and precipitation of pure Zn (hydr)oxide phases on mineral surfaces have been identified under laboratory conditions, however they are not common species in soils due to the instability and dissolution properties of the precipitates (Bochatay and Persson 2000, Waychunas et al. 2002, Nachtegaal and Sparks 2004, Terzano et al. 2007, Ha et al. 2009). In soils, Zn is more likely to be incorporated into a mixed metal surface precipitate, layered double hydroxide (LDH) with Al, Mg, or Mn, and/ or Zn-phyllosilicate (Ford and Sparks 2000, Trainor et al. 2000, Scheinost et al. 2002, Roberts et al. 2003, Nachtegaal and Sparks 2004, Jacquat et al. 2009b).

In the presence of Al (hydr)oxides and phyllosilicates, Zn will form either mixed metal LDH or Zn-phyllosilicates (Ford and Sparks 2000, Manceau et al. 2000, Trainor et al. 2000, Gaillard et al. 2001, Isaure et al. 2002, Roberts et al. 2002, Scheinost et al. 2002, Juillot et al. 2003, Manceau et al. 2003, Isaure et al. 2005, Manceau et al. 2005, Nachtegaal et al. 2005, Martinez et al. 2006, Jacquat et al. 2008, Jacquat et al. 2009a, b, c, Van Damme et al. 2010, Vespa et al. 2010). In single and binary mineral systems LDH precipitates form at elevated Zn concentrations in the presence of dissolved Al, Mg, Si and Mn at pH values greater than 6 (Ford and Sparks 2000, Trainor et al. 2000, Roberts et al. 2003, Nachtegaal and Sparks 2004, Jacquat et al. 2009b). The formation of Zn LDH is favored after available sorption sites have been saturated. As a result Zn adsorption is generally rapid initially (Zn-adsorption) followed by a slower kinetic uptake during precipitation (Table 11.3) (Ford and Sparks 2000, Nachtegaal and Sparks 2004).

Table 11.3 Review of previous XAFS studies documenting the formation of Zn precipitates.

Material	pH	Zn Concentration	Zinc Species Identified	Reference
			Precipitate	
Manganite	6–9.8	0.9–9.7 µmol m^{-2}	• Increasing pH resulted in the precipitation of a multinuclear hydrox complex or zinc hydroxide phase.	(Bochatay and Persson 2000)
Ferrihydrite	6.5, 6.7	0.02–2.7 mmol g^{-1}	• High surface coverage onset of a secondary precipitate with a Zn hydroxide like structure • Highest surface coverage-octahedrally coordinated secondary precipitate with a brucite like structure.	(Waychunas et al. 2002)
Goethite	7	~0.1–3.5 µmol m^{-2}	• pH 7-Octahedral inner-sphere-bidentate complex changing to a $Zn(OH)_2$ phase with time	(Nachtegaal and Sparks 2004)
Hematite	5.5	0.4–3.81 µmol m^{-2}	• Formation of a $Zn(OH)_{2(am)}$ precipitate at higher loading (≥ 3.4 µmol m^{-2})	(Ha et al. 2009)
Montmorillonite	7.0	1.5–2.7 µmol m^{-2}	• Octahedral multinuclear surface complex or precipitate	(Lee et al. 2004)
Hydroxy-interlayer montmorillonite	3.9	900 mg kg^{-1}	• Formation of a mixed Al-Zn hydroxide layer in hydroxy-interlayered montmorillonite	(Scheinost et al. 2002)
Hydroxy-interlayered smectite (HIS)	4.5–7	1615–8600 mg kg^{-1}	• Octahedral Zn incorporated into gibbsitic Al-polymers at low concentrations • Increase in Zn results in the addition of Zn adsorption to incomplete Al polymers.	(Jacquat et al. 2009b)
Gibbsite	6–7.5	0.3–3.8 µmol m^{-2}	• Low surface area gibbsite-Mixed metal Zn-Al layered double hydroxide precipitate	(Roberts et al. 2003)
Pyrophyllite	7.5	~0.1–1.6 µmol m^{-2}	• Mixed metal Zn-Al layered double hydroxide precipitate	(Ford and Sparks 2000)
γ-Al_2O_3	7–8.2	1.1–3.5 µmol m^{-2}	• High Zn-mixed metal Zn-Al layered double hydroxide precipitate, hydrotalcite type structure	(Trainor et al. 2000)

Table 11.3 contd....

Table 11.3 contd.

Material	pH	Zn Concentration	Zinc Species Identified	Reference
Precipitate				
Montmorillonite	7.3	500 µM	• Zn is structurally attached to edges of smectite platelets • Zn is present in both octahedral and tetrahedral coordination • High Si conc-nucleation and epitaxial growth of Zn-phyllosilicate	(Schlegel and Manceau 2006)
Goethite		+0.105 Zn:Goethite	• Up to 10.5% Zn substitution in the goethite crystal structure during precipitation	(Kaur et al. 2009)
Birnesite	4	0.128 Zn:Mn	• Zn adsorption onto birnessite resulted in the formation of chalophanite (Zn, Mn oxide) mineral structure	(Manceau et al. 2002)
Calcite	8.3	10 µM	• Zn incorporated into the calcite structure during precipitation	(Elzinga and Reeder 2002)

Fewer laboratory studies have examined the neoformation of Zn phyllosilicates (Table 11.3). Based on previous research, precipitation of Zn phyllosilicates begins with heterogeneous nucleation of Zn on clay surface and edges with either Zn, Al, or Mg (Schlegel et al. 2001, Lee et al. 2004, Juillot et al. 2006, Schlegel and Manceau 2006). Lee et al. (2004) utilized EXAFS to show the slow growth of a Zn phyllosilicate in the presences of montmorillonite (Fig. 11.9). The evolution of the oscillation at approximately 7.75 Å^{-1} in the χ function is indicative of a 2:1 Zn phyllosilicate (Schlegel et al. 2001). Schlegel et al. (2001) proposed a mechanism for the precipitation of Zn phyllosilicates using polarized EXAFS (p-EXAFS). They demonstrated that the formation of Zn phyllosilicates occurred through the initial adsorption of Zn onto the edges of hectorite (Mg-smectite). With the introduction of low concentrations of silica (Zn:Si = 8.5–17) small octahedrally coordinated Zn polymers, containing two to three Zn atoms formed and were surrounded by Mg and Si atoms in the hectorite structure. As Si concentrations increased (Zn:Si = 1) epitaxial growth of a Zn trioctahedral phyllosilicate occurred with only minimal inclusion of Mg in the octahedral layer. Thus, demonstrating the precipitation of a pure Zn phyllosilicate can occur at Si concentrations representative of natural conditions. In a follow-up study (Schlegel and

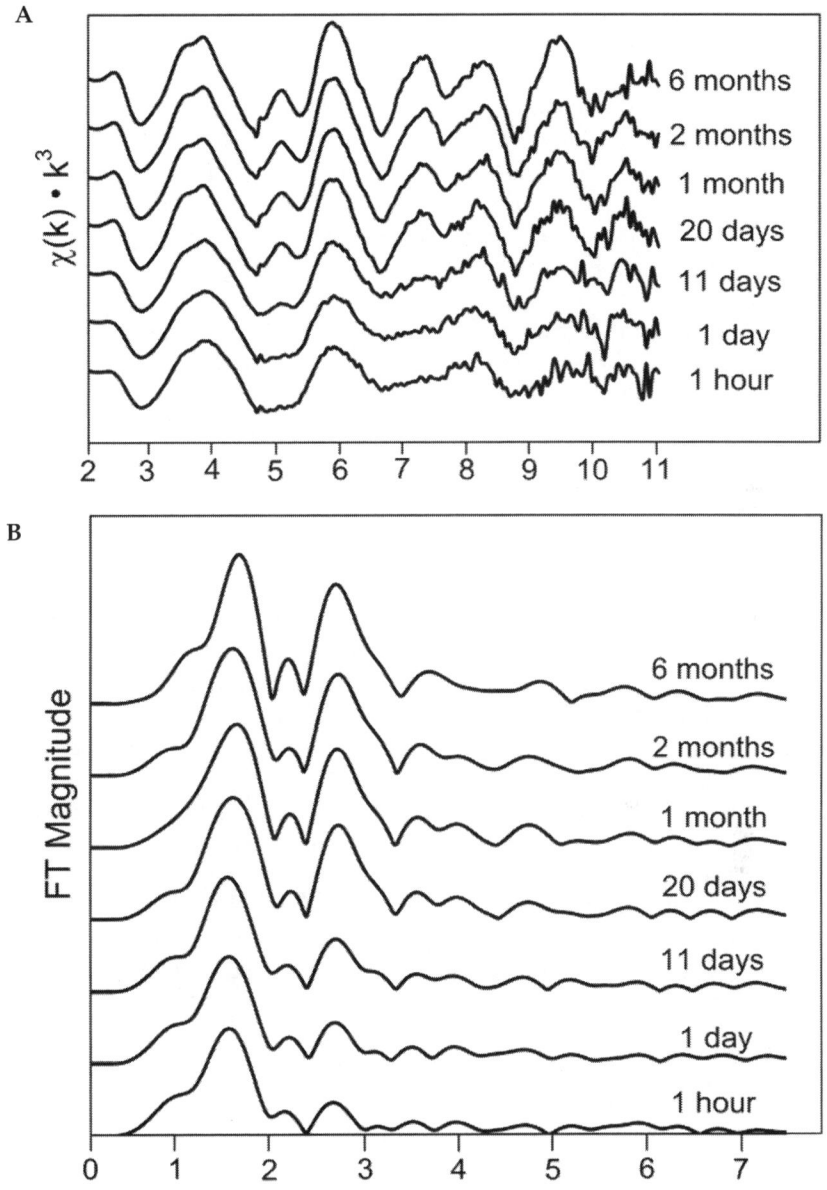

Figure 11.9 Changes in the Zn EXAFS spectrum for Zn adsorbed onto montmorillonite as a function of time. (A) χ-function and (B) radial distribution function. Changes in the shape of the χ-function and an increasing intensity for the peak located between 2 and 3 Å is associated with the precipitation of a secondary Zn(OH)$_2$ phase on montmorillonite. Figure reproduced with permission (Lee et al. 2004).

Manceau 2006) demonstrated a similar phenomenon with montmorillonite (Al-smectite). Interestingly, even with a dioctahedral mineral (Al-smectite), the Zn phyllosilicate phase retained its trioctahedral structure.

Several laboratory studies have focused on how Zn is incorporated into Fe, Mn and Al oxides, carbonate minerals and phyllosilicates through either diffusion into crystal lattice or mineral precipitation (Schlegel et al. 2001, Elzinga and Reeder 2002, Manceau et al. 2002, Lee et al. 2004, Manceau et al. 2005, Juillot et al. 2006, Jacquat et al. 2009b, Kaur et al. 2009). As expected, the chemical conditions and prevailing mineralogy govern the processes. For example, at low pH and in the presence of hydroxy interlayered minerals (vermiculite, smectite, lithiophorite), Zn will substitute for Al in the interlayer gibbsitic Al-polymers (Scheinost et al, 2002, Jacquat et al. 2009b). However, at higher pH Zn is more likely to precipitate as a LDH mineral (Jacquat et al. 2009c). Under basic conditions Zn will substitute for Ca in the calcite structure (Elzinga and Reeder 2002) and during iron (hydr)oxide precipitation Zn will substitute for Fe (Kaur et al. 2009). In the presence of phyllosilicate clays Zn will also diffuse into to the octahedral layer and substitute for Al or Mg (Juillot et al. 2003). The quantity of Zn substitution into the mineral lattice may be substantial. Kaur et al. (2009) reported that Zn substituted for up to 11 percent of the Fe in goethite. Zinc adsorption onto mineral surface and into interlayers has also resulted in the formation of new mineral phases. (Manceau et al. 2002) reported the formation of chalophanite (Zn, Mn oxide) after adsorption of Zn. Zinc replaced Mn^{3+} in the birnessite interlayers resulting in the formation of a mineral species.

The presence of mixed metal LDH, hydroxy interlayered minerals (HIM), Zn phyllosilicates, and Zn substitution in crystal lattices has been well documented in soils with anthropogenic or geogenic Zn sources (O'Day et al. 1998, Manceau et al. 2000, Isaure et al. 2002, Roberts et al. 2002, Scheinost et al. 2002, Voegelin et al. 2002, Juillot et al. 2003, Manceau et al. 2003, Isaure et al. 2005, Manceau et al. 2005, Nachtegaal et al. 2005, Voegelin and Kretzschmar 2005, Voegelin et al. 2005, Jacquat et al. 2008, Jacquat et al. 2009a, b, c, Van Damme et al. 2010, Vespa et al. 2010, Voegelin et al. 2011). Unlike the single mineral laboratory systems, Zn LDH have been identified at pH values near 5 in soils contaminated by Zn runoff (Fig. 11.10) (Jacquat et al. 2009b). The relative abundance of Zn LDH in the pedogenic fraction ranged from 10 to 45 percent, and was often the dominant pedogenic Zn species in loamy and sandy soils with pH values between 6 and 7 (Nachtegaal and Sparks 2004, Voegelin et al. 2005, Jacquat et al. 2009b, c). Zinc-HIM species in soils were first identified by Scheinost et al. (2002), and are typically found in soils with acidic pH and lower Zn concentrations (<2000 mg kg^{-1}). However, Jacquat et al. (2009c) identified Zn HIM in soils with Zn concentrations >100 mg kg^{-1} and with pH values

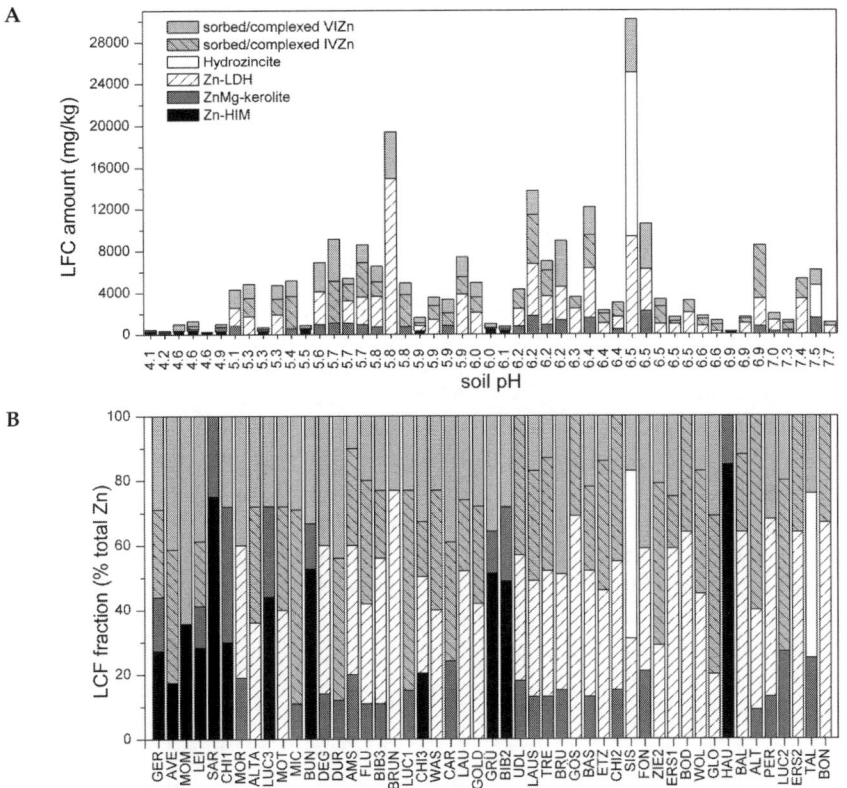

Figure 11.10 Zn speciation result from linear combination analysis of 49 different soils contaminated with Zn due to runoff from galvanized power line towers. (A) Total concentration of Zn in the soils, as determined by acid digestion, along with speciation data. (B) Normalized Zn speciation data for the 49 soils. Soils are arranged in order by pH. Figure reproduced with permission (Jacquat et al. 2009c). ZnMg kerolite is a Zn substituted 2:1 trioctahedral phyllosilicate used as a proxy for Zn-phyllosilicates.

up to 6.9 (Fig. 11.10). To better understand the environmental conditions under which Zn LDH or HIM phases precipitate, Voegelin et al. (2011) artificially contaminated three soils with ZnO. In the most acidic soil (pH 4) Zn-HIM was the most abundant species after 1 yr and remained so for the 4 yr experiment. At a soil pH of 5, the abundance of Zn LDH was slightly greater than Zn-HIM (30 vs. 20 percent), and at the two highest pH values (6.5 and 7.7) Zn-LDH phases were the most abundant with little to no detectable Zn-HIM. A similar trend, Zn-HIM forming in acidic soil and Zn-LDH in circum neutral and basic soils, was noted by Scheinost et al. (2002) and Jacquet et al. (2009b) (Fig. 11.10).

The majority of the literature on Zn speciation in soils has focused on the substitution of Zn in the phyllosilicate lattice and the precipitation of pure and enriched Zn-phyllosilicates. Both processes have been documented in soils with both geogenic and anthropogenic Zn sources, and phyllosilicates have been identified as the main long term sequestration mechanism in soils (Manceau et al. 2000, Juillot et al. 2003, Manceau et al. 2003, Isaure et al. 2005, Jacquat et al. 2009a, c, Van Damme et al. 2010, Vespa et al. 2010). Zinc phyllosilicate clays are found in wide range of mineralogical and pH environments. In a survey of 49 top soils, (Jacquat et al. 2009c) identified Zn phyllosilicates in 29 of the soils over pH range of 4.1 to 7.5 (Fig. 11.10). Zinc-phyllosilicates form in near neutral to basic pH environments where the ratio of available Zn and Si are saturated with respect to a specific phyllosilicate phase (Manceau et al. 2000, Isaure et al. 2002, Nachtegaal et al. 2005, Jacquat et al. 2008, Jacquat et al. 2009c). Jacquat et al. (2008, 2009c) noted that as available Si concentrations decreased the abundance of Zn-phyllosilicate species decreased. It was also noted that Zn-LDH species were inversely related to the quantity of free Si, suggesting that dissolved Si controls the precipitation of Zn-phyllosilicates. In a study comparing Zn-phyllosilicate species derived from geogenic and anthropogenic sources, Vespa et al. (2010) noted differences in the elemental composition of the phyllosilicate octahedral layer. In geogenic smectite clays, the octahedral layer was comprised of Zn and Al. However, the octahedral layers of smectites derived from anthropogenic sources of Zn were enriched in Fe. The enrichment is related to the weathering of Zn and Fe rich materials (franklinite; $ZnFe_2O_4$) deposited from smelting activities.

It should be noted that there has been considerable debate over the formation of and sequestration of Zn by either Zn-phyllosilicates or LDH minerals at near neutral to basic pH environments (Manceau et al. 2000, Isaure et al. 2002, Juillot et al. 2003, Nachtegaal et al. 2005, Panfili et al. 2005, Voegelin and Kretzschmar 2005, Van Damme et al. 2010, Vespa et al. 2010). Recently, Jacquat et al. (2009c) conducted a comprehensive analysis of how Zn speciation is influenced by soil properties for 49 soils with Zn concentrations ranging from background levels to highly contaminated soils (29 g kg^{-1}) and pH values between 4.1 and 7.7. The authors concluded that under near neutral to basic pH environments, Zn-LDH were the most common Zn precipitates. Further, the authors suggested that the quantity of available Si and Al may explain the preferential precipitation of one phase over another. It was suggested that soils with high Zn concentrations, Si may become limited and formation of Zn-LDH is preferred. At low to moderate Zn concentrations, Zn-phyllosilicate formation is favored over Zn-LDH assuming available Si is not limiting.

Zinc Speciation and Variability in Soil

Zinc speciation in soils is a function of numerous variables including: Zn source material, Zn concentration, soil chemical parameters (pH, oxidation-reduction potential) and mineralogy/organic content. Complicating the issue of solid phase speciation is that none of the parameters operate independently of each other and the interaction between specific variables can have a large impact on the Zn species present. The potential Zn species present in soils and the conditions under which they form and are stable in soils were outlined earlier. The following discussion will evaluate how speciation varies with soil properties and the Zn species likely to be present under those conditions.

Zinc source material and concentration

Zinc sources

The source of Zn from either geogenic or anthropogenic sources controls the concentration of Zn in soils. Further, the stability and dissolution chemistry of the source phase will dictate the free concentration of Zn for uptake and sequestration through pedogenic processes (Manceau et al. 2000, Manceau et al. 2003, Isaure et al. 2005, Manceau et al. 2005, Schuwirth et al. 2007, Jacquat et al. 2009b). In soils contaminated with smelting, and mining debris the fraction of Zn associated with the source material will generally exceed the pedogenic fraction and the total concentration of Zn in the soil will depend on the extent of contamination (Manceau et al. 2000, Isaure et al. 2002, Juillot et al. 2003, Isaure et al. 2005, Nachtegaal et al. 2005, Schuwirth et al. 2007, Donner et al. 2011, Williams et al. 2011). In general the Zn smelter contaminants will be associated with the dense (> 2.9 g cm^{-3}) sand fraction of the soil (> 50 µm) (Manceau et al. 2000, Isaure et al. 2002, Juillot et al. 2003, Isaure et al. 2005). However, it is not uncommon to find Zn minerals inherited from the smelting process in the soil clay fraction (Juillot et al. 2003).

As previously discussed a wide variety of Zn minerals may be associated with smelter particles. Therefore, understanding the mineral properties will aid in determining the quantity of free Zn in the soil available for pedological sequestration. Zinc minerals including franklinite, willemite, hemimorphite, Zn substituted magnetite, gahnite, and hardystonite are kinetically stable in oxic low temperature environments. This results in the slow release of Zn over time. Under these conditions the prevailing soil chemical properties will likely control the speciation of Zn. In contrast, Zn sulfide minerals (sphalerite, wurtzite, and marcasite) and Zn carbonates (smithsonite) are less kinetically stable with respect to other Zn minerals, and will readily undergo dissolution under specific environmental conditions (Isaure et al. 2002, Isaure et al. 2005, Schuwirth et al. 2007, Jacquat et al. 2009b, Van Damme et al. 2010, Voegelin et al. 2011). For example, Zn sulfides are inherently unstable in oxic environments and dissolution will proceed via an oxidative dissolution mechanism. A byproduct of sulfide dissolution is the acidification of the soil due to the production of sulfuric acid (Isaure et al. 2005, Van Damme et al. 2010). Acidification and subsequent lowering of the soil pH may inhibit precipitation of Zn LDH and phyllosilicates and promote Zn sequestration through HIM minerals and adsorption onto metal oxides present. Dissolution of Zn carbonates only occurs under acidic conditions, in alkaline soils the mineral is stable and will persist for an extended time (Van Damme et al. 2010). In acidic conditions pH < 6, where dissolution of Zn carbonates will occur, the speciation of Zn will be controlled by the prevailing soil chemical and mineralogical properties of the soil. Recently, Voegelin et al. (2011) demonstrated that the type of contaminant and the existing soil chemical properties can have a significant impact on pedogenic Zn speciation. They evaluated differences in Zn speciation for four different soils of varying acidity (pH 4, 5, 6.5, 7.7) artificially contaminated with zincite (ZnO) and sphalerite (ZnS). After 4 yr the most abundant pedogenic Zn species in the soils were Zn LDH for the ZnO, and Zn OSC and Zn-HIM for ZnS. It is also interesting to note that after 4 yr more than 90 percent of the ZnO in all soils had dissolved. In comparison between 26–75 percent of the ZnS material remained after 4 yr, thus illustrating the role of Zn sources on the total speciation of Zn in soils (Voegelin et al. 2011).

Soils contaminated by aqueous Zn^{2+} offer a unique insight to the pedogenic speciation. In soils contaminated with Zn runoff from power line towers, the principal contaminant is aqueous Zn^{2+}, and its concentration in soil is a function of the extent of metal corrosion and climate (average rainfall) (Jacquat et al. 2008, Jacquat et al. 2009c) as well as the prevailing soil chemical and mineralogical properties. Based on research from Degryse et al. (2011) and Jacquat et al. (2008, 2009c), the concentration of Zn varies by orders of magnitude (200–30,000 mg kg^{-1}). The specific sequestration mechanism for different soil properties may be determined based upon the length of time and the quantity of aqueous Zn^{2+} the soils have been exposed to (Jacquat et al. 2009c).

Zinc speciation introduced to soils through biosolid or manure applications will vary based on the condition of the organics (Donner et al. 2011). Fresh biosolids and manures are more likely to be enriched in ZnS phases while aged materials will be enriched in Zn substituted ferrihydrite, hopeite ($Zn_3(PO_4)_2$), and organic Zn complexes (Legros et al. 2010, Donner et al. 2011). Soils receiving biosolid applications are generally enriched in

Zn organic complexes and Zn associated with iron oxides, the two primary phases are preserved from the Zn species in the biosolids.

Geogenic sources of zinc vary widely depending on the local geology and geomorphology. The speciation of Zn in these soils will be governed by pedogenic transformations and Zn minerals in the parent material. The concentration of Zn in the soils is generally low compared to contaminated soils and likely close to or above the average crustal abundance of Zn (70 mg kg^{-1}) (Manceau et al. 2003, Manceau et al. 2005, Jacquat et al. 2009a). Jacquat et al. (2009a) examined changes in soil Zn speciation derived from Zn enriched limestone in three separate soils. The authors concluded that the source of Zn within the limestone was responsible for Zn speciation. In one soil, Zn was present as smithsonite and Zn species in the soil included Zn-HIS, Zn substituted kaolinite and adsorbed Zn. The difference in speciation in the parent material and soil combined with a decrease in the Zn content of the soil indicated much of the Zn was leached from the soil profile. In contrast a Zn substituted goethite (Zn-goethite) was the principal species in the limestone from a soil located nearby. In the second soil Zn was present predominantly as Zn substituted goethite within the soil. Based on the concentration of Zn in the parent material and soil the authors concluded the Zn-goethite was preserved during soil formation and not derived from pedogenic processes. Their results demonstrate how the Zn mineralogy, and principally the recalcitrant nature of Zn-goethite, plays a role in the speciation and mobility of Zn within soils with similar chemical and mineralogical properties.

Zinc concentration

The effect of Zn concentration on speciation is directly linked to the quantity of sorption sites (Jacquat et al. 2009c). As the ratio of Zn to sorption sites ([Zn]/ss) increases adsorption sites become limited promoting the formation of Zn precipitates. In contrast lower [Zn]/ss will favor the formation of Zn ISC and OSC depending on the pH and mineralogy.

Soil chemical properties: pH and oxidation reduction potential

Soil pH

Of the common chemical parameters measured in soils, pH is the most important variable controlling Zn speciation (Jacquat et al. 2009c). Under acidic soil conditions Zn-HIM, and Zn complexes (OSC and ISC) with OM and Fe/Al oxides are the dominant species (Roberts et al. 2002, Scheinost et al. 2002, Juillot et al. 2003, Manceau et al. 2005, Jacquat et al. 2009a, c). With increasing pH, Zn-HIM and Zn OSC become unstable and Zn hydrolysis promotes adsorption through ISC and precipitation of Zn-LDH and Zn-phyllosilicates (McBride and Blasiak 1979, Juillot et al. 2003, Isaure et al. 2005, Nachtegaal et al. 2005, Jacquat et al. 2009c, Voegelin et al. 2011). As previously stated the more abundant/stable species (Zn-LDH vs. Zn-phyllosilicates) is still a subject of debate. However, a broad survey of soils with varying Zn contents and pH demonstrated Zn-LDH was the more dominant phase at pH values exceeding 5.7 (Jacquat et al. 2009c). Under neutral and basic soil pH the dominant Zn species present include Zn-ISC, and Zn-LDH and Zn-phyllosilicates. However, in alkaline and carbonate rich soils soil hydrozincite ($Zn_5(CO_3)_2(OH)_6$) may form (Nachtegaal and Sparks 2004, Jacquat et al. 2008, Jacquat et al. 2009c).

How soil pH interacts with the ratio of [Zn]/ss is important in understanding the fate and mobility of Zn. Under acidic conditions where weakly held outer-sphere complexes and Zn-HIM species are favored, high [Zn]/ss ratios will likely result in the leaching of Zn from the soil profile due to the ready displacement of OSC by metal cations in the soil solution (Na, Ca, Mg, Al) and the unfavorable conditions for Zn precipitate formation (Sposito 1989, Roberts et al. 2002, Sparks 2003). Under similar conditions in near neutral to basic pH environments, formation of ISC and precipitation of Zn-LDH and phyllosilicates will sequester a large portion of dissolved Zn, reducing the quantity leached from soils (Sposito 1989, Roberts et al. 2002, Sparks 2003, Manceau et al. 2005, Voegelin et al. 2005, Jacquat et al. 2008, Jacquat et al. 2009a, Jacquat et al. 2009c, Voegelin et al. 2011).

Soil oxidation reduction potential

The majority of research regarding the speciation of Zn in soils has been conducted in well drained aerobic soils. Primarily due to the close proximity of these soils to anthropogenic sources of Zn. As such, Zn species identified in the previously (soil pH) represent the Zn pedogenic species found in oxic soil and sediment environments. However, changes in the redox environments will have a strong influence on the speciation of Zn. In reducing conditions, similar to Fe and other first row transition metals, neoformation of metal sulfides is possible, e.g., precipitation of pedogenic ZnS (Hesterberg et al. 1997b, Bostick et al. 2001, Peltier et al. 2003, Wilkin and Ford 2006, Donner et al. 2011). In basic and near neutral conditions in wetland soils and ground water (pH 6–7.8) the dominant Zn species present was ZnS (Hesterberg et al. 1997b, Bostick et al. 2001, Peltier et al. 2003). In addition to ZnS, Zn ISC/OSC, zinc carbonate and zinc carbonate hydroxide were identified. In slightly acidic (pH 5.5) fresh biosolids Zn was present as Zn substituted ferrihydrite > ZnS > Zn-OM ISC (Donner et al. 2011). The formation of the Zn-ferrihydrite is likely related to oxidation of the biosolids during dewatering. Interestingly, only minor quantities (> 15%) of Zn OSC and ISC were identified in the biosolids, wetland soils or ground water demonstrating the importance of Zn precipitates in reducing conditions. This point is further illustrated by Bostick et al. (2001) who evaluated Zn speciation in a wetland soil (pH 6) undergoing seasonal fluctuations in redox environments. The authors concluded that Zn cycled between ZnS and $ZnCO_3$ with a small recalcitrant fraction of Zn associated with metal (hydr)oxides, however, there was no conclusive evidence of a Zn ISC or OSC in the evaluation of the EXAFS data.

Summary

The commonality of Zn in the environment is derived from both anthropogenic and geogenic sources. While Zn is generally not the principal contaminant of concern for human health at multi-metal contaminated sites, its impact and toxicity on ecological receptors is of environmental concern. The partitioning, fate and transport of Zn in the environment are governed by its speciation in the aqueous and solid phase, which is a function of the chemical and mineralogical properties of the system.

Determining elemental speciation in the environment is not a trivial task. To identify and quantitatively determine Zn species in water, soil and sediments robust element specific techniques are required. This chapter has highlighted a few true speciation techniques: differential pulse cathodic stripping voltammetry (DPCSV), differential pulse anodic stripping voltammetry (DPASV) and X-ray absorption fine structure spectroscopy with complimentary synchrotron techniques. These methods have proven to be highly effective at elucidating Zn species present in aquatic, soil and sediment environments.

Differential pulse cathodic stripping voltammetry and differential pulse anodic stripping voltammetry have been used to accurately determine the presence and concentration of Zn species in aqueous environments utilizing rapid *in situ* quantification. The relative low cost required for this quantitative Zn speciation technique makes anodic/cathodic stripping voltammetry an attractive *in situ* technology.

X-ray absorption fine structure (XAFS) spectroscopy is an element specific technique for determining the *in situ* solid phase speciation of Zn. XAFS can determine the coordination environment for elements of interest in addition to oxidation state, coordination number, interatomic bond distances and identity of nearest-neighboring atoms. Coupling XAFS with other synchrotron techniques such as μXRF, which allows for spatial resolution and element co-association in environmental samples, and/or μXRD, for mineralogical identification, provides additional information on Zn speciation from multiple lines of evidence.

Five mechanisms have been identified for sequestration of Zn in soils and sediments: outer-sphere complex, inner-sphere complex, surface precipitation, incorporation/substitution in the mineral structure and neoformation of secondary phases. Zinc solid phase speciation in soils is governed by the prevailing chemical and mineralogical conditions. XAFS has been the preferred method in the literature for determining the solid phase speciation of Zn in soils and geologic environments.

Since ecoreceptors are the principal population of concern for Zn, it is imperative to utilize appropriate speciation techniques to adequately assess the risks of Zn in the environment. Regulators and stakeholders should consider Zn speciation to improve remediation strategies and risk assessments leading to better informed decision making and policy development.

References

Abbasi S., K. Khodarahmiyan and F. Abbasi. 2011. Simultaneous determination of ultra trace amounts of lead and cadmium in food samples by adsorptive stripping voltammetry. Food Chem. 128: 254–257.

Agbenin, J.O. 1998. Phosphate-induced zinc retention in a tropical semi-arid soil. Phosphate-induced zinc retention in a tropical semi-arid soil. Eur. J. Soil Sci. 49: 693–700.

Ankley, G.T., D.A. Benoit, R.A. Hoke, E.N. Leonard, C.W. West, G.L. Phipps, V.R. Mattson and L.A. Anderson. 1993. Development and evalaution of test methods for benthic invertebrates and sediments—effects of flow-rate and feeding on water-quality and exposure conditions. Arch. Environ. Con. Toxicol. 25: 12–19.

Ankley, G.T., D.M. DiToro, D.J. Hansen and W.J. Berry. 1996. Technical basis and proposal for deriving sediment quality criteria for metals. Environ. Toxicol. Chem. 15: 2056–2066.

Badora, A., G. Furrer, A. Grunwald and R. Schulin. 1998. Immobilization of zinc and cadmium in polluted soils by polynuclear Al-13 and Al-montmorillonite. J. Soil Con. 7: 573–588.

Bang, J.S. and D. Hesterberg. 2004. Dissolution of trace element contaminants from two coastal plain soils as affected by pH. J.Environ. Qual. 33: 891–901.

Barceloux, D.G. 1999. Zinc. Zinc. J. Toxicol-Clin. Toxic. 37: 279–292.

Barnes, R.M. 1990. Childhood soil ingestion—How much dirt do kids eat. Childhood soil ingestion—How much dirt do kids eat. Anal. Chem. 62: A1023–&.

Barrow, N.J. 1987. The effects of phosphate on zinc sorption by a soil. J. Soil Sci. 38: 453–459.

Basta, N.T., R. Gradwohl, K.L. Snethen and J.L. Schroder. 2001. Chemical immobilization of lead, zinc, and cadmium in smelter-contaminated soils using biosolids and rock phosphate. J. Environ. Qual. 30: 1222–1230.

Beckett, W.S., D.F. Chalupa, A. Pauly-Brown, D.M. Speers, J.C. Stewart, M.W. Frampton, M.J. Utell, L.-S. Huang, C. Cox, W. Zareba and G. Oberdörster. 2005. Comparing Inhaled Ultrafine versus Fine Zinc Oxide Particles in Healthy Adults. Am. J. Resp. Crit. Care 171: 1129–1135.

Binder, S., D. Sokal and D. Maughan. 1986. Estimating soil ingestion—the use of tracer elements in estimating the amount of soil ingested by young-children. Arch. Environ. Health. 41: 341–345.

Bochatay, L. and P. Persson. 2000. Metal ion coordination at the water-manganite (gamma-MnOOH) interface II. An EXAFS study of zinc(II). J. Colloid Interf. Sci. 229: 593–599.

Bostick, B.C., C.M. Hansel, M.J. La Force and S. Fendorf. 2001. Seasonal fluctuations in zinc speciation within a contaminated wetland. Environ. Sci. Technol. 35: 3823–3829.

Broos, K., M.S.J. Warne, D.A. Heemsbergen, D. Stevens, M.B. Barnes, R.L. Correll and M.J. McLaughlin. 2007. Soil factors controlling the toxicity of copper and zinc to microbial processes in Australian soils. Environ. Toxicol. Chem. 26: 583–590.

Brown, G.E., Jr. and G.A. Parks. 2002. Sorption of trace elements on mineral surfaces; modern perspectives from spectroscopic studies, and comments on sorption in the marine environment. International Book Series. 6: 69–179.

Brown, S.L., R.L. Chaney, J.S. Angle and J.A. Ryan. 1998. The phytoavailability of cadmium to lettuce in long-term biosolids-amended soils. J.Environ. Qual. 27: 1071–1078

.Brown, S., R. Chaney, J. Hallfrisch, J.A. Ryan and W.R. Berti. 2004. *In situ* soil treatments to reduce the phyto- and bioavailability of lead, zinc, and cadmium. J. Environ. Qual. 33: 522–531.

Brown, S., B. Christensen, E. Lombi, M. McLaughlin, S. McGrath, J. Colpaert and J. Vangronsveld. 2005. An inter-laboratory study to test the ability of amendments to reduce the availability of Cd, Pb, and Zn *in situ*. Environ. Pollut. 138: 34–45.

Brown, S., P. DeVolder, H. Compton and C. Henry. 2007. Effect of amendment C:N ratio on plant richness, cover and metal content for acidic Pb and Zn mine tailings in Leadville, Colorado. Environ. Pollut. 149: 165–172.

Buffle, J. and M.L. Tercier-Waeber. 2005. Voltammetric environmental trace-metal analysis and speciation: from laboratory to *in situ* measurements. Trends Anal. Chem. 24: 172–191.

Bunker, G. 2010. A Practical Guide to X-ray Absorption Fine Structure Spectroscopy Cambridge University Press.

Calabrese, E.J., E.J. Stanek and R. Barnes. 1997a. Soil ingestion rates in children identified by parental observation as likely high soil ingesters. J. Soil Con. 6: 271–279.

Calabrese, E.J., E.J. Stanek, R.C. James and S.M. Roberts. 1997b. Soil ingestion: A concern for acute toxicity in children. Soil ingestion: A concern for acute toxicity in children. Environ. Health Persp. 105: 1354–1358.

Cao, X., L.Q. Ma, D.R. Rhue and C.S. Appel. 2004. Mechanisms of lead, copper, and zinc retention by phosphate rock. Environ. Pollut. (Barking, Essex: 1987) 131: 435–444.

Casas, A.M. and E.A. Crecelius. 1994. Relationship between acid volatile sulfide and the toxicity of zinc, lead, and copper in marine-sediments. Environ. Toxicol.Chem. 13: 529–536.

Chaney, R.L. 1993. Zinc phytotoxicity. In: A.D. Robson [ed.]. Zinc in soils and plants. Kluwer Academic Publ, Dordrecht, The Netherlands. pp. 135–150.

Cigala, R.M., F. Crea, C.D. Stefano, G. Lando, D. Milea and S. Sammartano. 2010. Electrochemical Study on the Stability of Phytate Complexes with Cu^{2+}, Pb^{2+}, Zn^{2+}, and Ni^{2+}: A Comparison of Different Techniques†. J. Chem. Eng. Data. 55: 4757–4767.

Conklin, A.R. 2005. Introduction to Soil Chemistry: Analysis and Instrumentation John Wiley and Sons, Inc., Hoboken.

Cotter-Howells, J., J.M. Charnock, C. Winters, P. Kille, J.C. Fry and A.J. Morgan. 2005. Metal compartmentation and speciation in a soil sentinel: The earthworm, Dendrodrilus rubidus. Environ. Sci. Technol. 39: 7731–7740.

Cousins, R. Zinc. 2006. In: B. BA and R. RM [eds.]. Present Knowledge in Nutrition. ILSI Press, Washington DC. pp. 445–457.

D'Amore, J.J., S.R. Al-Abed, K.G. Scheckel and J.A. Ryan. 2005. Methods for speciation of metals in soils: A review. J. Environ. Qual. 34: 1707–1745.

Degryse, F., A. Voegelin, O. Jacquat, R. Kretzschmar and E. Smolders. 2011. Characterization of zinc in contaminated soils: complementary insights from isotopic exchange, batch extractions and XAFS spectroscopy. Eur. J. Soil Sci. 62: 318–330.

Donner, E., D.L. Howard, M.D. de Jonge, D. Paterson, M.H. Cheah, R. Naidu and E. Lombi. 2011. X-ray Absorption and Micro X-ray Fluorescence Spectroscopy Investigation of Copper and Zinc Speciation in Biosolids. Environ. Sci. Technol. 45: 7249–7257.

Elzinga, E.J. and R.J. Reeder. 2002. X-ray absorption spectroscopy study of Cu^{2+} and Zn^{2+} adsorption complexes at the calcite surface: Implications for site-specific metal incorporation preferences during calcite crystal growth. Geochim. Cosmochim. Ac. 66: 3943–3954.

Elzinga, E.J., J.J.M. VanGrinsven and F.A. Swartjes. 1999. General purpose Freundlich isotherms for cadmium, copper, and zinc in soils. Euro. J. Soil Sci. 50: 139–149.

Elzinga, E.J., A.A. Rouff and R.J. Reeder. 2006. The long-term fate of Cu^{2+}, Zn^{2+}, and Pb^{2+} adsorption complexes at the calcite surface: An X-ray absorption spectroscopy study. Geochim. Cosmochim. Ac. 70: 2715–2725.

Fomina, M., J. Charnock, A. Bowen and G.M. Gadd. 2007. X-ray absorption spectroscopy (XAS) of toxic metal mineral transformations by fungi. Environ. Microbiol. 9: 308–321.

Food and Nutrition Board Institute of Medicine. 2001. Zinc. Dietary reference intakes for vitamin A, vitamin K, arsenic, boron, chromium, copper, iodine, iron, manganese, molybdenum, nickel, silicon, vanadium, and zinc. National Academy Press, Washington DC. pp. 442–501.

Ford, R.G. and D.L. Sparks. 2000. The nature of Zn precipitates formed in the presence of pyrophyllite. Environ. Sci. Technol. 34: 2479–2483.

Fosmire, G. 1990. Zinc toxicity. Am. J. Clin.Nutr. 51:225–227.

Gaillard, J.F., S.M. Webb and J.P.G. Quintana. 2001. Quick X-ray absorption spectroscopy for determining metal speciation in environmental samples. J. Synchrotron Radiat. 8: 928–930.

Grafe, M. and D.L. Sparks. 2005. Kinetics of zinc and arsenate co-sorption at the goethite-water interface. Kinetics of zinc and arsenate co-sorption at the goethite-water interface. Geochim.Cosmochim. Ac. 69: 4573–4595.

Granero, S. and J.L. Domingo. 2002. Levels of metals in soils of Alcala de Henares, Spain: Human health risks. Levels of metals in soils of Alcala de Henares, Spain: Human health risks. Environ. Int. 28: 159–164.

Guine, V., L. Spadini, G. Sarret, M. Muris, C. Delolme, J.P. Gaudet and J.M.F. Martins 2006. Zinc sorption to three gram-negative bacteria: Combined titration, modeling, and EXAFS study. Environ. Sci. Technol. 40: 1806–1813.

Ha, J., T.P. Trainor, F. Farges and G.E. Brown, Jr. 2009. Interaction of Aqueous Zn(II) with Hematite Nanoparticles and Microparticles. Part 1. EXAFS Study of Zn(II) Adsorption and Precipitation. Langmuir. 25: 5574–5585.

Haghiri, F. 1974. Plant uptake of cadmium as influenced by cation exchange capacity, organic matter, zinc, and soil temperature. J. Environ. Qual. 3: 180–183.

Harter, R.D. 1983. Effect of soil pH on adsorption of lead, copper, zinc, and nickel. Soil Sci. Soc. Am. J. 47: 47–51.

He, G., G. Pan, M. Zhang and G.A. Waychunas. 2011. Coordination Structure of Adsorbed Zn(II) at Water–TiO2 Interfaces. Environ. Sci. Technol. 45: 1873–1879.

He, Z.L.L., X.E. Yang and P.J. Stoffella. 2005. Trace elements in agroecosystems and impacts on the environment. Trace elements in agroecosystems and impacts on the environment. J. Trace Elem. Med. Biol. 19: 125–140.

Hesterberg, D., D.E. Sayers, W. Zhou, G.M. Plummer and W. Robarge. 1997a. X-ray absorption spectroscopy of lead and zinc speciation in a contaminated groudwater aquifer. Environ. Sci. Technol. 31.

Hesterberg, D., D.E. Sayers, W.Q. Zhou, G.M. Plummer and W.P. Robarge. 1997b. X-ray absorption spectroscopy of lead and zinc speciation in a contaminated groundwater aquifer. Environ. Sci. Technol. 31: 2840–2846.

Hooda, P.S., C.J.K. Henry, T.A. Seyoum, L.D.M. Armstrong and M.B. Fowler. 2002. The potential impact of geophagia on the bioavailability of iron, zinc and calcium in human nutrition. Environ. Geochem. Health. 24: 305–319.

Hostynek, J.J., R.S. Hinz, C.R. Lorence, M. Price and R.H. Guy. 1993. Metals and the skin. Crit. Rev. Toxicol. 23: 171–235.

Ibekwe, A.M., J.S. Angle, R.L. Chaney and P. Vanberkum. 1995. Sewage sludge and heavy-metal effects on nodulation and nitrogen-fixation of legumes. J. Environ. Qual. 24: 1199–1204.

Isaure, M.P., A. Laboudigue, A. Manceau, G. Sarret, C. Tiffreau, P. Trocellier, G. Lamble, J.L. Hazemann and D. Chateigner. 2002. Quantitative Zn speciation in a contaminated dredged sediment by mu-PIXE, mu-SXRF, EXAFS spectroscopy and principal component analysis. Geochim. Cosmochim. Ac. 66: 1549–1567.

Isaure, M.P., A. Manceau, N. Geoffroy, A. Laboudigue, N. Tamura and M.A. Marcus. 2005. Zinc mobility and speciation in soil covered by contaminated dredged sediment using micrometer-scale and bulk-averaging X-ray fluorescence, absorption and diffraction techniques. Geochim. Cosmochim. Ac. 69: 1173–1198.

Jacquat, O., A. Voegelin, A. Villard, M.A. Marcus and R. Kretzschmar. 2008. Formation of Zn-rich phyllosilicate, Zn-layered double hydroxide and hydrozincite in contaminated calcareous soils. Geochim. Cosmochim. Ac. 72: 5037–5054.

Jacquat, O., A. Voegelin, F. Juillot and R. Kretzschmar. 2009a. Changes in Zn speciation during soil formation from Zn-rich limestones. Geochim. Cosmochim. Ac. 73: 5554–5571.

Jacquat, O., A. Voegelin and R. Kretzschmar. 2009b. Local coordination of Zn in hydroxy-interlayered minerals and implications for Zn retention in soils. Geochim. Cosmochim. Ac. 73: 348–363.

Jacquat, O., A. Voegelin and R. Kretzschmar. 2009c. Soil properties controlling Zn speciation and fractionation in contaminated soils. Geochim. Cosmochim. Ac. 73: 5256–5272.

Jacquat, O., C. Rambeau, A. Voegelin, N. Efimenko, A. Villard, K.B. Foellmi and R. Kretzchmar. 2011. Origin of high Zn contents in Jurassic limestone of the Jura mountain range and the Burgundy: evidence from Zn speciation and distribution. Swiss Journal of Geosciences. 104(3): 409–424.

Juillot, F., G. Morin, P. Ildefonse, T.P. Trainor, M. Benedetti, L. Galoisy, G. Calas and G.E. Brown. 2003. Occurrence of Zn/Al hydrotalcite in smelter-impacted soils from northern France: Evidence from EXAFS spectroscopy and chemical extractions. Am. Mineral. 88: 509–526.

Juillot, F., G. Morin, P. Ildefonse, G. Calas and G.E. Brown. 2006. EXAFS signature of structural Zn at trace levels in natural and synthetic trioctahedral 2:1 phyllosilicates. Am. Mineral. 91: 1432–1441.

Karlsson, T. and U. Skyllberg. 2007. Complexation of zinc in organic soils - EXAFS evidence for sulfur associations. Environ. Sci. Technol. 41: 119–124.

Kaur, N., M. Grafe, B. Singh and B. Kennedy. 2009. Simultaneous incorporation of Cr, Zn, Cd, and Pb in the goethite structure. Clay. Clay Miner. 57: 234–250.

Kelly, S.D., D. Hesterberg and B. Ravel. 2008. Analysis of Soils and Minerals Using X-ray Absorption Spectroscopy. Methods of Soil Analysis, part 5. Mineralogical Methods. SSSA, Madison, WI.

Korthals, G.W., M. Bongers, A. Fokkema, T.A. Dueck and T.M. Lexmond. 2000. Joint toxicity of copper and zinc to a terrestrial nematode community in an acid sandy soil. Ecotoxicology. 9: 219–228.

LaGoy, P. 1987. Estimated soil ingestion rates for use in risk assessment. Risk Anal. 7: 355–359.

Lee, S., P.R. Anderson, G.B. Bunker and C. Karanfil. 2004. EXAFS study of Zn sorption mechanisms on montmorillonite. Environ. Sci. Technol. 38: 5426–5432.

Legros, S., E. Doelsch, A. Masion, J. Rose, D. Borshneck, O. Proux, J.L. Hazemann, H. Saint-Macary and J.Y. Bottero. 2010. Combining Size Fractionation, Scanning Electron Microscopy, and X-ray Absorption Spectroscopy to Probe Zinc Speciation in Pig Slurry. J. Environ. Qual. 39: 531–540.

Li, X., G. Pan, Y. Qin, T. Hu, Z. Wu and Y. Xie. 2004. EXAFS studies on adsorption–desorption reversibility at manganese oxide–water interfaces: II. Reversible adsorption of zinc on δ-MnO2. J. Colloid Inter. Science. 271: 35–40.

Li, Y.M., R.L. Chaney, G. Siebielec and B.A. Kerschner. 2000. Response of four turfgrass cultivars to limestone and biosolids-compost amendment of a zinc and cadmium contaminated soil at Palmerton, Pennsylvania. J. Environ. Qual. 29: 1440–1447.

Lindsay, W.L. 1979. Chemical Equilibria in Soils John Wiley & Sons, New York, NY.

Lindsay, W.L. and W.A. Norvell. 1978. Development of a DTPA soil test for zinc, iron, manganese, and copper. Soil Sci. Soc. Am. J. 42: 421–428.

Lock, K. and C.R. Janssen. 2003.Comparative toxicity of a zinc salt, zinc powder and zinc oxide to Eisenia fetida, Enchytraeus albidus and Folsomia candida. Chemosphere. 53: 851–856.

Lock, K., C.R. Janssen and W.M. de Coen. 2000. Multivariate test designs to assess the influence of zinc and cadmium bioavailability in soils on the toxicity to Enchytraeus albidus. Environ. Toxicol. Chem. 19: 2666–2671.

Long, X.X., X.E. Yang, W.Z. Ni, Z.Q. Ye, Z.L. He, D.V. Calvert and J.P. Stoffella. 2003. Assessing zinc thresholds for phytotoxicity and potential dietary toxicity in selected vegetable crops. Commun. Soil Sci. Plan. 34: 1421–1434.

Lothenbach, B., G. Furrer, H. Scharli and R. Schulin. 1999. Immobilization of zinc and cadmium by montmorillonite compounds: Effects of aging and subsequent acidification. Environ. Sci. Technol. 33: 2945–2952.

Luo, X.S., D.M. Zhou, X.H. Liu and Y.J. Wang. 2006. Solid/solution partitioning and speciation of heavy metals in the contaminated agricultural soils around a copper mine in eastern Nanjing city, China. J. Hazard. Mater. 131: 19–27.

Magnier, A., G. Billon, Y. Louis, W. Baeyens and M. Elskens. 2011. On the lability of dissolved Cu, Pb and Zn in freshwater: Optimization and application to the Deûle (France). Talanta. 86: 91–98.

Maguire, M., J. Slavek, I. Vimpany, F.R. Higginson and W.F. Pickering. 1981. Influence of pH on copper and zinc uptake by soil clays. Influence of pH on copper and zinc uptake by soil clays. Aust. J. Soil Res. 19: 217–229.

Manceau, A., B. Lanson, M.L. Schlegel, J.C. Harge, M. Musso, L. Eybert-Berard, J.L. Hazemann, D. Chateigner and G.M. Lamble. 2000. Quantitative Zn speciation in smelter-contaminated soils by EXAFS spectroscopy. Am. J. Sci. 300: 289–343.

Manceau, A., B. Lanson and V.A. Drits. 2002. Structure of heavy metal sorbed birnessite, part III: Results from powder and polarized extended X-ray absorption fine structure spectroscopy. Geochim. Cosmochim. Ac. 66: 2639–2663.

Manceau, A., N. Tamura, R.S. Celestre, A.A. MacDowell, N. Geoffroy, G. Sposito and H.A. Padmore. 2003. Molecular-scale speciation of Zn and Ni in soil ferromanganese nodules from loess soils of the Mississippi Basin. Environ. Sci. Technol. 37: 75–80.

Manceau, A., C. Tommaseo, S. Rihs, N. Geoffroy, D. Chateigner, M. Schlegel, D. Tisserand, M.A. Marcus, N. Tamura and Z.S. Chen. 2005. Natural speciation of Mn, Ni, and Zn at the micrometer scale in a clayey paddy soil using X-ray fluorescence, absorption, and diffraction. Geochim. Cosmochim. Ac. 69: 4007–4034.

Marinsky, J.A., A. Wolf and K. Bunzl. 1979. The Binding of Trace Amounts of Lead(II), Copper(II), Cadmium(II), Zinc(II) and Calcium (II) to Soil Organic Matter. Talanta. 27: 461–468.

Martinez, C.E., M.B. McBride, M.T. Kandianis, J.M. Duxbury, S.J. Yoon and W.F. Bleam. 2002. Zinc-sulfur and cadmium-sulfur association in metalliferous peats evidence from spectroscopy, distribution coefficients, and phytoavailability. Environ. Sci. Technol. 36: 3683–3689.

Martinez, C.E., K.A. Bazilevskaya and A. Lanzirotti. 2006. Zinc coordination to multiple ligand atoms in organic-rich surface soils. Environ. Sci. Technol. 40: 5688–5695.

McBride, M.B. and J.J. Blasiak. 1979. Zinc and copper solubility as a function of pH in an acid soil. Soil Sci. Soc. Am. J. 43: 866–870.

McGrath, S.P., J.R. Sanders and M.H. Shalaby. 1988. The effects of soil organic-matter levels on soil solution concetrations and extractabilities of manganese, zinc, and copper. Geoderma. 42: 177–188.

McKone, T.E. and J.I. Daniels. 1991. Estimating human exposure through multiple pathways from air, water, and soil. Regul. Toxicol. Pharm. 13: 36–61.

Mench, M.J., A. Manceau, J. Vangronsveld, H. Clijsters and B. Mocquot. 2000. Capacity of soil amendments in lowering the phytoavailability of sludge-borne zinc. Agronomie. 20: 383–397.

Mertens, J., F. Degryse, D. Springael and E. Smolders. 2007. Zinc toxicity to nitrification in soil and soilless culture can be predicted with the same biotic ligand model. Environ. Sci. Technol. 41: 2992–2997.

Metwally, A.I., A.S. Mashhady, A.M. Falatah and M. Reda. 1993. Effect of pH on zinc adsorption and solubility in suspensions of different clays and soils. Z. Pflanzenernähr. Bodenk. 156: 131–135.

Msaky, J.J. and R. Calvet. 1990. Adsorption behavior of copper and zinc in soils: Influence of pH on adsorption characteristics. Soil Sci. 150: 513–522.

Nachtegaal, M. and D.L. Sparks. 2004. Effect of iron oxide coatings on zinc sorption mechanisms at the clay-mineral/water interface. J. Colloid Inter. Sci. 276: 13–23.

Nachtegaal, M., M.A. Marcus, J.E. Sonke, J. Vangronsveld, K.J.T. Livi, D. Van der Lelie and D.L. Sparks. 2005. Effects of *in situ* remediation on the speciation and bioavailability of zinc in a smelter contaminated soil. Geochim. Cosmochim. Ac. 69: 4649–4664.

Nagy, L., S. Yamashita, T. Yamaguchi, P. Sipos, H. Wakita and M. Nomura. 1998. The local structures of Cu(II) and Zn(II) complexes of hyaluronate. J. Inorg Biochem. 72: 49–55.

Nedeltcheva, T., M. Atanassova, J. Dimitrov and L. Stanislavova. 2005. Determination of mobile form contents of Zn, Cd, Pb and Cu in soil extracts by combined stripping voltammetry. Anal. Chim. Ac. 528: 143–146.

Newville, M. 2004. Fundementals of XAFS, Consortium for Advanced Raduiation Sources, University of Chicago, Chicago, IL.

Nohynek, G.J., J. Lademann, C. Ribaud and M.S. Roberts. 2007. Grey goo on the skin? Nanotechnology, cosmetic and sunscreen safety. Crit. Rev. Toxicol. 37: 251–277.

Noonan, C.W., S.J. Kathman, S.M. Sarasua and M.C. White. 2003. Influence of environmental zinc on the association between environmental and biological measures of lead in children. J. Expo. Anal. Environ. Epid. 13: 318–323.

Nriagu, J.O. 1980. Zinc in the Environment. Wiley, New York.

O'Day, P.A., S.A. Carroll and G.A. Waychunas. 1998. Rock-water interactions controlling zinc, cadmium, and lead concentrations in surface waters and sediments, US Tri-State Mining District. 1. Molecular identification using X-ray absorption spectroscopy. Environ. Sci. Technol. 32: 943–955.

Olsen, K.B., J. Wang, R. Setiadji and J. Lu. 1994. Field Screening of Chromium, Cadmium, Zinc, Copper, and Lead in Sediments by Stripping Analysis. Environ. Sci.Technol. 28: 2074–2079.

Pan, G., Y.W. Qin, X.L. Li, T.D. Hu, Z.Y. Wu and Y.N. Xie. 2004. EXAFS studies on adsorption-desorption reversibility at manganese oxides-water interfaces I. Irreversible adsorption of zinc onto manganite (γ-MnOOH). J. Colloid Inter. Sci. 271: 28–34.

Panfili, F.R., A. Manceau, G. Sarret, L. Spadini, T. Kirpichtchikova, V. Bert, A. Laboudigue, M.A. Marcus, N. Ahamdach and M.F. Libert. 2005. The effect of phytostabilization on Zn speciation in a dredged contaminated sediment using scanning electron microscopy, X-ray fluorescence, EXAFS spectroscopy, and principal components analysis. Geochim. Cosmochim. Ac. 69: 2265–2284.

Pearson, M.S., K. Maenpaa, G.M. Pierzynski and M.J. Lydy. 2000. Effects of soil amendments on the bioavailability of lead, zinc, and cadmium to earthworms. J. Environ. Qual. 29: 1611–1617.

Peltier, E.F., S.M. Webb and J.F. Gaillard. 2003. Zinc and lead sequestration in an impacted wetland system. Zinc and lead sequestration in an impacted wetland system. Adv. Environ. Res. 8: 103–112.

Provoost, J., C. Cornelis and F. Swartjes. 2006. Comparison of Soil Clean-up Standards for Trace Elements Between Countries: Why do they differ? J. Soil Sediment. 6: 173–181.

Pruvot, C., F. Douay, F. Herve and C. Waterlot. 2006. Heavy metals in soil, crops and grass as a source of human exposure in the former mining areas. J. Soil Sediment 6: 215–220.

Puls, R.W. and H.L. Bohn. 1988. Sorption of cadmium, nickel and zinc by kaolinite and montmorillonite suspensions. Soil Sci. Soc. Am. J. 52: 1289–1292.

Ravel, B. and M. Newville. 2005. ATHENA, ARTEMIS, HEPHAESTUS: data analysis for X-ray absorption spectroscopy using IFEFFIT. ATHENA, ARTEMIS, HEPHAESTUS: data analysis for X-ray absorption spectroscopy using IFEFFIT. J. Synchrotron Rad. 12: 537–541.

Renella, G., A.L.R. Ortigoza, L. Landi and P. Nannipieri. 2003. Additive effects of copper and zinc on cadmium toxicity on phosphatase activities and ATP content of soil as estimated by the ecological dose (ED50). Soil Biol. Biochem. 35: 1203–1210.

Roberts, D.R., A.C. Scheinost and D.L. Sparks. 2002. Zinc speciation in a smelter-contaminated soil profile using bulk and microspectroscopic techniques. Environ. Sci. Technol. 36: 1742–1750.

Roberts, D.R., R.G. Ford and D.L. Sparks. 2003. Kinetics and mechanisms of Zn complexation on metal oxides using EXAFS spectroscopy. J. Colloid Inter. Sci. 263: 364–376.

Rodda, D.P., B.B. Johnson and J.D. Wells. 1993. The Effect of Temperature and pH on the Adsorption of Copper(II), Lead(II), and Zinc(II) onto Goethite. J. Colloid. Inter. Sci. 161: 57–62.

Rohrs, L.C. 1957. Metal-Fume Fever from Inhaling Zinc Oxide. Metal-Fume Fever from Inhaling Zinc Oxide. AMA Arch. Intern. Med. 100: 44–49.

Roussel, H., C. Waterlot, A. Pelfrene, C. Pruvot, M. Mazzuca and F. Douay. 2010. Cd, Pb and Zn Oral Bioaccessibility of Urban Soils Contaminated in the Past by Atmospheric Emissions from Two Lead and Zinc Smelters. Arch. Environ.Con.Toxicol. 58: 945–954.

Ruyters, S., J. Mertens, D. Springael and E. Smolders. 2010. Stimulated activity of the soil nitrifying community accelerates community adaptation to Zn stress. Soil Biol. Biochem. 42: 766–772.

Sanna, G., M.I. Pilo, P.C. Piu, A. Tapparo and R. Seeber. 2000. Determination of heavy metals in honey by anodic stripping voltammetry at microelectrodes. Anal. Chim. Ac. 415: 165–173.

Sarret, G., A. Manceau, J.L. Hazemann, A. Gomez and M. Mench. 1997. EXAFS study of the nature of zinc complexation sites in humic substances as a function of Zn concentration. J. Phys. IV 7: 799–802.

Sarret, G., A. Manceau, L. Spadini, J.C. Roux, J.L. Hazemann, Y. Soldo, L. Eybert-Berard and J.J. Menthonnex. 1998. Structural determination of Zn and Pb binding sites in Penicillium chrysogenum cell walls by EXAFS spectroscopy. Enviro. Sci. Technol. 32: 1648–1655.

Saxe, J.K., C.A. Impellitteri, W. Peijnenburg and H.E. Allen. 2001. Novel model describing trace metal concentrations in the earthworm, Eisenia andrei. Environ. Sci. Technol. 35: 4522–4529.

Scheckel, K.G., R.L. Chaney, N.T. Basta and J.A. Ryan. 2009. Advances in Assessing Bioavailability of Metal(Loid)s in Contaminated Soils. In: D.L. Sparks [ed.]. Advances in Agronomy. Academic Press, Burlington. pp. 1–52.

Scheinost, A.C. and D.L. Sparks. 2000. Formation of Layered Single- and Double-Metal Hydroxide Precipitates at the Mineral/Water Interface: A Multiple-Scattering XAFS Analysis. J. Colloid Inter. Sci. 223: 167–178.

Scheinost, A.C., R. Kretzschmar, S. Pfister and D.R. Roberts. 2002. Combining selective sequential extractions, x-ray absorption spectroscopy, and principal component analysis for quantitative zinc speciation in soil. Environ. Sci. Technol. 36: 5021–5028.

Scheinost, A.C., A. Rossberg, M. Marcus, S. Pfister and R. Kretzschmar. 2005. Quantitative zinc speciation in soil with XAFS spectroscopy: evaluation of iterative transformation factor analysis. Physica Scripta (T115): 1038.

Schlegel, M.L. and A. Manceau. 2006. Evidence for the nucleation and epitaxial growth of Zn phyllosilicate on montmorillonite. Geochim. Cosmochim. Ac. 70: 901–917.

Schlegel, M.L., A. Manceau, L. Charlet, D. Chateigner and J.L. Hazemann. 2001. Sorption of metal ions on clay minerals. III. Nucleation and epitaxial growth of Zn phyllosilicate on the edges of hectorite. Geochim. Cosmochim. Ac. 65: 4155–4170.

Schuwirth, N., A. Voegelin, R. Kretzschmar and T. Hofmann. 2007. Vertical distribution and speciation of trace metals in weathering flotation residues of a zinc/lead sulfide mine. J. Environ. Qual. 36: 61–69.

Sedman, R.M. and R.J. Mahmood. 1994. Soil ingestion by children and adults reconsidered using the results of recent tracer studies. J. Air Waste Manage. Association. 44: 141–144.

Shuman, L.M. 1988. Effect of organic-matter on the distribution of manganese, copper, iron, and zinc in soil fractions. Soil Sci. 146: 192–198.

Shuman, L.M. 1999. Organic waste amendments effect on zinc fractions of two soils. J. Environ. Qual. 28: 1442–1447.

Smit, C.E. and C.A.M. VanGestel. 1996. Comparison of the toxicity of zinc for the springtail Folsomia candida in artificially contaminated and polluted field soils. App. Soil Ecol. 3: 127–136.

Smit, C.E. and C.A.M. Van Gestel. 1998. Effects of soil type, prepercolation, and ageing on bioaccumulation and toxicity of zinc for the springtail Folsomia candida. Environ. Toxicol. Chem. 17: 1132–1141.

Smolders, E., S.P. McGrath, E.Lombi, C.C. Karman, R. Bernhard, D. Cools, Van Den K. Brande, B.Van Os and N. Walrave. 2003. Comparison of toxicity of zinc for soil microbial processes between laboratory-contamined and polluted field soils. Environ. Toxicol. Chem. 22: 2592–2598.

Smolders, E., J. Buekers, I. Oliver and M.J. McLaughlin. 2004. Soil properties affecting toxicity of zinc to soil microbial properties in laboratory-spiked and field-contaminated soils. Environ. Toxicol. Chem. 23: 2633–2640.

Smolders, E., K. Oorts, P. van Sprang, I. Schoeters, C.R. Janssen, S.P. McGrath and M.J. McLaughlin. 2009. Toxicity of trace metals in soil as affected by soil type and aging after contamination: using calibrated bioavailability models to set ecological soil standards. Environ. Toxicol. Chem. 28: 1633–1642.

Sneddon, I.R., M. Orueetxebarria, M.E. Hodson, P.F. Schofield and E. Valsami-Jones. 2008. Field trial using bone meal amendments to remediate mine waste derived soil contaminated with zinc, lead and cadmium. App. Geochem. 23: 2414–2424.

Solomons, N. and R. Jacob. 1981. Studies on the bioavailability of zinc in humans: effects of heme and nonheme iron on the absorption of zinc. Am. J. Clin. Nutr. 34: 475–482.

Sparks, D.L. 1999. Soil Physical Chemistry CRC Press, Boca Raton.

Sparks, D.L. 2003. Environmental Soil Chemistry. 2nd edn. Academic Press, Amsterdam.

Sposito, G. 1989. The Chemistry of Soils Oxford University Press, New York.

Spurgeon, D.J. and S.P. Hopkin. 1996. Effects of variations of the organic matter content and pH of soils on the availability and toxicity of zinc to the earthworm Eisenia fetida. Pedobiologia. 40: 80–96.

Spurgeon, D.J., S.P. Hopkin and D.T. Jones. 1994. Effects of cadmium, copper, lead, and zinc on growth, reproduction, and survial, of the eathworm Eisenia-Foetida (Savigny) - Assessing the environmental-impact of point-source metal contamination in terrestrial ecosystems. Environ. Pollut. 84: 123–130.

Stahl, R.S. and B.R. James. 1991. Zinc sorption by manganese-oxide-coated sand as a function of pH. Soil Sci. Soc. Am. J. 55: 1291–1294.

Stanek, E.J. and E.J. Calabrese. 1995. Daily estimates of soil ingestion in children. Environ. Health Persp. 103: 276–285.

Stephan, C.H., F. Courchesne, W.H. Hendershot, S.P. McGrath, A.M. Chaudri, V. Sappin-Didier and S. Sauvé. 2008. Speciation of zinc in contaminated soils. Environ. Pollut. 155: 208–216.

Stevens, D.P., M.J. McLaughlin and T. Heinrich. 2003. Determining toxicity of lead and zinc runoff in soils: Salinity effects on metal partitioning and on phytotoxicity. Environ. Toxicol. Chem. 22: 3017–3024.

Stumm, W. and J.J. Morgan. 1996. Aquatic Chemistry Chemical Equilibria and Rates in Natural Waters. 3rd edn. John Wiley & Sons Inc., New York.

Sutton, S.R., P.M. Bertsch, M. Newville, M. Rivers, A. Lanzirotti and P. Eng. 2002. Microfluorescence and Microtomography Analyses of Heterogeneous Earth and Environmental Materials. Rev. Mineral. Geochem. 49: 429–483.

Teo, B.K. 1986. EXAFS: Basic Principals and Data analysis Springer-Verlag, Berlin.

Terzano, R., M. Spagnuolo, B. Vekemans, W. De Nolf, K. Janssens, G. Falkenberg, S. Flore and P. Ruggiero. 2007. Assessing the origin and fate of Cr, Ni, Cu, Zn, Ph, and V in industrial polluted soil by combined microspectroscopic techniques and bulk extraction methods. Environ. Sci. Technol. 41: 6762–6769.

Toner, B., A. Manceau, M.A. Marcus, D.B. Millet and G. Sposito. 2005. Zinc sorption by a bacterial biofilm. Enviro. Sci. Technol. 39: 8288–8294.

Trainor, T.P., G.E. Brown and G.A. Parks. 2000. Adsorption and precipitation of aqueous Zn(II) on alumina powders. J. Colloid Interf. Sci. 231: 359–372.

Trivedi, P., L. Axe and T.A. Tyson. 2001a. An analysis of zinc sorption to amorphous versus crystalline iron oxides using XAS. J. Colloid Interf. Sci. 244: 230–238.

Trivedi, P., L. Axe and T.A. Tyson. 2001b. XAS Studies of Ni and Zn Sorbed to Hydrous Manganese Oxide. Environ. Sci. Technol. 35: 4515–4521.

USEPA. 1997. Exposure Factors Handbook In: U. S. E. P. Agency [ed.].

USEPA. 2011a. National Priorities List (NPL).

USEPA. 2011b. Regional Screening Levels (RSL) for Chemical Contaminants at Superfund Sites.

Van Beelen, P. and P. Doelman. 1997. Significance and application of microbial toxicity tests in assessing ecotoxicological risks of contaminants in soil and sediment. Chemosphere 34: 455–499.

Van Damme, A., F. Degryse, E. Smolders, G. Sarret, J. Dewit, R. Swennen and A. Manceau. 2010. Zinc speciation in mining and smelter contaminated overbank sediments by EXAFS spectroscopy. Geochim. Cosmochim. Ac. 74: 3707–3720.

Vanwijnen, J.H., P. Clausing and B. Brunekreef. 1990. Estimated soil ingestion by children. Estimated soil ingestion by children. Environ. Res. 51: 147–162.

Velasquez, I.B., G.S. Jacinto and F.S. Valera. 2002. The speciation of dissolved copper, cadmium and zinc in Manila Bay, Philippines. Mar. Pollut. Bull. 45: 210–217.

Vespa, M., M.Lanson and A. Manceau. 2010. Natural Attenuation of Zinc Pollution in Smelter-Affected Soil. Environ. Sci. Technol. 44: 7814–7820.

Vicente-Martorell, J.J., M.D. Galindo-Riaño, M. García-Vargas and M.D. Granado-Castro. 2009. Bioavailability of heavy metals monitoring water, sediments and fish species from a polluted estuary. J. Hazard. Mat. 162: 823–836.

Voegelin, A. and R. Kretzschmar. 2005. Formation and dissolution of single and mixed Zn and Ni precipitates in soil: Evidence from column experiments and extended X-ray absorption fine structure spectroscopy. Environ. Sci. Technol. 39: 5311–5318.

Voegelin, A., A.C. Scheinost, K. Buhlmann, K. Barmettler and R. Kretzschmar. 2002. Slow formation and dissolution of Zn precipitates in soil—A combined column-transport and XAFS study. Environ. Sci. Technol. 36: 3749–3754.

Voegelin, A., K. Barmettler and R. Kretzschmar. 2003. Heavy metal release from contaminated soils: Comparison of column leaching and batch extraction results. J. Environ. Qual. 32: 865–875.

Voegelin, A., S. Pfister, A.C. Scheinost, M.A. Marcus and R. Kretzschmar. 2005. Changes in zinc speciation in field soil after contamination with zinc oxide. Environ. Sci. Technol. 39: 6616–6623.

Voegelin, A., O. Jacquat, S. Pfister, K. Barmettler, A.C. Scheinost and R. Kretzschmar. 2011. Time-Dependent Changes of Zinc Speciation in Four Soils Contaminated with Zincite or Sphalerite. Environ. Sci. Technol. 45: 255–261.

Wang, W.X., I. Stupakoff and N.S. Fisher. 1999. Bioavailability of dissolved and sediment-bound metals to a marine deposit-feeding polychaete. Mar. Ecol-Prog. Ser. 178: 281–293.

Waychunas, G.A., C.C. Fuller and J.A. Davis. 2002. Surface complexation and precipitate geometry for aqueous Zn(II) sorption on ferrihydrite I: X-ray absorption extended fine structure spectroscopy analysis. Geochim. Cosmochim. Ac. 66: 1119–1137.

Wilhelm, M., I. Lombeck and F.K. Ohnesorge. 1994. Cadmium, copper, lead, and zinc concentrations in hair and toenails of young-children and family members - A followup study. Sci. Total Environ. 141: 275–280.

Wilkin, R. and R. Ford. 2006. Arsenic solid-phase partitioning in reducing sediments of a contaminated wetland. Chem. Geol. 228: 156–174.

Williams, A.G.B., K.G. Scheckel, G. McDermott, D. Gratson, D. Neptune and J.A. Ryan. 2011. Speciation and bioavailability of zinc in amended sediments. Chem. Spec. Bioavailab. 23: 143–154.

Zachara, J.M. and J. Westall. 1999. Chemical Modeling of Ion Adsorption in Soils. *In:* D.L. Sparks [ed.]. Soil Physical Chemistry. CRC Press Boca Raton, FL (in press).

Zhang, H. 2004. *In situ* speciation of Ni and Zn in freshwaters: comparison between DGT measurements and speciation models.Environ. Sci. Technol. 38: 1421–1427.

Zvyagin, A.V., X. Zhao, A. Gierden, W. Sanchez, J.A. Ross and M.S. Roberts. 2008. Imaging of zinc oxide nanoparticle penetration in human skin *in vitro* and *in vivo*. J. Biomed. Opt. 13.

Zwonitzer, J.C., G.M. Pierzynski and G.M. Hettiarachchi. 2003. Effects of phosphorus additions on lead, cadmium, and zinc bioavailabilities in a metal-contaminated soil. Water Air Soil Pollut. 143: 193–209.

CHAPTER

12

Speciation Analysis of Tin in Environmental Samples

Valderi Luiz Dressler,[a],* Clarissa Marques Moreira dos Santos,[b] Fabiane Goldschmidt Antes,[c] Erico Marlon de Moraes Flores[d] and Dirce Pozebon[e]

Introduction

Tin is found naturally in divalent [Sn(II)] and tetravalent [Sn(IV)] states. While Sn(II) exists only in positive form, Sn(IV) has amphoteric properties and exists in both positive and negative forms. There are organic and inorganic compounds of tin, but only the organic compounds of the element have been widely investigated owing to their biocide properties and industrial application. The spectrum of organotin tin compounds (OTCs) is broad but very few of them exist in divalent state, having little importance (Rosenberg 2005).

The first syntheses of OTCs were carried out independently by Frankland in 1849 and by Löwig in 1852. Nevertheless, it was only in the 1940s that OTCs attracted major interest because the polymer industry

Federal University of Santa Maria, Chemistry Department, 97.105-900 – Santa Maria – RS – Brazil.
[a]Email: vdressler@gmail.com
[b]Email: clafarm_mm@yahoo.com.br
[c]Email: fabigold@gmail.com
[d]Email: emmflores@gmail.com
[e]Email: dircepoz@iq.ufrgs.br
*Corresponding author

discovered the stabilizing effect of polyvinyl chloride (PVC) by certain OTCs (Thoonen et al. 2004). Currently, about 70 percent of the production of mono- and dialkyl-substituted OTCs is used for PVC stabilization. Other important applications of OTCs were discovered since the 1940s and several aliphatic and aromatic OTCs have since been synthesized. The widespread use of OTCs has led to their influx into various ecosystems. Most of the OTCs are of industrial origin though inorganic tin may be biotransformed in the environment and generate other OTCs, mainly methyltin compounds. To sum up there exist about 800 OTCs that are man-made or produced in the environment (Hoch 2001).

Nowadays, OTCs are produced on an industrial scale via the following independent routes: 1) reaction of tin tetrachloride ($SnCl_4$) with Grignard reagent (RMgCl); 2) reaction of $SnCl_4$ with alkyl- or aryl-halide and elemental sodium (Wurtz reaction); 3) reaction of $SnCl_4$ with an aluminium alkyl compound, and 4) direct reaction of elemental Sn with an alkyl halide. Tin is in tetravalent state in all OTCs produced industrially (Gianguzza et al. 2012, Thoonen et al. 2004).

The OTCs are characterized by a central Sn atom covalently bonded to a variety of organic groups (methyl, ethyl, propyl, butyl, octyl, phenyl, etc.) through one or more carbon atoms. In general, OTCs are used as $R_nSnX_{(4-n)}$ (R = alkyl or aryl group; X = halide, nitrate, acetate, hydroxide, etc.; n = 1...4). The properties of OTCs are directly linked to the number (n) and nature of the organic group (R) covalently bound to the tin atom (Gianguzza et al. 2012, Thoonen et al. 2004).

The tin carbon bond (Sn-C) is stable up to 200°C and considered thermally stable under environmental conditions (Gianguzza et al. 2012). However, the degradation of OTCs is promoted by exposure to ultravilolet (UV) radiation, and biological and chemical action that cleavage the Sn-C bond. In general, the degradation of tetrasubstituted OTCs occurs stepwise, where the organic moieties are lost according the following scheme:

$$R_4Sn \rightarrow R_3SnX \rightarrow R_2SnX_2 \rightarrow RSnX_3 \rightarrow SnX_4$$

The degradation of OTCs in aqueous solution is very effective under UV radiation since its energy is sufficiently high (for instance, 300 kJ mol^{-1} at 300 nm) to cleave the Sn-C bond (around 220 kJ mol^{-1}) (Rosenberg 2005). Chemical cleavage of Sn-C bond can easily occur in presence of strong acids, halogens or alkali metals. However, these conditions are rare in the environment. Biodegradation of OTCs in the environment occurs as a result of the activity of various microalgae and bacteria. Nevertheless, biodegradation is limited by the high toxicity of OTCs.

Toxicity of OTCs

Inorganic tin compounds are generally considered nontoxic, contrary to OTCs, which have a complex pattern of toxicity. The OTCs are ubiquitous in all compartments of the environment and should be considered as global pollutants, in a similar way to polychlorinated biphenyls (PCBs), mercury and dioxins. The trisubstituted OTCs are active endocrine disrupters (interfere with the synthesis, secretion, transport, binding, action or elimination of natural hormones in the organism), even at low concentrations and they should be put at the top of the list of major pollutants (Donard et al. 2001).

The number and type of the organic radical bound to Sn influence on the toxicity of OTCs whereas the anion X (halide, nitrate, acetate, etc.) has low influence. For example, triethyltin (TEtT) is considered the most toxic tin compound for mammals while tributyltin (TBT) and triphenyltin (TPhT) exhibit high toxicity toward aquatic living organisms. The increase in chain length of the alkyl group R bound to Sn is accompanied by a decrease of biocidal activity of OTCs, making compounds such as octyltins (OcTs) essentially nontoxic. Thus, this compound is considered as relatively safe for use as PVC stabilizer for food packaging (Rosenberg 2005).

In general, the toxicity of the OTCs is in the order tetra- >tri-> di-> mono-substituted. The toxicity of OTCs for aquatic living organisms may be related to the solubilization degree of OTCs in water. The hydrophobicity and partition of OTCs between water and lipid phase in the organism are considered the main parameters of OTCs bioconcentration (Godoi et al. 2003). The physicochemical conditions of the aqueous medium influence the availability, the cell membrane permeability and consequently, the toxicity of OTCs in aqueous ecosystems (Looser et al. 1998). The effects of OTCs on aquatic organisms are well documented, however little data exists about the accumulation and effects of OTCs in higher trophic vertebrate predators (Rosenberg 2005).

Humans are at the top of the food web and could also be affected by contamination with OTCs. Even so, accumulation of OTCs in humans and respective effects are not well documented. The major routes of exposure are the ingestion of contaminated foodstuff and indirect exposure of OTCs present in household items. Levels of OTCs in human blood of American nationals were reported. The mean concentrations were 8 ng L^{-1} monobutyltin (MBT), 5 ng L^{-1} dibutyltin (DBT) and 8 ng L^{-1} tributyltin (TBT), while the level of total butyltins was up to 101 ng L^{-1}. It was assumed that the presence of OTCs in the human blood was due to exposure of the individuals to OTCs used as stabilizers or as biocides in household items. Much lower concentrations of OTCs were found in blood samples from the Environmental Specimen Bank/Human Specimen in Germany (Rosenberg 2005).

Use of OTCs

Organotin compounds have several industrial applications that are directly related to their chemical formulation. OTCs are mainly used as PVC stabilizers (76 percent of production of OTCs), catalysts (5 percent), antifouling biocides (5 percent) or agricultural biocides (8 percent) (Rosenberg 2005, Donard et al. 2001). Paints having TBT in their formulation when applied on the hulls of vessels provide protection towards organisms such as barnacles, mussels and algae. However, severe effects caused by TBT were observed in marine organisms, such as sterilization and changes in the sexual characteristics (imposex) (Okoro et al. 2011). Owing to the very dangerous environmental impact of OTCs, different countries have adopted a more restrictive policy to reduce their use. In particular, the use of TBT as antifouling agent in paints for ships was banned (AFS convention 2001, Antizar-Ladislao 2008) in many countries and new tin-free antifouling agents were proposed (Omae 2003a,b). Nevertheless, OTCs are still present in the marine environment and mainly originate from paints.

Trisubstituted OTCs, mostly TBT, TPhT and tricyclohexyltin (TcHT) comprises approximately 25 percent of the global OTCs production. Triphenyltin (present in commercial products such as Duter, Mertin and Brestan) has been used as fungicide in cultures of potato, celery, sugar beet, coffee and rice (Fent 1996). Although the dephenylation process is rapid in the environment, monophenyltin (MPhT) and especially diphenyltin (DPhT) are generated, which are also toxic. The fenbutatin oxide (bis[tris(2-methyl-2-phenylpropyl)]tin) has been used as acaricide in the culture of vegetables and fruits, including tomato, cucumber and banana. Tricyclohexyltin has been used as an agricultural pesticide, known commercially as Plictran (Donard et al. 2001).

The mono- and di-substituted OTCs, which exhibit relatively low toxicity, are frequently used as stabilizers for PVC and other plastics stabilizers, including packaging materials used for food (Okoro et al. 2011). Many domestic products (kitchen and toilet paper, sanitary towels and textile products such as socks) contain MBT and DBT or OcT. Complexes of disubstituted OTCs were investigated with the intention of exploiting their therapeutic activity (as anti-inflammatory, anti-microbial, anti-tuberculosis, anti-leukemic and anti-tumoural agents) (Nath et al. 2001, 2005, Basu Baul 2008, Kovala-Demertzi et al. 2009) and as stabilizers in lubricating oils (Piver 1973). The tetrasubstituted OTCs are mostly used as intermediate reagents in the synthesis of other OTCs species. They are, therefore, very often present as impurities in trisubstituted OTCs [e.g., tetrabutyltin (TeBT) in TBT] and may reach up to 10 percent in concentration of the final product. Similarly, trioctyltin (TOcT) is not directly used, but is often present in the commercially available dioctyltin (DOcT) as impurity, and hence can be found in various other products (Donard et al. 2001).

Tin in the Environment

Owing to the fairly high stability of OTCs, which consequently causes a very high accumulation in biota (Hsia and Liu 2003, Hassani et al. 2005, Strand and Jacobsen 2005) they are present in the environment mainly as residuals from previously widespread use and, probably, will still be present for a long time. Trisubstituted OTCs, which have extensively been used in antifouling paint formulations and fungicides in marine waters as well as in freshwater boating activities, contributed to the contamination of coastlines throughout the world. Besides the presence in different compartments of the environment, OTCs are also found in living aquatic organisms (Gibson and Wilson 2003, Murai et al. 2005, Zanon et al. 2009).

The behaviour of OTCs in the environment is complex and their distribution and transformation depend on hydrodynamic and biogeochemical conditions. Nontoxic inorganic tin may generate toxic OTCs by biomethylation in the environment.

Alkylation of tin occurs both in aerobic and anaerobic conditions, through a variety of bacterial substrates. Particularly, sulphate-reducing bacteria produce monomethyltin (MMeT) and dimethyltin (DMeT) from inorganic tin in sediment in anaerobic conditions. Abiotic methylation of tin in the aquatic environment is favoured at low pH and low ionic strength. Humic substances widely present in natural water and in sediments are important agents of methylation. In the case of anoxic sediments, butyltin species can be methylated to fully alkylated methylbutyltins (Gianguzza et al. 2012, Hoch 2001, Vella and Vassallo 2002). A possible mechanism of biomethylation of tin by reaction with methylcobalamin (CH_3-B_{12}) has been proposed. It requires one electron for oxidation of Sn(II) to Sn(III) radical, which can take place in presence of Fe(III). The Sn(III) radical can then react with CH_3-B_{12}[Co(III)] to produce (under conditions of high chloride concentration) CH_3-$SnCl_2$ and reduced cobalamin-Co(II). The biomethylation of tin was observed in a strain of *Pseudomonas* bacteria (Okoro et al. 2011, Hoch 2001).

The transformation of inorganic and organic tin species leads to widespread distribution of OTCs throughout the biogeochemical cycle (Antizar-Ladislao 2008). The atmosphere is the only exception; ethylated, butylated and propylated OTCs has been found in landfill gas (Mitra et al. 2005, Krupp et al. 2011, Vella and Vassallo 2002), nevertheless OTCs are virtually absent in the atmosphere, possibly due to their low stability in the presence of UV radiation from the sun light.

Due to their hydrophobic nature, tri-organotins are mostly associated with particulate matter in aquatic systems. As a consequence, tri-organotins would remain mainly accumulated on particulate matter or at the surface microlayer at the air-water interface (Rosenberg 2005, Antizar-Ladislao 2008).

Bottom sediments act as a reservoir of toxic pollutants where OTCs adsorb mainly onto sediment particles smaller than 63 μm. There, the OTCs are easily available to benthic organisms, in particular detritus feeders. Changes of physicochemical conditions in water and sediments, such as salinity, temperature and composition of organic matter, contribute to release of specific forms of OTCs from the sediment (Hoch and Schwesig 2004). Input of new OTCs into marine sediments may have origin from channels and basins in harbours and shipyards deepening. These conditions help OTCs to release and so they become bioavailable and harmful to organisms inhabiting the water column as well (Rüdel 2003).

Organotin compounds are stepwise dealkylated under UV radiation from sun light. However, those associated to sediments located in deep water can be stable for a long time due the absence of radiation - the UV radiation is strongly absorbed in natural water and does not reach bottom sediments.

Little is known about the occurrence and fate of OTCs in freshwater ecosystems but it is supposed that they follow the same trend as in marine ecosystems. However, there are differences basically that freshwater normally contains higher concentrations of humic substances to which OTCs preferentially interact and which is dependent on the physicochemical proprieties of the system and pH (Rosenberg 2005). The interaction with humic substances affects the amount of OTCs in soluble form and their stability. For example, the stability of phenyl- and cyclohexyl-tins in the environment is in general lower than that of butyltins.

Soil may also be contaminated with OTCs used as biocide in agriculture or from spreading activated sludge over cultivated fields. The stability of OTCs in the soil depends on several factors, whereas the carbon content plays a fundamental role. Photolytic degradation (mainly by UV radiation) of OTCs may take place although it is limited to the top layers of soil (Donard et al. 2001).

Organotin cations in aqueous solution can react with the naturally occurring and/or anthropogenic input components in the medium, leading to formation of other species that may influence the solubility of the OTCs (Inaba et al. 1995).

Natural aquatic systems, as well as wastewaters, generally contain several metal ions and ligands where formation of mixed species of tin is possible. Owing to their fairly high complexing capacity (according to the trend $RSn^{3+} > R_2Sn^{2+} > R_3Sn^+$), organotin cations are able to interact with different classes of organic ligands containing O-, N- and S- donor groups such as carboxilates, amines, aminoacids and peptides, policarboxilates, nucleotides and saccharides (Gianguzza et al. 2012). These ligands are present in algae, bacteria, humic substances and fulvic acids in the environment. The presence of natural organic matter in aquatic ecosystems, largely represented by humic substances and soluble fulvic acids, influences the bioavailability of tin ions and their mobility from water to sediments and vice-versa, and sorption/desorption kinetics of OTCs in sediments and also in soil (Hoch 2001, Staniszewska et al. 2008).

Mixed hydrolytic species of tin are often formed by reaction with organic and inorganic ligands, which depends on the pH of the medium. The formation of simple hydrolytic species prevails in alkaline pH conditions and in the absence of strong ligands. Alkyltin cations show a strong tendency to hydrolyze [equation (1)] in the following order: $RSn_3^+ > R_2Sn_2^+ > R_3Sn^+$ (Gianguzza et al. 2012). This trend of hydrolysis is for all the common species of mono-, di- and tri-alkyltin cations.

$$p\text{RxSn}^{(4-x)+} + q\text{H}_2\text{O} = (\text{RxSn})_p(\text{OH})_q^{(4p-4x-q)} + q\text{H}^+ \qquad \text{Equation (1)}$$

The stability of the species of mono-, di- and tri-alkyltin cations is a function of the stoichiometric coefficients p and q in equation (1). The number of alkyl groups bound to tin is the main factor affecting the hydrolysis, while the type of R has low influence.

The hydroxo complexes of methyl-, ethyl- and propyl-tin(IV) cations are soluble in water, including neutral species, in contrast with the most of metal cations (Gianguzza et al. 2012).

In the environment the OTCs are associated with complex matrixes and are present at low concentrations (at ng L^{-1} or ng g^{-1} levels) (Ebdon et al. 1998), making speciation analysis of OTCs a challenge.

Speciation Analysis of OTCs

Most investigations on Sn in the environment are related to speciation analysis of OTCs. Usually, mono-, di- and tri-OTCs are considered in the speciation analysis of tin in the environment.

Although methods that are based on selective extraction have been developed for the discrimination of OTCs from inorganic tin, most of the current methods for speciation of OTCs are based on the use of hyphenated techniques. Usually, a suitable separation technique is combined with a molecule- or element-specific detector. In this context, chromatographic techniques are commonly associated with molecular mass spectrometry (MS), inductively coupled plasma mass spectrometry (ICP-MS), microwave induced plasma optical emission spectrometry (MIP OES), flame photometric detector (FPD) and pulsed flame photometric detector (PFPD). Despite the use of highly selective and sensitive analytical techniques, the OTCs speciation analysis usually requires multistep procedures. In general, the OTCs speciation analysis involves sample collection, transport and storage, analyte separation from the matrix, derivatization, clean-up, preconcentration, species separation and quantification.

The scheme shown in Fig. 12.1 summarizes the main steps involved in the speciation analysis of OTCs, and production and application of OTCs. A typical gas chromatography-inductively coupled plasma mass spectrometry (GC-ICP-MS) chromatogram of OTCs in agricultural soil where Mertin (TPhT) was applied is shown in Fig. 12.2.

Techniques used for OTCs determination, limits of detection (LODs) and procedures of sample preparation are summarized in Table 12.1. The LODs cited may vary as function of the OTC, sample treatment, type of chromatografic column and detector used.

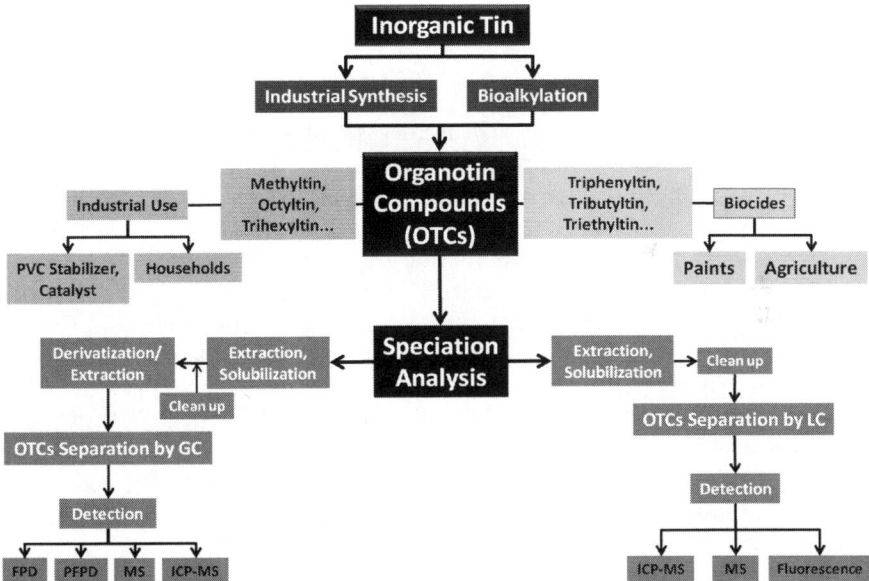

Figure 12.1 Scheme of production, application and speciation analysis of OTCs. PVC: polyvinyl chloride; GC: gas chromatography; LC: liquid chromatography; FPD: flame photometric detector: PFPD: pulsed flame photometric detector; ICP-MS: inductively coupled plasma mass spectrometry, and MS: mass spectrometry.

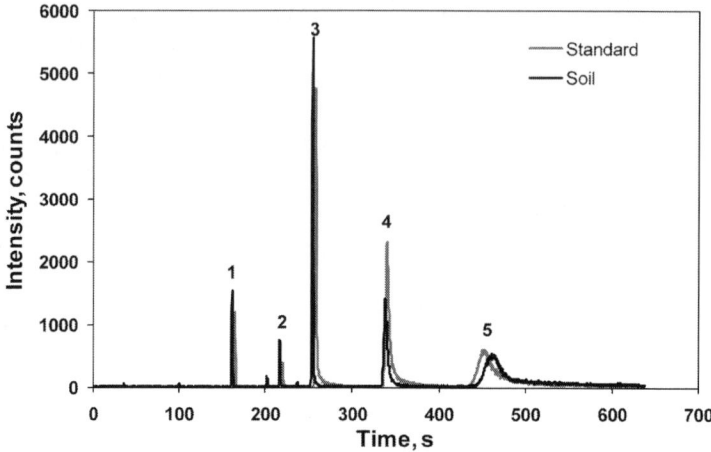

Figure 12.2 Chromatogram obtained for OTCs in soil treated with Mertin (TPhT) by using GC-ICP-MS. Grey line: standard solution (20 µg L⁻¹) containing MPhT, DPhT, and TPhT (as Sn); black line: soil sample. Peak 1, inorganic Sn; peak 2, TPrT (tripropyltin); peak 3, MPhT; peak 4, DPhT; peak 5, TPhT. Smaller peaks represent unknown OTCs.

Sampling and storage

One of the most difficult and critical task of environmental samples collection are the samples representativeness, because they are usually very heterogeneous in their chemical composition and an extensive screening of each sampling site is necessary. Sampling is always the first step of the analysis and its design and implementation has a decisive influence on the final analytical result. Environmental samples are not usually analyzed immediately after sampling and long-term storage can produce a significant alteration of the chemical species owing to volatilization, adsorption, chemical reactions between species, interaction with the container material and degradation due to microbial activity, temperature, pH and UV radiation (Donard et al. 2001).

For OTCs speciation analysis, solid and liquid samples should be stored in glass, polyethylene, polytetrafluorethylene, polycarbonate or aluminium containers. It is recommended to store the samples at low temperature (–20°C), in the dark and in the absence of air to avoid microbiological activities and physicochemical alterations. Under these conditions most of the OTCs are stable for a relatively long period of time. When samples need to be dried, the best way is by freeze-drying. Some OTCs, especially TPhT derivates, can decompose at temperature higher than 50°C (Smedes et al. 2000, Gomez-Ariza et al. 2001, Leroy et al. 1998).

Table 12.1 Speciation of OTCs using different procedures and techniques for detection.

Technique	LOD	Procedure	Reference
LC-ICP-MS	240 pg 1.8–2.8 ng g^{-1}	DBT, TBT, DPhT and TPhT in dried and powdered shellfish were extracted with methanol at pH 4.5 (acetic acid/sodium acetate buffer) under sonication; 65% acetonitrile in water provided the optimum separation by LC; 0.05% triethanolamine as ion-pair speeded up LC separation of the OTCs.	Yu et al. 2010
	2.8–16 pg	Inorganic tin, trimethyltin (TMeT), triethyltin (TEtT), tripropyltin (TPrT), TBT, and TPhT were separated by reversed phase LC; for sediment analysis, sodium 1-pentanesulphonate was used as ion pairing reagent, while 50% methanol and 5% acetic acid were used as organic modifier; HCl + tetrahydrofuran (1 + 11) and 0.1% tropolonebenzene followed by 25% NaCl and tropolonebenzene solution were used for extraction.	Yang et al. 1995
	0.7–2 ng g^{-1}	Pressurized liquid extraction with a ethanolic mixture of 0.5 mol L^{-1} acetic acid and 0.2% tropolone was used for extraction of MBT, DBT, TBT, MPhT, DPhT and TPhT from sediments; the use of 0.075% tropolone and 0.1% triethylamine in a mobile phase of methanol–acetic acid–water (72.5:6:21.5) allowed chromatographic separation of the OTCs.	Chiron et al. 2000
GC-ICP-MS	About 0.01 µg m^{-3}	Landfill gas samples were collected in Tedllar bags; sub-samples were taken and cryotrapped at - 80 °C, before analysis by capillary GC-ICP-MS; landfill gas was injected through a cartridge (filled with NaOH to reduce the trapped water and carbon dioxide) and cryotrapped; the cryotrapped gases were cryofocussed and then measured by GC-ICP-MS; dimethyldiethyltin, trimethylethyltin and propyltrimethyltin were detected.	Mitra et al. 2005
	0.6–20 pg L^{-1}	MBT, DBT, TBT, MPhT, DPhT, TPhT were determined in certified freeze-dried sediment and fish (PACS-2 and NIES 11, respectively): the sediment was extracted with glacial ethanoic acid under mechanical stirring; freeze-dried biological material was humidified with methanol under mechanical stirring before extraction with 0.12 mol L^{-1} HCl in methanol under sonication; TPrT was used as internal standard; the pH of the solutions was adjusted to 4.8 with NaOH; derivatization was carried out using 2% NaBEt$_4$ in presence of sodium ethanoate-ethanoic; a PDMS-SPME fibre was directly immersed into the derivatized sample under vigorous shaking; after analyte sorption, the fibre was directly placed into the injection port of the GC where the compounds were thermally desorbed.	Aguerre et al. 2001a

Table 12.1 contd....

Table 12.1 contd....

Technique	LOD	Procedure	Reference
	Not informed	OTCs (MBT, DBT and TBT) from dry certified sediment (PACS-2, CRM 462 and 646) were extracted with methanol-acetic acid and mechanical shaking; to increase the sensitivity the volumn of the hexane extract was reduce almost to dryness before injection into the GC-ICP-MS system; OTCs were determined by ID-ICP-MS after derivatization with NaBEt$_4$; double-spike ID was used to evaluate possible degradation of OTCs during solid-liquid extraction.	Alonso et al. 2002
GC-ICP OES	13–32 pg 1–42 ng L^{-1}	MBT, DBT, TBT, MPhT, DPhT and TPhT were determined in certified sediment and urban effluent; the sediment was extracted with glacial acetic acid and shaken at room temperature; TPrT was used as internal standard; the water effluent was extracted with of 0.1 mol L^{-1} ethanoate/ethanoic buffer (pH 4.8) and the OTCs were derivatized; by adding 1% NaBEt$_4$; a PDMS-SPME fibre was inserted into the sample (liquid phase) that was subsequently shaked; for classical liquid-liquid extraction, a similar procedure was performed: iso-octane was added to buffered sample solution together with 1% NaBEt$_4$ and the mixture submitted to mechanical shaken.	Aguerre et al. 2001b
	1–42 ng L^{-1}	The procedure was that used for OTCs speciation by GC-ICP-MS (see above).	Aguerre et al. 2001a
LC-ICP OES	200–1700 ng (ion pair) 450–1500 ng (ion exchange)	TMeT, TPhT and TBT were quantified in standard solutions using ion pair and ion exchange; solutions were prepared in methanol with the exception of TPhT that was firstly prepared in THF and then diluted with methanol to 50% THF; a Spherisorb ODS-2 column was used for OTCs separation.	Suyani et al. 1989
GC-PFPD	1.34–432 ng g^{-1}	MBT, DBT, TBT, MPhT, DPhT, TPhT were extracted from dried soil samples by using glacial acetic acid and mechanical shaking; for solid phase microextraction (SPME) extraction in headspace mode, derivatization with 2% NaBEt$_4$ at pH 4.8 (acetic acid/sodium acetate buffer) and shaking were used; after analyte sorption, the fibre was introduced into the GC injector port where the compounds were thermally desorbed; the accuracy was checked by recovery of OTCs spiked in the soil samples.	Zuliani et al. 2006
	1.0–1.4 ng g^{-1}	MPhT, DPhT and TPhT were quantified in soil; methanol + ethylacetate (1:1) and 0.1 mol L^{-1} HCl + 5% NaCl, or methanol + acetic acid (2 + 5) under mechanical shaking were tested for extraction; the extraction procedure was evaluated using spiked soil; TPrT was used as internal standard; the pH of the solution was adjusted to 4.9 (acetic acid/sodium acetate buffer) followed by addition of iso-octane and 2% NaBEt$_4$ for OTCs derivatization.	Antes et al. 2011

GC-FPD	0.3–1.3 ng g⁻¹	The stability of MBT, DBT, TBT, MPhT, DPhT, TPhT were evaluated; certified sediment samples were digested using water-HBr mixture (1 + 1) and extracted with 0.04% tropolone solution in dichloromethane under shaking; water samples were extracted using HBr + 0.08 tropolone in pentane under shaking; the organic phase was dried with anhydrous Na_2SO_4 and washed with hexane followed by addition of dichloromethane; prior to derivatization, the OTCs in water were submitted to solid phase extraction (SPE) and eluted with 1% HBr + 0.1% tropolone in methanol; after derivatization, a florisil column was used to extract the OTCs from sediment followed by elution with pentane; the OTCs were derivatized with 1 mol L⁻¹ pentylmagnesium bromide in diethyl ether at room temperature; certified sediment (CRM 462) and certified mussel (CRM 477) were used for validation of the procedures.	Gomez-Ariza et al. 1999
	4–10 pg	The procedure was that used for OTCs speciation by GC-ICP-MS (see above).	Aguerre et al. 2001(a)
QF AAS	0.4–2 ng	MMeT, DMeT, TMeT, diethyltin (DEtT), TEtT, MBT, DBT, TBT, and PhT in seawater and lake water were evaluated; water samples were transferred to a hydride generator, acidified with acetic acid and purged with He and the hydride trap cooled in liquid N_2; 4% $NaBH_4$ was injected in the reactor; the trapped hydrides were desorbed by heating, rising the temperature up to 80°C in a water bath.	Vernon et al. 1979
	4–7 pg	Extraction of MBT, DBT, TBT in certified sediment (BCR 462) were conducted using glacial acetic acid and shaking; quantification was carried out by standard additions where the extract and standards were mixed with glacial acetic acid; the solution was purged with He followed by addition of 2% $NaBH_4$; the hydrides were purged from the flask by a flow of He and trapped in a glass column packed with Chromosorb W HP SP 2100 immersed in liquid N_2; the trapped hydrides were desorbed by heating heating, rising rising the temperature up to 200°C.	Garcia et al. 1997

Table 12.1 contd....

Table 12.1 contd....

Technique	LOD	Procedure	Reference
GC-QF AAS	1–19 ng g^{-1}	Release of MBT, DBT and TBT from oyster, mussel and salmon tissue was obtained by solubilization of the samples with a mixture of enzymes in aqueous phosphate buffer; freeze-dried samples or wet samples were placed in a capped culture glass tube containing phosphate buffer (pH = 7.5), lipase and protease; the mixture was incubated in a thermostatic bath at 37°C under magnetic stirring; sub-samples of the extracts were directly introduced in the hydride generation reactor and analyzed.	Pannier et al. 1996
	6–25 pg	Harbour sediment (BCR-646) or oyster tissue (BCR-710) were wetted with methanol and equilibrated; TPrT was used as internal standard; HBr was added followed by sonication; after cooling, tropolone was added and the solution shaken and centrifuged the organic phase was transferred to a glass vial and dried by N$_2$; the residue was re-dissolved in toluene and NaBEt$_4$ added for derivatization; the solution was shaken followed by centrifuging; the toluene phase was used for TBT, DBT, MBT, TPhT, DPhT and MPhT determination.	Van et al. 2006b
	95–145 ng kg^{-1}	Non-certified and certified marine sediment (PACS-1 and NIES 12) were spiked with TBT and TPhT; 8-quinolinol, 6 mol L –1 HCl, 10% aqueous NaCl and toluene were added and the mixture mechanically shaken followed by centrifugation; extraction and centrifugation were repeated for the aqueous phase; the combined toluene extracts were dried with anhydrous Na$_2$SO$_4$, evaporated by flowing N$_2$; the extract was diluted with ethanol and passed through a Bondesil column under N$_2$ gas pressure; the adsorbed OTCs were eluted with methanol containing 1 mol L^{-1} HCl; 10% aqueous NaCl was added to the eluent that was extracted with hexane; the combined hexane phases were derivatized with NaBEt$_4$; the mixture was then washed with water; the hexane solution was injected in the GC-AAS system for TBT and TPhT determination.	Narasaki 2002
LC-MS	20–65 pg (SIM mode) 750–2000 pg (SCAN mode)	MBT, DBT, TBT, MPhT, DPhT and TPhT in certified sediment PACS-2 were determined; the sediment was extracted with glacial acetic acid/methanol (2:8) and mechanical shaking at room temperature; the supernatant was filtered and evaporated to dryness using a stream of N$_2$ at room temperature; the residue was redissolved in the mobile phase; an atmospheric pressure chemical ionization (APCI) ion source and Si-C$_{18}$ column were used; separation was accomplished using water/1% trifluoroacetic acid (TFA)/methanol gradient; pH was controlled by the relative fraction of H$_2$O (1% TFA) in the eluent.	Rosenberg et al. 2000

GC-MIP OES	0.5 µg g^{-1}	Mono-octyltin (MOcT), DOcT, DPhT, TPhT, dicyclohexyltin (DcHT) and TcHT were determined in certified (PACS-1) and non certified sediment; butyltin-free sediment was spiked with organotin mixed standard; the whole sediment sample was mixed and air-dried at room temperature; glacial acetic acid, water, NaCl and 0.3% tropolone in toluene were added and the mixture magnetically shaked; alternatively, HCl was used in place of glacial acetic acid; an aliquot of the toluene extract was removed and evaporated to dryness using a N$_2$ stream followed by hexane and ethylmagnesium bromide addition; the coloured substances derived from sediment and the excess of tropolone were removed by using a micro-column packed with silica gel and elution with hexane; the eluate volume was reduced to 1 mL using a N$_2$ stream and analyzed.	Chau et al. 1996 Ceulemans and
	0.15 ng L^{-1}	River water buffered to pH 5 was placed in a reaction vessel together with NaBEt$_4$; the sample was purged with He and the analytes were trapped in a fused-silica liner at –100°C; the trap was electrically heated to 200°C and the trapped TMeT, DMeT, DMeT and MMeT were released into the capillary column.	Adams 1996
GC-EI-MS	0.01–7.6 mg kg^{-1} DMTs and 0.001–0.63 mg kg^{-1} TMTs	Mono-, di-, tri- and tetramethyltin (TeMeT) in urban soil were extracted with 1 mol L^{-1} NH$_4$NO$_3$; speciation was performed by pH-gradient HG and purge-and-trap GC-ICP-MS; HG was commenced at pH 7 (adjusted by a citrate buffer) and gradually decreased to pH 1 whilst the NaBH$_4$ solution was added continuously. The structures of the species were confirmed by using combined GC-EI/MS and ICP-MS; total tin content determination was carried out by microwave digestion using reverse *aqua regia* followed by measurement by ICP-MS.	Duester et al. 2005

AAS: atomic absorption spectrometry; ID: isotope dilution; EI: Electron ionization; LC: liquid chromatography; GC: gas chromatography; ICP-MS: inductively coupled plasma mass spectrometry; ICP OES: inductively coupled plasma optical emission spectrometry; MIP: microwave induced plasma; MS: mass spectrometry; QF: quartz furnace.

A pilot inter-comparison for DBT and certification of a test sediment for TBT content revealed that gamma-irradiation, applied to preserve sediment sample, decreased the TBT content. The concentration of TBT in irradiated marine sediment was 13 percent lower than in non-irradiated sediment (Sturgeon et al. 2003).

Organotin compounds can be preserved in natural water for several months, but it depends on the storage conditions. The stability of OTCs was found to be dependent of the sample characteristic, temperature and pre-treatment applied for preservation. Good stability of OTCs in synthetic solutions properly stored has been reported by several authors. TPhT and butyltin species were found to be stable for 3 mon in HCl-acidified water stored in polyethylene bottles at 4°C in the dark and for at least 20 d in brown glass bottles stored at 25°C. However, OTCs in natural water samples were found to be unstable and filtering prior storage was advisable for turbid water (Gómez-Ariza et al. 1999, Staniszewska et al. 2008). One of the best ways to preserve OTCs present in water is the adsorption on silica-C_{18} cartridges where the adsorbed OTCs are stable even at room temperature.

Butyltins in neutral sweater stored in polycarbonate bottle at 4°C in the dark were stable for 7 mon. On the other hand, it was found that the concentration of TPhT decreased even in the first month of seawater storage. The stability of TPhT did not improve by acidification and storage in glass bottles. Better preservation of TPhT was obtained by its absorption on silica-C_{18} cartridge stored at room temperature, where no changes of TPhT concentration were observed during the first 2 mon (Gómez-Ariza et al. 1999).

The stability of OTCs in environmental samples such as sediment, soil and biological materials is critical. Freezing the sample followed by oven-drying for short periods of storage and freeze-drying for long periods of storage has been recommended to stabilize the OTCs. Freezing followed by oven-drying at 50°C has been shown to be suitable to preserve butyltins in sediments for 4 mon and both freezing and storage at 4°C preserve butyltins for 12 mon (Gómez-Ariza et al. 2000).

Significant transformation of butyltin species was observed in sediment samples submitted to air-drying under infrared radiation or oven-drying at 110°C. However, neither lyophilization nor drying in desiccators affected the stability of butyltin and phenyltin species. Butyltin species were stable for at least 1 yr after freezing, lyophilization and storage in a refrigerator or drying in desiccators, with storage at 4°C. Reduction of concentration of phenyltins was observed only after 3 mon of storage under these conditions, freezing and lyophilization being the most reliable (Gómez-Ariza et al. 1999).

Butyltin species in lyophilized sediment were stable over 5 mon but higher stability was observed for the wet sample stored at −20°C in the dark.

In this case, a reduction of about 14 percent in the TBT content was detected only after 9 mon, followed by an increase in DBT content. After this period of time, reduction of TBT and DBT levels and a corresponding increase of MBT were observed. After 18 mon, a general decrease of all butyltin compounds was found. This behaviour confirmed that TBT degraded by a stepwise debutylation mechanism to DBT, MBT and inorganic tin (Staniszewska et al. 2008).

Other critical aspects concerning OTCs speciation analysis in environmental samples are the low analyte concentration and the possibility of contamination and analyte losses. All materials that come in contact with the sample must be checked in advance to avoid the possible source of contamination or adsorption. In the case of water, the concentration of OTCs is in ng L^{-1} level and, therefore, it is recommended to collect a large volume of the sample for further OTCs preconcentration. However, the use of a preconcentration step or not depends on the technique used for OTCs detection (Donard et al. 2001).

Sample preparation

Sample preparation in speciation analysis usually involves several steps, which are dependent of the physicochemical condition of the sample and attention must be paid to preserve the species during sample handling. Appropriate sample preparation procedure must allow quantitative transfer of OTCs present in the sample into a solution suitable for further separation of the OTCs. In the easiest case, only solubilization has to be applied to bring the solid sample into solution. This can be achieved by acid extraction, alkaline hydrolysis using tetramethylammonium hydroxide, methanolic potassium hydroxide, methanolic sodium hydroxide or enzymatic hydrolysis (Dietz et al. 2007, Ceulemans et al. 2004, Monperrus et al. 2003a, Nagase and Hasebe 1993, Pannier et al. 1996, Pellegrino et al. 2000, Quevauviller and Morabito 2000). However, alkaline digestion or enzymatic hydrolysis is often efficient for biological tissues but rarely for soil and sediments. For such kind of samples, leaching is used widely in speciation analysis. Different solvents, combined or not with complexing agents and application of different forms of energy were proposed for OTCs leaching from solid samples (Staniszewska et al. 2008, Dietz et al. 2007, Leroy et al. 1998).

According to an inter-comparison study (Sturgeon et al. 2003), it was not possible to draw any firm conclusions regarding an optimum methodology for the extraction of TBT from marine sediments; a variety of approaches appeared appropriate, encompassing several solvent mixtures (acetic acid, hydrochloric acid, methanol, potassium hydroxide, tropolone, and hexane). Acceptable data within the uncertainty of the certified value were obtained when either microwave-assisted extraction or extraction with methanolic potassium hydroxide and heating were used.

Depending on the separation technique used and the characteristic of the extract, a clean-up procedure must be carried out on the extract and the OTCs derivatized in order to be adequately separated (Morabito et al. 2000). However, the greater the number of steps involved in the sample preparation, the greater the risk of analyte losses and contamination.

Extraction and Preconcentration. The characteristic of the OTCs, the type of sample and the technique used for separation/detection of the OTCs define the appropriate extraction procedures. The extraction of OTCs from environmental samples is considered critical because OTCs may interact strongly with several concomitants. As a consequence, extraction of OTCs from these samples is prone to (1) under-extraction (non-quantitative analyte recovery), (2) species inter-conversion, and (3) requirement for clean-up if numerous interfering species remain in the extract.

Solid samples are usually solubilized or leached by solvents (water, acid or organic solvent) and mechanical shaking, sonication, and microwave radiation or submitted to supercritical fluid extraction or pressurized liquid extraction (Staniszewska et al. 2008, Dietz et al. 2007, Leroy et al. 1998, Wasik and Ciesielski 2004, Wasik et al. 2007). OTCs are extracted or separated from water samples by using solid phase extraction (SPE) and liquid-liquid extraction (LLE) with non-polar solvents (mainly dichloromethane, hexane and iso-octane) in neutral or acidic conditions (Okoro et al. 2011).

Preconcentration and/or clean-up using LLE, SPE, solid phase microextraction (SPME) and stir bar sorptive extraction (SBSE) are techniques frequently used for OTCs speciation.

Liquid-liquid extraction allows transfer of OTCs to organic solvent (for instance, hexane and toluene) for subsequent determination. Mineral acids and mixtures of them, organic solvents with and without complexing agents (for instance, tropolone and sodium dithiocarbamate) have been used for OTCs extraction from water, sediments and biological samples. Acetic, hydrobromic and hydrochloric acids are among the most used for OTCs leaching from solid materials. Complexing agents are used to assist the extraction of lipophilic mono- and di-substituted OTCs, while organic solvents (for instance, hexane and pentane) are used for lipophilic tetra- and tri-substituted OTCs and complexed organotin species.

To improve the extraction efficiency, reducing the time required for extraction and amount of extracting reagents, application of microwave radiation, ultrasound radiation and supercritical fluid is recommended. Microwave radiation was applied to assist butyltin species extraction by means of a mixture of methanol and acetic acid. Accurate results for TBT (isotopically enriched TBT-[117]Sn was used) in certified sediment and biological materials were obtained by using microwave radiation during 2 min for the analyte extraction (Monperrus et al. 2003b). On the other hand, interconversion of phenyltins were observed when microwave or ultrasound irradiation were used for sample preparation (Van et al. 2008). Despite being more time consuming, the use of mechanical shaking generally leads to accurate results for phenyltin and butyltin, avoiding interconversion of them (Encinar et al. 2002, Van et al. 2008, Antes et al. 2011).

Liquid samples or the extract from solid samples may contain many components, such as sulphur and its derivatives, mineral oil, proteins, lipids and other organic and inorganic impurities. They may interfere in subsequent steps of derivatization and/or OTCs separation by chromatographic techniques, sometimes making it impossible to carry out the OTCs speciation analysis. In general, when chromatography is used for OTCs separation, interfering substances may reduce the separation capacity of the chromatographic column, leading to enhancement of the width of chromatographic peaks, peaks overlapping, appearance of ghost peaks, contamination and shortening of the lifetime of the column (Staniszewska et al. 2008). In this case, the sample solution or the extract must be purified (clean-up). Different procedures can be used for clean-up whereas SPE and purge-and-trap are commonly applied. Deactivated SiO_2, Al_2O_3, $Si-C_{18}$ or Florisil (activated magnesium silicate) are materials used for SPE. They are often used in the form of cartridges or as packed column. The retained OTCs are desorbed with an organic solvent such as hexane, which easily elutes non-polar OTCs derivates and separates them from more polar compounds (Smedes et al. 2000). However, SPE is not sufficiently efficient for sulphur species removal and high content of elemental sulphur and methylated derivates (natural or formed during derivatization) can make the determination of butyltin, particularly DBT, and phenyltin derivatives impossible. Activated copper, oxidation of sulphur compounds with dimethyldioxirane and elimination of the reaction products by adsorption on Al_2O_3, and crystallization of sulphur compounds after pressurized liquid extraction allows efficient separation of sulphur compounds prior to OTCs determination (Staniszewska et al. 2008, Bravo et al. 2004, Wasik et al. 2007).

Solid-phase extraction is very efficient for separation of interfering species, in addition to OTCs preconcentration that can be easily carried out on-line. With the use of SPE it is possible to reduce the volume of organic solvents used and contamination. The SPE consist of passing a liquid sample through a solid adsorbent in a column that retains the analyte by adsorption, complexation, ion-exchange or ion-pair interaction. Subsequently, the analyte is eluted with an appropriate solvent.

Solid phase microextraction, a solvent free technique, has been proposed as an alternative method to LLE. For SPME, the extracting phase is immobilized on a fused silica fibre substrate. The most commonly extracting phases used are polydimethylsiloxane (PDMS), PDMS-divinylbenzene (PDMS-DVB) and carboxen-PDMS. SPME is used in GC analysis where the OTCs adsorbed on the fibre are thermally desorbed directly into the injection port of the chromatograph. In general, SPME considerably improves the LODs of OTCs (Wahlen and Catterick 2003, Staniszewska et al. 2008, Dietz et al. 2007).

Solid phase microextraction can be used in direct mode (Aguerre et al. 2001b), where the fibre is immersed into the sample solution or in headspace mode (Vercauteren et al. 2000, Bravo et al. 2005) where the fibre is placed above the sample solution. The direct mode requires longer extraction time (up to 60 min) than the headspace mode (15–40 min) and is more liable to interferences, in particular in the analysis of biological matrices. The sensitivity usually decreases when environmental samples with a high content of organic matter (such as sewage sludge, sediments and soils) are analyzed. According to several authors (Wahlen and Catterick 2003, Staniszewska et al. 2008, Smedes et al. 2000, Dietz et al. 2007, Zuliani et al. 2006), matrix effects were reduced, memory effects lowered and the lifetime of the fibre increased by using SPME in headspace mode for OTCs speciation in environmental samples.

Stir bar sorptive extraction is also used to extract OTCs from aqueous solutions. A stir bar coated with PDMS is added to the aqueous sample for stirring and analyte extraction. The adsorbed OTCs are thermally desorbed into a gas chromatograph. Owing to the much larger surface area of the PDMS-phase, the extraction efficiency of SBSE is far superior to that of SPME (Wuilloud et al. 2004). The SBSE technique was used for TBT and TPhT speciation analysis in aqueous standard solutions where TPrT and TCyT were applied as internal standards. Detection limits about 0.1 pg L^{-1} were achieved (Vercauteren et al. 2000).

Purge-and-trap is an alternative clean-up technique for OTCs already derivatized to highly volatile species. The principle of purge-and-trap is the adsorption of a volatile species onto a solid sorbent followed by stripping of the species. The species are purged from the solution by the aid of an inert gas stream. Then, the adsorbed species is thermally stripped from the fibre (adsorbent) and directly introduced in the GC column. The advantage of purge-and-trap is the possibility to determine very low OTCs concentrations because the whole extracted species are introduced into the GC column. This clean-up technique can be applied after preliminary isolation and derivatization of OTCs from samples characterized by high content of inorganic compounds and large molecules of organic compounds as humic substances, without purification of the extract before the chromatographic analysis itself.

Due to the low concentration of the OTCs usually found in environmental samples a preconcentration step is quite often necessary. Solvent evaporation is usually used to enrich OTCs in the organic solvent extract. In this case, addition of small amount of iso-octane to the extract is recommended before its evaporation in order to avoid losses of the volatile derivatized OTCs. Purge-and-trap, SBSE, SPE, SPME and LLE are also used for OTCs preconcentration (Aguerre et al. 2001b, Vercauteren et al. 2000, Zachariadis and Rosenberg 2009, Girousi et al. 1997).

Derivatization

Derivatization is necessary to protect species integrity or to convert originally present species into a form suitable for extraction, separation and/or detection. Due to the low volatility of most of OTCs and in order to achieve better resolution, derivatization of OTCs is mandatory before their separation using gas chromatography (Morabito et al. 2000, Dietz et al. 2007). Ionic organotin species have to be extracted from the sample matrix and converted into their fully alkylated and more volatile forms by derivatization, which makes their separation possible. The derivatization usually leads to sharper peaks, better separation and higher sensitivity.

Low yields in derivatization as well as species degradation (especially phenyltins) can seriously affect the accuracy of the results. Independently of the derivatization method applied, the derivatization yield should allow quantitative results in any experimental conditions. However, the lack of commercially available derivatized organotin standards (pentylated, ethylated, etc.) may hinder the systematic quantitative evaluation of the derivatization yields.

The derivatization reactions for OTCs speciation analysis are usually hydride generation (HG) with $NaBH_4$, ethylation with $NaBEt_4$ and alkylation with Grignard reagents (R'MgX).

Hydride generation. Derivatization by means of HG is performed using an aqueous solution of $NaBH_4$ in acid media according to equation **(2)**.

$$R_nSn^{(4-n)+} + NaBH_4 + H^+ \rightarrow R_nSnH_{(4-n)} + H_2 \qquad \text{Equation (2)}$$

R = organic substituent

$n = 1, 2, 3$

The volatile hydrides generated can be separated from the sample matrix by using a stream of an inert gas. They can be retained in a purge-and-trap or in a cryogenic U-tube for subsequent separation and detection. By using HG combined with a U-tube separation device, inorganic tin and OTCs converted to the corresponding volatile hydrides are purged from the solution by an inerter gas stream and trapped in the U-tube, packed with suitable sorbent material maintained in liquid nitrogen. Next, the trap is gradually heated and the hydrides derivatives desorbed according to the respective boiling point. The concentration of the $NaBH_4$, pH and type of acid used of the reaction medium must be selected according to the sample matrix and the species considered (Diaz-Bone and Hitzke 2008). In specific conditions, $NaBH_4$ has sufficient reduction power to produce volatile species of most OTCs.

Although derivatization using HG is less expensive than the other derivatization methods, it has been only used for a few applications in OTCs speciation analysis. One reason is that the generated hydrides of OTCs are easily decomposed, leading to inaccurate and imprecise results. The sample matrix strongly affects the derivatization reaction because the substances present may react with the hydrides of the OTCs and cleave the Sn-H bond (Morabito et al. 2000). This matrix effect is critical for TBT speciation analysis (Cai et al. 1993) but not for methyltins derivatization where good results are obtained when HG is used. Hydridization of phenyltins compounds leads to low yields of derivatization and poor reproducibility and the species produced are not so volatile, making separation and detection more difficult.

When HG is used for derivatization prior to purge-and-trap there is a possibility to transfer water to the chromatographic column, which can damage the column. However, excessive water vapour can be removed from the purge gas by a condenser usually kept at –10 to –20°C, which requires the use of a cryostat or a methanol-water-CO_2 bath (Wasik et al. 1998, Ceulemans and Adams 1996).

Alkylation by Grignard reagents. Alkylation by Grignard reagents was the most frequently used method for derivatization of $R_nSn^{(4-n)+}$, carried out by methylation, ethylation, propylation, butylation, pentylation or hexylation in a suitable solvent, as schematized in equation **(3)** (Tao et al. 1999, Gui-Bin et al. 2000, Sutton et al. 2000). Alkylation by Grignard reagents allows formation of very stable tin derivatives that are suitable for GC separation.

$$\mathbf{R_nSn^{(4-n)+} + (4-n)R'MgX \rightarrow R_nSnR'_{(4-n)} + (4-n)Mg^{2+} + (4-n)X^-}$$ Equation **(3)**

R'= methyl, ethyl, propyl, *n*-butyl, pentyl

R = methyl, *n*-butyl, phenyl

n = 1, 2, 3

In view of the characteristics of the reagents, Grignard derivatization reaction has to be performed in dry organic solvent. Thus, the analyte present in environmental samples must be transferred to the organic solvent before the derivatization reaction. The derivatization reaction usually lasts 5 to 10 min and the excess of the Grignard reagent must be destroyed by acidifying the solution with sulphuric or hydrochloric acid.

In contrast to HG, Grignard derivatization leads to the determination of several species of OTCs (methyltins, butyltins, phenyltins, etc.) in different matrices, with higher derivatization yields and reproducibility. However, this derivatization approach involves additional steps because the mixed tetra-alkyltins produced have to be extracted back, increasing the risk of contamination and transformation of OTCs. It is also sensitive to traces of water present in the sample or sample extract, in addition to losses of the more volatile species such as methyltins and ethyltins.

Several derivatization procedures involving Grignard reagents were evaluated for determination of OTCs in river water and seawater (Tsunoi et al. 2002). Molar response, volatility of the derivatives and derivatization yields were evaluated whereas better derivatization was observed via pentylation. By using pentylation, the derivatization yields were higher but the volatility of derivatives was lower. Conversely, derivatization via methylation led to losses of OTCs species such as DBT and MPhT due to their high volatility.

Ethylation by sodium tetraethylborate. Reaction with $NaBEt_4$ to convert OTCs in more volatile compounds is currently the most frequently used derivatization method. This is attributed to the possibility to carry both derivatization and extraction in only one step. The OTCs derivatization with $NaBEt_4$ is conducted in aqueous solution, followed by extraction of the derivatized species into an appropriate organic solvent (Antes et al. 2011). This one step-derivatization and extraction reduces risks of contamination, losses of volatile OTCs compounds and makes the analysis faster than that involving Grignard reagent.

The derivatization reaction with $NaBEt_4$ is relatively fast and satisfactory yield is obtained within 5 min. The pH of the solution and concentration of $NaBEt_4$ are very important for the derivatization reaction. Usually, the pH of the sample solution is adjusted to 4.9 using acetate/acetic acid buffer and then an appropriate volume of a 2% (m/v) $NaBEt_4$ aqueous solution is added (Van et al. 2008).

The solvent used for extraction of ethylated tin compounds is added just before the addition of $NaBEt_4$ solution. After derivatization, solution phases are usually separated by centrifugation and the organic layer is taken for analysis by GC coupled to a suitable detection technique (Lespes et al. 2009).

The derivatized OTCs can also be extracted by SPME in headspace mode (Millán and Pawliszyn 2000, Carvalho et al. 2007) or by immersing the fibre into the sample solution (Aguerre et al. 2001a, b). The main advantage is that matrix interferences are strongly reduced and the analyte preconcentrated, improving the LOD.

Sodium tetraethylborate has wide applicability in OTCs speciation analysis in environmental matrices. However, despite the good chromatographic separation provided by $NaBEt_4$, this reagent cannot be used for samples containing ethylated Sn compounds and inorganic Sn, since it is not possible to distinguish between ethylated species and inorganic forms of Sn present in the sample; all of them are converted to tetraethyltin (Pinel-Raffaitin et al. 2007, Morabito et al. 2000). Another concern about the use of $NaBEt_4$ for derivatization is the stability of the reagent. It is recommended to prepare the $NaBEt_4$ solution daily (Lespes et al. 2009) or the prepared solution must be kept at low temperatures ($-20°C$) for use within 2 wk.

Techniques for OTCs separation and detection

Most of the analytical methods developed to quantify OTCs require hyphenated techniques, which are the on-line combination of a separation technique with a detection technique suitable for identification and/or quantification of a specific molecule or element. Alternatively, separations can be performed off-line, with species being separated and determined independently. Although a variety of separation techniques are used for OTCs, GC and LC are the most common.

Gas Chromatography. Gas chromatography is mostly used to separate OTCs and methods based on GC hyphenation have being extensively used. This can be attributed to the high resolving power and easy coupling of GC to sensitive and selective detectors such as FPD (Aguerre et al. 2001a), PFPD (Aguerre et al. 2001a), MS (Zachariadis and Rosenberg 2009), MIP OES (Zachariadis and Rosenberg 2009), ICP-MS (Aguerre et al. 2001a) and ICP OES (Aguerre et al. 2001b). The use of ICP-MS for OTCs speciation has increased and nowadays it is the detector of choice due to its excellent sensitivity, selectivity and the possibility of isotopic analysis. ICP-MS enables measurements of OTCs at ng L^{-1} and ng g^{-1} levels in environmental samples.

Although GC is a common analytical technique for several analytes separation, its use in the field of inorganic analysis has been relatively limited. This reflects the fact that most GC conventional detectors are not element specific and transferring the analyte from the end of the GC column to an atom/ion source may be problematic when atomic spectrometric based techniques are used for detection. The eluate from the chromatographic column is usually at an elevated temperature that must be maintained along the way to the atom/ion source. Failures in maintaining the temperature leads to cool spots and condensation of the analyte. A heated transferline is, therefore, mandatory. Ensuring that no cool spots exist in the transferline is relatively difficult. In addition, coupling GC with plasma based instruments may lead metallic transferlines acting like an "antenna" and thus coupling to the radiofrequency of the ICP. This is potentially dangerous and may cause instability of the plasma and instrument failure. Fortunately, several improvements have been made and feasible interfaces are now commercially available (McSheehy et al. 2007, Thermo Fisher Scientific 2009, Castro et al. 2012). Figure 12.3 shows a scheme of four interfaces used for GC hyphenation with ICP-MS or ICP OES.

The interface shown in Fig. 12.3(A) consists of a deactivated fused-silica (1.5 m length x 0.32 mm i.d.) capillary tube that is threaded through a 1 m flexible heated transferline and further through a 10 cm rigid transferline up to the end of the central channel of the ICP torch. The flexible part of the transferline consists of a stainless steel tube that is resistively heated and thermally insulated. Xenon gas can be added to the nebulizer gas (Ar)

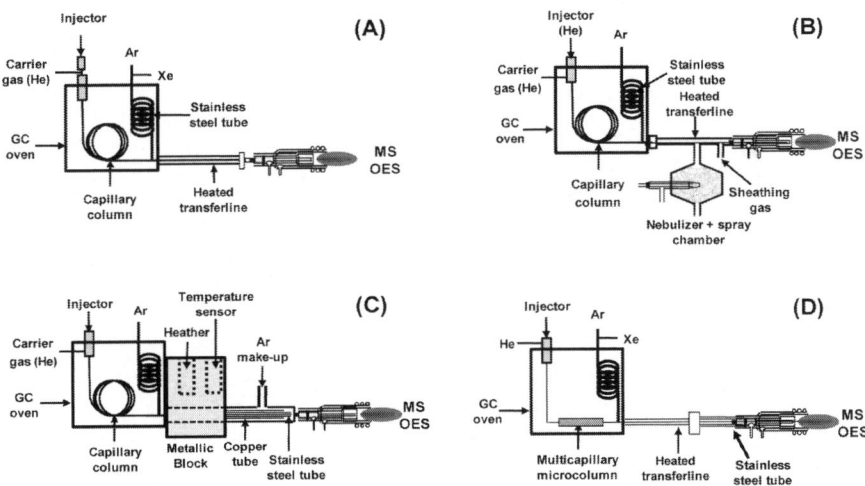

Figure 12.3 Scheme of interfaces for GC hyphenation with ICP-MS or ICP OES. Details of the interfaces are given in the text. (A) Adapted from Aguerre et al. 2001a (Reproduced by permission of The Royal Society of Chemistry); (B) Adapted from Aguerre et al. 2001b (Reproduced by permission of The Royal Society of Chemistry); (C) Adapted from Montes Bayón et al. 1999 (Reproduced by permission of The Royal Society of Chemistry); (D) Adapted with permission from Rodriguez et al. 1999. Copyrigth 1999 American Chemical Society.

via a T-piece. The make up gas (Ar with an admixture of Xe) is preheated by passing through a coil (1 m length, 1.58 mm i.d.) placed inside the chromatographic oven. The make up gas flows between the internal wall of the heated transferline and the external wall of the capillary tube and then merge with the GC carrier gas just before the plasma. The rigid part of the transferline encapsulates an additional heater and a thermocouple to measure the temperature. The heater extended to 5 cm from the end of the capillary. The end of the rigid part is placed inside the ICP torch in place of the conventional injector. No heating element is used in this part, but the stainless steel tube (1.58 mm) remains hot owing to its thermal conductivity (Rodriguez et al. 1999).

Figure 12.3(B) shows an interface that allows the continuous introduction of a nebulized standard solution into the plasma together with the gas flow eluting from the GC transferline. This type of interface allows the use of a continuous internal standard to correct for plasma variation effects. The transferline from GC to ICP consists of a deactivated stainless steel capillary (Silcosteel - 0.28 mm i.d., 0.53 mm o.d., 90 cm length) inserted into a flexible stainless steel tube (1.6 mm o.d., 1.02 mm i.d., 100 cm length). The Silcosteel capillary is connected to the end of the GC capillary column by means of a Silcosteel low dead-volume connector. The stainless steel tube is flushed with a heated Ar make up flow to promote the introduction of the analytes into the ICP. Before entering the transferline, the Ar make up gas is heated by passing through a loop made from stainless steel tubing (300 cm length, 1 mm i.d.) placed in the GC oven. Teflon-coated nichrome wire connected to a Variac, as power supply, is wrapped around the transfer line to allow the heating. The combination of the heated Ar make up gas flow and the heated transferline is suitable for maintaining the temperature about 160°C. The positions of the stainless steel tube in the injector tube of the torch and the Silcosteel capillary in the stainless steel tube are critical in terms of chromatographic performance, peak shape resolution and sensitivity (Aguerre et al. 2001b).

The interface shown in Fig. 12.3(C) consisits of a metallic T-piece connected to an assembly containing an outer copper tubing (8.91 cm length and 0.15 cm i.d.) and an internal stainless steel tubing (8.91 cm length and 0.06 cm i.d.) where the last 9 cm of the chromatographic column are placed and protected against breakage. The copper tubing is heated by a metallic block containing an electric heater and a temperature sensor, both controlled by the GC instrument. Part (1 cm) of the copper tubing passing through the metallic block is inserted into the oven wall of the GC instrument. The chromatographic column exiting the GC instrument reaches the top of the copper tubing and is immobilized there. The copper tubing is fixed to the metallic block by means of a screw drilled in its wall. Argon added as carrier make up gas is introduced through the side-arm of the T-piece and flows through a gap between the copper tubing and the T-piece and externally to the GC efluent. Pre-heating of the Ar make up gas prevents distortion of the less volatile TBT peak. This design leads to a high velocity intermediate sheathing Ar flow to transport the analytes from the GC column to the ICP-MS, avoiding condensation of tin species on the walls of the interface or in the connecting tubing. The other end of the T-piece is connected to a reducing union (6.35 mm x 3.2 mm) where a 1.5 mm i.d. PTFE tubing is directly attached and inserted into the central channel of the ICP torch using a rubber O-ring and standard glass union to obtain a gas tight connection (Montes Bayón et al. 1999).

The interface between GC and ICP-MS shown in Fig. 12.3(D) consists of a deactivated silica capillary tube (1.5 m lengh x 0.32 mm i.d.) that allows the transfer of the analytes from the end of the capillary column to the ICP-MS torch. This capillary is surrounded with a 0.16 mm i.d. stainless-steel tube. Argon used as auxiliary gas, previously heated in the oven, is passed through this tube. The heated argon ensures a sufficient and constant temperature all along the transferline. A high flow of the make up gas allows the analytes to be pushed from the end of the capillary column to the plasma without any loss and condensation. The flexible part of the transferline is also resistively heated at 250°C and thermally insulated. Xenon is added to the argon auxiliary gas at a constant flow rate, which is constantly measured in order to correct instrumental instabilities (Aguerre et al. 2001a).

Liquid chromatography. The use of liquid chromatography for the separation of OTCs has a number of differences in comparison to GC. These include the ease of interfacing the liquid flow from the column to the detector (a heated transferline is not necessary as in the case of using GC) and derivatization is not required to produce volatile OTCs. Thus, the analytical methodology is simplified and risks of contamination from the derivatization reagents eliminated. However, it is not possible to determine a broad range of OTCs differing in alkyl substituent (methyl, ethyl, etc.) and degree of substitution (mono-, di-, etc.) in one single chromatographic run because of differences of solubility of OTCs.

The time for a particular separation using LC is generally longer than that with GC, particularly when a solvent gradient is used. Because of its lower resolution and shortcoming in detection, the implementation of LC for OTCs separation is quite limited when compared to GC.

The separation of OTCs using LC can be performed by ion exchange, reversed phase, normal phase, ion-pair, size-exclusion, micelle and vesicle-mediated or supercritical fluid approaches (Ponce de León et al. 2002, González-Toledo et al. 2000, Ebdon et al. 1998). The detectors used in LC are non-element specific (spectrophotometric, voltametric or amperometric) or element specific (atomic absorption, atomic fluorescence, ICP OES and ICP-MS) (Ebdon et al. 1998, Wahlen and Catterick 2003).

Different strategies have been used for appropriate hyphenation of LC to detectors in order to improve the sensitivity of the overall system (for example, interfacing with HG) or make both techniques compatible (for example, addition of oxygen to the nebulizer gas in ICP-MS to avoid the deposition of carbon on the interface). Figure 12.4 shows the hyphenation of LC with ICP-MS or ICP OES. In general, the interface is simple and consists only of a capillary that connects the outlet of the chromatographic column to the nebulizer. However, when it is possible to generate volatile hydrides from the eluted OTCs, a HG system can be additionally used. In this case, an HG system is introduced between the column and the nebulizer or the HG system can be adapted directly to the ICP torch.

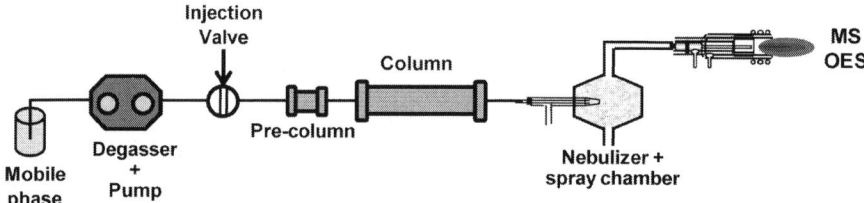

Figure 12.4 Scheme of hyphenation of LC with ICP-MS or ICP OES.

Quantification and Identification of OTCs

Quantification of OTCs is a difficult task due to their instability, low concentration levels and extraction complexity. No single methodology can be recommended for general use. Nowadays, ICP-MS is the technique of choice for OTCs detection and quantification. The most relevant advantage of ICP-MS is the high detection power (in the range of ng L^{-1}), linear range up to nine orders of magnitude and reduced interferences by co-elution of organic species. ICP-MS allows the measurement of isotopic ratios, which is useful in the evaluation of isotopic interferences, and quantification using isotope dilution (ID). In the case of LC-ICP-MS, the sensitivity is limited by the volume of sample that can be injected into the chromatographic column.

Calibration for OTCs quantification by GC-ICP-MS or LC-ICP-MS is usually carried out using tripropyltin as internal standard. Tripropyltin is not naturally present in the environment and is synthesized only for analytical purpose. External calibration is only appropriate when sample matrix has no effect on detection or derivatization. Otherwise, standard addition calibration (González-Toledo et al. 2003) or standard addition calibration with the use of an internal standard (Antes et al. 2011) is recommended. Tin has 10 naturally occurring isotopes, being ^{120}Sn (32.58

percent), ^{119}Sn (8.59 percent), ^{118}Sn (24.22 percent), ^{117}Sn (7.68 percent) and ^{116}Sn (14.54 percent) the most abundant. Detection of tin and respective OTCs by ICP-MS is usually performed by monitoring the signal of ^{120}Sn because this isotope is the most abundant and free of isobaric interference (González-Toledo et al. 2003).

The use of ID improves the accuracy and precision of measurements by ICP-MS. However, the application of ID for speciation analysis is limited by the commercial availability of isotopically enriched species species specific isotope dilution (SSID). Isotope dilution is based on the addition of a precise amount of an isotopically labelled form of the analyte to the sample. When the natural abundances of the analyte in the sample and in the spike are known, the concentration of the analyte in the sample can be calculated if a known amount of the spike is added and equilibrated with the analyte in the sample. If the equilibration of the spike and the analyte is fully achieved, ID allows correction of low derivatization yields, species interconvertion, signal intensity drift and other non-spectral interferences. However, incomplete species extraction may not be overcome. Contamination with inorganic tin and several OTCs were observed in MBT, DBT, TBT, DPhT and TPhT standards (Inagaki et al. 2007). Therefore, before using isotopically enriched OTCs for SSID the standards must be checked in relation to their purity and concentration of each species (Inagaki et al. 2007, Monperrus et al. 2003a,b).

Species-specific isotope dilution is recommended for monitoring and correcting transformation of species throughout the speciation analysis process and to compare different extraction procedures (Van et al. 2006a, Haas et al. 2001, Van et al. 2008, Encinar et al. 2002, Monperrus et al. 2003a). Significant degradation of TBT to DBT (and MBT), has been easily detected with use of enriched isotopic species (Encinar et al. 2002).

Species-specific isotope dilution was applied for OTCs determination in certified marine sediments (PACS-2 from NRCC and CRM 462 from IRMM) by using GC-ICP-MS. For DBT determination, Encinar et al. (2000) used ^{118}Sn-enriched DBT synthesized in their own laboratory. Monperrus et al. (2003a) used three different spiking procedures for TBT determination. 1) Enriched TBTCl (^{117}Sn abundance of 99.97%) was added to dry sediment followed by addition of methanol and subsequent mechanical shaking overnight in the dark. After methanol evaporation under nitrogen, the dried subsample was stirred for homogenization before microwave extraction with acetic acid. In this procedure, the spike and the extraction steps were successive whereas a long equilibration time and recomposition of the initial matrix was used to ensure full equilibration. 2) The dried sample was spiked with a known amount of enriched ^{117}SnTBTCl to which methanol and acetic acid were added and the mixture mechanically shaken overnight in the dark. Then, the mixture was centrifuged and the supernatant taken for analysis.

In this procedure, spiking and extraction were simultaneously carried out without waiting equilibration of the spike. 3) The dry sample was spiked with a known amount of the enriched [117]SnTBTCl followed by addition of methanol and acetic acid and the mixture submitted to microwave extraction. The results obtained showed both good accuracy (in accordance with certified TBT value) and good precision for the three different spiking procedures. The extraction with acetic acid was quantitative for TBT and no degradation of this species was observed during the microwave heating. Equilibration time was found not necessary for the enriched TBT. The equilibration between the spike and the analyte was rapidly achieved when using the microwave extraction.

An inter-comparison study on the determination of TBT in marine sediment was conducted (Sturgeon et al. 2003), involving National Metrology Institutes and "expert" laboratories. Eleven laboratories submitted results for TBT in prepared marine sediment. Two laboratories utilized a standard calibration approach based on a natural abundance TBT standard, whereas the others relied upon SSID for analyte quantification. A species specific [117]Sn-enriched TBT standard was used. A variety of approaches was adopted by the participants, including mechanical shaking, sonication, accelerated solvent extraction, microwave-assisted extraction and heating in combination with Grignard derivatization, ethylation and direct analysis. Detection techniques included GC-ICP-MS, LC-ICP-MS, GC-MS, GC OES and GC-FPD. Recovery of TBT from a control standard (PACS-2) averaged $93.5 \pm 2.4\%$. Satisfactory agreement amongst participants permitted scheduling of a follow-up key comparison for TBT, a pilot inter-comparison for DBT, and certification of test sediment for TBT content. Lower recovery of TBT from PACS-2 was obtained by those laboratories utilizing external calibration. Matrix effects occurred and the added internal standard (TPrT) served to only correct/compensate for volumetric factors. The major sources contributing to overall uncertainty identified by the participants were frequently related to the measurement of the reference-to-spike ratio in the blends. The results obtained were limited to the determination of TBT, and likely DBT.

References

Aguerre, S., G. Lespes, V. Desauziers and M. Potin-Gautier. 2001a. Speciation of organotins in environmental samples by SPME-GC: comparison of four specific detectors: FPD, PFPD, MIP-AES and ICP-MS. J. Anal. At. Spectrom. 16: 263–269.
Aguerre, S., C. Pécheryan, E. Krupp, O.F.X. Donard and M. Pontin-Gautier. 2001b. Optimisation of the hyphenation between solid-phase microextraction, capillary gas chromatography and inductively coupled plasma atomic emission spectrometry for the routine speciation of organotin compounds in the environment. J. Anal. At. Spectrom. 16: 1429–1433.

AFS Convention. 2001. International convention on the control of harmful anti-fouling systems on ships. London, UK. pp. 73–81.

Alonso, J.I.G., J.R. Encinar, P.R. González and A. Sanz-Medel. 2002. Determination of butyltin compounds in environmental samples by isotope-dilution GC-ICP-MS. Anal. Bioanal. Chem. 373: 432–440.

Antes, F.G., E. Krupp, E.M.M. Flores, V.L. Dressler and J. Feldmann. 2011. Speciation and degradation of triphenyltin in typical paddy fields and its uptake into rice plants. Environ. Sci. Technol. 45: 10524–10530.

Antizar-Ladislao, B. 2008. Environmental levels, toxicity and human exposure to tributyltin (TBT) contaminated marine environment. A review. Environ. Int. 34: 292–308.

Basu Baul, T.S. 2008. Antimicrobial activity of organotin(IV) compounds: a review. Appl. Organomet. Chem. 22: 195–204.

Bravo, M., G. Lespes, I. Gregori, H. Pinochet and M. Potin-Gautier. 2004. Identification of sulfur interferences during organotin determination in harbour sediment samples by sodium tetraethylborate ethylation and gas chromatography-pulsed flame photometric detection. J. Chromatogr. A 1046: 217–224.

Bravo, M., G. Lespes, I. Gregori, H. Pinochet and M. Potin-Gautier. 2005. Determination of organotin compounds by headspace solid-phase microextraction–gas chromatography–pulsed flame-photometric detection (HS-SPME-GC-PFPD). Anal. Bioanal. Chem. 383: 1082–1089.

Cai, Y., S. Rapsomamlus and M.O. Andreae. 1993. Determination of butyltin compounds in sediment using gas chromatography-atomic absorption spectrometry: comparison of sodium tetrahydroborate and sodium tetraethylborate derivatization methods. Anal. Chim. Acta. 274: 243–251.

Carvalho, P.N., L.F. Pinto, M.C.P. Basto and M.T.S.D. Vasconcelos. 2007. Headspace solid-phase micro-extraction and gas chromatography-ion trap tandem mass spectrometry method for butyltin analysis in sediments: Optimization and validation. Microchem. J. 87: 147–153.

Castro, J.E. Tessier, O.F.X. Donard and K. Neubauer. 2012. Mercury speciation in biological tissue and sediments by GC/ICP-MS. Perkin Elmer. Application note.

Ceulemans, M. and F.C. Adams. 1996. Integrated sample preparation and speciation analysis for the simultaneous determination of methylated species of tin, lead and mercury in water by purge-and-trap injection capillary gas chromatography–atomic emission spectrometry. J. Anal. At. Spectrom. 11: 201–206.

Ceulemans, M., C. Witte, R. Lobinski and F.C. Adams. 2004. Simplified sample preparation for GC speciation analysis of organotin in marine biomaterials. Appl. Organomet. Chem. 8: 451–461.

Chau, Y.K., F. Yang and R.J. Maguire. 1996. Improvement of extraction recovery for the monobutyltin species from sediment. Anal. Chim. Acta. 320: 165–169.

Chiron, S., S. Roy, R. Cottier and R. Jeannot. 2000. Speciation of butyl- and phenyltin compounds in sediments using pressurized liquid extraction and liquid chromatography-inductively coupled plasma mass spectrometry. J. Chromatograph. A 879: 137–145.

Diaz-Bone, R.A. and M. Hitzke. 2008. Multi-element organometal(loid) speciation by hydride generation-GC-ICP-MS: overcoming the problem of species-specific optima by using a pH-gradient during derivatisation. J. Anal. Atom. Spectrom. 23: 861–870.

Dietz, C., J. Sanz, E. Sanz, R. Munoz-Olivas and C. Cámara. 2007. Current perspectives in analyte extraction strategies for tin and arsenic speciation. J. Chromatogr. A 1153: 114–129.

Donard, O.F.X., G. Lespes, D. Arnouroux and R. Morabito. 2001. Organotin compounds in the environment: still a critical issue. *In:* L. Ebdon, L. Pitts, R. Cornelis, H. Crews, O.F.X. Donard and P. Quevauviller [eds.]. Trace Element Speciation for Environment, Food and Health. The Royal Society of Chemistry, Cambridge, UK. pp. 142–175.

Duester, L., R.A. Diaz-Bone, J. Kösters and A.V. Hirner. 2005. Methylated arsenic, antimony and tin species in soils. J. Environ. Monit. 7: 1186–1193.

Ebdon, L., S.J. Hill and C. Rivas. 1998. Organotin compounds in solid waste: a review of their properties and determination using high-performance liquid chromatography. Trends Anal. Chem. 17: 277–288.

Encinar, J.R., P.R. Gonzalez, J.I.G. Alonso and A. Sanz-Medel. 2000. Synthesis and application of isotopically labelled dibutyltin for isotope dilution analysis using gas chromatography-ICP-MS. J. Anal. Atom. Spectrom. 15: 1233–1239.

Encinar, J.R., P.R. Gonzalez, J.I.G. Alonso and A. Sanz-Medel. 2002. Evaluation of extraction techniques for the determination of butyltin compounds in sediments using isotope dilution-GC/ICP-MS with ^{118}Sn and ^{119}Sn-enriched species. Anal. Chem. 74: 5237–5242.

Feng, Y.-l. and H. Narasaki. 2002. Speciation of organotin compounds in marine sediments by capillary column gas chromatography-atomic absorption spectrometry coupled with hydride generation. Anal. Bioanal. Chem. 372: 382–386.

Fent, K. 1996. Ecotoxicology of organotin compounds. Crit. Rev. Toxicol. 26: 1–117.

Garcia, E.S., J.I.G. Alonso and A. Sanz-Medel. 1997. Determination of butyltin compounds in sediments by means of hydride generation/cold trapping gas chromatography coupled to inductively coupled plasma mass spectrometric detection. J. Mass Spectrom. 32: 542–549.

Gibson, C.P. and S.P. Wilson. 2003. Imposex still evident in eastern Australia 10 years after tributyltin restrictions. Mar. Environ. Res. 55: 101–112.

Gianguzza, A., O. Giuffrè, D. Piazzese and S. Sammartano. 2012. Aqueous solution chemistry of alkyltin(IV) compounds for speciation studies in biological fluids and natural waters. Coord. Chem. Rev. 256: 222–239.

Girousi, S., E. Rosenberg, A. Voulgaropoulos and M. Grasserbauer. 1997. Speciation analysis of organotin compounds in Thermaikos Gulf by GC-MIP-AED. Fresenius J. Anal. Chem. 358: 828–832.

Godoi, A.F.L., R. Favoreto and M. Santiago-Silva. 2003. Contaminação ambiental por compostos organoestânicos. Quím. Nova. 26: 708–716.

Gómez-Ariza, J.L., I. Giráldez, E. Morales, F. Ariese, W. Cofino and P. Quevauviller. 1999. Stability and storage problems in organotin speciation in environmental samples. J. Environ. Monit. 1: 197–202.

Gómez-Ariza, J.L., E. Morales, D. Sánchez-Rodas and I. Giráldez. 2000. Stability of chemical species in environmental matrices. Trends Anal. Chem. 19: 200–209

Gomez-Ariza, J.L., E. Morales, I. Giraldez and D. Sanchez-Rodas. 2001. Sample treatment and storage in speciation analysis. *In:* L. Ebdon, L. Pitts, R. Cornelis, H. Crews, O.F.X. Donard, and P. Quevauviller [eds.]. Trace Element Speciation for Environment, Food and Health. The Royal Society of Chemistry, Cambridge, UK. pp. 51–80.

González-Toledo, E., C. Leal, M. Granados, R. Companó and M.D. Prat. 2000. Separation of butyltin and phenyltin species by ion-exchange chromatography with complexing mobile phases. Chromatographia. 51: 443–449.

González-Toledo, E., R. Companó, M. Granados and M.D. Prat. 2003. Detection techniques in speciation analysis of organotin compounds by liquid chromatography. Trends Anal. Chem. 22: 26–33.

Gui-Bin, J., Z. Qun-Fang and H. Bin. 2000. Speciation of organotin compounds, total tin, and major trace metal elements in poisoned human organs by gas chromatography-flame hotometric detector and inductively coupled plasma-mass spectrometry. Environ. Sci. Technol. 34: 2697–2702.

Haas, K., J. Feldmann, R. Wennrich and H.-J. Stärk. 2001.Species-specific isotope-ratio measurements of volatile tin and antimony compounds using capillary GC–ICP–time-of-flight MS. Fresenius J. Anal. Chem. 370: 587–596.

Hassani, L.H., A.G. Frenich, J.L.M. Vidal, M.J.S. Muros and M.H. Benajiba. 2005. Study of the accumulation of tributyltin and triphenyltin compounds and their main metabolites in the sea bass, *Dicentrachus labrax*, under laboratory conditions. Sci. Tot. Environ. 348: 191–198.

Hoch, M. 2001. Organotin compounds in the environment—an overview. Appl. Geochem. 16: 719–743.

Hoch, M. and D. Schwesig. 2004. Parameters controlling the partitioning of tributyltin (TBT) in aquatic systems. Appl. Geochem. 19: 323–334.

Hsia, M.-P. and S.-M. Liu. 2003. Accumulation of organotin compounds in Pacific oysters, Crassostrea gigas, collected from aquaculture sites in Taiwan. Sci. Tot. Environ. 313: 41–48.

Inaba, K., H. Shiraishi and Y. Soma. 1995. Effects of salinity, pH and temperature on aqueous solubility of four organotin compounds. Water Res. 29: 1415–1417.

Inagaki, K., A. Takatsu, T. Watanabe, Y. Aoyagi, T. Yarita, K. Okamoto and K. Chiba. 2007. Certification of butyltins and phenyltins in marine sediment certified reference material by species-specific isotope-dilution mass spectrometric analysis using synthesized Sn-118-enriched organotin compounds. Anal. Bioanal. Chem. 387: 2325–2334.

Kovala-Demertzi, D., V. Dokorou, A. Primikiri, R. Vargas, C. Silvestru, U. Russo and M.A. Demertzis. 2009. Organotin meclofenamic complexes: Synthesis, crystal structures and antiproliferative activity of the first complexes of meclofenamic acid–Novel anti-tuberculosis agents. J. Inorg. Biochem. 103: 738–744.

Krupp, E.M., J.K. Merle, K. Haas, G. Foote, N. Maubec and J. Feldmann. 2011. Volatilization of organotin species from municipal waste deposits: novel species identification and modeling of atmospheric stability. Environ. Sci. Technol. 45: 943–950.

Leroy, M.J.F., P. Quevauviller, O.F.X. Donard and M. Astruc. 1998. Determination of tin species in environmental samples. Pure Appl. Chem. 70: 2051–2064.

Lespes, G., C. Marcic, J. Heroult, I. le Hecho and L. Denaix. 2009. Tributyltin and triphenyltin uptake by lettuce. J. Environ. Manag. 90: S60–S68.

Looser, P.W., S. Bertschi and K. Fent. 1998. Bioconcentration and bioavailability of organotin compounds: influence of pH and humic substances. Appl. Organomet. Chem. 12: 601–611.

McSheehy, S., M.J. Nash and W.M. Geiger. 2007. Handbook of Hyphenated ICP-MS Applications. Agilent, 1st. Ed. 68 p.

Millán, E. and J. Pawliszyn. 2000. Determination of butyltin species in water and sediment by solid-phase microextraction-gas chromatography-flame ionization detection. J. Chromatogr. 873: 63–71.

Mitra, S.K., K. Jiang, K. Haas and J. Feldmann. 2005. Municipal landfills exhale newly formed organotins. J. Environ. Monitor. 7: 1066–1068.

Monperrus, M., R.C.R. Martin-Doimeadios, J. Scancar, D. Amouroux and O.F.X. Donard. 2003a. Simultaneous sample preparation and species-specific isotope dilution mass spectrometry analysis of monomethylmercury and tributyltin in a certified oyster tissue. Anal. Chem. 75: 4095–4102.

Monperrus, M., O. Zuloaga, E. Krupp, D. Amouroux, R. Wahlen, B. Fairmanc and O.F.X. Donard. 2003b. Rapid, accurate and precise determination of tributyltin in sediments and biological samples by species specific isotope dilution-microwave extraction-gas chromatography-ICP mass spectrometry.J. Anal. At. Spectrom. 18: 247–253.

Montes Bayón, M., M.G. Camblor, J.I.G. Alonso and A. Sanz-Medel. 1999. An alternative GC-ICP-MS interface design for trace element speciation. J. Anal. At. Spectrom. 14: 1317–1322

Morabito, R., P. Massanisso and P. Quevauviller. 2000. Derivatization methods for the determination of organotin compounds in environmental samples. Trends Anal. Chem. 19: 113–119.

Murai, R., S. Takahashi, S. Tanabe and I. Takeuchi. 2005. Status of butyltin pollution along the coasts of western Japan in 2001, 11 years after partial restrictions on the usage of tributyltin. Mar. Pollut. Bull. 51: 940–949.

Nagase, M. and K. Hasebe. 1993. Determination of tributyltin and triphenyltin compounds in fish by gas chromatography with flame photometric detection. Anal. Sci. 9: 517–522.

Nath, M., S. Pokharia and R. Yadav. 2001. Organotin(IV) complexes of amino acids and Peptides. Coord. Chem. Rev. 215: 99–149.

Nath, M., S. Pokharia, G. Eng, X. Song and A. Kumar. 2005. Triphenyltin(IV) derivatives showed anti-inflammatory activity. Eur. J. Med. Chem. 40: 289–298.

Okoro, H.K., O.S. Fatoki, F.A. Adekola, B.J. Ximba and R.G. Snyman. 2011. Sources, environmental levels and toxicity of organotin in marine environment—A Review. Asian J. Chem. 23: 473–482.

Omae, I. 2003a. General aspects of tin-free antifouling paints. Chem. Rev. 103: 3431–3448.

Omae, I. 2003b. Organotin antifouling paints and their alternatives. Appl. Organomet. Chem. 17: 81–105.

Pannier, F., A. Astruc and M. Astruc. 1996. Determination of butyltin compounds in marine biological samples by enzymatic hydrolysis and HG-GC-QFAAS detection. Anal. Chim. Acta. 327: 287–293.

Pellegrino, C., P. Massanisso and R. Morabito. 2000. Comparison of twelve selected extraction methods for the determination of butyl- and phenyltin compounds in mussel samples. Trends Anal. Chem. 19: 97–105.

Pinel-Raffaitin, P., P. Rodríguez-González, M. Ponthieu, D. Amouroux, I. le Hecho, L. Mazeas, O.F.X. Donard and M. Potin-Gautier. 2007. Determination of alkylated tin compounds in landfill leachates using isotopically enriched tin species with GC-ICP-MS detection. J. Anal. At. Spectrom. 22: 258–266.

Piver, W.T. 1973. Organotin compounds: industrial applications and biological investigation. Environ. Health Perspect. 4: 61–79.

Ponce de León, C.A., M. Montes-Bayón and J.A. Caruso. 2002. Elemental speciation by chromatographic separation with inductively coupled plasma mass spectrometry detection. J. Chromatogr. A 974: 1–21.

Quevauviller, P. and R. Morabito. 2000. Evaluation of extraction recoveries for organometallic determinations in environmental matrices. Trends Anal. Chem. 19: 86–96.

Rosenberg, E. 2005. Speciation of tin. *In*: R. Cornelis, J. Caruso, H. Crews and K. Heumann. [eds.]. Handbook of Elemental Speciation II—Species in the Environment, Food, Medicine and Occupational Health.John Wiley & Sons Ltd., Chichester, England. pp. 422–463.

Rosenberg, E., V. Kmetov and M. Grasserbauer. 2000. Investigating the potential of high-performance liquid chromatography with atmospheric pressure chemical ionization-mass spectrometry as an alternative method for the speciation analysis of organotin compounds. Fresenius J. Anal. Chem. 366: 400–407

Rodriguez, I., S. Mounicou, R. Łobinski, V. Sidelnikov and Y. Patrushev. 1999. Species-selective analysis by microcolumn multicapillary gas chromatography with inductively coupled plasma mass spectrometric detection. Anal. Chem. 71: 4534–4543.

Rüdel, H. 2003. Case study: bioavailability of tin and tin compounds. Ecotox. Environ. Safe. 56: 180–189.

Smedes, M., A.S. Jong and I.M. Davies. 2000. Determination of (mono-, di- and) tributyltin in sediments. Analytical methods. J. Environ. Monit. 2: 54–549.

Staniszewska, M., B. Radke, J. Namiesnik and J. Bolalek. 2008. Analytical methods and problems related to the determination of organotin compounds in marine sediments. Int. J. Environ. Anal. Chem. 88: 747–774.

Strand, J. and J.A. Jacobsen. 2005. Accumulation and trophic transfer of organotins in a marine food web from the Danish coastal waters. Sci. Tot. Environ. 350: 72–85.

Sturgeon, R.E., R. Wahlen, T. Brandsch, B. Fairman, C. Wolf-Briche, J.I. Garcia Alonso, P.R. González, J.R. Encinar, A. Sanz-Medel, K. Inagaki, A. Takatsu, B. Lalere, M. Monperrus, O. Zuloaga, E. Krupp, D. Amouroux, O.F.X. Donard, H. Schimmel, B. Sejerøe-Olsen, P. Konieczka, P. Schultze, P. Taylor, R. Hearn, L. Mackai, R. Myors, T. Win, A. Liebich, R. Philipp, L. Yang and S. Willie. 2003. Determination of tributyltin in marine sediment: Comité Consultatif pour la Quantité de Matière (CCQM) pilot study P-18 international intercomparison. Anal. Bioanal. Chem. 376: 780–787.

Sutton, P.G., C.F. Harrington, Ben Fairman, Hywel Evans, L. Ebdon and T. Catterick. 2000. The small-scale preparation and NMR characterization of isotopically enriched organotin compounds. Appl. Organomet. Chem. 14: 691–700.

Suyani, H., J. Creed, T. Davidson and J. Caruso. 1989. Inductively coupled plasma mass spectrometry and inductively coupled plasma atomic emission spectrometry coupled to high performance liquid chromatography for speciation and detection of organotin compounds. J. Chromatogr. Sci. 27: 139–143.

Tao, H., R.B. Rajendran, C.R. Quetel, T. Nakazato, M. Tominaga and A. Miyazaki. 1999. Tin speciation in the femtogram range in open ocean seawater by gas chromatography/ inductively coupled plasma mass spectrometry using a shield torch at normal plasma conditions. Anal. Chem. 71: 4208–4215.

Thermo Fisher Scientific. 2009. Application note 30132. A novel GC-ICP-MS approach for speciation of sulfur in reformulated fuels.

Thoonen, S.H.L. B.J. Deelman and G. van Koten. 2004. Synthetic aspects of tetraorganotins and organotin(IV) halides. J. Organomet. Chem. 689: 2145–2157.

Tsunoi, S., T. Matoba, H. Shioji, L.T.H. Giang, H. Harino and M. Tanaka. 2002. Analysis of organotin compounds by grignard derivatization and gas chromatography-ion trap tandem mass spectrometry. J. Chromatogr. A 962: 197–206.

Van, D.N., Muppala, S.R.M., W. Frech and S. Tesfalidet. 2006a. Preparation, preservation and application of pure isotope-enriched phenyltin species. Anal. Bioanal. Chem. 386: 1505–1513.

Van, D.N., B. Radziuk and W. Frech. 2006b. A comparison between continuum- and line source AAS for speciation analysis of butyl- and phenyltin compounds. J. Anal. At. Spectrom.21: 708–711.

Van, D.N., T.T. Xuan and S. Tesfalidet. 2008. The transformation of phenyltin species during sample preparation of biological tissues using multi-isotope spike SSID-GC-ICP-MS. Anal. Bioanal. Chem. 392: 737–747.

Vella, A.J. and R. Vassallo. 2002. Emission to air of volatile organotins from tributyltin contaminated harbor sediments. Appl. Organomet. Chem. 16: 239–244.

Vercauteren, J., A. Meester, T. De Smaele, F. Vanhaecke, L. Moens, R. Dams and P. Sandra. 2000. Headspace solid-phase microextraction-capillary gas chromatography-ICP mass spectrometry for the determination of the organotin pesticide fentin in environmental samples. J. Anal. At. Spectrom. 15: 651-656.

Vernon F.H., S.L. Seidel and E.D. Goldberg. 1979. Determination of Tin(IV) and organotin compounds in natural waters, coastal sediments and macro algae by atomic absorption spectrometry. Anal. Chem. 51: 1256–1259.

Wahlen, R. and T. Catterick. 2003. Comparison of different liquid chromatography conditions for the separation and analysis of organotin compounds in mussel and oyster tissue by liquid chromatography-inductively coupled plasma mass spectrometry. J. Chromatogr. B 783: 221–229.

Wasik, A. and T. Ciesielski. 2004. Determination of organotin compounds in biological samples using accelerated solvent extraction, sodium tetraethylborate ethylation, and multicapillary gas chromatography–flame photometric detection. Anal. Bioanal. Chem. 378: 1357–1363.

Wasik, A., I.R. Pereiro and R. Lobinski. 1998. Interface for time-resolved introduction of gaseous analytes for atomic spectrometry by purge-and-trap multicapillary gas chromatography (PTMGC). Spectrochim. Acta. Part B. 53: 867–879.

Wasik, A., B. Radke, J. Bolałek and J. Namiesnik. 2007. Optimisation of pressurised liquid extraction for elimination of sulphur interferences during determination of organotin compounds in sulphur-rich sediments by gas chromatography with flame photometric detection. Chemosphere. 68: 1–9.

Wuilloud, J.C.A., R.G. Wuilloud, A.P. Vonderheide and J.A. Caruso. 2004. Gas chromatography/ plasma spectrometry—an important analytical tool for elemental speciation studies. Spectrochim. Acta. Part B. 59: 755–792.

Yang, H.-J., S.-J. Jiang, Y.-J. Yang and C.-J. Hwang. 1995. Speciation of tin by reversed phase liquid chromatography with inductively coupled plasma mass spectrometric detection. Anal. Chim. Acta. 312: 141–148.

Yu, Z.-H., J.-Q. Sun, M. Jing, X. Cao, F. Lee and X.-R. Wang. 2010. Determination of total tin and organotin compounds in shellfish by ICP-MS. Food Chem. 119: 364–367.

Zachariadis, G.A. and E. Rosenberg. 2009. Speciation of organotin compounds in urine by GC–MIP-AED and GC-MS after ethylation and liquid-liquid extraction. J. Chromatograph. B 877: 1140–1144.

Zanon, F., N. Rado, E. Centanni, N. Zharova and B. Pavoni. 2009. Time trend of butyl- and phenyl-tin contamination in organisms of the Lagoon of Venice (1999–2003). Environ. Monit. Assess. 152: 35–45.

Zuliani, T., G. Lespes, R. Milacic, J. Scancar and M. Potin-Gautier. 2006. Influence of the soil matrices on the analytical performance of headspace solid-phase microextraction for organotin analysis by gas chromatography-pulsed flame photometric detection. J. Chromatogr. A 1132: 234–240.

Speciation and Bioavailability of Iodine in Edible Seaweed

Vanessa Romaris Hortas,[a] Antonio Moreda Piñeiro[b] and
Pilar Bermejo Barrera[c],*

Introduction

Iodine is a naturally occurring non-metallic element of the periodic table's group VIIA. In nature it exists as a monovalent anion (iodide) in brines and in different molecular compounds, e.g., iodate or more complex organic iodinated compounds. It can exist in several oxidation statuses (–1, 0, +1, +3, +5 and +7); its stable natural isotope is ^{127}I; and it can be soluble in water or in organic solvents depending on its chemical species.

The average concentration on the Earth's crust is approximately 0.5 mg Kg^{-1}; it is between 45–60 µg l^{-1} in the oceans, and in the atmosphere the concentration ranges from 10 to 20 ng m^{-3} depending on the proximity to the sea and the soil type. The primary natural sources of iodine are the oceans. It accumulates in fish, shellfish and seaweed. In Fig. 13.1 we can see that the distribution, transport and transformation of iodine in the environment are

Department of Analytical Chemistry, Nutrition and Bromatology, Faculty of Chemistry, University of Santiago de Compostela, 15782 Santiago de Compostela, Spain.
[a] Email: vanessa.romaris@rai.usc.es
[b] Email: antonio.moreda@usc.es
[c] Email: pilar.bermejo@usc.es
*Corresponding author

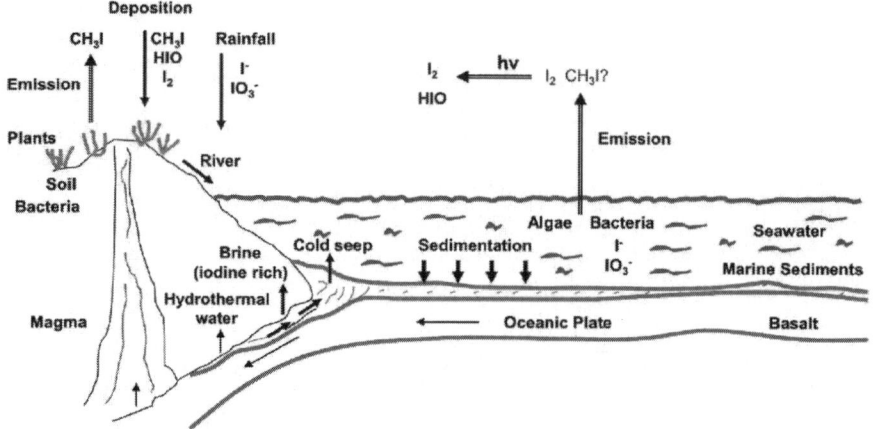

Figure 13.1 Global Cycle of Iodine.

Color image of this figure appears in the color plate section at the end of the book.

the products of complex physical, chemical and biological processes. This cycle is controlled by large exchanges from the oceans to the atmosphere in the marine boundary layer. Emissions of molecular iodine and volatile iodocarbons from different macroalgae and phytoplankton are the key processes in this cycle (Brownell 2010, Küpper et al. 2008, Lai et al. 2011). These processes are essential to life because the major part of iodine is not accessible to the living organisms; only small amounts can be liberated from the surface through the process of weathering and dissolution.

Iodine enters the environment in the form of sea-sprays or gases. Some authors (Laturnus 2001, Mäkelä et al. 2002), have proposed that organic iodine vapour is emitted by certain seaweed, but other authors (Ball et al. 2010, Huang et al. 2010, McFiggans et al. 2004, Saiz-Lopez et al. 2006) later proposed that molecular iodine is the main species emitted by seaweed. Recently, Kundel et al. 2012, using mass spectrometric techniques demonstrated that the species emitted by seaweed are molecular iodine (as the principal species) and some iodocarbon species. The total iodocarbon emission rates were one or two orders of magnitude lower than those of molecular iodine. These authors demonstrated that molecular iodine is the major iodine containing volatile species released by different types of seaweed. After the emission, molecular iodine and iodocarbons are probably involved in catalytic ozone destruction, in new particle formation and in cloud condensation (O'Dowd and Hoffmann 2005, Carperter 2003 and McFiggans et al. 2006). After the emission and due to photolytic dissociation, iodine can be transformed into methyl and iodine radicals producing different iodine forms such as HI, HIO, INO_2, $IONO_2$ or OIO (Chameides 1980, Cox 1999, Vogt 1999). These reactions are the principal cause of the transfer of iodine from the ocean to the atmosphere. The combustion of fossil fuels, which presents average iodine content of 4 mg Kg^{-1} in coal and 1 mg Kg^{-1} in petroleum, is another source of iodine in the atmosphere.

Deposition in soil is due to the weathering of rock, volcanic activity, decaying vegetation, rainfall and human activities. Microorganisms play an important role in the accumulation processes. The transfer of the iodine species to surface waters can be the product of ocean sprays and rainwater in coastal areas, or only rainwater in coastal and terrestrial areas alike (Baker et al. 2001a, b and Gilfedder et al. 2008). In drinking water, the oxidation of iodide to hypoiodous acid (HOI) is possible. This species is capable of reacting with some organic compounds that can cause taste and odour problems in tap water (Andersen et al. 2002, Bichsel and Von Gunten 1999). The transfer to the Earth's surface decreases with distance from the oceans and is also influenced by dry and wet deposition and by sorption of gases. In soil, iodine can react with some organic components such as humic substances, polyphenols and tyrosine residues (Whitehead 1984).

Iodine Species

The number of different iodine species in environmental and biological samples is very large (Table 13.1), and they are directly dependent on the type of matrix on which they are found.

The most abundant species are the inorganic ones such as iodide and iodate, which are present in environmental and biological materials, being especially abundant in surface waters, principally in seawater. But in this medium volatile organoiodine compounds are also present, commonly referred to as dissolved organic iodine (DOI) (Butler and Smith 1980, Edmons and Morita 1998). In seawater, the first specie identified was methyliodide. Other volatile iodinated species suh as ehyl-iodide, isopropyl-iodide, n-propyl-iodide chloroiodomethane, diiodomethane, 2-iodopropane, 1-iodobutane and 2 iodobutane were reported later. These species are responsible for the presence of iodine in marine atmospheres and in atmospheric particulate matter, Baker et al. 2000.

The origin of many volatile species in seawater and in the marine atmosphere is macroalgae (Gschend et al. 1985). Seaweed contains different types of iodinated compounds such as simple iodoalkanes, and other organoiodine compounds such as diiodoformaldehyde, 3-iodo-1-propanol, 1-bromo-3-iodo-2-propanone, 1-bromo-2,3-dichloro-1-iodoethane, bromoiodoacetic acid and diiodoacetamide (Edmons and Morita 1998, Gribble 2003). Low levels of other organic species with a more complex structure such as iodinated ketones, gamma-methylene lactones, iodo bromo aromatic sesquiterpenes, and iodine aminoacids, have also been reported, (Edmons and Morita 1998). Although some iodine amino acids

Table 13.1 Iodine species.

Name	Matrix	Structure
Iodide	Enviromental and biological samples	I^-
Iodate	Enviromental and biological samples	IO_3^-
Iodine	Enviromental and biological samples	I_2
Iodine monochloride	Enviromental and biological samples	ICl
Hypoiodous acid	Enviromental and biological samples	HIO
Methyl-iodide	Marine environment	CH_3I
Chloroiodomethane	Marine environment	CH_2ICl
Bromoiodomethane	Marine environment	CH_2IBr
Bromodiiodomethane	Marine environment	CHI_2Br
Dibromoiodomethane	Marine environment	$CHIBr_2$
Bromochloroiodomethane	Marine environment	$CHIClBr$
Diiodomethane	Marine environment	CH_2I_2
Triiodomethane	Marine environment	CHI_3
Ethyl-iodide	Marine environment	CH_3CH_2I
n-propyl-iodide	Marine environment	$CH_3 CH_2 CH_2I$
Isopropyl-iodide	Marine environment	$(CH_3)_2CHI$
2-iodopropane	Marine environment	$CH_3 CHI CH_3$
1-iodobutane	Marine environment	$CH_3 CH_2 CH_2 CH_2I$
2-iodobutane	Marine environment	$CH_3 CH_2 CHICH_3$
1-iodopentane	Marine environment	$CH_3 CH_2 CH_2 CH_2 CH_2I$
1-methyl-1-iodopropane	Seaweed	$CH_3 CH_2 CH(CH_3)I$
1-bromo-2-iodoethane	Seaweed	$CHI_2 CH_2Br$
1-bromo-2,3-dichloro-1-iodoethane	Seaweed	$CH_2Cl\ CHCl\ CHIBr$
Bromoiodoacetic acid	Seaweed	
Diiodoformaldehide	Seaweed	
3-iodo-1-propanol	Seaweed	

Table 13.1 contd....

Table 13.1 contd....

Name	Matrix	Structure
1-bromo-3-iodo-2-propanone	Seaweed	
Diiodoacetamide	Seaweed	
1-iodo-3,3-dibromo-2-heptanone	Enviromental and biological samples	
1-cloro, 3-iodoacetone	Enviromental and biological samples	
1,1-dibromo-, 3-iodoacetone	Enviromental and biological samples	
Ethyl bromoiodoacrylate	Enviromental and biological samples	
Ethyl iodoacrylate	Enviromental and biological samples	
Ethyl diiodoacrylate	Enviromental and biological samples	
Monoiodotyrosine (MIT)	Seaweed, sponges, mammalians	
Diiodotyrosine (DIT)	Seaweed, sponges, mammalians	

Table 13.1 contd....

Table 13.1 contd....

Name	Matrix	Structure
Thyroxine (T0)	Higher organisms	
Triiodothyronine (T3)	Higher organisms	
Reversed-triiodothyronine (rT3)	Higher organisms	
Diiodothyronine (T4)	Higher organisms	

such as 3-iodotirosine (MIT) and 3,5-diiodotyrosine (DIT) has been found in some seaweed and sponges, these iodine aminoacids are present principally in mammalians. They are key compounds for the biosynthesis of iodine containing hormones iodothyronines, such as 3,3',5,5'-tetraiodothyronine (thyroxine or T4), 3,3',5'-tri-iodothyronine (tri-iodothyronine or T3) and 3,5-di-iodothyronine (di-iodothyronine or T2), which play importants roles in the physical and intellectual developmnt of higher organisms (Underwood 1977).

The presence of complex organic iodine species in marine aerosols is worth mentioning (Baker et al. 2005, Gilfedder et al. 2008). Other species detected recently in particulate matter are from anthropogenic sources (Ruan et al. 2010); e.g., volatile neutral polyfluorinated iodine alkanes (PFIs) are compounds found as fluorinated additive materials in surfactants, lubricants, varnishes and pesticides.

Interest of Iodine Study

Iodine is an essential trace element in small amounts for normal physiological functions in humans. It is an essential substrate for the synthesis of thyroid hormones T4 and T3, which are necessary for the control of cellular metabolism, growth and development of body structures, neuronal function and development (Underwood 1977, Zimmermann et al. 2008). The human body cannot synthesize this element, obtaining it from different sources, with food being the most important. The WHO has recognized iodine deficiency as the most common preventable cause of brain damage in the world today. Therefore in 1986, the International Council for Control of Iodine Deficiency Disorders was created (Hetzel 2002). Although lack of iodine leads to iodine deficiency disorder, excessive iodine dietary intake can result in serious pathological problems. Recommended iodine intake has been established by several international organisms. The iodine intake recommended by the WHO is 150 μg /day for adults and adolescents over 13 yr of age; 200 μg/day for women during pregnancy and lactation; 120 μg/day for children 6–12 yr of age; and 90 μg/day for children 0–50 mon of age (WHO, 2001).

Bioavailability of Iodine in Seaweed

Due to the ability of seaweed to preconcentrate inorganic species from seawater, they have been recognized as a natural source of essential elements (Chapman and Chapman 1980). Although seaweed has been used as a food source in Asian countries since ancient times, the current interest on health food in the West has led to an increase of their presence in western markets. It is therefore necessary to ascertain the concentrations of essential and toxic elements. To know the possible effects of trace elements present in foods, it is not sufficient to know the total content, but also the bioavailable fraction of these elements. Bioavailability is an important factor in nutrition due to its variability among different foodstuffs and food components. It also depends on gastrointestinal conditions, and on several processes such as digestion, absorption, transport, utilization and elimination. To estimate the bioavailability of trace elements in humans, *in vivo* and *in vitro* methods have been proposed. *In vivo* methods using stable isotopes provide the best estimation of bioavailability, but the use of radioisotopes present problems in many laboratories, which is why *in vitro* methods have been preferred for investigating trace metal bioavailability. These methods are usually based on the simulation of gastric and intestinal digestion of food, and they measure the fraction of the element available for absorption. These methods are simple, rapid, inexpensive and easy to control. One of the methods used most often for studying bioavailability in foods is the one proposed by Miller et al. 1981.

Romaris-Hortas et al. 2011 performed a study on the bioavailability of iodine and bromine in edible seaweed using an *in vitro* method. This method was performed in two stages, the first stage simulating gastric digestion using pepsin, and the second stage simulating intestinal digestion using pancreatin with bile salts. Afterwards, the intestinal extract was passed through a dyalisis membrane with a molecular weight cut-off of 10 KDa. The concentration of the iodine in the solution within the dialysis membrane provided the dialysability data, and the dialysability percentage was calculated using the following equations:

% dialysability = $[I]_{dialysed}$ / $[I]_{seaweed}$. 100

($[I]_{dialysed}$ is the iodine concentration within the dialysis membrane, and $[I]_{seaweed}$ is the total iodine concentration in the seaweed).

In this chapter, nine different types of edible seaweed harvested in the Galician coast (northwestern Spain) obtained from a local manufacturer were studied. These samples are commercialized as dehydrated products: Dulse (*Palmaria palmata*) and Nori (*Prophyra umbilicalis*) as red seaweed; Kombu (*Laminaria ocholeuca*), Wakame (*Undaria pinatifida*) and Sea Spaghetti (*Himanthalia elongata*) as brown seaweed; Sea Lettuce (*Ulva rigida*) as green seaweed; the microalgae *Spirulina platensis*, often used in human and animal nutrition; and agar-agar, which is a hydrocolloid obtained from the red seaweed *Gelidiumm sesquipedale*. These samples were kept in an oven at 40°C to eliminate water traces before pulverization. Another sample studied is commercialized cooked and canned in brine, and consists of a mixture of two brown seaweed, Sea Spaguetti and Furbelows (*Saccorhiza polyschides*). Total iodine levels found in these seaweed were between 27.3 and 6138 µg/g. Kombu was the seaweed with the highest iodine concentration, while iodine was not detected in Spirulina. The highest concentrations of bioavailable iodine were found in the brown seaweed Kombu. The iodine dializability percentages were lower than 5 percent for all samples except for Kombu (17 ± 1.9%), Agar-agar (16 ± 1.8%) and Dulse (10 ± 1.5%) Correlations between the type of seaweed and dializability were not found, and the authors concluded that it is necessary to know the chemical iodine specie present in each seaweed to explain these results.

Speciation of Iodine in Seaweed

In addition to the total iodine content, the toxicity, mobility and availability of this element depends on the different chemical species present in foodstuffs. Moreover, in general, the bio-availability of elements from foodstuffs is also dependent on the specific chemical forms (Moreda-Piñeiro et al. 2011). Therefore, elemental speciation studies must be performed to evaluate the significance of the essential and non-essential elements.

Inorganic iodine (iodide and iodate) is frequently present in seaweed. But these aquatic plants contain numerous volatile iodoalkanes, mainly methyl-iodide as well as iodinated ketones, alcohols and carboxylic acids (McConnel and Fenical 1977, McConnel and Fenical 1980, Baker et al. 2001). Iodinated species of a more complex structure such as iodinated amino acids (3-iodotyrosine-MIT and 3,5-diiodotyrosine-DIT) have also been found in some seaweeds (Shah et al. 2005, Gomez-Jacinto et al. 2010).

The first studies on the speciation of iodine in some seaweeds were performed using neutron activation analysis (NAA), combined with a procedure of sequential extraction (Hou et al. 1997). These authors studied seven marine seaweeds: *Codium fragile, Ulva pertusa, Monostroma nitidum, Gracilaria confervoides, Sargassum kjellmanianum, Dictyopteris clivaricata and Laminaria japonica*. Using a sequential extraction they differentiated total iodine, water-soluble iodine, soluble organic iodide, iodide and iodate. The results indicated that the chemical species and contents of iodine are very different for different types of seaweeds. The highest iodine content of 734 mg/Kg was found in *Laminaria Japonica*, with 92 percent of the total iodine being water soluble. Other seaweeds present lower iodine content, with soluble iodine making up 16–41 percent of the total. In the aqueous extract, iodide is the main species, making up between 61–93 percent of the total soluble iodine. The percentage of organic iodine was between 5.5–37.4 percent, while the contents of iodate were found to be the lowest (1.4–4.5 percent). The same authors (Hou et al. 2000) studied the distribution of iodine in *Sargassum kjelmmanianum* using NAA combined with chemical and biochemical separation techniques. For the investigation of the leaching ratio of iodine by different chemical solvents they used ether, ethanol, 0,1M HCl and 0,1 M KOH. Iodine-bound protein was leached with Tris-HCl buffer and precipated in saturated ammonium sulphate solution. Fucoidin was leached with boiling water and precipitaed with alcohol. Algin was extracted with boiling desionized water, and polyphenols were extracted with ethanol and purified by column chromatography. The leaching rate of iodine is lowest in ether (8.9 percent) and highest in KOH (93.3 percent). Iodine is mainly bound to proteins. Another part is bound to pigments and polyphenols, and a smaller part to polysaccharides, such as algin fucoidan and cellulose.

More recent studies tend to perform iodine speciation using High Performance Liquid Chromatography (HPLC), mainly hyphenated with Inductively Coupled Plasma-Mass Spectrometry (ICP-MS). HPLC can function in size mode (size exclusion chromatogarphy, SEC) for assesing iodine bound to biomolecules; reverse phase mode (reverse phase chromatography, RPC) for determining organic iodinated forms, and anion exchange mode (anion exchange chromatography, AEC), which is mainly focused on separating inorganic iodine species but also suitable for some organic iodine species.

Shah et al. 2005, performed iodine speciation in commercially available samples of Kombu and Wakame, using a multidimensional chromatographic approach coupled with ICP-MS. The analysis of alkaline extracts by SEC-ICP-MS indicated the binding of iodine with high as well as low molecular weight fractions in Wakame; while in case of Kombu, only low molecular weight iodine species were found. Moreover, assocaition of iodine with protein as well as polyphenolic species was found for Wakame. The separation using Anion-exchange Chromatography coupled to ICP-MS after alkaline extraction confirmed that iodide is the predominant inorganic specie in both types of seaweeds. Moreover, after the enzymatic hydrolysis of the protein bound iodinated species, the presence of monoiodotyrosine and i-iodotyrosine were detected in Wakame.

Gomez-Jacinto et al. 2010, performed the characterization of iodine species in the microalgae *Chlorella vulgaris* after cultivation with different potassium iodide concentrations. They studied two iodine fractions—water-soluble and macromolecular fractions—using a sequential extraction scheme based on chemical reagents. Most iodine species separated from the water-soluble fraction with SEC-ICP-MS which are present in inorganic forms, IO_3^- (about 25 percent) and I^- (about 75 percent). The application of SEC-ICP-MS to the macromolecular fraction reveals the presence of four peaks from the void volume of about 67 KDa and a peak at 600 Kda. The iodine macromolecular fraction represents about 13 percent of total idine in the microalgae studied.

Recently, Romaris-Hortas et al. 2012 developed a new method using Anion Exchange Chromatography, AEC, for resolving simultaneously inorganic and organic iodine species, as well as bromine species. In this method, ICP-MS was used as a selective detector. This method was used to assess the bioavailable iodine species from edible Atlantic seaweed after an *in vitro* bioavailability approach previously studied by the same authors (Romaris-Hortas et al. 2011). The anion exchange column used was the Dionex AS7. A complete separation of iodide, iodate, MIT and DIT was obtained using a gradient program with a mobile phase of 175 mM ammonium nitrate, 15.0% (v/v) methanol at pH 3.8. With this column the presence of iodide, iodate, MIT and DIT was observed for a dialyzate from the seaweed Wakame. In order to confirm the presence of the organic iodine species, these authors used Reverse Phase Cromatography, RPC with an Agilent Zorbax XDB-C8 reverse phase column coupled to a Phenomenex C8 guard column. The mobile phase composition was 0.2 percent acetic acid in water (pH 2.7), and 2 percent in methanol (pH 2.3). With this column inorganic iodine elutes within the first few minutes, while MIT and DIT elute at approximately 14.9 and 17.5 min, respectively. With the use of both columns these authors performed iodine speciation in the dyalizable fraction of different types of seaweed. Iodide and MIT were observed in Nori and Dulse (red seaweed), and also in Sea lettuce (green seaweed), and in canned seaweed (a mixture of brown seaweeds). Iodide, MIT and DIT were observed in Wakame (brown seaweed), and also iodide and DIT (small signal) from Sea spaguetti (brown seaweed). For Spirulina, only a small signal of iodide was observed, while for NIES 09 Sargasso iodide, MIT, DIT, and also an unknown iodinated compound was observed. This unknown peak was also observed in Wakame and Kombu. Moreover, unresolved signals in the first part of the chromatograms for Kombu and agar-agar led to a poor resolution of MIT. Iodate was not observed in any dialyzate. High iodide concentrations were found in the dialyzates of Kombu, but this species was not detected in the Spirulina. Iodate was not detected in every seaweed study. For the MIT species, the highest concentration was observed in Nori and NIES 09 Sargasso. Kombu and Wakame were also present in significant amounts of this species. MIT was not detected in Spirulina. DIT was observed at higher levels in NIES 09 Sargasso, but was not detected in Dulse, Nori, Sea Lettuce, Sea Spaghetti, Spirulina or Agar-agar. In this chapter, the authors have concluded that further work is necessary to elucidate the structure of the unknown iodine species present in the dialyzates, (because these compounds are bioavailable for humans).

Conclusions

As can be seen in recent literature, the levels of iodine vary greatly between different types of seaweeds. In general brown seaweed, for example, Kombu, present very high iodine levels, whereas green or red seaweed present low levels. On the other hand, the bioavailability of iodine obtained using *in vitro* studies is generally low, less than 5 percent, but Kombu is the seaweed that presents the highest bioavailability, and this could be problematic due to the high iodine level present in this seaweed.

On the other hand, although in recent years some papers on the study of iodine speciation in seaweed have been developed, more studies are necessary in order to increase the knowledge of the iodine chemical species present in seaweed. It is necessary to improve the resolution of the chromatographic separations in some cases and to indentify the unknown peaks present in some seaweed.

Acknowledgments

The authors wish to thank the *Ministerio de Ciencia y Tecnología* (Project number AGL-2006-11034), and the *Xunta de Galicia* (Grupo de Referencia Competitiva 2007/000047-0) for financial support.

References

Andesern, S., B. Hvingel and P. Laurberg. 2002. Iodine content of traditional Greenlandic food items and tap water in East and West Greenland. Int. J. Circumpolar Health 61: 332–340.

Baker, A.R., D. Thompson, M.L.A.M. Campos, S.J. Parry and T.D. Jickells. 2000. Iodine concentration and availability in atmospheric aerosol. Atmos. Environ. 4 :4331–4336.

Baker, A.R., C. Tunnicliffe and T.D. Jickells. 2001a. Iodine speciation and deposition fluxes from the marine atmosphere. Journal of Geophysical Research. 106(D22): 28743–28749.

Baker, J.M., W.T. Sturges, J. Sugier, G. Sunnenberg, A.A. Lovett, C.E. Reeves, P.D. Nightingale and J.A. Penkett. 2001b. Emissions of CH_3Br, organochlorines and organoiodines from temperate macroalgae. Chemospher-Global Change Science. 3: 93–106.

Baker, A.R. 2005. Marine aerosol iodine chemistry: the importance of soluble organic iodine. Environ. Chem. 2: 295–298.

Ball, S.M., A.M. Hollingsworth, J. Humbles C. Leblanc, P. Potin and G. McFiggans. 2010. Spectroscopic studies of molecular iodine emitted into the gas phase by seaweed. Atmos. Chem. Phys. 10: 6237–6254.

Bichsel, Y. and U. Von Gunten. 1999. Oxidation of Iodide and Hypoiodous Acid in the Disinfection of Natural Waters. Environ. Sci. Tecnol.33: 4040–4045.

Brownell, D.K., R.M. Moore and J.J. Cullen. 2010. Production of methyl halides by Prochlorococcus and Synechococcus. Global Biogeochemical Cycles. 24, Gb2002.

Butler, E.C.V. and J.D. Smith. 1980. Iodine speciation in sea waters—the analytical use of ultraviolet photooxidation and differential pulse polarography. Deep-Sea Res. Part A. 27: 489–493.

Caperter, L.J. 2003. Iodine in the marine boundary layer. Chem. Rev. 103: 4953–4962.

Chameides, W.L. and D.D. Davis. 1980. Iodine: its possible role in tropospheric photochemistry. J. Geophys. Res. 85: 7383–7398.

Chapman, V.J. and D.J. Chapman. 1980. Sea Vegetables (algae as food for man). London, UK, Chapman & Hall.

Cox, R.A., W.J. Bloss, R.L. Jones and D.M. Fowley. 1999. OIO and the atmospheric cycle of iodine. Geophys. Res. Lett. 26: 1857–1860.

Edmons, J.S. and M. Morita. 1998. The determination of iodine species in environmental and biological samples. Pure and Applied Chem. 70: 1567–1584

Gilfedder, B.S., S.C. Lai, M. Petri, H. Biester and T. Hoffmann. 2008. Iodine speciation in rain, snow and aerosols. Atmospheric Chemistry and Physics. 8(20): 6069–6084.

Gómez-Jacinto,V., A. Arias-Borrego, T. García-Barrera, I. Garbayo, C. Vilchez and J.L. Gómez-Ariza. 2010. Iodine speciation in iodine-enriched microalgae Chlorella vulgaris. Pure and Applied Chemistry. 82: 473–481.

Gribble, G.W. 2003. The diversity of naturally produced organohalogens. Chemosphere 52: 289–297.

Gschwend, P.M., J.K. Macfarlane and K.A. Newman. 1985. Volatile halogenated organic compounds released to seawater from temperate marine macroalgae. Science. 227: 1033–1035.

Hetzel, B.S. 2002. Eliminating iodine deficiency disorders, the role of the International Council in the Global partnership. Bull.W.H.O. 80: 410–414.

Hou, X., C. Chai, Q. Qian, X. Xan and X. Fan. 1997. Determination of chemical species of iodine in some seaweeds (I). The Science of the Total Environment. 204: 215–221

Hou, X., X. Yan and C. Chai. 2000. Chemical Species of iodine in some seaweeds II. Iodine-Bound Biological Macromolecules. Journal of Radioanal. Nucl. 245: 461–467.

Huang, R.J., K. Seitz, J. Buxmann, D. Pöhler, K.E. Hornsby, L.J. Carperter, U. Platt and T. Hoffmann. 2010. *In situ* measurements of molecular iodine in the marine boundary layer: the link to macroalgae and the implications for O_3, IO, OIO and NOx. Atmos. Chem. Phys. 10: 4823–4833.

Kundel, M., U.T. Thorenz, J.H. Petersen, R. Huang, N.H. Bings and T. Hoffmann. 2012. Application of mass spectrometric technique for the trace analysis of short-lived iodine-containig volatiles emitted by seaweed. Anal. Bioanal. Chem. 402: 3345–3357.

Küpper, F.C., L.J. Carpenter, G.B. McFiggans, C.J. Palmer, T.J. Waite, E.M. Boneberg, S. Woitsch, M. Weiller, R. Abela, D. Grolimund, P. Potin, A. Butler, G.W. Luther, P.M.H. Kroneck, W. Meyer-Klaucke and M.C. Feiters. 2008. Iodide accumulation provides kelp with an inorganic antioxidant impacting atmospheric chemistry. Proceedings of the National Academy of Sciences of the United States of America. 105(19): 6954–6958.

Lai, S.C., J. Williams, S.R. Arnold, E.L. Atlas, S. Gebhardt and T. Hoffmann. 2011. Iodine containing species in the remote marine boundary layer: A link to oceanic phytoplankton. Geophysical Research Letters. 38, L20801.

Laturnus, F. 2001. Marine macroalgae in polar regions as natural sources for volatile orgobohalogens. Environ. Sci. Pollut. Res. 8: 103–108.

Mäkelä, J.M., T. Hoffmann, C. Holzke, M. Vakeva, T. Suni, T. Mattila, P.P. Aalto, U. Tapper, E.I. Kauppinen and C.D. O'Dowd. 2002. Biogenic iodine emissions and identification of end-products in coastal ultrafine particles during nucleation bursts. J. Geophys. Res. 107(D19), PAR14/1-PAR14/14.

McConnell, O. and W. Fenical. 1977. Halogen Chemistry of the Red Alag Asparagopsis, Phytochemistry. 16: 367–374.

McConnell, O. and W. Fenical. 1980. Halogen Chemistry of the Red Alaga Bonnemaisonia, Phytochemistry. 19: 233–247.

McFiggans, G., H. Coe, R. Burgess, J. Allan, M. Cubison, M.R. Alfarra, R. Saunders, A. Saiz-Lopez, J.M.C. Plane, D.J. Wevill, L.J. Carpenter, A.R. Rickard and P.S. Monks. 2004. Direct evidence for coastal iodine particles from Laminaria macroalgae—linkage to emissions of molecular iodine. Atmospheric Chemistry and Physics. 4: 701–713.

McFiggans, G., P. Artaxo, U. Baltensperger, H. Coe, M.C. Facchini, G. Feingold, S. Fuzzi, M. Gysel, A. Laaksonen, U. Lohmann, T.F. Mentel, D.M. Murphy, C.D. O'Dowd J.R. Snider and E. Weingartner. 2006. The effect of physical and chemical aerosol properties on warm cloud droplet activation. Atmos. Ohys. 6: 2593–2649.

Miller, J.C., B.R. Schricker, R.R. Rasmussen and D. Van Campen. 1981. An *in vitro* estimation of iron availability from meals. The American Journal of Clinical Nutrition 34: 2248–2256.

Moreda-Piñeiro, J., A. Moreda-Piñeiro, V. Romaris-Hortas, P. Moscoso-Pérez, P. Lopéz-Mahia, S. Muniategui-Lorenzo, P. Bermejo-Barrera and D. Prada-Rodríguez. 2011. *In vivo* and *in vitro* testing to assess the bioaccessibility and the bioavailability of arsenic, selenium and mercury species in food samples. Trends. Anal. Chem. 30: 324–345.

O'Dowd, C.D. and T. Hoffmann. 2005.Coastal new particle formation: a review of the current state-of-the-art. Environ Chem. 2: 245–255.

Romarís-Hortas, V., C. García-Sartal, M.C. Barciela-Alonso, R. Domínguez-García, A. Moreda-Piñeiro and P. Bermejo-Barrera. 2011. Bioavailability study using an *in vitro* method of iodine and bromine in edible seaweed. Food Chemistry. 124: 1747–1752.

Romarís-Hortas, V., P. Bermejo-Barrera and P. Moreda-Piñeiro. 2012. Development of anion-exchange/reversed-phase high performance liquid chromatography-inductively coupled plasma-mass spectrometry methods for the speciation of bio-available iodine and bromine from edible seaweed. Journal of Chromatography. A 1236: 164–176.

Ruan T., Y. Wang, Q. Zhang, L. Ding, P. Wang, G. Qu, C. Wang, T. Wang and G. Jiang. 2010. Trace determination of airborne polyfluorinated iodine alkanes using multisorbent thermal desorption/gas chromatography/high resolution mass spectrometry, J. Chromatog. A 1217: 4439–4447.

Saiz-Lopez, A., J.M.C. Plane, G. McFiggans, P.I. Williams, S.M. Ball, M. Bitter, R.L. Jones, C.Hongwei and T. Hoffmann. 2006. Modelling molecular iodine emissions in a coastal marine environment: the link to new particle formation. Atmospheric Chemistry and Physics. 6: 883–895.

Shah, M., R.G. Wuilloud, S.S. Kannamkumarath and J.A. Caruso. 2005. Iodine speciation studies in commercially available seaweed by coupling different chromatographic techniques with UV and ICP-MS detection. J. Anal. Atom. Spectrometry. 20: 176–182.

Underwood, E.J. 1977. Trace Elements in Human and Animal Nutrition, Academic Press, New York.

Vogt, R., R. Sander, R.Von Glasow and P.J. Crutzen. 1999. Iodine chemistry and its role in halogen activation and ozone loss in the marine boundary layer: a model study. J. Atmos. Chem. 32: 375–395.

Whitehead, D.C. 1984. The distribution and transformations of iodine in the environment. Environ. Inter. 10: 321–339.

World Health Organization, UNICEF and International Council for the Control of Iodine Deficiency Disorder. 2001. WHO, WHO/Euro/NUT. Geneva. pp. 1–107.

Zimmermann, M.B., P.L. Jooste and C.S. Pnadav. 2008. Iodine-deficiency disorders. Lancet. 372: 1251–1262.

Speciation and Determination of Tellurium in Water, Soil, Sediment and other Environmental Samples

M.S. El-Shahawi,[1,a],* H.M. Al-Saidi,[2] E.A. Al-Harbi,[3] A.S. Bashammakh[1] and A.A. Alsibbai[1]

1. Introduction

In the past, the determination of total element concentrations was considered to be enough for clinical and environmental analyses. However, the knowledge of chemical speciation has primary importance because the toxicity, mobility, bioavailability and bioaccumulation of elements largely depend on their chemical species (Ebdon et al. 2001). Chemical species are specific forms of an element defined as isotopic composition, electronic or

[1]Department of Chemistry, Faculty of Science, King Abdulaziz University, P.O.Box 80203, Jeddah 21589, Saudi Arabia.
[a]Email: malsaeed@kau.edu. sa, mohammad_el_shahawi@yahoo.co.uk
[2]Department of Chemistry, University College, Umm Al Qura University, Makkah, Saudi Arabia.
[3]Department of Applied Chemistry, College of Applied Science, Taibah University, Al-Madinah Al-Munawarah, P.O. Box: 3193, Saudi Arabia.
*Corresponding author

oxidation state and/or complex or molecular structure (Templeton et al. 2000). Tellurite, Te (IV) is 10 times more toxic than tellurate, Te (VI) (Huang and Hu 2008). Therefore, speciation studies are important for the accurate evaluation of pollution levels in a wide variety of samples such as water, soil and sediment.

Speciation analyses have been widely used to identify the metal species that have adverse effects on living organisms (Cava-Montesinos et al. 2004). The interaction of metal ions with biota is highly dependent on their oxidation state and/or organic or inorganic structure rather than to their total concentration. Due to its wide applications as alloying additives in steel, copper, lead alloys, vulcanizing agent and accelerator in the processing of rubber, and the component of catalysts for synthetic fiber production, the level of exposure of tellurium has largely increased in the last years (Jain and Ali 2000). Therefore, an overview on distribution, toxicity and chemical speciation of tellurium will be presented in this chapter. The recent developments in the analytical techniques frequently used for chemical speciation of tellurium will be discussed in detail.

2. Occurrence, Uses and Toxicity of Tellurium

Tellurium was discovered in 1782 by the Austrian mineralogist Baron Franz Joseph Muller (Nordberg et al. 2007, Bragnall 1966). The element seldom occurs in its pure state in nature. However, it is usually found as a compound in ores of gold, silver, copper, lead, mercury or bismuth (Cava-Montesinos et al. 2004, Jain and Ali 2000). The most well-known ores containing tellurium are the ores of gold and silver, e.g., Calaverite ($AuTe_2$), Sylvanite ($AgAuTe_4$) (Tsai and Jan 1993, Emsley 2002). Tellurium is considered one of the rarest stable solid elements on the Earth's crust even in comparison with some lanthanides and it is usually associated with selenium at trace and ultra trace levels (Cooper 1971, Sadeh 1987). The level of Te in soil ranges from 0.05 to 30 µg kg^{-1}, while in sea Te level is about 0.15 ng L^{-1} (Emsley 2002).

Tellurium exists in the environment in elemental (Te), inorganic–telluride (Te^{2-}), Te (IV), Te (VI) and organic (dimethyl telluride (CH_3TeCH_3)) forms (Cooper 1971), oxyanions are more stable and common than the elemental state (Summers and Jacoby 1977). Many Te compounds are redox-active, with the formal oxidation states of Te in these compounds range from −2 to +6 (Yosef et al. 2007, Cunha et al. 2009). Chemical and physical properties of tellurium are largely similar to those of selenium (Cooper 1971). At physiological pH, Te is less soluble than Se and its oxidation states (IV), e.g., TeO_3^{2-}, and (VI), e.g., TeO_4^{2-} or TeO_2 occur easily. Tellurium has many stable isotopes involving [130]Te, 34.48%; [128]Te, 31.79%; [126]Te, 18.71%; [125]Te, 6.99%; [124]Te, 4.61%; [122]Te, 2.46%; [123]Te, 0.87%; [120]Te, 0.089% (CRC Handbook 1975).

Tellurium has been used in metallurgic industry to improve the mechanical properties of steels and other ferrous alloys. It is also employed in the form of alloys with copper and lead, used in welding and chemical equipment. Tellurium is utilized in the rubber industry to improve heat resistance and as a catalyst in many industrial applications. Thermo electrical devices used in power plants and refrigeration are based on the see beck effect, for current production when heating up a junction of two different metals, and in the Peltier effect that consists of heat transfer by means of an electrical current flow through a metallic junction (Geological Survey 2011). Due to relatively high conductivity, Te is used in semiconductors and other electronic devices. Tellurium also finds application in daylight lamps and ceramics. Photographic and pharmaceutical industries also utilize some quantities of this element (Kabata-Pendias 2011). Tellurium found historical applications in the treatment of microbial infections prior to the discovery of antibiotics. Early documentation in 1926 reports its use in the treatment of syphilis and leprosy (De Meio and Henriques 1947).

The oxyanion tellurite has been used in microbiology since the 1930s when Alexander Fleming reported its antibacterial properties (Fleming 1932, Fleming and Young 1940). In 1984, it was suggested that TeO_2^{3-} could be a potential antisickling agent of red blood cells in the treatment of sickle cell anemia (Asakura et al. 1984). While in 1988, tellurium-containing immuno modulating drugs were proposed as treatment agents for AIDS (Jacobs 1989). On the other hand, Te compounds, especially organo-tellurium compounds, have been found to exhibit probable antioxidative and anticancer properties (Bagnall 1966, Jacob et al. 2000, Engman et al. 2002).

The toxicology of tellurium has received less attention compared to selenium. However, toxicity of tellurium compounds largely depends on the chemical form and the quantity of the element consumed. Tellurium showed acute toxicity in young children when ingestion of metal-oxidizing solutions containing significant concentrations of Te (Yarema and Curry 2005, Taylor 1996). Tellurium can be accumulated in the kidney, heart, liver, spleen and muscle and its content in liver and kidney as an example in excess of 0.002 g kg^{-1} (Chai and Zhu 1994). The occurrence of Te in mammalian tissues is at the level of 20 µg kg^{-1} (Chai and Zhu 1994). Tellurium slowly leaches out and its half-time is estimated to be 600 d (Hollins 1969). Tellurium level in human urine concentration is reported < 1 µg l^{-1} (Reimann and Caritat 1998), whereas in blood it is less than 0.3 µg l^{-1}. Tellurium toxicity is associated with impaired neurotransmission that affects saliva and sweat secretion in humans. The daily dietary intake of Te by adults from food has been estimated at around 0.1 mg kg^{-1} BW (Kobayashi 2004).

3. Analytical Methods for Chemical Speciation and Determination of Tellurium

Many analytical techniques such as voltammetry, flame atomic absorption spectrometry (FAAS), electrothermal atomic absorption spectrometry (ET–AAS), inductively coupled plasma-mass spectroscopy (ICP–MS), inductively coupled plasma-optical emission spectroscopy (ICP–OES), X-ray fluorescence, spectrophotometry, and electrophoresis have been used for Te determination in various environmental matrices (Table 14.1). Among these techniques, AAS and ICP coupled with a suitable hydride generation system have been widely used for the chemical speciation of Te in various samples due to their high sensitivity. However, the direct determination of Te using most detection techniques mentioned above is still a difficult task due to its occurrence in trace and ultra trace levels and strong interference of the sample matrices. Therefore, a preconcentration step becomes necessary to get accurate results. Below, we will discuss some different analytical techniques widely used for the estimation of Te in detail.

3.1 Determination of Te in water

3.1.1 Speciation of Te using hydride generation

Hydride generation (HG) using tetrahydroborate (THB) at present is considered to be one of the most widespread derivatization procedures for the determination of trace and ultra-trace amounts of chemical vapor forming elements, in combination with various atomic spectrophotometric techniques (Xi et al. 2010). Tellurium (IV) is most often determined by HG followed by atomic absorption spectrometry (AAS) or atomic fluorescence (AFS) (El-Hadri et al. 2005), while, tellurium (VI) concentration is obtained by the difference between the total tellurium and Te (IV) contents. Thus, both HG-AAS and GH-AFS techniques have found wide applications in chemical speciation of Te (IV) and Te (VI) in various matrices.

3.1.2 Determination of Te using liquid-phase microextraction

Atomic spectrometric techniques, e.g., AAS, ICP–OES, atomic fluorescence spectrometry and ICP–MS in combination with an efficient pre-concentration technique have been successfully used for the chemical speciation of tellurium (El-Hadri et al. 2005). Most of these methods are based upon selective extraction of the Te (IV) complexes. Therefore, Te (II) or Te (0) have to be oxidized to Te (IV), while, Te (VI) must be reduced to Te (IV) for determination of total Te species (Cava-Montesinos et al. 2004).

Table 14.1 List of some analytical methods used for tellurium determination in variety of samples.*

Technique	Sample type	Pre-concentration technique	Analytical features	Ref.
ICP-MS	Freshwater	Non polar silica-based octadecyl (C18) sorbent	LOD= 3 ng L^{-1}; Linearity =0–5.0 μgL^{-1}	Yu et al. 2003
LC–HG–AFS	Wastewater	Anion-exchange-complexing agent	LOD = 600 ng L^{-1} (Te^{4+}), 700 ng L^{-1} (Te^{6+}); Linearity = 2–100 μgL^{-1}	Viñas et al. 2005
ICP-MS	Seawater	γ-MPTMS	LOD = 0.079 ng L^{-1}	Huang and Hu 2008
GF-AAS	Tapwater	Dowex X8 and XAD resins	(Dowex 1X8) LOD = 7 ng L^{-1}; Linearity (23.0 –400 ng L^{-1}); (XAD) LOD = 66 ng L^{-1} Linearity (220–1000 ng L^{-1})	Pedro et al. 2008
GF-AAS	Water	Voltammetry	LOD = 2.0 ng L^{-1}	Ghasemi et al. 2009
GF-AAS	Natural water	HF-LPME	LOD = 4.0 ng L^{-1} Linearity = 0.040–40 μgL^{-1}	Ghasemi et al. 2010
GF-AAS	Natural water	DLLME	LOD = 4.0 ng L^{-1}; Linearity = 0.015–1.0 ngmL^{-1}	Najafi et al. 2010
HG-CL	Natural water		LOD = 2000.0 ng L^{-1}; Linearity = 10–200 μg L^{-1}	Luo et al. 2011
ICP-MS	water	Modified silica sorbents	LOD < 10 μg L^{-1}	Urbánková et al. 2011
GF-AAS	Natural water	USAE-SFODME	LOD = 3.0 ng L^{-1}; Linearity= 0.01–0.24 ng mL^{-1}	Fathirad et al. 2012
FAAS	Geological and sediment	MIBK extraction	—	Hubert and Chao 1985)
GF-AAS	Ores and sediments	Iron collection and xanthate extraction	LOD = 0.1 μg g^{-1}	Donaldon and Leaver 1990
HG-ICP-MS	Geological matrices		LOD = 1.0 mg Kg^{-1}	Hall and Pelchat 1997

Table 14.1 contd....

Table 14.1 contd.

Technique	Sample type	Pre-concentration technique	Analytical features	Ref.
CSV	Geological matrices	SPE by chelating resin	LOD < μg Kg^{-1}	Ferri et al. 1998
HG–AFS	Air	Extraction with aqua regia	LOD= 6 × 10^{-3}–0.2 ng m^{-3}; Linearity =1.04–9.7 ng ml^{-1}	Moscoso-Pérez et al. 2004
HG–AFS	Tea leaves	-	LOD= 0.0022 μg g^{-1}	Zhang et al. 2011
HG-IAT-FAAS	Fly ash; Garlic		LOD = 900.0 ng L^{-1}; The precision, as RSD, 7.0% (n = 6)	Matusiewicz and Krawczyk 2007
HPLC/ICP-MS	Water; urine; fish and soil		LOD = 0.41μg L^{-1}	Lindemann et al. 2000
D-HGAAS	Urine		LOD = 260.0 ng L^{-1}	Ha et al. 2001
HG–AFS	Milk	Leaching of slurries by sonication	LOD = 0.57 ng g^{-1}; Linearity= 0.0–0.5 ng mL^{-1}	Ródenas-Torralba et al. 2005
Differential oscillopolarography	Heart, Liver, Kidney, Spleen and Lung	Solvent extraction	LOD = 7.0 ngL^{-1}; Linearity=0.02–2.0ng mL^{-1}; RSD = 0.54%–2.10%, Recovery 98.4%–95.7%	Li et al. 2000

*Abbreviations: CSV = Cathodic stripping voltammetry; D-HGAAS = Hydride generation atomic absorption spectrometry; DLLME = dispersive liquid–liquid microextraction; HF-LPME = Hollow fiber-based liquid-phase microextraction; HG-AFS = hydride generation atomic fluorescence spectrometry; HG-CL = Hydride generation chemiluminescence; HG-IAT-FAAS = Hydride generation with *in situ* trapping flame atomic absorption spectrometry; HG-ICP-MS = Hydride Generation Inductively Coupled Plasma Mass Spectrometry; HPLC/ICP-MS = High-performance liquid chromatography/Inductively coupled plasma mass spectrometry; GF–AAS= Graphite furnace atomic absorption spectrometry; FAAS = Flame atomic absorption spectrometry; LC–HG–AFS=liquid chromatography/hydride-generation/atomic fluorescence spectrometry; MPTMS= mercaptopropyltrimethoxysilane; ICP-MS = Inductively coupled plasma mass spectrometry; USAE-SFODME = Ultrasound-assisted emulsification solidified floating organic drops microextraction.

There are many studies describing the use of liquid-phase microextraction techniques in combination with AAS for the chemical speciation of Te in different environmental samples.

Recently, excellent pre-concentration techniques before the measurement step by AAS have been reported for the chemical speciation and total determination of various tellurium species in complicated matrices (Ghasemi et al. 2010, Najafi et al. 2010, Fathirad et al. 2012). The method of Ghasemi et al. 2010 was based upon extraction of Te (IV) complex species with ammonium pyrolidine dithiocarbamate (APDC) by hollow fiber liquid phase microextraction (HF-LPME), while dispersive liquid–liquid microextraction (DLLME) and ultrasound-assisted emulsification solidified floating organic drops microextraction (USAE-SFODME) have been reported by Najafi et al. 2010 and Fathirad et al. 2012, respectively. In these methods Te (VI) was first reduced to Te (IV), and then the microextraction technique was then applied before AAS determination of total Te.

3.1.3 Determination of Te using Solid Phase Extraction

The applications of solid–phase extraction (SPE) were more common than those of liquid phase extraction for Te speciation analysis. In general, SPE is based on the utilization of a major constituent as the bonded stationary phase immobilized different ligand or functional group (Marahel et al. 2011). Mercapto-modified silica microcolumn was used as an effective stationary phase for developing a novel and sensitive method for the determination of trace amounts of Te (IV) in waters by HG-AAS (Körez et al. 2000). This method offered the limit of detection of less than 0.04 ng ml^{-1} in sea water.

Yu et al. 2003 have described a procedure for the speciation of tellurium (IV, VI) by SPE-ICP-MS. In this method, Te (IV) complex with APDC was completely retained on a nonpolar silica-based octadecyl (C18) sorbent-containing SPE cartridge, while the uncomplexed Te (VI) passed through the cartridge and remained as a free species in the solution prior to its determination by ICP-MS. The Te (IV) concentration was calculated as the difference between total tellurium obtained by ICP–MS and Te(VI) concentrations. Simultaneous determination of Te, As, Sb, and Se by ICP–MS after pre-concentration of analytes using modified silica sorbent was describe Urbánková et al. (Urbánková et al. 2011). The analytes were retained on modified silica in the form of ion associates with cationic surfactants in the presence of the presence of 4-(2-pyridylazo) resorcinol, 2-pyrrolidinecarbodithioate and 8-hydroxyquinoline-5-sulfonic acid as chelating agents. The quantitative retention occurred at pH 7 ± 0.2 and the mixture of acetone with ethanol in ratio 1:1 in the presence of 0.1 mol L^{-1} HCl was used for the quantitative elution. Organic solvents and the excess of acid were removed by evaporation prior to the determination by ICP-MS.

Tellurium is present in seawater in the form of tellurium (IV) and (VI) at trace level (ng L^{-1}), on the other hand, the high concentration of matrix ions in sea water causes instrumental drift, signal suppression and clogging of the sample introduction system of instrument. Therefore, pre-concentration and extraction of Te from the matrix of seawater becomes an important step prior to measurement. Huang and Hu used the magnetic nanoparticles (MNPs) as SPE adsorbent for the separation of inorganic tellurium species from seawaters prior to their determination by ICP–MS (Huang and Hu 2008). Within the pH range of 2–9, Te (IV) could be quantitatively adsorbed on γ-mercaptopropyltrimethoxysilane (γ–MPTMS) modified silica-coated magnetic nanoparticles (MNPs), while the Te (VI) was not retained and remained in solution. Like other magnetic separation techniques, tellurite loaded on MNPs could be separated easily from the aqueous solution by applying external magnetic field without filtration or centrifugation. Under the optimal conditions, the LOD obtained for Te (IV) was 0.079 ng/L, while the precision was 7.0 percent.

3.1.4 Determination of Te by chemiluminescence

HG technique coupled with chemiluminescence (CL) has been successfully used to minimize matrix interference during determination of tellurium (IV) in natural water (Luo et al. 2011). The developed method was based upon the strong CL emission obtained by the reaction between hydrogen telluride and luminol in basic medium. Under the optimized condition, excellent linear dynamic range of CL intensity vs. Te (IV) concentrations (10–200 μg L^{-1}) was achieved. The limits of detection and quantification by this method were found excellent compared to the reported methods in the literature.

3.1.5 On-line pre-concentration of Te in water

Two flow-injection graphite furnace atomic absorption spectrometry (FI-GFAAS) systems for the determination of Te in tap water have been reported by Pedro et al. (Pedro et al. 2008). Because Te (VI) is hydride inactive, this procedure is suitable for Te (IV) and if Te (VI) needs to be detected, the pre-reduction step is necessary. The first approach was based on the on-line pre-concentration of the analyte onto a strong anionic resin (Dowex X8) using as packaging material of a micro-column inserted in the flow system. The second approach was based on the co-precipitation of tellurium with La(OH)$_3$ followed by retention onto adsorbent fillings (XAD resins). Although, the use of Dowex 1X8 as material filling fulfills the requirements for the on-line operation of the FI-GFAAS system, the low selectivity of the anionic exchange resin constrains its use in complex matrix like saline samples. On the other hand, amberlite XAD shows a higher level of tolerance to the presence of dissolved ions. But neither automatic operation nor long-term use of the adsorbent resin is possible due to its severe compaction during the default sequence of washing of the GF auto sampler.

Pre-concentration of Te using selective electrodeposition with mercury-coated electrode was proposed for the chemical speciation of Te (IV) and Te (VI) in water samples by GF–AAS (Ghasemi et al. 2009). The method is based on the selective reduction of the Te (IV) at a controlled applied potential (2.0 V) on the mercury-coated electrode. In acidic media (1.0 mol L^{-1} HCl solution), only Te (IV) can be electrodeposited onto the mercury electrode surface and separated, while, Te (VI) remained in solution. The electrode was withdrawn from the solution and the spent electrolyte containing Te (VI) was measured by GFAAS. Tellurium (IV) was calculated from the difference between the measured total Te and Te (VI) content.

3.1.6 The chromatographic determination of Te

According to our knowledge, the applications of chromatography for the determination of Te species are limited. As an example, Liquid Chromatography–Hydride Generation coupled Atomic Fluorescence Spectrometry (LC–HG–AFS) was used for speciation of Te (IV) and Te (VI) in wastewater (Viñas et al. 2005). Anion-exchange LC with multidentate complexing agents, e.g., EDTA and potassium hydrogen phthalate (KHP) in the mobile phase was used to improve column efficiency because such agents have very high complexing capacity and can transform the positive metal ion into negatively charged complexes.

3.1.7 Spectrophotometric methods of Te determination in water

Direct determination of Te using spectrophotometric methods is rarely used due to the low sensitivity of such methods. However, survey of the literature reveals that a few spectrophotometric methods were proposed for the determination of Te in the environmental samples. One of the most recent spectrophotometric methods was proposed for direct determination of Te in water, plant materials and soil (Prasad et al. 2007). The method is based on the oxidation of leuco methylene green (LMG) to its blue form by Te (IV) in acetate buffer medium (pH 3.0 to 5.0). Measurements were linear over the concentration range 0.4–2.5 µg mL^{-1}. The extractive spectrophotometric method with Bismuthiol II is highly sensitive and selective for the determination of Te in water and other environmental samples (Toshida et al. 1966). In acid media, the complexing agent bismuthiol II reacts with Te (IV) to form a neutral 1:4 (Te: reagent) complex which is extractable into $CHCl_3$. The unreacted (free) bismuthiol II is also extracted into chloroform and it absorbs in the same region with its Te (IV) complex. Excess bismuthiol II in the organic phase was stripped into the aqueous phase by shaking with buffer solution of pH $\approx$ 8. However, this method suffers from serious interference of Hg^{2+}, Se^{4+} Fe^{3+} and Sb^{3+}. From the view of sustainable methods, the method is not recommend because of the use of chloroform.

3.2 Tellurium in plants, soil and geological samples

3.2.1 Tellurium in plants

Te concentrations in plants varied from < 0.013 to 0.35 mg kg^{-1} and do not exceed 1 mg kg^{-1} in contaminated systems (Moreno et al. 2007). For uncontaminated systems Te concentrations less than 0.001 mg kg^{-1} have been reported by Fathirad et al. 2012. Some plants, such as tea, accumulate Te because Te displaces Se and inhibits biological activities (Nordberg et al. 2007). Therefore, simultaneous determination of both Te and Se in plants samples is a worthwhile task. Studies concerned with development of new methods for the determination of Te in plants are still limited. Zhang et al. 2011 have designed a new multi-channel hydride generation atomic fluorescence spectrometry (HG–AFS) for determination of As, Bi, Se, and Te in tea plants (Zhang et al. 2011). After sample pre-treatment using HNO_3 and H_2O_2 and under optimal conditions, the detection limits for As, Bi, Te and Se in tea leaves were 0.0152, 0.0080, 0.0022, 0.0068 µg g^{-1}, respectively.

3.2.2 Tellurium in soils

Limited data are available on the occurrence of tellurium in soils. As a representative example, Te content in some types in USA is in the range 0.02 and 0.69 mg kg^{-1} (Govindaraju 1994). Trace and ultra trace levels of tellurium in soil make the direct determination of Te element a very difficult task. Chemical speciation of Te and Se was carried out simultaneously using chelating resin combined with cathodic stripping voltammetry (Ferri et al. 1998). The developed method was found suitable for the determination of Te in soil at µg kg^{-1} with good precision. Pre-treatment by microwave digest in the acid mixture of HCl, HNO_3, and HF has been reported as an effective digestion approach for Te in soil (Hubert and Chao 1985). Determination of Au, In, Te, and Th in geological materials by flame atomic-absorption spectrophotometry has been reported by Hubert and Chao 1985. In this method, the sample was decomposed by a mixture of hydrofluoric acid, aqua regia, and hydrobromic acid-bromine solution. The analytes were separated and pre-concentrated by two steps–MIBK extraction at two concentrations of HBr. First, gold and thallium were extracted from HBr (0.1 mol L^{-1}) medium, while Te and In were separated by HBr (3.0 mol L^{-1}).

Total concentrations of tellurium in dried mine tailings and in downstream waters and sediments have been reported (Wray 1998, Moreno et al. 2007). Moreno et al. 2007 found that Te and other base metals are mainly associated with the particulate inhalable fraction of the dry tailings, while, elevated levels of these trace elements have been reported in waters and sediments downstream from the same tailings deposits (Kyle et al. 2011). Harada and Takahashi 2009 studied the distribution and speciation of Te and Se between the solid and aqueous phases in synthetic soil. Under oxen conditions both elements are mainly associated with iron (III) hydroxides, and Te (IV) and (VI) species were both found to inner-sphere complexes. Under reducing conditions, tellurium (0) species was formed in batch studies. Tellurium distribution between soil and water was much lower than Se under a wide range of redox conditions, likely due to higher affinities of Te (IV) and (VI) for iron (III) hydroxides.

3.2.3 Geological samples

Donaldson and Leaver described a method for quantifying Te down to ~ 0.01 μg g^{-1} in ores, rocks, soils and sediments (Donaldson and Leaver 1990). After sample decomposition and evaporation of the solution to incipient dryness, Te is separated from > 300 μg of copper by co-precipitation with hydrous ferric oxide from an ammonia solution and the precipitate is dissolved in HCl (1.0 mol L^{-1}). Tellurium in the resultant solutions is reduced to Te (VI) by heating and separated from iron, lead and various other elements by a single cyclohexane extraction of its xanthate complex from HCl (~ 9.5 mol L^{-1}) in the presence of thiosemicarbazide as a complexing agent for copper. After washing with HCl (10.0 mol L^{-1}) hydrochloric acid followed by water is used to remove residual iron, chloride and soluble salts, tellurium is stripped from the extract with nitric acid (16.0 mol L^{-1}) and finally determined in a 2% v/v nitric acid medium, by graphite-furnace atomic-absorption spectrometry at 214.3 nm in the presence of nickel as a matrix modifier. Small amounts of gold and palladium, which are partly co-extracted as xanthates if the iron-collection step is omitted, do not interfere. The method is directly applicable, without the co-precipitation step, to most rocks, soils and sediments.

The influence of pre-treatment type on Te determination in some geological samples by GH–ICP–MS was critically investigated (Hall and Pelchat 1997). The study was based on the change of HCl concentration in the mixture of pre-treatment (aqua regia and HF-HClO$_4$-HNO$_3$-HCl).

Arsenic (III and V) were also examined as a potential interference and it was found that As (V) severely suppressed the Te signal in both the 2 and 4 mol L^{-1} HCl experiments. As (III) did not interfere and Ge, Sn and Pb were not measured, because formation of their hydrides at the concentration of HCl used (2–4 mol L^{-1}) would be negligible. 4 mol L^{-1} HCl was determined to be the preferred acid medium because of reduced interferences and a shorter wash-out time between samples (i.e., decreased memory effect).

Hydride generation combined with an integrated atom trap (HG-IAT) atomizer for flame AAS was proposed for the determination of Te in reference material (GBW 07302 Stream Sediment), coal fly ash and garlic (Matusiewicz and Krawczyk 2007). The results confirm that the present hyphenated technique using a continuous mode hydride generation gas phase *in situ* trapping on an integrated silica tube trap, followed by atomization in acetylene–air flame with simultaneous direct thermal heating of the collector (atomizer), can be used for the determination of trace amounts of Te in samples and reference material. Following the trapping stage, the performance of the device and related problems are quite similar to the case of hydride generation-electrothermal atomization (*in situ* trapping) HG-GF-AAS. The detection limit of this HG-IATFAAS system for Te is considerably improved compared with those reported for measurements of Te by any flame AAS approach. High-performance liquid chromatography (HPLC) coupled with inductively coupled plasma mass spectrometry (ICP-MS) was presented for speciation analysis of Te in a variety of samples involving soil (Lindemann et al. 2000). Tellurium was almost quantitatively extracted from samples using CH_3OH and H_2O, while, recoveries after extraction with water and sulfuric acid (0.01 mol L^{-1}) were below 20 percent.

3.3 Tellurium in food, biological samples and air

3.3.1 Tellurium in food

Tellurium (VI) is hydride inactive; therefore, this oxidation state must be quantitatively reduced to Te (IV). The reduction is performed mostly by HCl or HBr at various concentrations, temperatures and reaction times (Ulivo 1997). The main problem with the reduction procedures is the low selectivity, since an undetermined fraction of other inorganic and organic species may be converted into Te (IV) during the reduction step, causing an over-estimation of Te (VI). Thus, reduction of Te (VI) to Te (IV) has been carried out in a microwave oven for Te speciation in milk using hydride generation, atomic fluorescence (HG-AFS) (Ródenas-Torralba et al. 2005). Batch leaching of Te has been performed by sonication at room temperature for 10 min using aqua regia (Ródenas-Torralba et al. 2005). The extract was treated by $NaBH_4$ in HCl medium to form the corresponding hydrides and finally AFS measurements were processed in front of external calibrations prepared and measured in the same way as samples. The proposed method provided a high sampling frequency of 24 hr^{-1} for the determination of both, free Te (IV) ions and total Te, in a same sample. Hydride generation atomic fluorescence spectrometry (HG-AFS) after the microwave-assisted sample digestion procedure established for tellurium determination in milk. The method provides sensitivity values of 1591 and 997 fluorescence unit's ng ml^{-1} with detection limits of 0.015 ng ml^{-1} for Te.

Application of the methodology to the analysis of cow milk samples in the Spanish market gave evidence of the presence of concentration ranges from 1.04 to 9.7 ng ml^{-1} for Te having found a good comparability with data obtained after dry-ashing of samples (Cava-Montesinos 2003). Trace Te from some food samples typically consumed by the French population was determined during the second French Total Diet Study (TDS). Among the main trace element analyzed by inductively coupled plasma-mass spectrometry (ICP-MS) after microwave-assisted digestion. The contents were compared using data from worldwide total diet studies. Data for tin in canned food and beverages were compared with European guidelines. The food groups with the highest levels of Te were "Sweeteners, honey and confectionery" with dark chocolate, "Fish and fish products" and particularly "Shellfish" (Te (0.003 mg kg^{-1}) and "Fat and oil" (Milloura 2012).

3.3.2 Tellurium in biological samples

Aggarwal et al. 1994 have developed a method for the determination of Te in urine. The method has been based upon conversion of Te(IV) into volatile compounds using a Grignard reagent [4-fluorophenyl) magnesium bromide] in dry diethyl ether to give the final product [$Te(FC_6H_4)_2$]. After removing the excess of Grignard reagent, Te derivatives were extracted by toluene. The extract was evaporated to dryness and the residue was reconstituted in methylene chloride. The method was found suitable for GC–MS isotope dilution determination of Te in urine using [120]Te as an internal standard.

Hydride generation atomic absorption spectrometry with derivative signal processing (D–HG-AAS) has been proposed for Te determination in urine (Ha et al. 2001). The method offered limit of detection less than 0.2 µg L^{-1}, recovery of 98 percent and excellent sensitivity and selectivity for Te estimation in urine samples. Differential oscillopolarography has been used for speciation analysis of Te in biological samples, e.g., heart, liver, kidney, spleen and lung (Li et al. 2000). The method was based upon the activation effect of Te (IV) on the slow Pd (II)-catalyzed reaction between sodium hypophosphite and methyl red and the analyte was extracted from samples by liquid–liquid extraction using MIBK.

3.3.3 Tellurium in air samples

The concentration of Te in air ranges from 0.35 to 50 ng m^{-3} (Kobayashi 2004). Because of the low level of Te in various matrices, few methods were found suitable for Te detection in air. The most widely used technique for the determination of Te in air is hydride generation coupled with AAS or AFS. A highly sensitive and simple method, based on HG–AFS has been developed for the determination of Te (IV) in aqua regia extracts from atmospheric particulate matter samples (Moscoso-Pérez et al. 2004). Atmospheric particulates were collected on glass fiber filters using a medium volume sampler (PM10 particulate matter). Two-level factorial designs have been used to optimize the hydride generation atomic fluorescence spectrometry (HG–AFS) procedure. The effects of several parameters affecting the hydride generation efficiency (hydrochloric acid, sodium tetrahydroborate and potassium iodide concentrations and flow rates) were evaluated using a Plackett–Burman experimental design. The parameters affecting the hydride measurement, e.g., delay, analysis and memory times were also investigated. The significant parameters on using sodium tetrahydroborate and sodium tetrahydroborate at selected flow rate for Te(IV)) were optimized by using 2^n + star central composite design. Using a univariate approach these parameters were optimized. The accuracy of methods were verified using several certified reference materials: SRM 1648 (urban particulate matter) and CRM 1649a (urban dust). Detection limits in the range of 6×10^{-3} to 0.2 ng m^{-3} was achieved. The developed methods were applied to several atmospheric particulate matter samples corresponding to A Coruña city (NW Spain).

4. Conclusion

Determination of different chemical forms of tellurium in different matrices (water, food, biological and air samples) is important because of their various toxicological effects. In analysis of tellurium in environmental samples, the techniques used should be sensitive and selective. The inherent advantages and disadvantages of each method can be used to select the most appropriate method based on the type of sample matrix to be analyzed and the tellurium species and concentration levels to be determined.

References

Aggarwal, S.K., M. Kinter, J. Nicholson and D.A. Herold. 1994. Determination of tellurium in urine by isotope dilution gas chromatography/mass spectrometry using (4-fluorophenyl) magnesium bromide as a derivatizing agent and a comparison with electrothermal atomic absorption spectrometry. Anal. Chem. 66: 1316–1322.

Asakura, T., Y. Shibutani, M.P. Reilly and R.H. DeMeio. 1984. Anti stickling effect of tellurite: apotent membrane acting agent *in vitro*. Blood. 64: 305–307.

Bragnall, K.W. 1966. The Chemistry of Selenium, Tellurium, and Polonium. Elsevier Publishing Co., London, UK.

Cava-Montesinos, P., A. de la Guardia, C. Teutsch, M.L. Cervera and M. de la Guardia. 2004. Speciation of selenium and tellurium in milk by hydride generation atomic fluorescence spectrometry.J. Anal. Atom. Spectrom. 19: 696–699.

Chai, Z.F. and H.M. Zhu. 1994. Introduction to Trace Element Chemistry, Atomic Energy Press. Peking, China.

Cooper, W.C. 1971. Tellurium. Van Nostrand Reinhod Co., New York. USA.

CRC Handbook of Chemistry and Physics. 1975. CRC Press. Cleveland, OH, US.

Cunha, R.L., I.E. Gouvea and L. Juliano. 2009. A glimpse on biological activities of tellurium compounds. An Acad Bras Cienc. 81: 393–407.

De Meio, R.H. and F.C. Henriques. 1947.Tellurium IV, excretion and distribution in tissues studied with a radioactive isotope. J. Biol. Chem 169: 609–623.

Donaldson, E.M. and M.E. Leaver. 1990. Determination of tellurium in ores, concentrates and related materials by graphite-furnace atomic-absorption spectrometry after separations by iron collection and xanthate extraction. Talanta. 37: 173–183.

Ebdon, L., L. Pitts, R. Cornelis, H. Crews, O.F.X. Donard and P. Quevauviller. 2001. Trace Element Speciation for Environment, Food and Health, The Royal Society of Chemistry, Cambridge, UK.

Emsley, J. 2002. Nature's Building Blocks: An A-Z Guide to the Elements; Oxford University Press, New York, USA.

El-Hadri, F., A. Morales–Rubio and M. de la Guardia. 2005. Determination of tellurium by continuous hydride generation and atomic fluorescence spectrometry. Ciencia. 13(2): 218–227.

Engman, L., N. Al-Maharik, M. McNaughton, A. Birmingham and G. Powis. 2002. Thioredoxin reductase and cancer cell growth inhibition by organotellurium antioxidants. Anticancer Drugs. 14: 153–161.

Fathirad, F., D. Afzali, A. Mostafavi and M. Ghanbarian. 2012. Ultrasound-assisted emulsification solidified floating organic drops microextraction of ultra trace amount of Te (IV) prior to graphite furnace atomic absorption spectrometry determination. Talanta. 88: 759–764.

Ferri, T., S. Rossi and P. Sangiorgio. 1998. Simultaneous determination of the speciation of selenium and tellurium in geological matrices by use of an iron(III)-modified chelating resin and cathodic stripping voltammetry. Anal. Chim. Acta. 361: 113–123.

Fleming, A. 1932. On the specific antibacterial properties of penicillin and potassium tellurite. J. Pathol. Bacteria. 35: 831–842.

Fleming, A. and M.Y. Young. 1940. The inhibitor action of Potassium tellurite oncoliform bacteria. J. Pathol. Bacteriol. 51: 29–35.

Ghasemi, E., N.M. Najafi, S. Seidi, F. Raofie and A. Ghassempour. 2009. Speciation and Determination of Trace Inorganic Tellurium in Environmental Samples by Electrodeposition-Electrothermal Atomic Absorption Spectroscopy. J. Anal. At. Spectrom. 24: 1446–1451.

Ghasemi, E., N.M. Najafi, F. Raofie and A. Ghassempour. 2010. Simultaneous speciation and preconcentration of ultra traces of inorganic tellurium and selenium in environmental samples by hollow fiber liquid phase microextraction prior to electrothermal atomic absorption spectroscopy determination. J. Hazard. Mater. 181: 491–496.

Govindaraju, K. 1994. Compilation of working values and sample description for 383 geostandards. Geostand Newsletters. (Special Issue) 18: 1–158.

Ha, J., H.W. Sun, J.M. Sun, D.Q. Zhang and L.L. Yang. 2001. Determination of tellurium in urine by hydride generation atomic absorption spectrometry with derivative signal processing. Anal. Chim. Acta. 448: 145–149.

Hall, G.E.M. and J.-C. Pelchat. 1997. Analysis of Geological Materials for Bismuth, Antimony, Seleniumand Tellurium by Continuous Flow Hydride Generation Inductively Coupled Plasma Mass Spectrometry. J. Anal. At. Spectrom. 12 : 97–102.

Harada, T. and Y. Takahashi. 2009. Origin of the difference in the distribution behavior of tellurium and selenium in a soil–water system. Geochim. Cosmochim. Acta. 72: 1281–1294.

Hollins, J.G. 1969. The metabolim of tellurium in rats. Health Phys. 17: 495–505.

Huang, C. and B. Hu. 2008. Speciation of inorganic tellurium from seawater by ICP-MS following magnetic SPE separation and Preconcentration. J. Sep. Sci. 31: 760–767.

Hubert, A.E. and T.T. Chao. 1985. Determination of gold, indium, tellurium and thallium in the same sample digest of geological materials by atomic-absorption spectroscopy and two-step solvent extraction. Talanta. 32: 568–570.

Jacob, C., G.E. Arteel, T. Kanda, L. Engman and H. Sies. 2000. Water-soluble organotellurium compounds: catalytic protection against peroxynitrite and release of zinc from metallothionein. Chem. Res. Toxicol. 13: 3–9.

Jacobs, J.L. 1989. Immunologic developments in AIDS. Year Immunol. 4: 276–285.

Jain, C.K. and I. Ali. 2000. Arsenic: occurrence, toxicity and speciation techniques. Water Res. 34: 4304 –4312.

Kabata-Pendias, A. 2011. 21 Elements of Group 16 (Previously Group VIa). *In:* Trace Elements in Soils and Plants. Taylor and Francis Group, LLC, CRC Press New York, USA. p. 381.

Kobayashi, R. 2004. Tellurium. *In:* E. Merian, M. Anke, M. Ihnat and M. Stoeppler [eds.]. Elements and their compounds in the environment. 2nd edn. Wiley-VCH. Weinheim. pp. 1407–1414.

Körez, A., A.E. Eroğlu, M. Volkan and O.Y. Ataman. 2000. Speciation and preconcentration of inorganic tellurium from waters using a mercapto silica microcolumn and determination by hydride generation atomic absorption spectrometry. J. Anal. At. Spectrom. 15: 1599–1605.

Kyle, J.H., P.L. Breuer, K.G. Bunney and R. Pleysier. 2011. Review of trace toxic elements (Pb, Cd, Hg, As, Sb, Bi, Se, Te) and their deportment in gold processing: Part II: Deportment in gold ore processing by cyanidation. Hydrometallurgy. 111–112: 10–20.

Li,L., P. He and Y. Fang. 2000. Catalytic determination of ultra trace amounts of tellurium by 2.5-order differential oscillopolarography. Fresenius J. Anal. Chem. 366: 239–243.

Lindemann, T., A. Prange, W. Dannecker and B. Neidhart. 2000. Stability studies of arsenic, selenium, antimony and tellurium species in water, urine, fish and soil extracts using HPLC/ICP-MS. Fresenius J. Anal. Chem. 368: 214–220.

Luo, L., Y. Tang, M. Xi, W. Li, Y. Lv and K. Xu. 2011. Hydride generation induced chemiluminescence for the determination of tellurium (IV). Microchem. J. 98: 51–55.

Marahel, F., M. Ghaedi, M. Montazerozozhori, M.N. Biyareh, S.N. Kokhdan and M. Soylak. 2011. Solid Phase Extraction and Determination of Trace Amount of Some Metal Ions on Duolite XAD 761 Modified with a New Schiff Base as Chelating Agent in Some Food Samples. Food and Chemical Toxicology. 49: 208–214.

Matusiewicz, H. and M. Krawczyk. 2007. Determination of tellurium by hydride generation with in situ trapping flame atomic absorption spectrometry. Spectrochim. Acta. Part B. 62: 309–316.

Milloura, S., L. Noël, R. Chekri, C. Vastel, A. Kadar, V. Sirot, J. Leblanc and T. Guérin. 2012. Strontium, silver, tin, iron, tellurium, gallium, germanium, barium and vanadium levels in foodstuffs from the Second French Total Diet Study, J. Food Comp. Anal. 25: 108–129.

Moreno, O., A., I. MacDonald and W. Gibbons. 2007. Preferential fractionation of trace metals–metalloids into PM10 resuspended from contaminated gold mine tailings at Rodalquilar, Spain. Water, Air, and Soil Pollution. 179: 93–105.

Moscoso-Pérez, C., J. Moreda-Piñeiro, P. López-Mahía, S. Muniategui-Lorenzo, E. Fernández-Fernández and D. Prada-Rodríguez. 2004. Hydride generation atomic fluorescence spectrometric determination of As, Bi, Sb, Se(IV) and Te(IV) in aqua regia extracts from atmospheric particulate matter using multivariate optimization. Anal. Chim. Acta. 526: 185–192.

Najafi, N.M., H. Tavakoli, R. Alizadeh and S. Seidi. 2010. Speciation and determination of ultratraceamounts of inorganic tellurium in environmental water samples by dispersive liquid–liquid microextraction and electrothermal atomic absorption spectrometry. Anal. Chim. Acta. 670: 18–23.

Nordberg, G., B. Fowler, M. Nordberg and L. Friberg. 2007. Handbook on the Toxicology of Metals. 3rd edn. Elsevier. San Diego Califorina, USA.

Pedro, J., J. Stripekis, A. Bonivardi and M. Tudino. 2008. Determination of tellurium at ultra-trace levels in drinking water by on-line solid phase extraction coupled to graphite furnace atomic absorption spectrometer. Spectrochim. Acta Part B: Atomic Spectroscopy. 63: 86–91.

Prasad, P.R., J.D. Kumar, B.K. Priya, P. Subrahmanyam, S. Ramanaiah and P. Chiranjeevi. 2007. Determination of tellurium(IV) in various environmental samples with spectrophotometry. E-J. Chem. 4: 354–362.

Reimann, C. and P. de Caritat. 1998. Chemical elements in the environment. Springer-Verlag, Berlin Heidelberg.

Ródenas-Torralba, E., A. Morales-Rubio and M. de la Guardia. 2005. Multicommutation hydride generation atomic fluorescence determination of inorganic tellurium species in milk. Food Chem. 91: 181–189.

Sadeh, T. 1987. *In*: S. Patai [ed.]. Biological and biochemical aspects of tellurim derivatives. The Chemistry of Organic Selenium and Tellurium Compounds. Wiley. Chichester. pp. 367–376.

Stewart, D.A., P.M. Tyroler and S. Stupavsky. 1985. The Removal of selenium and tellurium from copper electrolyte at INCO's copper refinery electrowinning department. *In*: A.J. Oliver [ed.]. 15th Annual Hydrometallurgical Meeting. Vancouver, Canada. CIM Metallurgical Society. Paper No. 18.

Summers, A.O. and G.A. Jacoby. 1977. Plasmid mediated resistance to tellurium compounds. J. Bacteriol. 129: 276– 281.

Taylor, A. 1996. Biochemistry of tellurium. Biol. Trace Elem Res. 55: 231–239.

Templeton, D.M., F. Ariese, R. Cornelis, L.G. Danielsson, H. Muntau, H.P. van Leenwen and R. Lobinski. 2000. *Guidelines for terms related to chemical speciation and fractionation of elements. Definitions, structural aspects, and methodological approaches.* Pure Appl. Chem. 72: 1453–1470.

Toshida, H., M. Taga and S. Hikime. 1966. Spectrophotometric determination of small amounts of tellurium with bismuthiol II. Talanta. 13: 185–191.

Tsai, S.J. and C.C. Jan. 1993. Determination of trace amounts of thallium and tellurium in nickel-base alloys by electrothermal-atomic absorption spectroscopy. Analyst. 118: 1183–1191.

U.S. Geological Survey. 2011. Mineral Commodity Summaries. Reston, Virginia.

Ulivo, A.D. 1997. Determination of Selenium and Tellurium in Environmental Samples. Analyst. 122: 117R–144R.

Urbánková, K., M. Moos, J. Machát and L. Sommer. 2011. Simultaneous determination of inorganic arsenic, antimony, selenium and tellurium by ICP-MS in environmental waters using SPE preconcentration on modified silica. Int. J. Environ. Anal. Chem. 91: 1077–1087 .

Viñas, P., G.I. López-García, B. Merino-Meroño and M. Hernández-Córdoba. 2005. Ion chromatography-hydride generation-atomic fluorescence spectrometry speciation of tellurium. Appl. Organometal. Chem. 19: 930–934.

Wray, D.S. 1998. The impact of unconfined mine tailings and anthropogenic pollution on a semi-arid environment—an initial study of the Rodalquilar mining district, south east Spain. Environ. Geochem. Health. 20: 29–38.

Xi, M., R. Liu, P. Wu, K. Xu, X. Hou and Y. Lv. 2010. Atomic absorption spectrometric determination of trace tellurium after hydride trapping on platinum-coated tungsten coil. Microchem. J. 95: 320–325.

Yarema, M.C. and S.C. Curry. 2005. Acute tellurium toxicity from ingestion of metal-oxidizing solutions. Pediatrics. 116: 319–321.

Yosef, S., M. Brodsky, B. Sredni, A. Albeck and M. Albeck. 2007. Octa-O-bis-(R,R)-tartarate ditellurane (SAS)—a novel bioactive organotellurium(IV) compound: synthesis, characterization, and protease in hibitory activity. Chem. Med. Chem. 767: 1601–6. 768.

Yu, C.H., Q.T. Cai, Z.X. Guo, Z.G. Yang and S.B. Khoo. 2003. Speciation analysis of tellurium by solid-phase extraction in the presence of ammonium pyrrolidine dithiocarbamate and inductively coupled plasma mass spectrometry. Anal. Bioanal. Chem. 376: 236–242.

Zhang, N., N. Fu, Z. Fang, Y. Feng and L. Ke. 2011. Simultaneous multi-channel hydride generation atomic fluorescence spectrometry determination of arsenic, bismuth, tellurium and selenium in tea leaves. Food Chem. 124: 1185–1188.

Trace Elements and Human Health

Mehrdad Gholami[1],* and Hojatollah Yamini[2]

Introduction

Elements which occur in the organism in very small quantities, less than 0.01 percent, are generally referred to as trace elements (Oliver 1997). Trace elements play very important roles in nutrition and health. There have been several reports pointing to the correlation of diseases linked to mineral deficiency. For instance, in the 1930s, occurrence of goiter in the goiter belt of the United States was associated with insufficient iodine in plants and soil (Zimmermann 2011, Anderson et al. 2010, White and Zasoski 1999). Another example is the Keshan disease, widespread and endemic in some areas of China, leading to myocardial fibrosis and necrosis. In the 1970s, selenium deficiency was discovered to be one of the main determinant factors of Keshan Disease through two main observations: firstly, crops in the area had an extremely low selenium content, and secondly administration of selenium before establishment of the disease had a salutary effect on the health of the patients (Hurst 2011, Burk and Lane 1983, Tan et al. 2002). On the other

[1]Department of Chemistry, Islamic Azad University, Marvdasht Branch, Marvdasht, Iran. Email: m.gholami@miau.ac.ir
[2]Department of English, Islamic Azad University, Marvdasht Branch, Marvdasht, Iran. Email: arashyamini@yahoo.com
*Corresponding author

hand, for many decades, human beings have contaminated the environment by dumping and emission of trace element species because they thought this would be harmless due to eventual dilution of contaminations in the environment. This was a naive idea which overlooked the transformations of trace element species in the biosphere which can totally alter their behavior. Trace element species without any doubt affect the food chain. However this was a grave mistake with lethal consequences since human beings are contaminated by various harmful trace element species. These species accumulate in the flora and fauna and their concentrations increasing at accelerating rate and ascending order in the food chain as a result of a process known as **bioaccumulation**. As an example, in 1956, the subtle and serious consequences of methyl mercury exposure became evident in Minamata, Japan. Minamata Bay was contaminated with mercury and methyl mercury from a factory manufacturing chemical acetaldehyde. Mercury was used in the manufacturing process, both mercury and methyl mercury being discharged into Minamata Bay. The fish in the bay accumulated increasing amounts of methyl mercury, which was subsequently passed to the fish-consuming residents of the area. This was one of the first modern lessons of the consequences of the bioaccumulation of methyl mercury (Chan 2011, Walker 2010, Gilbert 2005). Another tragic incident in Iraq clearly documented the fatal effects of maternal methyl mercury exposure. During the winter of 1971, some 73,000 tons of wheat and 22,000 tons of barley were imported into Iraq. This grain, pink-colored mercury-coated seed grain, intended for planting, was treated with various organic mercurials. A severe drought in Iraq resulted in a loss of seed grain as people struggling with malnutrition consumed the seed grains. Unfortunately, the illiterate local population could not realize the English writings and warnings on the seed bags nor recognize the pink seeds as hazardous. Bread made from these seeds was pink, tasty, but toxic, particularly for the growing children, resulting in the hospitalization of some 6530 and death of 459 people at the time of the study (Bakir et al. 1973, Syversen and Kaur 2012).

Trace elements can be categorized as i) essential elements with physiological relevance (microelements or micronutrients), ii) elements which can become toxic at high concentrations and iii) elements that are intrinsically toxic (Reinhold 1975). Since different organisms have different nutritional requirements, some elements can be essential for one organism but toxic to other ones. Calcium, cobalt, chromium (III), copper, fluorine, iodine, iron, magnesium, manganese, molybdenum, potassium, selenium, sodium, and zinc are *essential elements*. Currently, iron, zinc, copper, chromium, iodine, cobalt, molybdenum, and selenium are considered *essential* for nutrition by the WHO (WHO 2002). Boron, nickel, silicon, manganese, and vanadium are *elements with possible beneficial effect*. Aluminum, antimony, arsenic, barium, beryllium, cadmium, lead, mercury, silver, strontium, thallium, and tin are elements *without any beneficial effects.* This list includes elements that can be found in the environment and to which human exposure should be limited. If the absence or deficiency of an element brings abnormalities, which can be connected to specific biochemical changes reversed by supplying the element, that element is considered to be essential. Most of these elements act primarily as catalysts for enzymes, and all of them can be toxic at high concentrations. In tissues and fluids, metals are mostly present as complexes with organic compounds: amino acids, proteins and peptides, organic acids and glutathione. Transporters of metals and transporters of their complexes are major players in homoeostasis and in mediating effects of toxic metals because some of them are not highly specific and interact with multiple metals (Ballatori 2002). One significant feature, when considering health effects of trace elements, is their slow accumulation in tissues even at low doses. Therefore, acute effects are reported very rarely, whereas chronic exposure can lead to gathering of higher concentrations and outbreak of diseases.

Figure 15.1 A shows the safe range of intake and dietary recommendations for healthy individuals. However, in the population there may be individuals

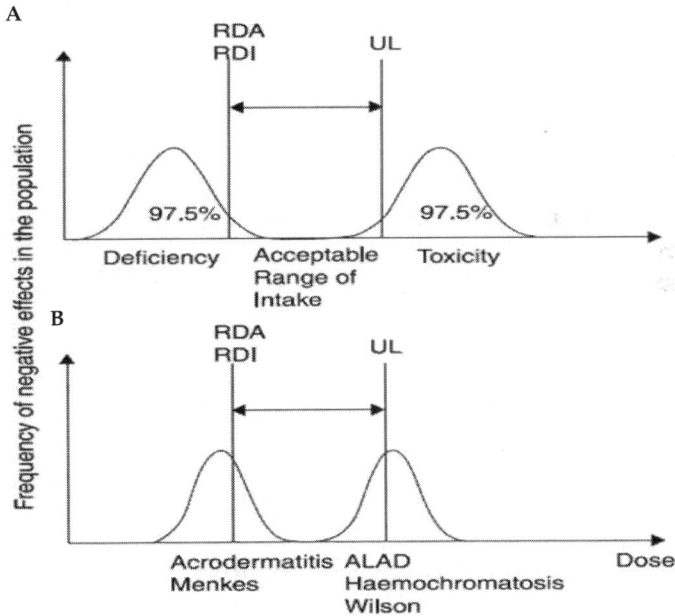

Figure 15.1 (A) shows the safe range of intake and dietary recommendations for healthy individuals. RDA, Recommended Daily Allowance; RDI, Recommended Nutrient Intake; UL, Upper Limit. (A) RDA and RDI are sufficient for avoiding deficiency symptoms in 97.5 percent of healthy individuals, whereas UL is sufficient to avoid toxicity symptoms in 97.5 percent of healthy individuals. (B) Susceptible populations may manifest deficiency symptoms at levels above RDA-RDI and may show toxicity symptoms below the UL.

who show hypersensitivity or hyper-susceptibility to some elements. For these persons toxicity may occur at levels below the Upper Tolerable Limit (UTL), or deficiency symptoms may appear even at the Recommended Daily Allowance (RDA) (Fig. 1B) (Prasad 2008).

1. Antimony

1.1 Introduction

Antimony (Sb) is vastly applied to make alloys, pigments, opacifying agent, medicine, coloring matter, cable covering, ammunitions, bearings, polyethylene terephthalate, brake linings and flame retardants (wilson et al. 2012, Westerhoff et al. 2008, Smichowski 2007). A vast amount of antimony has been emitted into the environment during the sentential application of antimony in many industries. Important natural sources of antimony pollution are volcanic emission, rock weathering and soil erosion (Wilson 2012, Westerhoff 2008, Shotyk et al. 2004, Smichowski 2007). Antimony is not an essential element and it is potentially toxic at even low concentrations (Smichowski 2007). Antimony in the form of Sb_2O_3 has been approved to be carcinogenic (International Agency for Research on Cancer 1998). Antimony is listed as one of the most important pollutants by the US Environmental Protection Agency (1979) and Council of the European Communities (1976) due to its potential carcinogenicity. A serious antimony pollution poses a great threat to food safety, and to human and animal health. It was reported that antimony has toxicological effects on plant growth such as root growth and sprout (He and Yang 1999). Milligrams per liter of Antimony prevents bacteria and soil microbial enzymatic activities such as urease (An and Kim 2009, Tylenda and Fowler 2007).

1.2 Human health

Information regarding acute and chronic toxicity of antimony essentially originated from results found in connection with industrial atmospheric exposure. Treatment of tropical diseases with antimony compounds and its side-effects are discussed only briefly (Sundar and Chakravarty 2010). Similar to arsine (AsH_3), the inhalation of Stibine SbH_3 damages the red blood cells and causes hemolysis. Symptoms of shock and hemoglobinuria are among the signs of this kind of poisoning (Stokinger 1963).

1.2.1 Gastrointestinal effects

Acute antimony poisoning, exhibited as vomiting, nausea and diarrhea, was reported in 150 children who drank a contaminated lemon drink (about 30 mg/L) (Werrin 1963). In order to induce vomiting, in the past, wine and other drinks stored in antimony cups were used (Fairhall and Hyslop 1947, McCallum 1977) for adults, the emetic dose is in the range of 30–60 mg (Martindale 1941). Long time exposure to airborne antimony trichloride (Taylor 1966), antimony trisulfide (Brieger et al. 1954) and/or antimony oxide (Renes 1953) have been reported to cause abdominal pain, diarrhea, vomiting and ulcers.

1.2.2 Respiratory effects

Inhalational exposure to several antimony compounds, e.g., antimony trioxide, stibine (antimony hydride), antimony trisulfide, antimony pentoxide, antimony trichloride, antimony pentasulfide, metallic antimony, etc., affects health. The antimony absorption from the respiratory tract is a function of particle size. Aerosols containing larger particles with high water solubility composed of antimony compounds (e.g., particles of antimony tartrate) are retained in the lungs for a shorter period of time than those containing small particles with low water solubility (e.g., particles of antimony oxides) (Thomas et al. 1973, Felicetti et al. 1974). Irritation and soreness of the upper respiratory tract in seven workers caused by acute respiratory exposure to antimony trichloride (73 mg/m^3) (Taylor 1966). Cordasco and Stone (1973) reported three cases, two of which dead of severe pulmonary edema caused by antimony pentachloride. Renes examined 78 workers, present in smelting processes, for periods exceeding 2 wk. Concentrations of exposure ranged from 4.7–11.8 mg/m^3, 20 percent of the workers suffered from rhinitis, 8 percent from pharyngitis, 5.5 percent from pneumonitis and 1 percent from tracheitis. Nose bleeding and soreness in the nose were reported by more than 70 percent. The workers were, however, also exposed to arsenic at concentrations around 0.7 mg/m^3 (ACGIH 1983, Tylenda and Fowler 2007).

1.2.3 Dermal effects

"Persons working with antimony and antimony salts suffer from antimony spots", pustular skin eruptions. These eruptions are short-living and mainly affect skin areas where sweating occurs and those areas exposed to heat (McCallum 1963, Paschoud 1964, Potkonjak and Pavlovich 1983, Renes 1953, Stevenson 1965, Sunder and Chakravarty 2010). Transferring the worker to a cooler environment often results in the rash clearing up within 3–14 days.

1.2.4 Cardiovascular effects

Hypertension and altered ECG (Electro Cardiogram) T-wave abnormalities were observed in workers exposed to 2.15 mg antimony/m³ as antimony trisulfide from 8 mon to 2 yr. Nevertheless, these workers had also been exposed to phenol formaldehyde resin (Brieger et al. 1954, Tylenda and Fowler 2007).

1.2.5 Reproductive effects

Spontaneous abortions and disturbances in menstruation were reported in women working at an antimony metallurgical plant as compared to a control group. Women were exposed to a mixture of antimony trioxide, antimony pentasulfide and metallic antimony (Belyaeva 1967, Tylenda and Fowler 2007).

1.2.6 Genotoxicity

The sister chromatid test was used to assess the genotoxicity of extracts of soil containing high concentrations of antimony and arsenic, both in the pentavalent state revealed low genotoxicity and partial antagonism (Gebel et al. 1997). Five antimony compounds were examined for genotoxicity. The study showed that stibine and trimethylstibine were genotoxic whereas potassium antimony tartrate, potassium hexahydroxyantimonate and trimethyl antimony dichloride were not (Andrewes et al. 2004, Tylenda and Fowler 2007).

1.2.7 Carcinogenicity

Since the environmental distribution of antimony is low, there is not enough evidence for carcinogenicity of antimony trioxide and trisulphide in human beings (Gebel 1997, Hayes 1997). "Antimony trioxide is categorized as possibly carcinogenic to human beings (Group 2B) by the International Agency for Research on Cancer" (International Agency for Research on Cancer 1989, as cited in Tylenda and Fowler 2007).

1.2.8 Effects of stibine gas

Stibine (SbH_3) gas has been observed to produce hemolysis and hemoglobinuria; however, it seems to be less acutely toxic than arsine (AsH_3) gas (Gallichio et al. 2001). Compared to arsine gas, information on the carcinogenic or other toxic potential of stibine is limited (Andrewes et al. 2004, Gallicchio et al. 2001, Tylenda and Fowler 2007).

1.2.9 Antimony in therapeutics

It is well known that antimony is panacea of all kinds of diseases in the Middle Ages. Since the last century, antimonials have been used for the treatment of two parasitic diseases schistosomiasis and leishmaniasis (Farid et al. 1968, Davis 1968, Cook 1993, Tylenda and Fowler 2007).

2. Arsenic

2.1 Introduction

The 33rd element of the Periodic Table of the chemical elements is Arsenic. Although it is classified as a metalloid, it is also referred to as a metal, and in the context of toxicology as a heavy metal (Mandal and Suzuki 2002). The oxidation numbers of arsenic are commonly +5, +3, and −3. This element is able to form inorganic and organic compounds both within the human body and the in environment (Orloff et al. 2009). When arsenic is combined with hydrogen and carbon, the element is referred to as "organic arsenic". However, when combined with other elements such as oxygen, sulfur and chlorine, it is called "inorganic arsenic". Since most arsenic compounds are colorless and without smell, the presence of arsenic cannot easily be detected in food, water or air. As a result, they pose a serious threat to human health. Indeed, the history of arsenic is synonymous with poison (Mandal and Suzuki 2002). Arsenic can be found everywhere in nature, and its abundance ranks 20th on the Earth's crust, 14th in seawater and 12th in the human body (Jomova et al. 2010).

2.2 Sources and routes of exposure to arsenic

Arsenic trioxide (As_2O_3) is the most widespread inorganic arsenical found in air, while different kinds of inorganic arsenical such as, arsenates (AsO_4^{3-}) or arsenites (AsO_2^-) occur in water, soil or food (Magalhaes 2002, Chou et al. 2007). Gallium arsenide (GaAs) is an inorganic arsenic compound which may also have an adverse effect on human health, due to its prevalent use in the microelectronics industry. Food, especially seafood, rice, mushroom and poultry, which are the main dietary forms, contains the largest source of arsenic and other metals. (Jones 2007, Petroczi and Nepusz et al. 2009, Smedley and Kinniburgh 2002). During the time when there is more arsenic per Se in seafood, it is mostly in an organic form called arsenobetaine which is much less harmful than others. Arsenic poisoning mainly occurs through industrial exposure, from contaminated wine or moonshine, or by malicious administration. It has been reported that traditional Chinese herbal products, deliberately fortified with arsenic for therapeutic purposes, may cause serious health problems (Martena et al. 2010). Cosmetic products such as eye-shadows color pigments, which are advertised, sold and used frequently, contain toxic elements, including arsenic (Sainio et al. 2000). The skin of the eyelids is very sensitive, and the use of eye-shadows may lead to eczemas. Moreover, arsenic particles can be water soluble and may be absorbed through the wet skin at high concentrations, as a result of which arsenic may present a potential risk of carcinogenesis. Cosmetic products should contain less than 5 ppm of metal impurities (Jomova et al. 2010). Ingested arsenic, through arsenic-contaminated water, soil or food, may quickly enter the human body. Breathing air, containing arsenic dusts, results in the settling of dust particles on the lining of the lungs (Chen et al. 2006). The absorbed organic and inorganic forms of arsenic is depleted in urine after several days, although some amount may remain for a few months or even longer (Aposhian et al. 2000a, b). The majority of organic arsenic is rejected more rapidly and usually within several days. Groundwater, contaminated by arsenic and other metals, can severely impact the health of the populations of various regions in the world. Some of the most serious examples of arsenic contamination occur in Bangladesh and West Bengal, India, where almost 43 million people were found to have been drinking water loaded with arsenic (Chowdhury et al. 2000). To place this in perspective, the WHO recommended limit for arsenic in water is 10 µg l^{-1} (WHO factsheet no. 210 May 2001).

2.3. Toxicity of arsenic

Inorganic arsenic includes arsenite, As(III) and arsenate As(V) and can be either methylated to form monomethylarsonic acid MMA(V) or dimethylated as in dimethylarsinic acid DMA(V). Inorganic arsenic is far more toxic than organic arsenic (Shi et al. 2004, Valko et al. 2005). Arsenic is toxic to most of the body organs, the kidneys being the most sensitive ones (Cohen et al. 2006). The degree of arsenic poisoning depends on different factors such as dose, individual susceptibility to arsenic and the age of the affected individuals. Chronic arsenic exposure impacts the vascular system and causes hypertension and cardiovascular diseases. Acute arsenic toxicity may produce cardiomyopathy and hypotension. The most common neurological effect of long-term arsenic toxicity is minor neuropathy, and gastrointestinal effects are revealed by toxic hepatitis followed by increased levels of liver enzymes. Trivalent arsenic prevents a lot of cellular enzymes through sulfydryl group binding. It also prevents the uptake of glucose into cells, leading to gluconeogenesis, fatty acid oxidation and further production of acetyl CoA. Trivalent arsenic also prevents the production of glutathione, which protects cells against oxidative damage (Miller et al. 2002). The toxicity of pentavalent inorganic arsenic is partly due to its conversion to trivalent arsenic, from which other toxic effects proceed.

2.4 Arsenic and human health

The integrity of cellular genetic constituents involves an alteration by oxidants or free radical species, which induce genotoxicity by arsenic, "Arsenic-induced". Experimental evidence prove that arsenic-induced generation of free radicals and oxidative stress cause cell damage and cell death through activation of oxidative sensitive signaling pathways (De Vizcaya-Ruiz et al. 2009). Exposure to arsenic causes various types of cancer (Miller et al. 2002), cardiovascular diseases (Navas-Acien et al. 2005), diabetes (Díaz-Villaseñor et al. 2007), neurological disorders (Vahidnia et al. 2007) and dermal problems (Cohen et al. 2006). Chronic exposure to arsenic leads to dermal diseases, including hyperkeratosis and hyper-pigmentation, often used as diagnostic criteria for arsenicosis (McCarty et al. 2007). Exposure to arsenic causes dermal effects, including symptoms of the early stages of arsenic poisoning. Exposure of workers to inorganic arsenic in the air causes dermatitis and mild dermal irritation. Similar dermal effects (hyperkeratosis and hyper-pigmentation) have been observed among farmers in Taiwan who had been drinking arsenic-contaminated well water (Tseng 1977). Keratosis was observed in female workers of a chemical plant; they had been exposed to arsanilic acid 0.065 mg m^{-3} (Chou et al. 2007).

2.4.1 Cancer

Arsenic is a destructive environmental carcinogen, and leads mainly to cancers of the skin; however, there is epidemiological evidence for various types of cancers of body organs like the lung, the bladder, the liver and the kidneys, being caused by exposure to arsenic (Rossman 2003). The mechanism by which these cancers originate may involve the elevation of oxidative stress by arsenic compounds, in which the antioxidant capacity of the living organism is overcome by Reactive Oxygen Species (ROS), causing molecular damage to proteins, lipids and most significantly DNA (Liu et al. 2001). Trivalent arsenic shows a greater toxicity than the corresponding pentavalent forms, as well as a far more distinct ability to release iron from the iron storage protein ferritin (Salnikow and Zhitkovich 2008). There are several studies reporting the fact that inhalation exposure to inorganic arsenic increases the risk of lung cancer. This is more serious for workers exposed primarily to arsenic trioxide dust (Wall 1980, Welch et al. 1982); nevertheless, lung cancers have also been observed among workers mainly exposed to arsenate (Bulbulyan et al. 1996).

It was revealed that the occurrence of lung cancer correlates highly with increasing arsenic exposure in both smokers and nonsmokers (Järup and Pershagen 1991). Histological studies indicate that arsenic does not specifically increase the occurrence of one particular type of lung cancer. Moreover, other minor types of nonrespiratory cancers, associated with inhalation exposure to inorganic arsenic, have been reported. Increasing risk of stomach cancers, among workers who had been exposed to the highest concentrations of arsenic, has also been reported (Jomova et al. 2010). Enterline et al. (1995) reported a significant increase in the mortality rates due to cancers of the large intestine and the bones. However, it should be noted that the apparent enhancement in the risk of bone cancer was based on a very small number of observations. Enhancement in non-melanoma skin cancers as a result of exposure from a Slovakian coal-burning power plant have also been reported (Pesch et al. 2002).

Inhalation and ingestion of inorganic arsenic have been reported to cause cancers of the bladder and the kidneys, especially urothelial cancers (Chen et al. 1985, 1986, 1997b, IARC 2004, NRC 1999, WHO 2001). Long-term exposure to ingested and inhaled arsenic has been associated with both hepatic angiosarcoma and hepatocellular carcinoma (Bates et al. 1992, Chen et al. 1997b, Falk et al. 1981a, b, IARC 2004, NRC 1999, WHO 1981, 2001). Inhalation and ingestion of inorganic arsenic is associated with an increased risk of gastrointestinal cancers, hematolymphatic malignancies, and malignant neoplasm of the nervous system (Chen et al. 1997b, IARC 2004, NRC 1999, WHO 1981, 2001).

2.4.2 Cardiovascular effects

Oral exposure to arsenic causes serious and unfavorable effects on the cardiovascular system; besides, there is some proof from epidemiological studies that the cardiovascular system may also be affected by inhaled inorganic arsenic (Navas-Acien et al. 2005, States et al. 2009). Wang et al. (2003) found an enhanced rate of disease in the blood vessels in Taiwanese communities living in areas with arsenic-polluted well water (> 0.35 mg L^{-1}). Ecological studies conducted in the USA show a significant increase in the number of deaths from arteriosclerosis, aneurysm and other related diseases in the areas in which the drinking water contained arsenic concentrations > 20 µg L^{-1} (Engel and Smith 1994). Another disorder associated with increased arsenic exposure is hypertension (Yang et al. 2007).

2.4.3 Gastrointestinal disturbances

Clinical signs of gastrointestinal irritation, including nausea, vomiting, diarrhea and abdominal pain are observed in all cases of short-term high dose, and longer-term lower dose exposures to inorganic arsenic (Uede and Furukawa 2003, Vantroyen et al. 2004). The gastrointestinal tract seems to be the crucial target of toxicity following oral exposure to MMA (Lee et al. 1995).

2.4.4 Liver disease

Several studies showed symptoms of hepatic injury after oral exposure of human beings to inorganic arsenic. Clinical experiments also revealed liver damage (Liu et al. 2002) and blood tests exhibited elevated levels of hepatic enzymes.

2.4.5 Renal disease

Inorganic arsenicals do not show any important renal injury in humans. Elevated levels of creatinine or bilirubin have been reported (Moore et al. 1994).

2.4.6 Neurological disorders

Inorganic arsenic can create serious neurological effects after both inhalation (Calderon et al. 2001, Lagerkvist and Zetterlund 1994) and oral exposure (Uede and Furukawa 2003). A possible association between arsenic in drinking water and neurobehavioral alterations in children has been reported (Tsai et al. 2003). Peripheral neuropathy which may last for several years is the most typical neurological feature of arsenic neurotoxicity (Mathew et al. 2010).

3. Chromium

3.1 Introduction

Chromium (Cr) is a fragile, hard, and shiny metal. Chromium is silver-gray and can be highly polished. It does not lose its shine in air; when heated, it burns and forms the green chromic oxide. In oxygen, Cr is unstable and immediately produces a thin oxide layer that is impermeable to oxygen and preserves the metal below. Crocoite, $PbCrO_4$, ore with Cr in the trivalent Cr (III) form is the main source of Chromium (Cr) in nature. Chromite ore is applied for manufacturing Cr metal as well as monochromates, dichromates, chromic acid and Cr pigments. Hexavalent, Cr(VI), is absorbed through the airways and the digestive tract much quicker than in the Cr(III) state (Langard and Costa 2007).

3.2 Chromium and human health

Chromium exists in all human organs of adults and newborns. The concentration of chromium is generally highest in lung tissue and it increases with age due to inhalation and retention of Cr compounds with low water solubility. Chromium is discharged through the urine and feces, essentially through the urine (Langard and Costa 2007). Trivalent chromium may play a role in the metabolism of glucose. Nevertheless, the mechanism still is not quite clear. The main source of acute and long-term adverse effects of Cr results from Cr (VI) compounds. Allergic reaction of the skin most frequently observed as the adverse effect for Cr, and Cr (VI)-induced ulcers of the skin and ulcerations of the mucosa of the nasal septum are still observed in developing countries. Prolonged inhalation exposure to various Cr (VI) compounds has been reported to result in a high risk of carcinomas of the respiratory organs. Compounds of Cr (VI) are reported to cause mutations, chromosomal aberrations, DNA damage in the form of single-strand breaks, DNA–protein and DNA–DNA cross-links. Cr (III) is believed to be responsible for much of the DNA damage and mutations induced by reducing Cr (VI). Food is the main source of Cr (III) compounds in the general population; however, no adverse effects of Cr have been reported on the basis of Cr in food. Cr compounds have been classified as carcinogenic to human beings (IARC 1990).

3.2.1 Chrome ulcers

Corrosive action of Cr (VI) compounds generally induces chrome ulcers. A deeply penetrating round hole of the skin may develop as a result of deposition of chromic acid, dichromate compounds or other Cr (VI) compounds on broken skin surface (Dewirtz 1929). Acute Irritative Dermatitis disease reported in the Cr industry, especially, among workers in contact with Cr (VI) compounds (Langard and Costa 2007). Cement eczema can be induced by traces of chromate compounds in the cement (Engebrigtsen 1952, Freget 1981).

3.2.2 Airways

Bronchial asthma is reported as a result of inhalation of dust containing chromate or chromic acid fumes (Card 1935, Joules 1932, Williams 1969). The dust particles stay in the lung airbags and produce the side-effects.

3.2.3 The kidneys

Tubular necrosis of kidneys contributes to death in fatal intoxications with chromates. Approximately 5 g of chromate ingestion may result in toxic effects within 12 hr. One to four days after ingestion of smaller amounts of chromate (i.e., 1 or 2 g in adults) may result in tubular necrosis (Ellis et al. 1982, Fristedt et al. 1965, Kaufman et al. 1970).

3.2.4 The liver

Ingestion of chromates may cause necrosis of the liver. These adverse effects seem to be most important in cases where the ingested dose is 1.5–2 g or less (Ellis et al. 1982, Fristedt et al. 1965, Kaufman et al. 1970).

3.2.5 Cardiovascular system

Bleeding from the digestive tract and massive loss of fluid may take place, and death may occur following ingestion of 5 g or more of chromates. Hemorrhagic diathesis may be a predominant symptom in nonfatal cases. "It is not clear whether chromates may exert direct effects on the heart or if the cardiovascular symptoms are secondary to untoward effects in other organs" (Brieger 1920, Ellis et al. 1982, Fristedt et al. 1965, Partington 1950, Schiffl et al. 1982, Schlatter and Kissling 1973).

3.2.6 Cancer

There are several studies whose findings confirm that some Cr (VI) compounds have the potential to enhance the risk of cancer in the respiratory organs (Costa 1997, De Flora 2000, IARC 1990, Langård 1990). A possible connection between exposure to chromates and enhanced risk of cancer of the intestines is reported after observing five digestive tract cancer cases among 44 deceased chromate workers. The elevated risks of cancer of the stomach and digestive tract and lung cancer on the relationship between exposure to Cr (VI) compounds are reported (Costa 1997, De Flora 2000, IARC 1990, Langård 1990, Langard and Costa 2007).

4. Iodine

4.1 Introduction

An essential component of the hormones produced by the thyroid gland is iodine. So, iodine and thyroid hormones are essential for human life. Marine and Kimball proved that endemic goiter (thyroid enlargement) was caused by Iodine Deficiency (ID), which could be prevented by iodine supplementation (Marine and Kimball 1990). Iodine (as iodide) is vastly but asymmetrically distributed on the Earth's environment. In many regions, iodide in top soil is depleted during glaciations, flooding and erosion of soils. Plants grown in these iodine-depleted regions have low levels of iodine; as a result, animals and human beings consuming food grown in these soils become iodine-deficient. Iodine deficiency has multiple harmful effects on growth and development in animals and human beings. These are usually termed iodine deficiency disorders which are listed in Table 15.1 (Hetzel 1983, Zimmermann et al. 2008). Lack of sufficient iodine results in inadequate thyroid hormone production.

In 1980, WHO reported that up to 60 percent of the world's population, most of whom are from developing countries, are affected by Iodine Deficiency. The consequences of ID could be prevented by a low-cost intervention, Universal Salt Iodization (USI) (Zimmermann et al. 2008). Since then, the number of households around the world using iodized salt has increased from < 20% to > 70% resulting dramatically to reduced ID (Andersson et al. 2010). In spite of this progress, in 2007, WHO estimated nearly two billion individuals, one-third of all school-aged children are among them, still suffer from ID (De Benoist et al. 2008).

Table 15.1 The iodine deficiency disorders, by age group (Hetzel 1983, Zimmermann et al. 2008).

Age groups	Health consequences of iodine deficiency
All ages	Goiter
Fetus	-Increased susceptibility of the thyroid gland to nuclear radiation -Abortion -Stillbirth -Congenital anomalies -Perinatal mortality
Neonate	-Infant mortality -Endemic cretinism
Child and Adolescent	-Impaired mental function -Delayed physical development
Adults	-Impaired mental function -Reduced work productivity -Toxic nodular goiter -Iodine-induced hyperthyroidism -Increased occurrence of hypothyroidism in moderate-to-severe iodine deficiency -Decreased occurrence of hypothyroidism in mild-to-moderate iodine deficiency

4.2 Iodine and human health

4.2.1 Iodine deficiency and thyroid metabolism

Absorption of dietary iodide from iodized salt and seafood occurs rapidly and nearly completely (> 90%) in the stomach and duodenum (Alexander et al. 1967, Nath et al. 1992). Iodine in the body of a healthy adult contains up to 20 mg, of which 70–80 percent is in the thyroid (Fisher and Oddie 1969). The iodine content of the thyroid may fall to < 20 µg in chronic ID. The adult thyroid keeps 60–80 µg of iodine/day in iodine-sufficient areas to balance losses and maintain thyroid hormone synthesis (Degroot 1966, Stanbury 1954). Thyroid hormone synthesis in chronic severe ID is gradually reduced and causes hypothyroidism and its consequences. During pregnancy the iodine requirement is increased by ≥ 50% (Glinoer 1997). In case of pregnancy of a chronically iodine deficient woman, goiter and hypothyroidism occur, resulting in unfavorable effects on maternal and fetal health.

4.2.2 Effects of hypothyroxinemia on the brain

Thyroxine (T4) and small amounts of triiodothyronine (T3) are two thyroid hormones which are secreted by the thyroid. Nevertheless, T3 is biologically more active, and is formed mainly in the brain by deiodination of circulating T4 (Bernal 2005). Most of the biological activities of T3 in the Central Nervous System (CNS) are mediated by Thyroid hormone Receptors (TRs). In the nucleus, TRs bind to thyroid hormone-responsive elements in the promoter region of their target genes to regulate transcription (Flamant and Samarut 2003). The thyroid hormone has several important roles in the developing brain such as accelerated myelination and improved cell migration, differentiation and maturation (Bernal 2005, Flamant and Samarut 2003, Davis et al. 2005, Zoeller and Rovet 2004). The role of iodine in human growth and development has recently been reviewed by Zimmermann (Zimmermann 2011).

4.2.3 Iodine toxicity

Long-term ingestion of iodine more than dietary requirements may lead to *iodism*, which is defined as "poisoning caused by sensitivity to or overuse of iodine or its compounds." The direct acute toxicity of iodine results from its irritant properties (NAS 1980). Elemental iodine (I2) in excessive amounts is corrosive, and irritates tissues through all routes of exposure (i.e., inhalation, ingestion and skin contact). Airborne iodine causes irritation of the respiratory system, eyes and skin, and may have unfavorable effects on the central nervous and cardiovascular systems (Genium 1999).

4.2.4 Acute effects

Short-term exposure to a high concentration or a large amount of iodine may result in acute iodine toxicity. An average fatal dose in adults of ingested iodine crystals or powder has been reported to be 2–4 grams. Ingestion causes burns in the mouth, vomiting, abdominal pain and diarrhea. In serious intoxication, headache, delirium, and a drop in blood pressure may occur. Inhalation results in eye and nose irritation and also chest tightness. Skin contact may result in burns, irritation and rashes (Genium 1999).

4.2.5 Chronic effects

Repeated ingestion of iodine in amounts that is more than dietary requirements results in a toxic syndrome called *iodism*. An unpleasant brassy taste, burning of the mouth and throat, and soreness of the teeth and gums are initial symptoms of iodism. Moreover, increased salivation, inflammation of mucous membranes of the nose (rhinitis), eye and mouth, sneezing, laryngitis, bronchitis and skin rashes are frequently reported (Hardman et al. 1996).

4.2.6 Skin and eye contact

Contact with strong iodine solutions result in burns (AIHA 2002). Small amounts of iodine may be absorbed via the skin (NAS 1980). Eye contact with saturated iodine vapor causes brown staining of the outermost cell layer of the cornea or the corneal epithelium and spontaneous loss of these cells. Nevertheless, recovery is observed in 2–3 d (AIHA 2002).

5. Mercury

5.1 Introduction

Mercury exists in different forms of inorganic mercury and organic mercury with very different properties that is generated from mercury and accumulates in some commonly consumed species of fish. Mercury is industrially useful and potentially harmful. The history of human beings' use of mercury shows our struggle to balance and understand the usefulness of this compound, and prevent its harmful effects to human beings and the environment (Gilbert 2003).

Mercury leaves a variety of impacts on human health and the environment throughout the world. Mercury and its compounds are highly toxic to the developing nervous system. This toxicity toward humans depends on the chemical form, the amount, the pathway of exposure and the vulnerability of the exposed person. Human beings are exposed to mercury through a variety of pathways: consumption of fish (Chan 2011, Benefice et al. 2010, Jewett and Duffy 2007, Agusa 2005), occupational and household applications (Nance 2012, Spiegel and Veiga 2010, Eckley et al. 2011, Hylander 2011, Shimshack and Ward 2010, Risher et al. 2003, Junghans 1983), dental amalgams (Ye et al. 2009, Bates 2011, Richardson et al. 2011), mercury-coated seed grains and mercury-containing vaccines (Dórea et al. 2011), and chlor-alkali industry (Grangeon et al. 2012, Reis et al. 2009, Montuori et al. 2006, Soto et al. 2011). Because of the high toxicity of mercury, special attention has been paid to its prevention (Ibáñez-Palomino et al. 2012). Especially, methyl mercury (MeHg), the most toxic form of mercury, causes severe neurological damage to human beings and wildlife (Grandjean et al. 1999, Clarkson et al. 2003).

5.2 Mercury and human health

5.2.1 Elemental mercury

Elemental mercury is applied industrially in electric lamps, switches, gauges and controls (e.g., thermometers, barometers and thermostats), battery production, nuclear weapons production and the specialty chemical industry such as the production of caustic soda. Since elemental mercury has a high affinity for gold and silver, it has been, and continues to be, used in precious metal extraction from ore. For over a century, elemental mercury has been used in mercury–silver amalgam preparations to repair dental cavities.

The Chinese applied cinnabar, red mercury(II) sulfide (HgS), is the common ore of mercury, to make red ink about 1000 BC, and in cosmetics, soaps, and laxatives. Inorganic mercury, as mercury nitrate, was applied in the felting industry to aid in matting felt. A study of 25 hat factories in Connecticut demonstrated evidence of chronic mercurialism among 59 of 534 hat workers (Gilbert 2003).

Dental amalgams were applied as early as the seventh century, and the first commercial mercury dental amalgam was used in the 1930s in New York. Chronic mercury exposure among dentists and dental assistants is a well-recognized occupational hazard (Gilbert 2003, Ye et al. 2009, Richardson et al. 2011). Concerns over the public health risks of mercury amalgam fillings have also been raised in the scientific literature; however, it is highly controversial. Recent studies indicate that the amount of mercury in urine is related to the number of dental amalgams, and a similar relationship exists for mercury excretion in mothers' breast milk (Gilbert 2003, Ye et al. 2009, Richardson et al. 2011).

5.2.2 Toxicity of elemental mercury

When mercury vapor is formed, elemental mercury is inhaled; it is readily and rapidly absorbed into the bloodstream, and easily crosses the blood–brain barrier and the placenta. However, the ingestion of elemental mercury, due to poor absorption in the gut, is far less hazardous than inhalation of mercury vapor. Mercury is oxidized after entering the brain, and will not move back across the blood–brain barrier; thus, continued exposure to mercury vapor will result in mercury accumulation in the nervous system. Acute, high level exposure to mercury vapor can result in respiratory, cardiovascular, neurological and gastrointestinal malfunctioning, and even death (Gilbert 2003).

5.2.3 Inorganic mercury

Inorganic mercury exits in the form of salts as either monovalent (Hg^+, mercuric) or Calomel (Hg_2Cl_2), divalent (Hg^{2+}, mercurous) or (mercuric chloride) sublimate are two major mercury chloride salts that were produced in the Middle Ages. Inorganic mercury-based skin creams were first applied during this period for the treatment of syphilis. During the early 1900s inorganic mercury was used as a clinical diuretic (Berlin et al. 2007).

5.2.4 Toxicity of inorganic mercury

Several studies examining the effects of oral exposure to inorganic mercury salts have shown renal toxicity in human beings as a result of acute oral exposures. Kidney effects (e.g., heavy albuminuria, hypoalbuminemia, edema and hypercholesterolemia) have been reported after therapeutic administration of inorganic mercury (Basic Information I Mercury I US EPA 2012).

5.2.5 Organic mercury

Although there are various synthetic organic mercury compounds, the most important one is the naturally-occurring form, methyl mercury (MeHg). In the environment, inorganic mercury is primarily biotransformed to MeHg through microbial methylation in sediments of freshwater and seawater. Once MeHg is formed, it readily enters the seafood chain and bioaccumulates in tissues of sea organisms. Since MeHg is stored throughout the life of sea organisms, it is transferred up the food chain and results in the highest concentrations in larger, long-lived, predatory species in oceans, such as swordfish, pike and tuna. Mercury is stored in fish muscles in the form of MeHg. MeHg concentration in fish depends on the age and trophic level of the particular fish, and can be quite substantial (> 1000 µg/kg). For example, the total mercury in the edible tissues of shark and swordfish averages 1200 µg/kg. Organomercurials used to be applied as fungicides, paint preservatives and in medicinal applications, though these uses have ceased as a result of their recognized neurotoxicity. Therefore, fish and marine mammals are regarded as the primary sources of human MeHg exposure, and to a lesser degree research applications of MeHg and other organomercurials (Gilbert 2003, Benefice et al. 2010, Jewett and Duffy 2007, Agusa et al. 2005). Moderate consumption of fish (with low mercury levels) is not likely to result in exposures of concern. Nevertheless, people who consume higher amounts of fish or marine mammals may be highly exposed to mercury and are, therefore, at higher risk (Hightower and Moore 2003).

The most dramatic case of severe MeHg poisoning is from Minamata Bay, Japan, in the late 1950s the subtle and serious consequences of methyl mercury exposure became evident in Minamata, Japan. Initially, early signs of uncoordinated movement and numbness around the lips and extremities followed by constriction in visual fields in fishermen and their families, baffled health experts. Developmental effects were clearly evident in infants who exhibited subtle to severe disabilities. This spectrum of adverse effects was finally related to methyl mercury exposure from consumption of contaminated fish. Minamata Bay was contaminated with mercury and methyl mercury from a factory where mercury was used as a catalyst in manufacturing the chemical acetaldehyde. The amount of mercury discharged from 1932 to 1968 was estimated about 456 tons. As a result, hundreds of people died, and thousands were negatively affected with permanent damage (Harada 1995, Akagi et al. 1998). Mercury was used in the manufacturing process, which also resulted in both mercury and methyl mercury being discharged into Minamata Bay. The fish in the bay accumulated increasing amounts of methyl mercury, which was subsequently passed to the fish-consuming residents of the area. This was one of the first modern cases of the consequences of the bioaccumulation of methyl mercury. Recent studies in the Brazilian Amazon demonstrated increased exposure to MeHg among the local people because of their consumption of fish contaminated by upstream gold-mining activities (Lebel et al. 1998, Grandjean et al. 1999, Dolbec et al. 2000).

5.2.6 Toxicity and effects on humans

Once MeHg is dispersed in the blood stream throughout the body and enters the brain, it may cause structural damage. The most vital target for MeHg toxicity is the central nervous system. The physical lesions may lead to tingling and numbness in fingers and toes, loss of coordination, difficulty in walking, generalized weakness, impairment of hearing and vision, tremors, and finally loss of consciousness and death. The developing fetus may be at particular risk from MeHg exposure. Infants whose mothers were exposed to MeHg during pregnancy exhibit a variety of developmental neurological abnormalities, including the following: delayed onset of walking and talking, cerebral palsy, altered muscle tone and deep tendon reflexes, and reduced neurological test scores. Maternal toxicity may or may not have been present during pregnancy for those offsprings exhibiting adverse effects. For the general population, the observed critical effects following MeHg exposure are multiple central nervous system effects including ataxia and paresthesia (Prasad 2008).

6.2.7 Regulatory standards

6.2.7.1 Inorganic mercury

Metallic mercury is poorly absorbed after oral ingestion; thus, this is much less hazardous than inhalation. Below are some of the advisories on mercury vapor inhalation. The liquid silver inorganic mercury evaporates into the atmosphere. When inhaled, mercury easily crosses into the blood and then, to the brain; thus, the primary hazard concern is from inhalation (Gilbert 2005, (Basic Information | Mercury | US EPA 2012).

- Agency for Toxic Substances and Disease Registry, ATSDR—minimal risk level (MRL) – 0.2 $\mu g/m^3$
- American Conference of Governmental Industrial Hygienists, ACGIH —threshold limit value (TLV)-TWA – 0.05 mg/m^3

6.2.7.2 Organic methyl mercury

The primary human exposure to methyl mercury is from consumption of contaminated fish. The most sensitive population is the developing fetus or infant due to the effects of methyl mercury on the nervous system (neurotoxic) and developmental effects. Exposure limits and fish consumption advisories are directed at pregnant women, women of childbearing age, and children. All agencies also recognize that fish consumption has many nutritional benefits and is an important part of people's diet. Nevertheless, the widespread distribution of mercury and subsequent bioaccumulation of methyl mercury requires that engaged agencies should develop recommendation for levels of mercury in fish. A list of some of these recommendations is provided below, but it is very important to consult the local fish consumption advisories.

- FDA – 1 ppm in commercially harvested fish (i.e., tuna fish).
- FDA – action level – 0.47 $\mu g/kg$ per day.

- ATSDR – minimal risk levels (MRLs) – 0.30 μg/ kg per day.
- Washington State – total daily intake – 0.035–0.08 μg/ kg per day.
- EPA – Reference Dose (RfD) – 0.1 μg/ kg per day (In 1997, the EPA estimated that 7 percent of the women of childbearing age in the United States exceeds the established RFD of 0.1 μg/ kg per day).
- 41 states have issued over 2000 fish consumption advisories related to mercury.

The federal government develops regulations and recommendations to protect public health. Regulations can be enforced by law. Federal agencies that develop regulations for toxic substances include the Environmental Protection Agency (EPA), the Occupational Safety and Health Administration (OSHA), and the Food and Drug Administration (FDA). Recommendations, on the other hand, provide valuable guidelines to protect public health, but cannot be enforced by law. Federal organizations that develop recommendations for toxic substances include the Agency for Toxic Substances and Disease Registry (ATSDR) and the National Institute for Occupational Safety and Health (NIOSH) (Gilbert 2005, Basic Information | Mercury | US EPA 2012).

6.2.8. Mercury and cancer

There is only a few available research studies on the risk of cancer of the exposition to Hg (Boffetta 1993). However, EPA has determined that mercuric chloride ($HgCl_2$) and methyl mercury (CH_3HgCl) are possible human carcinogens. Mercuric chloride has also been found to cause increases in several types of tumors in rats and mice, and methyl mercury caused kidney tumors in male mice. Different groups of workers professionally exposed have been studied. The researchers' main finding is that Hg did not produce cancer in the cases evaluated. Although, for instance, in three cases reported (Barregard et al. 1990, Ellingsen et al. 1993, Boffetta et al. 1998) different explanations are argued, the consistent fact is that an increment (although small) of the lung and liver cancer is produced (Dolbec et al. 2000).

7. Selenium

7.1 Introduction

Interest in selenium and health was focused primarily on the potentially toxic effects of high intakes in human beings (Smith et al. 1936). There has been growing interest in selenium in relation to Keshan disease (an endemic cardiomyopathy) and also possible protective effects against cancer and other chronic diseases (Clark et al. 1996). There is a relatively narrow margin between selenium intakes that results in deficiency or toxicity. Moreover, the species of selenium is another determinant of its health effect (Fairweather-Tait et al. 2011).

7.2 Selenium and human health

Globally, total soil selenium concentrations typically lie within the range 0.01–2.0 mg/kg with an overall mean of 0.4 mg/kg (Fordyce 2005).

Selenium uptake of crop is influenced greatly by the availability and chemical species of selenium in soil. Inorganic selenium occurs in three soil-phases (fixed, adsorbed, and soluble), and only adsorbed/soluble forms of selenium are believed to be available for plant uptake (Stroud et al. 2010).

Selenium amount in the diet largely depends on the place where crops are grown and cultivated, the soil/fodder to which animals are exposed and the actual foods consumed. The main food groups providing selenium in the diet are bread and cereals, meat, fish, eggs and milk/dairy products are shown in Fig. 15.2.

Sodium selenate, sodium selenite and sodium hydrogen selenite are the only permitted species of selenium added to foods for particular nutritional use in Europe, including baby formula milk and total parental nutrition foods, whereas the predominant selenium species in most natural foods is selenomethionine (Flynn et al. 2009).

As a long-term biomarker, the analysis of selenium in hair and toenails is useful (Ashton et al. 2009) because both tissues are easy to access and are noninvasive, they are also suitable for fieldwork, but samples must be

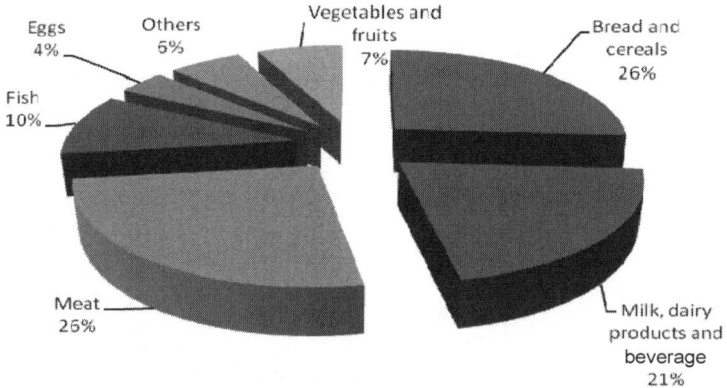

Figure 15.2 Contribution of each food group to total population dietary exposure in the UK. Adapted from data presented in the UK Food Standards Agency document (Stroud et al. 2010, Clark et al. 1996, FSA 2009).

Color image of this figure appears in the color plate section at the end of the book.

prepared with care to avoid contamination. For example, hair samples can be affected by selenium containing shampoo residues (Satia et al. 2006).

Selenium has three functions in the human body: in thyroid hormone metabolism, in antioxidant defense and oxidative metabolism system, and in the immune system (Lu and Holmgren 2009).

Regarding thyroid hormone metabolism, the redox-protective effects of selenoproteins may be of particular importance in the thyroid gland, whose long-lived cells generate H_2O_2 required for the synthesis of thyroid hormones (Kryukov et al. 2003).

Table 15.2 gives the details about known selenoproteins that carry out nutritional functions of selenium effective in antioxidant defense and oxidative metabolism system (Sunde 1997, Diplock 1994, Behne et al. 2000, Behne et al. 1997, Fairweather-Tait et al. 2011, Rayman 2000).

Limited data in human beings show that when intakes of selenium are sub-optimal, selenium supplements can enhance immune responses (Hoffmann and Berry 2008).

7.3 Clinical disorders

Clinical disorders can be investigated under two categories: A) Deficiency, and B) Toxicity.

7.3.1 Deficiency

1. The name Keshan disease originates from a severe outbreak in Keshan County, Heilongjiang Province, China, in 1935. Keshan disease is an endemic cardiomyopathy observed in selenium-deficient areas of China. The main clinical features of Keshan disease are acute or chronic episodes of a heart disorder characterized by cardiogenic shock and/ or congestive heart failure. Keshan disease can be clinically classified into four types: acute, sub-acute, chronic and latent. Dilatation of the heart is commonly observed (Ge and Yang 1993).

2. Role of selenium in Kashin-Beck Disease (KBD) is mainly found in a diagonal belt from northeast to southwest China, and also in Mongolia, Siberia, and North Korea. S·. Kashin-Beck Disease (KBD) is an endemic, chronic, degenerative osteoarthropathy that is present in selenium-deficient areas in the world, and it was first described in 1848 by Nickolay Kashin in the Bajkal area of Russia (Kashin 1859) and later in 1906 by Eugene Beck (Beck 1906).

Table 15.2 Known selenoproteins that carry out nutritional functions of selenium (Fairweather-Tait et al. 2011, Rayman 2000).

Selenoprotein	Function
Glutathione peroxidases (GPx1, GPx2, GPx3, GPx4)	Antioxidant enzymes: remove hydrogen peroxide, and lipid and phospholipid hydroperoxides (thereby maintaining membrane integrity, modulating eicosanoid synthesis, modifying synthesis, modifying inflammation and likelihood of propagation of further oxidative damage to biomolecules such as lipids, lipoproteins, and DNA) (Sunde 1997, Diplock 1994, Allan et al. 1999, Spallholz et al. 1990)
(Sperm) mitochondrial capsule selenoproteins	Form of glutathione peroxidase (GPx4): shields developing sperm cells from oxidative damage and later polymerizes into structural protein required for stability/motility of mature sperm (Ursini et al. 1999)
Iodothyronine deiodinases (three isoforms) Thioredoxin reductases (probably three isoforms)	Production and regulation of level of active thyroid hormone, T_3, from thyroxine, T_4 (Sunde 1997). Thioredoxin reductases (probably three isoforms) Reduction of nucleotides in DNA synthesis; regeneration of antioxidant systems; (Allan et al. 1999) maintenance of intracellular redox state, critical for cell viability and proliferation; (Allan et al. 1999) regulation of gene expression by redox control of binding of transcription factors to DNA (Allan et al. 1999).
Selenophosphate synthetase, SPS2	Required for biosynthesis of selenophosphate, the precursor of selenocysteine, and therefore for selenoproteins Synthesis (Allan et al. 1999).
Selenoprotein P	Found in plasma and associated with endothelial cells. Appears to protect endothelial cells against damage from Peroxynitrite (Sunde 1997, Diplock 1994, Allan et al. 1999, Spallholz et al. 1990, Arteel et al. 1999)
Selenoprotein W Prostate epithelial selenoproteins (15kDa)	Needed for muscle function (Sunde 1997, Diplock 1994, Allan et al. 1999, Arteel et al. 1999) Found in epithelial cells of ventral prostate. Seems to have redox function (resembles GPx4), perhaps protecting secretory cells against development of carcinoma (Behne et al. 1997).
DNA-bound spermatid selenoprotein (34 kDa) 18 kDa selenoprotein	Glutathione peroxidase-like activity. Found in stomach and in nuclei of spermatozoa. May protect developing sperm (Behne et al. 1997). Important selenoprotein, found in kidney and large number of other tissues. Preserved in selenium deficiency (Behne et al. 2000).

7.3.2 Toxicity

Selenium toxicity, much less common than selenium deficiency, can affect individuals as a result of oversupplementation (MacFarquhar et al. 2010), accidental or deliberate (suicidal) ingestion of very high doses (Lech

2002), or through high levels in the food supply. Brittle hair and brittle, thickened, stratified nails, leading to loss in some cases, along with an odor of garlic on the breath and skin (MacFarquhar et al. 2010, Rayman 2008) are characteristic features of *selenosis* which occur in population groups exposed to unusually high levels of dietary selenium. Additional symptoms, including vomiting and pulmonary edema, are a feature of more acute selenium poisoning (Rayman 2009).

7.3.2.1 Cardiovascular disease (CVD)

Selenium is essential for selenium-dependent antioxidant enzymes such as GPxs, TXNRD, SePP and other selenoproteins. Because of the antioxidant properties of selenium and/or selenoenzymes, it has been concluded that selenium may prevent (CVD). Many observational studies investigating the association of low selenium concentrations with cardiovascular outcomes and randomized trials investigating whether selenium supplements prevent coronary heart disease (CHD) have been inconclusive, so, the observational evidence that low selenium concentrations are associated with cardiovascular risk should be treated as suggestive but not definitive. There is uncertainty about cause and effect; therefore, time-resolved and prospective studies are needed in different pathological settings. Further, when investigating the relationship between selenium and disease risk, future studies are needed to determine not only selenium status but also genotype in relation to selenoproteins and related pathways (Rayman 2009).

7.3.2.2 Cancer

There are many studies investigating the effect of selenium on cancer; several recent reviews focus on mainly the animal model, the potential mechanisms of action using evidence from *in vitro* cell culture studies and *in vivo* studies (Zhuo and Diamond 2009, Selenius et al. 2010, Lu and Jiang 2005, Jackson and Combs 2008). Proposed mechanisms of the effects of selenium on cancer are summarized in Fig. 15.3. They include regulation of cell cycle and apoptosis, antioxidant effect through the action of selenoproteins, in particular, GPx1, GPx4, Sep15, SEPP1, and TXNRD1 (Zhuo and Diamond 2009, Rayman 2008), modulation of angiogenesis (Lu and Jiang 2005), and the extracellular matrix (Hurst et al. 2008, Carlson et al. 2009), histone deacetylase inhibition (Pinto et al. 2010), carcinogen detoxification, induction of GSTs, alteration of DNA damage and repair mechanisms, and also immune system modulation (Zhuo and Diamond 2009, Selenius et al. 2010, Lu and Jiang 2005, Jackson and Combs 2008). However, the effects

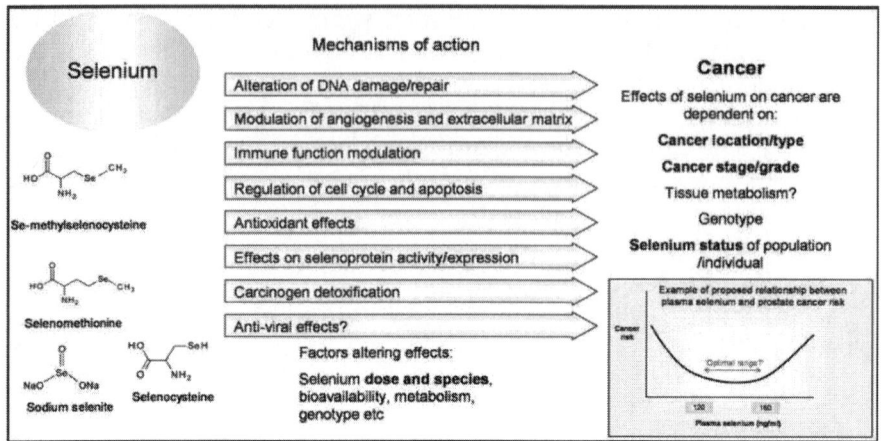

Figure 15.3 Mechanisms of action of selenium against cancer and key factors modulating the effect of selenium (Fairweather-Tait et al. 2011).

of selenium on cancer are species-specific, dose-specific, and cancer type-specific and may be affected by genotype and the bioavailability of selenium (Fig. 15.3), discussed in more detail below.

Association between selenium and cancer was investigated by Willett et al. (Willett et al. 1983). They reported that the relative risk (RR) of cancer was higher in individuals with low plasma selenium concentrations (<115 ng/mL compared with 128 to > 154 ng/mL). There have been several reports about the association between selenium and various cancers such as, gastrointestinal cancers (Bjelakovic et al. 2008, Blot et al. 1995, Li et al. 2004, Wang et al. 1994), prostate cancer (Brinkman et al. 2006, Duffield-Lillico et al. 2003, Etminan et al. 2005, Meyer et al. 2005), lung cancer (Zhuo et al. 2004), skin cancer (Clark et al. 1997, Zhuo et al. 2004) and liver cancer (Yu et al. 1997).

In summary, although direct comparisons of odd ratios, hazard ratios (HR), and relative risks for many studies are not possible because the results are study-specific, there is a consistent trend throughout several of the human studies demonstrating potential protective effects with plasma/serum selenium between (120–160 ng/mL) and reduced risk of some types of cancer when compared with the low plasma selenium status, namely < 120 ng/mL. Above 160 ng/mL the cancer protective effect is likely to diminish and the risk perhaps increases for some types of cancer (Fairweather-Tait et al. 2011). Literature from the 1950s and 1960s indicates that an inappropriately high dose range of selenium may actually increase the incidence of certain types of cancer in animal samples, and, as a result, selenium used to be classified as a carcinogen in animals when used at high exposure (Shapiro 1972). Therefore, a careful balance ensuring selenium intakes and selenium status fall in the relatively narrow base of the U-shaped risk-response curve is critical for potential modulation of certain cancer-type specific risk profiles (Fairweather-Tait et al. 2011).

7.3.2.3 Diabetes

Variable evidence supporting an effect of selenium on the risk of diabetes exists, occasionally conflicting, and limited to very few human studies. Following a trial, investigating the effect of selenium supplementation (200 mg/day) on skin cancer, subsequent analysis showed that there was an increased risk of developing type 2 diabetes in the supplemented group (Stranges et al. 2007). Both low and high selenium intakes could influence the risk of diabetes, and this relationship requires further investigation (Stranges et al. 2007).

7.3.2.3 Inflammation and inflammatory disorders

Selenium has anti-inflammatory properties and the underlying mechanisms have recently been reviewed elsewhere (Duntas 2009). Selenium has also been accepted as a potential therapy for patients suffering rheumatoid arthritis (RA). Like many other inflammatory disorders, selenium status appears to be lower in RA patients than in control subjects, summarized in Tarp (Tarp 1995).

7.3.2.4 Fertility

The role of selenium in male spermatogenesis and semen quality (e.g., sperm count, semen volume, motility and morphology) has been investigated, but links have also been made to female reproductive issues such as pre-eclampsia and miscarriage (Rayman 2000). The evidence supports a limited role for selenium in female fertility although data suggests that women with unexplained infertility may have lower selenium levels in the follicular fluid than those with explained infertility (Paszkowski et al. 1995). Couples were assessed over a period of 5 yr and it was found that the pregnancy rate was greatest in the mid-range of selenium status (Bleau et al. 1984); however, status was only measured in the semen of the men, so, these findings require cautious interpretation, and more investigation, as the exposure of both partners would not necessarily be similar.

7.3.3 Dietary reference intakes

There is a relatively narrow range of intake between which selenium deficiency and toxicity occurs. Current estimates suggest that intakes below 30 μg/day are inadequate and those exceeding 900 μg/day are potentially harmful (Yang and Zhou 1994, WHO 1996).

Possible relationships between blood selenium concentration and selenium function or health effects are summarized in Fig. 15.4. Nevertheless, the relationship between selenium intake/status and risk of disease is complex, cancer type specific (location of tumor and grade/severity of disease), and specific to populations or individuals, all of which being dependent on baseline selenium.

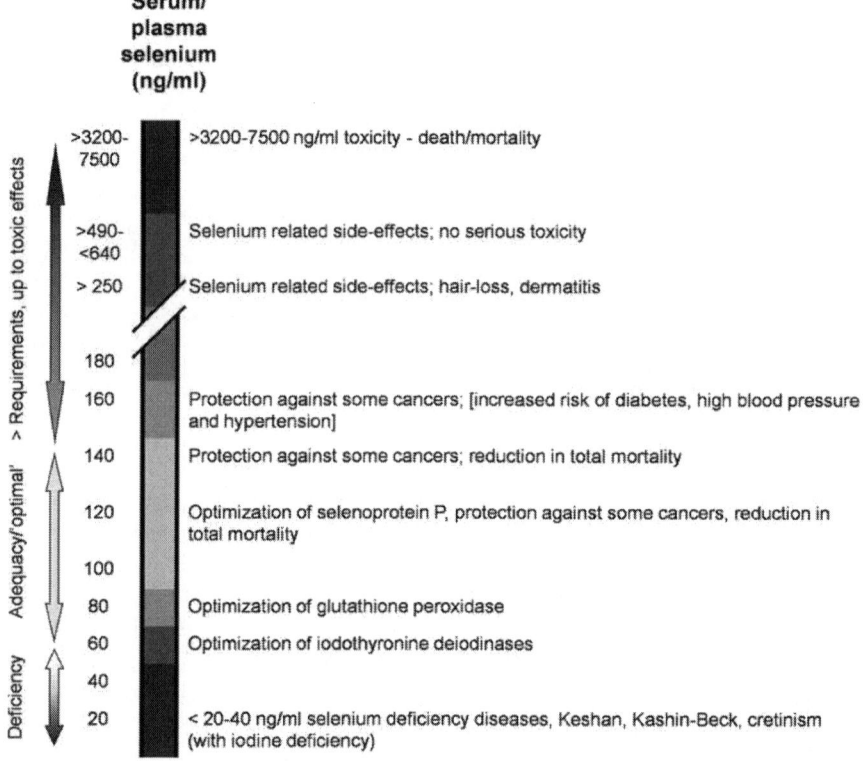

Figure 15.4 Range of blood selenium concentrations with possible related health effects from deficiency to toxicity (Fairweather-Tait et al. 2011).

8. Tellurium

8.1 Introduction

Tellurium occurs in two allotropic forms of a silvery white crystalline metal and as a black amorphous powder. Tellurium has 21 isotopes, with atomic masses ranging from 115–135. Tellurium chemically is similar to selenium and sulfur. As a result of burning in air a greenish blue flame produces and forms tellurium dioxide, which has low solubility in water. Tellurium has compounds in oxidation states −2, +2, +4, and +6. Elemental tellurium, hydrogen telluride, tellurium hexafluoride are toxicologically important. Tellurium dioxide, tellurous and telluric acids, and sodium and potassium tellurites and tellurates, which are soluble in water are also toxicologically important. Several complex compounds of tellurium and organometallic have been introduced (e.g., aryl-, diaryl-, and triaryl- tellurium (IV) chlorides). Preparation of a series of organotellurium compounds has been reported (e.g., bis2-hydroxymethylphenyltelluride, bis 2-hydroxymethylphenyl ditelluride and bis 4- hydroxymethylphenyl telluride) (Al-Rubaie and Al-Jadaan 2002). Tellurium is regarded as a non-essential trace element. However, a typical human body holds >0.5 g of Te, which is comparable to the levels of all other trace elements in human beings, except for iron, zinc and rubidium (Cohen 1984, Schroeder et al. 1967). Tellurium is found to be an essential element in a manner similar to selenium (Markert 1992, Markert 1994, Chasteen 1993). Although tellurium atom is generally regarded as a toxic metalloid, it plays a role in some biological systems (Taylor 1996). Yosef et al. have synthesized a group of based compounds with varied Te valences. The compounds exerting the most pronounced biological activities are AS101 [ammonium trichloro(dioxoethylene-O,O') tellurate] and SAS [Octa-O-bis-(R,R)-Tartarate Ditellurane] (Yosef et al. 2007).

8.2 Tellurium and human health

Exposures to tellurium vapor and hydrogen telluride may cause irritation of the respiratory tract (Izrael'son 1973, Popova et al. 1965), bronchitis and pneumonia (Lewis 1996). Exposure to tellurium hexafluoride causes dermatitis and blue-black skin discoloration. Ingestion of about 40 µg of soluble tellurium may cause the unusual breath odor. Formation of organic tellurium compounds (e.g., dimethyl telluride) may be responsible for the garlic-like odor in breathing Taylor 1996, Blackadder and Manderson 1975). Only a few cases of non-occupational poisoning have been reported so far. Three cases of accidental tellurium poisoning have been reported by Keall et al. (Keall et al. 1946). They had a solution of sodium tellurite by mistake. They showed symptoms such as vomiting, renal pain, stupor, loss of consciousness, irregular breathing, and cyanosis, and after approximately 6 hr, two of the patients died. During the autopsy of all tissues a strong garlic odor emitted. In the mucosa of the bladder and of the ureter, deposition of black tellurium was found. In the lungs, liver, spleen, and kidney, congestion was also detected. Distinct symptoms during tellurium intoxication are: Loss of appetite, dryness of the mouth, suppression of sweating, a metallic taste in the mouth, and, most notably, a sharp garlic odor of the breath, sweat and urine (Muller et al. 1989).

Ammonium trichloro (dioxoethylene-O, O′) tellurate (AS101) and Octa-O-bis-(R, R)-tartarate ditellurane (SAS), two inorganic tellurium complexes, have been thoroughly investigated (Sredni 2012), as well as their anti-cancer properties and explaining their mechanism of action. AS101 is probably the most vastly studied synthetic tellurium compound from the point of view of its biological activity. It is an effective immunomodulator (both *in vitro* and *in vivo*) with different potential therapeutic applications. It is most likely the only tellurium compound to be tested in phase I/II clinical studies in cancer patients. AS101 or SAS may have promising anti-cancer therapeutic potential coupled with the excellent safety profile for tellurium compounds (Sredni 2012).

9. Thallium

9.1 Introduction

Thallium exists in two oxidation states, Tl^+ and Tl^{3+}, however, Tl^+ generally forms most stable compounds (Shand et al. 1998). Tl^+ is stable and has properties similar to potassium; Tl^{3+}, which resembles aluminum, has strong oxidizing properties and is slowly converted to a monovalent state (Delvalls et al. 1999). Table 15.3 presents data on the occurrence of thallium in the environment.

Thallium exposure can happen through air, water and food. However, the levels of thallium in the air and water are very low. The greatest exposure occurs by eating food, mostly homegrown fruits and green vegetables contaminated by thallium. Small amounts of thallium are released into the air from coal-burning power plants, cement factories and smelting operations (ATSDR 1992).

Thallium salts were extensively used to treat syphilis (since 1883), and to reduce night sweats in tuberculosis patients (Delvalls et al. 1999). Application of thallium salts as poisons for rodents and later, as insecticides began in 1920 and for the following 45 yr it was the principal use for this element (Nriagu 1998). In the past, thallium was extensively used

Table 15.3 Thallium in the environment (Twidwell and Beam 2002).

Source	μg/L (ppb)	μg/kg (ppb)
Drinking water	7.2	
Groundwater, river water, groundwater	20–24	
Deep-sea sediments		200–5700
Deep-sea manganese nodules		to 100,000
Coal power plants (worldwide), kg/year		599,000
Lanmuchang thallium ores, Hg–Tl		720,000–3,800,000
Lanmuchang thallium ores, Hg–Tl Water (well, spring, stream, mine water)	0.4–2.7	
Nanhua thallium ores, As–Tl		960–1,900
Nanhua thallium ores, As–Tl Water (spring, mine water, slag water)	0.1–16.5	
Sulfide minerals: chalcopyrite		To 5%
Galena		1400–20,000
Sphalerite		8000–45,000
Pyrite		5000–23,000
Production from sulfides (worldwide), kg/year		15,500 in 1998
Entering, Pb, Zn, Cu smelting (worldwide), kg/year		> 160,000
Production from iron smelting (US), kg/year		> 140,000
Total world mobilization, kg/year		> 2,000,000

for medical purposes. In the treatment of ringworm of the scalp, it was given to children to produce hair loss, thallium was also used widely in the treatment of venereal diseases, tuberculosis and malaria (Kazantzis 2000). Thallium salts were marketed to treat scalp ringworm as they rapidly caused alopecia (baldness), enabling ointments to be applied more effectively to control the fungal infection. They have also been used as a depilatory. Among other uses of thallium and its salts are: manufacture of fake jewelry, low-temperature thermometers, ceramic semiconductor material, scintillation counters for radioactivity quantitation and inoptical lenses, to which it confers a high refractive index (Arzate and Santamaria 1998). Thallium is also applied for oxidation of hydrocarbons and olefins, for polymerization and for epoxidation. Thallium compounds are used in infra-red spectrometers, crystals, other optical systems, and for coloring glass (Kazantzis 2007). Thallium is mainly applied in alloys, electronic devices and specialized glass (Zitko 1975). The greatest application of thallium is in specialized electronic research equipment (EPA 2012). Among the growing uses of thallium, the following are mentioned: laser industry, fiber (optical) glass, scintillographic imaging, superconductivity and as a molecular probe to emulate the biological function of alkali-metal ions. In addition, other various applications of thallium compounds are: fireworks (thallium nitrate), pigments (thallium chromate) and dyes, impregnation of wood and leather against fungi and bacteria and in mineralogical separation (IPCS 1996). Due to excellent Nuclear Magnetic Resonance (NMR) properties of thallium ions, they have been used as a probe to emulate the biological functions of alkali metal ions, eespecially K^+ and Na^+ (Kazantzis 2000).

9.2 Thallium and human health

9.2.1 Thallium toxicity

Thallium is highly toxic and it is an Environmental Protection Agency (EPA) priority pollutant element, also it is more acutely toxic than Hg, Cd, Pb, Zn and Cu in mammals (Cheam 2001). It is absorbed through mucous membranes and skin. Thallium is widely distributed throughout the body. It accumulates in bones, renal medulla and eventually, in the central nervous system. Thallium has the biological half-life of 3–8 d. It is excreted mainly in the urine (Zitko 1975). Thallium salts are colorless and odorless and have been used for homicidal purposes, and for illegal abortion (Kazantzis 2000).

9.2.2 Acute toxicity

The effects of large amounts of thallium ingestion over a short period of time includes: vomiting, diarrhea, temporary hair loss, and effects on the nervous system, lungs, heart, liver and kidneys, and finally death. The effects of ingesting low levels of thallium over a long time are not known (EnviroTools 2002). The symptoms of acute thallium poisoning in human beings are gastroenteritis, polyneuropathy and alopecia (Kazantzis 2007). Dermic signs may encompass palmarerythema, acne, anhydrosis and dry scaly skin which are caused by the toxic effect of thallium on sweat and sebaceous glands (Mulkey and Oehme 1993). With acute intoxication, the exposure to a large single dose, there is usually an initial hypotension and bradycardia owing to direct effects of the sinus node and cardiac muscle, followed by hypertension and tachycardia thought to be due to vagal nerve degeneration (Arzate and Santamaria 1998). The central and peripheral nervous system is the main critical organ in thallium intoxication (Kemper and Bertram 1991). Symptoms of acute toxicity depend on age, route and dose of administration (IPCS 1996).

9.2.3 Chronic toxicity

Symptoms of thallium poisoning include anorexia, headache and pains in abdomen, upper arms and thighs and all over the body. In extreme cases, alopecia, blindness, and eventually death may occur (Zhang et al. 1998). Some studies indicate that there is no thallium mutagenicity or teratogenicity (Kemper and Bertram 1991). If the intake of thallium occurrs over a long period, white streaks (Mee's lines) may be seen on fingernails and toenails (Arzate and Santamaria 1998). Limited data are available on the effects of thallium on human reproductive system. Menstrual cycle, libido and male potency may be adversely affected. Sperms are known to be affected following chronic intoxication (IPCS 1996).

John Peter at al. reviewed various treatment options and removal technologies in order to protect the environment from thallium toxicity (John Peter and Viraraghavan 2005).

10. Tin

10.1 Introduction

Tin (Sn) belongs to Group IV A of the Periodic Table. Tin compounds are found in the environment with an oxidation state of + 2 or + 4; however, the trivalent forms are unstable. So, the stannous compounds, SnX_2 of bivalent tin, and the stannic compounds, SnX_4 of tetravalent tin, are two main types. Anionic stannite and stannate are insoluble and stable whereas cationic Sn^{2+} and Sn^{4+} are stable. Although tin is found everywhere, it is one of the least abundant elements on the Earth's crust (Cima 2011).

10.2 Tin and human health

10.2.1 Inorganic tin and human health

There are several reports on occurrence of food poisoning from tin contamination in canned food. Large numbers of people in Kuwait, after consuming formulated orange juice and apple juice containing 250–390 mg/kg of tin, suffered gastrointestinal disturbances, such as vomiting and diarrhea (Benoy et al. 1971). Acute gastroenteritis, a condition that causes irritation and inflammation of the stomach and intestines, was reported after ingestion of canned tomato juice containing tin at concentration of 131–405 mg/kg (Barker and Runte 1972).

Canned peaches containing 563 mg tin/kg had also caused gastrointestinal symptoms (Nehring 1972). Prolonged suppression of DNA synthesis in human lymphocytes has been reported as a result of inorganic stannous and stannic chloride at a concentration of 10–50 µmol/L (McLean et al. 1983).

Abdominal pain, anemia, and liver and kidney problems occurred as a result of high intakes of inorganic tin compounds (Williams et al. 1999). Inhalation of tin hydride is neurotoxic, causing nerve damage. Although it is similar to arsine gas, the effects are less serious because it does not cause hemolysis (Baldwin and Marshall 1999). Skin rashes, nausea, stomach upsets, diarrhea, vomiting, abdominal pain, palpitations, and headache are symptoms of high levels of tin poisoning. Fatigue, depression, low adrenals, shortness of breath, asthma, headaches, low cardiac output (left) and insomnia are caused by low levels of tin exposure. Tin exposures show an epidemiological relationship with ischemic heart disease, a term used to describe patients whose heart can no longer pump enough blood throughout their body due to coronary artery disease, and the metabolism of cholesterol (Klevay 2000). Tin can increase cholesterol in plasma (Klevay 1984). Risk enhancements of chronic renal failure are also reported because of increasing exposure to tin (Nuyts et al. 1995).

10.2.2 Organotin and human health

Since organic tin compounds inhibit synthesis of heme oxygenase, they are toxic and can also be genotoxic. Because organic tin compounds are absorbed through the skin, they cause serious irritation and burning of the skin. When these compounds are inhaled or ingested, they cause systemic toxicity, mainly anemia and renal and hepatocellular damage. They may also lead to hyperglycemia, changes in blood pressure and damage to the immune system. Alkyl and aromatic tin compounds have the potential to be neurotoxins. Encephalopathy and cerebral edema can be caused by trialkyltin compounds, such as triethyltin. Extreme industrial exposure to triethyltin produces symptoms of visual defects, headaches and nausea (Prull and Rompel 1970). Organic tin compounds (trimethyl and triethyl derivatives) develop psychomotor tremors, disturbances, hallucinations, convulsions, and psychotic behavior (Hu 1998). Inhalation of fungicide powder, containing 60 percent triphenyltin, causes dizziness, nausea and photophobia (Manzo et al. 1981). There are no neurobehavioral studies of children whose mothers had been exposed to organotins during pregnancy (Sokas 1998). Workers exposed to liquid tri- and dibutyltin-chlorides because of leaking gloves or failure to wear hand protection complained of skin burns and severe lesions on the hands (Lyle 1958).

The main toxic effects of organotins causing environmental pollution are summarized in Table 15.4 (Cima 2011). Chapter 42 of Handbook on the Toxicology of Metals (Third Edition) (Ostrakhovitch and Cherian 2007) is recommended for further reading.

Table 15.4 Human toxicity of organotins (Cima 2011).

Toxic· effects	Clinical findings	Compound	Abbreviations	
Local effects Skin irritation and lesion	Erythematous eruption Contact dermatitis Ulceration	All	ACT DBT DET DMT	azocyclotin dibutyltin diethyltin dimethyltin
Eye irritation	Profuse lacrimation Eye inflammation Photophobia Blurred vision	All	FBTO MBT MET MMT	fenbutatin oxide monobutyltin monoethyltin monomethyltin
Respiratory Damage	Breath shortness Coughing Wheezing	All	TBT TBT·MMA	tributyltin tributyltinmethyl methacrylate copolymer
Gastrointestinal Irritation	Pharyngitis Nausea Vomiting Hematemesis Inappetence Abdominal pain	TBT, TPTA TBT, TPTA TBT, TPTA TBT, TPTA TBT, TPTA TBT, TPTA	TBTO TCHT TET TMr TPT TPTA TPTC TPTH	tributyltin oxide tricyclohexyltin triethylti n trimethyltin triphenyltin triphenyltin· acetate triphenyltin chloride triphenyltin hydroxide

Table 15.4 contd....

Table 15.4 contd.

Toxic· effects	Clinical findings	Compound	Abbreviations
Systemic effects	Smell disturbance	TBT	
Neurological Effects	Hearing loss	TMT	
	Visual disturbance	TMT	
	Seizures	TMT	
	Neural necrosis	TMT, TET	
	Encephalopathy	TPT	
	Cerebral edema	TPT	
	Involuntary hand movements	TPT	
	Facial twitching	TPT	
	Crying	TPT	
	Diplopia	TPT	
	Drowsiness	TPT	
	Giddiness	TPT	
	Bidirectional nystagmus	TPT	
	Calculation ability impairment	TPT	
	Time and placedisorientation	TPT	
	Amnesia	TMT, TPTA	
	Headache	TET, TBT	
	Delayed sensomotor polyneuropathy	TPTA	
Cardiovascular Effects	Vasodilatation	TBTO	
	Hypotension	TBTO	
	Heart failure	TBTO	
Hepatic Effects	Elevated hepatic enzymes	TPT	
	Elevated C-reactive protein	TPTA	
	Hepatomegaly	TPTA	
	Hepatic inflammation	TPTA	
Genitourinary Effects	Acute renal failure	TPTA	
Hematological effects	Hemolysis	TBT> TPT> TET>DBT> TMT= MBT	
	Leucopoenia	TPTA, TPT	

11. Zinc

11.1 Introduction

Zinc (Zn) is a trace element which can be found almost everywhere. It is one of the most significant trace elements in the body, and it is essential to the growth and development of microorganisms, plants and animals. Zinc exists in all body tissues and secretions in relatively high concentrations, 85 percent of the whole body with zinc being stored in muscles and bones, 11 percent in the skin and the liver, and the remaining in other tissues; the highest concentrations are in the prostate and parts of the eye. In the adult body, the average amount of Zn is about 1.4–2.3 g (Calesnick and Dinan 1988, Stefanidou et al. 2006, Prasad 2009, Bhowmik et al. 2010). After iron, zinc is the second most abundant transition metal ion in living organisms; however, ignoring hemoglobin-bound iron, zinc becomes the most abundant transition metal (Vasak and Hasler 2000).

   Unlike other transition metal ions, such as copper and iron, zinc does not undergo redox reactions since the d shell of zinc element is filled. Zinc is a biologically significant trace element and essential for cell growth, development, differentiation, homeostasis, connective tissue growth and maintenance, DNA synthesis, RNA transcription, cell division and cell activation. It is also essential in taste acuity, immune system function, prostaglandin production, bone mineralization, wound healing, proper thyroid function, blood clotting, cognitive functions, fetal growth and sperm production. It adjusts body fluid pH; it elevates the formation of collagen to make hair, skin, nails; it helps to enhance memory and improves mental development; it keeps the normal function of the prostate; and it has a significant role in testosterone secretion (Bhowmik et al. 2010). Moreover, zinc is required in immune responses, apoptosis, oxidative stress, and aging (Stefanidou et al. 2006, Murakami and Hirano 2008, Prasad 2009, Plum et al. 2010).

11.2 Zinc and human health

11.2.1 Zinc deficiency

"Even a small deficiency of zinc leads to a disaster. Lack of zinc causes smell and taste failure, loss of appetite anorexia", and may affect the immune system, triggering arteriosclerosis and anemia. Zinc deficiency causes impaired hemostasis because of defective platelet aggregation, a decrease in T-cell number and a decreased response of T-lymphocytes to phytomitogens. Zinc deficiencies also accompany a lot of diseases such as gastrointestinal disorders, renal disease, sickle cell anemia, alcoholism, some cancer types,

AIDS, burns and aging (Keen and Gershwin 1990, Mocchegiani and Fabris 1995, Fraker et al. 2000, Mocchegiani and Muzzioli 2000).

Zinc deficiency in pregnant women may decrease fetal brain cells, and may influence their development. Zinc deficiency in children may detain intellectual development, normal growth and reproductive system health. Zinc deficiency in adult males may result in prostatic hyperplasia, influencing the reproductive function and fertility. However, zinc supplementation is a powerful therapeutic tool in managing a long list of illnesses (Bhowmik et al. 2010). The recommended daily dietary zinc intake is estimated at 15 mg/day (Tapiero and Tew 2003, MacDonald 2000), and the recommended tolerable upper intake level of zinc is 25 mg/day (SCF 2003). The efficiency of zinc supply is highly related to the dose and length of the treatment. Long treatment or high doses of zinc may raise zinc accumulation with subsequent damage on immune efficiency (Mocchegiani et al. 2001). Modest immune modifications are observed in old people treated with high doses of zinc for short periods, as well as with physiological zinc doses (12 mg Zn^{++}/day) (USDA 1976) for long periods of up to 1 yr. Zinc accumulation in both conditions may result in toxic effects. After zinc physiological supplementation, immune recovery has been observed in the elderly, in infections, in cancer, and in patients with sickle cell anemia (Mocchegiani and Muzzioli 2000). Zinc therapy has been very successful in patients with acrodermatitis enteropathica and Wilson's disease (Prasad 2009). However, since excess zinc is toxic for the cell and may cause acute and chronic zinc poisoning (Pagani et al. 2007), the cellular level of zinc must be controlled within a satisfactory range, which is normally between 0.1 and 0.5 mM (Eide 2006).

11.2.2 Zinc and the immune system

Zinc is considered critical for immune responses. It effects and interacts especially with components of the immune system, a highly proliferative system (Wellinghausen and Rink 1998).

11.2.3 Zinc and depression

Zinc deficiency causes impaired brain zinc homeostasis and leads to change in behavior, mental function, learning and susceptibility to epileptic convulsions. It was claimed that human depression might be accompanied with lower serum zinc concentrations in subjects suffering from depression (Takeda 2000, Nowak et al. 2005). Previous research with human beings and animals showed that dietary intake of zinc may regulate symptoms of depression. Moreover, clinical studies demonstrated the benefit of zinc supplementation in antidepressant therapy in major depression (Nowak et al. 2003, Whittle et al. 2009). Amani et al. (2010) showed a positive correlation between dietary intake of zinc and serum levels of zinc.

11.2.4 Zinc and oxidative stress

Cells in all living systems require adequate levels of antioxidant defenses to avoid the harmful effect of an excessive production of Reactive Oxygen Species (ROS), and to hinder damage to the immune cells (Puertollano et al. 2011). Zinc deficiency is associated with increased levels of oxidative damage, including increased lipid, protein and DNA oxidation (Prasad 2009, Jomovaa and Valko 2011).

11.2.5 Zinc and cardiovascular diseases

Cardiovascular Disease (CVD) is believed to be the major cause of death worldwide. Cardiovascular cells have interactions with zinc. Zinc ions are rapidly taken up by endothelial cells, possibly by endocytosis of albumin-bound zinc, so changes in the dietary zinc have the potential to change endothelial cell levels of zinc. The levels of zinc in plasma decrease with age and have a strong relation with increasing CVD; thus, there is a strong relation between zinc deficiency and increased CVD (Little et al. 2010).

11.2.6 Zinc and ageing

Aging is an unavoidable biological process associated with successive and automatic biochemical and physiological changes and increased vulnerability to diseases. In aging, loss of immunological responses may have various origins, among them decline in neuroendocrine function and increase in apoptosis regulated by zinc deficiency that leads to low zinc ion bioavailability and high MTs levels (Mocchegiani et al. 2000, Mocchegiani and Muzzioli 2000). In slowing down aging, zinc supplementation is considered necessary since this causes improvement of immune functions (Mocchegiani et al. 2011).

11.2.7 Zinc and Alzheimer's disease

There are research findings which prove that zinc metabolism is varied in Alzheimer's disease and other neurodegenerative diseases (Aschner 1996, Wang et al. 2010). The presence of extracellular b-amyloid (Ab) plaques in the brain is one of the pathological signs of Alzheimer's Disease (AD). Abnormal zinc homeostasis is a factor which leads to Ab aggregation, and alteration of zinc homeostasis may become a possible way for AD therapeutic (Wang et al. 2010). However, it was found that a high level of zinc also increased Ab generation through altering the expression levels of APP (Amyloid Precursor Protein) and APP cleavages enzymes *in vivo* and *in vitro* (Wang et al. 2010).

11.2.8 Zinc and diabetes

It is well known that a physical chemical relationship exists between insulin and zinc. It was clear that the addition of zinc to insulin would change the time course of the effect of a given dose of insulin (Dodson and Steiner 1998) and, hence adequate insulin amounts could be stored in pancreatic b-cells to allow enough release after a meal (Carroll et al. 1988). It appears to be a complex interrelationship between zinc and Type I and/or Type II diabetes. Diabetes may be related to increased intracellular oxidants and free radicals associated with decreases in intracellular zinc and in zinc-dependent antioxidant enzymes. Investigations revealed that the islet-restricted zinc transporter, ZnT8, is a likely to play a role in the control of insulin secretion and the risk of developing Type II diabetes, and this transporter represents an exciting therapeutic target for intervention in Type II diabetes (Chimienti et al. 2006, Rutter 2010).

11.2.9 Zinc and Wilson's disease

Wilson's disease, an inherited autosomal recessive disorder of copper balance, causes hepatic damage and neurological disturbance of variable degree. The defective gene ATP7B, which plays a key role in human copper metabolism, encodes a hepatic copper-transporting protein. Many clinical findings are related to copper accumulation predominantly in the liver and the brain and include hepatic disease ranging from mild hepatitis to acute liver failure or cirrhosis and/or neurological symptoms such as dystonia, tremors, dysarthria, and psychiatric disturbances (Prasad 2009). Zinc treatment in Wilson's disease does not aggravate the patients' clinical signs and/or laboratory findings. Nevertheless, it improves some clinical symptoms of the patients. The administration of zinc has some side effects as well; however, none of them is very serious. So, for long term management of patients with Wilson's disease, zinc acetate is a recommended therapy (Huster 2010, Shimizu et al. 2010). A review article prepared by Chasapis et al. is recommended for further information (Chasapis et al. 2012).

11.2.10 Zinc and cancer

Zinc is believed to contribute in cancer prevention, so it plays an essential role in host defense against the initiation and promotion of several malignancies (Dhawan and Chadha 2010). Zinc levels in serum and malignant tissues of patients with different types of cancer are abnormal, confirming the involvement of zinc in cancer development (John et al. 2010). Studies have shown that zinc levels in serum are reduced in patients with cancer of breast (Schlag et al. 1978), gallbladder (Gupta et al. 2005), lung (Issell et al. 2006), colon, head and neck (Bu¨ntzel et al. 2007), and bronchus (Issell et al. 2006, Büntzel et al. 2007, Chakravarty et al. 1985).

References

ACGIH. 1983. Supplemental documentation 1983. American Conference of Governmental Industrial Hygienists. Cincinnati, OH.

Agusa, T., T. Kunito and H. Iwata. 2005. In Monirith, Touch Seang Tana, Annamalai Subramanian, Shinsuke Tanabe, Mercury contamination in human hair and fish from Cambodia: levels, specific accumulation and risk assessment. Environmental Pollution. 134(1): 79–86.

AIHA. 2002. American Industrial Hygiene Association. The AIHA 2002 Emergency Response Planning Guidelines and Workplace Environmental Exposure Level Guides Handbook. American Industrial Hygiene Association. Fairfax, VA.

Akagi, H., P. Grandjean, Y. Takizawa and P. Weihe. 1998. Methylmercury dose estimation frommumbilical cord concentrations in patients with Minamata disease. Environ. Res. 77: 98–103.

Alexander, W., R. Harden, M. Harrison and J. Shimmins. 1967. Some aspects of the absorption and concentration of iodide by the alimentary tract in man. Proc. Nutr. Soc. 26: 62–6.

Alimonti, A., G. Forte and S. Spezia. 2005. Rapid Commun. Mass Spectrom. 19: 3131–3138.

Allan, C.B., G.M. Lacourciere and T.C. Stadtman. 1999. Responsiveness of selenoproteins to dietary selenium. Ann. Rev. Nutr. 19: 1–16.

Al-Rubaie, A.Z. and S.A.S. Al-Jadaan. 2002. Appl. Organometallic Chem. 16: 649–654.

Amani, R., S. Saeidi, Z. Nazari and S. Nematpour. 2010. Correlation between dietary zinc intakes and its serum levels with depression scales in young female students. Biol. Trace Elem. Res. 137(2): 150–158.

An, Y.J. and M.J. Kim. 2009. Effect of antimony on the microbial growth and the activities of soil enzymes. Chemosphere. 74: 654–659.

Andersson, M., B. De Benoist and L. Rogers. 2010. Epidemiology of iodine deficiency: salt iodisation and iodine status. Best Pract. Res. Clin. Endocrinol. Metab. 24(1): 1–11.

Andrewes, P., K.T. Kitchin and K. Wallace. 2004. Toxicol. Appl. Pharmacol. 194: 41–48.

Aposhian, H.V., B. Zheng, M.M. Aposhian, X.C. Le, M.E. Cebrian, W. Cullen, R.A. Zakharyan, M. Ma, R.C. Dart, Z. Cheng, P. Andrewes, L. Yip, G.F. O'Malley, R.M. Maiorino, W. Van Voorhies, S.M. Healy and A. Titcomb. 2000a. DMPS arsenic challenge test. II. Modulation of arsenic species, including monomethylarsonous acid (MMA(III)), excreted in human urine. Toxicol. Appl. Pharmacol. 165: 74–83.

Aposhian, H.V., E.S. Gurzau, X.C. Le, A. Gurzau, S.M. Healy, X. Lu, M. Ma, L. Yip, R.A. Zakharyan, R.M. Maiorino, R.C. Dart, M.G. Tircus, D. Gonzalez-Ramirez, D.L. Morgan, D. Avram and M.M. Aposhian. 2000b. Occurrence of monomethylarsonous acid in urine of humans exposed to inorganic arsenic. Chem. Res. Toxicol. 13: 693–697.

Arteel, G.A, K. Briviba and H. Sies. 1999. Protection against peroxynitrite. FEBS Lett. 445: 226–30.

Arzate, S.G. and A.Santamaria. 1998. Thallium toxicity. Toxicol. Lett. 99: 1–13.
Aschner, M. 1996. The functional significance of brain. Metallothionein. FASEB. J. 10: 1129–1136.
Ashton, K., L. Hooper, L.J. Harvey, R. Hurst, A. Casgrain and S.J. Fairweather-Tait. 2009. Methods of assessment of selenium status in humans: a systematic review. Am. J. Clin. Nutr. 89: 2025S–2039S.
ATSDR (Agency for Toxic Substances and Disease Registry). 1992. Toxicological profile for thallium. U.S. Public Health Service.
Bakir, F., S.F. Damluji, L. Amin-Zaki, M. Murtadha, A. Khalidi, N.Y. Al-Rawi, S. Tikriti, H.I. Dhahir, T.W. Clarkson, J.C. Smith and R.A. Doherty. 1973. Methylmercury Poisoning in Iraq, Science. 181(4096): 230–241.
Baldwin, D.R and W.J. Marshall. 1999. Heavy metal poisoning and its laboratory investigation. Ann. Clin. Biochem. 36 (3): 267–300.
Ballatori, N. 2002. Transport of toxic metals by molecular mimicry. Environ. Health Persp. 110(suppl 5): 689–694.
Barker, W.H. Jr. and V. Runte. 1972. Tomato juice-associated gastroenteritis, Washington and Oregon, 1969. Am. J. Epidemiol. 96(3): 219–226.
Barnes, J.M. and P.N. Magee. 1958. J. Pathol. Bacteriol. 75: 267–279.
Barregard, L., G. Sallsten and B. Jarvholm. 1990. Mortality and cancer incidence in chloralkali workers exposed to inorganic mercury. Br. J. Ind. Med. 47: 99–104.
Basic Information | Mercury | US EPA". 2012. http://www.epa.gov/mercury/about.htm.
Bates M.N. 2011. Dental Amalgam Fillings: A Source of Mercury Exposure, Encyclopedia of Environmental Health. pp. 11–20.
Bates, M.N., A.H. Smith and C. Hopenhayn-Rich. 1992. Am. J. Epidemiol. 135: 462–476.
Beck, E.B. 1906. To the problem of disforming endemic osteoarthritis in the Baikal area. Russk Vrach. Russian Physician. 3: 74–75.
Behne, D., A. Kyriakopoulos, M. Kalcklosh et al. 1997. Two new selenoproteins found in the prostatic glandular epithelium and the spermatid nuclei. Biomed. Environ. Sci. 10: 340–45.
Behne, D., H. Pfiefer, D. Rothlein and A. Kyriakopoulos. 2000. Cellular and subcellular distribution of selenium and selenoproteins. *In:* A.M. Roussel, A. Favier and R.A. Anderson [eds.]. Trace elements in man and animals 10: Proceedings of the Tenth International Symposium on Trace Elements in Man and Animals. Plenum Press. New York. pp. 29–33.
Belyaeva, A.P. 1967. The effect of antimony on reproduction. Gig. Truda. Prof. Zabol. 11: 32.
Benefice, E., S. Luna-Monrroy and R. Lopez-Rodriguez. 2010. Fishing activity, health characteristics and mercury exposure of Amerindian women living alongside the Beni River (Amazonian Bolivia), International Journal of Hygiene and Environmental Health. 213(6): 458–464.
Benoy, C.J., P.A. Hooper and R. Schneider. 1971. Food Cosmet. Toxicol. 9: 645–656.
Berlin, M., R.K. Zalups and B.A. Fowler. 2007. Mercury. *In:* Handbook on the Toxicology of Metals 3rd edn. Academic Press, Burlington, MA 01803, USA, pp. 675–729.
Bernal J. 2005. Thyroid hormones and brain development. Vitam Horm. 71: 95–122.
Bhowmik, D., K.P. Chiranjib and K.P.S. Kumar. 2010. A potential medicinal importance of zinc in human health and chronic disease. Int. J. Pharm. Biomed Sci. 1(1): 05–11.
Bjelakovic, G., D. Nikolova, R.G. Simonetti and C. Gluud. 2008. Antioxidant supplements for preventing gastrointestinal cancers. Cochrane Database Syst. Rev: CD004183.
Blackadder, E.S. and W.G. Manderson. 1975. Br. J. Ind. Med. 32: 59–61.
Bleau, G., J. Lemarbre, G. Faucher, K.D. Roberts and A. Chapdelaine. 1984. Semen selenium and human fertility. Fertil Steril. 42: 890–894.
Blot, W.J., J.Y. Li, P.R. Taylor, W. Guo, S.M. Dawsey and B. Li. 1995. The Linxian trials: mortality rates by vitamin-mineral intervention group. Am. J. Clin. Nutr. 62: 1424S–1426S.
Boffetta, P., E. Merler and H. Vainio. 1993. Carcinogenicity of mercury and mercury compounds. Scand. J. Work Environ. Health. 19: 1–7.
Boffetta, P., M. Garcia-Gomez and V. Pompe-Kirn. 1998. Cancer occurrence among European mercury miners. Cancer Causes & Control. 9: 591–599.

Brieger, H. 1920. Z. Exp. Pathol. Ther. 21: 393–408.

Brieger, H., C.W. Semisch, J. Stasney and D.A. Piatnek. 1954. Industrial antimony poisoning. Ind. Med. Surg. 23: 521–523.

Brinkman, M., R.C. Reulen, E. Kellen, F. Buntinx and M.P. Zeegers. 2006. Are men with low selenium levels at increased risk of prostate cancer? Eur. J. Cancer. 42: 2463–2471.

Bu¨ntzel, J., F. Bruns, M. Glatzel, A. Garayev, R. Mu¨cke, K. Kisters, U. Scha¨fer, K.Scho¨nekaes and O. Micke. 2007. Zinc concentrations in serum during head and neck cancer progression. Anticancer Res. 27(4A): 1941–1943.

Bulbulyan, M.A., N.J. Jourenkova, P. Boffetta, S.V. Astashevsky, A.F. Mukeria and D.G. Zaridze. 1996. Mortality in a cohort of Russian fertilizer workers. Scand. J. Work Environ. Health. 22: 27–33.

Burk, R.F. and J.M. Lane. 1983. Modification of chemical toxicity by selenium deficiency. Fundamental and Applied Toxicology. 3(4): 218–221.

Calderon, J., M.E. Navarro, Jimenez M.E. Capdeville, M.A. Santos Diaz, A. Golden, Rodriguez I. Leyva, V. Borja Aburto and F. Díaz Barriga. 2001. Exposure to arsenic and lead an neuropsychological development in Mexican children. Environ. Res. 85: 69–76.

Calesnick, B. and AM. Dinan. 1988. Zinc deficiency and zinc toxicity. Am. Fam. Physician. 37: 267–270.

Card, W.I. 1935. Lancet. 2: 1348–1349.

Carlson, B.A., M.H. Yoo, Y. Sano, A. Sengupta, J.Y. Kim, R. Irons, V. N. Gladyshev, D.L. Hatfield and J.M. Park. 2009. Selenoproteins regulate macrophage invasiveness and extracellular matrixrelated gene expression. BMC Immunol. 10: 57.

Carroll, R.J., R.E. Hammer, S.J. Chan, H.H. Swift, A.H. Rubenstein and D.F. Steiner. 1988. A mutant human proinsulin is secreted from islets of Langerhans in increased amounts via an unregulated pathway. Proc. Natl. Acad. Sci. USA. 85: 8943–8947.

Chakravarty, P.K., A. Ghosh and J.R. Chowdhury. 1985. Zinc in human malignancies. Neoplasma. 33: 85–90.

Chan, H.M. 2011. Mercury in Fish: Human Health Risks, Encyclopedia of Environmental Health. 697–704.

Chasapis, C.T., A.C. Loutsidou, C.A. Spiliopoulou and M.E. Stefanidou. 2012. Zinc and human health: an update, Arch. Toxicol. 86: 521–534.

Chasteen, T.G. 1993. Confusion between dimethyl selenenyl sulfide and dimethyl selenone released by bacteria. Appl. Organomet. Chem. 7: 335–42.

Cheam, V. 2001. Thallium contamination of water in Canada. Water Qual. Res J. Can. 36(4): 851–77.

Chen, B., C.T. Burt, P.L. Goering et al. 1986. Biochem. Biophys. Res. Comm. 139: 228–34.

Chen, C.J., Y.-C. Chuang, T.M. Lin et al. 1985. Cancer Res. 45: 5895–5899.

Chen, C.J., Y.M. Hsueh, H.Y. Chiou et al. 1997b. *In:* C.O. Abernathy, R.L. Calderon and W.R. Chappell [eds.]. Arsenic: Exposure and Health Effects. Chapman & Hall, London. pp. 232–242.

Chen, W., J. Yang, J. Chen and J. Bruch. 2006. Exposures to silica mixed dust and cohort mortality study in tin mines: exposure–response analysis and risk assessment of lung cancer. Am. J. Ind. Med. 49: 67–76.

Chimienti, F., S. Devergnas, F. Pattou, F. Schuit, R. Garcia-Cuenca, B. Vandewalle et al. 2006. *In vivo* expression and functional characterization of the zinc transporter ZnT8 in glucose-induced insulin secretion. J. Cell. Sci. 119: 4199–4206.

Chou, S., C. Harper, L. Ingerman, F. Llados, J. Colman, L. Chappell, M. Osier, M. Odin and G. Sage. 2007. Toxicological Profile for Arsenic. US Department of Health and Human Service, Agency for Toxic Substances and Disease Registry.

Chowdhury, U.K., B.K. Biswas, T.R. Chowdhury, G. Samanta, B.K. Mandal, G.C. Basu, C.R. Chanda, D. Lodh, K.C. Saha, S.K. Mukherjee, S. Roy, S. Kabir, Q. Quamruzzaman and D. Chakraborti. 2000. Groundwater arenic contamination in Bangladesh and West Bengal, India. Environ. Health Perspect. 108: 393–397.

Cima, F. 2011. Tin: Environmental Pollution and Health Effects, Encyclopedia of Environmental Health. 351–359.

Clark, L.C, G.F. Jr. Combs, B.W. Turnbull, E.H. Slate, D.K. Chalker, J. Chow, L.S. Davis, R.A. Glover, G.F. Graham, E.G. Gross, A. Krongrad, J.L. Jr. Lesher, H.K. Park, B.B. Jr. Sanders, C.L. Smith and J.R. Taylor. 1996. Effects of selenium supplementation for cancer prevention in patients with carcinoma of the skin. A randomized controlled trial. Nutritional Prevention of Cancer Study Group. JAMA. 276: 1957–1963.

Clarkson, T.W., L. Magos and G.J. Myers. 2003. Human exposure to mercury: The three modern dilemmas. J. Trace Elem. Exp. Med. 16: 321–343.

Cohen, B.L. 1984. Anomalous behavior of tellurium abundances. Geochim Cosmochim Acta. 48: 203–5.

Cohen, S.M., L.L. Arnold, M. Eldan, A.S. Lewis and B.D. Beck. 2006. Methylated arsenicals: the implications of metabolism and carcinogenicity studies in rodents to human risk assessment. Crit. Rev. Toxicol. 36: 99–133.

Cook, G.C. 1993. Leishmaniasis: some recent developments in chemotherapy. J. Antimicrob. Chemother. 31: 327–330.

Cordasco, E.M. and F.D. Stone. 1973. Pulmonary edema of environmental origin. Chest. 64: 182–185.

Costa, M. 1997. Toxicity and carcinogenicity of Cr(VI) in animal models and humans. CRC Crit. Rev. Toxicol. 27: 431–442.

Davis, A. 1968. Comparative trials of antimonial drugs on urinary schistosomiasis. Bull. World Health Organ. 38: 197–227.

Davis, P.J., F.B. Davis and V. Cody. 2005. Membrane receptors mediating thyroid hormone action. Trends Endocrin Met. 16: 429–35.

De Benoist, B., E. McLean, M. Andersson and L. Rogers. 2008. Iodine deficiency in 2007: global progress since 2003. Food Nutr. Bull. 29: 195–202.

De Flora, S. 2000. Carcinogenesis. 21: 533–541.

De Vizcaya Ruiz, A., O. Barbier, R. Ruiz Ramos and M.E. Cebrian. 2009. Biomarkers of oxidative stress and damage in human populations exposed to arsenic. Mutat. Res. 674: 85–92.

Degroot, L.J. 1966. Kinetic analysis of iodine metabolism. J. Clin. Endocr. Metab. 26: 149–73.

Delvalls T.A., V. Saenz, A.M. Arias and J. Blasco. 1999. Thallium in the marine environment: first ecotoxicological assessments in the Guadalquivir estuary and its potential adverse effect on the Don̄ana European natural reserve after the Aznalcollar mining spill. Cienc Mar. 25(2): 161–175.

Dewirtz, A.P. 1929. Dermatol. Chromate of Lime Ulcers. Dermatol. Wochenschr. 89: 1801–1818.

Dhawan, D.K. and V.D. Chadha. 2010. Zinc: a promising agent in dietary chemoprevention of cancer. Indian J. Med. Res. 132(6): 676–682.

Díaz Villaseñor, A., A.L. Burns, M. Hiriart, M.E. Cebrián and P. Ostrosky Wegman. 2007. Arsenic induced alteration in the expression of genes related to type 2 diabetes mellitus. Toxicol. Appl. Pharmacol. 225: 123–133.

Diplock, A.T. 1994. Antioxidants and disease prevention. Mol Aspects Med. 15: 293–376.

Dodson, G. and D. Steiner. 1998. The role of assembly in insulin's biosynthesis. Curr. Opin. Struct. Biol. 8: 189–194.

Dolbec, J., D. Mergler, C.J.S. Passos, S.S. De Morais and J. Lebel. 2000. Methylmercury exposure affects motor performance of a riverine population of the Tapajos river, Brazilian Amazon. Int Arch Occup Environ Health. 73: 195–203.

Dórea, J.G., V. Lucia V.A. Bezerra, V. Fajon and M. Horvat. 2011. Speciation of methyl- and ethyl-mercury in hair of breastfed infants acutely exposed to thimerosal-containing vaccines, Clinica Chimica Acta. 412(17–18): 1563–1566.

Duffield-Lillico, A.J., B.L. Dalkin, M.E. Reid, B.W. Turnbull, E.H. Slate, E.T. Jacobs, J.R. Marshall and L.C. Clark. 2003. Selenium supplementation, baseline plasma selenium status and incidence of prostate cancer: an analysis of the complete treatment period of the Nutritional Prevention of Cancer Trial. BJU Int. 91: 608–612.

Duntas, L.H. 2009. Selenium and inflammation: underlying antiinflammatorymechanisms. Horm. Metab. Res. 41: 443–447.

Eckley, C.S., M. Gustin, F. Marsik and M.B. Miller. 2011. Measurement of surface mercury fluxes at active industrial gold mines in Nevada (USA), Science of The Total Environment. 409(3): 514–522.

Eide, D.J. 2006. Zinc transporters and the cellular trafficking of zinc. Biochim. Biophys. Acta. 1763(7): 711–722.

Ellingsen, D.G., A. Andersen, H.P. Nordhagen, J. Efskind and H. Kjuus. 1993. Incidence of cancer and mortality among workers exposed to mercury vapour in the Norwegian chloralkali industry. Br. J. Ind. Med. 50: 875–880.

Ellis, E.N., B.H. Brouhard, R.E. Lynch et al. 1982. J. Toxicol. Clin. Toxicol. 19: 249–258.

Engebrigtsen, J.K. 1952. Some investigations on hypersensitiveness to bichromate in cement workers. Acta Derm.-Venereol. 32: 462–468

Engel, R.R. and A.H. Smith. 1994. Arsenic in drinking water and mortality from vascular disease: An ecologic analysis in 30 counties in the United States. Arch. Environ. Health. 49: 418–427.

Enterline, P.E., R. Day and G.M. Marsh. 1995. Cancers related to exposure to arsenic at a copper smelter. Occup. Environ. Med. 52: 28–32.

EnviroTools factsheets. Factsheets on Thallium. (available at http://www.envirotools.org/factsheets/contaminants/thallium.shtml); 2002.

Etminan, M., J.M. FitzGerald, M. Gleave and K. Chambers. 2005. Intake of selenium in the prevention of prostate cancer: a systematic review and meta-analysis. Cancer Causes Control. 16: 1125–1131.

Fairhall, L.T. and F. Hyslop. 1947. The toxicology of antimony. Public Health Rep. Suppl. 195.

Fairweather-Tait, S.J., Y. Bao, M.R. Broadley, R. Collings, D. Ford, J.E. Hesketh and R. Hurst. 2011. Selenium in Human Health and Disease, Antioxidants and Redox signaling. 14(7): 1337–1383.

Falk, H., G.G. Caldwell, K.G. Ishak, L.B. Thomas and H. Popper. 1981. Arsenic related hepatic angiosarcoma. American Journal of Industrial Medicine. 2: 43–50.

Falk, H., J.T. Herbert, L. Edmonds et al. 1981. Review of four cases of childhood hepatic angiosarcoma—elevated environmental arsenic exposure in one case. Cancer 47: 382–391.

Farid, Z., S. Bassily, D.C. Kent, A. Hassan, M.F. Abdel-Wahab and J. Wissa. 1968. Urinary schistosomiasis treated with sodium antimony tartrate a quantitative evaluation. Brit. Med. J. 3: 713–714.

Felicetti, S.W., R.G.Thomas and R.O. McClellan. 1974. Retention of inhaled antimony-124 in the beagle dog as a function of temperature of aerosol formation. Health Phys. 26: 525–531.

Fisher, D.A. and T.H. Oddie. 1969. Thyroid iodine content and turnover in euthyroid subjects—validity of estimation of thyroid iodine accumulation from shortterm clearance studies. J. Clin. Endocr. Metab. 29: 721–7.

Flamant, F. and J. Samarut. 2003. Thyroid hormone receptors: lessons from knockout and knock-in mutant mice. Trends Endocrin Met. 14: 85–90.

Flynn, A., T. Hirvonen, G.B. Mensink, M.C. Ocke, L. Serra-Majem, K. Stos, L. Szponar, I. Tetens, A. Turrini, R. Fletcher and T. Wildemann. 2009. Intake of selected nutrients from foods, from fortification and from supplements in various European countries. Food Nutr. Res. 53.

Fordyce, F. 2005. Selenium deficiency and toxicity in the environment. *In:* O. Selinus, B. Alloway, J. Centeno, R. Finkelman R. Fugeu. Lindh and P. Smedley [eds.]. Essentials of Medical Geology. Elsevier. London. pp. 373–415.

Fraker, P.J., L.E. King, T. Laakko and T.L. Vollmer. 2000. The dynamic link between the integrity of the immune system and zinc status. J. Nutr. 130(5S Suppl): 1399S–1406S.

Freget, S. 1981. Chromium valencies and cement dermatitis. Br. J. Dermatol. 105, Suppl. 21: 7–9.

Fristedt, B., A. Linquist, A. Schütz and P. Ovrum. 1965. Survival in a case of acute chromic acid poisoning with acute renal failure treated by haemodialysis. Acta medica scandinavica. 177: 153–159.

FSA, Food Standards Agency. 2009. Survey on measurement of the concentrations of metals and other elements from the 2006 UK total diet study. Food Survey Information Sheet 01/09, 2009. 16–17, 33: 37–45. Food Standards Agency, London.

Gallicchio, L., B.A. Fowler and E.F. Madden. 2001. *In:* E. Bingham, B. Cohrssen and C.H. Powell [eds.]. Patty's Toxicology, vol. 2, 5th edn. John Wiley & Sons, Inc. New York. pp. 747–800.

Ge, K. and G. Yang. 1993. The epidemiology of selenium deficiency in the etiological study of endemic diseases in China. Am. J. Clin. Nutr. 57: 259S–263S.

Gebel, T., S. Christensen and H. Dunkelberg. 1997. Anticancer Res. 17: 2603–2607.

Genium. 1999. Genium Publishing Corporation. Iodine. Genium's Handbook of Safety, Health, and Environmental Data for Common Hazardous Substances. McGraw-Hill Companies, Inc., Schenectady, NY.

Gilbert, S.G. 2003. A Small Dose of Toxicology. Taylor & Francis Ltd.

Gilbert, S.G. 2005. A small dose of toxicology, Taylor & Francis e-Library.

Glinoer, D. 1997. The regulation of thyroid function in pregnancy: pathways of endocrine adaptation from physiology to pathology. Endocr Rev. 18: 404–33.

Grandjean, P., R.F. White, A. Nielsen, D. Cleary and E.C.D Santos. 1999. Methylmercury neurotoxicity in Amazonian children downstream from gold mining. Environ. Health Perspect. 107: 587–591.

Grangeon, S., S. Guédron, J. Asta, G. Sarret and L. Charlet. 2012. Lichen and soil as indicators of an atmospheric mercury contamination in the vicinity of a chlor-alkali plant Grenoble, France. Ecological Indicators. 13(1): 178–183.

Gupta, S.K., S.P. Singh and V.K. Shukla. 2005. Copper, zinc, and Cu/Zn ratio in carcinoma of the gallbladder. J. Surg. Oncol. 91: 204–208.

Harada, M. 1995. Minamata Disease—Methylmercury Poisoning in Japan Caused by Environmental-Pollution. Crit. Rev. Toxicol. 25: 1–24.

Hardman, J.G., L.E. Limbird, P.B. Molinoff, R.W. Ruddon and A.G. Gilman. 1996. Goodman and Gilman's The Pharmacological Basis of Therapeutics. 9th edn. McGraw-Hill Companies, Inc., New York, NY.

Hayes, R.B. 1997. The carcinogenicity of metals in humans. Cancer Causes Control. 8: 371–385.

He, M.C. and J.R.Yang. 1999. Effects of different forms of antimony on rice during the period of germination and growth and antimony concentration in rice tissue. Science of Total Environment. 243–244, 149–155.

Hetzel, B.S. 1983. Iodine deficiency disorders (IDD) and their eradication. Lancet. 2: 1126–9.

Hightower, J.M. and D. Moore. 2003. Mercury levels in high-end consumers of fish. Environ Health Perspect. 111: 604–608.

Hoffmann, P.R. and M.J. Berry. 2008. The influence of selenium on immune responses. Mol. Nutr. Food Res. 52: 1273–1280.

Hu, H. 1998. Heavy metal poisoning. *In:* A.S. Fauci, E. Braunwald, K.J. Isselbacher, J.D. Wilson, J.B. Martin, D.L. Kasper, S.L. Hauser and D.L. Longo [eds.]. Harrison's Principles of Internal Medicine. 14th ed. McGraw-Hill, New York. pp. 2564–2569.

Hurst, R., R.M. Elliott, A.J. Goldson and S.J. Fairweather-Tait. 2008. Se methylselenocysteine alters collagen gene and protein expression in human prostate cells. Cancer Lett. 269: 117–126.

Hussain, S.A., J.D.E. Ane and P.V. Taberner. 1998. Lack of inhibition of human plasma cholinesterase and red cell acetylcholinesterase by antimony compounds including stibine. Hum. Exp. Toxicol. 17: 140–143.

Huster, D. 2010. Wilson disease. Best Pract Res. Clin. Gastroenterol. 24(5): 531–539.

Hylander, L.D. 2011. Gold and Amalgams: Environmental Pollution and Health Effects, Encyclopedia of Environmental Health. 1015–1026.

IARC Monographs. 2004. Evaluation of Carcinogenic Risks to Humans: Some Drinking-water Disinfectants and Contaminants, including Arsenic, vol. 84. International Agency for Research on Cancer, Lyon.

Ibáñez-Palomino, C., J.F. López-Sánchez and A. Sahuquillo. 2012. Certified reference materials for analytical mercury speciation in biological and environmental matrices: Do they meet user needs?; a review, Analytica Chimica Acta, vol. 720, pp. 9–15.

International Agency for Research on Cancer. 1998. http://www.inchem.org/documents/iarc/vol47/47-11.html.

Issell, B.F., B.V. Macfadyen, E.T. Gum, M. Valdivieso, S.J. Dudrick and G.P. Bodey. 2006. Serum zinc levels in lung cancer patients. Cancer. 47: 1845–1848.

Izrael'son, Z.I. 1973. *In:* Z.I. Izrael'son, O.J. Mogilevskaja and S.V. Suvorov [eds.]. Problems of Industrial Hygiene and Occupational Diseases in Work with Rare Metals. Medicina, Moscow (in Russian). pp. 258–266.

Jackson, M.I. and G.F. Jr. Combs. 2008. Selenium and anticarcinogenesis: underlying mechanisms. Curr. Opin. Clin. Nutr. Metab. Care. 11: 718–726.

Järup, L. and G. Pershagen. 1991. Arsenic exposure, smoking, and lung cancer in smelter workers—a case–control study. Am. J. Epidemiol. 134: 545–551.

Jewett, S.C. and L.K. Duffy. 2007. Mercury in fishes of Alaska, with emphasis on subsistence species, Science of The Total Environment. 387(1–3) 3–27.

John Peter, A.L. and T. Viraraghavan. 2005. Thallium: a review of public health and environmental concerns, Environment International. 31: 493–501.

John, E., T.C. Laskow, W.J. Buchser and B.R. Pitt. 2010. Zinc in innate and adaptive tumor immunity. J. Transl. Med. 8: 118.

Jomova, K., Z. Jenisova, M. Feszterova, S. Baros, J. Liska, D. Hudecova, C.J. Rhodesd and M. Valko. 2010. Arsenic: toxicity, oxidative stress and human disease. J. Appl. Toxicol. 31: 95–107.

Jomovaa, K. and M. Valko. 2011. Advances in metal-induced oxidative stress and human disease. Toxicology. 283(2–3): 65–87.

Jones, F.T. 2007. A broad view of arsenic. Poultry Sci. 86: 2–14.

Joules, H. 1932. Lancet. 2: 182–183.

Junghans, R.P. 1983. A review of the toxicity of methylmercury compounds with application to occupational exposures associated with laboratory uses, Environmental Research. 31(1): 1–31.

Kashin, N.I. 1859. The description of endemic and other disease, prevailing in the Urov-river area. *In:* The Records of Physico-Medical Scientific Society Attached to the Moscow University. Moscow.

Kaufman, D.B., W. DiNicola and R. McIntosh. 1970. Am. J. Dis. Child. 119: 374–376.

Kazantzis, G. 2000. Thallium in the environment and health effects. Environ Geochem Health. 22: 275– 280.

Kazantzis, G. 2007. Thallium. *In:* Handbook on the Toxicology of Metals. 3rd edn. Burlington. pp. 827–837.

Keall, J.H.H., N.H., Martin and R.E. Tunbridge. 1946. Br. J. Ind. Med. 3: 175–176.

Keen, C.L. and ME. Gershwin. 1990. Zinc deficiency and immune function. Annu. Rev. Nutr. 10: 415–431.

Kemper, F.H. and H.P. Bertram. 1991. Thallium. Metals and their compounds in the environment: occurrence, analysis, and biological relevance. Weinheim. New York. 7. pp. 1227– 41.

Klevay, L.M. 1984. *In:* O.M. Rennert and W.Y. Chan [eds.]. Metabolism of Trace Metals in Man. CRC Press, Boca Raton, FL. pp. 129–157.

Klevay, L.M. 2000. *In:* J.D. Bogden and L.M. Klevay [eds.]. Clinical Nutrition of the Essential Trace Elements and Minerals: The Guide for Health Professionals. Humana Press Inc., Totowa, N.J. pp. 251–271.

Kryukov, G.V., S. Castellano, S.V. Novoselov, A.V. Lobanov, O. Zehtab, R. Guigo and V.N. Gladyshev. 2003. Characterization of mammalian selenoproteomes. Science. 300: 1439–1443.

Lagerkvist, B.J. and B. Zetterlund. 1994. Assessment of exposure to arsenic among smelter workers: a five year follow up. Am. J. Ind. Med. 25: 477–488.

Langård, S. 1990. One hundred years of chromium and cancer: A review of epidemiological evidence and selected case reports. Am. J. Ind. Med. 17: 189–215.

Langard, S. and M. Costa. 2007. Chromium. *In:* Handbook on the Toxicology of Metals. 3rd edn. Burlington. pp. 487–510.

Lebel, J., D. Mergler, F. Branches, M. Lucotte, M. Amorim, F. Larribe and J. Dolbec. 1998. Neurotoxic effects of low-level methylmercury contamination in the Amazonian Basin. Environ Res. 79: 20–32.

Lech, T. 2002. Suicide by sodium tetraoxoselenate(VI) poisoning. Forensic Sci Int. 130: 44–48.

Lee, D.C., J.R. Roberts, J.J. Kelly and S.M. Fishman. 1995. Whole-bowel irrigation as an adjunct in the treatment of radiopaque arsenic. Am. J. Emerg. Med. 13: 244–245.

Lewis, R.J. 1996. Sax's Dangerous Properties of Industrial Materials. 9th edn. Wiley 8: Sons, New York.

Li, H., H.Q. Li, Y. Wang, H.X. Xu, W.T. Fan, M.L. Wang, P.H. Sun and Xie XY. 2004. An intervention study to prevent gastric cancer by micro-selenium and large dose of allitridum. Chin. Med J. (Engl). 117: 1155–1160.

Little, P.J., R. Bhattacharya, A.E. Moreyra and I.L. Korichneva. 2010. Zinc and cardiovascular disease. Nutrition. 26: 1050–1057.

Liu, J., B. Zheng, H.V. Aposhian, Y. Zhou, M.L. Chen, A. Zhang and M.P. Waalkes. 2002. Chronic arsenic poisoning from burning high arsenic containing coal in Guizhou, China. Environ. Health Perspect. 110: 119–122.

Liu, S.X., M. Athar, I. Lippai, C. Waldren and T.K. Hei. 2001. Induction of oxyradicals by arsenic: implication for mechanism of genotoxicity. Proc. Natl Acad. Sci. USA. 98: 1643–1648.

Lu, J. and A. Holmgren. 2009. Selenoproteins. J. Biol. Chem. 284: 723–727.

Lu, J. and C. Jiang. 2005. Selenium and cancer chemoprevention: hypotheses integrating the actions of selenoproteins and selenium metabolites in epithelial and non-epithelial target cells. Antioxid Redox Signal. 7: 1715–1727.

Lyle, W.H. 1958. Lesions of the Skin in Process Workers Caused by Contact with Butyl Tin Compounds. Br. J. Ind. Med. 15: 193–196.

MacDonald, R.S. 2000. The role of zinc in growth and cell proliferation. J. Nutr. 130: 1500S–1508S.

MacFarquhar, J.K., D.L. Broussard, P. Melstrom, R. Hutchinsonm, A. Wolkin, C. Martin, R.F. Burk, J.R. Dunn, A.L. Green, R. Hammond, W. Schaffner and T.F. Jones. 2010. Acute selenium toxicity associated with a dietary supplement. Arch. Intern. Med. 170: 256–261.

Magalhaes, M.C.F. 2002. Arsenic. An environmental problem limited by solubility. Pure Appl. Chem. 74: 1843–1850.

Mandal, B.K. and K.T. Suzuki. 2002. Arsenic round the world: a review. Talanta. 58: 201–235.

Manzo, L., P. Richelmi, E. Sabbioni, R. Pietra, F. Bono and L. Guardia. 1981. Poisoning by triphenyltin acetate. Report of two cases and determination of tin in blood and urine by neutron activation analysis. Clin. Toxicol. 18: 1343–1353.

Marine, D. and O.P. Kimball. 1990. The prevention of simple goiter in man. A survey of the incidence and types of thyroid enlargements in the schoolgirls of Akron (Ohio), from the 5th to the 12th grades, inclusive—the plan of prevention proposed, 1917. J. Lab. Clin. Med. 115: 128–36.

Markert, B. 1992. Presence and signifi cance of naturally occurring chemical elements of the periodic system in the plant organism and consequences for future investigations on inorganic environmental chemistry in ecosystems. Plant Ecol. 103: 1–30.

Markert, B. 1994. The biological system of the elements (BSE) for terrestrial plants (Glycophytes). Sci. Total Environ. 155: 221–8.

Martena, M.J., J.C. Van Der Wielen, I.M. Rietjens, W.N. Klerx, H.N. De Groot and E.J. Konings. 2010. Monitoring of mercury, arsenic, and lead in traditional Asian herbal preparations on the Dutch market and estimation of associated risks. Food Addit. Contam. Part A Chem. Anal. Control Expos. Risk Assess. 27: 190–205.

Martindale, W. 1977. *In:* The Extra Pharmacopoeia. 27th ed. A. Wade and J.E.F. Reynolds (eds.). The Pharmaceutical Press, London: 1721–1723.

Mathew, L., A. Vale and J.E. Adcock. 2010. Arsenical peripheral neuropathy. Pract. Neurol. 10: 34

McCallum, R.I. 1977. Martindale: The Extra Pharmacopoeia. 1941. 22nd edn. Pharmaceutical Press, London as cited in Proc. R. Soc. Med. 70: 756–763.

McCarty, K.M., Y.C. Chen, Q. Quamnuzzaman, M. Rahman, G. Mahiuddin, Y.M. Hsueh, L. Su, T. Smith, L. Ryan and D.C. Christiani. 2007. Arsenic methylation, GSTT1, GSTM1, GSTP1 polymorphisms and skin lesions. Environ. Health Perspect. 115: 341–345.

McLean, J.R., H.C. Birnboim, R. Pontefact et al. 1983. The effect of tin chloride on the structure and function of dna in human white blood cells. Chem. Biol. Interact. 46: 189–200.

Meyer, F., P. Galan, P. Douville, I. Bairati, P. Kegle, S. Bertrais, C. Estaquio and S. Hercberg. 2005. Antioxidant vitamin and mineral supplementation and prostate cancer prevention in the SU.VI.MAX trial. Int. J. Cancer. 116: 182–186.

Miller, Jr., W.H., H.M. Schipper, J.S. Lee, J. Singer and S. Waxman. 2002. Mechanisms of action of arsenic trioxide. Cancer Res. 62: 3893–3903.

Mocchegiani, E. and M. Muzzioli. 2000. Zinc, metallothioneins, immune responses, survival and ageing. Biogerontology. 1: 133–143.

Mocchegiani, E. and N. Fabris. 1995. Age-related thymus involution: zinc reverses *in vitro* the thymulin secretion effect. Int. J. Immunopharmacol. 17: 745–749.

Mocchegiani, E., L. Costarelli, R. Giacconi, F. Piacenza, A. Basso and M. Malavolta. 2011. Zinc, metallothioneins and immunosenescence: effect of zinc supply as nutrigenomic approach. Biogerontology. [Epub ahead of print] PMID: 21503725.

Mocchegiani, E., M. Muzzioli and R. Giacconi. 2000. Zinc and immunoresistance to infections in ageing: new biological tools. Trends Pharmacol. Sci. 21: 205–208.

Mocchegiani, E., R. Giacconi, C. Cipriano, M. Muzzioli, P. Fattoretti, C. Bertoni-Freddari, G. Isani, P. Zambenedetti and P. Zatta. 2001. Zinc-bound metallothioneins as potential biological markers of ageing. Brain Res. Bull. 55: 147–153.

Montuori, P., E. Jover, S. Díez, N. Ribas-Fitó, J. Sunyer, M. Triassi, J. and M. Bayona. 2006. Mercury speciation in the hair of pre-school children living near a chlor-alkali plant, Science of The Total Environment. 369(1–3): 51–58.

Moore, M.M., K. Harrington Brock and C.L. Doerr. 1994. Genotoxicity of arsenic and its methylated metabolites. Environ. Geochem. Health. 16: 191–198.

Mulkey, J.P. and F.W. Oehme. 1993. A review of thallium toxicity. Vet Hum Toxicol. 35: 445–53.

Muller, R., W. Zschiesche, H.M. Steffen et al. 1989. Tellurium-Intoxication. Klin. Wochenschr. 67: 1152–1155.

Murakami, M. and T. Hirano. 2008. Intracellular zinc homeostasis and zinc signaling. Cancer Sci. 99: 1515–1522.

Nance, P., J. Patterson, A. Willis, N. Foronda and M. Dourson. 2012. Human health risks from mercury exposure from broken compact fluorescent lamps (CFLs), Regulatory Toxicology and Pharmacology. 62(3): 542–552.

NAS. 1980. National Academy of Sciences. 3: Drinking Water and Health. National Research Council, Safe Drinking Water Committee. National Academy Press; Washington, D.C.

Nath, S.K., B. Moinier, F. Thuillier, M. Rongier and J.F. Desjeux. 1992. Urinary excretion of iodide and fluoride from supplemented food grade salt. Int. J. Vitam Nutr. Res. 62: 66–72.

Navas Acien, A., A.R. Sharrett, E.K. Silbergeld, B.S. Schwartz, K.E. Nachman, T.A. Burke and E. Guallar. 2005. Arsenic exposure and cardiovascular disease: a systematic review of the epidemiologic evidence. Am. J. Epidemiol. 162: 1037–1049.

Nehring, P. 1972. Ind. Obst. Gemusseverwert. 57: 489–492.

Nepusz, T., A. Petroczi and D.P. Naughton. 2009. Food alert patterns for metal contamination analyses in seafoods: longitudinal and geographical perspectives. Environ. Int. 35: 1030–1033.

Nowak, G., B. Szewczyk and A. Pilc. 2005. Zinc and depression. An update. Pharmacol. Rep. 57: 713–718.

Nowak, G., M. Siwek, D. Dudek, A. Zieˆba and A. Pilc. 2003. Effect of zinc supplementation on antidepressant therapy in unipolar depression: a preliminary placebo-controlled study. Pol. J. Pharmacol. 55: 1143–1147

NRC. 1999. Arsenic in Drinking Water. National Research Council. National Academy Press, Washington, DC.

Nriagu, J.O. 1998. Thallium in the Environment. Wiley Series *In:* Advances in Environmental Science and Technology, Vol. 29. John Wiley and Sons, New York.

Nuyts, G.D., E. Van Vlem, J. Thys et al. 1995. New occupational risk factors for chronic renal failure. Lancet. 346: 7–11.

O'Callaghan, J.P. and D.B. Miller. 1988. Acute exposure of the neonatal rat to triethyltin results in persistent changes in neurotypic and gliotypic proteins. J. Pharmacol. Exp. Ther. 246: 394–402.

Oliver, M.A. 1997. Soil and human health: A review. Eur. J. Soil Sci. 48: 573–592.

Orloff, K., K. Mistry and S. Metcalf. 2009. Biomonitoring for environmental exposures to arsenic. J. Toxicol. Environ. Health B. Crit. Rev. 12: 509–524.

Ostrakhovitch, E.A. and M. Ge. Cherian. 2007. Tin. *In:* Handbook on the Toxicology of Metals. 3rd edn. Academic Press. Burlington. pp. 839–859.

Pagani, A., L. Villarreal, M. Capdevila and S. Atrian. 2007. The saccharomyces cerevisiae Crs5 metallothionein metal-binding abilities and its role in the response to zinc overload. Mol. Microbiol. 1: 256–269.

Partington, C.N. 1950. Acute Poisoning with Potassium Bichromate. BM J. 2: 1097–1098.

Paschoud, J.M. 1964. Notes cliniques au sujet des eczémas de contact professionnels par l'arsenic et l'antimoine. Dermatologica. 129: 410–415.

Paszkowski, T., A.I. Traub, S.Y. Robinson and D. McMaster. 1995. Selenium dependent glutathione peroxidase activity in human follicular fluid. Clin. Chim. Acta. 236: 173–180.

Pesch, B., U. Ranft, P. Jakubis, M.J. Nieuwenhuijsen, A. Hergemöller, K. Unfried, M. Jakubis, P. Miskovic and T. Keegan. 2002. Environmental arsenic exposure from a coalburning power plant as a potential risk factor for nonmelanoma skin carcinoma: Results from a case–control study in the district of Prievidza, Slovakia. Am. J. Epidemiol. 155: 798–809.

Petroczi, A. and D.P. Naughton. 2009. Mercury, cadmium and lead contamination in seafood: a comparative study to evaluate the usefulness of Target Hazard Quotients. Food Chem. Toxicol. 47: 298–302.

Pinto, J.T., J.I. Lee, R. Sinha, M.E. Macewan and A.J. Cooper. 2010. Chemopreventive mechanisms of alpha-keto acid metabolites of naturally occurring organoselenium compounds. Amino Acids [Epub ahead of print].

Plum, L.M, L. Rink and H. Haase.2010. The essential toxin: impact of zinc on human health. Int. J. Environ. Res. Public Health 7:1342–1365.

Popova, T.B., M.N. Ryzkova and K.V. Glotova. 1965. *In:* A.A. Letavet [ed.]. Occupational Diseases in Chemical Industry. Medicina, Moscow (in Russian). pp. 246–250.

Potkonjak, V. and M. Pavlovich. 1983. Antimoniosis: A particular form of pneumoconiosis. I. Etiology, clinical and X-ray findings. Int. Arch. Occup. Environ. Health. 51: 199–207.

Prasad, A.S. 2009. Zinc: role in immunity, oxidative stress and chronic inflammation. Curr. Opinion. Clin. Nutr. Metab. Care. 12: 646–652.

Prasad, M.N.V. 2008. Trace Elements as Contaminants and Nutrients, Consequences in Ecosystems and Human Health. John Wiley & Sons, Chichester, New York, Tokyo. pp. 343–367.

Prull, G. and K. Rompel. 1970. EEG changes in acute poisoning with organic tin compounds. Electroencephalogr. Clin. Neurophysiol. 29: 215.

Puertollano, M.A., E. Puertollano, G.A'. de Cienfuegos and M.A. de Pablo. 2011. Dietary antioxidants: immunity and host defense. Curr. Top. Med. Chem. 11(14): 1752–1766.

Rayman, M.P. 2000. The importance of selenium to human health. Lancet. 356: 233–241.

Rayman, M.P. 2008. Food-chain selenium and human health: emphasis on intake. Br. J. Nutr. 100: 254–268.

Rayman, M.P. 2009. Selenoproteins and human health: insights from epidemiological data. Biochim. Biophys. Acta. 1790: 1533–1540.

Reinhold, J.G. 1975. Trace elements—A selective survey. Clin. Chem. 21: 476–500.

Reis, A.T., S.M. Rodrigues, C. Araújo, J.P. Coelho, E. Pereira and A.C. Duarte. 2009. Mercury contamination in the vicinity of a chlor-alkali plant and potential risks to local population, Science of The Total Environment. 407(8): 2689–2700.

Renes, L.E. 1953. Antimony poisoning in industry. Arch. Ind. Hyg. 7: 99–108.

Richardson, G.M., R. Wilson, D. Allard, C. Purtill, S. Douma and J. Gravière. 2011. Mercury exposure and risks from dental amalgam in the US population, post-2000, Science of The Total Environment. 409(20): 4257–4268.

Risher, J.F., R.A. Nickle and S.N. Amler. 2003. Elemental mercury poisoning in occupational and residential settings, International Journal of Hygiene and Environmental Health. 206(4-5): 371–379.

Rossman, T.G. 2003. Mechanism of arsenic carcinogenesis: an integrated approach. Mutat. Res. 533: 37-65.

Rutter, G.A. 2010. Think zinc, new roles for zinc in the control of insulin secretion. Islets. 2(1): 49–50.

Rycroft, R.J.G. 2006. Occupational Dermatoses from Warm Dry Air. British Journal of Dermatology 105, no. s21: 29–34.

Sainio, E.L., R. Jolanki, E. Hakala and L. Kanerva. 2000. Metals and arsenic in eye shadows. Contact Dermat. 42: 5–10.

Salnikow, K. and A. Zhitkovich. 2008. Genetic and epigenetic mechanisms in metal carcinogenesis and cocarcinogenesis: nickel, arsenic, and chromium. Chem. Res. Toxicol. 21: 28–44.

Satia, J.A., I.B. King, J.S. Morris, K. Stratton and E. White. 2006. Toenail and plasma levels as biomarkers of selenium exposure. Ann. Epidemiol. 16: 53–58.

SCF. 2003. Opinion of the Scientific Committee on Food on the tolerable upper intake level of zinc. European Commission Schiffl, H., P. Weidmann, M. Weiss et al. 1982. Miner. Electrolyte Metab. 7: 28–35.

Schiffl , H., Weidmann, P., Weiss, M., et al. 1982. Miner. Electrolyte Metab. 7: 28–35.

Schlag, P., W. Seeling, P. Merkle and M. Betzler. 1978. Changes of serumzinc in breast cancer. Langenbecks. Arch. Chir. 2: 129–133.

Schlatter, C. and U. Kissling. 1973. Beitr. Gerichtl. Med. 30: 382–388.

Schroeder, H.A., J. Buckman and J.J. Balassa. 1967. Abnormal trace elements in man: tellurium. J. Chronic. Dis. 20: 147–61.

Selenius, M., A.K. Rundlof, E. Olm, A.P. Fernandes and M. Bjornstedt. 2010. Selenium and the selenoprotein thioredoxin reductase in the prevention, treatment and diagnostics of cancer. Antioxid Redox Signal. 12: 867–880.

Shand, P., W.M. Edmunds and J. Ellis. 1998. The hydrogeochemistry of thallium in natural waters. Water–rock interaction: proceedings of the 9th International symposium on water–rock interaction. New Zealand Taupo. 75–78.

Shapiro, J.R. 1972. Selenium and carcinogenesis: a review. Ann. N. Y. Acad. Sci. 192: 215–219.

Shi, H., X. Shi and K.J. Liu. 2004. Oxidative mechanism of arsenic toxicity and carcinogenesis. Mol. Cell. Biochem. 255: 67–78.

Shimizu, N., J. Fujiwara, S. Ohnishi, M. Sato, H. Kodama, T. Kohsaka, A. Inui, T. Fujisawa, H. Tamai S. Ida, S. Itoh, M. Ito, N. Horiike, M. Harada, M. Yoshino and T. Aoki. 2010. Effects of long-term zinc treatment in Japanese patients with Wilson disease: efficacy stability, and copper metabolism. Transl. Res. 156(6): 350–357.

Shimshack, J.P. and M.B. Ward. 2010. Mercury advisories and household health trade-offs. Journal of Health Economics. 29(5): 674–685.

Shotyk, W., M. Krachler and B. Chen. 2004. Antimony in recent, ombrotrophic peat from Switzerland and Scotland: comparison with natural background values (5,320 to 8,020 14C yr BP) and implications for the global atmospheric Sb cycle. Global Biogeochemical Cycles,18, GB1016.1–GB1016.13.

Smedleym, P.L. and D.G. Kinniburgh. 2002. A review of the source, behavior and distribution of arsenic in natural waters. Appl. Geochem. 17: 517–568.

Smichowski, P. 2007. Antimony in the environment as a global pollutant: a review on analytical methodologies for its determination in atmospheric aerosols. Talanta. 75: 2–14.

Smith, M., K.W. Franke and B.B. Westfall. 1936. The selenium problem in relation to public health. A preliminary survey to determine the possibility of selenium intoxication in the rural population living in seleniferous soil. US Public Health Report. 1496–1505.

Sokas, R.K. 1998. Reproductive Hazards of the Workplace. John Wiley and Sons, Inc., New York.

Soto, D.X., R. Roig, E. Gacia and J. Catalan. 2011. Differential accumulation of mercury and other trace metals in the food web components of a reservoir impacted by a chlor-alkali plant (Flix, Ebro River, Spain): Implications for biomonitoring, Environmental Pollution. 159(6): 1481–1489.

Spallholz, J.E., L.M. Boylan and H.S. Larsen. 1990. Advances in understanding selenium's role in the immune system. Ann. N.Y. Acad. Sci. 587: 123–39.

Spiegel, S.J. and M.M. Veiga. 2010. International guidelines on mercury management in small-scale gold mining. Journal of Cleaner Production. 18(4): 375–385.

Sredni, B. 2012. Immunomodulating tellurium compounds as anti-cancer agents. Seminars in Cancer Biology. 22: 60–69.

Stanbury, J.B. 1954. The adaptation of man to iodine deficiency. Harvard University Press. Cambridge, MA. pp. 1–209.

States, J.C., S. Srivastava, Y. Chen and A. Barchowsky. 2009. Arsenic and cardiovascular disease. Toxicol. Sci. 107: 312–323.

Stefanidou, M., C. Maravelias, A. Dona and C. Spiliopoulou. 2006. Zinc: a multipurpose trace element. Arch Toxicol. 80(1): 1–9 Review.

Stevenson, C.J. 1965. Antimony spots. Trans. St. John's Hosp. Dermatol. Soc. 51: 40–45.

Stokinger, H.F. 1963. *In:* A. Patty and D.D. Irish [eds.]. Industrial Hygiene and Toxicology. 2nd edn. 1963. Interscience Publishers, New York. pp. 993–998.

Stranges, S., J.R. Marshall, R. Natarajan, R.P. Donahue, M. Trevisan, G.F. Combs, F.P. Cappuccio, A. Ceriello and M. E. Reid. 2007. Effects of long-term selenium supplementation on the incidence of type 2 diabetes: a randomized trial. Ann. Intern. Med. 147: 217–223.

Stroud, J.L., M.R. Broadley, I. Foot, S.J. Fairweather-Tait, D.J. Hart, R. Hurst, P. Knott, H. Mowat, K. Norman, P. Scott, M. Tucker, P.J. White, S.P. McGrath and F.-J. Zhao. 2010. Soil factors affecting selenium concentration in wheat grain and the fate and speciation of Se fertilisers applied to soil. Plant Soil. 332: 19–30.

Summaries & Evaluations. Antimony Trioxide and Antimony Trisulfide; International Agency for Research on Cancer: Lyon, France. 1989. Available online: http://www.inchem.org/documents/iarc/vol47/47-11.html (accessed on 29 January 2010).

Sundar, S. and J. Chakravarty. 2010. Antimony Toxicity Int. J. Environ. Res. Public Health. 7: 4267–4277.

Sunde, R.A. 1997. Selenium. *In:* B.L. O'Dell and R.A. Sunde [eds.]. Handbook of nutritionally essential mineral elements. Marcel Dekker Inc. New York. pp. 493–556.

Syversen, T. and P. Kaur. 2012. The toxicology of mercury and its compounds, Journal of Trace Elements in Medicine and Biology. 26(4): 215–226.

Takeda, A. 2000. Movement of zinc and its functional significance in the brain. Brain Res. Bull. 34: 137–148.

Tan, J., W. Zhu, W. Wang, R. Li, S. Hou, D. Wang and L. Yang. 2002. Selenium in soil and endemic diseases in China. Sci. Total Environ. 284: 227–235.

Tapiero, H. and K.D. Tew. 2003. Trace elements in human physiology and pathology: zinc and metallothioneins. Biomed Pharmacother 57: 399–411.

Tarp, U. 1995. Selenium in rheumatoid arthritis. A review. Analyst. 120: 877–81.

Taylor, A. 1996. Biochemistry of Tellurium. Biological Trace Element Research 55, no. 3 (December): 231–239. http://pubget.com/paper/9096851.

Taylor, P.J. 1966. Acute intoxication from antimony trichloride. Br. J. Ind. Med. 23: 318–321.

Thomas, R.G., S.W. Felicetti, R.V. Lucchino and R.O. McClellan. 1973. Retention patterns of antimony in mice following inhalation of particles formed at different temperatures. Proc. Exp. Biol. Med. 144: 544–550.

Tsai, S., H. Chou, H. The, C.M. Chen and C.J. Chen. 2003. The effects of chronic arsenic exposure from drinking water on the neurobehavioral development in adolescence. Neurotoxicology. 24: 747–753.

Tseng, W. 1977. Effects and dose–response relationships of skin cancer and blackfoot disease with arsenic. Environ. Health Perspect. 19: 109–119.

Twidwell, L.G. and C.W. Beam. 2002. Potential technologies for removing thallium from mine and process wastewater: an abbreviated annotation of literature. Eur. J. Miner Process Environ Prot. 2(1): 1–10.

Tylenda, C.A. and B.A. Fowler. 2007. Antimony. *In:* Handbook on the Toxicology of Metals 3rd edn Academic Press, Burlington, MA 01803, USA. 353–365.

Uede, K. and F. Furukawa. 2003. Skin manifestations in acute arsenic poisoning from the Wakayama curry poisoning incident. Br. J. Dermatol. 149: 757–762.

Ursini, F., S. Heim, M. Kiess et al. 1999. Dual function of the selenoprotein PHGPx during sperm maturation. Science. 285: 1393–96.

Vahidnia, A., G.B. van der Voet and F.A. de Wolff. 2007. Arsenic neurotoxicity—a review. Hum. Exp. Toxicol. 26: 823–832.

Valko, M., H. Morris and M.T. Cronin. 2005. Metals, toxicity and oxidative stress. Curr. Med. Chem. 12: 1161–1208.

Vantroyen, B., J.F. Heilier, A. Meulemans, A. Michels, J.P. Buchet, S. Vanderschueren, V. Haufroid and M. Sabbe. 2004. Survival after a lethal dose of arsenic trioxide. J. Toxicol. Clin. Toxicol. 42: 889–895.

Vasak, M. and D.W. Hasler. 2000. Metallothioneins: new functional and structural insights. Curr. Opin. Chem. Biol. 4: 177–183.

Walker, B.L. 2010. Toxic Archipelago: A History of Industrial Disease in Japan. University of Washington Press. ISBN 0-295-98954-8.

Wall, S. 1980. Survival and mortality pattern among Swedish smelter workers. Int. J. Epidemiol. 9: 73– 87.

Wang, C.Y., T. Wang, W. Zheng, B.L. Zhao, G. Danscher, Y. Chen and Z. Wang. 2010. Zinc overload enhances APP cleavage and Ab deposition in the Alzheimer mouse brain. PLoS One. 5(12): e15349.

Wang, G.Q., S.M. Dawsey, J.Y. Li, P.R. Taylor, B. Li, W.J. Blot, W.M. Weinstein, F.S. Liu, K.J. Lewin, H. Wang et al. 1994. Effects of vitamin/mineral supplementation on the prevalence of histological dysplasia and early cancer of the esophagus and stomach: results from the General Population Trial in Linxian, China. Cancer Epidemiol Biomarkers Prev. 3: 161–166.

Wang, S.L., J.M. Chiou, C.J. Chen, C.H. Tseng, W.L. Chou, C.C. Wang, T.N. Wu and L.W. Chang. 2003. Prevalence of non insulin dependent diabetes mellitus and related vascular diseases in southwestern arseniasisendemic and nonendemic areas in Taiwan. Environ. Health Perspect. 111: 155–159.

Welch, K., I. Higgins M. Oh and C. Burchfiel. 1982. Arsenic exposure, smoking, and respiratory cancer in copper smelter workers. Arch Environ Health. 37: 325–335.

Wellinghausen, N. and L. Rink. 1998. The significance of zinc for leukocyte biology. J. Leukoc Biol. 64: 571–577.

Werrin, M. 1963. Hussock. Q. Bull. Assoc. Food Drug Officials US. 27: 38–45.

Wester, P.O. 1973. Acta Med. Scand. 194: 505–512.

Westerhoff P., P. Prapaipong, E. Shock and A. Hillaireau. 2008. Antimony leaching from polyethylene terephthalate (PET) plastic used for bottled drinking water, Water Research. 42(3): 551–556.

White, J.G. and R.J. Zasoski. 1999. Mapping soil micronutrients. Field Crop Res. 60: 11–26.

Whittle, N., G. Lubec and N. Singewald. 2009. Zinc deficiency induces enhanced depression-like behaviour and altered limbic activation reversed by antidepressant treatment in mice. Amino Acids. 36(1): 147–158.

WHO, World Health Organization. 1996. Trace Elements in Human Nutrition and Health. World Health Organisation. Geneva.

WHO. 1981. Environmental Health Criteria, Arsenic. World Health Organization, Geneva. pp. 1–174.

WHO. 2001. Environmental Health Criteria, 224, Arsenic and Arsenic Compounds. 2nd edn. World Health Organization, Geneva. pp. 385–392.

Willett, W.C., B.F. Polk, J.S. Morris, M.J. Stampfer, S. Pressel, B. Rosner, J.O. Taylor, K. Schneider and C.G. Hames. 1983. Prediagnostic serum selenium and risk of cancer. Lancet. 2: 130–134.

Williams, C.D. 1969. Asthma related to chromium compounds. Report of two cases and review of the literature on chromate diseases. N.C. Med. J. 30: 482–490.

Williams, F., R. Robertson and M. Roworth. 1999. Scottish Centre for Infection and Environmental Health In: Detailed Profile of 25 Major Organic and Inorganic Substances. 1st edn. SCEIH, Glascow. Ann. Clin. Biochem. 36(Pt 3): 267–300.

Wilson N., J. Webster-Brown and K. Brown. 2012. The behaviour of antimony released from surface geothermal features in New Zealand. Journal of Volcanology and Geothermal Research. 247–248: 158–167.

World Health Organization. 2002. Environmental Health Criteria 228—Principles and Methods for the Assessment of Risk from Essential Trace Elements. WHO. Geneva

Yang, G. and R. Zhou. 1994. Further observations on the human maximum safe dietary selenium intake in a seleniferous area of China. J Trace Elem Electrolytes Health Dis. 8: 159–165.

Yang, H.T., H.J. Chou, B.C. Han and S.Y. Huang. 2007. Lifelong inorganic arsenic compounds consumption affected blood pressure in rats. Food Chem. Toxicol. 45: 2479–2487.

Ye, X., H. Qian, P. Xu, L. Zhu, M.P. Longnecker and H. Fu. 2009. Nephrotoxicity, neurotoxicity, and mercury exposure among children with and without dental amalgam fillings. International Journal of Hygiene and Environmental Health. 212(4): 378–386.

Yosef, S., M. Brodsky and B. Sredni. 2007. Amnon Albeck, and Michael Albeck. Octa-O-bis-(R,R)-Tartarate Ditellurane (SAS)—a Novel Bioactive Organotellurium(IV) Compound: Synthesis, Characterization, and Protease Inhibitory Activity. Chem. Med. Chem. 2: 1601–1606.

Yu, S.Y., Y.J. Zhu and W.G. Li. 1997. Protective role of selenium against hepatitis B virus and primary liver cancer in Qidong. Biol. Trace Elem Res. 56: 117–124.

Zhang, Z., B. Zhang, J. Long, X. Zhang and G. Chen. 1998. Thallium pollution associated with mining of thallium deposits. Sci. China, Ser. D. 41(1): 75–81.

Zhuo, H., A. H. Smith and C. Steinmaus. 2004. Selenium and lung cancer: a quantitative analysis of heterogeneity in the current epidemiological literature. Cancer Epidemiol Biomarkers Prev. 13: 771–778.

Zhuo, P. and A.M. Diamond. 2009. Molecular mechanisms by which selenoproteins affect cancer risk and progression. Biochim. Biophys. Acta. 1790: 1546–1554.

Zimmermann, M.B. 2011. The role of iodine in human growth and development. Seminars in Cell & Developmental Biology. 22: 645–652.

Zimmermann, M.B., P.L. Jooste and C.S. Pandav. 2008. Iodine-deficiency disorders. Lancet. 372: 1251–62.

Zitko, V. 1975. Toxicity and pollution potential of thallium. Sci. Total Environ. 4: 185–92.

Zoeller, R.T. and J. Rovet. 2004. Timing of thyroid hormone action in the developing brain: clinical observations and experimental findings. J. Neuroendocrinol. 16: 809–18.

Index

A

Adsorptive Stripping Voltammetry 439
Airborne particulate matter 343–346
Analytical Methods 278, 288, 393, 424, 530, 531
Anion Exchange Chromatography 522
Anodic Stripping Voltammetry 439–441, 467
Antimony 341–362
Arsenic 363–389
Ashes 344, 346, 353–356, 359
Atmospheric aerosols 341–362
Atomic absorption spectroscopy 319

B

Bioaccumulation 546, 563, 564
Bioavailability 264, 265, 268, 269, 272, 273, 277, 278, 293, 294
Bioavailability of Iodine 513–526
Biomarker 566
Bioremediation 274, 276

C

Cancer 548, 550, 553, 554, 557, 565, 569–572, 574, 581, 582, 585
Capillary electrophoresis 368, 376, 377
Carcinogenicity 548, 550
Certified reference materials 5, 11, 319
Chemical speciation 527, 528, 530, 533, 535, 536
Chemical vapor generation 13
Chi-function 444
Chromatographic methods 393, 407, 408, 414, 415, 418, 423
Chromatographic separation 11, 12
Chromium 306–324
Chronic effects 560
Chronic toxicity 548, 577
Cloud point extraction 61, 62, 67, 68, 70, 100, 123, 158
Coupled techniques 205

D

Determination 531–544
Dispersive liquid-liquid micro-extraction 20, 22, 23, 31, 93, 100, 101, 104, 158

E

Edible seaweed 513–526
Electrophoresis 204, 205, 216, 218, 219, 222, 233, 246, 249, 257
Environmental analysis 220
Environmental Legislation 277
Environmental pollution 578
Environmental samples 242–262, 390–432, 478–512
Enzymatic hydrolysis 17
Enzymatic hydrolysis methods 147, 149, 150, 155, 156
Essential elements 546
Ethylation 497–499, 506
Extended X-Ray Adsorption Fine Structure 443
Extraction 5, 10, 11, 264, 279, 286–288, 290, 291, 347, 349, 350, 352, 354–356, 358, 368, 370, 371, 374–376, 382, 393–400, 408, 409, 413, 414, 416–419, 423, 484, 487–490, 493–497, 499, 504–506

F

Fly ashes 346, 354, 356
Food and biological samples 538
Formation constant 328, 336, 338, 339
Fractionation 4, 341–362

G

Gas chromatography (GC) 204, 205, 210, 214, 247, 249, 250–253, 407–412, 422, 423, 485, 491, 497, 500
Genotoxicity 550, 553
Geological samples 536, 537
Global cycle of iodine 514

H

Health 263–265, 278
Hollow fiber liquid phase micro-extraction
    22, 31, 109–111, 159
HPLC 247–249, 252, 256, 257, 369, 370, 376,
    411–414, 423
Human health 545–598
Hydride generation 367, 374, 377
Hyphenated systems 13
Hyphenated techniques 242–262

I

ICP-MS 204, 205, 208, 209, 214, 216, 219,
    220, 223, 224, 226, 228–232, 397, 402,
    403, 405, 409–411, 413, 414, 417, 419,
    423
Immune system 567, 569, 578, 581, 582
Inductively Coupled Plasma Mass Spec-
    trometry 484, 485, 491, 522
Inner-Sphere Adsorption Complex 453, 454
Inorganic Selenium 297–304
Inorganic Tin 479, 480, 482, 484, 487, 493,
    598, 505
Interference 11, 369, 370
Iodine 513–526
Iodine chemical species 523
Iodine intake recommendation 519
Iodine speciation in seaweed 523
Ion chromatography 256, 259

L

Lanmuchangite 326
Leaching 310, 312, 314–320
Liquid chromatography 204–206, 210, 485,
    491, 503
Liquid-liquid extraction 20–23, 26–28, 31,
    33, 97, 100, 109, 110, 116, 118, 160, 161,
    393–395, 397, 399
Lorandite 326

M

Mass spectrometry 203, 204, 210, 214, 220,
    232, 246–249, 252, 255–258
Matrix solid phase dispersion 21, 156, 161
Mercury 390–432
Metalloids 243, 250–252, 256
Metals 243, 244, 248, 250, 251
Method validation 14, 320
Methylated arsenic species 366, 377
Micro X-ray Fluorescence Spectroscopy 446

Microporous membrane liquid liquid
    extraction 21, 109, 118, 160
Microwave assisted extraction 23, 121, 123,
    160

N

Neutron Activation Analysis 521
Non-chromatographic methods 393, 423
Non-chromatographic speciation 11, 12
Non-chromatographic techniques 367

O

Organomethalic Selenium 269, 288
Organotin Compounds 481, 483, 492
Outer-Sphere Adsorption Complex 450,
    452
Oxidation state 307, 308, 311

P

Plants 529, 536
Precipitation and co-precipitation 24–26
Preconcentration 530
Pressurized liquid extraction 21–23, 121,
    141, 142, 161
Purge and Trap 24, 34–38, 40, 44, 161

R

Radial distribution function 445, 446, 459
Recommended Daily Allowance 547, 548
Recommended Nutrient Intake 547
Reproductive effects 550
Road dust 343, 344, 350, 356–359

S

Sample preparation 2, 10, 14, 15, 204–206,
    221, 245, 258, 485, 493–495
Sediments 263–305, 308–310, 319, 320, 363,
    368, 374, 381, 382
Selenium biogeochemistry 271
Separation techniques 202–241
Single-drop micro-extraction 22, 31, 94–96,
    98, 160
Size Exclusion Chromatography 522
Soil 306–324, 363–389, 527–544
Solid matrices 309, 320
Solid phase extraction 20–22, 45, 46, 48, 50,
    55, 56, 162, 396, 398, 399, 416
Solid phase micro-extraction 20–23, 72, 73,
    76, 79, 158, 160, 162
Solidified drop liquid phase micro-extrac-
    tion 105, 162

Solid-liquid extraction 119, 120, 123, 124, 156
solubility of thallium 325, 326, 330, 339
Solubility product constant 326, 328–330, 332, 334, 335
Solvent bar micro-extraction 21, 109, 115, 116, 161
Speciation 1–18, 263–324, 341–477
Speciation analysis 203–206, 212, 213, 215–218, 220–223, 233, 242–262, 478–512
Specific interaction theory (SIT) 326, 328, 329, 332, 334, 335, 339, 340
Spike 11
Stir bar sorptive extraction 20, 72, 87, 88, 91, 115, 159, 161
Storage 5, 8, 9, 10, 14
Supercritical fluid extraction 162, 400

T

Tellurium 527–544
Tellurium Occurrence 528–530, 536
Thallium speciation 340
Thallium species 325, 326
Thallium chemistry 326
Tin 478–512
Tolerable weekly intake 366
Toxicity 263–265, 268, 269, 271, 274–278, 306–308, 364, 365, 382, 527–533, 547, 548, 550, 552, 553, 555, 559–565, 567, 568, 572, 576–579
Traffic related elements 344

U

Ultrasound assisted extraction 22, 131, 134
Upper Tolerable Limit 548

V

Vanadium 306–324

W

water 3, 6–13, 25, 27–30, 32, 34, 36–39, 42–48, 52, 56, 60, 62, 63, 70, 71, 80, 82, 83, 85, 89, 90, 94, 96–99, 101–106, 108, 109, 111, 112, 114, 117, 118, 120–124, 127, 130–137, 139, 141, 144, 145, 147, 149–154, 157, 158, 161, 203, 205, 207, 211, 213, 214, 218, 222, 223, 225–229, 231, 233, 243–245, 250, 251, 253, 256–260, 264–266, 271, 273, 274, 276, 279–281, 283, 284, 286–288, 290–293, 307, 308, 310, 311, 313, 314, 316, 317, 320, 342, 343, 347–350, 355, 364, 366, 370, 371, 373–376, 379, 382, 392, 395, 396, 398, 402, 406–408, 410–417, 419–421, 435, 437, 438, 440, 466, 467, 480, 482–484, 487–494, 498, 499, 513, 515, 520–523, 527, 528, 530–535, 537, 538, 540, 549, 551–556, 573–575

X

X-Ray Adsorption Fine Structure Spectroscopy 443, 449, 467
X-Ray Adsorption Near Edge Structure 443

Z

Zinc 433–477
Zinc-Anthropogenic sources 434, 462, 466
Zinc-Exposure Pathways 435

Color Plate Section

Chapter 11

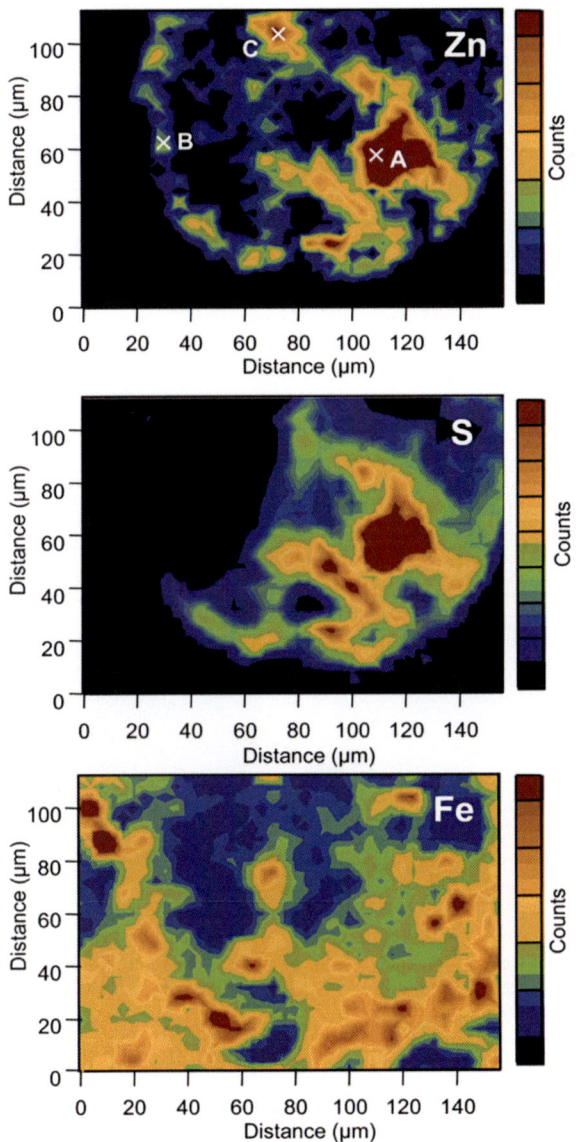

Figure 11.6 X-ray fluorescence maps for zinc, sulfur and iron from a porous black slag material. The scale bars to the right of the images denote intensity of the fluoresce signal which is related to abundance of the element. Maps show the co-localization of zinc with sulfur and iron. Figure reproduced with permission (Isaure et al. 2002).

Chapter 13

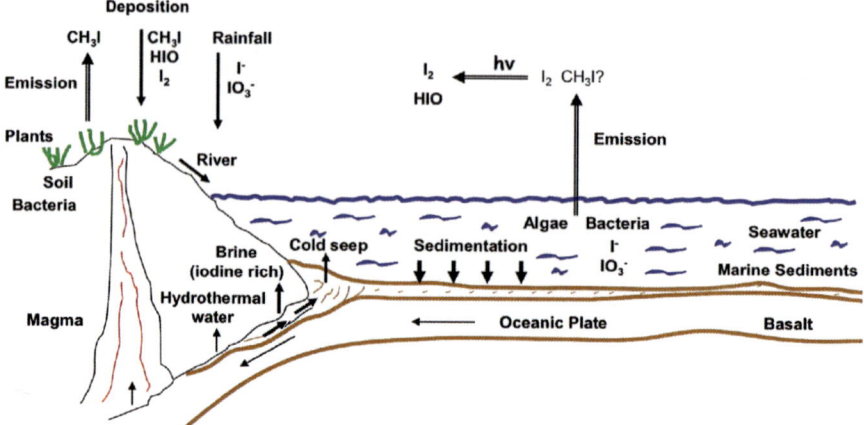

Figure 13.1 Global Cycle of Iodine.

Chapter 15

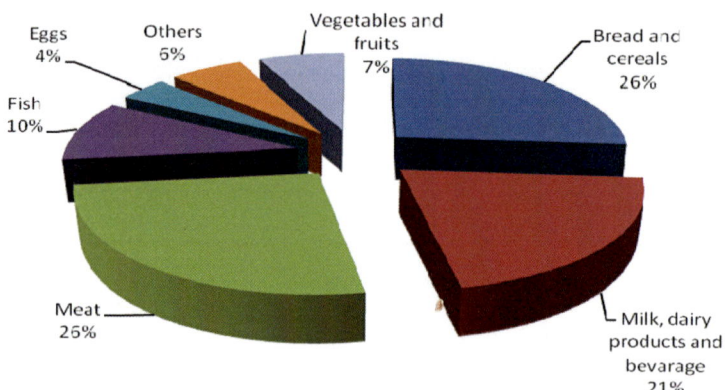

Figure 15.2 Contribution of each food group to total population dietary exposure in the UK. Adapted from data presented in the UK Food Standards Agency document (Stroud et al. 2010, Clark et al. 1996, FSA 2009).